Lecture Notes in Artificial Intelligence   8190

Subseries of Lecture Notes in Computer Science

LNAI Series Editors

Randy Goebel
University of Alberta, Edmonton, Canada
Yuzuru Tanaka
Hokkaido University, Sapporo, Japan
Wolfgang Wahlster
DFKI and Saarland University, Saarbrücken, Germany

LNAI Founding Series Editor

Joerg Siekmann
DFKI and Saarland University, Saarbrücken, Germany

Hendrik Blockeel   Kristian Kersting
Siegfried Nijssen   Filip Železný (Eds.)

Machine Learning and Knowledge Discovery in Databases

European Conference, ECML PKDD 2013
Prague, Czech Republic, September 23-27, 2013
Proceedings, Part III

 Springer

Volume Editors

Hendrik Blockeel
KU Leuven, Department of Computer Science
Celestijnenlaan 200A, 3001 Leuven, Belgium
E-mail: hendrik.blockeel@cs.kuleuven.be

Kristian Kersting
University of Bonn, Fraunhofer IAIS, Department of Knowledge Discovery
Schloss Birlinghoven, 53754 Sankt Augustin, Germany
E-mail: kristian.kersting@iais.fraunhofer.de

Siegfried Nijssen
Universiteit Leiden, LIACS, Niels Bohrweg 1, 2333 CA Leiden, The Netherlands
and KU Leuven, Department of Computer Science, 3001 Leuven, Belgium
E-mail: snijssen@liacs.nl

Filip Železný
Czech Technical University, Department of Computer Science and Engineering
Technicka 2, 16627 Prague 6, Czech Republic
E-mail: zelezny@fel.cvut.cz

Cover image: © eleephotography

ISSN 0302-9743   e-ISSN 1611-3349
ISBN 978-3-642-40993-6   e-ISBN 978-3-642-40994-3
DOI 10.1007/978-3-642-40994-3
Springer Heidelberg New York Dordrecht London

Library of Congress Control Number: 2013948101

CR Subject Classification (1998): I.2.6, H.2.8, I.5.2, G.2.2, G.3, I.2.4, I.2.7, H.3.4-5, I.2.9, F.2

LNCS Sublibrary: SL 7 – Artificial Intelligence

Typesetting: Camera-ready by author, data conversion by Scientific Publishing Services, Chennai, India

Printed on acid-free paper

Springer is part of Springer Science+Business Media (www.springer.com)

Preface

These are the proceedings of the 2013 edition of the European Conference on Machine Learning and Principles and Practice of Knowledge Discovery in Databases, or ECML PKDD for short. This conference series has grown out of the former ECML and PKDD conferences, which were Europe's premier conferences on, respectively, Machine Learning and Knowledge Discovery in Databases. Organized jointly for the first time in 2001, these conferences have become increasingly integrated, and became one in 2008. Today, ECML PKDD is a world–leading conference in these areas, well–known in particular for having a highly diverse program that aims at exploiting the synergies between these two different, yet related, scientific fields.

ECML PKDD 2013 was held in Prague, Czech Republic, during September 23–27. Continuing the series' tradition, the conference combined an extensive technical program with a variety of workshops and tutorials, a demo track for system demonstrations, an industrial track, a nectar track focusing on particularly interesting results from neighboring areas, a discovery challenge, two poster sessions, and a rich social program.

The main technical program included five plenary talks by invited speakers (Rayid Ghani, Thorsten Joachims, Ulrike von Luxburg, Christopher Re and John Shawe-Taylor) and a record–breaking 138 technical talks, for which further discussion opportunities were provided during two poster sessions. The industrial track had four invited speakers: Andreas Antrup (Zalando), Ralf Herbrich (Amazon Berlin), Jean-Paul Schmetz (Hubert Burda Media), and Hugo Zaragoza (Websays). The demo track featured 11 software demonstrations, and the nectar track 5 talks. The discovery challenge, this year, focused on the task of recommending given names for children to soon–to–be–parents. Twelve workshops were held: Scalable Decision Making; Music and Machine Learning; Reinforcement Learning with Generalized Feedback; Languages for Data Mining and Machine Learning; Data Mining on Linked Data; Mining Ubiquitous and Social Environments; Tensor Methods in Machine Learning; Solving Complex Machine Learning Problems with Ensemble Methods; Sports Analytics; New Frontiers in Mining Complex Pattern; Data Analytics for Renewable Energy Integration; and Real–World Challenges for Data Stream Mining. Eight tutorials completed the program: Multi–Agent Reinforcement Learning; Second Order Learning; Algorithmic Techniques for Modeling and Mining Large Graphs; Web Scale Information Extraction; Mining and Learning with Network–Structured Data; Performance Evaluation of Machine Learning Algorithms; Discovering Roles and Anomalies in Graphs: Theory and Applications; and Statistically Sound Pattern Discovery.

The conference offered awards for distinguished papers, for the paper from ECML / PKDD 2003 with the highest impact after a decade, and for the best demonstration. In addition, there was the novel Open Science Award. This award was installed in order to promote reusability of software, data, and experimental setups, with the aim of improving reproducibility of research and facilitating research that builds on other authors' work.

For the first time, the conference used a mixed submission model: work could be submitted as a journal article to Machine Learning or Data Mining and Knowledge Discovery, or it could be submitted for publication in the conference proceedings. A total of 182 original manuscripts were submitted to the journal track, and 447 to the proceedings track. Of the journal submissions, 14 have been published in the journal, as part of a special issue on ECML PKDD 2013, and 14 have been redirected to the proceedings track. Among the latter, 13 were accepted for publication in the proceedings. Finally, of the 447 submissions to the proceedings track, 111 have been accepted. Overall, this gives a record number of 629 submissions, of which 138 have been scheduled for presentation at the conference, making the overall acceptance rate 21.9%.

The mixed submission model was introduced in an attempt to improve the efficiency and reliability of the reviewing process. Reacting to criticism on the conference–based publication model that is so typical for computer science, several conferences have started experimenting with multiple reviewing rounds, continuous submission, and publishing contributions in a journal instead of the conference proceedings. The ECML PKDD model has been designed to maximally exploit the already existing infrastructure for journal reviewing. For an overview of the motivation and expected benefits of this new model, we refer to *A Revised Publication Model for ECML PKDD*, available at arXiv:1207.6324.

These proceedings of the 2013 European Conference on Machine Learning and Principles and Practice of Knowledge Discovery in Databases contain full papers of work presented at the main technical track, abstracts of the journal articles and invited talks presented there, and short papers describing the demonstrations and nectar papers. We thank the chairs of the demo track (Andreas Hotho and Joaquin Vanschoren), the nectar track (Rosa Meo and Michèle Sebag), and the industrial track (Ulf Brefeld), as well as the proceedings chairs Yamuna Krishnamurthy and Nico Piatkowski, for their help with putting these proceedings together. Most importantly, of course, we thank the authors for their contributions, and the area chairs and reviewers for their substantial efforts to guarantee and sometimes even improve the quality of these proceedings. We wish the reader an enjoyable experience exploring the many exciting research results presented here.

July 2013

Hendrik Blockeel
Kristian Kersting
Siegfried Nijssen
Filip Železný

Organization

ECML PKDD 2013 was organized by the Intelligent Data Analysis Research Lab, Department of Computer Science and Engineering, of the Czech Technical University in Prague.

Conference Chair

Filip Železný   Czech Technical University in Prague,
                                Czech Republic

Program Chairs

Hendrik Blockeel   KU Leuven, Belgium & Leiden University,
                               The Netherlands
Kristian Kersting   University of Bonn & Fraunhofer IAIS,
                               Germany
Siegfried Nijssen   Leiden University, The Netherlands & KU
                               Leuven, Belgium
Filip Železný   Czech Technical University in Prague,
                               Czech Republic

Local Chair

Jiří Kléma   Czech Technical University in Prague,
                               Czech Republic

Publicity Chair

Élisa Fromont   Université Jean Monnet, France

Proceedings Chairs

Yamuna Krishnamurthy   TU Dortmund, Germany
Nico Piatkowski   TU Dortmund, Germany

Workshop Chairs

Niels Landwehr   University of Potsdam, Germany
Andrea Passerini   University of Trento, Italy

Tutorial Chairs

Kurt Driessens   Maastricht University, The Netherlands
Sofus A. Macskassy   University of Southern California, USA

Demo Track Chairs

Andreas Hotho   University of Würzburg, Germany
Joaquin Vanschoren   Leiden University, The Netherlands

Nectar Track Chairs

Rosa Meo   University of Turin, Italy
Michèle Sebag   Université Paris-Sud, France

Industrial Track Chair

Ulf Brefeld   TU Darmstadt, Germany

Discovery Challenge Chair

Taneli Mielikäinen   Nokia, USA

Discovery Challenge Organizers

Stephan Doerfel   University of Kassel, Germany
Andreas Hotho   University of Würzburg, Germany
Robert Jäschke   University of Kassel, Germany
Folke Mitzlaff   University of Kassel, Germany
Jürgen Müller   L3S Research Center, Germany

Sponsorship Chairs

Peter van der Putten   Leiden University & Pegasystems,
                                   The Netherlands
Albert Bifet   University of Waikato & Yahoo, New Zealand

Awards Chairs

Bart Goethals   University of Antwerp, Belgium
Peter Flach   University of Bristol, UK
Geoff Webb   Monash University, Australia

Open-Science Award Committee

Tias Guns   KU Leuven, Belgium
Christian Borgelt   European Center for Soft Computing, Spain
Geoff Holmes   University of Waikato, New Zealand
Luis Torgo   University of Porto, Portugal

Web Team

Matěj Holec, Webmaster   Czech Technical University in Prague,
                               Czech Republic
Radomír Černoch   Czech Technical University in Prague,
                               Czech Republic
Filip Blažek   Designiq, Czech Republic

Software Development

Radomír Černoch   Czech Technical University in Prague,
                               Czech Republic
Fabian Hadiji   University of Bonn, Germany
Thanh Le Van   KU Leuven, Belgium

ECML PKDD Steering Committee

Fosca Gianotti, Chair   Universitá di Pisa, Italy
Jose Balcazar   Universitat Polytécnica de Catalunya, Spain
Francesco Bonchi   Yahoo! Research Barcelona, Spain
Nello Cristianini   University of Bristol, UK
Tijl De Bie   University of Bristol, UK
Peter Flach   University of Bristol, UK
Dimitrios Gunopulos   University of Athens, Greece
Donato Malerba   Universitá degli Studi di Bari, Italy
Michèle Sebag   Université Paris-Sud, France
Michalis Vazirgiannis   Athens University of Economics and Business,
                               Greece

Area Chairs

Henrik Boström   Stockholm University, Sweden
Jean-François Boulicaut   University of Lyon, France
Carla Brodley   Tuft University, USA
Ian Davidson   University of California, Davis, USA
Jesse Davis   KU Leuven, Belgium
Tijl De Bie   University of Bristol, UK
Janez Demšar   University of Ljubljana, Slovenia
Luc De Raedt   KU Leuven, Belgium

Pierre Dupont   UC Louvain, Belgium
Charles Elkan   University of California, San Diego, USA
Alan Fern   Oregon State University, USA
Johannes Fürnkranz   TU Darmstadt, Germany
Joao Gama   University of Porto, Portugal
Thomas Gärtner   University of Bonn and Fraunhofer IAIS,
                                  Germany
Aristides Gionis   Aalto University, Finland
Bart Goethals   University of Antwerp, Belgium
Geoff Holmes   Waikato University, New Zealand
Andreas Hotho   University of Würzburg, Germany
Eyke Hüllermeier   Philipps-Universität Marburg, Germany
Manfred Jaeger   Aalborg University, Denmark
Thorsten Joachims   Cornell University, USA
George Karypis   University of Minnesota, USA
Stefan Kramer   University of Mainz, Germany
Donato Malerba   University of Bari, Italy
Dunja Mladenic   Jožef Stefan Institute, Slovenia
Marie-Francine Moens   KU Leuven, Belgium
Bernhard Pfahringer   University of Waikato, New Zealand
Myra Spiliopoulou   Magdeburg University, Germany
Hannu Toivonen   University of Helsinki, Finland
Marco Wiering   University of Groningen, The Netherlands
Stefan Wrobel   University of Bonn and Fraunhofer IAIS,
                                  Germany

Program Committee

Leman Akoglu   Andras Benczur
Mohammad Al Hasan   Bettina Berendt
Aris Anagnostopoulos   Michael Berthold
Gennady Andrienko   Indrajit Bhattacharya
Annalisa Appice   Albert Bifet
Cedric Archambeau   Mario Boley
Marta Arias   Francesco Bonchi
Hiroki Arimura   Gianluca Bontempi
Ira Assent   Christian Borgelt
Martin Atzmüller   Zoran Bosnic
Chloe-Agathe Azencott   Abdeslam Boularias
Antonio Bahamonde   Kendrick Boyd
James Bailey   Pavel Brazdil
Jose Balcazar   Ulf Brefeld
Christian Bauckhage   Björn Bringmann
Roberto Bayardo   Wray Buntine
Aurelien Bellet   Robert Busa-Fekete

Toon Calders
Andre Carvalho
Francisco Casacuberta
Michelangelo Ceci
Loic Cerf
Duen Horng Chau
Sanjay Chawla
Weiwei Cheng
Fabrizio Costa
Sheldon Cooper
Vitor Costa
Bruno Cremilleux
Tom Croonenborghs
Boris Cule
Tomaz Curk
James Cussens
Martine De Cock
Colin de la Higuera
Juan del Coz
Francois Denis
Jana Diesner
Wei Ding
Janardhan Doppa
Devdatt Dubhashi
Ines Dutra
Sašo Džeroski
Tina Eliassi-Rad
Tapio Elomaa
Seyda Ertekin
Floriana Esposito
Ines Faerber
Fazel Famili
Hadi Fanaee Tork
Elaine Faria
Ad Feelders
Stefano Ferilli
Carlos Ferreira
Jordi Fonollosa
Antonino Freno
Elisa Fromont
Fabio Fumarola
Patrick Gallinari
Roman Garnett
Eric Gaussier
Ricard Gavalda

Pierre Geurts
Rayid Ghani
Fosca Giannotti
David Gleich
Vibhav Gogate
Michael Granitzer
Dimitrios Gunopulos
Tias Guns
Jiawei Han
Daniel Hernandez Lobato
Frank Hoeppner
Thomas Hofmann
Jaako Hollmen
Arjen Hommersom
Vasant Honavar
Tamás Horváth
Dino Ienco
Elena Ikonomovska
Robert Jäschke
Frederik Janssen
Szymon Jaroszewicz
Ulf Johansson
Alipio Jorge
Kshitij Judah
Hachem Kadri
Alexandros Kalousis
U Kang
Panagiotis Karras
Andreas Karwath
Hisashi Kashima
Samuel Kaski
Latifur Khan
Angelika Kimmig
Arno Knobbe
Levente Kocsis
Yun Sing Koh
Alek Kolcz
Andrey Kolobov
Igor Kononenko
Kleanthis-Nikolaos Kontonasios
Nitish Korula
Petr Kosina
Walter Kosters
Georg Krempl
Sergei Kuznetsov

Helge Langseth
Pedro Larranaga
Silvio Lattanzi
Niklas Lavesson
Nada Lavrač
Gregor Leban
Chris Leckie
Sangkyun Lee
Ping Li
Juanzi Li
Edo Liberty
Jefrey Lijffijt
Jessica Lin
Francesca Lisi
Corrado Loglisci
Eneldo Loza Mencia
Peter Lucas
Francis Maes
Michael Mampaey
Giuseppe Manco
Stan Matwin
Michael May
Mike Mayo
Wannes Meert
Ernestina Menasalvas
Rosa Meo
Pauli Miettinen
Bamshad Mobasher
Joao Moreira
Emmanuel Müller
Mohamed Nadif
Alex Nanopoulos
Balakrishnan Narayanaswamy
Sriraam Natarajan
Aniruddh Nath
Thomas Nielsen
Mathias Niepert
Xia Ning
Niklas Noren
Eirini Ntoutsi
Andreas Nürnberger
Gerhard Paass
David Page
Rasmus Pagh
Spiros Papadimitriou

Panagiotis Papapetrou
Andrea Passerini
Mykola Pechenizkiy
Dino Pedreschi
Jian Pei
Nikos Pelekis
Ruggero Pensa
Marc Plantevit
Pascal Poncelet
Aditya Prakash
Kai Puolamaki
Buyue Qian
Chedy Raïssi
Liva Ralaivola
Karthik Raman
Jan Ramon
Huzefa Rangwala
Umaa Rebbapragada
Jean-Michel Renders
Steffen Rendle
Achim Rettinger
Fabrizio Riguzzi
Celine Robardet
Marko Robnik Sikonja
Pedro Rodrigues
Juan Rodriguez
Irene Rodriguez-Lujan
Ulrich Rückert
Stefan Rüping
Jan Rupnik
Yvan Saeys
Alan Said
Lorenza Saitta
Antonio Salmeron
Scott Sanner
Raul Santos-Rodriguez
Sam Sarjant
Claudio Sartori
Taisuke Sato
Lars Schmidt-Thieme
Christoph Schommer
Michèle Sebag
Marc Sebban
Thomas Seidl
Giovanni Semeraro

Junming Shao
Pannaga Shivaswamy
Jonathan Silva
Kevin Small
Koen Smets
Padhraic Smyth
Carlos Soares
Mauro Sozio
Eirini Spyropoulou
Ashwin Srinivasan
Jerzy Stefanowski
Benno Stein
Markus Strohmaier
Mahito Sugiyama
Einoshin Suzuki
Sandor Szedmak
Andrea Tagarelli
Nima Taghipour
Nikolaj Tatti
Matthew Taylor
Maguelonne Teisseire
Evimaria Terzi
Ljupco Todorovski
Luis Torgo
Panagiotis Tsaparas
Vincent Tseng
Grigorios Tsoumakas
Antti Ukkonen
Athina Vakali

Guy Van den Broeck
Matthijs van Leeuwen
Joaquin Vanschoren
Michalis Vazirgiannis
Shankar Vembu
Celine Vens
Sicco Verwer
Enrique Vidal
Herna Viktor
Christel Vrain
Jilles Vreeken
Byron Wallace
Fei Wang
Xiang Wang
Takashi Washio
Jörg Wicker
Gerhard Widmer
Aaron Wilson
Chun-Nam Yu
Jure Zabkar
Gerson Zaverucha
Bernard Zenko
Min-Ling Zhang
Elena Zheleva
Arthur Zimek
Albrecht Zimmermann
Indre Zliobaite
Blaz Zupan

Demo Track Program Committee

Alan Said
Albert Bifet
Andreas Nürnberger
Bettina Berendt
Christian Borgelt
Daniela Stojanova
Gabor Melli
Geoff Holmes
Gerard de Melo
Grigorios Tsoumakas
Jaako Hollmen

Lars Schmidt-Thieme
Michael Mampaey
Mikio Braun
Mykola Pechenizkiy
Omar Alonso
Peter Reutemann
Peter van der Putten
Robert Jäschke
Stephan Doerfel
Themis Palpanas

Nectar Track Program Committee

Maria Florina Balcan
Christian Böhm
Toon Calders
Luc De Raedt
George Karypis

Hugo Larochelle
Donato Malerba
Myra Spiliopoulou
Vicenc Torra

Additional Reviewers

Rohit Babbar
Aubrey Barnard
Christian Beecks
Alejandro Bellogin
Daniel Bengs
Souhaib Ben Taieb
Mansurul Bhuiyan
Sam Blasiak
Patrice Boizumault
Teresa Bracamonte
Janez Brank
George Brova
David C. Anastasiu
Cécile Capponi
Annalina Caputo
Jeffrey Chan
Anveshi Charuvaka
Claudia d'Amato
Xuan-Hong Dang
Ninh Dang Pham
Lucas Drumond
Wouter Duivesteijn
François-Xavier Dupé
Ritabrata Dutta
Pavel Efros
Dora Erdos
Pasqua Fabiana Lanotte
Antonio Fernandez
Georg Fette
Manoel França
Sergej Fries
Atsushi Fujii
Patrick Gabrielsson
Esther Galbrun
Michael Geilke

Christos Giatsidis
Robby Goetschalckx
Boqing Gong
Michele Gorgoglione
Tatiana Gossen
Maarten Grachten
Xin Guan
Massimo Guarascio
Huan Gui
Amaury Habrard
Ahsanul Haque
Marwan Hassani
Kohei Hayashi
Elad Hazan
Andreas Henelius
Shohei Hido
Patricia Iglesias Sanchez
Roberto Interdonato
Baptiste Jeudy
Hiroshi Kajino
Yoshitaka Kameya
Margarita Karkali
Mehdi Kaytoue
Fabian Keller
Mikaela Keller
Eamonn Keogh
Umer Khan
Tushar Khot
Benjamin Kille
Dragi Kocev
Jussi Korpela
Domen Kosir
Hardy Kremer
Tanay K. Saha
Gautam Kunapuli

Guest Editorial Board (Journal Track)

Luc De Raedt	KU Leuven, Belgium
Luis Torgo	University of Porto, Portugal
Marie-Francine Moens	KU Leuven, Belgium
Matthijs van Leeuwen	KU Leuven, Belgium
Michael May	Fraunhofer IAIS, Germany
Michael R. Berthold	Universität Konstanz, Germany
Nada Lavrač	Jožef Stefan Institute, Slovenia
Nikolaj Tatti	University of Antwerp, Belgium
Pascal Poupart	University of Waterloo, Canada
Pierre Dupont	UC Louvain, Belgium
Prasad Tadepalli	Oregon State University, USA
Roberto Bayardo	Google Research, USA
Soumya Ray	Case Western Reserve University, USA
Stefan Wrobel	University of Bonn and Fraunhofer IAIS, Germany
Stefan Kramer	University of Mainz, Germany
Takashi Washio	Osaka University, Japan
Tamás Horváth	Fraunhofer IAIS, Germany
Tapio Elomaa	Tampere University of Technology, Finland
Thomas Gärtner	University of Bonn and Fraunhofer IAIS, Germany
Tijl De Bie	University of Bristol, UK
Toon Calders	Eindhoven University of Technology, The Netherlands
Willem Waegeman	Ghent University, Belgium
Wray Buntine	NICTA, Australia

Additional Reviewers (Journal Track)

Babak Ahmadi	Emma Brunskill
Amr Ahmed	Michelangelo Ceci
Leman Akoglu	Sanjay Chawla
Mohammad Al Hasan	Weiwei Cheng
Massih-Reza Amini	KyungHyun Cho
Bart Baesens	Tom Croonenborghs
Andrew Bagnell	Florence d'Alché-Buc
Arindam Banerjee	Bhavana Dalvi
Christian Bauckhage	Kurt De Grave
Yoshua Bengio	Bolin Ding
Albert Bifet	Chris Ding
Andrew Bolstad	Jennifer Dy
Byron Boots	Sašo Džeroski
Karsten Borgwardt	Alan Fern
Kendrick Boyd	Luis Ferre
Ulf Brefeld	Daan Fierens

Arik Friedman
Mark Gales
Joao Gama
Roman Garnett
Rainer Gemulla
Marek Grzes
Tias Guns
Gregor Heinrich
Katherine Heller
Francisco Herrera
McElory Hoffmann
Andreas Hotho
Christian Igel
Mariya Ishteva
Manfred Jaeger
Kshitij Judah
Ata Kaban
Lars Kaderali
Alexandros Kalousis
Panagiotis Karras
Hendrik Blockeel
Kristian Kersting
Siegfried Nijssen
Filip Železný
Marius Kloft
Hans-Peter Kriegel
Gautam Kunapuli
Niels Landwehr
Ni Lao
Agnieszka Lawrynowicz
Honglak Lee
Nan Li
Yu-Feng Li
James Robert Lloyd
Manuel Lopes
Haibing Lu
Michael Mampaey
Dragos Margineantu
Stan Matwin
Francisco Melo
Pauli Miettinen
Alessandro Moschitti
Uwe Nagel
Mirco Nanni
Sriraam Natarajan

Mathias Niepert
John Paisley
Ankur Parikh
Srinivasan Parthasarathy
Alessandro Perina
Jan Peters
Massimiliano Pontil
Foster Provost
Ricardo Prudencio
Tao Qin
Novi Quadrianto
Ariadna Quattoni
Chedy Raïssi
Alexander Rakhlin
Balaraman Ravindran
Jesse Read
Chandan Reddy
Achim Rettinger
Peter Richtarik
Volker Roth
Céline Rouveirol
Cynthia Rudin
Stefan Rüping
Scott Sanner
Lars Schmidt-Thieme
Jeff Schneider
Stephen Scott
Michèle Sebag
Mathieu Serrurier
Sohan Seth
Jude Shavlik
Le Song
Peter Stone
Peter Sunehag
Charles Sutton
Johan Suykens
Nima Taghipour
Emmanuel Tapia
Nikolaj Tatti
Graham Taylor
Manolis Terrovitis
Grigorios Tsoumakas
Tinne Tuytelaars
Antti Ukkonen
Laurens van der Maaten

Marcel van Gerven
Martijn van Otterlo
Lieven Vandenberghe
Joaquin Vanschoren
Michalis Vazirgiannis
Aki Vehtari
Byron Wallace
Thomas J. Walsh
Chao Wang
Pu Wang
Shaojun Wang
Randy Wilson

Han-Ming Wu
Huan Xu
Zhao Xu
Jieping Ye
Yi-Ren Yeh
Shipeng Yu
Dengyong Zhou
Shenghuo Zhu
Arthur Zimek
Albrecht Zimmermann
Indre Zliobaite

Sponsors

Gold Sponsor
Winton Capital   http://wintoncapital.com

Silver Sponsors
Cisco Systems, Inc.   http://www.cisco.com
Deloitte Analytics   http://www.deloitte.com
KNIME   http://www.knime.com
Yahoo! Labs   http://www.yahoo.com

Bronze Sponsors
CSKI   http://www.cski.cz
Definity Systems   http://www.definity.cz
DIKW Academy   http://dikw-academy.nl
Google   http://research.google.com
Xerox Research Centre Europe   http://www.xrce.xerox.com
Zalando   http://www.zalando.de

Prize Sponsors
Data Mining and Knowledge Discovery   http://link.springer.com/
                                        journal/10618

Deloitte Analytics   http://www.deloitte.com
Google   http://research.google.com
Machine Learning   http://link.springer.com/
                                        journal/10994

Yahoo! Labs   http://www.knime.com

Abstracts of Invited Talks

Using Machine Learning Powers for Good

Rayid Ghani

The past few years have seen increasing demand for machine learning and data mining—both for tools as well as experts. This has been mostly motivated by a variety of factors including better and cheaper data collection, realization that using data is a good thing, and the ability for a lot of organizations to take action based on data analysis. Despite this flood of demand, most applications we hear about in machine learning involve search, advertising, and financial areas. This talk will talk about examples on how the same approaches can be used to help governments and non-prpofits make social impact. I'll talk about a summer fellowship program we ran at University of Chicago on social good and show examples from projects in areas such as education, healthcare, energy, transportation and public safety done in conjunction with governments and non-profits.

Biography

Rayid Ghani was the Chief Scientist at the Obama for America 2012 campaign focusing on analytics, technology, and data. His work focused on improving different functions of the campaign including fundraising, volunteer, and voter mobilization using analytics, social media, and machine learning; his innovative use of machine learning and data mining in Obama's reelection campaign received broad attention in the media such as the New York Times, CNN, and others. Before joining the campaign, Rayid was a Senior Research Scientist and Director of Analytics research at Accenture Labs where he led a technology research team focused on applied R&D in analytics, machine learning, and data mining for large-scale & emerging business problems in various industries including healthcare, retail & CPG, manufacturing, intelligence, and financial services. In addition, Rayid serves as an adviser to several start-ups in Analytics, is an active organizer of and participant in academic and industry analytics conferences, and publishes regularly in machine learning and data mining conferences and journals.

Learning with Humans in the Loop

Thorsten Joachims

Machine Learning is increasingly becoming a technology that directly interacts with human users. Search engines, recommender systems, and electronic commerce already heavily rely on adapting the user experience through machine learning, and other applications are likely to follow in the near future (e.g., autonomous robotics, smart homes, gaming). In this talk, I argue that learning with humans in the loop requires learning algorithms that explicitly account for human behavior, their motivations, and their judgment of performance. Towards this goal, the talk explores how integrating microeconomic models of human behavior into the learning process leads to new learning models that no longer reduce the user to a "labeling subroutine". This motivates an interesting area for theoretical, algorithmic, and applied machine learning research with connections to rational choice theory, econometrics, and behavioral economics.

Biography

Thorsten Joachims is a Professor of Computer Science at Cornell University. His research interests center on a synthesis of theory and system building in machine learning, with applications in language technology, information retrieval, and recommendation. His past research focused on support vector machines, text classification, structured output prediction, convex optimization, learning to rank, learning with preferences, and learning from implicit feedback. In 2001, he finished his dissertation advised by Prof. Katharina Morik at the University of Dortmund. From there he also received his Diplom in Computer Science in 1997. Between 2000 and 2001 he worked as a PostDoc at the GMD Institute for Autonomous Intelligent Systems. From 1994 to 1996 he was a visiting scholar with Prof. Tom Mitchell at Carnegie Mellon University.

Unsupervised Learning with Graphs:
A Theoretical Perspective

Ulrike von Luxburg

Applying a graph–based learning algorithm usually requires a large amount of data preprocessing. As always, such preprocessing can be harmful or helpful. In my talk I am going to discuss statistical and theoretical properties of various preprocessing steps. We consider questions such as: Given data that does not have the form of a graph yet, what do we loose when transforming it to a graph? Given a graph, what might be a meaningful distance function? We will also see that graph–based techniques can lead to surprising solutions to preprocessing problems that a priori don't involve graphs at all.

Biography

Ulrike von Luxburg is a professor for computer science/machine learning at the University of Hamburg. Her research focus is the theoretical analysis of machine learning algorithms, in particular for unsupervised learning and graph algorithms. She is (co)–winner of several best student paper awards (NIPS 2004 and 2008, COLT 2003, 2005 and 2006, ALT 2007). She did her PhD in the Max Planck Institute for Biological Cybernetics in 2004, then moved to Fraunhofer IPSI in Darmstadt, before returning to the Max Planck Institute in 2007 as a research group leader for learning theory. Since 2012 she is a professor for computer science at the University of Hamburg.

Making Systems That Use Statistical Reasoning Easier to Build and Maintain over Time

Christopher Re

The question driving my work is, how should one deploy statistical data–analysis tools to enhance data–driven systems? Even partial answers to this question may have a large impact on science, government, and industry—each of whom are increasingly turning to statistical techniques to get value from their data.

To understand this question, my group has built or contributed to a diverse set of data–processing systems: a system, called GeoDeepDive, that reads and helps answer questions about the geology literature; a muon filter that is used in the IceCube neutrino telescope to process over 250 million events each day in the hunt for the origins of the universe; and enterprise applications with Oracle and Pivotal. This talk will give an overview of the lessons that we learned in these systems, will argue that data systems research may play a larger role in the next generation of these systems, and will speculate on the future challenges that such systems may face.

Biography

Christopher Re is an assistant professor in the department of Computer Sciences at the University of Wisconsin-Madison. The goal of his work is to enable users and developers to build applications that more deeply understand and exploit data. Chris received his PhD from the University of Washington, Seattle under the supervision of Dan Suciu. For his PhD work in the area of probabilistic data management, Chris received the SIGMOD 2010 Jim Gray Dissertation Award. Chris's papers have received four best papers or best–of–conference citations (best paper in PODS 2012 and best–of–conference in PODS 2010, twice, and one in ICDE 2009). Chris received an NSF CAREER Award in 2011.

Deep–er Kernels

John Shawe-Taylor

Kernels can be viewed as shallow in that learning is only applied in a single (output) layer. Recent successes with deep learning highlight the need to consider learning richer function classes. The talk will review and discuss methods that have been developed to enable richer kernel classes to be learned. While some of these methods rely on greedy procedures many are supported by statistical learning analyses and/or convergence bounds. The talk will highlight the trade–offs involved and the potential for further research on this topic.

Biography

John Shawe-Taylor obtained a PhD in Mathematics at Royal Holloway, University of London in 1986 and joined the Department of Computer Science in the same year. He was promoted to Professor of Computing Science in 1996. He moved to the University of Southampton in 2003 to lead the ISIS research group. He was Director of the Centre for Computational Statistics and Machine Learning at University College, London between July 2006 and September 2010. He has coordinated a number of European wide projects investigating the theory and practice of Machine Learning, including the PASCAL projects. He has published over 300 research papers with more than 25000 citations. He has co-authored with Nello Cristianini two books on kernel approaches to machine learning: "An Introduction to Support Vector Machines" and "Kernel Methods for Pattern Analysis".

Abstracts of Industrial
Track Invited Talks

ML and Business: A Love–Hate Relationship

Andreas Antrup

Based on real world examples. the talk explores common gaps in the mutual understanding of the business and the analytical side; particular focus shall be on misconceptions of the needs and expectations of business people and the resulting problems. It also touches on some approaches to bridge these gaps and build trust. At the end we shall discuss possibly under–researched areas that may open the doors to a yet wider usage of ML principles and thus unlock more of its value and beauty.

Bayesian Learning in Online Service: Statistics Meets Systems

Ralf Herbrich

Over the past few years, we have entered the world of big and structured data— a trend largely driven by the exponential growth of Internet–based online services such as Search, eCommerce and Social Networking as well as the ubiquity of smart devices with sensors in everyday life. This poses new challenges for statistical inference and decision–making as some of the basic assumptions are shifting:

- The ability to optimize both the likelihood and loss functions
- The ability to store the parameters of (data) models
- The level of granularity and 'building blocks' in the data modeling phase
- The interplay of computation, storage, communication and inference and decision–making techniques

In this talk, I will discuss the implications of big and structured data for Statistics and the convergence of statistical model and distributed systems. I will present one of the most versatile modeling techniques that combines systems and statistical properties—factor graphs—and review a series of approximate inference techniques such as distributed message passing. The talk will be concluded with an overview of real–world problems at Amazon.

Machine Learning in a Large diversified Internet Group

Jean-Paul Schmetz

I will present a wide survey of the use of machine learning techniques across a large number of subsidiaries (40+) of an Internet group (Burda Digital) with special attention to issues regarding (1) personnel training in state of the art techniques, (2) management buy–in of complex non interpretable results and (3) practical and measurable bottom line results/solutions.

Some of the Problems and Applications of Opinion Analysis

Hugo Zaragoza

Websays strives to provide the best possible analysis of online conversation to marketing and social media analysts. One of the obsessions of Websays is to provide "near–man–made" data quality at marginal costs. I will discuss how we approach this problem using innovative machine learning and UI approaches.

Abstracts of Journal Track Articles

The full articles have been published in *Machine Learning* or *Data Mining and Knowledge Discovery*.

Fast sequence segmentation using log–linear models
Nikolaj Tatti
Data Mining and Knowledge Discovery
DOI 10.1007/s10618-012-0301-y

Sequence segmentation is a well–studied problem, where given a sequence of elements, an integer K, and some measure of homogeneity, the task is to split the sequence into K contiguous segments that are maximally homogeneous. A classic approach to find the optimal solution is by using a dynamic program. Unfortunately, the execution time of this program is quadratic with respect to the length of the input sequence. This makes the algorithm slow for a sequence of non–trivial length. In this paper we study segmentations whose measure of goodness is based on log–linear models, a rich family that contains many of the standard distributions. We present a theoretical result allowing us to prune many suboptimal segmentations. Using this result, we modify the standard dynamic program for 1D log–linear models, and by doing so reduce the computational time. We demonstrate empirically, that this approach can significantly reduce the computational burden of finding the optimal segmentation.

ROC curves in cost space
Cesar Ferri, Jose Hernandez-Orallo and Peter Flach
Machine Learning
DOI 10.1007/s10994-013-5328-9

ROC curves and cost curves are two popular ways of visualising classifier performance, finding appropriate thresholds according to the operating condition, and deriving useful aggregated measures such as the area under the ROC curve (AUC) or the area under the optimal cost curve. In this paper we present new findings and connections between ROC space and cost space. In particular, we show that ROC curves can be transferred to cost space by means of a very natural threshold choice method, which sets the decision threshold such that the proportion of positive predictions equals the operating condition. We call these new curves rate–driven curves, and we demonstrate that the expected loss as measured by the area under these curves is linearly related to AUC. We show that the rate–driven curves are the genuine equivalent of ROC curves in cost space, establishing a point–point rather than a point–line correspondence. Furthermore, a decomposition of the rate–driven curves is introduced which separates the loss due to the threshold choice method from the ranking loss (Kendall

τ distance). We also derive the corresponding curve to the ROC convex hull in cost space: this curve is different from the lower envelope of the cost lines, as the latter assumes only optimal thresholds are chosen.

A framework for semi–supervised and unsupervised optimal extraction of clusters from hierarchies

Ricardo J.G.B. Campello, Davoud Moulavi, Arthur Zimek and Jörg Sander

Data Mining and Knowledge Discovery
DOI 10.1007/s10618-013-0311-4

We introduce a framework for the optimal extraction of flat clusterings from local cuts through cluster hierarchies. The extraction of a flat clustering from a cluster tree is formulated as an optimization problem and a linear complexity algorithm is presented that provides the globally optimal solution to this problem in semi–supervised as well as in unsupervised scenarios. A collection of experiments is presented involving clustering hierarchies of different natures, a variety of real data sets, and comparisons with specialized methods from the literature.

Pairwise meta–rules for better meta–learning–based algorithm ranking

Quan Sun and Bernhard Pfahringer

Machine Learning
DOI 10.1007/s10994-013-5387-y

In this paper, we present a novel meta–feature generation method in the context of meta–learning, which is based on rules that compare the performance of individual base learners in a one–against–one manner. In addition to these new meta–features, we also introduce a new meta–learner called Approximate Ranking Tree Forests (ART Forests) that performs very competitively when compared with several state–of–the–art meta–learners. Our experimental results are based on a large collection of datasets and show that the proposed new techniques can improve the overall performance of meta–learning for algorithm ranking significantly. A key point in our approach is that each performance figure of any base learner for any specific dataset is generated by optimising the parameters of the base learner separately for each dataset.

Block coordinate descent algorithms for large–scale sparse multiclass classification

Mathieu Blondel, Kazuhiro Seki and Kuniaki Uehara
Machine Learning
DOI 10.1007/s10994-013-5367-2

Over the past decade, ℓ_1 regularization has emerged as a powerful way to learn classifiers with implicit feature selection. More recently, mixed–norm (e.g., ℓ_1/ℓ_2) regularization has been utilized as a way to select entire groups of features. In this paper, we propose a novel direct multiclass formulation specifically designed for large–scale and high–dimensional problems such as document classification. Based on a multiclass extension of the squared hinge loss, our formulation employs ℓ_1/ℓ_2 regularization so as to force weights corresponding to the same features to be zero across all classes, resulting in compact and fast–to–evaluate multiclass models. For optimization, we employ two globally–convergent variants of block coordinate descent, one with line search (Tseng and Yun in Math. Program. 117:387423, 2009) and the other without (Richtrik and Tak in Math. Program. 138, 2012a, Tech. Rep. arXiv:1212.0873, 2012b). We present the two variants in a unified manner and develop the core components needed to efficiently solve our formulation. The end result is a couple of block coordinate descent algorithms specifically tailored to our multiclass formulation. Experimentally, we show that block coordinate descent performs favorably compared to other solvers such as FOBOS, FISTA and SpaRSA. Furthermore, we show that our formulation obtains very compact multiclass models and outperforms ℓ_1/ℓ_2–regularized multiclass logistic regression in terms of training speed, while achieving comparable test accuracy.

A comparative evaluation of stochastic–based inference methods for Gaussian process models

Maurizio Filippone, Mingjun Zhong and Mark Girolami
Machine Learning
DOI 10.1007/s10994-013-5388-x

Gaussian process (GP) models are extensively used in data analysis given their flexible modeling capabilities and interpretability. The fully Bayesian treatment of GP models is analytically intractable, and therefore it is necessary to resort to either deterministic or stochastic approximations. This paper focuses on stochastic–based inference techniques. After discussing the challenges associated with the fully Bayesian treatment of GP models, a number of inference strategies based on Markov chain Monte Carlo methods are presented and rigorously assessed. In particular, strategies based on efficient parameterizations and efficient proposal mechanisms are extensively compared on simulated and real data on the basis of convergence speed, sampling efficiency, and computational cost.

Probabilistic topic models for sequence data

Nicola Barbieri, Antonio Bevacqua, Marco Carnuccio, Giuseppe Manco and Ettore Ritacco

Machine Learning
DOI 10.1007/s10994-013-5391-2

Probabilistic topic models are widely used in different contexts to uncover the hidden structure in large text corpora. One of the main (and perhaps strong) assumptions of these models is that the generative process follows a bag–of–words assumption, i.e. each token is independent from the previous one. We extend the popular Latent Dirichlet Allocation model by exploiting three different conditional Markovian assumptions: (i) the token generation depends on the current topic and on the previous token; (ii) the topic associated with each observation depends on topic associated with the previous one; (iii) the token generation depends on the current and previous topic. For each of these modeling assumptions we present a Gibbs Sampling procedure for parameter estimation. Experimental evaluation over real–word data shows the performance advantages, in terms of recall and precision, of the sequence–modeling approaches.

The flip–the–state transition operator for restricted Boltzmann machines

Kai Brügge, Asja Fischer and Christian Igel

Machine Learning
DOI 10.1007/s10994-013-5390-3

Most learning and sampling algorithms for restricted Boltzmann machines (RBMs) rely on Markov chain Monte Carlo (MCMC) methods using Gibbs sampling. The most prominent examples are Contrastive Divergence learning (CD) and its variants as well as Parallel Tempering (PT). The performance of these methods strongly depends on the mixing properties of the Gibbs chain. We propose a Metropolis–type MCMC algorithm relying on a transition operator maximizing the probability of state changes. It is shown that the operator induces an irreducible, aperiodic, and hence properly converging Markov chain, also for the typically used periodic update schemes. The transition operator can replace Gibbs sampling in RBM learning algorithms without producing computational overhead. It is shown empirically that this leads to faster mixing and in turn to more accurate learning.

Differential privacy based on importance weighting

Zhanglong Ji and Charles Elkan

Machine Learning
DOI 10.1007/s10994-013-5396-x

This paper analyzes a novel method for publishing data while still protecting privacy. The method is based on computing weights that make an existing dataset, for which there are no confidentiality issues, analogous to the dataset that must be kept private. The existing dataset may be genuine but public already, or it may be synthetic. The weights are importance sampling weights, but to protect privacy, they are regularized and have noise added. The weights allow statistical queries to be answered approximately while provably guaranteeing differential privacy. We derive an expression for the asymptotic variance of the approximate answers. Experiments show that the new mechanism performs well even when the privacy budget is small, and when the public and private datasets are drawn from different populations.

Activity preserving graph simplification

Francesco Bonchi, Gianmarco De Francisci Morales, Aristides Gionis and Antti Ukkonen
Data Mining and Knowledge Discovery
DOI 10.1007/s10618-013-0328-8

We study the problem of simplifying a given directed graph by keeping a small subset of its arcs. Our goal is to maintain the connectivity required to explain a set of observed traces of information propagation across the graph. Unlike previous work, we do not make any assumption about an underlying model of information propagation. Instead, we approach the task as a combinatorial problem.
We prove that the resulting optimization problem is **NP**–hard. We show that a standard greedy algorithm performs very well in practice, even though it does not have theoretical guarantees. Additionally, if the activity traces have a tree structure, we show that the objective function is supermodular, and experimentally verify that the approach for size–constrained submodular minimization recently proposed by Nagano et al (2011) produces very good results. Moreover, when applied to the task of reconstructing an unobserved graph, our methods perform comparably to a state–of–the–art algorithm devised specifically for this task.

ABACUS: frequent pattern mining based community discovery in multidimensional networks

Michele Berlingerio, Fabio Pinelli and Francesco Calabrese
Data Mining and Knowledge Discovery
DOI 10.1007/s10618-013-0331-0

Community Discovery in complex networks is the problem of detecting, for each node of the network, its membership to one of more groups of nodes, the communities, that are densely connected, or highly interactive, or, more in general, similar, according to a similarity function. So far, the problem has been widely studied in monodimensional networks, i.e. networks where only one connection between two entities may exist. However, real networks are often multidimensional, i.e., multiple connections between any two nodes may exist, either reflecting different kinds of relationships, or representing different values of the same type of tie. In this context, the problem of Community Discovery has to be redefined, taking into account multidimensional structure of the graph. We define a new concept of community that groups together nodes sharing memberships to the same monodimensional communities in the different single dimensions. As we show, such communities are meaningful and able to group nodes even if they might not be connected in any of the monodimensional networks. We devise ABACUS (frequent pAttern mining–BAsed Community discoverer in mUltidimensional networkS), an algorithm that is able to extract multidimensional communities based on the extraction of frequent closed itemsets from monodimensional community memberships. Experiments on two different real multidimensional networks confirm the meaningfulness of the introduced concepts, and open the way for a new class of algorithms for community discovery that do not rely on the dense connections among nodes.

Growing a list
Benjamin Letham, Cynthia Rudin and Katherine A. Heller
Data Mining and Knowledge Discovery
DOI 10.1007/s10618-013-0329-7

It is easy to find expert knowledge on the Internet on almost any topic, but obtaining a complete overview of a given topic is not always easy: Information can be scattered across many sources and must be aggregated to be useful. We introduce a method for intelligently growing a list of relevant items, starting from a small seed of examples. Our algorithm takes advantage of the wisdom of the crowd, in the sense that there are many experts who post lists of things on the Internet. We use a collection of simple machine learning components to find these experts and aggregate their lists to produce a single complete and meaningful list. We use experiments with gold standards and open–ended experiments without gold standards to show that our method significantly outperforms the state of the art. Our method uses the ranking algorithm Bayesian Sets even when its underlying independence assumption is violated, and we provide a theoretical generalization bound to motivate its use.

What distinguish one from its peers in social networks?
Yi-Chen Lo, Jhao-Yin Li, Mi-Yen Yeh, Shou-De Lin and Jian Pei
Data Mining and Knowledge Discovery
DOI 10.1007/s10618-013-0330-1

Being able to discover the uniqueness of an individual is a meaningful task in social network analysis. This paper proposes two novel problems in social network analysis: how to identify the uniqueness of a given query vertex, and how to identify a group of vertices that can mutually identify each other. We further propose intuitive yet effective methods to identify the uniqueness identification sets and the mutual identification groups of different properties. We further conduct an extensive experiment on both real and synthetic datasets to demonstrate the effectiveness of our model.

Spatio–temporal random fields: compressible representation and distributed estimation

Nico Piatkowski, Sangkyun Lee and Katharina Morik
Machine Learning
DOI 10.1007/s10994-013-5399-7

Modern sensing technology allows us enhanced monitoring of dynamic activities in business, traffic, and home, just to name a few. The increasing amount of sensor measurements, however, brings us the challenge for efficient data analysis. This is especially true when sensing targets can interoperate—in such cases we need learning models that can capture the relations of sensors, possibly without collecting or exchanging all data. Generative graphical models namely the Markov random fields (MRF) fit this purpose, which can represent complex spatial and temporal relations among sensors, producing interpretable answers in terms of probability. The only drawback will be the cost for inference, storing and optimizing a very large number of parameters—not uncommon when we apply them for real–world applications.

In this paper, we investigate how we can make discrete probabilistic graphical models practical for predicting sensor states in a spatio–temporal setting. A set of new ideas allows keeping the advantages of such models while achieving scalability. We first introduce a novel alternative to represent model parameters, which enables us to compress the parameter storage by removing uninformative parameters in a systematic way. For finding the best parameters via maximum likelihood estimation, we provide a separable optimization algorithm that can be performed independently in parallel in each graph node. We illustrate that the prediction quality of our suggested method is comparable to those of the standard MRF and a spatio–temporal k–nearest neighbor method, while using much less computational resources.

Table of Contents – Part III

Ensembles

Statistical Learning

Semi-supervised Learning

Unsupervised Learning

Subgroup Discovery, Outlier Detection and Anomaly Detection

Privacy and Security

Data Mining and Constraint Solving

Evaluation

Applications

Medical Applications

Nectar Track

Demo Track

AR-Boost: Reducing Overfitting by a Robust Data-Driven Regularization Strategy

Baidya Nath Saha[1], Gautam Kunapuli[2], Nilanjan Ray[3],
Joseph A. Maldjian[1], and Sriraam Natarajan[4]

[1] Wake Forest School of Medicine, USA
{bsaha,maldjian}@wakehealth.edu
[2] University of Wisconsin-Madison, USA
kunapuli@wisc.edu
[3] University of Alberta, Canada
nray1@ualberta.ca
[4] Indiana University, USA
natarasr@indiana.edu

Abstract. We introduce a novel, robust data-driven regularization strategy called Adaptive Regularized Boosting (AR-Boost), motivated by a desire to reduce overfitting. We replace AdaBoost's hard margin with a regularized soft margin that trades-off between a larger margin, at the expense of misclassification errors. Minimizing this regularized exponential loss results in a boosting algorithm that *relaxes the weak learning assumption further*: it can use classifiers with error greater than $\frac{1}{2}$. This enables a natural extension to multiclass boosting, and further reduces overfitting in both the binary and multiclass cases. We derive bounds for training and generalization errors, and relate them to AdaBoost. Finally, we show empirical results on benchmark data that establish the robustness of our approach and improved performance overall.

1 Introduction

Boosting is a popular method for improving the accuracy of a classifier. In particular, AdaBoost [1] is considered the most popular form of boosting and it has been shown to improve the performance of base learners both theoretically and empirically. The key idea behind AdaBoost is that it constructs a strong classifier using a set of weak classifiers [2,3]. While AdaBoost is quite powerful, there are two major limitations: (1) if the base classifier has a misclassification error of greater than 0.5, generalization decreases, and (2) it suffers from overfitting with noisy data [4,5].

The first limitation can become severe in multiclass classification, where the error rate of random guessing is $\frac{C-1}{C}$, where C is the number of classes [6]. AdaBoost requires weak classifiers to achieve an error rate less than 0.5, which can be problematic in multiclass classification. The second limitation of overfitting occurs mainly because weak classifiers are unable to capture "correct" patterns inside noisy data. Noise can be introduced into data by two factors – (1) mislabeled data, or (2) limitation of the hypothesis space of the base classifier [7]. During training, AdaBoost concentrates on learning difficult data patterns accurately,

H. Blockeel et al. (Eds.): ECML PKDD 2013, Part III, LNAI 8190, pp. 1–16, 2013.

and potentially distorts the optimal decision boundary. AdaBoost maximizes the "hard margin", namely the smallest margin of those noisy data patterns and consequently *the margin of other data points may decrease significantly*. Different regularization strategies such as early stopping, shrinking the contribution of the individual weak classifiers, and soft margins, have been proposed [2,4,5,7,8,9,10] to combat this issue.

AdaBoost's use of a hard margin increases the penalty exponentially for larger negative margins; this further increases error due to outliers. We propose an approach that combines early convergence with a soft margin by introducing a regularization term inside the exponential loss function. In every boosting round, the regularization term vanishes only if the weak classifier chosen at the current stage classifies the observations correctly. We derive a modified version of the AdaBoost algorithm by minimizing this regularized loss function and this leads to *Adaptive Regularized Boosting* (**AR-Boost**).

We show that choosing optimal values of a data-driven regularized penalty translates to the selection of optimal weights of the misclassified samples at each boosting iteration. These optimal weights force the weak classifiers to correctly label misclassifications in the previous stage. Consequently, AR-Boost *converges faster* than AdaBoost, and is also *more robust* to outliers. Finally, the proposed regularization allows boosting to employ weak classifiers even if their *error rate is greater than* 0.5. This is especially *suited to the multiclass setting*, where the permissible error is $\frac{C-1}{C} > \frac{1}{2}$. This serves as another significant motivation for the development of this approach.

Many properties that motivate this approach are *controlled by the user through tuning* a single regularization parameter $\rho > 1$, and this parameter determines how much differently AR-Boost behaves, compared to AdaBoost. The parameter ρ softens the margin, making our approach more robust to outliers. This is because it does not force classification of outliers according to their (possibly) incorrect labels, and thus does not distort the optimal decision boundary. Instead, it allows *a larger margin at the expense of some misclassification error*. To better understand this, consider the example presented in Figure 1. When the data is noisy, AdaBoost will still aim to classify the noisy example into one of the classes; our approach instead avoids this, leading to a more robust decision boundary. This added robustness allows for better generalization (shown in the bottom row of Figure 1). In addition to an empirical demonstration of this approach's success, we also derive theoretical bounds on the training and generalization error.

The rest of the paper is organized as follows. After reviewing existing work on boosting in Section 2, we describe binary AR-Boost in Section 3, and provide justification for our choice of regularization. In Section 4, we investigate the theoretical properties of our approach by deriving training and generalization error bounds. We describe the multiclass extension of AR-Boost in Section 5. In Section 6, we investigate the empirical properties of binary and multiclass AR-Boost, and compare their performance to some well-known regularized boosting approaches, and conclude in Section 7.

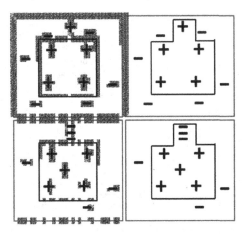

Fig. 1. Decision boundary made by the decision stumps (linear thresholds) used as weak classifiers in AR-Boost (*left column*) and AdaBoost (*right column*) on training (*top row*) and test (*bottom row*) dataset.

2 Background and Related Work

For training data $(\mathbf{x}_i, y_i)$, $i = 1, \ldots, n$, we assume that $\mathbf{x}_i \in \mathbb{R}^p$, $y_i \in \{-1, 1\}$ for binary classification, and $y_i \in \{1, \ldots, C\}$, for C-class classification. AdaBoost learns a strong classifier $f(\mathbf{x}) = \mathsf{sign}\left(\sum_{t=1}^{T} \alpha_t h_t(\mathbf{x})\right)$, by combining weak classifiers in an iterative manner [2]. Here, α_t is the weight associated with the weak classifier $h_t(\cdot)$. The value of α_t is derived by minimizing an exponential loss function: $L(y, f(\mathbf{x})) = \exp(-y f(\mathbf{x}))$.

AdaBoost is prone to overfitting and several strategies were developed to address this issue. Mease and Wyner [4] experimentally demonstrated that boosting often suffers from overfitting run for a large number of rounds. Model selection using Akaike or Bayesian information criteria (AIC/BIC) [11,12] achieved moderate success in addressing overfitting. Hastie *et al.*, [2] proposed ϵ-Boost where they regularize by shrinking the contribution of each weak classifier: $f(\mathbf{x}) = \mathsf{sign}\left(\sum_{t=1}^{T} \nu \alpha_t h_t(\mathbf{x})\right)$. More shrinkage (smaller ν) increases training error over AdaBoost for the same number of rounds, but reduces test error.

Jin *et al.*, [7] proposed Weight-Boost, which uses input-dependent regularization that combines the weak classifier with an instance-dependent weight factor: $f(\mathbf{x}) = \mathsf{sign}\left(\sum_{t=1}^{T} \exp(-|\beta f_{t-1}(\mathbf{x})|)\alpha_t h_t(\mathbf{x})\right)$. This trades-off between the weak classifier at the current iteration and the meta-classifier from previous iterations. The factor $\exp(-|\beta f_{t-1}(\mathbf{x})|)$ only considers labels provided by $h_t(\mathbf{x})$ when the previous meta-classifier f_{t-1} is not confident on its decision. Xi *et al.*, [13] minimized an L_1-regularized exponential loss $L(y, f(\mathbf{x})) = \exp(-y f(\mathbf{x}) + \beta \|\alpha\|_1)$, $\beta > 0$, which provides sparse solutions and early stopping. Rätsch *et al.*, [5,9] proposed a weight-decay method, in which they softened the margin by introducing a *slack variable* $\xi = \left(\sum_{t=1}^{T} \alpha_t h_t(\mathbf{x})\right)^2$ in the exponential loss function

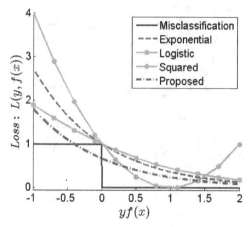

Fig. 2. Common loss functions for binary classification compared with the proposed loss $L(y, f(\mathbf{x})) = \exp(-yf(\mathbf{x}) - \lambda|y - h_t(\mathbf{x})|)$, where h_t is the most recent weak learner.

$L(y, f(\mathbf{x})) = \exp(-yf(\mathbf{x}) - C\xi)$, $C \geq 0$. They found that the asymptotic margin distribution for AdaBoost with noisy data is very similar to that of SVMs [14]; analogous to SVMs, "easy" examples do not contribute to the model and only "difficult" patterns with small margins are useful.

3   AR-Boost for Binary Classification

For any condition π, let $\delta[\![\pi]\!] = 1$, if π holds, and 0 otherwise. AdaBoost minimizes an exponential loss function, $L(y, f(\mathbf{x})) = \exp(-yf(\mathbf{x}))$. The misclassification loss, $L(y, f(\mathbf{x})) = \delta[\![yf(\mathbf{x}) < 0]\!]$ penalizes only the misclassified examples (with $yf(\mathbf{x}) < 0$) with an exact penalty of 1. Other loss functions (see Figure 2) attempt to overestimate the discontinuous misclassification loss with continuous/differentiable alternatives. Of these, the squared loss, $L(y, f(\mathbf{x})) = (y - f(\mathbf{x}))^2$ does not decrease monotonically with increasing margin $yf(\mathbf{x})$. Instead, for $yf(\mathbf{x}) > 0$ it increases quadratically, with increasing influence from observations that are correctly classified with increasing certainty. This significantly reduces the relative influence misclassified examples.

While exponential loss is monotonically decreasing, it penalizes larger misclassified margins exponentially, is exponentially large for these outliers and leads to worse misclassification rates [2]. This motivates the novel loss function,

$$L(y, f(\mathbf{x})) = \exp(-yf(\mathbf{x}) - \lambda|y - h_t(\mathbf{x})|), \tag{1}$$

where $\lambda > 0$ and $h_t(\cdot)$ is the weak classifier chosen at the current step, t[1]. The additional term in the loss function $|y - h_t(\mathbf{x})|$ acts as a regularizer, in

---

[1] As the loss function incorporates the weak classifier from the last round h_t, it should be written $L_t(y, f(\mathbf{x}))$; we drop the subscript t from L to simplify notation, as the dependence of the loss on t is apparent from the context.

Algorithm 1. AR-Boost for Binary Classification

---

input: $\lambda = \frac{1}{2}\log\rho$ {select $\rho > 1$ such that $\lambda > 0$}

$w_i^1 = \frac{1}{n}$, $i = 1, \ldots, n$ {initialize example weight distribution uniformly}

for $t = 1$ **do**

    $h_t = \mathsf{WeakLearner}\left((\mathbf{x}_i, y_i, w_i^t)_{i=1}^n \right)$ {train weak classifier h_t using weights w_i^t}

    $\varepsilon_t = \sum_{i=1}^n w_i^t \delta[\![\, y_i \neq h_t(\mathbf{x}_i)\,]\!]$ {sum of weights of examples misclassified by h_t}

    if $\varepsilon_t \geq \frac{\rho}{\rho+1}$ **then**

        $T = t - 1$; **break.**

    else

        $\alpha_t = \frac{1}{2}\log\frac{\rho(1-\varepsilon_t)}{\varepsilon_t}$ {update α_t with adaptive regularization parameter ρ}

        $w_i^{t+1} = \frac{w_i^t \exp\left(2\alpha_t\, \delta[\![\, y_i \neq h_t(\mathbf{x}_i)\,]\!]\right)}{Z_t}$ {update weights with normalization Z_t}

    end if

end for

output: $f(\mathbf{x}) = \mathsf{sign}\left(\sum_{t=1}^T \alpha_t h_t(\mathbf{x}) \right)$ {final classifier}

---

conjunction with the margin term $yf(\mathbf{x})$. This term does not resemble typical norm-based regularizations that control the structure of the hypothesis space, such as ℓ_1 or ℓ_2-norms. It behaves like a regularization term because, it controls the hypothesis space by relaxing the weak learning assumption in order to admit hypotheses that have error greater than $\frac{1}{2}$ into the boosting process.

At iteration t, the proposed loss function is the same as AdaBoost's loss function if the misclassification error is zero. However, the penalty associated with this loss is less than that of AdaBoost's loss if an example is misclassified (Figure 2). AdaBoost maximizes the hard margin, $\gamma = yf(\mathbf{x})$ without allowing any misclassification error, $\mathsf{e_{tr}} = \frac{1}{n}\sum_{i=1}^n \delta[\![\, y_i f(\mathbf{x}_i) < 0\,]\!]$. Inspired by SVMs [14], our function maximizes a soft margin, $\gamma = yf(\mathbf{x}) + \lambda|y - h_t(\mathbf{x})|$. Instead of enforcing outliers to be classified correctly, this modification allows for a larger margin *at the expense of some misclassification errors*, $\mathsf{e'_{tr}} = \frac{1}{n}\sum_{i=1}^n \delta[\![\, y_i f(\mathbf{x}_i) + \lambda|y_i - h_t(\mathbf{x}_i)| < 0\,]\!]$, and tries to avoid overfitting.

We derive a modified AdaBoost algorithm that we call *Adaptive Regularized Boosting* (AR-Boost); the general procedure of AR-Boost for binary classification is shown in Algorithm 1. The derivation of the updates is shown in Appendix A. AR-Boost finds the hypothesis weight, $\alpha_t = \frac{1}{2}\log\rho\frac{(1-\varepsilon_t)}{\varepsilon_t}$, with $\lambda = \frac{1}{2}\log\rho > 0$. When $\rho = 1$, AR-Boost is the same as AdaBoost. As mentioned earlier, the WeakLearner is capable of learning with classifiers with an error rate $\varepsilon_t > 0.5$. The extent to which this error is tolerated is further discussed in the next section. For all learners, we have $\alpha_t = \lambda + \frac{1}{2}\log\frac{(1-\varepsilon_t)}{\varepsilon_t}$. This is equivalent to computing $\alpha_t^{\mathsf{AR\text{-}Boost}} = \lambda + \alpha_t^{\mathsf{AdaBoost}}$.

One additional advantage of this regularized loss function is that the penalty for negative margins can be adjusted after observing the classifier performance. Accordingly, we determine the value of λ or ρ through cross validation, by choosing the parameter for which the average misclassification error, $\mathsf{e'_{tr}} = \frac{1}{n}\sum_{i=1}^n \delta[\![\, -y_i f(\mathbf{x}_i) - \lambda|y_i - h_t(\mathbf{x}_i)| < 0\,]\!]$ is smallest. For instance, in Figure 3

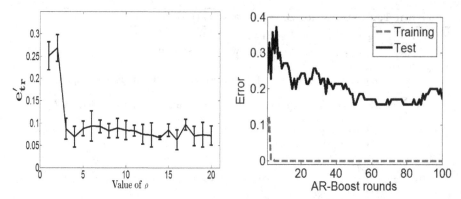

Fig. 3. PIMA Indian Diabetes data set (**left**) 5-fold cross validation with smallest cross-validation error for $\rho = 4$; (**right**) training and test errors over a number of boosting iterations for $\rho=4$

(left), we show AR-Boost's 5-fold cross-validation error for the PIMA Indian Diabetes data set [7]. The best value is $\rho = 4$, and the corresponding training and test error curves are shown in Figure 3 (right) for this choice of ρ. This behavior is similar to AdaBoost, in that even when the training error has been minimized, the test error continues to decrease.

We derive the multiclass version of AR-Boost in Section 5, which takes advantage of AR-Boost's ability to handle weaker classifiers. Before proceeding, we further analyze AR-Boost's ability to relax the weak-learner assumption.

3.1   Relaxing the Weak Learning Assumption of AdaBoost

At the t-th iteration, $\alpha_t = \frac{1}{2} \log \frac{\rho(1-\varepsilon_t)}{\varepsilon_t}$, $\rho > 1$. The hypothesis weight $\alpha_t > 0$ only when $\frac{\rho(1-\varepsilon_t)}{\varepsilon_t} > 1$. From this, it is immediately apparent that

$$\varepsilon_t < \frac{\rho}{1+\rho}, \tag{2}$$

and when $\rho = 1$, we have that $\varepsilon_t < 0.5$; this is the standard weak learning assumption that is used in AdaBoost. As we start increasing the value of $\rho > 1$, we can see that $\frac{\rho}{\rho+1} \to 1$ and AR-Boost is able to accommodate classifiers with $\varepsilon_t \in [0.5, \frac{\rho}{\rho+1})$. Thus, AR-Boost is able to *learn with weaker hypotheses than afforded by the standard weak learning assumption*; how weak these learners can be is controlled by the choice of ρ. This can be seen in Figure 4, which shows the AR-Boost objective values, and their minima plotted for various weak-learner errors ε. AR-Boost can handle weaker classifiers than AdaBoost, and assigns them increasingly lower weights $\alpha \to 0$, the weaker they are.

3.2   How Does Relaxing the Weak Learning Assumption Help?

Similar to Zhu *et al.*, [6] who illustrate that AdaBoost fails in the multiclass setting, we show how this can also happen in binary classification. We conduct

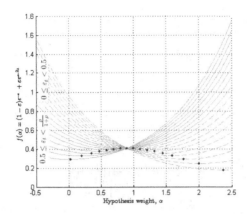

Fig. 4. Objective values of AR-Boost ($f(\alpha) = (1-\varepsilon)e^{-\alpha}+\varepsilon e^{\alpha-2\lambda}$) plotted as a function of α for different values of ε. The minimum of each objective is also shown by dot. The curves in dashed green represent classifiers that satisfy the weak learner assumption: $0 \le \varepsilon < 0.5$, while the curves in orange represent classifiers that exceed it: $0.5 \le \varepsilon < \frac{\rho}{\rho+1}$. These curves were plotted for $\rho = 6$.

an experiment with simple two-class data, where each example $\mathbf{x} \in \mathbb{R}^{10}$, and $x_{ij} \sim \mathcal{N}(0,1)$. The two classes are defined as,

$$c = \begin{cases} 1, & \text{if } 0 \le \sum x_j^2 < \chi^2_{10,1/2}, \\ -1 & \text{otherwise}, \end{cases}$$

where $\chi^2_{10,1/2}$ is the $(1/2)100\%$ quantile of the χ^2_{10} distribution. The training and test set sizes were 2000 and 10,000, with class sizes being approximately equal. We use decision stumps (single node decision tree) as weak learners.

Figure 5 (top row) demonstrates how AdaBoost sometimes fails in binary classification. Training and test errors remain unchanged over boosting rounds (Figure 5 top left). The error ε_t and the AdaBoost weights α_t for each round t are shown in Figure 5 top center, and right. The value of ε_t starts below $\frac{1}{2}$, and after a few iterations, it overshoots $\frac{1}{2}$ ($\alpha_t < 0$), then is quickly pushed back down to $\frac{1}{2}$ (Figure 5 top center). Now, once ε_t is equal to $\frac{1}{2}$, the weights of subsequent examples *are no longer updated* ($\alpha_t = 0$). Thus, no new classifiers are added to $f(\mathbf{x})$, and the *overall error rate remains unchanged*.

Unlike AdaBoost, AR-Boost *relaxes the weak learning assumption further*: it can use classifiers with error greater than $\frac{1}{2}$ as shown in bottom row of Figure 5. Here, both training and test error decrease with boosting iterations, which is what we would expect to see from a successful boosting algorithm. AR-Boost can successfully incorporate weak classifiers with error as large as $\varepsilon_t < \rho/(\rho+1)$ for binary classification as shown in Algorithm 1. Similar behavior holds for the C-class case, which can incorporate classifiers with error up to $\varepsilon < \rho(C-1)/(\rho(C-1)+1)$ as we show below, in Algorithm 2. This limiting value of ε_t is not artificial, it follows naturally by minimizing the proposed novel regularized

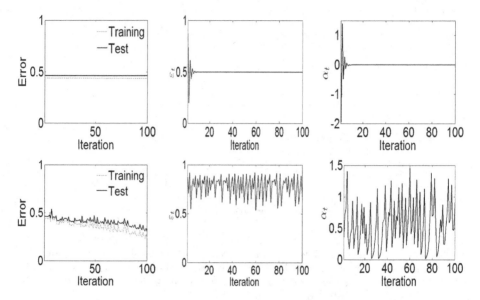

Fig. 5. Comparing AdaBoost and AR-Boost on a simple two-class simulated example, with decision stumps (single node decision trees) used as weak learners. The top row (AdaBoost) and bottom row (AR-Boost) show (**left**) training and test errors; (**center**) weak classifier error at round t, ε_t; and (**right**) example weight at round t, α_t respectively for AdaBoost and AR-Boost. For AR-Boost, $\rho = 5$.

exponential loss function. It provides softer margins and smaller penalties for larger negative margins than AdaBoost (Figure 2).

4   Analysis of AR-Boost

We now analyze the behavior of AR-Boost via upper bounds on the training and generalization error and compare these bounds with AdaBoost.

4.1   Training Error

We can formally analyze the behavior of the algorithm by deriving an upper bound on the training error. To do so, we first state the following.

Lemma 1. *At the t-th iteration, define the* goodness *γ_t of the current weak learner $h_t(\mathbf{x})$ as how much better it does than the worst allowable error: $\varepsilon_t = \frac{\rho}{\rho+1} - \gamma_t$. The normalization Z_t of the weights w_i^{t+1} (in Algorithm 1) can be bounded by*

$$Z_t \leq \exp\left(-\frac{(\rho+1)^2}{2\rho}\gamma_t^2 + \frac{\rho^2-1}{2\rho}\gamma_t\right). \tag{3}$$

This Lemma is proved in Appendix C. Now, we can state the theorem formally.

Theorem 1. *If the goodness of weak learners at every iteration is bounded by $\gamma_t \geq \gamma$, the training error of AR-Boost, e_{tr}, after T rounds is bounded by*

$$e_{tr} \leq \prod_{t=1}^{T} \exp\left(-\frac{(\rho+1)^2}{2\rho}\gamma^2 + \frac{\rho^2-1}{2\rho}\gamma\right). \tag{4}$$

If $\gamma \geq \frac{\rho-1}{\rho+1}$, the training error exponentially decreases.

Proof. After T iterations, the example weights w_i^{T+1} can be computed using step 9 of Algorithm 1. By recursively unraveling this step, and recalling that $w_i^1 = \frac{1}{n}$, we have

$$w_i^{T+1} = \frac{e^{-y_i f(\mathbf{x}_i)}}{n \prod_{t=1}^{T} Z_t}. \tag{5}$$

The training error is $e_{tr} = \frac{1}{n}\sum_{i=1}^{n}\delta[\![y_i \neq f(\mathbf{x}_i)]\!]$. For all misclassified examples, we can bound the training error by

$$e_{tr} \leq \frac{1}{n}\sum_{i=1}^{n}e^{-y_i f(\mathbf{x}_i)} = \prod_{t=1}^{T}Z_t,$$

where we use (5) and the fact that $\sum_{i=1}^{n} w_i^{t+1} = 1$. The bound follows from Lemma 1 and the fact that $\gamma_t \geq \gamma$.   □

First, note that when $\rho = 1$, the training error bound is exactly the same as that of AdaBoost. Next, to understand the behavior of this upper bound, consider Figure 6. The bound of AdaBoost is shown as the dotted line, while the remaining curves are the AR-Boost training error for various values of $\rho > 1$. It is evident that it is possible to *exponentially shrink the training error* for increasing T, as long as the *goodness of the weak learners* is at least $\gamma = \frac{\rho-1}{\rho+1}$, which means that the error at each iteration, $\varepsilon_t \leq \frac{1}{\rho+1}$.

4.2 Generalization Error

Given a distribution D over $X \times \{\pm 1\}$ and a training sample S drawn i.i.d. from D, Schapire *et al.*, [15] showed that the upper bound on the generalization error of AdaBoost is, with probability $1 - \delta$, $\forall \theta > 0$,

$$\Pr_D\left[\![yf(\mathbf{x}) \leq 0]\!\right] \leq \Pr_S\left[\![yf(\mathbf{x}) \leq \theta]\!\right] + O\left(\frac{1}{\sqrt{n}}\left(\frac{d\log^2(\frac{n}{d})}{\theta^2} + \log\frac{1}{\delta}\right)^{\frac{1}{2}}\right), \tag{6}$$

where d is the Vapnik-Chervonenkis (VC) dimension of the space of base classifiers. This bound depends on the training error $e_{tr}^{\theta} = \Pr_S\left[\![yf(\mathbf{x}) \leq \theta]\!\right]$ and is *independent* of the number of boosting rounds T.

Schapire *et al.*, explained AdaBoost's ability to avoid overfitting using the margin, $m = yf(\mathbf{x})$, the magnitude of which represents the measure of confidence

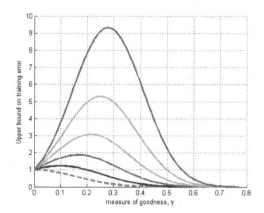

Fig. 6. The upper bound on the training error for various choices of ρ, as the goodness of the weak learners improves. AdaBoost ($\rho = 1$) is the dashed line. The behavior of AR-Boost for various values of $\rho > 1$ is similar to AdaBoost if $\gamma > \frac{\rho-1}{\rho+1}$, that is, the training error decreases exponentially as $T \to \infty$.

on the predictions of the base classifiers. The bound above shows that a large margin on the training set results in a superior bound on the generalization error. Since boosting minimizes the exponential loss $L(y, f(\mathbf{x})) = e^{-yf(\mathbf{x})}$, the margin is maximized. The result below shows that AR-Boost also behaves similarly to maximize the margin, $m = yf(\mathbf{x}) + \lambda|y - h_t(\mathbf{x})|$, where $\lambda > 0$.

Theorem 2. *At every iteration* $t = 1, \ldots, T$, *let the base learner produce classifiers with training errors* ε_t. *Then, for any* $\theta > 0$, *we have*

$$\mathbf{e}_{tr}^{\theta} = Pr_S[\![yf(\mathbf{x}) \le \theta]\!] \le \left(\sqrt{\rho^{1+\theta}} + \frac{1}{\sqrt{\rho^{1-\theta}}}\right)^T \prod_{t=1}^{T}\sqrt{\varepsilon_t^{1-\theta}(1 - \varepsilon_t)^{1+\theta}}. \quad (7)$$

Proof. If $yf(\mathbf{x}) \le \theta$, then $y\sum_{t=1}^{T}\alpha_t h_t(\mathbf{x}) \le \theta\sum_{t=1}^{T}\alpha_t$ and $\exp(-y\sum_{t=1}^{T}\alpha_t h_t(\mathbf{x}) + \theta\sum_{t=1}^{T}\alpha_t) \ge 1$. Using this, we have that

$$Pr_S[\![yf(\mathbf{x}) \le \theta]\!] \le E_S\left[\!\left[\exp\left(-y\sum_{t=1}^{T}\alpha_t h_t(\mathbf{x}) + \theta\sum_{t=1}^{T}\alpha_t\right)\right]\!\right]$$

$$= \frac{1}{n}\exp\left(\theta\sum_{t=1}^{T}\alpha_t\right)\sum_{i=1}^{n}\exp\left(-y_i f(\mathbf{x}_i)\right).$$

Using the value of α_t, and equations (5) and (10) gives us the result. $\qquad\square$

As before, we immediately note that when $\rho = 1$, this bound is exactly identical to the bound derived by Schapire *et al.* [15]. To further analyze this bound, assume that we are able to produce classifiers with $\varepsilon_t \le \frac{\rho}{\rho+1} - \gamma$, with some goodness $\gamma > 0$. We know from Theorem 1 that the upper bound of training reduces exponentially with T. Then, we can simplify the upper bound in (7) to

$$Pr_S[\![yf(\mathbf{x}) \le \theta]\!] \le \left(\frac{1}{\rho} - \frac{(\rho+1)}{\rho}\gamma\right)^{\frac{(1-\theta)T}{2}}(1 + (\rho+1)\gamma)^{\frac{(1+\theta)T}{2}}$$

Algorithm 2. AR-Boost for multiclass Classification

input: $\lambda = \frac{C-1}{2} \log \rho$ {select $\rho > 1$ such that $\lambda > 0$}
$w_i^1 = \frac{1}{n}$, $i = 1, \ldots, n$ {initialize example weight distribution uniformly}
for $t = 1$ **do**
    $H_t = \mathsf{WeakLearner}\left((\mathbf{x}_i, \mathbf{y}_i, w_i^t)_{i=1}^n \right)$ {train weak classifier H_t using weights w_i^t}

$$\varepsilon_t = \sum_{i=1}^{n} w_i^t \delta[\![c_i \neq H_t(\mathbf{x}_i)]\!] \text{ {sum of weights of examples misclassified by } } H_t \}$$

    if $\varepsilon_t \geq \frac{\rho(C-1)}{\rho(C-1)+1}$ **then**
      $T = t - 1$; **break.**
    else

$$\alpha_t = \log \frac{\rho(1-\varepsilon_t)}{\varepsilon_t} + \log(C-1) \text{ {update } } \alpha_t \text{ with adaptive parameter } \rho\}$$

$$w_i^{t+1} = \frac{w_i^t \exp\left(\alpha_t \, \delta[\![c_i \neq H_t(\mathbf{x}_i)]\!] \right)}{Z_t} \text{ {update weights with normalization } } Z_t\}$$

    end if
end for
output: $f(\mathbf{x}) = \arg\max_k \left(\sum_{t=1}^{T} \alpha_t \delta[\![H_t(\mathbf{x}) = k]\!] \right)$ {final classifier}

Hence, $\forall \theta < \frac{1}{2}\frac{\rho-1}{\rho+1}$, and $\forall \gamma > \frac{\rho-1}{\rho+1}$, we have that $\mathsf{Pr}_S[\![yf(\mathbf{x}) \leq \theta]\!] \to 0$ as $T \to \infty$. This suggests that $\lim_{T\to\infty} \min_i y_i f(\mathbf{x}_i) \geq \gamma$, showing that better weak hypotheses, with greater γ, provide larger margins.

5 AR-Boost for Multiclass Classification

In the C-class classification case, each data point can belong to one of C classes i.e., the label of the i-th data point $c_i \in 1, \ldots, C$. For this setting, we can recode the output as a C-dimensional vector $\mathbf{y}_i$ [6,16] whose entries are such that $y_i^k = 1$, if $c_i = k$; else $y_i^k = -\frac{1}{C-1}$, if $c_i \neq k$. The set of C possible output vectors for a C-class problem is denoted by Υ. Given the training data, we wish to find a C-dimensional vector function $\mathbf{f}(\mathbf{x}) = \left(f^1(\mathbf{x}), \ldots, f^C(\mathbf{x}) \right)'$ such that

$$\mathbf{f}(\mathbf{x}) = \begin{array}{c} \arg\min\limits_{\mathbf{f}} \ \sum_{i=1}^{n} L(\mathbf{y}_i, \mathbf{f}(\mathbf{x}_i)) \\ \text{subject to } \sum_{k=1}^{C} f^k(\mathbf{x}) = 0. \end{array}$$

We consider $\mathbf{f}(\mathbf{x}) = \sum_{t=1}^{T} \alpha_t \mathbf{h}_t(\mathbf{x})$, where $\alpha_t \in \mathbb{R}$ are coefficients, and $\mathbf{h}_t(\mathbf{x})$ are basis functions. These functions $\mathbf{h}_t(\mathbf{x}) : X \to \Upsilon$ are required to satisfy the symmetric constraint: $\sum_{k=1}^{C} h_t^k(\mathbf{x}) = 0$. Finally, every $\mathbf{h}_t(\mathbf{x})$ is associated with a multiclass classifier $H_t(\mathbf{x})$ as, $h_t^k(\mathbf{x}) = 1$, if $H_t(\mathbf{x}) = k$; else $h_t^k(\mathbf{x}) = -\frac{1}{C-1}$, if $H_t(\mathbf{x}) \neq k$, such that solving for $\mathbf{h}_t$ is equivalent to finding the multiclass classifier $H_t : X \to \{1, \ldots, C\}$; in turn, $H_t(\mathbf{x})$ can generate $h_t(\mathbf{x})$ resulting in a one-to-one correspondence.

    The proposed multiclass loss function for AR-Boost is $L(\mathbf{y}, \mathbf{f}(\mathbf{x})) = \exp\left(-\frac{1}{C}\mathbf{y}'\mathbf{f}(\mathbf{x}) - \frac{\lambda}{C}\|\mathbf{y} - \mathbf{h}_t(\mathbf{x})\|_1 \right)$, which extends the binary loss function to the multiclass case discussed in Section 3. This loss is the natural generalization of the exponential loss for binary classification proposed by Zhu et al., [6] as

Table 1. Description of binary and multiclass data sets used in the experiments

DATASET	#EXAMPLES	#TRAIN	#TEST	#FEATURES	#CLASSES
ionosphere	351	238	113	341	2
german	1000	675	325	20	2
diabetes	768	531	237	8	2
wpbc	198	141	57	30	2
wdbc	569	423	146	30	2
spambase	4601	3215	1386	58	2
vowel	990	528	462	13	11
pen digits	10992	7494	3498	16	10
letter	20000	16000	4000	16	26
thyroid	215	160	55	5	3
satimage	6435	4435	2000	36	7
segmentation	2310	210	2100	19	7

Stage-wise Additive Modeling that uses a multiclass Exponential loss function (SAMME). The general procedure of multiclass AR-Boost is shown in Algorithm 2 and the details of the derivation are shown in Appendix B. AR-Boost finds the feature weight, $\alpha_t = \log \rho \frac{1-\varepsilon_t}{\varepsilon_t} + \log(C-1)$, with $\rho > 1$. When $\rho = 1$, the AR-Boost algorithm becomes the SAMME algorithm.

6   Experimental Results

We compare AR-Boost with AdaBoost and four other regularized boosting algorithms: ϵ-boost [2], L_1-regularized boost [13], AdaBoost$_{reg}$ [5] and Weight-Boost [7]. We chose 12 data sets (6 binary and 6 multiclass problems) (see Table 1) from the UCI machine learning repository [17] that have been previously used in literature [6,7]. For all the algorithms, the maximum training iterations is set to 100. We also use classification and regression trees (CART) [2] as the baseline algorithm. We compared multiclass AR-Boost discussed in Section 5 to two commonly used algorithms: AdaBoost.MH [18] and SAMME [6]. The parameter ρ was tuned through cross validation.

Figure 7 (left) shows the results of binary classification across 6 binary classification tasks. The baseline decision tree has the worst performance and AdaBoost improves upon trees. Using regularization, however, gives different levels of improvement over AdaBoost. Our AR-Boost approach yields further improvements compared to the other regularized boosting methods on all data sets except spambase. On spambase, AR-Boost produces test error of 4.91%, while Weight Decay and WeightBoost give errors of 4.5% and 4.2% respectively. These results demonstrate that performance is significantly improved for smaller data sets (for example, the improvement is nearly 35% for wpbc). This shows that AR-Boost is able to reduce overfitting, significantly at times, and achieves better generalization compared to state-of-the-art regularized boosting methods on binary classification problems.

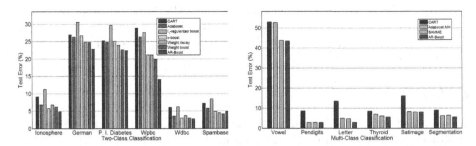

Fig. 7. Misclassification errors of binary AR-Boost (**left**) and multiclass AR-Boost (**right**) compared with other AdaBoost algorithms on UCI data sets (see Table 1).

Similar results are observed in the multiclass setting (Figure 7, right). Boosting approaches are generally an order of magnitude better than the baseline; AR-Boost is comparable (in 3 tasks) or better (in 3 tasks) than SAMME, the best of the other methods. The most interesting result is found on the `vowel` dataset; both AR-Boost and SAMME achieve around 40% test error, which is almost 15% better than AdaBoost.MH. This demonstrates that our approach can seamlessly extend to the multiclass case as well. Again, similar to the binary case, AR-Boost improves robustness to overfitting, especially for smaller data sets (for example, nearly 33% improvement for the `thyroid` data set).

Finally, we investigate an important property of AR-Boost: robustness to outliers, which is a prime motivation of this approach. In this experiment, we introduced different levels of label noise (10%, 20%, 30%) in the binary classification tasks, and compared AR-Boost to the baseline and AdaBoost. We randomly flip the label to the opposite class for random training examples for the benchmark data. Increasing levels of noise: 10%, 20% and 30% were introduced, with those probabilities of flipping a label. AR-Boost exhibits superior performance (Figure 8) at all noise levels. The key result that needs to be emphasized is that at higher noise levels, the difference becomes more pronounced. This suggests that AR-Boost is reasonably *robust* to increasing noise levels, while performance decreases for other approaches, sometimes drastically. Thus, AR-Boost can learn successfully in various noisy settings, and also with small data sets.

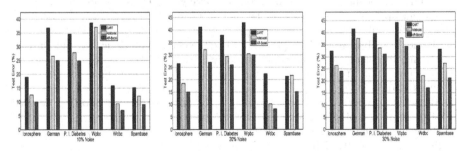

Fig. 8. Misclassification errors of CART, AdaBoost and AR-Boost on UCI datasets with 10%, 20% ad 30% label noise.

These results taken together demonstrate that AR-Boost addresses two limitations in AdaBoost, and other regularized boosting approaches to date, which are also two motivating objectives: robustness to noise, and ability to effectively handle multiclass classification.

7   Conclusion

We proposed Adaptive Regularized Boosting (AR-Boost) that appends a regularization term to AdaBoost's exponential loss. This is a data-driven regularization method, which softens the hard margin of AdaBoost by assigning a smaller penalty to misclassified observations at each boosting round. Instead of forcing outliers to be labelled correctly, AR-Boost allows a larger margin; while this comes at the cost of some misclassification errors, it improves robustness to noise. Compared to other regularized AdaBoost algorithms, AR-Boost uses weaker classifiers, and thus can be used in the multiclass setting. The upper bound of training and generalization error of AR-Boost illustrate that the error rate decreases exponentially with boosting rounds. Extensive experimental results show that AR-Boost outperforms state-of-the-art regularized AdaBoost algorithms for both binary and multiclass classification. It remains an interesting future direction to understand the use of such an approach in other problems such as semi-supervised learning and active learning.

Acknowledgements. Sriraam Natarajan (SN) gratefully acknowledges the support of Defense Advanced Research Projects Agency (DARPA) DEFT Program under Air Force Research Laboratory (AFRL) prime contract FA8750-13-2-0039. SN and Gautam Kunapuli gratefully acknowledge the support of Defense Advanced Research Projects Agency (DARPA) Machine Reading Program under Air Force Research Laboratory (AFRL) prime contract FA8750-09-C-0181. Any opinions, findings, and conclusion expressed in this material are those of the authors and do not necessarily reflect the view of the DARPA, AFRL, or the US government.

References

1. Freund, Y., Schapire, R.E.: A decision-theoretic generalization of on- line learning and an application to boosting. Journal of Computer and System Sciences 55, 119–139 (1997)
2. Hastie, T., Tibshirani, R., Friedman, J.: The Elements of Statistical Learning: Data Mining, Inference, and Prediction, 2nd edn. Springer (2009)
3. Schapire, R.E., Freund, Y.: Boosting Foundations and Algorithms, 1st edn. MIT Press (2012)
4. Mease, D., Wyner, A.: Evidence contrary to the statistical view of boosting. Journal of Machine Learning Research, 131–156 (1998)
5. Rätsch, G., Onoda, T.: Müller, K.R.: An improvement of AdaBoost to avoid overfitting. In: Proc. ICONIP, pp. 506–509 (1998)
6. Zhu, J., Zhou, H., Rosset, S., Hastie, T.: Multi-class AdaBoost. Statistics and Its Inference 2, 349–360 (2009)

7. Jin, R., Liu, Y., Si, L., Carbonell, J., Hauptmann, A.G.: A new boosting algorithm using input-dependent regularizer. In: Proc. ICML, pp. 615–622 (2003)
8. Friedman, J., Hastie, T., Tibshirani, R.: Additive logistic regression: a statistical view of boosting. The Annals of Statistics 28, 2000 (1998)
9. Rätsch, G., Onoda, T., Müller, K.R.: Soft margins for AdaBoost. In: Machine Learning, pp. 287–320 (2000)
10. Sun, Y., Li, J., Hager, W.: Two new regularized AdaBoost algorithms. In: Proc. ICMLA (2004)
11. Bühlmann, P., Hothorn, T.: Boosting algorithms: Regularization, prediction and model fitting. Statistical Science, 131–156 (2007)
12. Kaji, D., Watanabe, S.: Model selection method for adaBoost using formal information criteria. In: Köppen, M., Kasabov, N., Coghill, G. (eds.) ICONIP 2008, Part II. LNCS, vol. 5507, pp. 903–910. Springer, Heidelberg (2009)
13. Xi, Y.T., Xiang, Z.J., Ramadge, P.J., Schapire, R.E.: Speed and sparsity of regularized boosting. In: Proc. AISTATS, pp. 615–622 (2009)
14. Vapnik, V.: The Nature of Statistical Learning Theory. Springer (2000)
15. Schapire, R.E., Freund, Y., Bartlett, P., Lee, W.S.: Boosting the margin: A new explanation for the effectiveness of voting methods (1998)
16. Lee, Y., Lin, Y., Wahba, G.: Multicategory support vector machines, theory, and application to the classification of microarray data and satellite radiance data. Journal of the American Statistical Association 99, 67–81 (2004)
17. Bache, K., Lichman, M.: UCI Machine Learning Repository (2013), http://archive.ics.uci.edu/ml
18. Schapire, R.E., Singer, Y.: Improved boosting algorithms using confidence-rated predictions. In: Machine Learning, vol. 37, pp. 297–336 (1999)

Appendix

A   Derivation of Binary AR-Boost

At the t-th iteration, the loss function is $L(y, f(\mathbf{x})) = \exp\left(-yf(\mathbf{x}) - \lambda|y - h_t(\mathbf{x})|\right)$. Let $f_t(\mathbf{x}) = f_{t-1}(\mathbf{x}) + \alpha_t h_t(\mathbf{x})$ be the strong classifier composed of first t classifiers. We have that $\alpha_t = \arg\min_\alpha \sum_{i=1}^n \exp\left(-y_i f_{t-1}(x_i) - \alpha y_i h_t(\mathbf{x}_i) - \lambda|y_i - h_t(\mathbf{x}_i)|\right)$. Using the fact that $w_i^t = \exp\left(-y_i f_{t-1}(x_i)\right)$, we have

$$\alpha_t = \arg\min_\alpha \sum_{i=1}^n w_i^t \exp\left(-\alpha y_i h_t(\mathbf{x}_i) - \lambda|y_i - h_t(\mathbf{x}_i)|\right)$$

$$= \arg\min_\alpha \sum_{i:y_i=h_t(\mathbf{x}_i)} w_i^t e^{-\alpha} + \sum_{i:y_i \neq h_t(\mathbf{x}_i)} w_i^t e^{\alpha-2\lambda}.$$

Letting $\varepsilon_t = \sum_{i=1}^n w_i^t \delta[\![\, y_i \neq h_t(\mathbf{x}_i)\,]\!]$, and observing that w_i^t are normalized ($\sum_{i=1}^n w_i^t = 1$), we have $\alpha_t = \arg\min_\alpha e^{-\alpha}(1 - \varepsilon_t) + e^{\alpha-2\lambda}\varepsilon_t$. This gives us

$$\alpha_t = \frac{1}{2}\log\frac{\rho(1 - \varepsilon_t)}{\varepsilon_t}, \tag{8}$$

where we use $\lambda = \frac{1}{2}\log\rho$. Now, observing from above that $w_i^{t+1} = \exp\left(-y_i f_t(x_i)\right) = w_i^t \exp\left(-y_i \alpha_t h_t(\mathbf{x}_i)\right)$ and using the fact that $-y_i h_t(\mathbf{x}_i) = 2\delta[\![\, y_i \neq h_t(\mathbf{x}_i)\,]\!] - 1$, we have

$$w_i^{t+1} = w_i^t \exp\left(2\alpha_t \delta[\![\, y_i \neq h_t(\mathbf{x}_i)\,]\!]\right). \tag{9}$$

The term $e^{-\alpha_t}$ is dropped as it appears in $w_i^{t+1} \; \forall i$ and cancels during normalization. Then w_i^{t+1} is expressed in terms of w_i^t, y, α_t and h_t. Subsequently, the summation breaks into two parts: $y = h_t$ and $y \neq h_t$ and finally it uses (8), to get the final expression for Z_t:

$$Z_t = \sum_{i=1}^{n} w_i^{t+1} = \left(\frac{1}{\sqrt{\rho}} + \sqrt{\rho}\right)\sqrt{\varepsilon_t(1 - \varepsilon_t)}. \tag{10}$$

B Derivation of Multiclass AR-Boost

At the t-th iteration, the loss function is $L(\mathbf{y}, \mathbf{f}(\mathbf{x})) = \exp\left(-\frac{1}{C}\mathbf{y}'\mathbf{f}(\mathbf{x}) - \frac{\lambda}{C}\|\mathbf{y} - \mathbf{h}_t(\mathbf{x})\|_1\right)$. Let $\mathbf{f}_t(\mathbf{x}) = \mathbf{f}_{t-1}(\mathbf{x}) + \alpha_t\mathbf{h}_t(\mathbf{x})$ be the strong classifier composed of first t classifiers. We need $\alpha_t = \arg\min_\alpha \sum_{i=1}^{n} \exp\left(-\frac{1}{C}\mathbf{y}_i'(\mathbf{f}_{t-1}(\mathbf{x}_i) + \alpha_t\mathbf{h}_t(\mathbf{x}_i)) + \frac{\lambda}{C}\|\mathbf{y}_i - \mathbf{h}_t(\mathbf{x}_i)\|_1\right)$. Analogous to the two class case, we have $w_i^t = \exp\left(-\frac{1}{C}\mathbf{y}_i'\mathbf{f}_{t-1}(\mathbf{x}_i)\right)$. Recall, that solving for $\mathbf{h}_t(\mathbf{x})$ is encoded as finding the multiclass classifier $H_t(\mathbf{x})$ that yields $\mathbf{h}_t(\mathbf{x})$. Thus, we have

$$\alpha_t = \arg\min_\alpha \sum_{i=1}^{n} w_i^t \exp\left(-\frac{\alpha}{C}\mathbf{y}_i'\mathbf{h}_t(\mathbf{x}_i) - \frac{\lambda}{C}\|\mathbf{y}_i - h_t(\mathbf{x}_i)\|_1\right)$$

$$= \arg\min_\alpha \sum_{i:c_i = H_t(\mathbf{x}_i)} w_i^t e^{-\frac{\alpha}{C-1}} + \sum_{i:c_i \neq H_t(\mathbf{x}_i)} w_i^t e^{\frac{\alpha}{(C-1)^2} - \frac{2\lambda}{C-1}}.$$

As before, we set $\varepsilon_t = \sum_{i=1}^{n} w_i^t \delta[\![c_i \neq H_t(\mathbf{x}_i)]\!]$ and we get $\hat{\alpha}_t = \frac{(C-1)^2}{C}\alpha_t$

$$\alpha_t = \log\frac{\rho(1 - \varepsilon_t)}{\varepsilon_t} + \log(C - 1), \tag{11}$$

where $\lambda = \frac{C-1}{2}\log\rho$. This allows us to write

$$w_i^{t+1} = \begin{cases} w_i^t e^{-\frac{C-1}{C}\alpha_t}, & \text{if } c_i = H_t(\mathbf{x}_i), \\ w_i^t e^{\frac{1}{C}\alpha_t} , & \text{if } c_i \neq H_t(\mathbf{x}_i). \end{cases} \tag{12}$$

After normalization, this weight above is equivalent to the weight used in Algorithm 2. Finally, it is simple to show that the output after T iterations, $\mathbf{f}_T(\mathbf{x}) = \arg\max_k (f_T^1(\mathbf{x}), \dots, f_T^k(\mathbf{x}), \dots, f_T^C(\mathbf{x}))'$ and is equivalent to $\mathbf{f}_T(\mathbf{x}) = \arg\max_k \sum_{t=1}^{T} \alpha_t\delta[\![H_t(\mathbf{x}) = k]\!]$.

C Proof for Lemma 1

At the t-th iteration, setting $\varepsilon_t = \frac{\rho}{\rho+1} - \gamma_t$ in (10) and simplifying gives us

$$Z_t = \frac{\rho+1}{\sqrt{\rho}}\sqrt{\left(\frac{\rho}{\rho+1} - \gamma_t\right)\left(\frac{1}{\rho+1} + \gamma_t\right)} = \sqrt{1 - \frac{(\rho+1)^2}{\rho}\gamma_t^2 + \frac{\rho^2 - 1}{\rho}\gamma_t}$$

Now, using $1 + x \leq e^x$ gives (3).   □

Parallel Boosting with Momentum

Indraneel Mukherjee[1], Kevin Canini[1], Rafael Frongillo[2], and Yoram Singer[1]

[1] Google Inc.
indraneelenator@gmail.com, {canini,singer}@google.com
[2] Computer Science Division, University of California Berkeley
raf@cs.berkeley.edu

Abstract. We describe a new, simplified, and general analysis of a fusion of Nesterov's accelerated gradient with parallel coordinate descent. The resulting algorithm, which we call BOOM, for *boo*sting with *m*omentum, enjoys the merits of both techniques. Namely, BOOM retains the momentum and convergence properties of the accelerated gradient method while taking into account the curvature of the objective function. We describe a *distributed* implementation of BOOM which is suitable for massive high dimensional datasets. We show experimentally that BOOM is especially effective in large scale learning problems with rare yet informative features.

Keywords: accelerated gradient, coordinate descent, boosting.

1 Introduction

Large scale supervised machine learning problems are concerned with building accurate prediction mechanisms from massive amounts of high dimensional data. For instance, Bekkerman et al. [15] gave as an example the problem of training a Web search ranker on billions of documents using user-clicks as labels. This vast amount of data also comes with a plethora of user behavior and click patterns. In order to make accurate predictions, different characteristics of the users and the search query are extracted. As a result, each example is typically modeled as a very high-dimensional vector of binary features. Yet, within any particular example only a few features are non-zero. Thus, many features appear relatively rarely in the data. Nonetheless, many of the infrequent features are both highly informative and correlated with each other. Second order methods take into account the local curvature of the parameter space and can potentially cope with the skewed distribution of features. However, even memory-efficient second order methods such as L-BFGS [12] cannot be effective when the parameter space consists of 10^9 dimensions or more. The informativeness of rare features attracted practitioners who crafted domain-specific feature weightings, such as TF-IDF [13], and learning theorists who devised stochastic gradient and proximal methods that adapt the learning rate per feature [4,6]. Alas, our experiments with state-of-the-art stochastic gradient methods, such as AdaGrad [4], underscored the inability of stochastic methods to build very accurate predictors

H. Blockeel et al. (Eds.): ECML PKDD 2013, Part III, LNAI 8190, pp. 17–32, 2013.

which take into account the highly correlated, informative, yet rare features. The focus of this paper is the design and analysis of a *batch* method that is highly parallelizable, enjoys the fast convergence rate of modern proximal methods, and incorporates a simple and efficient mechanism that copes with what we refer to as the *elliptical* geometry of the feature space.

Our algorithm builds and fuses two seemingly different approaches. Due to the scale of the learning problems, we have to confine ourselves to first order methods whose memory requirements are *linear* in the dimension. One of the most effective approaches among first order optimization techniques is Nesterov's family of accelerated gradient methods. Yuri Nesterov presented the core idea of accelerating gradient-based approaches by introducing a momentum term already in 1983 [8]. The seemingly simple modification to gradient descent obtains optimal performance for the complexity class of first-order algorithms when applied to the problem of minimization of smooth convex functions [7]. Nesterov and colleagues presented several modifications in order to cope with non-smooth functions, in particular composite (sum of smooth and non-smooth) functions. The paper of the late Paul Tseng provides an excellent, albeit technically complex, unified analysis of gradient acceleration techniques [16]. This paper is also the most related to the work presented here as we elaborate in the sequel.

Both Nesterov himself as well as the work of Beck and Teboulle [1], who built on Nesterov's earlier work, studied accelerated gradient methods for composite objectives. Of the two approaches, the latter is considered more efficient to implement as it requires storing parameters from only the last two iterations and a single projection step. Beck and Teboulle termed their approach FISTA for Fast Iterative Shrinkage Thresholding Algorithm. We start our construction with a derivation of an alternative analysis for FISTA and accelerated gradient methods in general. Our analysis provides further theoretical insights and distills to a broad framework within which accelerated methods can be applied. Despite their provably fast convergence rates, in practice accelerated gradient methods often exhibit slow convergence when there are strong correlations between features, amounting to elliptical geometry of the feature space. Putting the projection step aside, first order methods operate in a subspace which conforms with the span of the examples and as such highly correlated rare features can be overlooked. This deficiency is the rationale for incorporating an additional component into our analysis and algorithmic framework.

Coordinate descent methods [17] have proven to be very effective in machine learning problems, and particularly in optimization problems of elliptical geometries, as they can operate on each dimension separately. However, the time complexity of coordinate descent methods scale linearly with the dimension of the parameters and thus renders them inapplicable for high-dimensional problems. Several extensions that perform mitigated coordinate descent steps in parallel for multiple coordinates have been suggested. Our work builds specifically on the parallel boosting algorithm from [2]. However, parallel boosting algorithms on their own are too slow. See for instance [14] for a primal-dual analysis of the rate of convergence of boosting algorithms in the context of loss minimization.

Our approach combines the speed of accelerated gradient methods with the geometric sensitivity of parallel coordinate descent, and enjoys the merits of both approaches. We call the resulting approach BOOM, for *boosting with momentum*. As our experiments indicate, BOOM clearly outperforms both parallel boosting and FISTA over a range of medium- to large-scale learning problems. The aforementioned paper by Tseng also marries the two seemingly different approaches. It is based on extending the so called 1-memory version of accelerated gradient to inner-product norms, and indeed by employing a matrix-based norm Tseng's algorithm can model the elliptical geometry of the parameter space. However, our analysis and derivation are quite different than Tseng's. We take a more modular and intuitive approach, starting from the simplest form of proximal methods, and which may potentially be used with non-quadratic families of proximal functions.

The rest of the paper is organized as follows. We describe the problem setting in Sec. 2 and review in Sec. 3 proximal methods and parallel coordinate descent for composite objective functions. For concreteness and simplicity of our derivations, we focus on a setting where the non-smooth component of the objective is the 1-norm of the vector of parameters. In Sec. 4 we provide an alternative view and derivation of accelerated gradient methods. Our derivation enables a unified view of Nesterov's original acceleration and the FISTA algorithm as special cases of an abstraction of the accelerated gradient method. This abstraction serves as a stepping stone in the derivation of BOOM in Sec. 5. We provide experiments that underscore the merits of BOOM in Sec. 6. We briefly discuss a loosely-coupled distributed implementation of BOOM which enables its usage on very large scale problems. We comments on potential extensions in Sec. 7.

2   Problem Setting

We start by first establishing the notation used throughout the paper. Scalars are denoted by lower case letters and vectors by boldface letters, e.g. $\mathbf{w}$. We assume that all the parameters and instances are vectors of fixed dimension n. The inner product of two vectors is denoted as $\langle \mathbf{w}, \mathbf{v} \rangle = \sum_{j=1}^{n} w_j v_j$. We assume we are provided a labeled dataset $\{(\mathbf{x}_i, y_i)\}_{i=1}^{m}$ where the examples have feature vectors in $\mathbb{R}^n$. The unit vector whose j'th coordinate is 1 and the rest of the coordinates are 0 is denoted $\mathbf{e}_j$. For concreteness, we focus on binary classification and linear regression thus the labels are either in $\{-1, +1\}$ or real-valued accordingly. A special case which is nonetheless important and highly applicable is when each feature vector is sparse, that is only a fraction of its coordinates are non-zero. We capture the sparsity via the parameter κ defined as the maximum over the 0-norm of the examples, $\kappa = \max_i |\{j : x_{i,j} \neq 0\}|$. The convex optimization problem we are interested in is to associate an importance weight with each feature, thus find a vector $\mathbf{w} \in \mathbb{R}^n$ which minimizes:

$$\mathcal{L}(\mathbf{w}) = \sum_{i=1}^{m} \ell(\langle \mathbf{w}, \mathbf{x}_i \rangle, y_i) + \lambda_1 \|\mathbf{w}\|_1 = \mathcal{F}(\mathbf{w}) + \lambda_1 \|\mathbf{w}\|_1 ,$$

where $\mathcal{F}$ denotes the smooth part of $\mathcal{L}$. As mentioned above, the non-smooth part of $\mathcal{L}$ is fixed to be the 1-norm of $\mathbf{w}$. Here $\ell : \mathbb{R} \times \mathbb{R} \to \mathbb{R}_+$ is a prediction loss function. In the concrete derivations presented in later sections we focus on two popular loss functions: the squared error $\ell(\hat{y}, y) = \frac{1}{2}(\hat{y} - y)^2$ for real-valued labels, and the logistic loss $\ell(\hat{y}, y) = 1 + e^{-y\hat{y}}$ for binary labels, where $\hat{y}$ is the prediction and y is the true label. Throughout the paper $\mathbf{w}^*$ denotes the point minimizing $\mathcal{L}$.

3   Proximal Methods

We begin by reviewing gradient descent for smooth objective functions and then show how to incorporate non-smooth 1-norm regularization. We review how the same minimization task can be carried out using parallel coordinate descent.

Gradient Descent. Let us first assume that $\lambda_1 = 0$, hence $\mathcal{L} = \mathcal{F}$. We denote by L the maximum curvature of $\mathcal{F}$ in any direction, so that $\nabla^2 \mathcal{F} \preceq \mathsf{L}I$. This property of $\mathcal{F}$ coincides with having a Lipschitz-continuous gradient of a Lipschitz constant L. We can locally approximate $\mathcal{F}$ around any point $\mathbf{v}$ using the following quadratic upper bound: $\mathcal{F}(\mathbf{w}+\boldsymbol{\delta}) \leq \mathcal{F}(\mathbf{w}) + \langle \nabla \mathcal{F}(\mathbf{v}), \boldsymbol{\delta} \rangle + \frac{1}{2} \underbrace{\boldsymbol{\delta}^{\dagger}(\mathsf{L}I)\boldsymbol{\delta}}_{\mathsf{L}\|\boldsymbol{\delta}\|^2}$.

In each iteration, the new weight $\mathbf{w}_{t+1}$ is chosen to minimize the above bound anchored at the previous iterate $\mathbf{w}_t$, which amounts to,

$$\mathbf{w}_{t+1} = \mathbf{w}_t - (1/\mathsf{L})\nabla \mathcal{F}(\mathbf{w}_t) . \tag{1}$$

For this update, recalling $\mathcal{L} = \mathcal{F}$, simple algebra yields the following drop in the loss,

$$\mathcal{L}(\mathbf{w}_t) - \mathcal{L}(\mathbf{w}_{t+1}) \geq \|\nabla \mathcal{L}(\mathbf{w}_t)\|_2^2 / (2\mathsf{L}) . \tag{2}$$

The guarantee of (2), yields that it takes $O\big(\mathsf{L}\|\mathbf{w}_0 - \mathbf{w}^*\|^2/\epsilon\big)$ iterations to obtain an approximation that is ϵ-close in the objective value.

Incorporating 1-norm regularization. When $\lambda_1 > 0$, the local approximation has to explicitly account for the ℓ_1 regularization, we have the following local approximation:

$$\mathcal{L}(\mathbf{w} + \boldsymbol{\delta}) \leq \mathcal{F}(\mathbf{w}) + \langle \boldsymbol{\delta}, \nabla \mathcal{F}(\mathbf{w}) \rangle + (\mathsf{L}/2)\|\boldsymbol{\delta}\|^2 + \lambda_1 \|\mathbf{w} + \boldsymbol{\delta}\|_1 .$$

We can decompose the above Taylor expansion in a coordinate-wise fashion

$$\mathcal{L}(\mathbf{w} + \boldsymbol{\delta}) \leq \mathcal{L}(\mathbf{w}) + \sum_{j=1}^n f_j^{\mathsf{L}}(\delta_j) , \tag{3}$$

where

$$f_j^{\mathsf{L}}(\delta) \triangleq g_j \delta + (1/2)\mathsf{L}\delta^2 + \lambda_1 |w_j + \delta| - \lambda_1 |w_j| , \tag{4}$$

where g_j denotes $\nabla \mathcal{F}(\mathbf{w})_j$. Multiple authors (see e.g. [5]) showed that the value δ_j^* minimizing f_j satisfies $w_j + \delta_j^* = \mathcal{P}_{\mathsf{L}}^{g_j}(w_j)$, where

$$\mathcal{P}_{\mathsf{L}}^g(w) \triangleq \operatorname{sign}(w - g/\mathsf{L}) \left[|w - g/\mathsf{L}| - \lambda_1/\mathsf{L} \right]_+ . \tag{5}$$

In fact, with this choice of δ_j^* we can show the following guarantee using standard arguments from calculus.

Lemma 1. *Let f_j^L be defined by (4) and δ_j^* chosen as above, then*

$$f_j^L(0) - f_j^L(\delta_j^*) \geq (L/2)\delta_j^{*2} .$$

Gradient descent with ℓ_1 regularization iteratively chooses $\mathbf{w}_{t+1}$ to minimize the local approximation around $\mathbf{w}_t$, which amounts to

$$\forall j : w_{t+1,j} = \mathcal{P}_L^{g_j}(w_{t,j}) , \tag{6}$$

and using (3) and Lemma 1 we obtain the following guarantee:

$$\mathcal{L}(\mathbf{w}_t) - \mathcal{L}(\mathbf{w}_{t+1}) \geq (L/2)\|\mathbf{w}_t - \mathbf{w}_{t+1}\|^2 . \tag{7}$$

This guarantee yields the same convergence rate of $O(L\|\mathbf{w}_0 - \mathbf{w}^*\|^2/\epsilon)$ as before.

Coordinate Descent. In coordinate descent algorithms, we take into account the possibility that the curvature could be substantially different for each coordinate. Let L_j denotes the maximum curvature of $\mathcal{F}$ along coordinate j, then as we show in the sequel that parallel coordinate descent achieves a convergence rate of $O(\sum_{j=1}^n \kappa L_j(w_{0,j} - w_j^*)^2/\epsilon)$, where κ is the aforementioned sparsity parameter. Note that when the dataset is perfectly spherical and all the coordinate wise curvatures are equal to L_0, convergence rate simplifies to $O(\kappa L_0\|\mathbf{w}_0 - \mathbf{w}^*\|^2/\epsilon)$, and we approximately recover the rate of gradient descent. In general though, the coordinate-specific rate can yield a significant improvement over the gradient descent bounds, especially for highly elliptical datasets.

Let us describe a toy setting that illustrates the advantage of coordinate descent over gradient descent when the feature space is elliptical. Assume that we have a linear regression problem where all of the labels are 1. The data matrix is of size $2(n{-}1)\times n$. The first $n{-}1$ examples are all the same and equal to the unit vector $\mathbf{e}_1$, that is, the first feature is 1 and the rest of the features are zero. The next $n{-}1$ examples are the unit vectors $\mathbf{e}_2,\ldots,\mathbf{e}_n$. The matrix $\nabla^2\mathcal{F}$ is diagonal where the first diagonal element is $n-1$ and the rest of the diagonal elements are all 1. The optimal vector $\mathbf{w}^*$ is $(1,\ldots,1)$ and thus its squared 2-norm is n. The largest eigen value of $\nabla^2\mathcal{F}$ is clearly $n-1$ and thus $L = n-1$. Therefore, gradient descent converges at a rate of $O(n^2/\epsilon)$. The rest of the eigen values of $\nabla^2\mathcal{F}$ are 1, namely, $L_j = 1$ for $j \geq 2$. Since exactly one feature is "turned on" in each example $\kappa = 1$. We therefore get that coordinate descent converges at a rate of $O(n/\epsilon)$ which is substantially faster in terms of the dimensionality of the problem. We would like to note in passing that sequential coordinate descent converges to the optimal solution in exactly n steps.

To derive the parallel coordinate descent update, we proceed as before via a Taylor approximation, but this time have a separate approximation per coordinate: $\mathcal{F}(\mathbf{w}+\theta\mathbf{e}_j) \leq \mathcal{F}(\mathbf{w})+\theta\nabla\mathcal{F}(\mathbf{w})_j+(L_j/2)\theta^2$. In order to simultaneously step in each coordinate, we employ the sparsity parameter κ and Jensen's inequality (see e.g. [3]) to show

$$\mathcal{F}(\mathbf{w} + \boldsymbol{\theta}/\kappa) \leq \mathcal{F}(\mathbf{w}) + (1/\kappa)\sum_{j=1}^n \left(\theta_j\nabla\mathcal{F}_j(\mathbf{w}) + (L_j/2)\theta_j^2\right)$$
$$= \mathcal{F}(\mathbf{w}) + \sum_{j=1}^n \left(g_j\theta_j/\kappa + (1/2)L_j\kappa\left(\theta_j/\kappa\right)^2\right) .$$

By replacing θ_j / κ by δ_j, we get

$$\mathcal{F}(\mathbf{w} + \boldsymbol{\delta}) \leq \mathcal{F}(\mathbf{w}) + \sum_{j=1}^{n} \left(g_j \delta_j + (\kappa \mathsf{L}_j / 2) \delta_j^2 \right). \tag{8}$$

Introducing the ℓ_1 regularization terms on both sides, we have

$$\mathcal{L}(\mathbf{w} + \boldsymbol{\delta}) \leq \mathcal{L}(\mathbf{w}) + \sum_{j=1}^{n} f_j^{\kappa \mathsf{L}_j}(\delta_j) , \tag{9}$$

where $f_j^{\kappa \mathsf{L}_j}$ is as in (4). From our earlier discussions, we know the optimal choice $\boldsymbol{\delta}^*$ minimizing the previous expression satisfies $\delta_j^* = \mathcal{P}_{\kappa \mathsf{L}_j}^{g_j}(w_j)$. Accordingly, the parallel coordinate descent update is

$$w_{t+1,j} = \mathcal{P}_{\kappa \mathsf{L}_j}^{g_j}(w_{t,j}) , \tag{10}$$

and using (9) and Lemma 1, we have

$$\mathcal{L}(\mathbf{w}_t) - \mathcal{L}(\mathbf{w}_{t+1}) \geq \sum_{j=1}^{n} (\kappa \mathsf{L}_j / 2) \left(w_{t,j} - w_{t+1,j} \right)^2 . \tag{11}$$

As before, with this guarantee on the progress of each iteration we can show that the convergence rate is $O(\sum_{j=1}^{n} \kappa \mathsf{L}_j (w_{0,j} - w_j^*)^2 / \epsilon)$.

4   Accelerated Gradient Methods

In this section we give a new alternative derivation of the accelerated gradient method (AGM) that underscores and distills the core constituents of the method. Our view serves as a stepping stone towards the extension that incorporates parallel boosting in the next section. Accelerated methods take a more general approach to optimization than gradient descent. Instead of updating the parameters using solely the most recent gradient, AGM creates a sequence of auxiliary functions $\phi_t : \mathbb{R}^n \to \mathbb{R}^n$ that are increasingly accurate approximations to the original loss function. Further, the approximating functions uniformly converge to the original loss function as follows,

$$\phi_{t+1}(\mathbf{w}) - \mathcal{L}(\mathbf{w}) \leq (1 - \alpha_t)(\phi_t(\mathbf{w}) - \mathcal{L}(\mathbf{w})) , \tag{12}$$

where each $\alpha_t \in (0,1)$ and the entire sequence determines the rate of convergence. At the same time, AGM produces weight vectors $\mathbf{w}_t$ such that the true loss is lower bounded by the minimum of the auxiliary function as follows,

$$\mathcal{L}(\mathbf{w}_t) \leq \min_{\mathbf{w}} \phi_t(\mathbf{w}) . \tag{13}$$

The above requirement yields directly the following lemma.

Lemma 2. *Assume that (12) and (13) hold in each iteration, then after t iterations, $\mathcal{L}(\mathbf{w}_t) - \mathcal{L}(\mathbf{w}^*)$ is be upper bounded by,*

$$\left(\prod_{k<t} (1 - \alpha_k) \right) (\phi_0(\mathbf{w}^*) - \phi_0(\mathbf{w}_0) + \mathcal{L}(\mathbf{w}_0) - \mathcal{L}(\mathbf{w}^*)) .$$

The inequality (12) is achieved by choosing a linear function $\hat{\mathcal{L}}_t$ that lower bounds the original loss function, $\hat{\mathcal{L}}_t(\mathbf{w}) \leq \mathcal{L}(\mathbf{w})$, and then squashing ϕ_t towards the linear function by a factor α_t,

$$\phi_{t+1} = (1 - \alpha_t)\phi_t + \alpha_t\hat{\mathcal{L}}_t. \tag{14}$$

Nesterov chose the initial auxiliary function to be a quadratic function centered at $\mathbf{v}_0 = \mathbf{w}_0$ (an arbitrary point vector), with curvature $\gamma_0 = \mathsf{L}/2$, and intercept $\phi_0^* = \mathcal{L}(\mathbf{w}_0)$, $\phi_0(\mathbf{w}) = \gamma_0\|\mathbf{w} - \mathbf{v}_0\|^2 + \phi_0^* = \frac{\mathsf{L}}{2}\|\mathbf{w} - \mathbf{w}_0\|^2 + \mathcal{L}(\mathbf{w}_0)$. Then, using an inductive argument, each ϕ_t is also a quadratic function with a center $\mathbf{v}_t$, curvature $\gamma_t = \gamma_0\prod_{t'<t}(1 - \alpha_{t'})$, and intercept ϕ_t^*

$$\phi_t(\mathbf{w}) = \gamma_t\|\mathbf{w} - \mathbf{v}_t\|^2 + \phi_t^* . \tag{15}$$

Moreover, if the linear function $\hat{\mathcal{L}}_t$ has slope $\boldsymbol{\eta}_t$, algebraic manipulations yield:

$$\mathbf{v}_{t+1} = \mathbf{v}_t - (\alpha_t/2\gamma_{t+1})\boldsymbol{\eta}_t \tag{16}$$

$$\begin{aligned}\phi_{t+1}^* &= \phi_{t+1}(\mathbf{v}_t) - (\phi_{t+1}(\mathbf{v}_t) - \phi_{t+1}(\mathbf{v}_{t+1})) \\ &= (1 - \alpha_t)\phi_t^* + \alpha_t\hat{\mathcal{L}}_t(\mathbf{v}_t) - \gamma_{t+1}\|\mathbf{v}_{t+1} - \mathbf{v}_t\|^2 \\ &= (1 - \alpha_t)\phi_t^* + \alpha_t\hat{\mathcal{L}}_t(\mathbf{v}_t) - (\alpha_t^2/4\gamma_{t+1})\|\boldsymbol{\eta}_t\|^2 . \end{aligned} \tag{17}$$

The last two equalities follow from (14), (15), and (16). To complete the algorithm and proof, we need to choose $\hat{\mathcal{L}}_t$, α_t and $\mathbf{w}_{t+1}$ so as to satisfy (13). Namely, $\mathcal{L}(\mathbf{w}_{t+1}) \leq \phi_{t+1}^*$. All acceleration algorithms satisfy these properties by tackling the expression in (17) in two steps. First, an intermediate point $\mathbf{y}_{t+1} = (1 - \alpha_t)\mathbf{w}_t + \alpha_t\mathbf{v}_t$ is chosen. Note by linearity of $\hat{\mathcal{L}}_t$ we have,

$$\hat{\mathcal{L}}_t(\mathbf{y}_{t+1}) = (1 - \alpha_t)\hat{\mathcal{L}}_t(\mathbf{w}_t) + \alpha_t\hat{\mathcal{L}}_t(\mathbf{v}_t) \leq (1 - \alpha_t)\mathcal{L}(\mathbf{w}_t) + \alpha_t\hat{\mathcal{L}}_t(\mathbf{v}_t) .$$

Then, a certain proximal step is taken from $\mathbf{y}_{t+1}$ in order to reach $\mathbf{w}_{t+1}$, making sufficient progress in the process that satisfies,

$$\hat{\mathcal{L}}_t(\mathbf{y}_{t+1}) - \mathcal{L}(\mathbf{w}_{t+1}) \geq (\alpha_t^2/4\gamma_{t+1})\|\boldsymbol{\eta}_t\|^2 . \tag{18}$$

Combining the above two inequalities and inductively assuming that $\mathcal{L}(\mathbf{w}_t) \leq \phi_t^*$, it can be shown that $\mathcal{L}(\mathbf{w}_{t+1})$ is at most ϕ_{t+1}^* as given by (17).

The acceleration method in its abstract and general form is described in Algorithm 1. Further, based on the above derivation, this abstract algorithm ensures that (12) and (13) hold on each iteration. Consequently Lemma 2 holds as well as the following theorem.

Theorem 1. *The optimality gap $\mathcal{L}(\mathbf{w}_t) - \mathcal{L}(\mathbf{w}^*)$ when $\mathbf{w}_t$ is constructed according to Algorithm 1 is at most,*

$$\left(\prod_{k<t}(1 - \alpha_k)\right)\mathsf{L}\|\mathbf{w}_0 - \mathbf{w}^*\|^2 . \tag{19}$$

Proof. It suffices to show that the bound of Lemma 2 can be distilled to the bound of (19). Using the definition of ϕ_0 we have

$$\phi_0(\mathbf{w}^*) - \phi_0(\mathbf{w}_0) = (\mathsf{L}/2)\|\mathbf{w}^* - \mathbf{w}_0\|^2 . \tag{20}$$

Further, since the maximum curvature is L, the function has a Lipschitz-continuous gradient with Lipschitz constant L, namely, for any two vectors $\mathbf{x}, \mathbf{x}'$ the following inequality holds $\mathcal{L}(\mathbf{u}) - \mathcal{L}(\mathbf{u}') \leq (L/2)\|\mathbf{x} - \mathbf{x}'\|^2$ and in particular for $\mathbf{u} = \mathbf{w}_0$ and $\mathbf{u}' = \mathbf{w}^*$. Summing the term from inequality (20) with the Lipschitz-continuity bound completes the proof.

We next present two concrete instantiations of Algorithm 1. Each variant chooses a lower bounding function $\hat{\mathcal{L}}_t$, a proximal step for reaching $\mathbf{w}_{t+1}$ from $\mathbf{y}_{t+1}$, and finally α_t so that (18) holds.

Nesterov's Acceleration [9]. In this setting there is no 1-norm regularization, $\lambda_1 = 0$, so that $\mathcal{L} = \mathcal{F}$. Here $\hat{\mathcal{L}}_t$ is the tangent plane to the surface of $\mathcal{L}$ at $\mathbf{y}_{t+1}$, $\boldsymbol{\eta}_t = \nabla\mathcal{L}(\mathbf{y}_{t+1})$ and $\hat{\mathcal{L}}_t(\mathbf{w}) = \mathcal{L}(\mathbf{y}_{t+1}) + \langle\boldsymbol{\eta}_t, \mathbf{w} - \mathbf{y}_{t+1}\rangle$, which by definition lower bounds $\mathcal{L}$. Further, let $\mathbf{w}_{t+1}$ be obtained from $\mathbf{y}_{t+1}$ using the proximal step in (1),

$$\mathbf{w}_{t+1} = \mathbf{y}_{t+1} - (1/L)\nabla\mathcal{L}(\mathbf{y}_{t+1}) . \tag{21}$$

Then, we have the same guarantee as in (2),

$$\mathcal{L}(\mathbf{y}_{t+1}) - \mathcal{L}(\mathbf{w}_{t+1}) \geq 1/(2L)\|\nabla\mathcal{L}(\mathbf{y}_{t+1})\|^2 .$$

By definition of $\boldsymbol{\eta}$, (18) holds if we choose α_t to satisfy

$$1/(2L) = \alpha_t^2/(4\gamma_{t+1}), \tag{22}$$

which by expanding γ_k and using $\gamma_0 = L/2$, simplifies to

$$\alpha_t^2/(1 - \alpha_t) = \prod_{k<t}(1 - \alpha_k) . \tag{23}$$

From the above recurrence, the following inverse quadratic upper bound holds.

Lemma 3. *Assume that (23) holds, then* $\prod_{s<t}(1 - \alpha_s) \leq \frac{2}{(t+1)^2}$

Lemma 3 with Thm. 1 yields a rate of convergence of $O(L\|\mathbf{w} - \mathbf{w}^*\|^2/\sqrt{\epsilon})$.

FISTA [1]. In FISTA, $\mathbf{w}_{t+1}$ is set from $\mathbf{y}_{t+1}$ using (6), namely, $w_{t+1,j} = \mathcal{P}_L^{g_j}(y_{t+1,j})$. With this choice of $\mathbf{w}_{t+1}$, $\hat{\mathcal{L}}_t$ is constructed as follows,

$$\boldsymbol{\eta}_t = L\left(\mathbf{y}_{t+1} - \mathbf{w}_{t+1}\right), \hat{\mathcal{L}}_t(\mathbf{w}) = \mathcal{L}(\mathbf{w}_{t+1}) + (1/2L)\|\boldsymbol{\eta}\|^2 + \langle\boldsymbol{\eta}, \mathbf{w} - \mathbf{y}_{t+1}\rangle . \tag{24}$$

The fact that $\hat{\mathcal{L}}_t$ lower bounds $\mathcal{L}$ is not obvious as was shown in [1]. We provide a more general proof later in Lemma 4. Note that the definition (24) implies that, $\hat{\mathcal{L}}_t(\mathbf{y}_{t+1}) - \mathcal{L}(\mathbf{w}_{t+1}) = (1/2L)\|\boldsymbol{\eta}\|^2$. Thus (18) holds when α_t is set as in (22) so as to satisfy the recurrence (23). Once again invoking Lemma 3 and Theorem 1, we obtain the same convergence rate of $O(L\|\mathbf{w}_0 - \mathbf{w}^*\|^2)/\sqrt{\epsilon})$. The resulting algorithm may appear different than the original FISTA algorithm, but can be shown to be equivalent.

Algorithm 1. Accelerated Gradient	Algorithm 2. Boosting with Momentum

1: **inputs:** loss $\mathcal{L} : \mathbb{R}^n \to \mathbb{R}$,
     curvature $\mathsf{L} \in \mathbb{R}_+$.
	1: **inputs:** loss $\mathcal{L}$,
	     sparsity κ and $\mathsf{L}_1, \ldots, \mathsf{L}_n$.

2: **initialize:** $\mathbf{w}_0 = \mathbf{v}_0 \in \mathbb{R}^n$,
     $\gamma_0 \leftarrow \mathsf{L}/2$.
	2: **initialize:** $\mathbf{w}_0 = \mathbf{v}_0 \in \mathbb{R}^n$,
	     $\forall j : \gamma_{0,j} \leftarrow \kappa \mathsf{L}_j/2$.

3: **for** $t = 0, 1, \ldots,$ **do**	3: **for** $t = 0, 1, \ldots,$ **do**
4: Pick $\alpha_t \in (0,1)$	4: Pick $\alpha_t \in (0,1)$
5: Set $\gamma_{t+1} = \gamma_0 \prod_{k \le t}(1 - \alpha_k)$.	5: Set $\gamma_{t+1,j} = \gamma_{0,j} \prod_{k \le t}(1 - \alpha_k)$.
6: $\mathbf{y}_{t+1} \leftarrow (1 - \alpha_t)\mathbf{w}_t + \alpha_t \mathbf{v}_t$.	6: $\mathbf{y}_{t+1} \leftarrow (1 - \alpha_t)\mathbf{w}_t + \alpha_t \mathbf{v}_t$.

7:     Choose linear function
         $\hat{\mathcal{L}}_t \le \mathcal{L}$ with slope $\boldsymbol{\eta}_t$
	7: Choose linear function
	         $\hat{\mathcal{L}}_t \le \mathcal{L}$ with slope $\boldsymbol{\eta}_t$

8:     Choose $\mathbf{w}_{t+1}$ using $\mathbf{y}_{t+1}$
         so that (18) holds
	8: Choose $\mathbf{w}_{t+1}$ using $\mathbf{y}_{t+1}$
	         so that (28) holds

9: $\mathbf{v}_{t+1} \leftarrow \mathbf{v}_t - (\alpha_t/2\gamma_{t+1})\boldsymbol{\eta}_t$	9: $\forall j : v_{t+1,j} \leftarrow v_{t,j} - (\alpha_t/2\gamma_{t+1,j})\eta_{t,j}$.
10: **end for**	10: **end for**

5   BOOM: A Fusion

In this section we use the derivation of the previous section in a more general setting in which we combine the momentum-based gradient descent with parallel coordinate decent. As mentioned above we term the approach BOOM as it fuses the parallel *boo*sting algorithm from [2] with *m*omentum accelerated gradient methods [9,10,11].

The structure of this section will closely mirror that of Section 4, the only difference being the details for handling per-coordinate curvature. We start by modifying the auxiliary functions to account for the different curvatures L_j of $\mathcal{F}$ in each direction, starting with the initial function,

$$\phi_0(\mathbf{w}) = \sum_{j=1}^n \gamma_{0,j}(w_j - v_{0,j})^2 + \phi_0^*.$$

The $\gamma_{0,j}$ are initialized to $\kappa \mathsf{L}_j/2$ for each coordinate j, and ϕ_0^* is set to $\mathcal{L}(\mathbf{w}_0)$:

$$\phi_0(\mathbf{w}) = \sum_{j=1}^n (\mathsf{L}_j/2)(w_j - v_{0,j})^2 + \mathcal{L}(\mathbf{w}_0).$$

So instead of a spherical quadratic, the new auxiliary function is elliptical, with curvatures matching those of the smooth part $\mathcal{F}$ of the loss function. As before, we will choose a linear function $\hat{\mathcal{L}}_t \le \mathcal{L}$ in each round and squash ϕ_t towards it to obtain the new auxiliary function. Therefore (14) continues to hold, and we can again inductively prove that ϕ_t continues to retain an elliptical quadratic form:

$$\phi_t(\mathbf{w}) = \sum_{j=1}^n \gamma_{t,j} \mathsf{L}_j (w_j - v_{t,j})^2 + \phi_t^*, \tag{25}$$

where $\gamma_{t,j} = \gamma_{0,j} \prod_{k < j}(1 - \alpha_k)$. In fact, if $\hat{\mathcal{L}}_t$ has slope $\boldsymbol{\eta}_t$, then we can show as before:

$$v_{t+1,j} = v_{t,j} - (\alpha_t/2\gamma_{t+1,j})\eta_{t,j} \tag{26}$$

$$\begin{aligned}
\phi_{t+1}^* &= \phi_{t+1}(\mathbf{v}_t) - (\phi_{t+1}(\mathbf{v}_t) - \phi_{t+1}(\mathbf{v}_{t+1})) \\
&= (1-\alpha_t)\phi_t^* + \alpha_t\hat{\mathcal{L}}_t(\mathbf{v}_t) - \sum_{j=1}^t \gamma_{t+1,j}(v_{t+1,j} - v_{t,j})^2 \\
&= (1-\alpha_t)\phi_t^* + \alpha_t\hat{\mathcal{L}}_t(\mathbf{v}_t) - \sum_{j=1}^n (\alpha_t^2/4\gamma_{t+1,j})\eta_{t,j}^2.
\end{aligned} \tag{27}$$

By picking $\mathbf{y}_{t+1} = (1-\alpha_t)\mathbf{w}_t + \alpha_t\mathbf{v}_t$ as before and arguing similarly we can inductively show the same invariant $\mathcal{L}(\mathbf{w}_t) \leq \phi_t^*$, except $\mathbf{w}_{t+1}$ has to satisfy the following guarantee instead of (18):

$$\hat{\mathcal{L}}_t(\mathbf{y}_{t+1}) - \mathcal{L}(\mathbf{w}_{t+1}) \geq \sum_{j=1}^n (\alpha_t^2/4\gamma_{t+1,j})\eta_{t,j}^2. \tag{28}$$

The algorithm in this abstract form is given in Algorithm 2. We have now established that (12) and (13) hold in each iteration, and hence using Lemma 2 and arguing as before, we have the following theorem.

Theorem 2. *If Algorithm 2 outputs $\mathbf{w}_t$ in iteration t, then the suboptimality $\mathcal{L}(\mathbf{w}_t) - \mathcal{L}(\mathbf{w}^*)$ can be upper bounded by $\left(\prod_{k<t}(1-\alpha_k)\right)\sum_{j=1}^n \kappa L_{0,j}(w_{t,j} - w_j^*)^2$.*

In order to make Algorithm 2 concrete, we must choose $\{\alpha_t\}$, $\hat{\mathcal{L}}_t$, and $\mathbf{w}_{t+1}$ to satisfy the required constraints. Our selection will be analogous to the choices made by FISTA, but incorporating different curvatures. We first select $\mathbf{w}_{t+1}$ from $\mathbf{y}_{t+1}$ in a way similar to (10), $\forall j : w_{t+1,j} = \mathcal{P}_{\kappa L_j}^{g_j}(y_{t+1,j})$, where $g_j = \nabla \mathcal{F}(\mathbf{y}_{t+1})_j$. Based on this choice, we select $\hat{\mathcal{L}}_t$ as follows:

$$\eta_{t,j} = \kappa L_j (y_{t+1,j} - w_{t+1,j}) \tag{29}$$

$$\hat{\mathcal{L}}_t(\mathbf{w}) = \mathcal{L}(\mathbf{w}_{t+1}) + \sum_{j=1}^n (\eta_j^2/2\kappa L_j) + \langle \boldsymbol{\eta}_t, \mathbf{w} - \mathbf{y}_{t+1} \rangle. \tag{30}$$

By extending Lemma 2.3 in [1] we can show $\hat{\mathcal{L}}_t \leq \mathcal{L}$.

Lemma 4. *If $\hat{\mathcal{L}}_t$ is defined as in (30), then $\forall \mathbf{w} : \hat{\mathcal{L}}_t(\mathbf{w}) \leq \mathcal{L}(\mathbf{w})$.*

The proof relies on optimality properties of the choice $\mathbf{w}_{t+1}$ and involves some subgradient calculus. We defer it to the supplementary materials. In addition to the lower bounding property $\hat{\mathcal{L}}_t \leq \mathcal{L}$, from the definition (30), we also have

$$\hat{\mathcal{L}}_t(\mathbf{y}_{t+1}) - \mathcal{L}(\mathbf{w}_{t+1}) = \sum_{j=1}^n (1/2\kappa L_j)\eta_j^2.$$

Then the constraint (28) will follow if we set α_t to satisfy:

$$\alpha_t^2/(4\gamma_{t,j}) = 1/(2\kappa L_j). \tag{31}$$

Expanding out $\gamma_{t,j}$ and using $\gamma_{0,j} = L_j/2$ we obtain $\frac{\alpha_t^2}{1-\alpha_t} = \frac{2\gamma_{0,j}}{\kappa L_j}\prod_{k<t}(1-\alpha_k) = \prod_{k<t}(1-\alpha_k)$, which is identical to (23).

We have now defined a particular instantiation of Algorithm 2, which satisfies the required conditions by the above discussion. We dub this instantiation *BOOM*, and give the full procedure in Algorithm 3 for completeness. Applying Theorem 2 and once again invoking Lemma 3, we have the following theorem, which yields a $O(\sum_{j=1}^n \kappa L_j(w_{0,j} - w_j^*)^2/\sqrt{\epsilon})$ convergence rate.

Theorem 3. *The* $\mathbf{w}_t$ *output by Algorithm 3 satisfies the bound*

$$\mathcal{L}(\mathbf{w}_t) - \mathcal{L}(\mathbf{w}^*) \leq (2/(t+1)^2) \sum_{j=1}^{n} \kappa L_j (w_{0,j} - w_j^*)^2.$$

As examples, consider linear and logistic loss. When $\mathcal{L}$ uses the linear loss, the curvature parameters are given by $L_j = \sum_i x_{i,j}^2$ where $\mathbf{x}_i \in \mathbb{R}^n$ is the feature vector for the ith example in the training set. With logistic loss, the curvature can be bounded by $L_j = (1/4) \sum_i x_{i,j}^2$ [5].

6 Experiments

Algorithm 3. BOOM

1: **inputs:** loss $\mathcal{L}$, regularizer λ_1,
2: **parameters:** κ and $L_1, \ldots, L_n$
3: **initialize:** $\mathbf{w}_0 = \mathbf{v}_0 = \mathbf{0}$, $\gamma_{0,j} \leftarrow \kappa L_j/2$
4: **for** $t = 0, 1, \ldots,$ **do**
5:   Pick α_t satisfying (23)
6:   Set $\gamma_{t+1,j} \leftarrow \gamma_{0,j} \alpha_t^2/(1 - \alpha_t)$.
7:   $\mathbf{y}_{t+1} \leftarrow (1 - \alpha_t)\mathbf{w}_t + \alpha_t\mathbf{v}_t$.
8:   $\mathbf{g} \leftarrow \nabla\mathcal{L}(\mathbf{y}_{t+1})$
9:   $w_{t+1,j} \leftarrow \mathcal{P}_{\kappa L_j}^{g_j}(y_{t+1,j})$
10:   $v_{t+1,j} \leftarrow v_{t,j} - \dfrac{\alpha_t \kappa L_j}{2\gamma_{t+1,j}}(y_{t+1,j} - w_{t+1,j})$
11: **end for**

We tested the performance of four algorithms: (1) parallel boosting, discussed in Section 3, (2) FISTA, discussed in Section 4, (3) BOOM, shown in Algorithm 3, and (4) sequential boosting. Note that for the first three algorithms, within each iteration, each coordinate in the feature space $\{1, \ldots, n\}$ could be assigned a separate processing thread that could carry out all the computations for that coordinate: e.g., the gradient, the step-size, and the change in weight for that coordinate. Therefore by assigning enough threads, or for the case of massively high dimensional data, implementing these algorithms on a distributed architecture and allocating enough machines, the computation time could remain virtually constant even as the number of dimensions grows. However, in sequential boosting a single iteration consists of choosing n coordinates uniformly at random with replacement, then making optimal steps along these coordinates, one by one in order. Therefore, in terms of computational time, one iteration of the sequential algorithm is actually on the order of n iterations of the parallel algorithms. The goal in including sequential boosting was to get a sense for how well the parallel algorithms can compete with a sequential algorithm, which in some sense has the best performance in terms of number of iterations. In all the experiments, when present, the curve corresponding to parallel boosting is shown in solid red lines, Fista in dashed blue, BOOM in solid lightgreen, and sequential boosting in dashed black.

We next describe the datasets. The synthetic datasets were designed to test how algorithms perform binary classification tasks with varying levels of sparsity, ellipticity, and feature correlations. We generated 9 synthetic datasets for binary classification, and 9 for linear regression. Each dataset has 1000 examples (split into train and test in a 2:1 ratio) and 100 binary features. Each feature is sparse or dense, occurring in 5% or 50%, resp. of the examples. The fraction of sparse features is 0, 0.5, or 1.0. Note that with a 0.5 fraction of sparse features, the ellipticity is higher than when the fraction is 0 or 1. For each of these settings, either 0, 50, or 100 percent of the features are grouped into equal blocks of identical features. The labels were generated by a random linear combination of the features, and contain 10% white label noise for binary classification, or 10% multiplicative Gaussian noise for the linear regression datasets. We ran each of the four algorithms for 100 iterations on each dataset.

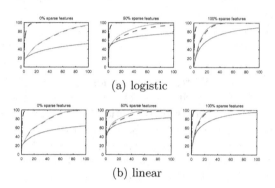

(a) logistic

(b) linear

For each algorithm, in each iteration we measure progress as the drop in loss since the first iteration, divided by the best loss possible, expressed as a percentage. For the training set we considered regularized loss, and for the test set unregularized loss. We partition the datasets based on whether the fraction of sparse features, is 0, 0.5 or 1. For each partition, we plot the progress on the training loss of each algorithm, averaged across all the datasets in the partition.

The results are tabulated in Figure 1 separately for logistic and linear regression.

Fig. 1. (Synthetic data) The top and bottom rows represent logistic and linear regression experiments, resp. The columns correspond to partitions of the datasets based on fraction of sparse features. The x-axis is iterations and the y-axis is the progress after each iteration, averaged over the datasets in the partition.

BOOM outperforms parallel boosting and FISTA uniformly across all datasets (as mentioned above, the sequential boosting is shown only as a benchmark). Against FISTA, the difference is negligible for the spherical datasets (where the fraction of sparse features is 0 or 1), but more pronounced for elliptical datsets

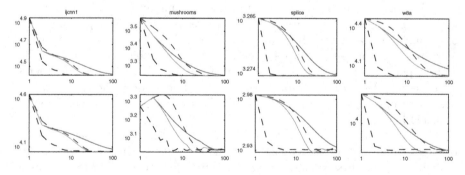

Fig. 2. (Medium-size real data) The top and bottom rows correspond to training and test data, resp. The x-axis measures iterations, and the y-axis measures loss with and without regularization for the training and test sets, resp.

(where the fraction of sparse features is 0.5). The plots on the test loss have the same qualitative properties, so we omit them.

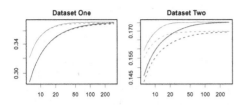

Fig. 3. (Large scale data) The x-axis is iterations and the y-axis is r-squared of the loss. The solid curves correspond to r-squared over training sets, and the dashed curve over the test-sets.

We next ran each algorithm for 100 iterations on four binary classification datasets: ijcnn1, splice, w8a and mushrooms. The criteria for choosing the datasets were that the number of examples exceeded the number of features, and the size of the datasets were reasonable. The continuous features were converted to binary features by discretizing them into ten quantiles. We made random 2:1 train/test splits in each dataset. The datasets can be found at http://www.csie.ntu.edu.tw/~cjlin/libsvmtools/datasets/binary.html. The log-log plot of train and test losses against the number of iterations are shown respectively in the left and right columns of Figure 2. BOOM significantly outperforms both FISTA and parallel boosting on both train and test loss on each dataset, suggesting that most real datasets are elliptical enough for the coordinate-wise approach of BOOM to make a noticeable improvement over FISTA.

Finally, we report large-scale experiments over two anonymized proprietary datasets. Both are linear regression datasets, and we report R-squared errors for the training set in solid lines, and test set in dashed. Dataset One contains 7.964B and 80.435M examples in the train and test sets, and 24.343M features, whereas Dataset Two contains 5.243B and 197.321M examples in the train and test sets, and 712.525M features. These datasets have very long tails, and the sparsity of the features varies drastically. In both datasets 90% of the features occur in less than 7000 examples, whereas the top hundred or so features occur in more than a hundred million examples each. Because of this enormous ellipticity, FISTA barely makes any progress, performing miserably even compared to parallel boosting, and we omit its plot. Sequential boosting is infeasibly slow to implement, and therefore we only show plots for BOOM and parallel boosting on these datasets. The results are shown in Figure 3, where we see that BOOM significantly outperforms parallel boosting on both datasets.

7   Conclusions

We presented in this paper an alternative abstraction of the accelerated gradient method. We believe that our more general view may enable new methods that can be accelerated via a momentum step. Currently the abstract accelerated gradient method (Algorithm 1) requires an update scheme which provides a guarantee on the drop in loss. Just as the choices of step size and slope were be abstracted, we suspect that this gradient-based condition can be relaxed, resulting in a potentially non-quadratic family of proximal functions. Thus, while the present paper focuses on accelerating parallel coordinate descent, in principle the techniques could be applied to other update schemes with a provable guarantee on the drop in loss.

References

1. Beck, A., Teboulle, M.: Fast iterative shrinkage-thresholding algorithm for linear inverse problems. SIAM Journal of Imaging Sciences 2, 183–202 (2009)
2. Collins, M., Schapire, R.E., Singer, Y.: Logistic regression, AdaBoost and Bregman distances. Machine Learning 47(2/3), 253–285 (2002)
3. Cover, T.M., Thomas, J.A.: Elements of Information Theory. Wiley (1991)
4. Duchi, J., Hazan, E., Singer, Y.: Adaptive subgradient methods for online learning and stochastic optimization. Journal of Machine Learning Research, 2121–2159 (2011)
5. Duchi, J., Singer, Y.: Boosting with structural sparsity. In: Proceedings of the 26th International Conference on Machine Learning (2009)
6. McMahan, H.B., Streeter, M.: Adaptive bound optimization for online convex optimization. In: Proceedings of the Twenty Third Annual Conference on Computational Learning Theory (2010)
7. Nemirovski, A., Yudin, D.: Problem complexity and method efficiency in optimization. John Wiley and Sons (1983)
8. Nesterov, Y.: A method of solving a convex programming problem with convergence rate $o(1/k^2)$. Soviet Mathematics Doklady 27(2), 372–376 (1983)
9. Nesterov, Y.: Introductory Lectures on Convex Optimization. Kluwer Academic Publishers (2004)
10. Nesterov, Y.: Smooth minimization of nonsmooth functions. Mathematical Programming 103, 127–152 (2005)
11. Nesterov, Y.: Primal-dual subgradient methods for convex problems. Mathematical Programming 120(1), 221–259 (2009)
12. Nocedal, J., Wright, S.: Numerical Optimization, 2nd edn. Springer Series in Operations Research and Financial Engineering (2006)
13. Salton, G., Buckley, C.: Term weighting approaches in automatic text retrieval. Information Processing and Management 24(5) (1988)
14. Shalev-Shwartz, S., Singer, Y.: On the equivalence of weak learnability and linear separability: new relaxations and efficient algorithms. In: Proceedings of the Twenty First Annual Conference on Computational Learning Theory (2008)
15. Svore, K., Burges, C.: Large-scale learning to rank using boosted decision trees. In: Bekkerman, R., Bilenko, M., Langford, J. (eds.) Scaling Up Machine Learning. Cambridge University Press (2012)
16. Tseng, P.: On accelerated proximal gradient methods for convex-concave optimization. Submitted to SIAM Journal on Optimization (2008)
17. Tseng, P., Yun, S.: A coordinate gradient descent method for nonsmooth separable minimization. Mathematical Programming Series B 117, 387–423 (2007)

Appendix

In this appendix we provide technical proofs for theorems and lemmas whose proof were omitted from the body of the manuscript.

Proof of Lemma 1 Let ∂f_j^L denotes a subgradient of f_j^L. By convexity of f_j^L and optimality of δ_j^*, ∂f_j^L is increasing, in the range $(0, \delta_j^*)$. Further, by optimality of δ_j^*, there f_j^L has a zero subgradient at the point δ_j^*. The form of (4) implies that ∂f_j^L increases at the rate of at least L_j. Therefore, we get that

$$f_j^L(0) - f_j^L(\delta_j^*) = \int_{\delta_j^*}^{0} \partial f^L(z) dz \geq \int_{0}^{\delta_j^*} L z \, dz = \tfrac{L}{2}\delta_j^{*2} . \quad \square$$

Proof of Lemma 2 The proof amounts to application of (13) and (12) as follows,

$$\forall \mathbf{w}^* : \mathcal{L}(\mathbf{w}_t) - \mathcal{L}(\mathbf{w}^*) \leq \phi_t(\mathbf{w}^*) - \mathcal{L}(\mathbf{w}^*) \quad [\text{using (13)}]$$
$$\leq \left(\prod_{k<t}(1 - \alpha_k)\right)(\phi_0(\mathbf{w}^*) - \mathcal{L}(\mathbf{w}^*)) \quad [\text{recursively applying (12)}]$$
$$= \left(\prod_{k<t}(1 - \alpha_k)\right)(\phi_0(\mathbf{w}^*) - \phi_0(\mathbf{w}_0) + \phi_0(\mathbf{w}_0) - \mathcal{L}(\mathbf{w}^*))$$
$$= \left(\prod_{k<t}(1 - \alpha_k)\right)(\phi_0(\mathbf{w}^*) - \phi_0(\mathbf{w}_0) + \mathcal{L}(\mathbf{w}_0) - \mathcal{L}(\mathbf{w}^*)) . \quad \square$$

Proof of Lemma 3 We have $\alpha_t^2 = \prod_{s \leq t}(1 - \alpha_s)$. Notice that this implies an implicit relation $\alpha_{t+1}^2 = (1 - \alpha_{t+1})\alpha_t^2$. To finish the proof, we show that $\alpha_{t-1} \leq 2/t$. We have

$$\frac{1}{\alpha_{t+1}} - \frac{1}{\alpha_t} = \frac{\alpha_t - \alpha_{t+1}}{\alpha_t \alpha_{t+1}} = \frac{\alpha_t^2 - \alpha_{t+1}^2}{\alpha_t \alpha_{t+1}(\alpha_t + \alpha_{t+1})} = \frac{\alpha_t^2 - \alpha_t^2(1 - \alpha_{t+1})}{\alpha_t \alpha_{t+1}(\alpha_t + \alpha_{t+1})},$$

where the last equality follows from the implicit relation. Now, using the fact that $\alpha_t > \alpha_{t+1}$ we get that $\frac{1}{\alpha_{t+1}} - \frac{1}{\alpha_t} \geq \frac{\alpha_t^2 \alpha_{t+1}}{\alpha_t \alpha_{t+1} 2\alpha_t} = \frac{1}{2}$. We also have $\alpha_0^2/(1 - \alpha_0) = 1$, and thus $\alpha_0 = (\sqrt{5} - 1)/2 < 1$. Therefore, we get $1/\alpha_{t-1} \geq (t-1)/2 + 1/\alpha_0 \geq (t+1)/2$. This in turn implies that $\alpha_{t-1}^2/2 \leq 2/(t+1)^2$, and the proof is completed. $\quad \square$

Proof of Lemma 4 The proof is very similar to the proof of Lemma 2.3 in [1]. The proof essentially works by first getting a first order expansion of the loss $\mathcal{L}$ around the point $\mathbf{w}_{t+1}$, and then shifting the point of expansion to $\mathbf{y}_{t+1}$.

In order to get the expansion around $\mathbf{w}_{t+1}$, we will separately get expansions for the smooth part $\mathcal{F}$ and the 1-norm parts of the loss. For the smooth part, we first take the first order expansion around $\mathbf{y}_{t+1}$:

$$\mathcal{F}(\mathbf{w}) \geq \mathcal{F}(\mathbf{y}_{t+1}) + \langle \nabla \mathcal{F}(\mathbf{y}_{t+1}), \mathbf{w} - \mathbf{y}_{t+1} \rangle. \tag{32}$$

We will combine this with the expansion in (8) around the point $\mathbf{y}_{t+1}$ to get an upper bound for the point $\mathbf{w}_{t+1}$. We have:

$$\mathcal{F}(\mathbf{y}_{t+1} + \boldsymbol{\delta}) \leq \mathcal{F}(\mathbf{y}_{t+1}) + \sum_{j=1}^{n}\left(g_j \delta_j + (\kappa L_j/2)\delta_j^2\right),$$

where $g_j = \nabla \mathcal{F}(\mathbf{y}_{t+1})_j$. Substituting $\mathbf{w}_{t+1}$ for $\mathbf{y}_{t+1} + \boldsymbol{\delta}$ we get

$$\mathcal{F}(\mathbf{w}_{t+1}) \leq \mathcal{F}(\mathbf{y}_{t+1}) + \langle \nabla \mathcal{F}(\mathbf{y}_{t+1}), \mathbf{w}_{t+1} - \mathbf{y}_{t+1} \rangle + \sum_{j=1}^{n} (\kappa \mathsf{L}_j/2)(w_{t+1,j} - y_{t+1,j})^2.$$

Subtracting the previous equation from (32) and rearranging

$$\mathcal{F}(\mathbf{w}) \geq \mathcal{F}(\mathbf{w}_{t+1}) + \langle \nabla \mathcal{F}(\mathbf{y}_{t+1}), \mathbf{w} - \mathbf{w}_{t+1} \rangle - \sum_{j=1}^{n} (\kappa \mathsf{L}_j/2)(w_{t+1,j} - y_{t+1,j})^2.$$

Next we get an expansion for the non-smooth 1-norm part of the loss. If $\boldsymbol{\nu}$ is a subgradient for the $\lambda_1 \|\cdot\|_1$ function at the point $\mathbf{w}_{t+1}$, then we have the following expansion: $\lambda_1 \|\mathbf{w}\|_1 \geq \lambda_1 \|\mathbf{w}_{t+1}\|_1 + \langle \boldsymbol{\nu}, \mathbf{w} - \mathbf{w}_{t+1} \rangle$. Combining with the expansion of the smooth part we get

$$\mathcal{L}(\mathbf{w}) \geq \mathcal{L}(\mathbf{w}_{t+1}) + \langle \nabla \mathcal{F}(\mathbf{y}_{t+1}) + \boldsymbol{\nu}, \mathbf{w} - \mathbf{w}_{t+1} \rangle - \sum_{j=1}^{n} (\kappa \mathsf{L}_j/2)(w_{t+1,j} - y_{t+1,j})^2.$$

We will carefully choose the subgradient vector $\boldsymbol{\nu}$ so that the jth coordinate of $\boldsymbol{\nu} + \nabla \mathcal{F}(\mathbf{y}_{t+1})$, i.e., $\nu_j + g_j$ satisfies

$$\nu_j + g_j = \kappa \mathsf{L}_j (w_{t+1,j} - y_{t+1,j}), \tag{33}$$

matching the definition of $\eta_{t,j}$ in (29). We will show how to satisfy (33) later, but first we show how this is sufficient to complete the proof. Using this we can write the above expansion as

$$\mathcal{L}(\mathbf{w}) \geq \mathcal{L}(\mathbf{w}_{t+1}) + \langle \boldsymbol{\eta}_t, \mathbf{w} - \mathbf{w}_{t+1} \rangle - \sum_{j=1}^{n} (\kappa \mathsf{L}_j/2)(w_{t+1,j} - y_{t+1,j})^2.$$

By shifting the base in the inner product term to $\mathbf{y}_{t+1}$ we can write it as

$$\langle \boldsymbol{\eta}_t, \mathbf{w} - \mathbf{w}_{t+1} \rangle = \langle \boldsymbol{\eta}_t, \mathbf{w} - \mathbf{y}_{t+1} \rangle + \langle \boldsymbol{\eta}_t, \mathbf{y}_{t+1} - \mathbf{w}_{t+1} \rangle$$
$$= \langle \boldsymbol{\eta}_t, \mathbf{w} - \mathbf{y}_{t+1} \rangle + \sum_{j=1}^{n} \kappa \mathsf{L}_j (w_{t+1,j} - y_{t+1,j})^2.$$

Substituting this in the above, we get

$$\mathcal{L}(\mathbf{w}) \geq \mathcal{L}(\mathbf{w}_{t+1}) + \langle \boldsymbol{\eta}_t, \mathbf{w} - \mathbf{y}_{t+1} \rangle + \sum_{j=1}^{n} (\kappa \mathsf{L}_j/2)(w_{t+1,j} - y_{t+1,j})^2.$$

Notice that the right side of the above expression is $\hat{\mathcal{L}}_t(\mathbf{w})$.

To complete the proof we need to show how to select $\boldsymbol{\nu}$ so as to satisfy (33). We will do so based on the optimality properties of $\mathbf{w}_{t+1}$. Recall that by choice $w_{t+1,j} = \mathcal{P}_{\kappa \mathsf{L}_j}^{g_j}(y_{t+1,j}) = y_{t+1,j} + \delta_j^*$, where δ_j^* minimizes the function $f_j^{\kappa \mathsf{L}_j}$ defined as in (4): $f_j^{\kappa \mathsf{L}_j}(\delta) = g_j \delta + \frac{1}{2} \kappa \mathsf{L}_j \delta^2 + \lambda_1 |y_{t+1,j} + \delta| - \lambda_1 |y_{t+1,j}|$, By optimality conditions for the convex function $f_j^{\kappa \mathsf{L}_j}$, we have $g_j + \kappa \mathsf{L}_j \delta_j^* + \nu_j = 0$, for some ν_j that is a subgradient of the $\lambda_1 |y_{t+1,j} + \cdot|$ function at the point δ_j^*, or in other words, a subgradient of the function $\lambda_1 |\cdot|$ at the point $\mathbf{w}_{t+1,j}$. Therefore there exists a subgradient $\boldsymbol{\nu}$ of the $\lambda_1 \|\cdot\|_1$ function at the point satisfying (33), completing the proof.

Inner Ensembles: Using Ensemble Methods Inside the Learning Algorithm

Houman Abbasian[1], Chris Drummond[2], Nathalie Japkowicz[1],
and Stan Matwin[3,4]

[1] School of Electrical Engineering and Computer Science
University of Ottawa,
Ottawa, Ontario,Canada, K1N 6N5
habba057@uottawa.ca, nat@site.uottawa.ca
[2] National Research Council of Canada,
Ottawa, Ontario, Canada, K1A 0R6
Christopher.Drummond@nrc-cnrc.gc.ca
[3] Dalhousie University, Halifax, Canada
stan@cs.dal.ca
[4] Institute for Computer Science, Polish Academy of Sciences, Poland

Abstract. Ensemble Methods represent an important research area
within machine learning. Here, we argue that the use of such methods
can be generalized and applied in many more situations than they have
been previously. Instead of using them only to combine the output of
an algorithm, we can apply them to the decisions made inside the learning algorithm, itself. We call this approach Inner Ensembles. The main
contribution of this work is to demonstrate how broadly this idea can
applied. Specifically, we show that the idea can be applied to different
classes of learner such as Bayesian networks and K-means clustering.

Keywords: Inner Ensembles, Bayesian Network, K-means, Comprehensibility.

1 Introduction

The idea of the *wisdom of crowds* is that decisions made by groups of people
are often more accurate, and more robust, than those made by individuals. An
important sub-field in machine learning, ensemble methods, has exploited this
idea very effectively, particularly in producing substantial performance gains.
However, we argue that there is considerable room to extend it further. We
believe our work is just the beginning of a much wider use of ensemble methods.
Here, instead of combining the output of models, we apply ensemble methods
to the decisions used to generate the models. We call this idea *Inner Ensembles*
as the *wisdom of crowds* is applied inside the learning algorithm. Although this
idea has been applied to decision trees [1,2], it is in fact much more general
and has broader benefits than previously thought. Here we argue that like more
traditional ensemble methods, *Inner Ensembles* define a broad framework that
has the potential to impact all kinds of algorithms other than just decision trees.

H. Blockeel et al. (Eds.): ECML PKDD 2013, Part III, LNAI 8190, pp. 33–48, 2013.
© Springer-Verlag Berlin Heidelberg 2013

Using this framework, we can realize many of the advantages of traditional ensemble methods while restoring the more intuitive models produced by the base algorithms. Many of us have worked extensively with such models and we have a clear sense of what has been learned. This is particularly important when the task is knowledge discovery rather than prediction. On a more practical level, Inner Ensembles produce models with a number of clear advantages: comprehensibility, stability, simplicity, fast classification and small memory footprint. Certainly, many of these are problematical for traditional ensemble methods [3]. However, we recognize that these advantages must often be traded off against absolute performance. The work reported here shows that much of the improved performance can be maintained. Continued refinement of the approach should lead to further improvement.

Comprehensibility, how understandable a model is to users, is essential in many real-world problems: medicine, fraud detection in insurance companies, loan concession in financial environments [2]. Comprehensibility acts as a validation tool in some domains such as medical diagnosis; users are confident in a system only when they understand how it arrives at decisions [4]. A comprehensible model helps in identifying important hidden feature relationships. It may suggest better representations, improving an algorithm's generalization power [5]. Finally, comprehensibility may help to refine "approximately-correct"' domain theories [6]. Comprehensibility is an important feature of inner-ensembles, but not the only one. Stability is the property of being robust to small changes in the underlying data [7]. Robust models are important because they evoke more confidence that the underlying concept has been truly captured and their accuracy is less susceptible to noise. Simplicity is an important property in its own right, typically motivated by Occam's razor [8]. The closely related concept of over-fitting avoidance has been an important issue within machine learning for some years. It is not without controversy though and the exact reasons for the desirability are open to question [9]. Simplicity in terms of the actual features used also leads to another two desirable properties: fast classification and small memory footprint. In many on-line applications, the speed of determining membership, either in classification or clustering, is an important practical consideration [10].

There has been quite a bit of work addressing the shortcomings of the standard ensemble method, particularly the lack of comprehensibility. There are generally two directions that have been followed. The first uses an ensemble as part of the predictive model gaining the comprehensibility through the simplified high-level structure [11,12]. The second uses a standard ensemble as a guide to growing a new and simpler model that is comprehensible [13,14,15]. However, our approach, Inner Ensembles, is quite different, using the ensemble to chose the parts of the model. To illustrate the point, standard ensemble learning is analogous to a management meeting in a company where decisions are made by voting; Inner Ensembles learning is analogous to one manager being selected by voting to make the decision on behalf of the rest of the group. To the best of our knowledge there are two pieces of work that can be considered as Inner Ensembles [1,2]. However, these two were focused solely on decision trees and then just on choosing the right feature to split. What we are doing is generalizing this idea to work for many different kinds of algorithms. Certainly, our framework offers additional benefits. To support this claim we introduce general guidelines for using Inner Ensembles which we have applied to two categories of learning: supervised and unsupervised. For the former we use Bayesian networks, for the latter K-means clustering. We use ensemble methods similar to bagging. In the future work section, we discuss the other potential ways of extending this idea especially using boosting instead of bagging and using Inner Ensembles for other kinds of algorithms. In the rest of the paper, we begin by explaining how to apply the framework for Inner Ensembles to existing algorithms. Next, we present experiments that show the efficacy of our framework. Finally, we will draw conclusions and suggest how future work will explore new applications for Inner Ensembles.

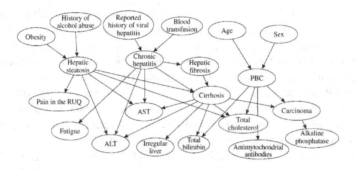

Fig. 1. An example of a liver disorder diagnosis network [16]

2   Inner Ensembles in Practice

In this section, we describe general guidelines for applying Inner Ensembles to any algorithms. Then using these guidelines, we describe in detail how our framework is applied to two quite different algorithms: Bayesian networks and K-means clustering. By choosing both a supervised and unsupervised algorithm, we aim to demonstrate the broad applicability of our framework. As the structure of each algorithm is different, it is difficult to define a precise method for applying Inner Ensembles in every case. However, with the knowledge of how each algorithm works, we can follow some general guidelines. We need to identify decision points, where choices are made based on a measure. Then, we need to locate the input and output of that decision maker. By manipulating the input, say by sampling the data, we produce different outputs. We combine these outputs and apply the result to the decision being made at that point inside the algorithm. This procedure is shown in algorithm 1.

Algorithm 1. General guidelines for using Inner Ensembles

1: **Locating a decision maker inside the algorithm.**
2: **Finding a measure based on which that decision maker works.**
3: **Indicating the input and output of the measure.**
4: **Applying the ensemble on the measure:**
5:        Changing the input of the measure in different ways. {It can be sampling of data or feature based modifications or other methods.}
6:        Generating an output based on each input.
7:        Combining the outputs
8: **Applying the output of the ensemble for decision making.**

2.1 Learning Bayesian Network Structure

Bayesian networks have been used for many applications: medicine, expert systems [17] and path finder systems [18]. One of the major advantages is their comprehensibility [19]. This advantage would clearly be lost when multiple networks form an ensemble. Let us illustrate this point using the real world application of liver disorder diagnosis [16]. The network, shown in figure 1, consists of the risk factors and symptoms for several related disorders and it is important that it be understandable for a clinician. For example, the network shows that *Alcohol Abuse* and *Obesity* are risk factors for *Hepatic Steatosis*, a fatty liver. *Hepatic Steatosis*, itself, may produce *Pain* and is linked to *Cirhosis*. When using an ensemble of Bayesian networks, we would have many such networks with different numbers of arcs which may represent quite different relationships; it would be difficult for clinician to understand such a model. By using Inner Ensembles, we gain many of the benefits of the traditional ensemble method while keeping the comprehensibility of the base algorithm.

Bayesian networks specify a set of conditional independence assumptions; this is captured in the structure of the network. To completely specify the network, we also need conditional probability tables [20]. Therefore, for Bayesian networks, there are two distinct learning problems, we focus on the former. Specifically, we apply our framework to build the structure of the network during the learning phase of the algorithm. Bayesian network algorithms search for the best network structure among all possible ones. The popular K2 algorithm, shown in Algorithm 2, uses a scoring function to determine the better of two networks and a given ordering of nodes to determine the sequence in which they are processed [21]. For all the nodes in the ordered list (line 2), the algorithm begins with no parents for each node (line 3), π_i is the set of parents of the ith node. At each step (line 4), one parent is added to the node, and the score, $g(i, \pi_i \cup \{z\})$, of the network structure is calculated (line 5), $Prec(x_i)$ in algorithm 2 denote the nodes that precede node x_i in the ordered list. The parent that increases the score the most is added to the node's list of parents π_i. Adding parents to the node stops if the addition does not increase the score of the structure (lines 7-11). On completion, the algorithm produces a final structure.

We follow the general guidelines presented in algorithm 1 to apply Inner Ensembles to the Bayesian network algorithm. First we locate a decision maker.

Algorithm 2. K2 algorithm

1: **Input:** A set of n nodes, An ordering of the nodes, u maximum number of parents for each node, A database D containing N instances.
2: **for** i=1 to n **do**
3:     $\pi_i = \phi$, $P_{old} = g(i, \pi_i)$, OKTOProceed = true.
4:     **while** OKTOProceed **and** $|\pi_i| < u$ **do**
5:         z is the node in $Pred(x_i)$ - π_i that maximizes $g(i, \pi_i \cup \{z\})$
6:         $P_{new} = g(i, \pi_i \cup \{z\})$
7:         **if** $P_{new} > P_{old}$ **then**
8:             $P_{old} = P_{new}$, $\pi_i = \pi_i \cup \{z\}$
9:         **else**
10:            OKTOProceed = false
11:        **end if**
12:    **end while**
13:    **write** (Node: x_i , Parents of this node: , π_i)
14: **end for**

Algorithm 3. LOO global score "g" for K2 algorithm.

1: $Acc = 0$.
2: $D = D_1, ... D_N$. {Instances}
3: **for** $i = 1$ to N **do**
4:     Estimate conditional probability for network using existing structure and D-D_i.

5:     $Acc = Acc + PredictAccuracy(D_i)$
6: **end for**
7: score is: $\frac{Acc}{N}$.

According to algorithm 2, it decides if a node can be added as a parent. The next step is to find a score function for this decision maker, $g(i, \pi_i \cup \{z\})$. There are two types: local and global [22]. We use the global score function g (line 3 algorithm 2) calculated as shown in algorithm 3. Here, using leave-one-out cross validation (LOO), the algorithm extracts one instance for validation, the rest of the data forming the training set (line 4). At each iteration, the network is built using the training set and tested on the single instance. The final score of the network is the average of the scores across all splits of the data (lines 5,7). The nest step according to algorithm 1 is to indicating the input and the output of the scoring function (LOO global score). The input is the training data and the output is the score. Then, line 5 of algorithm 1, we change the input of the measure by generating E sampling of the data for which E is the ensemble size. For each of those E samplings, the output of the LOO global score is calculated (line 6 algorithm 1) and finally the E outputs are combined by averaging (line 7). Using this procedure, the global score is redefined as shown in algorithm 4, we call this the $g_{Ensemble}$ score. Sampling of the data in algorithm 4 can be any kind of sampling. It can be with or without replacement. It can also be a different size with respect to the original dataset. Depending on the type of the sampling, we have different ensemble methods. For example, in the case of sampling with

Algorithm 4. Ensemble LOO global score "$g_{Ensemble}$" for K2 algorithm.

1: E {Ensemble Size}
2: $D = D_1, ... D_N$. {Instances}
3: $FinalScore = 0$.
4: **for** $k = 1$ to E **do**
5:    $D_S = Sample(D, Size)$ {$D_S = D_{S1}, ... D_{SSize}$}
6:    $FinalScore = FinalScore + LOOScore(D_S)$.
7: **end for**
8: score is: $\frac{FinalScore}{E}$

---

replacement and sample size of 100%, we have bagging. Once the best structure has been selected, the algorithm proceeds in the conventional fashion calculating the conditional probability tables necessary for the complete Bayesian network.

2.2   K-means Clustering

For unsupervised learning, we apply Inner Ensembles to K-means clustering, a very popular clustering method. One advantage of such a method is that it characterizes each cluster in terms of a single point, called a prototype. The prototype is a representative of all samples inside that cluster and thus the meaning of that cluster. The idea that a prototype is an important way of representing a particular concept is central to certain theories in Cognitive Science [23]. For our purposes, what is important about prototypes is that they help in the human understanding of the structure of a particular problem or domain. In medical diagnosis, a prototype might be a vector of numerical results from tests or measurements. A clinician assessing an individual's risk for a heart problems will combine measurements of blood pressure, cholesterol and weight. A prototype, identified in a clustering over many subjects, representing a high risk patient would be clearly separated from one of low risk. Prototypes offer other advantages such as use in compression and for efficient finding of the nearest neighbor [24]. Using a traditional ensemble of K-means clustering, we loose the prototypes which are the cluster centers for K-means. Using Inner Ensembles we keep the cluster prototype, and thus comprehesibility, while gaining the performance advantage of the ensemble method.

Algorithm 5 details the steps in K-means clustering [25]. It starts by initializing the cluster centers (line 1). Then, over several iterations, it assigns instances to the closest cluster center and updates the centers (lines 4-7). Finally, the algorithm stops when there are no more changes in the cluster centers (line 9). Following the lines of general guidelines, algorithm 1, we need to find a decision maker inside K-means algorithm that assigns each instance to the closest cluster center (line 1). Next we find a score function, the Euclidean distance in this case (line 2). Next we find the input and output of the score function (line 3). The input of the Euclidean distance is the cluster centers and the data; the output is the distance of the data to each cluster center. Thus we can change the input of the measure that is the data by generating E different sampling

---

Algorithm 5. K-means clustering algorithm.

1: Initialize K cluster centers $\mu_1, \mu_2, ..., \mu_K$
2: Data: $X = \{X_1, X_2, ..., X_N\}$
3: **repeat**
4:    **for** All data instances X_i **do**
5:       $m = argmin_K\{d(X_i, \mu_K)\}$ {Closest cluster center to each instance}
6:       Assign instance X_i to cluster center μ_m
7:    **end for**
8:    Update cluster centers.
9: **until** (No assignment change or max iterations)
10: **Return** The clusters.

Algorithm 6. Inner K-means clustering algorithm

1: Initialize K cluster centers $\mu_1, \mu_2, ..., \mu_K$
2: Data: $X = \{X_1, X_2, ..., X_N\}$
3: **repeat**
4:    **for** All data instances X_i **do**
5:       $m = \{m_1, ..., m_N\}$.
6:       **for** j= 1 to Ensemble size **do**
7:          $S = Sample(FeatureSpace)$
8:          $I = argmin_E\{d(X_{Sj}, \mu_E)\}$ {Closest cluster center to each instance}
9:          $m_I = m_I + 1$
10:       **end for**
11:       $L = argmax_e(m_e)$ {Voting}
12:       Assign instance X_i to cluster center μ_L
13:    **end for**
14:    Update cluster centers.
15: **until** (No assignment change or max iterations)
16: **Return** The clusters.

---

of the data (line 5). Although we initially used sampling of the instances, our experimental results were poor because it has little impact on the location of the centers. Thus we lose diversity that is very important for ensemble methods. To improve diversity, we use random subsets of features for each ensemble member. The algorithm repeatedly selects a random subset of features. Next the outputs of each generated input are calculated (line 6). In this case based on the feature subset, the closest cluster center is found using Euclidean distance. Therefore for each data instance, we have a set of candidate cluster centers according to different feature subsets. In the combining step (line 7), the cluster center with the most number of votes is selected for that particular data instance. This is shown in algorithm 6.

3   Experimental Results

In this section, we run several experiments for our new versions of the K2 Bayesian network and K-means clustering algorithms. We believe that Inner

Ensembles attain many of the benefits of traditional ensemble through similar means, i.e.. reducing the variance in the bias-variance trade-off [26]. So, we expect that whenever the ensemble methods work, Inner Ensembles will. Thus, the experiments test two different hypotheses: that our new versions are superior in performance to the base algorithms and that whenever the traditional ensemble method improves the performance, Inner Ensembles improves it too. For Bayesian networks, we compare the results with bagging because the Inner Ensembles sampling method we used is similar to bagging. For Inner K-means, we compare the results with several cluster ensemble methods in terms of different cluster validation measures. For both experiments, we use UCI repository datasets [27]. As the statistical test, we used Friedmans test and for post-hoc we use Nemenys test both with $\alpha = 0.05$ in all of the experiments. We report the results of a sensivity analysis study in the course of which different parameters of the algorithm are considered. These parameters include:

- *Ensemble Size:* 10 different ensemble sizes 10-100.
- *Sample Size:*10 different sampling sizes 10%-100%
- *Resampling Mode: With* or *without* replacement.

3.1 Inner Ensemble K2

In addition to the experiments noted above, for Bayesian networks we run another to confirm that in terms of classification time as well as in terms of comprehensibility our method is superior to bagging. We use 14 datasets from the UCI repository with different number of classes, features and instances. In all of the tables, BN is the original network, IEBN the Inner Ensemble Bayesian Network and BGBN the bagging of Bayesian networks.

For each of the parameters, we used 10 fold cross validation 10 times. For each run, the average error of the network is calculated. Table 1 shows the comparison of the accuracy for different ranges of parameters. In this table IEBN No Rep means sampling without replacement. Also S-Size and En-Size list the range of parameters for the sample size and the ensemble size for which IEBN performs better than BN. For better understanding, we include the percentage of parameters that results in better performance and we list the average and the best results for those parameters. For simplicity of presentation, we only report the results obtained for sampling without replacement because in this sampling regimen, IEBN works better than BN on 62% of the parameters while for sampling with replacement it only works on 46% of the parameters.

We begin by comparing the average performances of IEBN and BGBN with original BN. The null hypothesis for Friedman's test is BN , IEBN and BGBN perform similarly. The Friedman statistic is 26.14 for IEBN and BGBN while the critical value for Friedman's test is 6.00 for the 3 models. Thus we can reject the null hypothesis, there is at least one classifier with significant difference in performance. We use Nemenyi's test to rank the classifiers according to their performance. The critical value for Nemenyi's test is equal to 2.34 for the 3 models and the q-values for BN-IEBN and IEBN-BGBN are 2.83 and 2.27 respectively

Table 1. The Accuracy of the Various Algorithms

Dataset	BN	IEBN No Rep			IEBN No Rep		BGBN	
		S-Size	En Size	Perc	Avrg	Best	Avrg	Best
Breast-w	97.1	[10-90]	[10-100]	70%	97.2	97.31	97.18	97.18
Credit-a	85	[10-90]	[10-100]	99%	85.62	86.06	86.24	86.48
Ecoli	80.77	[10-90]	[10-100]	50%	81.03	81.7	85.35	85.92
Glass	70.84	[10-90]	[10-50]	4%	71.1	71.31	74.21	75.14
Heart-c	80.76	[10-90]	[10-100]	96%	81.63	82.74	82.56	82.94
Heart-S	82.19	[10-90]	[10-100]	32%	82.48	82.96	82.71	83.19
Hepatitis	82.39	[10-60]	[40-100]	10%	82.75	83.55	82.88	83.03
Iris	92.87	[10-90]	[10-100]	81%	93.24	93.8	94.58	94.93
Labor	88.25	[10-90]	[10-100]	99%	90.64	92.81	93.04	94.04
Liver	56.93	[10-90]	[10-100]	90%	57.41	58.41	63.75	64.58
Lymph	81.96	[10-90]	[10-100]	97%	83.19	85.14	84.66	85.41
Pima	74.66	[10-90]	[10-100]	97%	75.35	76.05	75.77	75.98
Tic-tac	92.44	[50-90]	[10-100]	18%	92.58	92.73	96.87	97.18
Vote	96.05	[30-90]	[10-100]	26%	96.12	96.25	96.15	96.21
Average	83			62%	83.59	84.34	85.42	85.87

which shows that the performance of BN on all the datasets is significantly different from that of IEBN. On the other hand the performance of IEBN and BGBN is not significantly different. We also use Friedman's test to compare the BN with the best results of IEBN and the best results of BGBN. Here the Friedman statistic is 22.29 and the critical value for Friedman's test is 6.00 for the 3 models. Thus we can reject the null hypothesis. By performing Nemenyi's test, the q-value for BN-IEBN and IEBN-BGBN are 3.4 and 1.13 which confirm that IEBN is significantly different from BN but not from BGBN. Generally we can see that with sampling without replacement, on 62% of the entire parameters IEBN works significantly better than BN on all of the datasets.

Generally, as expected, IEBN improves the performance over the original network whenever Bagging does but the improvement is smaller. Therefore we are not expecting improvements on Breast-w and Vote dataset because bagging does not. For individual datasets on which bagging improves the performance , except for Tic-tac-toe, the performance improves over that of the base algorithm. However the gain for Ecoli and Glass datasets are small. Future work on the way of creating ensembles inside the Bayesian network, will determine how closely our method can approach bagging in terms of performance. At present, our method is just applied to learning the structure of the network. One possible avenue for future work is to apply Inner Ensembles for the second part of Bayesian network learning, estimating conditional probabilities.To compare the classification times of Inner Ensembles and bagging, we ran the experiment on all of the datasets for different ensemble sizes, and averaged over all of the datasets. Figure 2 shows the average time for different ensemble size for BGBN versus IEBN. As we can see from this figure, the classification time is a lot faster for Inner Ensembles than bagging. An increase in the ensemble size in the case of Inner Ensembles

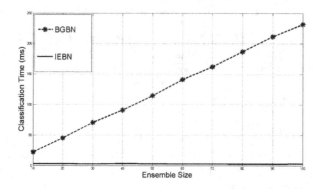

Fig. 2. Classification time (mS) for BGBN vs IEBN

has little effect on classification time because the number of models remains the same.

So in this section, we have shown how our method achieves some of the performance improvement of bagging while maintaining the comprehensibility, classification performance and fast classification time of the original Bayesian network.

3.2   Inner Ensemble K-means

To test our new version of K-means, we compare results on 15 UCI datasets, each with the necessary numerical attributes used by such a clustering system. Among possible different performance measures for clustering, we use Normalized Mutual Information that determines the shared information between the clusterings [28] and Purity that measures the coherence of a cluster [29]. Equation 1 shows NMI for which $I(X,Y)$ is the mutual information between two random variables X,Y; $H(X)$ is the entropy of X; X is the result of the clustering; Y contains the true labels. Equation 2 shows the cluster Purity. In this equation, n is the total number of instances, m the number of clusters and P_j is the fraction of the cluster to which the largest class of objects is assigned. For both measures, the larger the value the better the results.

$$NMI(X,Y) = \frac{I(X,Y)}{\sqrt{H(X)H(Y)}} \qquad (1)$$

$$Purity = \sum_{j=1}^{m} \frac{n_j}{n} P_j \qquad (2)$$

For these experiments, we compare the performance of inner K-means with two different types of cluster ensemble method. The first is a graph-based approach (G-Based) that includes three different methods CSPA, MCLA and HGPA [28]. The second is based on pair-wise similarity. The methods of this latter group first build a similarity matrix, called a co-association matrix and then uses it for different forms of hierarchical clustering. This group is also called

Hierarchical Agglomerative Clustering Consensus (HAC) [30]. There are three different methods for HAC based on the hierarchical clustering: single, average and complete linkage.

For simplicity, we do not report the results of all 6 ensemble methods. Instead, for the graph based methods, we report the best performance of CSPA, MCLA and HGPA. For the HAC methods, we report the best performance among the different linkage types. For each of the parameters (ensemble size, sample size), we run Inner Ensemble, cluster ensemble and original K-means 30 times with 30 different initial cluster centers and the average performance is calculated. For a fair comparison, we run original K-means with the same cluster centers as used for Inner Ensembles. As the Inner Ensembles uses different subsets of features, we generate the members for cluster ensemble the similar way. If the number of instances are larger than 500, we sample the data without replacement until 500 is reached.

In table 2, we report the results of a sensitivity analysis using a range of parameters. We only report the results obtained with sampling with replacement, using this resampling strategy, Inner Ensemble K-means (IEK-m) works better than K-means on 55% of the parameters while for sampling without replacement it just works on 53% of the parameters. From this table, we can see that IEK-m generally outperforms K-means, the graph-based and HAC methods in terms of the NMI measure. For the Heart-s, Pima and Sonar datasets, the NMI is close to zero, the clusters that where extracted from these datasets are nothing like the true classes. For Sonar and Vehicles and Yeast the cluster ensemble methods do not improve the performance and, as we expected, IEK-m does not improve it either. IEK-m improves the performance on the rest of the datasets except for Ecoli for which one of the ensemble methods decreases the performance. In comparison, HAC improves the performance on just 8 out of 15 datasets, does not improve it on 7 .G-based methods improve the performance on 9 out of 15, do not improve it on 3 and decrease it on 3 datasets.

To compare the performances all datasets, first we compare the average performance of K-means, IEK-m, G-Based and HAC. The Friedman statistics for the average case for K-means, IEK-m, G-Based and HAC is 9.66, larger than the critical value 7.5 for 4 models. The q-value for K-means and IEK-m for NMI is 2.75, larger than the critical value which is 2.57 for 4 models. This shows that IEK-m is significantly different from K-means on all datasets in the average case. Also for the NMI measure, the q-values for HAC and G-Based are 0.14 and 1.06 respectively which are both less than 2.57 showing that that for the average case, IEK-m is not significantly different from G-Based and HAC. To compare the best results, the Friedman statistics for the best case is 16.5 which is larger than the critical value 7.5 for 4 models. The q-value for K-means and IEK-m for NMI is 3.88 which is larger than the critical value which is 2.57 for 4 models. This shows that for the best case IEK-m is significantly different from the K-means on the entire data. Also for the NMI measure in the best case, the q-values for HAC and G-Based are 1.06 and 2.12 which both are less than 2.57 . In the best case , IEK-m is not significantly different from G-Based and HAC.

Table 2. Comparison of K-means, Inner Ensemble K-means and Ensemble K-means (G-Based, HAC) in terms of NMI

Dataset	K-means	IEK-m Rep S-Size	En Size	Perc	IEK-m Rep Avrg	Best	G-Based Avrg	Best	HAC Avrg	Best
Ecoli	0.59	[-]	[-]	0%	0.59	0.59	0.57	0.57	0.61	0.62
Glass	0.38	[40-90]	[10-100]	49%	0.4	0.42	0.34	0.34	0.38	0.38
Heart-s	0.02	[20-90]	[10-100]	88%	0.11	0.31	0.04	0.05	0.02	0.02
Iris	0.71	[50-90]	[10-100]	48%	0.75	0.8	0.77	0.82	0.77	0.79
Letter	0.44	[30-90]	[10-100]	63%	0.45	0.46	0.47	0.47	0.47	0.48
Optdig	0.72	[20-90]	[10-100]	86%	0.74	0.75	0.75	0.76	0.75	0.77
Pendig	0.67	[40-90]	[10-100]	54%	0.68	0.7	0.68	0.69	0.69	0.7
Pima	0.02	[30-90]	[10-100]	68%	0.05	0.09	0.11	0.11	0.02	0.02
Segment	0.54	[20-90]	[10-100]	67%	0.59	0.63	0.56	0.58	0.61	0.62
Sonar	0.01	[10-90]	[10-100]	66%	0.01	0.02	0.01	0.01	0.02	0.02
Vehicle	0.19	[20-90]	[10-100]	80%	0.19	0.2	0.19	0.19	0.19	0.2
Vowel	0.2	[20-80]	[10-100]	34%	0.24	0.29	0.21	0.21	0.21	0.21
Wavef	0.36	[20-90]	[10-100]	53%	0.37	0.38	0.36	0.37	0.36	0.37
Wine	0.43	[20-90]	[10-100]	69%	0.55	0.72	0.45	0.51	0.43	0.43
Yeast	0.28	[90-90]	[70-70]	1%	0.28	0.28	0.27	0.27	0.28	0.28
Average	0.37			55%	0.4	0.44	0.39	0.4	0.39	0.39

Table 3. Comparison of K-means, Inner Ensemble K-means and Ensemble K-means (G-Based, HAC) in terms of purity

Dataset	K-means	IEK-m Rep S-Size	En Size	Perc	IEK-m Rep Avrg	Best	G-Based Avrg	Best	HAC Avrg	Best
Ecoli	0.81	[90-90]	[30-90]	6%	0.81	0.82	0.8	0.8	0.81	0.82
Glass	0.57	[30-90]	[10-100]	49%	0.58	0.6	0.6	0.62	0.57	0.57
Heart-s	0.59	[20-90]	[10-100]	88%	0.68	0.81	0.62	0.63	0.59	0.59
Iris	0.83	[50-90]	[10-100]	56%	0.86	0.9	0.91	0.94	0.9	0.92
Letter	0.32	[20-90]	[10-100]	76%	0.34	0.35	0.34	0.35	0.34	0.35
Optdig	0.74	[10-90]	[10-100]	96%	0.77	0.79	0.81	0.82	0.78	0.8
Pendig	0.69	[40-90]	[10-100]	62%	0.71	0.73	0.72	0.73	0.7	0.71
Pima	0.66	[30-80]	[10-100]	48%	0.67	0.7	0.68	0.69	0.66	0.66
Segment	0.57	[20-90]	[10-100]	79%	0.61	0.64	0.62	0.66	0.64	0.65
Sonar	0.55	[10-90]	[10-100]	66%	0.56	0.58	0.55	0.56	0.55	0.56
Vehicle	0.45	[20-90]	[10-100]	84%	0.46	0.47	0.47	0.47	0.47	0.47
Vowel	0.23	[20-80]	[10-100]	46%	0.26	0.31	0.24	0.24	0.23	0.23
Wavef	0.56	[10-90]	[10-100]	40%	0.56	0.59	0.55	0.55	0.56	0.56
Wine	0.7	[20-90]	[10-100]	71%	0.81	0.92	0.75	0.78	0.7	0.7
Yeast	0.52	[50-90]	[50-80]	8%	0.52	0.52	0.52	0.52	0.54	0.54
Average	0.59			58%	0.61	0.65	0.61	0.62	0.6	0.61

Table 3 shows the results of the clustering in terms of the purity measure. We report the results obtained for sampling with replacement because this sampling regimen, IEK-m works better than BN on 58% of the parameters while for sampling without replacement it only works on 53% of the parameters. IEK-m generally outperforms both K-means and ensemble methods. For the average case IEK-m improves the performance on most i.e. 12 out of 15 datasets and for the best case 14 out of 15 datasets. Also in the average case , the HAC methods improve the performance on 7 out of 15 datasets and on the best case 10 out of 15, but not on the rest. The average case for the G-based methods improves the performance on 11 out of 15, decrease the performance of Waveform and Ecoli and with no improvement on the rest and for the best case,G-Based improves the performance on 12 out of 15 dataset and it is otherwise to the average case.

The Friedman statistics for the average case is 11.5 which is larger than the critical value 7.5 for 4 models. On average, the q-value for K-means and IEK-m for Purity is 2.89 which is larger than the critical value which is 2.57 for 4 models. This shows that IEK-m performs significantly better than K-means on all datasets. For the Purity measure, in the average case, the q-values for HAC and G-Based are 0.84 and 0.07 respectively which are both larger than 2.57. Thus IEK-m is not significantly different from G-Based and HAC. The Friedman statistics for the best case is 17.00 larger than the critical value 7.5 for 4 models. In the best case, the q-value for K-means and IEK-m for Purity is 3.81 which is larger than the critical value of 2.57. Thus IEK-m performs significantly better than K-means on all datasets. Also for the best case, the q-values for HAC and G-Based are 1.55 and 0.56 both larger than 2.57. Thus IEK-m is not significantly different from G-Based and HAC.

To sum up, our experiments show that our framework is broadly applicable on different types of algorithms. More specifically, we apply our framework on two groups of supervised and un-supervised learning. Our experiments generally show that: (1) As we expected, wherever ensemble methods work, our framework improves the performance on both the supervised and the un-supervised cases. (2) Our framework improves the performance on a large portion of the parameters. (3) For both supervised and un-supervised learning, our framework improves the performance significantly over the original method. But it does not improve the performance over ensemble methods for supervised learning. However, for un-supervised learning, our results shows that Inner Ensembles works comparably or are superior to the regular ensembles.

4   Limitations and Future Work

One limitation of the work is that the reported results are based on the parameters for which Inner Ensembles improve the performance. A better way of doing the experiments is to find the best parameters prior to evaluating performance. To address this issue, we have run some initial experiments using cross-validation to find the best parameters before running the algorithm on the test data. So far we have only done this for K-means where we obtained very similar results to those presented earlier. To validate the stability, we run IEK-m on different noisy data. We add random noise to the most 50% significant features of the datasets from 20%, 40%, 60% and the primary results shows that the IEK-m is more stable than original K-means and generally comparable to cluster ensembles on 20% and 40% noise. On 60% some of the cluster ensemble methods are more stable than IEK-m but still IEK-m is more stable than original K-means.

So far, we do the experiments to confirm the performance of the Inner Ensembles. But as we mentioned in the introduction there are other advantages of using Inner Ensembles such as stability, classification time and small memory usage. We need further experiments to confirm the stability of the IEBN and a discussion about the memory usage of the Inner Ensembles and speed of IEK-m which we thing can be faster in online clustering. For future work, we want to explore other classification and clustering algorithms. Nor are we restricted solely to those situations, we can apply this idea to a variety of other algorithms, such as the type of search used in learning or the pruning method for decision trees or other groups of methods such as feature selection. One important factor that affects the performance of the ensemble method is diversity. For future work the effect of diversity on the performance of Inner Ensembles needs to be investigated too. The idea can be extended in other ways. First we can improve the specific methods discussed in this paper by using different kind of samplings, such as weighted ones. So far we just used voting, akin to bagging. However, we intend to investigate different kinds of ensemble methods, such as boosting.

5 Conclusion

The main objective of this paper was to extend the possible ways that the machine learning community makes use of ensemble methods. Our particular approach is called Inner Ensembles. By using this new approach on supervised and unsupervised learning algorithms, we showed that this idea is broadly applicable. For supervised learning, we applied our method to the learning of the structure of Bayesian networks, a popular classification algorithm; for unsupervised learning we used it for K-means clustering, a popular clustering method. In the case of Bayesian network, Inner Ensembles work generally better than original Bayesian network but worse than bagging. On the other hand, for K-means, inner K-means performs better than original K-means and comparable to two families of clustering ensemble. We introduced this idea as a framework that has the potential of of being applicable in many different ways other than those we have discussed in this paper.

References

1. Geurts, P., Wehenkel, L.: Investigation and Reduction of Discretization Variance in Decision Tree Induction. In: Lopez de Mantaras, R., Plaza, E. (eds.) ECML 2000. LNCS (LNAI), vol. 1810, pp. 162–170. Springer, Heidelberg (2000)

2. Prez, J.M., Muguerza, J., Arbelaitz, O., Gurrutxaga, I.: A New Algorithm to Build Consolidated Trees: Study of the Error Rate and Steadiness. In: Proceedings of the International Intelligent Information Processing and Web Mining Conference. Advances in Soft Computing, pp. 79–88 (2004)

3. Ferri, C., Hernández-Orallo, J., Ramírez-Quintana, M.J.: From Ensemble Methods to Comprehensible Models. In: Proceedings of the 5th International Conference on Discovery Science, pp. 165–177 (2002)

4. Wolberg, W.H., Street, W.N., Mangasarian, O.L.: Machine Learning Techniques to Diagnose Breast Cancer from Image-Processed Nuclear Features of Fine Needle Aspirates. Cancer Letters 77, 163–171 (1994)

5. Flann, N.S., Dietterich, T.G.: Selecting Appropriate Representations for Learning from Examples. In: Proceedings of the 5th National Conference on Artificial Intelligence, pp. 460–466 (1986)

6. Craven, M.W., Shavlik, J.W.: Extracting Comprehensible Concept Representations from Trained Neural Networks. In: Working Notes on the International Joint Conference on AI Workshop on Comprehensibility in Machine Learning, pp. 61–75 (1995)

7. Dwyer, K., Holte, R.: Decision Tree Instability and Active Learning. In: Kok, J.N., Koronacki, J., Lopez de Mantaras, R., Matwin, S., Mladenič, D., Skowron, A. (eds.) ECML 2007. LNCS (LNAI), vol. 4701, pp. 128–139. Springer, Heidelberg (2007)

8. Blumer, A., Ehrenfeucht, A., Haussler, D., Warmuth, M.K.: Occam's Razor. Information Processing Letters 24(6), 377–380 (1987)

9. Domingos, P.: Occam's Two Razors: The Sharp and the Blunt. In: Knowledge Discovery and Data Mining, pp. 37–43 (1998)

10. Williams, N., Zander, S., Armitage, G.: A Preliminary Performance Comparison of Five Machine Learning Algorithms for Practical IP Traffic Flow Classification. Special Interest Group on Data Communication 36(5), 5–16 (2006)

11. Zimmermann, A.: Ensemble-Trees: Leveraging Ensemble Power Inside Decision Trees. In: Proceedings of the 11th International Conference on Discovery Science, pp. 76–87 (2008)

12. Lou, Y., Caruana, R., Gehrke, J.: Intelligible Models for Classification and Regression. In: Proceedings of the 18th ACM SIGKDD International Conference on Knowledge Discovery and Data Mining, pp. 150–158 (2012)

13. Van Assche, A., Blockeel, H.: Seeing the Forest through the Trees: Learning a Comprehensible Model from a First Order Ensemble. In: Kok, J.N., Koronacki, J., Lopez de Mantaras, R., Matwin, S., Mladenič, D., Skowron, A. (eds.) ECML 2007. LNCS (LNAI), vol. 4701, pp. 418–429. Springer, Heidelberg (2007)

14. Zhou, Z.H., Jiang, Y., Chen, S.F.: Extracting Symbolic Rules from Trained Neural Network Ensembles. AI Communications 16(1), 3–15 (2003)

15. Domingos, P.: Knowledge Discovery Via Multiple Models. Intelligent Data Analysis 2, 187–202 (1998)

16. Pourret, O., Marcot, B., Naim, P.: Bayesian Networks: A Practical Guide to Applications. Statistics in Practice. Wiley, Chichester (2008)

17. Desmedt, J.: Computer-Aided Electromyography and Expert Systems. Clinical Neurophysiology Updates. Elsevier, Amsterdam (1989)

18. Heckerman, D.E., Nathwani, B.N.: Towards Normative Expert Systems: Part ii, Probability-Based Representations for Efficient Knowledge Acquisition and Inference Methods of Information in Medicine. In: Methods of Information in Medicine, pp. 106–116 (1992)

19. Vomlel, J.: Two Applications of Bayesian Networks. In: Proceedings of Conference Znalosti, pp. 73–82 (2002)

20. Mitchell, T.M.: Machine Learning, 1st edn. McGraw-Hill, Inc., New York (1997)
21. Cooper, G.F., Herskovits, E.: A Bayesian Method for the Induction of Probabilistic Networks from Data. Machine Learning 9(4), 309–347 (1992)
22. Bouckaert, R.: Bayesian Network Classifiers in Weka. Working paper series. Department of Computer Science, University of Waikato (2004)
23. Lakoff, G.: Women, Fire and Dangerous Things. University of Chicago Press, Chicago (1987)
24. Tan, P.N., Steinbach, M., Kumar, V.: Introduction to Data Mining, 1st edn. Addison-Wesley Longman Publishing Co., Inc., Boston (2005)
25. Duda, R.O., Hart, P.E.: Pattern Classification and Scene Analysis. John Willey & Sons, New York (1973)
26. Breiman, L.: Bias, Variance, and Arcing Classifiers. Technical report, Statistics Department, University of California at Berkeley (1996)
27. Asuncion, A., Newman, D.J.: UCI Machine Learning Repository (2007)
28. Strehl, A., Ghosh, J., Cardie, C.: Cluster Ensembles - A Knowledge Reuse Framework for Combining Multiple Partitions. Journal of Machine Learning Research 3, 583–617 (2002)
29. Rendón, E., Abundez, I.M., Gutierrez, C., Zagal, S.D., Arizmendi, A., Quiroz, E.M., Arzate, H.E.: A Comparison of Internal and External Cluster Validation Indexes. In: Proceedings of the American Conference on Applied Mathematics and the 5th International Conference on Computer Engineering and Applications, pp. 158–163 (2011)
30. Nguyen, N., Caruana, R.: Consensus Clusterings. In: Proceedings of the 7th International Conference on Data Mining, pp. 607–612 (2007)

Learning Discriminative Sufficient Statistics Score Space for Classification

Xiong Li[1,2], Bin Wang[1], Yuncai Liu[1], and Tai Sing Lee[2]

Department of Automation, Shanghai Jiao Tong University, Shanghai 200240, China
Computer Science Department, Carnegie Mellon University, Pittsburgh 15213, USA
flit.lee@gmail.com, {binwang,whomliu}@sjtu.edu.cn, tai@cs.cmu.edu

Abstract. Generative score spaces provide a principled method to exploit generative information, e.g., data distribution and hidden variables, in discriminative classifiers. The underlying methodology is to derive measures or score functions from generative models. The derived score functions, spanning the so-called score space, provide features of a fixed dimension for discriminative classification. In this paper, we propose a simple yet effective score space which is essentially the sufficient statistics of the adopted generative models and does not involve the parameters of generative models. We further propose a discriminative learning method for the score space that seeks to utilize label information by constraining the classification margin over the score space. The form of score function allows the formulation of simple learning rules, which are essentially the same learning rules for a generative model with an extra posterior imposed over its hidden variables. Experimental evaluation of this approach over two generative models shows that performance of the score space approach coupled with the proposed discriminative learning method is competitive with state-of-the-art classification methods.

Keywords: generative score space, sufficient statistics, discriminative learning, classification.

1 Introduction

Probabilistic generative models and discriminative models are two complementary [1] and important paradigms in machine learning. Generative models are designed to model data distribution, particularly good at dealing with missing data and structured data, e.g., tree structure data or sequences with variable length. They seek to explain data in terms of hierarchical models with hidden variables. These hidden variables encode higher order information related to observed data that could be informative in the identification of data samples. Further, generative models can be used to construct classifier by means of the maximum a posteriori (MAP) decision rule, resulting in naive Bayes or MAP classifier. However, generative models in general are inferior to discriminative classifiers [2,3] which are designed to directly capture the decision boundaries among different classes. Discriminative classifiers can adapt to complex data

H. Blockeel et al. (Eds.): ECML PKDD 2013, Part III, LNAI 8190, pp. 49–64, 2013.
© Springer-Verlag Berlin Heidelberg 2013

using furnished or learned kernel similarity. The feature spaces underlying the kernels are generally implicit.

To integrate the capabilities of generative and discriminative models, several schemes [4,5,6,7] have been proposed. Among them, generative score spaces [8,9,10,11,12] provide necessary explicit feature mappings required in many practical applications [13,8] and is the focus of this paper. These explicit feature mappings or score functions are derived from the generative models of the data distribution. Their values, i.e., features, are then delivered to discriminative classifiers to perform classification. While score spaces have shown promising performance in a variety of challenging applications [14,8,15], discriminative learning approaches [16,6,17,18,19] which can exploit label information in general perform better and still furnish state-of-the-art performance.

In this paper, we propose a score space method with an effective score space and a discriminative learning approach. The score space is spanned by the sufficient statistics of an adopted generative model, and is called sufficient statistics score space (SS). Its score function is a function over random variables, which is distinct from earlier methods [8,10,11] in which the scores are functions over random variables and model parameters. We propose a discriminative learning approach to learn the score space by subjecting the classifier over score space to margin constraints. The simple form of the score function results in simple learning rules, which are the same as those for the generative models but with a discriminative posterior imposed over the hidden variables. This posterior in fact introduces a mechanism to generate a more suitable score space for classification.

Further, we will establish the following properties of the score space: (1) the classification error of a zero-loss linear classifier over the score space is at least as low as that of a MAP classifier; (2) the MAP estimation of the linear classifier weights implied in our discriminative learning approach results in an expression of classifier weights that are equal to the weights of the linear SVMs classifiers over the discriminative score space; (3) the discriminative learning approach favors generative models with less hidden variables.

2   Related Works

2.1   Generative Score Spaces

Generative score space [11,12,20,14,8,10] is a class of methods developed to exploit information provided by generative models for discriminative classification. Score functions or feature mappings are functions defined over the observed data, and the hidden variables and parameters of the generative models. The spaces spanned by the score functions are called score spaces or feature spaces.

The score functions generally are measures over generative models. Fisher score (FS) [11] derives score functions by measuring how model parameters affect the log likelihood. Let $\mathbf{x} \in \mathbb{R}^D$ be the observed variable and $P(\mathbf{x} \,|\, \theta)$ be its marginal distribution parameterized by a vector θ, the i-th component of FS is the differential with respect to the parameter θ_i,

$$\Phi_i(\mathbf{x}, \theta) = \nabla_{\theta_i} \log P(\mathbf{x} \,|\, \theta)$$

Table 1. Summary of related discriminative learning approaches

Methods	Feature Mapping	Dis. Learn. Criterion
FKL [19]	"partially" explicit	1-NN
LM-HMM [6]	-	large-margin
disHMM [16]	-	min. hinge-loss
Med-LDA [17]	topic variable	max-margin
disLDA [18]	-	conditional max. likelihood

Free energy score space (FESS) [8] is based on the measures on how well a data point fits random variables. The resulting score functions are the summation terms of the log likelihood function. Posterior divergence (PD) [10] derives a set of comprehensive measures that are connected to both FS and FESS. Another variant class of these methods derives the score function based on class-conditional models, with a model trained for each class, seeking to utilize the label information. The score functions in [20] are log likelihood functions. TOP kernel (TK) [12] extends FS to operate on the MAP discriminant function instead of the log likelihood function. FS was operated on class-conditional models in [14]. These score spaces, working with classifiers, combine and integrate the capabilities from generative and discriminative models, with competitive results in a variety of challenging tasks [14,8,15] such as image recognition. However, these methods learn score spaces and the classifier separately, and might not fully exploit and utilize the label information.

2.2 Discriminative Learning

Several discriminative learning approaches [16,6,17,18,19] have been proposed to exploit the capabilities of generative models and discriminative models simultaneously. Gales et al. [21] comprehensively reviewed the discriminative learning approaches for speech recognition. Table 1 provides a summary of these approaches. Although several discriminative learning criteria are involved, margin based criteria [6,17] exhibit highly competitive performance.

Fisher kernel learning (FKL) [19] is most related to our approach. It proposed a discriminative learning method for Fisher kernel by minimizing the error rate of 1-nearest neighbor (1-NN) classifiers. We observed that, when the learned kernel or score space working with SVMs or its variants, the potential of this method can be further exploited. A potential improvement for this method is to replace the error measure of the 1-NN classifier with the error measure or some criteria of a classifier that will be used to perform classification.

3   Sufficient Statistics Score Space

We here describe how to formulate the sufficient statistics score space, starting from the variational lower bound of generative models. The idea is to decompose the log likelihood into parameter-based parts and variable-based parts.

3.1   Variational Inference of Exponential Family

We consider a general case where $P(\mathbf{x}; \theta)$ is a hierarchical generative model. Let $P(\mathbf{x}, \mathbf{h}; \theta)$ be its joint distribution with a set of hidden variables $\mathbf{h}$ and the parameter vector θ. In this case, it is usually difficult to obtain the close form of $P(\mathbf{x}; \theta)$ since the integration is usually intractable. A practical method is to resort to the lower bound of $\log P(\mathbf{x}; \theta)$. We here use the lower bound given by variational inference [22], for sample $\mathbf{x}^t$,

$$\log P(\mathbf{x}^t; \theta) \geq \mathrm{KL}(Q(\mathbf{h}^t) \,\|\, P(\mathbf{x}^t, \mathbf{h}^t; \theta)) = \mathcal{F}^t(Q, \theta), \tag{1}$$

where $\mathbf{h}^t$ indicates that it depends on $\mathbf{x}^t$ [8]; $Q(\mathbf{h}^t)$ is the approximate distribution of the real posterior $P(\mathbf{h}^t \,|\, \mathbf{x}^t, \theta)$; $\mathcal{F}^t(Q, \theta)$ is the negative free energy function or the lower bound of $\log P(\mathbf{x}^t; \theta)$. It is worth noting that, the approximation of the real posterior $P(\mathbf{h}^t \,|\, \mathbf{x}^t, \theta)$ using $Q(\mathbf{h}^t)$ and of the the real log likelihood $\log P(\mathbf{x}^t; \theta)$ using the lower bound $\mathcal{F}^t(Q, \theta)$ is often satisfied. In fact, the approximation error can be zero since $Q(\mathbf{h}^t)$ exactly equals to $P(\mathbf{h}^t \,|\, \mathbf{x}^t, \theta)$ and $\mathcal{F}^t(Q, \theta)$ exactly equals to $\log P(\mathbf{x}^t; \theta)$ when using exact inference. Learning generative models based on the variational lower bound can be expressed as,

$$\max_{Q, \theta} \sum_t \mathcal{F}^t(Q, \theta) = \max_{Q, \theta} \sum_t -\mathrm{KL}(Q(\mathbf{h}^t) \,\|\, P(\mathbf{x}^t, \mathbf{h}^t; \theta)) \tag{2}$$

An assumption here, as also made in most probabilistic generative models [23], is that the joint distribution $P(\mathbf{x}, \mathbf{h}; \theta)$ of a generative model belongs to the exponential family, written as [23],

$$P(\mathbf{x}, \mathbf{h}; \theta) = \exp\{\alpha(\theta)^T T(\mathbf{x}, \mathbf{h}) + A(\theta)\} \tag{3}$$

where $\alpha(\theta)$ is a vector-valued function; $T(\mathbf{x}, \mathbf{h})$ is the vector of sufficient statistics over $\mathbf{x}$ and $\mathbf{h}$; $A(\theta)$ is a scalar function. Since $P(\mathbf{x}, \mathbf{h}) = P(\mathbf{x} \,|\, \mathbf{h}) P(\mathbf{h})$, $P(\mathbf{h})$ also belongs to exponential family, $P(\mathbf{h}; \theta_h) = \exp\{\alpha(\theta_h)^T T(\mathbf{h}) + A(\theta_h)\}$. As was done in [24], we assume that, for a sample $\mathbf{x}^t$, the approximate posterior $Q(\mathbf{h}^t)$ shares the same form as $P(\mathbf{h}; \theta_h)$, but with different parameters,

$$Q(\mathbf{h}^t) = \exp\{\alpha(\theta_h^t)^T T(\mathbf{h}^t) + A(\theta_h^t)\} \tag{4}$$

where θ_h^t is a vector of parameters and depends on the sample $\mathbf{x}^t$. Substituting Eqs. (3) and (4) into Eq. (1), it can be verified that,

$$
\begin{aligned}
\mathcal{F}^t(Q, \theta) &= \mathrm{E}_{Q(\mathbf{h}^t)}[\alpha(\theta)^T T(\mathbf{x}^t, \mathbf{h}^t) + A(\theta) - \alpha(\theta_h^t)^T T(\mathbf{h}^t) - A(\theta_h^t)] \\
&= \mathrm{E}_{Q(\mathbf{h}^t)}[\alpha(\theta)^T T(\mathbf{x}^t, \mathbf{h}^t) - \mathbf{1}^T \mathrm{diag}(\alpha(\theta_h^t)) T(\mathbf{h}^t) - A(\theta_h^t) + A(\theta)] \\
&= \alpha(\theta)^T \mathrm{E}_{Q(\mathbf{h}^t)}[T(\mathbf{x}^t, \mathbf{h}^t)] - \mathbf{1}^T \mathrm{diag}(\alpha(\theta_h^t)) \mathrm{E}_{Q(\mathbf{h}^t)}[T(\mathbf{h}^t)] - A(\theta_h^t) + A(\theta) \\
&= \eta^T \mathrm{E}_{Q(\mathbf{h}^t)}[\phi(\mathbf{x}^t, \mathbf{h}^t)] = \eta^T \varPhi(\mathbf{x}^t) \tag{5}
\end{aligned}
$$

where $\eta = (\alpha(\theta)^T, -\mathbf{1}^T, -1, A(\theta))^T$ only depends on parameter θ; $\phi^t(\mathbf{x}^t, \mathbf{h}^t)$ is a function over $\mathbf{x}^t$, $\mathbf{h}^t$ and θ_h^t, depending on $\mathbf{x}^t$,

$$\phi(\mathbf{x}^t, \mathbf{h}^t) = (T(\mathbf{x}^t, \mathbf{h}^t)^T, (\mathrm{diag}(\alpha(\theta_h^t)) T(\mathbf{h}^t))^T, A(\theta_h^t), 1)^T \tag{6}$$

Note that $\mathbf{h}^t$ and θ_h^t depend on the specific sample $\mathbf{x}^t$. Therefore they reflect some attributes or encode some information related to $\mathbf{x}^t$. $\Phi(\mathbf{x}^t)$ is the score function or feature mapping, taking the following form,

$$\Phi(\mathbf{x}^t) = \mathrm{E}_{Q(\mathbf{h}^t)}[\phi(\mathbf{x}^t, \mathbf{h}^t)] \tag{7}$$

The function $\Phi(\mathbf{x}^t)$ is termed as sufficient statistics score function since its main components are sufficient statistics $T(\mathbf{x}, \mathbf{h})$ and $T(\mathbf{h})$. $\mathcal{F}^t(Q, \theta)$ is decomposed into the linear combination of η which depends on all training samples and the score function $\Phi(\mathbf{x}^t)$ which depends on the sample $\mathbf{x}^t$.

The above formulation is based on the variational inference in Eq. (1) and the approximate posterior in Eq. (4). The approximation works well when the real log likelihood are intractable [8,10], and equals to the real log likelihood exactly when using exact inference. The derived score function in Eq. (7) is compatible with other inference methods as we can estimate the posterior (Eq. (4)) using the outputs of those methods, e.g., using the samples drawn by Gibbs sampling.

3.2   Error Rate Comparison with MAP Classification

As score spaces typically work with linear classifiers, [12] proposed a method to analyze the classification error of a linear classifier $y = \mathrm{sign}(\mathbf{w}^T \Phi(\mathbf{x}) + b)$ where $\mathbf{w} \in \mathbb{R}^d$ is the weight and $b \in \mathbb{R}$ is the bias. We assume that $\mathbf{w}$ and b are learned by an optimal learning algorithm on a sufficiently large training set. Letting $\Psi(a)$ be the zero-one loss function that outputs 1 if $a > 0$ and 0 otherwise, the classification error can be expressed as,

$$R(\Phi) = \min_{\mathbf{w},b} \mathrm{E}_{\mathbf{x},y} \Psi[-y(\mathbf{w}^T \Phi(\mathbf{x}) + b)],$$

where $\mathrm{E}_{\mathbf{x},y}$ denotes the expectation over the true distribution. Note that $R(\Phi)$ is exactly the test error if the test set and the training set share the same distribution. We assume this condition holds, as was done in [12,8,10].

Previous works [12,8,10] have shown that, in the case that the model is trained using samples from the positive class and the log likelihood $\log P(\mathbf{x} \,|\, y = +1)$ is available, the error rate $R(\Phi)$ of a linear classifier operating on the score space is at least as low as the error rate $R(\lambda)$ of the MAP classifier,

$$R(\lambda) = \mathrm{E}_{\mathbf{x},y} \Psi[-y(P(y = +1 \,|\, \mathbf{x}) - \frac{1}{2})] = \mathrm{E}_{\mathbf{x},y} \Psi[-y(\log P(y = +1 \,|\, \mathbf{x}) - \log \frac{1}{2})]$$

$$= \mathrm{E}_{\mathbf{x},y} \Psi[-y(\log P(\mathbf{x} \,|\, y = +1) - \log \frac{1}{2})] = \mathrm{E}_{\mathbf{x},y} \Psi[-y(\eta^T \Phi(\mathbf{x}) - \log \frac{1}{2})]$$

$$\geq \min_{\mathbf{w},b} \mathrm{E}_{\mathbf{x},y} \Psi[-y(\mathbf{w}^T \Phi(\mathbf{x}) + b)] = R(\Phi)$$

In the case that the exact log likelihood might be intractable, as shown in Eq. (5), we resort to the lower bound $\mathcal{F}_{+1}(\mathbf{x})$ and $\mathcal{F}_{-1}(\mathbf{x})$ for a pair of models θ_{+1} and θ_{-1} that are respectively trained using the positive samples and negative samples, and accordingly resort to the free energy test [8]. That is,

$\hat{y} = \text{sign}(\mathcal{F}_{+1}(\mathbf{x}) - \mathcal{F}_{-1}(\mathbf{x}))$. Applying the formulation in Eq. (5), then we have $\mathcal{F}_{+1}(\mathbf{x}) = \eta_{+1}^T \Phi_{+1}(\mathbf{x})$ and $\mathcal{F}_{-1}(\mathbf{x}) = \eta_{-1}^T \Phi_{-1}(\mathbf{x})$. We accordingly define the score function over a pair of models as $\Phi(\mathbf{x}) = (\Phi_{+1}(\mathbf{x})^T, \Phi_{-1}(\mathbf{x})^T)^T$. The above inequality $R(\Phi) \leq R(\lambda)$ still holds,

$$\begin{aligned}
R(\lambda) &= \mathrm{E}_{\mathbf{x},y}\Psi[-y(\mathcal{F}_{+1}(\mathbf{x}) - \mathcal{F}_{-1}(\mathbf{x}))] \\
&= \mathrm{E}_{\mathbf{x},y}\Psi[-y(\eta_{+1}^T \Phi_{+1}(\mathbf{x}) - \eta_{-1}^T \Phi_{-1}(\mathbf{x}))] \\
&\geq \min_{\mathbf{x},y} \mathrm{E}_{\mathbf{x},y}\Psi[-y(\mathbf{w}^T \Phi(\mathbf{x}) + b)] = R(\Phi)
\end{aligned}$$

The above justifications also hold for [11,12,8,10] because $\mathcal{F}(\mathbf{x}, \theta)$ can be expressed as a linear combination of any of the score functions.

4 Learning Discriminative Score Space

To exploit label information, we propose a discriminative learning method that learns score space as well as generative models under the classification margin constraints of a linear classifier in the score space.

4.1 The Learning Problem

First we will use a probabilistic classifier because of its compatibility with probabilistic generative models. Let $\mathbf{x}$ be the input data and $y \in \{-1, +1\}$ be the output label; $\mathcal{S} = \{(\mathbf{x}^t, y^t)\}_t$ be the training set whose samples are indexed by t. Let $\mathbf{x}$ be the augmented sample $(\mathbf{x}^T, 1)^T$; $\mathbf{w}$ be the weight including the bias; γ^t be the desired margin for the sample $\mathbf{x}^t$. The classifier subject to margin constraint is given by [4],

$$\min_{Q(\mathbf{w})Q(\gamma^t)} \mathrm{KL}(Q(\mathbf{w})Q(\gamma^t) \,\|\, P(\mathbf{w})P(\gamma^t)) \tag{8}$$

$$\text{s.t. } \mathrm{E}_{Q(\mathbf{w})}[y^t \mathbf{w}^T \mathbf{x}^t] \geq \mathrm{E}_{Q(\gamma^t)}[\gamma^t], \; \forall \, t, \tag{9}$$

where $P(\mathbf{w})$ and $Q(\mathbf{w})$ are the prior and posterior for the weight respectively; $P(\gamma^t)$ and $Q(\gamma^t)$ are the prior and posterior for the margin respectively. The margin γ^t is specified for $\mathbf{x}^t$. This formulation allows for a tunable and flexible margin, which functions in a way similar to the soft margin in SVMs.

Now we have shown the objective functions of generative models (Eq. (1)) and the classifier (Eqs (8) and (9)). Learning discriminative score space subject to margin constrains means we need to maximize Eq. (1) and minimize Eq. (8) simultaneously, subject to Eq. (9). The learning problem can be expressed as,

$$\min_{\mathcal{Q},\theta} \sum_t \underbrace{\mathrm{KL}(Q(\mathbf{h}^t) \,\|\, P(\mathbf{x}^t, \mathbf{h}^t; \theta))}_{\mathrm{KL}_\theta \text{ (generative)}} + \xi \underbrace{\mathrm{KL}(Q(\mathbf{w})Q(\gamma^t) \,\|\, P(\mathbf{w})P(\gamma^t))}_{\mathrm{KL}_\mathbf{w} + \mathrm{KL}_\gamma \text{ (discriminative)}} \tag{10}$$

$$\text{s.t. } \mathrm{E}_{\mathcal{Q}}[y^t \mathbf{w}^T \phi(\mathbf{x}^t, \mathbf{h}^t) - \gamma^t] \geq 0, \; \forall \, t \tag{11}$$

where $\mathcal{Q} = \{Q(\mathbf{h}^t), Q(\gamma^t), Q(\mathbf{w})\}$. The first term in Eq. (10) is the objective function for the generative model as in Eq. (2), where $P(\mathbf{x}, \mathbf{h}; \theta)$ is the joint distribution and $Q(\mathbf{h}^t)$ is the approximate posterior. The second term of Eq. (10) and the constraint Eq. (11) form the objective function of the classifier, where $P(\gamma^t)$ and $P(\mathbf{w})$ are priors on the margins and the weights respectively. $\xi > 0$ is a weight that tunes the balance between the generative model and the classifier.

4.2   Inference and Parameter Estimation

The quantities to be estimated in the objective function Eq. (10) and Eq. (11) include $Q(\mathbf{h}^t)$, $Q(\gamma^t)$, $Q(\mathbf{w})$ and θ. To estimate these quantities, we first specify the priors $P(\mathbf{w})$ and $P(\gamma^t)$. Similar to that in [4], we set the priors,

$$P(\mathbf{w}) = \mathcal{N}(0, \mathbf{I}), \tag{12}$$

$$P(\gamma^t) = ce^{-c(a-\gamma^t)} \text{ for } \gamma^t \le a. \tag{13}$$

where a, c are two parameters to be specified. The learning problem in Eq. (10) and Eq. (11) takes the exact form of posterior regularization [25], and in principle can be solved using EM-like procedures [25,26]. In optimization [26], to estimate $Q(\mathbf{h}^t)$, $Q(\gamma^t)$, $Q(\mathbf{w})$ and θ, we alternatively solve sub-problems with respect to some of these quantities while keeping the others fixed in each pass. The solution of θ will benefit from the form of $\Phi(\mathbf{x})$ because $\Phi(\mathbf{x})$ and the constraints Eq. (11) are not related to θ.

Posterior $Q(\mathbf{h}^t)$ of Hidden Variables. By fixing quantities $Q(\gamma^t)$ and θ, the solution of $Q(\mathbf{h}^t, \mathbf{w})$ takes the following form [25,4],

$$Q(\mathbf{h}^t, \mathbf{w}) \propto P(\mathbf{x}^t, \mathbf{h}^t; \theta) P(\mathbf{w}) \cdot \exp\left\{ \sum_t \lambda^t \left[y^t \mathbf{w}^T \phi^t - E_{Q(\gamma^t)}[\gamma^t] \right] \right\}, \tag{14}$$

where $\phi^t = \phi(\mathbf{x}^t, \mathbf{h}^t)$, and λ^t is the Lagrange multiplier for the t-th inequality of Eq. (11). Note that $\mathbf{w}$ follows a normal prior in Eq. (12), making the integration $Q(\mathbf{h}^t) = \int Q(\mathbf{h}^t, \mathbf{w}) d\mathbf{w}$ tractable,

$$Q(\mathbf{h}^t) \propto \underbrace{P(\mathbf{x}^t, \mathbf{h}^t; \theta)}_{\propto P(\mathbf{h}^t | \mathbf{x}^t, \theta)} \underbrace{\exp\left\{ \sum_t \lambda^t E_{Q(\gamma^t)}[\gamma^t] - \frac{1}{2} \sum_{t,t'} \lambda^t \lambda^{t'} y^t y^{t'} (\phi^t)^T \phi^{t'} \right\}}_{\propto \text{discriminative posterior}} \tag{15}$$

This formula shows that the posterior of hidden variables is proportional to the product of (1) the joint distributions of the naive generative model and (2) an exponential term that is derived from the classifier and favors large margin. When $\mathbf{h}$ is a set of discrete variables, the posterior and $E_{Q(\mathbf{h}^t)}[\phi^t]$ are straightforward to compute; when $\mathbf{h}$ is a set of continuous variables, without an analytical solution in most cases, we resort to estimate the expectation $E_{Q(\mathbf{h}^t)}[\phi^t]$ by,

$$E_{Q(\mathbf{h}^t)}[\phi(\mathbf{x}^t, \mathbf{h}^t)] \approx \frac{1}{n} \sum_{i=1}^n \phi(\mathbf{x}^t, \mathbf{h}_i^t), \tag{16}$$

where $\mathbf{h}_i^t$ is the i-th sample of all the n samples drawn from the posterior $Q(\mathbf{h}^t)$. Gibbs-rejection sampling [27] can be very effective in drawing samples from Eq. (15). A sample $\mathbf{h}_i^t$ drawn from $P(\mathbf{x}^t, \mathbf{h}^t; \theta)$ will be accepted or rejected based on the exponential term.

Posterior $Q(\gamma^t)$ of Margins. By fixing the quantities $Q(\mathbf{h}^t)$ and θ, the posterior $Q(\gamma^t, \mathbf{w})$ can be solved, in the same way as in the solution of $Q(\mathbf{h}^t, \mathbf{w})$. We compute the posterior $Q(\gamma^t) = \int Q(\gamma^t, \mathbf{w})d\mathbf{w}$ as,

$$Q(\gamma^t) \propto \int P(\gamma^t) \exp\left\{\lambda^t \mathbf{E}_{Q(\mathbf{h}^t)}\left[y^t \mathbf{w}^T \phi^t - \gamma^t\right]\right\} d\mathbf{w}$$

$$\propto \exp\left\{-\left(c - \lambda^t\right)\left(a - \gamma^t\right)\right\}, \tag{17}$$

For the exponential distribution $P(\gamma^t) = ce^{-c\gamma^t}$ with $\gamma^t \geq 0$ (Eq. (13)), the mean of γ^t is $\mathbf{E}_{P(\gamma^t)}[\gamma^t] = c^{-1}$. The expected margin can be similarly derived,

$$\mathbf{E}_{Q(\gamma^t)}[\gamma^t] = a - \left(c - \lambda^t\right)^{-1}. \tag{18}$$

which adapts to samples, for example, by taking negative values for incorrect classification, which essentially implements a soft-margin.

Lagrange Multipliers $\lambda = \{\lambda^1, \lambda^2, ..., \lambda^N\}$. Every Lagrange multiplier here corresponds to an inequality constraint. Fixing $Q(\mathbf{h}^t)$, $Q(\gamma^t)$ and θ leads to,

$$Q(\mathbf{w}) = \frac{1}{Z(\lambda)} P(\mathbf{w}) \exp\left\{\sum_t \lambda^t \mathbf{E}_{Q(\mathbf{h}^t, \gamma^t)}\left[y^t \mathbf{w}^T \phi^t - \gamma^t\right]\right\},$$

where $Z(\lambda) = \int Q(\mathbf{w})d\mathbf{w}$ is the partition function. Then $\lambda \geq 0$ is obtained by maximizing the objective function $J_\lambda = -\log Z(\lambda)$. Using the same integration in Eq. (15), we have,

$$J_\lambda = \sum_t \lambda^t \mathbf{E}_{Q(\gamma^t)}[\gamma^t] - \frac{1}{2}\sum_{t,t'} \lambda^t \lambda^{t'} y^t y^{t'} \mathbf{E}_{Q(\mathbf{h}^t)}[\phi^t]^T \mathbf{E}_{Q(\mathbf{h}^{t'})}[\phi^{t'}]. \tag{19}$$

This is a standard quadratic programming problem, which can be efficiently solved. It differs from the dual form of SVMs because of the extra weight $\mathbf{E}_Q[\gamma^t]$.

Parameters θ of Generative Models. In the objective function Eq. (10) and the constraint Eq. (11), only the term KL_θ depend on the parameter θ. So minimizing the objective function with respect to θ equals to minimizing KL_θ with respect to θ, not subjecting to any inequality constraint. The resulting update rules for θ are the *same* as those for the original generative models.

The learning procedure of the proposed method is summarized in Algorithm 1. The output is the parameter θ of a generative model. Given the generative model trained by Algorithm 1, we are now equipped to compute score functions for test samples. The procedure constructing discriminative score space is summarized in Algorithm 2.

4.3   Classifier Learning Rules

Given the discriminatively learned generative models and score spaces, there are two ways to obtain classifiers over the score spaces: (1) train classifiers on the

---

Algorithm 1. Discriminative learning of generative models

1: **input:** training data set $\mathcal{S} = \{(\mathbf{x}^t, y^t)\}_{t=1}^N$
2: initialize parameters $\hat{\theta}, \mathbf{u}, \lambda$
3: **repeat**
4:   **for** $t = 1$ to N **do**
5:     sample $\{\mathbf{h}_i^t\}_i$ from $Q(\mathbf{h}^t)$ (Eq. (15))
6:     estimate $\mathrm{E}_{Q(\mathbf{h}^t)}[\phi(\mathbf{x}^t, \mathbf{h}^t)]$ (Eq. (16))
7:     compute $\mathrm{E}_{Q(\gamma^t)}[\gamma^t]$ (Eq. (18))
8:   **end for**
9:   update λ (Eq. (19))
10:   update $\hat{\theta}$ with $\{\mathbf{h}_i^t\}_{ti}$ using the rules of the original generative models
11: **until** convergence
12: **output:** $\hat{\theta}$

Algorithm 2. Construct discriminative sufficient statistics score spaces

1: **input:** generative model $\hat{\theta}$ and input data set $\{(\mathbf{x}^t, y^t)\}_{t=1}^{N_t}$
2: **for** $t = 1$ to N_t **do**
3:   sample $\{\mathbf{h}_i^t\}_i$ from $Q(\mathbf{h}^t)$ (Eq. (15))
4:   estimate $\Phi(\mathbf{x}^t) = \mathrm{E}_{Q(\mathbf{h}^t)}[\phi(\mathbf{x}^t, \mathbf{h}^t)]$ (Eq. (16))
5: **end for**
6: **output:** $\{\Phi(\mathbf{x}^t)\}_{t=1}^{N_t}$

---

score spaces using any standard method; (2) estimate SVMs like classifiers using the quantities produced by Algorithm 1. We will now present the details of (2).

The learning problem in Eq. (10) and Eq. (11) already includes a linear classifier with the following decision rule,

$$\hat{y} = \mathrm{sign}(\mathrm{E}_{Q(\mathbf{w})}[\mathbf{w}^T \Phi(\mathbf{x})])$$

To estimate a classifier based on the quantities produced by Algorithm 1, we just need to estimate the weight $\mathbf{w}$. First, we specify the posterior of the classifier $\mathbf{w}$ to be a Gaussian distribution with unit covariance matrix [28],

$$Q_s(\mathbf{w}) = N(\mathbf{u}, \mathrm{I}). \tag{20}$$

where $\mathbf{u}$ is the mean to be estimated from training data. Considering the above specification for $Q_s(\mathbf{w})$ and the specification for $P(\mathbf{w})$ (Eq. (12)), it can be verified that $\mathrm{KL}_{\mathbf{w}} = \mathrm{KL}(Q(\mathbf{w}) \| P(\mathbf{w})) = \frac{1}{2}\mathbf{u}^T\mathbf{u}$. This means that minimizing $\mathrm{KL}_{\mathbf{w}}$ in Eq. (10) encourages $\mathbf{u}$ to have a short length. Under the above specifications, the solution of $\mathbf{w}$ takes the following form.

Proposition 1. *Let $\Phi(\mathbf{x}) = \mathrm{E}_{Q(\mathbf{h})}[\phi(\mathbf{x}, \mathbf{h})]$ be the score function derived from (Algorithm 2) the discriminatively trained generative models (Algorithm 1). With the specification in Eq. (20), the maximum a posteriori (MAP) estimation of $\mathbf{w}$ in Eq. (10) takes the same form as the solution of the linear SVMs equipped with the score function $\Phi(\mathbf{x})$.*

Proof. The solution of $Q(\mathbf{w})$ can be expressed as,

$$Q(\mathbf{w}) = \frac{1}{Z}P(\mathbf{w})\exp\left\{\sum_t \lambda^t E_{Q(\mathbf{h}^t,\gamma^t)}\left[y^t\mathbf{w}^T\phi^t - \gamma^t\right]\right\}$$

$$= \frac{1}{Z}P(\mathbf{w})\exp(\alpha^T\mathbf{w} - \beta), \qquad (21)$$

where $Z = \int P(\mathbf{w})\exp(\alpha^T\mathbf{w} + \beta)d\mathbf{w}$ is the partition function to ensure $Q(\mathbf{w})$ being a probabilistic distribution; $\alpha = \sum_t \lambda^t y^t E_{Q(\mathbf{h}^t)}[\phi^t]$ and $\beta = E_{Q(\gamma^t)}[\gamma^t]$. Considering the specification $Q_s(\mathbf{w})$ in Eq. (20), the MAP estimation $\hat{\mathbf{w}}$ of $\mathbf{w}$ satisfies $\hat{\mathbf{w}} = E_{Q(\mathbf{w})}[\mathbf{w}] = \mathbf{u}$ and can be determined by minimizing the I-projection between the specified posterior Eq. (20) and the derived posterior Eq. (21),

$$\min_{\mathbf{u}} \mathrm{KL}\left[Q_s(\mathbf{w}) \| \frac{1}{Z}P(\mathbf{w})\exp(\alpha^T\mathbf{w} + \beta)\right]$$

$$= \min_{\mathbf{u}} E_{Q_s(\mathbf{w})}\left[\log Q_s(\mathbf{w}) - \log\frac{1}{Z}P(\mathbf{w})\exp(\alpha^T\mathbf{w} + \beta)\right]$$

$$= \min_{\mathbf{u}} E_{Q_s(\mathbf{w})}\left[\mathbf{w}^T\mathbf{u} - \frac{1}{2}\mathbf{u}^T\mathbf{u} - (\alpha^T\mathbf{w} + \beta)\right] + \log Z$$

$$= \min_{\mathbf{u}} \left[\frac{1}{2}\mathbf{u}^T\mathbf{u} - \alpha^T\mathbf{u} - \beta\right] + \log Z.$$

where Z does not depend on $\mathbf{u}$. Letting $\frac{\partial \mathrm{KL}}{\partial \mathbf{u}} = 0$, we has an analytical solution,

$$\hat{\mathbf{w}} = \mathbf{u} = \alpha = \sum_t \lambda^t y^t E_{Q(\mathbf{h}^t)}[\phi(\mathbf{x}^t, \mathbf{h}^t)]. \qquad (22)$$

This is equivalent to the solution of linear SVMs [2].

5   Experiments

We experimented with two generative models in the proposed framework in the context of classification. As shown in Section 4.2 and Algorithm 1, we only need to specify the feature mapping Φ for each adopted generative model. In each experiment, we compare (1) the proposed sufficient statistics (SS) score space which learns the score spaces (including generative models) and the discriminative classifiers separately, under no discriminative constraint; (2) the discriminative learning of SS subject to margin constraints (MSS), as proposed in Section 4; (3) Fisher score (FS) method [11]; (4) free energy score space (FESS) method [8] and other state-of-the-art methods. Here, we omit the comparison with other hybrid methods [5,10] due to the space limitation. For each problem, we repeatedly test 20 rounds. In each round, training and test sets are formed by random sampling from the dataset.

The MSS approach is proposed in the setting of binary classification. It is straightforward to extend it to multi-class classification problems, by splitting each multi-class problem into several binary problems and combine the MSS features separately learned from each of the binary classification problems. The

Table 2. Summary of classification accuracy (%) on sequence datasets. Discrete HMMs are used to model the distribution of sequences. SS is the baseline version of the proposed method without using discriminative learning.

Class	C	LM-HMM	FKL [19]	FS [11]	FESS [8]	SS	MSS
Character	20	94.26	**95.71**	95.20	93.99	93.55	95.62
Hill Valley	2	58.71	54.00	63.60	55.41	53.39	**65.68**
Jap. Vowel	9	92.26	**96.16**	88.93	90.63	91.26	93.40
Hand Move.	15	78.10	75.22	79.11	79.89	78.00	**82.22**
Promoter Gene	2	67.92	69.81	63.38	65.77	65.35	**74.23**
Junction Gene	3	58.64	57.05	58.71	58.78	58.86	**65.37**
Protein Kinase	3	72.18	74.15	73.24	72.65	73.53	**78.53**
SCOP Protein	7	**64.75**	60.96	64.17	64.12	64.24	64.64
Chicken Shape	5	77.64	79.83	79.63	80.26	79.58	**83.36**

parameters ($\xi = 1$ (Eq. (10)), $a = 1$ and $c = 6$ (Eq. (13)), the number of topics M of LDA (Section 5.2), the number of hidden states K of HMMs (Section 5.1)) used in the following experiments are chosen through an *offline* cross validation method, i.e., the parameters are chosen using cross validation on a dataset and then applied to all datasets. The reasons of using offline rather that online method are that (1) online cross validation for 5 parameters are computationally very expensive; (2) offline method produces satisfied performance.

For score spaces FS, FESS and SS, we use the same scheme as [8], i.e., train a generative model for each class and combine the features obtained from these models. This scheme is empirically validated to be more effective than the score space derived from one generative model of all samples. For all score space methods (FS, FESS, SS, MSS), we use linear SVMs (libsvm toolbox [29]) as the classifier. For localized multiple kernel learning (LMKL) [3], Fisher kernel learning (FKL) [19] and FESS, we use the authors' implementations, which can be downloaded from their websites. FS-HMMs, FS-LDA, LM-HMMs [6] and the proposed methods are implemented by ourselves.

5.1   Sequence Recognition: Hidden Markov Models

In the first experiment, we learn the score space for sequence recognition with hidden Markov models (HMMs) [30] as the generative model. Let $\mathbf{x}$ be the sequence with length $L_{\mathbf{x}}$. We here consider the discrete case where $\mathbf{x}^l$ is a vector of binary indicators of states at position l along the sequence, i.e., $x_k^l = 1$ if the k-th of the K possible observed states is selected at position l. $\mathbf{q}^l$ is the binary indicator for hidden states, where $q_i^l = 1$ if the i-th of the M possible hidden states is selected at position l. The joint distribution is given by,

$$P(\mathbf{x}, \mathbf{q}; \theta) = \prod_{i=1}^{M} \pi_i^{q_i^0} \cdot \prod_{l=0}^{L_{\mathbf{x}}-1} \prod_{i,j=1}^{M,M} a_{ij}^{q_i^l q_j^{l+1}} \cdot \prod_{l=0}^{L_{\mathbf{x}}} \prod_{i,k=1}^{M,K} b_{ik}^{q_i^l x_k^l}$$

where $\theta = \{\pi_i, a_{ij}, b_{ik}\}_{ijk}$. Let $\hat{\pi} = \{\hat{\pi}_i\}_i$, $\hat{\mathbf{a}} = \{\hat{a}_{ij}\}_{ij}$ and $\hat{\mathbf{b}} = \{\hat{b}_{ik}\}_{ik}$ respectively be the initial, state transition and emission probabilities of the approximate posterior. The score function is $\Phi(\mathbf{x}) = E_{Q(\mathbf{q})}[\phi(\mathbf{x}, \mathbf{q})]$, where,

$$\phi(\mathbf{x}, \mathbf{q}) = \text{vec}\left(\left\{ q_i^0, \sum_{l=0}^{L_\mathbf{x}-1} q_i^l q_j^{l+1}, \sum_{l=0}^{L_\mathbf{x}} q_i^l x_k^l, \right.\right.$$
$$\left.\left. q_i^0 \log \hat{\pi}_i, \sum_{l=0}^{L_\mathbf{x}-1} q_i^l q_j^{l+1} \log \hat{a}_{ij}, \sum_{l=0}^{T_\mathbf{x}} q_i^l x_k^l \log \hat{b}_{ik} \right\}_{i,k} \right).$$

Given the hidden states of the input sequence inferred with the Baum-Welch algorithm [31], it is easy to estimate the posterior probabilities, i.e. initial, transition, and emission probabilities conditioned on $\mathbf{x}$. Using the sampling distribution in Eq. (15), we are able to draw examples of hidden states and re-estimate their posterior. The quantity $E_{Q(\mathbf{z})}[\cdot]$ can be computed effectively since $\mathbf{z}$ is a discrete variable.

We compare the performance of SS and MSS against that of FS, FESS, FKL [19] and large margin HMM (LM-HMM) [6]. The number of hidden states M is set to be $M = 3$ for MSS and $M = 10$ for FS, FESS and SS based on cross-validation performance as shown in Fig. (1). For FKL and LM-HMM, we chose M from $2, 5, 10$ using offline cross validation. We randomly select 50% samples for training and the rest for testing. The learned score space is evaluated on 9 sequence datasets where SCOP protein is obtained from ASTRAL database with similar sequences reduced by a E-value threshold of 10^{-25}; the chicken piece shape dataset is collected by [32]; the rest are obtained from UCI database. For FS, FESS, SS and MSS, the datasets with continuous values are quantized to state sequences for the discrete HMMs, i.e., 8 states for chicken piece shape and 20-40 states for other datasets. For FKL, we use continuous HMMs for continuous data and discrete HMMs which is implemented by configuring the graph of MRF for state sequence data.

The results are reported in Table 2. Our SS's performance is competitive against FS and FESS, even though it does not utilize the parameters of generative models as was done in FS and FESS. Our MSS outperforms other methods in 8 of the 13 experiments. The improvement of MSS over SS is brought about by the discriminative learning paradigm. The comparison between MSS against LM-HMM or FKL is particularly worth noting because both LM-HMM and FKL are methods that learn generative models and discriminative models jointly. We should also note that the data representation used in FKL is slightly different with that used in MSS. That is, FKL uses continuous data on the first 4 datasets while MSS quantities them into discrete data, and thus its performance might suffer from this quantization. Further, MSS is effective for a small number of hidden states and thus is more efficient to train, even with limited samples. We will discuss these issues more in Section 5.3.

5.2   Image Recognition: Latent Dirichlet Allocation

We also evaluate the framework when Latent Dirichlet Allocation (LDA) [34] is used as its generative model in the context of image scene recognition. In this

Table 3. Classification accuracy (%) on OT, Scene-15 and UIUC-sports datasets

Dataset	C	PHOW [33]	Med-LDA[17]	FS [11]	FESS [8]	SS	MSS
OT	8	87.21	89.50	86.42	88.89	88.25	**90.36**
Scene-15	15	79.83	81.05	78.68	81.92	79.25	**83.64**
UIUC-sports	8	80.04	82.79	79.91	82.18	80.34	**84.67**

task, visual words are used to represent images as is typically done in computer vision. The LDA version in [35] is used to model the distribution of visual words, with each topic associated with a particular distribution. We sample topic variables using collapsed Gibbs sampling [35], and reject examples according to the rule stipulated in Eq. (15). Differing from [35], we update model parameters α and β in each iteration. In order to make it compatible with [35], we only use word and topic to construct feature mapping $\Phi(w^d_{\cdot})$,

$$\phi = \text{vec}\left(\left\{z^k_{dn}, w^d_n z^k_{dn}, 1\right\}_{n,k}\right),$$

where w and z denote word and topic respectively. d, n and k index image, word and topic respectively. For FS [11] and FESS [8], we extract features from the trained LDA model and use them with linear SVM.

The OT scene dataset, Scene-15 dataset and UIUC-sports dataset are used for evaluation. They contains 8, 15 and 8 categories respectively. For each image, dense SIFT descriptors [36] are extracted from 20×20 grid patches over 4 scales. The descriptors are clustered (using K-mean on randomly selected descriptors) into 200 visual words in a code book. An image is represented by a histogram of the frequency of the observed visual words. The number of topics is set to $K = 10$ for MSS and $K = 50$ for FS, FESS and SS throughout the experiment. For OT, Scene-15 and UIUC-sports datasets, 5%, 100, 70 images per category are randomly selected as training set and the rest as test set in each test.

The evaluation results are reported in Table 3. PHOW [33] is a state-of-the-art feature descriptor for scene recognition, and Med-LDA [17] is a discriminative learning approach for LDA that has been shown to be superior to disLDA [18]. MSS outperforms all compared methods on all three datasets. The performance of SS is slightly inferior to FESS but slightly better than FS, which indicates that even though SS does not fully exploit the model parameters, it still captures rich generative information. We also evaluate the feature mapping as a function of the number of samples and topics, as shown in Fig. 2, and show that MSS works well with few topics (hidden variables).

5.3 Computational Efficiency

The proposed discriminative learning method is an iterative process, involving the inference step and the parameter estimation step, where the parameter estimation is slower because it needs to solve a quadratic programming and update the generative model. Learning can be greatly sped up in the following way. Instead of cycling through the steps T times, we can pre-train the generative models (e.g. with T iterations) and then launch Algorithm 1 for a few iterations, about 10 iterations empirically.

The performance of a score space, to some extent, depends on the number of hidden variables (e.g., the number of topics in LDA and the number of hidden states in HMMs) in the generative model used [10,19]. We investigate this dependency by evaluating the method's performance as a function of the number of hidden variables. We evaluate the score space using two classification schemes: (1) multi-class classification; (2) splitting the multi-class problem into a group of binary classification problems and averaging their results. We find that the two methods of evaluation share very similar trend, and report only experimental results based on (2) in Fig. 1 (HMMs), and Fig. 2 (LDA). Overall, we found the proposed method works well with generative models with a few hidden variables, fewer than other methods required, which also makes our method more efficient. In addition, as shown in Fig. 1 and Fig. 2, MSS's performance over other methods is robust against the percentage of total samples used as training samples.

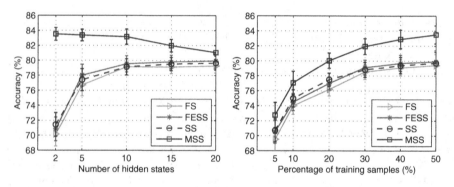

Fig. 1. Accuracy (%) w.r.t. the number of hidden states (HMMs) and the percentage of training samples on Chicken data

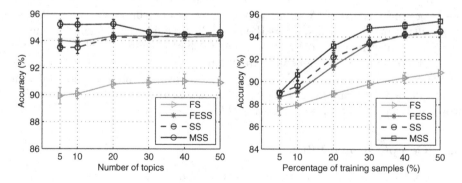

Fig. 2. Accuracy (%) w.r.t. the number of topics (LDA) and the percentage of training samples on OT (City vs rest)

Empirically, The MSS-HMM and MSS-LDA can be faster than MAP-HMM and MAP-LDA in part because the number of topics or hidden variables are smaller.

6 Conclusions

In this paper, we derive a new score space (SS) by decomposing the lower bound of the log likelihood into a linear combination of two parts. The first part is related to model parameters while the second part is related data samples. The second part, based mainly on sufficient statistics, provides the score functions to span the score space. This decomposition allows us to develop a computationally tractable method to learn score space discriminatively, subject to margin constraints of a classifier over the score space. We provide an EM-like algorithm for inference and learning, where the posterior introduced by discriminative factors (margin constraints) feed-back discriminative information to tune the score space. This method works well with a small number of hidden variables, which makes inference and learning fast and efficient. We show that this approach is competitive against other state-of-the-art methods in a variety of datasets.

Acknowledgement. Thanks NSF CISE IIS 0713206, 973 Program 2011CB302203, NSFC 60833009, NSFC 60975012 for support, and to Liwei Wang, Jiaya Jia and Zhuowen Tu for helpful discussions. This work was done when X. Li visited the Computer Science Department at CMU with support from the Chinese Scholarship Council.

References

1. Ng, A., Jordan, M.: On discriminative vs. generative classifiers: A comparison of logistic regression and naive bayes. In: NIPS (2002)
2. Vapnik, V.: The nature of statistical learning theory. Springer (2000)
3. Gönen, M., Alpaydin, E.: Localized multiple kernel learning. In: ICML (2008)
4. Jaakkola, T., Meila, M., Jebara, T.: Maximum entropy discrimination. TR AITR-1668, MIT (1999)
5. Raina, R., Shen, Y., Ng, A., McCallum, A.: Classification with hybrid generative/discriminative models. In: NIPS (2004)
6. Sha, F., Saul, L.: Large margin hidden markov models for automatic speech recognition. In: NIPS (2007)
7. Li, X., Zhao, X., Fu, Y., Liu, Y.: Bimodal gender recognition from face and fingerprint. In: CVPR (2010)
8. Perina, A., Cristani, M., Castellani, U., Murino, V., Jojic, N.: Free energy score spaces: using generative information in discriminative classifiers. IEEE Trans. on PAMI 34(7), 1249–1262 (2012)
9. Li, X., Lee, T., Liu, Y.: Stochastic feature mapping for pac-bayes classification. arXiv:1204.2609 (2012)
10. Li, X., Lee, T., Liu, Y.: Hybrid generative-discriminative classification using posterior divergence. In: CVPR (2011)
11. Jaakkola, T., Haussler, D.: Exploiting generative models in discriminative classifiers. In: NIPS (1999)

12. Tsuda, K., Kawanabe, M., Ratsch, G., Sonnenburg, S., Muller, K.: A new discriminative kernel from probabilistic models. Neural Computation 14(10), 2397–2414 (2002)
13. Vedaldi, A., Zisserman, A.: Efficient additive kernels via explicit feature maps. IEEE Trans. on PAMI 34(3), 480–492 (2012)
14. Holub, A.D., Welling, M., Perona, P.: Hybrid generative-discriminative visual categorization. International Journal of Computer Vision 77(1-3), 239–258 (2008)
15. Chatfield, K., Lempitsky, V., Vedaldi, A., Zisserman, A.: The devil is in the details: an evaluation of recent feature encoding methods. In: BMVC (2011)
16. Collins, M.: Discriminative training methods for hidden markov models: theory and experiments with perceptron algorithms. In: EMNLP, pp. 1–8 (2002)
17. Zhu, J., Ahmed, A., Xing, E.P.: Medlda: maximum margin supervised topic models for regression and classification. In: ICML, pp. 1257–1264 (2009)
18. Lacoste-Julien, S., Sha, F., Jordan, M.: Disclda: Discriminative learning for dimensionality reduction and classification. In: NIPS (2008)
19. der Maaten, L.V.: Learning discriminative fisher kernels. In: ICML, pp. 217–224 (2011)
20. Smith, N., Gales, M.: Speech recognition using svms. NIPS 14, 1197–1204 (2001)
21. Gales, M.J.F., Watanabe, S., Fosler-Lussier, E.: Structured discriminative models for speech recognition: an overview. IEEE Signal Processing Magazine 29(6), 70–81 (2012)
22. Neal, R., Hinton, G.: A new view of the em algorithm that justifies incremental, sparse and other variants. Learning in Graphical Models, 355–368
23. Wainwright, M.J., Jordan, M.I.: Graphical Models, Exponential Families, and Variational Inference. Now Publishers Inc., Hanover (2008)
24. Jordan, M., Ghahramani, Z., Jaakkola, T., Saul, L.: An introduction to variational methods for graphical models. Machine Learning 37(2), 183–233 (1999)
25. Graça, J., Ganchev, K., Taskar, B.: Expectation maximization and posterior constraints. In: NIPS (2007)
26. Friedman, J., Hastie, T., Tibshirani, R.: The Elements of Statistical Learning. Spinger (2008)
27. Gilks, W.R., Wild, P.: Adaptive rejection sampling for gibbs sampling. Applied Statistics, 337–348 (1992)
28. Langford, J.: Tutorial on practical prediction theory for classification. JMLR 6(1), 273–306 (2006)
29. Chang, C.C., Lin, C.J.: Libsvm: a library for support vector machines. ACM Transactions on Intelligent Systems and Technology 2(3), 27 (2011)
30. Rabiner, L.: A tutorial on hidden Markov models and selected applications in speech recognition. Proceeding of the IEEE 77(2), 257–286 (1989)
31. Baum, L., Petrie, T., Soules, G., Weiss, N.: A maximization technique occurring in the statistical analysis of probabilistic functions of markov chains. The Annals of Mathematical Statistics 41(1), 164–171 (1970)
32. Mollineda, R., Vidal, E., Casacuberta, F.: Cyclic sequence alignments: approximate versus optimal techniques. IJPRAI 16, 291–299 (2002)
33. Vedaldi, A., Gulshan, V., Varma, M., Zisserman, A.: Multiple kernels for object detection. In: ICCV (2009)
34. Blei, D., Ng, A., Jordan, M.: Latent dirichlet allocation. JMLR 3, 993–1022 (2003)
35. Griffiths, T., Steyvers, M.: Finding scientific topics. PNAS 101(suppl. 1), 5228–5235 (2004)
36. Lowe, D.: Distinctive image features from scale-invariant keypoints. IJCV 60(2), 91–110 (2004)

The Stochastic Gradient Descent for the Primal L1-SVM Optimization Revisited

Constantinos Panagiotakopoulos and Petroula Tsampouka

School of Technology, Aristotle University of Thessaloniki, Greece
costapan@eng.auth.gr, petroula@gen.auth.gr

Abstract. We reconsider the stochastic (sub)gradient approach to the unconstrained primal L1-SVM optimization. We observe that if the learning rate is inversely proportional to the number of steps, i.e., the number of times any training pattern is presented to the algorithm, the update rule may be transformed into the one of the classical perceptron with margin in which the margin threshold increases linearly with the number of steps. Moreover, if we cycle repeatedly through the possibly randomly permuted training set the dual variables defined naturally via the expansion of the weight vector as a linear combination of the patterns on which margin errors were made are shown to obey at the end of each complete cycle automatically the box constraints arising in dual optimization. This renders the dual Lagrangian a running lower bound on the primal objective tending to it at the optimum and makes available an upper bound on the relative accuracy achieved which provides a meaningful stopping criterion. In addition, we propose a mechanism of presenting the same pattern repeatedly to the algorithm which maintains the above properties. Finally, we give experimental evidence that algorithms constructed along these lines exhibit a considerably improved performance.

1 Introduction

Support Vector Machines (SVMs) [1,20,5] have been extensively used as linear classifiers either in the space where the patterns originally reside or in high dimensional feature spaces induced by kernels. They appear to be very successful at addressing the classification problem expressed as the minimization of an objective function involving the empirical risk while at the same time keeping low the complexity of the classifier. As measures of the empirical risk various quantities have been proposed with the 1- and 2-norm loss functions being the most widely accepted ones giving rise to the optimization problems known as L1- and L2-SVMs [4]. SVMs typically treat the problem as a constrained quadratic optimization in the dual space. At the early stages of SVM development their efficient implementation was hindered by the quadratic dependence of their memory requirements on the number of training examples a fact which rendered prohibitive the processing of large datasets. The idea of applying optimization only to a subset of the training set in order to overcome this difficulty resulted in the development of decomposition methods [16,9]. Although such methods led

H. Blockeel et al. (Eds.): ECML PKDD 2013, Part III, LNAI 8190, pp. 65–80, 2013.

to improved convergence rates, in practice their superlinear dependence on the number of examples, which can be even cubic, can still lead to excessive runtimes when dealing with massive datasets. Recently, the so-called linear SVMs [10,7,8,13] taking advantage of linear kernels in order to allow parts of them to be written in primal notation succeeded in outperforming decomposition SVMs.

The above considerations motivated research in alternative algorithms naturally formulated in primal space long before the advent of linear SVMs mostly in connection with the large margin classification of linearly separable datasets a problem directly related to the L2-SVM. Indeed, in the case that the 2-norm loss takes the place of the empirical risk an equivalent formulation exists which renders the dataset linearly separable in a high dimensional feature space. Such alternative algorithms ([14,15] and references therein) are mostly based on the perceptron [17], the simplest online learning algorithm for binary linear classification, with their key characteristic being that they work in the primal space in an online manner, i.e., processing one example at a time. Cycling repeatedly through the patterns they update their internal state stored in the weight vector each time an appropriate condition is satisfied. This way, due to their ability to process one example at a time, such algorithms succeed in sparing time and memory resources and consequently become able to handle large datasets.

Since the L1-SVM problem is not known to admit an equivalent maximum margin interpretation via a mapping to an appropriate space fully primal large margin perceptron-like algorithms appear unable to deal with such a task.[1] Nevertheless, a somewhat different approach giving rise to online algorithms was developed which focuses on the minimization of the regularized 1-norm soft margin loss through stochastic gradient descent (SGD). Notable representatives of this approach are the pioneer NORMA [12] (see also [21]) and Pegasos [18,19] (see also [2,3]). SGD gives rise to a kind of perceptron-like update having as an important ingredient the "shrinking" of the current weight vector. Shrinking always takes place when a pattern is presented to the algorithm with it being the only modification suffered by the weight vector if no loss is incurred. Thus, due to lack of a meaningful stopping criterion the algorithm without user intervention keeps running forever. In that sense the algorithms in question are fundamentally different from the mistake-driven large margin perceptron-like classifiers which terminate after a finite number of updates. There is no proof even for their asymptotic convergence when they use as output the final hypothesis but they do exist probabilistic convergence results or results in terms of the average hypothesis.

In the present work we reconsider the straightforward version of SGD for the primal unconstrained L1-SVM problem assuming a learning rate inversely proportional to the number of steps. Therefore, such an algorithm can be regarded

---

[1] The Margin Perceptron with Unlearning (MPU) [13] addresses the L1-SVM problem by keeping track of the number of updates caused by each pattern in parallel with the weight vector which is updated according to a perceptron-like rule. In that sense MPU uses dual variables and should rather be considered a linear SVM which, however, possesses a finite time bound for achieving a predefined relative accuracy.

either as NORMA with a specific dependence of the learning rate on the number of steps or as Pegasos with no projection step in the update and with a single example contributing to the (sub)gradient ($k = 1$). We observe here that this algorithm may be transformed into a classical perceptron with margin [6] in which the margin threshold increases linearly with the number of steps. The obvious gain from this observation is that the shrinking of the weight vector at each step amounts to nothing but an increase of the step counter by one unit instead of the costly multiplication of all the components of the generally non-sparse weight vector with a scalar. Another benefit arising from the above simplified description is that we are able to demonstrate easily that if we cycle through the data in complete epochs the dual variables defined naturally via the expansion of the weight vector as a linear combination of the patterns on which margin errors were made satisfy automatically the box constraints of the dual optimization. An important consequence of this unexpected result is that the relevant dual Lagrangian which is expressed in terms of the total number of margin errors, the number of complete epochs and the length of the current weight vector provides during the run a lower bound on the primal objective function and gives us a measure of the progress made in the optimization process. Indeed, by virtue of the strong duality theorem the dual Lagrangian and the primal objective coincide at optimality. Therefore, assuming convergence to the optimum an upper bound on the relative accuracy involving the dual Lagrangian may be defined which offers a useful and practically achievable stopping criterion. Moreover, we may now provide evidence in favor of the asymptotic convergence to the optimum by testing experimentally the vanishing of the duality gap. Finally, aiming at performing more updates at the expense of only one costly inner product calculation we propose a mechanism of presenting the same pattern repeatedly to the algorithm consistently with the above interesting properties.

The paper is organized as follows. Section 2 describes the algorithm and its properties. In Section 3 we give implementational details and deliver our experimental results. Finally, Section 4 contains our conclusions.

2   The Algorithm and Its Properties

Assume we are given a training set $\{(\boldsymbol{x}_k, l_k)\}_{k=1}^m$, with vectors $\boldsymbol{x}_k \in \mathbb{R}^d$ and labels $l_k \in \{+1, -1\}$. This set may be either the original dataset or the result of a mapping into a feature space of higher dimensionality [20,5]. By placing $\boldsymbol{x}_k$ in the same position at a distance ρ in an additional dimension, i.e., by extending $\boldsymbol{x}_k$ to $[\boldsymbol{x}_k, \rho]$, we construct an embedding of our data into the so-called augmented space [6]. The advantage of this embedding is that the linear hypothesis in the augmented space becomes homogeneous. Following the augmentation, a reflection with respect to the origin of the negatively labeled patterns is performed allowing for a uniform treatment of both categories of patterns. We define $R \equiv \max_k \|\boldsymbol{y}_k\|$ with $\boldsymbol{y}_k \equiv [l_k \boldsymbol{x}_k, l_k \rho]$ the k-th augmented and reflected pattern.

Let us consider the regularized empirical risk

$$\frac{\lambda}{2} \|w\|^2 + \frac{1}{m} \sum_{k=1}^{m} \max\{0, 1 - w \cdot y_k\}$$

involving the 1-norm soft margin loss $\max\{0, 1 - w \cdot y_k\}$ for the pattern y_k and the regularization parameter $\lambda > 0$ controlling the complexity of the classifier w. For a given dataset of size m minimization of the regularized empirical risk with respect to w is equivalent to the minimization of the objective function

$$\mathcal{J}(w, C) \equiv \frac{1}{2} \|w\|^2 + C \sum_{k=1}^{m} \max\{0, 1 - w \cdot y_k\} \ ,$$

where the "penalty" parameter $C > 0$ is related to λ as

$$C = \frac{1}{\lambda m} \ .$$

This is the L1-SVM problem expressed as an unconstrained optimization.

The algorithms we are concerned with are classical SGD algorithms. The term stochastic refers to the fact that they perform gradient descent with respect to the objective function in which the empirical risk $(1/m) \sum_{k=1}^{m} \max\{0, 1 - w \cdot y_k\}$ is approximated by the instantaneous risk $\max\{0, 1 - w \cdot y_k\}$ on a single example. The general form of the update rule is then

$$w_{t+1} = w_t - \eta_t \nabla_{w_t} \left[\frac{1}{2} \|w_t\|^2 + \frac{1}{\lambda} \max\{0, 1 - w_t \cdot y_k\} \right] \ ,$$

where η_t is the learning rate and ∇_{w_t} stands for a subgradient with respect to w_t since the 1-norm soft margin loss is only piecewise differentiable ($t \geq 0$). We choose a learning rate $\eta_t = 1/(t+1)$ which satisfies the conditions $\sum_{t=0}^{\infty} \eta_t^2 < \infty$ and $\sum_{t=0}^{\infty} \eta_t = \infty$ usually imposed in the convergence analysis of stochastic approximations. Then, noticing that $w_t - \frac{1}{t+1} w_t = \frac{t}{t+1} w_t$, we obtain the update

$$w_{t+1} = \frac{t}{t+1} w_t + \frac{1}{\lambda(t+1)} y_k \tag{1}$$

whenever

$$w_t \cdot y_k \leq 1 \tag{2}$$

and

$$w_{t+1} = \frac{t}{t+1} w_t \tag{3}$$

otherwise. In deriving the above update rule we made the choice $w_t - \lambda^{-1} y_k$ for the subgradient at the point $w_t \cdot y_k = 1$ where the 1-norm soft margin loss is not differentiable. We assume that $w_0 = 0$. We see that if $w_t \cdot y_k > 1$ the update consists of a pure shrinking of the current weight vector by the factor $t/(t+1)$.

The update rule may be simplified considerably if we perform the change of variable

$$w_t = \frac{a_t}{\lambda t} \tag{4}$$

for $t > 0$ and $w_0 = a_0 = 0$ for $t = 0$. In terms of the new weight vector a_t the update rule becomes

$$a_{t+1} = a_t + y_k \tag{5}$$

whenever

$$a_t \cdot y_k \leq \lambda t \tag{6}$$

and

$$a_{t+1} = a_t \tag{7}$$

otherwise.[2] This is the update of the classical perceptron algorithm with margin in which, however, the margin threshold in condition (6) increases linearly with the number of presentations of patterns to the algorithm independent of whether they lead to a change in the weight vector a_t. Thus, t counts the number of times any pattern is presented to the algorithm which corresponds to the number of updates (including the pure shrinkings (3)) of the weight vector w_t. Instead, the weight vector a_t is updated only if (6) is satisfied meaning that a margin error is made on y_k.

In the original formulation of Pegasos [18] the update is completed with a projection step in order to enforce the bound $\|w_t\| \leq 1/\sqrt{\lambda}$ which holds for the optimal solution. We show now that this is dynamically achieved to any desired accuracy after the elapse of sufficient time. In practice, however, it is in almost all cases achieved after one pass over the data.

Proposition 1. *For $t > 0$ the norm of the weight vector w_t is bounded from above as follows*

$$\|w_t\| \leq \frac{1}{\sqrt{\lambda}} \sqrt{1 + \left(\frac{R^2}{\lambda} - 1\right)\frac{1}{t}} \; . \tag{8}$$

Proof. From the update rule (5) taking into account condition (6) under which the update takes place we get

$$\|a_{t+1}\|^2 - \|a_t\|^2 = \|y_k\|^2 + 2a_t \cdot y_k \leq R^2 + 2\lambda t \; .$$

Obviously, this is trivially satisfied if (6) is violated and (7) holds. A repeated application of the above inequality with $a_0 = 0$ gives

$$\|a_t\|^2 \leq R^2 t + 2\lambda \sum_{k=0}^{t-1} k = R^2 t + \lambda t(t-1) = (R^2 - \lambda)t + \lambda t^2$$

from where using (4) and taking the square root we obtain (8). $\qquad\square$

---

[2] For $t = 0$ (6) becomes $a_0 \cdot y_k \leq 0$ instead of $a_0 \cdot y_k \leq 1$ which is obtained from (2) with $w_0 = a_0$. Since both are satisfied with $a_0 = 0$ (6) may be used for all t.

**The Stochastic Gradient Descent Algorithm
with random selection of examples**

Input: A dataset $S = (\boldsymbol{y}_1, \ldots, \boldsymbol{y}_k, \ldots, \boldsymbol{y}_m)$
with augmentation and reflection assumed
Fix: C, $t_{\max}$
Define: $\lambda = 1/(Cm)$
Initialize: $t = 0$, $\boldsymbol{a}_0 = \boldsymbol{0}$
while $t < t_{\max}$ **do**
    Choose $\boldsymbol{y}_k$ from S randomly
    if $\boldsymbol{a}_t \cdot \boldsymbol{y}_k \leq \lambda t$ **then**
        $\boldsymbol{a}_{t+1} = \boldsymbol{a}_t + \boldsymbol{y}_k$
    else
        $\boldsymbol{a}_{t+1} = \boldsymbol{a}_t$
    $t \leftarrow t + 1$
$\boldsymbol{w}_t = \boldsymbol{a}_t/(\lambda t)$

Combining (8) with the initial choice $\boldsymbol{w}_0 = \boldsymbol{0}$ we see that for all t the weaker bound $\|\boldsymbol{w}_t\| \leq R/\lambda$ previously derived in [19] holds.

SGD gives naturally rise to online algorithms. Therefore, we may choose the examples to be presented to the algorithm at random. However, the L1-SVM optimization task is a batch learning problem which may be better tackled by online algorithms via the classical conversion of such algorithms to the batch setting. This is done by cycling repeatedly through the possibly randomly permuted training dataset and using the last hypothesis for prediction. This traditional procedure of presenting the training data to the algorithm in complete epochs has in our case, as we will see shortly, the additional advantage that there exists a lower bound on the optimal value of the objective function to be minimized which is expressed in terms of quantities available during the run. The existence of such a lower bound provides an estimate of the relative accuracy achieved by the algorithm.

Proposition 2. *Let us assume that at some stage the whole training set has been presented to the algorithm exactly T times. Then, it holds that*

$$\mathcal{J}_{\text{opt}}(C) \equiv \min_{\boldsymbol{w}} \mathcal{J}(\boldsymbol{w}, C) \geq \mathcal{L}^T \equiv C\frac{M}{T} - \frac{1}{2}\|\boldsymbol{w}^T\|^2 \,, \tag{9}$$

where M is the total number of margin errors made up to that stage and $\boldsymbol{w}^T \equiv \boldsymbol{w}_{(mT)}$ the weight vector at $t = mT$ with m being the size of the training set.

Proof. Let I_k^t denote the number of margin errors made on the pattern $\boldsymbol{y}_k$ up to time t such that $\boldsymbol{a}_t = \sum_k I_k^t \boldsymbol{y}_k$. Obviously, it holds that

$$0 \leq I_k^{(mT)} \leq T \tag{10}$$

since $\boldsymbol{y}_k$ up to time $t = mT$ has been presented to the algorithm exactly T times. Then, taking into account (4) we see that at time t the dual variable α_k^t associated with $\boldsymbol{y}_k$ is $\alpha_k^t = I_k^t/(\lambda t)$ and consequently the dual variable $\alpha_k^{(mT)}$ after T complete epochs is given by

$$\alpha_k^{(mT)} = \frac{I_k^{(mT)}}{\lambda mT} = C\frac{I_k^{(mT)}}{T} \,. \tag{11}$$

With use of (10) we readily conclude that the dual variables after T complete epochs automatically satisfy the box constraints

$$0 \leq \alpha_k^{(mT)} \leq C \,. \tag{12}$$

From the weak duality theorem it follows that

$$\mathcal{J}(\boldsymbol{w}, C) \geq \mathcal{L}(\boldsymbol{\alpha}) = \sum_k \alpha_k - \frac{1}{2} \sum_{i,j} \alpha_i \alpha_j \boldsymbol{y}_i \cdot \boldsymbol{y}_j \; ,$$

where $\mathcal{L}(\boldsymbol{\alpha})$ is the dual Lagrangian[3] and the variables α_k obey the box constraints $0 \leq \alpha_k \leq C$. Thus, setting $\alpha_k = \alpha_k^{(mT)}$ in the above inequality, noticing that $\sum_k \alpha_k^{(mT)} = (C/T) \sum_k I_k^{(mT)} = CM/T$ and substituting $\sum_k \alpha_k^{(mT)} \boldsymbol{y}_k$ with $\boldsymbol{w}^T$ we obtain $\mathcal{J}(\boldsymbol{w}, C) \geq \mathcal{L}^T$ which is equivalent to (9). $\qquad \square$

In the course of proving Proposition 2 we saw that although the algorithm is fully primal the dual variables α_k^t defined through the expansion $\boldsymbol{w}_t = \sum_k \alpha_k^t \boldsymbol{y}_k$ of the weight vector $\boldsymbol{w}_t$ as a linear combination of the patterns on which margin errors were made obey after T complete epochs automatically the box constraints (12) encountered in dual optimization.[4] This surprising result allows us to construct the dual Lagrangian $\mathcal{L}^T$ which provides a lower bound on the optimal value $\mathcal{J}_{\text{opt}}$ of the objective $\mathcal{J}$ and assuming $\mathcal{L}^T > 0$ to obtain an upper bound $\mathcal{J}/\mathcal{L}^T - 1$ on the relative accuracy $\mathcal{J}/\mathcal{J}_{\text{opt}} - 1$ achieved as the algorithm keeps running. Thus, we have for the first time a primal SGD algorithm which may use the relative accuracy as stopping criterion.[5] It is also worth noticing that $\mathcal{L}^T$ involves only the total number M of margin errors and does not require that we keep the values of the individual dual variables during the run.

Although the automatic satisfaction of the box constraints by the dual variables is very important it is by no means sufficient to ensure vanishing of the duality gap and consequently convergence to the optimal solution. To demonstrate convergence to the optimum relying on dual optimization theory we must make sure that the Karush-Kuhn-Tucker (KKT) conditions [20,5] are satisfied. Their approximate satisfaction demands that the only patterns which have a substantial loss be the ones which have dual variables equal or at least extremely close to C (bound support vectors) and moreover that the patterns which have zero loss and margin considerably larger than $1/\|\boldsymbol{w}^T\|$ should have vanishingly small dual variables. Patterns with margin very close to $1/\|\boldsymbol{w}^T\|$ may

---

[3] Maximization of $\mathcal{L}(\boldsymbol{\alpha})$ subject to the constraints $0 \leq \alpha_k \leq C$ is the dual of the primal L1-SVM problem expressed as a constrained minimization.

[4] We expect that the dual variables will also satisfy the box constraints in the limit $t \to \infty$ if the patterns presented to the algorithm are selected randomly with equal probability since asymptotically they will all be selected an equal number of times.

[5] It is, of course, computationally expensive to evaluate at the end of each epoch the exact primal objective. Thus, an approximate calculation of the loss using the value that the weight vector had the last time each pattern was presented to the algorithm is preferable. This way we exploit the already computed inner product $\boldsymbol{a}_t \cdot \boldsymbol{y}_k$ which is needed in order to decide whether condition (6) is satisfied. If this approximate calculation gives a value of the relative accuracy which is not larger than f times the one set as stopping criterion we proceed to a proper calculation of the primal objective. The comparison coefficient f is given empirically a value close to 1.

The Stochastic Gradient Descent Algorithm
with relative accuracy ϵ

Input: A dataset $S = (\boldsymbol{y}_1, \ldots, \boldsymbol{y}_k, \ldots, \boldsymbol{y}_m)$
with augmentation and reflection assumed
Fix: C, ϵ, f, $T_{\max}$
Define: $q_k = \|\boldsymbol{y}_k\|^2$, $\lambda = 1/(Cm)$, $\epsilon' = f\epsilon$
Initialize: $t = 0$, $T = 0$, $M = 0$, $r_0 = 0$, $\boldsymbol{a}_0 = \boldsymbol{0}$
while $T < T_{\max}$ do
  Permute(S)
  $L = 0$
  for $k = 1$ to m do
    $p_{tk} = \boldsymbol{a}_t \cdot \boldsymbol{y}_k$
    $\theta_t = \lambda t$
    if $p_{tk} \leq \theta_t$ then
      $\boldsymbol{a}_{t+1} = \boldsymbol{a}_t + \boldsymbol{y}_k$
      $r_{t+1} = r_t + 2p_{tk} + q_k$
      $M \leftarrow M + 1$
      if $t > 0$ then
        $L \leftarrow L + 1 - p_{tk}/\theta_t$
      else
        $L \leftarrow L + 1$
    else
      $\boldsymbol{a}_{t+1} = \boldsymbol{a}_t$
      $r_{t+1} = r_t$
    $t \leftarrow t + 1$
  $T \leftarrow T + 1$
  $\theta = \lambda t$
  $w2 = r_t/(2\theta^2)$
  $\mathcal{J} = w2 + CL$
  $\mathcal{L} = CM/T - w2$
  if $\mathcal{J} - \mathcal{L} \leq \epsilon'\mathcal{L}$ then
    $L = 0$
    for $k = 1$ to m do
      $p_k = \boldsymbol{a}_t \cdot \boldsymbol{y}_k$
      if $p_k < \theta$ then
        $L \leftarrow L + \theta - p_k$
    $L \leftarrow L/\theta$
    $\mathcal{J} = w2 + CL$
    if $\mathcal{J} - \mathcal{L} \leq \epsilon\mathcal{L}$ then
      break
$\boldsymbol{w}_t = \boldsymbol{a}_t/(\lambda t)$

---

have dual variables with values between 0 and C and play the role of the non-bound support vectors. From (11) we see that the dual variable associated with the k-th pattern is equal to CT_k/T where $T_k \equiv I_k^{(mT)}$ is the number of epochs for which the k-th pattern was found to be a margin error. It is apparent that if there exists a number of epochs no matter how large it may be after which a pattern is consistently found to be a margin error then in the limit $T \to \infty$ we will have $(T_k/T) \to 1$ and the dual variable associated with it will asymptotically approach C. In contrast, if a pattern after a specific number of epochs is never found to be a margin error then $(T_k/T) \to 0$ and its dual variable will tend asymptotically to zero reflecting the accumulated effect of the shrinking that the weight vector suffers each time a pattern is presented to the algorithm. Therefore, the algorithm has the necessary ingredients for asymptotic satisfaction of the KKT conditions for the vanishing of the duality gap. The potential danger remains, however, that they may exist patterns with margin not very close to $1/\|\boldsymbol{w}^T\|$ which do not belong to any of the above categories and occasionally either become margin errors although most of the time are not or become classified with sufficiently large margin despite of the fact that they are most of the time margin errors. The hope is that with time the changes in the weight vector $\boldsymbol{w}_t$ will become smaller and smaller and such events will become more and more rare leading eventually to convergence to the optimal solution.

The above discussion cannot be regarded as a formal proof of the asymptotic convergence of the algorithm. We believe, however, that it does provide a convincing argument that assuming convergence (not necessarily to the optimum) the duality gap will eventually tend to zero and the lower bound $\mathcal{L}^T$ on the primal objective $\mathcal{J}$ given in Proposition 2 will approach the optimal primal objective $\mathcal{J}_{\mathrm{opt}}$, thereby proving that convergence to the optimum has been achieved. If, instead, we make the stronger assumption of convergence to the optimum then, of course, the vanishing of the duality gap follows from the strong duality theorem. In any case the stopping criterion exploiting the upper bound $\mathcal{J}/\mathcal{L}^T - 1$ on the relative accuracy $\mathcal{J}/\mathcal{J}_{\mathrm{opt}} - 1$ is a meaningful one.

Our discussion so far assumes that in an epoch each pattern is presented only once to the algorithm. We may, however, consider the option of presenting the same pattern $\boldsymbol{y}_k$ repeatedly ℓ times to the algorithm[6] aiming at performing more updates at the expense of only one calculation of the costly inner product $\boldsymbol{a}_t \cdot \boldsymbol{y}_k$. Proposition 2 and the analysis following it will still be valid on the condition that all patterns in each epoch are presented exactly the same number ℓ of times to the algorithm. Then, such an epoch should be regarded as equivalent to ℓ usual epochs with single presentations of patterns to the algorithm and will have as a result the increase of t by an amount equal to $m\ell$.

It is, of course, important to be able to decide in terms of just the initial value of $\boldsymbol{a}_t \cdot \boldsymbol{y}_k$ how many, let us say ℓ_+, out of these ℓ consecutive presentations of the pattern $\boldsymbol{y}_k$ to the algorithm will lead to a margin error, i.e., to an update of $\boldsymbol{a}_t$, with each of the remaining $\ell_- = \ell - \ell_+$ presentations necessarily corresponding to just an increase of t by 1 which amounts to a pure shrinking of $\boldsymbol{w}_t$.

Proposition 3. *Let the pattern $\boldsymbol{y}_k$ be presented at time t repeatedly ℓ times to the algorithm. Also let*

$$P = \boldsymbol{a}_t \cdot \boldsymbol{y}_k - \lambda t .$$

Then, the number ℓ_+ of times that $\boldsymbol{y}_k$ will be found to be a margin error is given by the following formula

$$\text{if } P > (\ell - 1)\lambda \qquad \ell_+ = 0 ,$$

$$\text{if } P \le (\ell - 1)\lambda \qquad \ell_+ = \min\left\{\ell, \left[\frac{(\ell-1)\lambda - P}{\max\{\|\boldsymbol{y}_k\|^2, \lambda\}}\right] + 1\right\} . \qquad (13)$$

Here $[x]$ denotes the integer part of $x \ge 0$.

Proof. For the sake of brevity we call a plus-step a presentation of the pattern $\boldsymbol{y}_k$ to the algorithm which leads to a margin error and a minus-step a presentation which does not. If at time t a plus-step takes place $\boldsymbol{a}_{t+1} \cdot \boldsymbol{y}_k - \lambda(t+1) = (\boldsymbol{a}_t \cdot \boldsymbol{y}_k - \lambda t) + (\|\boldsymbol{y}_k\|^2 - \lambda)$ while if a minus-step takes place $\boldsymbol{a}_{t+1} \cdot \boldsymbol{y}_k - \lambda(t+1) = (\boldsymbol{a}_t \cdot \boldsymbol{y}_k - \lambda t) - \lambda$. Thus, a plus-step adds to P the quantity $\|\boldsymbol{y}_k\|^2 - \lambda$ while a minus-step the quantity $-\lambda$. Clearly, after ℓ consecutive presentations of $\boldsymbol{y}_k$ to the algorithm it holds that $\boldsymbol{a}_{t+\ell} \cdot \boldsymbol{y}_k - \lambda(t+\ell) = P + \ell_+(\|\boldsymbol{y}_k\|^2 - \lambda) - (\ell - \ell_+)\lambda$.

---

[6] Multiple updates were introduced in [13,14]. A discussion in a context related to the present work is given in [11]. However, a proper SGD treatment in the presence of a regularization term for the 1-norm soft margin loss was not provided. Instead, a "forward-backward splitting" approach was adopted in which a multiple update in the absence of the regularizer is followed by ℓ pure regularizer-induced $\boldsymbol{w}_t$ shrinkings.

If $P > (\ell - 1)\lambda$ it follows that $P - (\ell - 1)\lambda > 0$ which means that after $\ell - 1$ consecutive minus-steps condition (6) is still violated and an additional minus-step must take place. Thus, $\ell_- = \ell$ and $\ell_+ = 0$.

For $P \le (\ell - 1)\lambda$ we first treat the subcase $\max\{\|\boldsymbol{y}_k\|^2, \lambda\} = \lambda$. If $\|\boldsymbol{y}_k\|^2 \le \lambda$ and $P \le 0$ condition (6) is initially satisfied and will still be satisfied after any number of plus-steps since the quantity $\|\boldsymbol{y}_k\|^2 - \lambda$ that is added to P with a plus-step is non-positive. Thus, $\ell_+ = \ell$. This is in accordance with (13) since $((\ell - 1)\lambda - P)/\lambda \ge \ell - 1$ or $[((\ell - 1)\lambda - P)/\lambda] + 1 \ge \ell$ leading to $\ell_+ = \ell$. It remains for $\|\boldsymbol{y}_k\|^2 \le \lambda$ to consider P in the interval $0 < P \le (\ell - 1)\lambda$ which can be further subdivided as $(\ell_1 - 1)\lambda < P \le \ell_1 \lambda$ with the integer ℓ_1 satisfying $1 \le \ell_1 \le \ell - 1$. For P belonging to such a subinterval condition (6) is initially violated and will still be violated after $\ell_1 - 1$ minus-steps while after one more minus-step will be satisfied. It will still be satisfied after any number of additional plus-steps because the quantity $\|\boldsymbol{y}_k\|^2 - \lambda$ that is added to P with a plus-step is non-positive. Thus, $\ell_- = \ell_1$ and $\ell_+ = \ell - \ell_1$. This is in accordance with (13) since $(\ell - \ell_1 - 1)\lambda \le (\ell - 1)\lambda - P < (\ell - \ell_1)\lambda$ leads to $[((\ell - 1)\lambda - P)/\lambda] + 1 = \ell - \ell_1$.

The subcase $\|\boldsymbol{y}_k\|^2 > \lambda$ of the case $P \le (\ell - 1)\lambda$ is far more complicated. If $\|\boldsymbol{y}_k\|^2 > \lambda$ with $P \le -(\ell - 1)(\|\boldsymbol{y}_k\|^2 - \lambda)$ condition (6) is initially satisfied and will still be satisfied after $\ell - 1$ plus-steps since $P + (\ell - 1)(\|\boldsymbol{y}_k\|^2 - \lambda) \le 0$. Thus, $\ell_+ = \ell$. This is consistent with (13) because $(\ell - 1)\lambda - P \ge (\ell - 1)\|\boldsymbol{y}_k\|^2$ or $[((\ell - 1)\lambda - P)/\|\boldsymbol{y}_k\|^2] + 1 \ge \ell$ leading to $\ell_+ = \ell$. It remains to be examined the case $\|\boldsymbol{y}_k\|^2 > \lambda$ with P in the interval $-(\ell - 1)(\|\boldsymbol{y}_k\|^2 - \lambda) < P \le (\ell - 1)\lambda$. The above interval can be expressed as a union of subintervals $(\ell - \ell_1 - 1)\lambda - \ell_1(\|\boldsymbol{y}_k\|^2 - \lambda) < P \le (\ell - \ell_1)\lambda - (\ell_1 - 1)(\|\boldsymbol{y}_k\|^2 - \lambda)$ with the integer ℓ_1 satisfying $1 \le \ell_1 \le \ell - 1$. Let P belong to such a subinterval. Let us also assume that the pattern $\boldsymbol{y}_k$ has been presented $\kappa \le \ell$ consecutive times to the algorithm as a result of which κ_+ plus-steps and κ_- minus-steps have taken place and the quantity $\kappa_+(\|\boldsymbol{y}_k\|^2 - \lambda) - \kappa_-\lambda$ has been added to P. Then $P_{\kappa_+,\kappa_-} \equiv P + \kappa_+(\|\boldsymbol{y}_k\|^2 - \lambda) - \kappa_-\lambda$ satisfies $(\ell - \ell_1 - 1 - \kappa_-)\lambda - (\ell_1 - \kappa_+)(\|\boldsymbol{y}_k\|^2 - \lambda) < P_{\kappa_+,\kappa_-} \le (\ell - \ell_1 - \kappa_-)\lambda - (\ell_1 - 1 - \kappa_+)(\|\boldsymbol{y}_k\|^2 - \lambda)$. As κ increases either κ_+ will first reach the value ℓ_1 with $\kappa_- < \ell - \ell_1$ or κ_- will first reach the value $\ell - \ell_1$ with $\kappa_+ < \ell_1$. In the former case $0 \le (\ell - \ell_1 - 1 - \kappa_-)\lambda < P_{\kappa_+,\kappa_-}$. This means that condition (6) is violated and will continue being violated until the number of minus-steps becomes equal to $\ell - \ell_1 - 1$ in which case one more minus-step must take place. Thus, all steps taking place after κ_+ has reached the value ℓ_1 are minus-steps. In the latter case $P_{\kappa_+,\kappa_-} \le -(\ell_1 - 1 - \kappa_+)(\|\boldsymbol{y}_k\|^2 - \lambda) \le 0$. This means that condition (6) is satisfied and will continue being satisfied until the number of plus-steps becomes equal to $\ell_1 - 1$ in which case one more plus-step must take place. Thus, all steps taking place after κ_- has reached the value $\ell - \ell_1$ are plus-steps. In both cases $\ell_+ = \ell_1$. This is again in accordance with (13) because $(\ell_1 - 1)\|\boldsymbol{y}_k\|^2 \le (\ell - 1)\lambda - P < \ell_1 \|\boldsymbol{y}_k\|^2$ or $[((\ell - 1)\lambda - P)/\|\boldsymbol{y}_k\|^2] + 1 = \ell_1$. $\square$

With ℓ_+ given in Proposition 3 the update of multiplicity ℓ of the weight vector $\boldsymbol{a}_t$ is written formally as

$$\boldsymbol{a}_{t+\ell} = \boldsymbol{a}_t + \ell_+ \boldsymbol{y}_k \ . \tag{14}$$

3   Implementation and Experiments

We implement three types of SGD algorithms[7] along the lines of the previous section. The first is the plain algorithm with random selection of examples, denoted SGD-r, which terminates when the maximum number $t_{\max}$ of steps is reached. Its pseudocode is given in Section 2. The dual variables in this case do not satisfy the box constraints as a result of which relative accuracy cannot be used as stopping criterion. The SGD algorithm with relative accuracy ϵ, the pseudocode of which is also given in Section 2, is denoted SGD-s where s designates that in an epoch each pattern is presented a single time to the algorithm. It terminates when the relative deviation of the primal objective $\mathcal{J}$ from the dual Lagrangian $\mathcal{L}^T$ just falls below ϵ provided the maximum number $T_{\max}$ of full epochs is not exhausted. A variation of this algorithm, denoted SGD-m, replaces in the T-th epoch the usual update with the multiple update (14) of multiplicity $\ell = 5$ only if $0 < T \bmod 9 < 5$. For both SGD-s and SGD-m the comparison coefficient takes the value $f = 1.2$ unless otherwise explicitly stated.

Algorithms performing SGD on the primal objective are expected to perform better if linear kernels are employed. Therefore the feature space in our experiments will be chosen to be the original instance space. As a consequence, our algorithms should most naturally be compared with linear SVMs. Among them we choose SVM^{perf} [8] [10], the first cutting-plane algorithm for training linear SVMs, the Optimized Cutting Plane Algorithm for SVMs[9] (OCAS) [7], the Dual Coordinate Descent[10] (DCD) algorithm [8] and the Margin Perceptron with Unlearning[11] (MPU) [13]. We also include in our study Pegasos[12] ($k = 1$). The SGD algorithms of [2,3] implemented in single precision are not considered.

The datasets we used for training are the binary Adult and Web datasets as compiled by Platt,[13] the training set of the KDD04 Physics dataset[14] (with 70 attributes after removing the 8 columns containing missing features), the Real-sim, News20 and Webspam (unigram treatment) datasets,[15] the multiclass Covertype UCI dataset[16] and the full Reuters RCV1 dataset.[17] Their number of instances and attributes are listed in Table 1. In the case of the Covertype dataset we study the binary classification problem of the first class versus rest while for the RCV1 we consider both the binary text classification tasks of the C11 and CCAT classes versus rest. The Physics and Covertype datasets were rescaled by

---

[7] Sources available at http://users.auth.gr/costapan
[8] Source (version 2.50) available at http://svmlight.joachims.org
[9] Source (version 0.96) available at http://cmp.felk.cvut.cz/~xfrancv/ocas/html
[10] Source available at http://www.csie.ntu.edu.tw/~cjlin/liblinear. We used the slightly faster older liblinear version 1.7 instead of the latest 1.93.
[11] Source available at http://users.auth.gr/costapan
[12] Source available at http://ttic.uchicago.edu/~shai/code
[13] http://research.microsoft.com/en-us/projects/svm/
[14] http://osmot.cs.cornell.edu/kddcup/datasets.html
[15] http://www.csie.ntu.edu.tw/~cjlin/libsvmtools/datasets
[16] http://archive.ics.uci.edu/ml/datasets.html
[17] http://www.jmlr.org/papers/volume5/lewis04a/lyrl2004_rcv1v2_README.htm

Table 1. The number T of complete epochs required in order for the SGD-s algorithm to achieve $(\mathcal{J} - \mathcal{L}^T)/\mathcal{L}^T \leq 10^{-5}$ for $C = 0.1$

data set	#instances	#attributes	SGD-s $\epsilon = 10^{-5}$ $C = 0.1$		
			T	$\mathcal{J}$	$\mathcal{L}^T$
Adult	32561	123	208174	1149.904	1149.893
Web	49749	300	16849	755.1139	755.1064
Physics	50000	70	13668	4995.139	4995.089
Realsim	72309	20958	4209	1437.315	1437.301
News20	19996	1355191	2178	902.5611	902.5521
Webspam	350000	254	27680	8284.781	8284.698
Covertype	581012	54	712648	36427.52	36427.16
C11	804414	47236	5670	5174.432	5174.381
CCAT	804414	47236	7987	12114.29	12114.17

multiplying all the features with 0.001. The experiments were conducted on a 2.5 GHz Intel Core 2 Duo processor with 3 GB RAM running Windows Vista. The C++ codes were compiled using the g++ compiler under Cygwin.

First we perform an experiment aiming at demonstrating that our SGD algorithms are able to obtain extremely accurate solutions. More specifically, with the algorithm SGD-s employing single updating we attempt to diminish the gap between the primal objective $\mathcal{J}$ and the dual Lagrangian $\mathcal{L}^T$ setting as a goal a relative deviation $(\mathcal{J} - \mathcal{L}^T)/\mathcal{L}^T \leq 10^{-5}$ for $C = 0.1$. In the present and in all subsequent experiments we do not include a bias term in any of the algorithms (i.e., in our case we assign to the augmentation parameter the value $\rho = 0$). In order to keep the number T of complete epochs as low as possible we increase the comparison coefficient f until the number of epochs required gets stabilized. This procedure does not entail, of course, the shortest training time but this is not our concern in this experiment. In Table 1 we give the values of both $\mathcal{J}$ and $\mathcal{L}^T$ and the number T of epochs needed to achieve these values. If multiple updates are used a larger number of epochs is, in general, required due to the slower increase of $\mathcal{L}^T$. Thus, SGD-s achieves, in general, relative accuracy closer to ϵ than SGD-m does. This is confirmed by subsequent experiments.

In our comparative experimental investigations we aim at achieving relative accuracy $(\mathcal{J} - \mathcal{J}_{\text{opt}})/\mathcal{J}_{\text{opt}} \leq 0.01$ for various values of the penalty parameter C assuming knowledge of the value of $\mathcal{J}_{\text{opt}}$. For Pegasos and SGD-r we use as stopping criterion the exhaustion of the maximum number of steps (iterations) $t_{\max}$ which, however, is given values which are multiples of the dataset size m. The ratio $t_{\max}/m$ may be considered analogous to the number T of epochs of the algorithm SGD-s since equal values of these two quantities indicate identical numbers of $\boldsymbol{w}_t$ updates. The input parameter for SGD-s and SGD-m is the (upper bound on) the relative accuracy ϵ. For MPU we use the parameter $\epsilon = \delta = \delta_{\text{stop}}$, where δ is the before-run relative accuracy and δ_{stop} the stopping threshold for the after-run relative accuracy. For SVM^{perf} and DCD we use as input their parameter ϵ while for OCAS the primal objective value $q = 1.01\mathcal{J}_{\text{opt}}$ (not given

Table 2. Training times of SGD algorithms to achieve $(\mathcal{J} - \mathcal{J}_{\mathrm{opt}})/\mathcal{J}_{\mathrm{opt}} \leq 0.01$ for $C = 1$

data set	Pegasos		SGD-r		SGD-s			SGD-m		
	$t_{\max}/m$	s	$t_{\max}/m$	s	ϵ	T	s	ϵ	T	s
						$C = 1$				
Adult	181	4.4	116	0.55	0.105	111	0.56	0.33	50	0.27
Web	53	1.0	46	0.34	0.054	26	0.20	0.1	14	0.11
Physics	2	0.20	6	0.09	2.1	1	0.03	0.14	3	0.06
Realsim	66	3.4	70	2.0	0.046	20	0.58	0.061	16	0.47
News20	89	10.2	88	7.5	0.023	39	3.1	0.029	25	2.1
Webspam	8	3.4	9	2.1	0.068	14	3.0	0.21	5	1.2
Covertype	-	-	62	10.9	0.264	65	8.7	1.12	18	2.5
C11	41	31.4	39	21.1	0.05	16	7.9	0.136	8	4.2
CCAT	37	32.7	36	19.8	0.055	16	8.1	0.163	7	4.1

Table 3. Training times of linear SVMs to achieve $(\mathcal{J} - \mathcal{J}_{\mathrm{opt}})/\mathcal{J}_{\mathrm{opt}} \leq 0.01$ for $C = 1$

data set	SVM^{perf}		OCAS	DCD		MPU	
	ϵ	s	s	ϵ	s	ϵ	s
			$C = 1$				
Adult	0.7	1.5	0.08	2.8	0.16	0.02	0.09
Web	0.2	0.33	0.30	6	0.06	0.01	0.05
Physics	1.0	0.30	0.02	23	0.06	0.06	0.06
Realsim	0.08	0.80	0.62	0.7	0.22	0.06	0.23
News20	0.14	12.8	6.0	0.4	0.64	0.03	1.5
Webspam	0.5	7.3	4.2	2.5	1.4	0.1	0.98
Covertype	4.2	45.8	3.4	6.5	9.1	0.1	6.2
C11	0.09	12.8	9.0	1.4	3.5	0.09	2.5
CCAT	0.25	19.3	12.9	1.6	3.6	0.1	3.2

in the tables) with the relative tolerance taking the default value $r = 0.01$. Any difference in training time between Pegasos and SGD-r for equal values of $t_{\max}/m$ should be attributed to the difference in the implementations. Any difference between $t_{\max}/m$ for SGD-r and T for SGD-s is to be attributed to the different procedure of choosing the patterns that are presented to the algorithm. Finally, the difference in the number T of epochs between SGD-s and SGD-m reflects the effect of multiple updates. It should be noted that in the runtime of SGD-s and SGD-m several calculations of the primal and the dual objective are included which are required for checking the satisfaction of the stopping criterion. If SGD-s and SGD-m were using the exhaustion of the maximum number $T_{\max}$ of epochs as stopping criterion their runtimes would certainly be shorter.

Tables 2 and 3 contain the results of the experiments involving the SGD algorithms and linear SVMs for $C = 1$. We observe that, in general, there is a progressive decrease in training time as we move from Pegasos to SGD-m through SGD-r and SGD-s due to the additive effect of several factors. These factors are the more efficient implementation of our algorithms exploiting the change of variable given by (4), the presentation of the patterns to SGD-s and SGD-m in

Table 4. Training times of SGD algorithms to achieve $(\mathcal{J} - \mathcal{J}_{\mathrm{opt}})/\mathcal{J}_{\mathrm{opt}} \leq 0.01$ for $C = 10$

data set	Pegasos		SGD-r		SGD-s			SGD-m		
	$t_{\max}/m$	s	$t_{\max}/m$	s	ϵ	T	s	ϵ	T	s
Adult	-	-	1146	5.3	0.098	1172	5.8	0.35	455	2.4
Web	338	6.5	330	2.4	0.049	220	1.7	0.105	99	0.76
Physics	12	1.1	51	0.78	0.203	17	0.27	0.223	19	0.33
Realsim	746	30.5	738	21.1	0.0162	487	12.9	0.027	261	7.0
News20	796	76.1	797	63.0	0.0104	719	53.0	0.012	279	20.6
Webspam	-	-	64	14.5	0.125	62	12.3	0.4	26	5.7
Covertype	-	-	472	82.8	0.343	462	60.2	0.77	269	36.3
C11	446	332.1	441	238.7	0.0415	178	82.0	0.085	116	53.9
CCAT	-	-	387	212.2	0.0471	170	79.2	0.112	98	46.7

Table 5. Training times of linear SVMs to achieve $(\mathcal{J} - \mathcal{J}_{\mathrm{opt}})/\mathcal{J}_{\mathrm{opt}} \leq 0.01$ for $C = 10$.

data set	SVM^{perf}		OCAS	DCD		MPU	
	ϵ	s	s	ϵ	s	ϵ	s
Adult	0.6	38.0	0.33	2.6	1.2	0.04	0.62
Web	0.23	1.4	0.55	8	0.20	0.02	0.08
Physics	1.3	2.9	0.09	23	0.30	0.05	0.20
Realsim	0.031	4.4	2.6	0.25	0.45	0.02	0.41
News20	0.019	147.0	40.7	0.2	1.5	0.02	2.1
Webspam	0.36	37.1	6.8	2.2	5.3	0.2	2.5
Covertype	2.1	52.0	17.3	6.1	90.8	0.08	38.5
C11	0.079	39.6	25.0	0.65	9.2	0.02	5.7
CCAT	0.14	72.0	35.2	0.85	11.7	0.02	7.9

complete epochs (see also [3,19]) and the use by SGD-m of multiple updating. The overall improvement made by SGD-m over Pegasos is quite substantial. DCD and MPU are certainly statistically faster but their differences from SGD-m are not very large especially for the largest datasets. Moreover, SGD-s and SGD-m are considerably faster than SVM^{perf} and statistically faster than OCAS. Pegasos failed to process the Covertype dataset due to numerical problems.

Tables 4 and 5 contain the results of the experiments involving the SGD algorithms and linear SVMs for $C = 10$. Although the general characteristics resemble the ones of the previous case the differences are magnified due to the intensity of the optimization task. Certainly, the training time of linear SVMs scales much better as C increases. Moreover, MPU clearly outperforms DCD and OCAS for most datasets. SGD-m is still statistically faster than SVM^{perf} but slower than OCAS. Finally, Pegasos runs more often into numerical problems.

In contrast, as C decreases the differences among the algorithms are alleviated. This is apparent from the results for $C = 0.05$ reported in Tables 6 and 7. SGD-r, SGD-s and SGD-m all appear statistically faster than the linear SVMs. Also Pegasos outperforms SVM^{perf} for the majority of datasets with preference for the largest ones. Seemingly, lowering C favors the SGD algorithms.

Table 6. Training times of SGD algorithms to achieve $(\mathcal{J} - \mathcal{J}_{\text{opt}})/\mathcal{J}_{\text{opt}} \leq 0.01$ for $C = 0.05$

data set	Pegasos		SGD-r		SGD-s			SGD-m		
	$t_{\max}/m$	s	$t_{\max}/m$	s	ϵ	T	s	ϵ	T	s
Adult	7	0.17	12	0.06	0.07	10	0.05	0.15	6	0.03
Web	4	0.11	4	0.03	0.06	2	0.02	0.06	2	0.02
Physics	1	0.09	1	0.02	0.11	1	0.02	0.06	1	0.02
Realsim	3	0.27	3	0.09	0.09	1	0.06	0.14	1	0.06
News20	4	0.72	3	0.33	0.08	1	0.20	0.12	1	0.20
Webspam	1	0.50	2	0.47	0.4	1	0.44	0.18	1	0.44
Covertype	5	4.1	4	0.72	0.25	4	0.64	0.27	5	0.80
C11	2	1.6	3	1.6	0.03	2	1.5	0.1	1	1.0
CCAT	2	2.1	2	1.1	0.12	1	1.0	0.12	1	1.0

Table 7. Training times of linear SVMs to achieve $(\mathcal{J} - \mathcal{J}_{\text{opt}})/\mathcal{J}_{\text{opt}} \leq 0.01$ for $C = 0.05$

data set	SVM^{perf}		OCAS	DCD		MPU	
	ϵ	s	s	ϵ	s	ϵ	s
Adult	1.1	0.14	0.05	3	0.03	0.01	0.03
Web	0.3	0.09	0.09	4	0.03	0.01	0.03
Physics	1.1	0.08	0.02	12	0.03	0.02	0.02
Realsim	0.7	0.23	0.20	0.7	0.16	0.2	0.16
News20	0.9	1.3	0.73	3	0.23	0.2	0.56
Webspam	1.1	2.8	2.3	1	1.1	0.2	0.72
Covertype	2.9	53.7	1.6	6	1.2	0.3	1.4
C11	0.11	4.7	2.9	4	1.4	0.1	1.4
CCAT	0.25	7.2	5.7	1	2.7	0.1	1.6

4   Conclusions

We reexamined the classical SGD approach to the primal unconstrained L1-SVM optimization task and made some contributions concerning both theoretical and practical issues. Assuming a learning rate inversely proportional to the number of steps a simple change of variable allowed us to simplify the algorithmic description and demonstrate that in a scheme presenting the patterns to the algorithm in complete epochs the naturally defined dual variables satisfy automatically the box constraints of the dual optimization. This opened the way to obtaining an estimate of the progress made in the optimization process and enabled the adoption of a meaningful stopping criterion, something the SGD algorithms were lacking. Moreover, it made possible a qualitative discussion of how the KKT conditions will be asymptotically satisfied provided the weight vector $\boldsymbol{w}_t$ gets stabilized. Besides, we showed that in the limit $t \to \infty$ even without a projection step in the update it holds that $\|\boldsymbol{w}_t\| \leq 1/\sqrt{\lambda}$, a bound known to be

obeyed by the optimal solution. On the more practical side by exploiting our simplified algorithmic description and employing a mechanism of multiple updating
we succeeded in substantially improving the performance of SGD algorithms.
For optimization tasks of low or medium intensity the algorithms constructed
are comparable to or even faster than the state-of-the-art linear SVMs.

References

1. Boser, B., Guyon, I., Vapnik, V.: A training algorithm for optimal margin classifiers. In: COLT, pp. 144–152 (1992)
2. Bordes, A., Bottou, L., Gallinari, P.: SGD-QN: Careful quasi-Newton stochastic
   gradient descent. JMLR 10, 1737–1754 (2009)
3. Bottou, L.: Stochastic gradient descent examples (Web Page),
   http://leon.bottou.org/projects/sgd
4. Cortes, C., Vapnik, V.: Support vector networks. Mach. Learn. 20, 273–297 (1995)
5. Cristianini, N., Shawe-Taylor, J.: An Introduction to Support Vector Machines.
   Cambridge University Press, Cambridge (2000)
6. Duda, R.O., Hart, P.E.: Pattern Classsification and Scene Analysis. Wiley, Chichester (1973)
7. Frank, V., Sonnenburg, S.: Optimized cutting plane algorithm for support vector
   machines. In: ICML, pp. 320–327 (2008)
8. Hsieh, C.-J., Chang, K.-W., Lin, C.-J., Keerthi, S.S., Sundararajan, S.: A dual coordinate descent method for large-scale linear SVM. In: ICML, pp. 408–415 (2008)
9. Joachims, T.: Making large-scale SVM learning practical. In: Advances in Kernel
   Methods-Support Vector Learning. MIT Press, Cambridge (1999)
10. Joachims, T.: Training linear SVMs in linear time. In: KDD, pp. 217–226 (2006)
11. Karampatziakis, N., Langford, J.: Online importance weight aware updates. In:
    UAI, pp. 392–399 (2011)
12. Kivinen, J., Smola, A., Williamson, R.: Online learning with kernels. IEEE Trans-
    actions on Signal Processing 52(8), 2165–2176 (2004)
13. Panagiotakopoulos, C., Tsampouka, P.: The margin perceptron with unlearning.
    In: ICML, pp. 855–862 (2010)
14. Panagiotakopoulos, C., Tsampouka, P.: The margitron: A generalized perceptron
    with margin. IEEE Transactions on Neural Networks 22(3), 395–407 (2011)
15. Panagiotakopoulos, C., Tsampouka, P.: The perceptron with dynamic margin. In:
    Kivinen, J., Szepesvári, C., Ukkonen, E., Zeugmann, T. (eds.) ALT 2011. LNCS,
    vol. 6925, pp. 204–218. Springer, Heidelberg (2011)
16. Platt, J.C.: Sequential minimal optimization: A fast algorithm for training support
    vector machines. Microsoft Res. Redmond WA, Tech. Rep. MSR-TR-98-14 (1998)
17. Rosenblatt, F.: The perceptron: A probabilistic model for information storage and
    organization in the brain. Psychological Review 65 (6), 386–408 (1958)
18. Shalev-Schwartz, S., Singer, Y., Srebro, N.: Pegasos: Primal estimated sub-gradient
    solver for SVM. In: ICML, pp. 807–814 (2007)
19. Shalev-Schwartz, S., Singer, Y., Srebro, N., Cotter, A.: Pegasos: Primal estimated
    sub-gradient solver for SVM. Mathematical Programming 127(1), 3–30 (2011)
20. Vapnik, V.: Statistical learning theory. Wiley, Chichester (1998)
21. Zhang, T.: Solving large scale linear prediction problems using stochastic gradient
    descent algorithms. In: ICML (2004)

Bundle CDN: A Highly Parallelized Approach for Large-Scale ℓ_1-Regularized Logistic Regression

Yatao Bian, Xiong Li, Mingqi Cao, and Yuncai Liu

Institute of Image Processing and Pattern Recognition,
Shanghai Jiao Tong University, Shanghai 200240, China
{bianyatao,lixiong,caomingqi,whomliu}@sjtu.edu.cn

Abstract. Parallel coordinate descent algorithms emerge with the growing demand of large-scale optimization. In general, previous algorithms are usually limited by their divergence under high degree of parallelism (DOP), or need data pre-process to avoid divergence. To better exploit parallelism, we propose a coordinate descent based parallel algorithm without needing of data pre-process, termed as Bundle Coordinate Descent Newton (BCDN), and apply it to large-scale ℓ_1-regularized logistic regression. BCDN first randomly partitions the feature set into Q non-overlapping subsets/bundles in a Gauss-Seidel manner, where each bundle contains P features. For each bundle, it finds the descent directions for the P features in parallel, and performs P-dimensional Armijo line search to obtain the stepsize. By theoretical analysis on global convergence, we show that BCDN is guaranteed to converge with a high DOP. Experimental evaluations over five public datasets show that BCDN can better exploit parallelism and outperforms state-of-the-art algorithms in speed, without losing testing accuracy.

Keywords: parallel optimization, coordinate descent newton, large-scale optimization, ℓ_1-regularized logistic regression.

1 Introduction

High dimensional ℓ_1-regularized models arise in a wide range of applications, such as sparse logistic regression [12] and compressed sensing [10]. Various optimization methods such as coordinate minimization [4], stochastic gradient [15] and trust region [11] have been developed to solve ℓ_1-regularized models, among which coordinate descent newton (CDN) is proven to be promising [17].

The growing demand of scalable optimization along with the stagnant CPU speed impels people to design computers with more cores and heterogeneous computing frameworks, such as generous purpose GPU (GPGPU). To fully utilize these kinds of devices, parallel algorithms pop up like mushrooms in various areas, such as parallel annealed particle filter for motion tracking by Bian et al [1] and parallel stochastic gradient descent by Niu et al [13]. While works in [7,19] perform parallelization over samples, there are often much more features than

H. Blockeel et al. (Eds.): ECML PKDD 2013, Part III, LNAI 8190, pp. 81–95, 2013.

samples in ℓ_1-regularized problems. Bradley et al [2] proposed Shotgun CDN for ℓ_1-regularized logistic regression by directly parallelizing the updates of features. However, Shotgun CDN is easily affected by interference among parallel updates, which limits its DOP. To get more parallelism, Scherrer et al [14] proposed to conduct feature clustering, which would introduce extra computing overhead.

To better exploit parallelism, we propose a new globally convergent algorithm, Bundle Coordinate Descent Newton (BCDN), without needing of data pre-process. In each outer iteration, BCDN randomly partitions the feature index set $\mathcal{N}$ into Q subsets/bundles with size of P in a Gauss-Seidel manner. In each inner iteration it first parallelly finds the descent directions for P features in a bundle and second, it conducts P-dimensional Armijo line search to find the stepsize. A set of experiments demonstrate its remarkable properties: a highly parallelized approach with strong convergence guarantee. Experimental results with different bundle size P (DOP) indicate that it could run with high DOP (large bundle size P). Also, its high parallelism ensures good scalability on different parallel computing frameworks (e.g. multi-core, cluster, heterogeneous computing).

The contributions of this paper are mainly threefold: (1) proposing a highly parallelized coordinate descent based algorithm, BCDN; (2) giving strong convergence guarantee by theoretical analysis; (3) applying BCDN to large-scale ℓ_1-regularized logistic regression.

For readability, we here briefly summarize the mathematical notations as follows. s and n denote the number of training samples and the number of features respectively. $\mathcal{N} = \{1, 2, \cdots, n\}$ denotes the feature index set. $(\mathbf{x}_i, y_i), i = 1, \cdots, s$ denote the sample-label pairs, where $\mathbf{x}_i \in \mathbb{R}^n, y_i \in \{-1, +1\}$. $\mathbf{X} \in \mathbb{R}^{s \times n}$ denotes the design matrix, whose i^{th} row is $\mathbf{x}_i$. $\mathbf{w} \in \mathbb{R}^n$ is the unknown vector of model weights; $\mathbf{e}_j$ denotes the indicator vector with only the jth element equaling 1 and others 0. $\|\cdot\|$ and $\|\cdot\|_1$ denote the 2-norm and 1-norm, respectively.

The remainder of this paper is organized as follows. We first briefly review two related algorithms for ℓ_1-regularized logistic regression in Section 2, then describe BCDN and its high ideal speedup in Section 3. We give the theoretical analysis for convergence guarantee of BCDN in Section 4 and present implementation and datasets details in Section 5. Experimental results will be reported in Section 6.

2   Algorithms for ℓ_1-Regularized Logistic Regression

Consider the following unconstrained ℓ_1-regularized optimization problem:

$$\min_{\mathbf{w} \in \mathbb{R}^n} F_c(\mathbf{w}) \equiv c \sum_{i=1}^{s} \varphi(\mathbf{w}; \mathbf{x}_i, y_i) + \|\mathbf{w}\|_1, \tag{1}$$

where $\varphi(\mathbf{w}; \mathbf{x}_i, y_i)$ is a non-negative and convex loss function; $c > 0$ is the regularization parameter. For logistic regression, the overall training losses can be expressed as follows:

$$L(\mathbf{w}) \equiv c \sum_{i=1}^{s} \varphi(\mathbf{w}; \mathbf{x}_i, y_i) = c \sum_{i=1}^{s} \log(1 + e^{-y_i \mathbf{w}^T \mathbf{x_i}}). \tag{2}$$

A number of solvers are available for this problem. In this section, we focus on two effective solvers: CDN [17] and its parallel variant, Shotgun CDN [2].

2.1 Coordinate Descent Newton (CDN)

Yuan et al [17] have demonstrated that CDN is a very efficient solver for large-scale ℓ_1-regularized logistic regression. It is a special case of coordinate gradient descent (CGD) proposed in [16]. The overall procedure of CDN is summarized in Algorithm 1.

Given the current model $\mathbf{w}$, for the selected feature $j \in \mathcal{N}$, $\mathbf{w}$ is updated along the descent direction $\mathbf{d}^j = d(\mathbf{w}; j)\mathbf{e}_j$, where,

$$d(\mathbf{w}; j) \equiv \arg\min_d \{\nabla_j L(\mathbf{w})d + \frac{1}{2}\nabla^2_{jj} L(\mathbf{w})d^2 + |w_j + d|\}. \tag{3}$$

Armijo rule is adopted based on [3] to determine the stepsize for the line search procedure. Let $\alpha = \alpha(\mathbf{w}, \mathbf{d})$ be the determined stepsize, where,

$$\alpha(\mathbf{w}, \mathbf{d}) \equiv \max_{t=0,1,2,\cdots} \{\beta^t \mid F_c(\mathbf{w} + \beta^t \mathbf{d}) - F_c(\mathbf{w}) \le \beta^t \sigma \Delta\}, \tag{4}$$

where $0 < \beta < 1$, $0 < \sigma < 1$, β^t denotes β to the power of t, $\Delta \equiv \nabla L(\mathbf{w})^T \mathbf{d} + \|\mathbf{w} + \mathbf{d}\|_1 - \|\mathbf{w}\|_1$. This rule requires only function evaluations. According to [16], larger stepsize will be accepted if we choose σ near 0.

Algorithm 1. Coordinate Descent Newton (CDN) [17]

1: Set $\mathbf{w}^1 = \mathbf{0} \in \mathbb{R}^n$.
2: **for** $k = 1, 2, 3, \cdots$ **do**
3:    **for all** $j \in \mathcal{N}$ **do**
4:        Obtain $\mathbf{d}^{k,j} = d(\mathbf{w}^{k,j}; j)\mathbf{e}_j$ by solving Eq. (3).
5:        Find the stepsize $\alpha^{k,j} = \alpha(\mathbf{w}^{k,j}, \mathbf{d}^{k,j})$ by solving Eq. (4).   // 1-dimensional line search
6:        $\mathbf{w}^{k,j+1} \leftarrow \mathbf{w}^{k,j} + \alpha^{k,j}\mathbf{d}^{k,j}$.
7:    **end for**
8: **end for**

2.2 Shotgun CDN (SCDN)

Shotgun CDN (SCDN) [2] simply updates $\bar{P}$ features in parallel, where each feature update corresponds to one inner iterations in CDN, so its DOP[1] is $\bar{P}$. However, its parallel updates might increase the risk of divergence, which comes from the correlation among features. Bradley et al [2] provided a problem-specific

---

[1] DOP is a metric indicating how many operations can be or being simultaneously executed by a computer.

measure for SCDN's potential of parallelization: the spectral radius ρ of $X^T X$. With this measure, an upper bound is given to $\bar{P}$, i.e., $\bar{P} \leq n/\rho + 1$ to achieve speedups linear in $\bar{P}$. However, ρ can be very large for most large-scale datasets, e.g. $\rho = 20,228,800$ for dataset gisette with $n = 5000$, which limits the parallel ability of SCDN. Therefore, algorithms with high parallelism are desired to deal with large-scale problems. The details of SCDN can be found in Algorithm 2.

Algorithm 2. Shotgun CDN (SCDN) [2]

1: Choose the number of parallel updates $\bar{P} \geq 1$.
2: Set $\mathbf{w} = \mathbf{0} \in \mathbb{R}^n$.
3: **while** not converged **do**
4:     **In parallel** on $\bar{P}$ processors
5:         Choose $j \in \mathcal{N}$ uniformly at radom.
6:         Obtain $\mathbf{d}^j = d(\mathbf{w}; j)\mathbf{e}_j$ by solving Eq. (3).
7:         Find the stepsize $\alpha^j = \alpha(\mathbf{w}, \mathbf{d}^j)$ by solving Eq. (4).   //1-dimensional line
    search
8:         $\mathbf{w} \leftarrow \mathbf{w} + \alpha^j \mathbf{d}^j$.
9: **end while**

---

3   Bundle Coordinate Descent Newton (BCDN)

SCDN places no guarantee on its convergence when the number of features to be updated in parallel is greater than a threshold, i.e., $\bar{P} > n/\rho + 1$. This is because the 1-dimensional line search (step 7 in Algorithm 2) inside each parallel loop of SCDN cannot ensure the descent of $F_c(\mathbf{w})$ for all the $\bar{P}$ parallel feature updates. Motivated by this observation and some experimental results, we propose to perform high dimensional line search to ensure the descent of $F_c(\mathbf{w})$. The proposed method is termed as Bundle Coordinate Descent Newton (BCDN) whose overall procedure is summarized in Algorithm 3.

In each outer iteration, BCDN first randomly[2] partitions the feature index set $\mathcal{N}$ into Q subsets/bundles $\mathcal{B}^1, \mathcal{B}^2, \cdots, \mathcal{B}^Q$ in a Gauss-Seidel manner,

$$\bigcup_{q=1}^{Q} \mathcal{B}^q = \mathcal{N} \text{ and } \mathcal{B}^p \bigcap_{p \neq q} \mathcal{B}^q = \emptyset, \ \forall \ 1 < p, q < Q. \tag{5}$$

For simplicity, in practice, all bundles are set to have the same size P, then the number of bundles $Q = \lceil \frac{n}{P} \rceil$. Note that in the following theoretical analysis, the bundles can have different sizes.

Then in each inner iteration, BCDN first finds 1-dimensional descent directions (step 7) for features in $\mathcal{B}^q$ in parallel, then performs P-dimensional line search (step 10) to get the stepsize along the descent direction.

---

[2] The randomness is conducted by a random permutation of the feature index.

Algorithm 3. Bundle CDN (BCDN)

1: Choose the bundle size $P \in [1, n]$.
2: Set $\mathbf{w}^1 = \mathbf{0} \in \mathbb{R}^n$.
3: **for** $k = 1, 2, 3, \cdots$ **do**
4:     Randomly partition $\mathcal{N}$ to $\mathcal{B}^1, \mathcal{B}^2, \cdots, \mathcal{B}^Q$ satisfying Eq. (5).
5:     **for all** $\mathcal{B}^q \subseteq \mathcal{N}$ **do**
6:        **for all** $j \in \mathcal{B}^q$ **in parallel do**
7:           Obtain $\mathbf{d}^{k,j} = d(\mathbf{w}^{k,\mathcal{B}^q}; j)\mathbf{e}_j$ by solving Eq. (3).
8:           $\mathbf{d}^{k,\mathcal{B}^q} \leftarrow \mathbf{d}^{k,\mathcal{B}^q} + \mathbf{d}^{k,j}$.
9:        **end for**
10:       Find the stepsize $\alpha^{k,\mathcal{B}^q} = \alpha(\mathbf{w}^{k,\mathcal{B}^q}, \mathbf{d}^{k,\mathcal{B}^q})$ by solving Eq. (4).//*P-dimensional line search*
11:       $\mathbf{w}^{k,\mathcal{B}^{q+1}} \leftarrow \mathbf{w}^{k,\mathcal{B}^q} + \alpha^{k,\mathcal{B}^q} \mathbf{d}^{k,\mathcal{B}^q}$.
12:    **end for**
13: **end for**

Obviously, CDN is a special case of BCDN with the setting $Q = n$. That is, $\mathcal{B}^q = \{q\}, q = 1, 2, \cdots, n$.

3.1   High Ideal Speedup[3] of BCDN

In the following part, we will demonstrate that the ideal speedup of BCDN is the bundle size P, compared to CDN.

First, in the computing procedure for descent direction (step 7 in Algorithm 3), the computing of 1-dimensional descent directions for each feature is independent of each other. Therefore, the DOP is P and the ideal speedup also is P. Then, we argue that, in BCDN, the P-dimensional line search (step 10 in each outer iteration in Algorithm 3) also has the ideal speedup of P, in comparison with CDN. In each outer iteration, BCDN runs $Q = \lceil \frac{n}{P} \rceil$ times of P-dimensional line search (step 10 in Algorithm 3), while CDN runs n times of 1-dimensional line search (step 5 in Algorithm 1). However, the P-dimensional line search in BCDN costs about the same computing time as the 1-dimensional line search in CDN, which will be shown as follows.

First, each P-dimensional line search in BCDN will terminate roughly within the same finite number of steps, with respect to CDN. This will be proven in Theorem 1 and verified by experiments in Section 6.1. Second, the time costs of one step of line search in BCDN and CDN are equivalent. In our BCDN implementation, we maintain both $\mathbf{d}^T \mathbf{x}_i$ and $e^{\mathbf{w}^T \mathbf{x}_i}, i = 1, \cdots, s$ and follow the

---

[3] Here we introduce a notation "ideal speedup" to measure the speedup ratio for a parallel algorithm on an ideal computation platform. The ideal platform is assumed to have unlimited computing resources, and have no parallel schedule overhead.

Algorithm 4. Armijo Line Search Details

1: Given $\beta, \sigma, \nabla L(\mathbf{w}), \mathbf{w}, \mathbf{d}$ and $e^{\mathbf{w}^T \mathbf{x}_i}, i = 1, \cdots, s$.
2: $\Delta \leftarrow \nabla L(\mathbf{w})^T \mathbf{d} + \|\mathbf{w} + \mathbf{d}\|_1 - \|\mathbf{w}\|_1$.
3: *Compute $\mathbf{d}^T \mathbf{x}_i, i = 1, \cdots, s$.
4: **for** $t = 0, 1, 2, \cdots$ **do**
5:    **if** Eq. (6) is satisfied **then**
6:       $\mathbf{w} \leftarrow \mathbf{w} + \beta^t \mathbf{d}$.
7:       $e^{\mathbf{w}^T \mathbf{x}_i} \leftarrow e^{\mathbf{w}^T \mathbf{x}_i} e^{\beta^t \mathbf{d}^T \mathbf{x}_i}, i = 1, \cdots, s$.
8:       break
9:    **else**
10:       $\Delta \leftarrow \beta \Delta$.
11:       $\mathbf{d}^T \mathbf{x}_i \leftarrow \beta \mathbf{d}^T \mathbf{x}_i, i = 1, \cdots, s$.
12:    **end if**
13: **end for**

---

implementation technique of Fan et al (see Appendix G of [5]). That is, the sufficient decrease condition in Eq. 4 is computed using the following form:

$$f(\mathbf{w} + \beta^t \mathbf{d}) - f(\mathbf{w})$$
$$= \|\mathbf{w} + \beta^t \mathbf{d}\|_1 - \|\mathbf{w}\|_1 + c\left(\sum_{i=1}^{s} \log\left(\frac{e^{(\mathbf{w} + \beta^t \mathbf{d})^T \mathbf{x}_i} + 1}{e^{(\mathbf{w} + \beta^t \mathbf{d})^T \mathbf{x}_i} + e^{\beta^t \mathbf{d}^T \mathbf{x}_i}}\right) + \beta^t \sum_{i: y_i = -1} \mathbf{d}^T \mathbf{x}_i\right) \quad (6)$$
$$\leq \sigma \beta^t (\nabla L(\mathbf{w})^T \mathbf{d} + \|\mathbf{w} + \mathbf{d}\|_1 - \|\mathbf{w}\|_1)$$

It is worth noting that BCDN and CDN share some steps in Algorithm 4: (1) compute Δ using the pre-computed value $\nabla L(\mathbf{w})$ (step 2 in Algorithm 4); (2) in each line search step, they both check if Eq. (6) is satisfied. The only difference is the rule of computing $\mathbf{d}^T \mathbf{x}_i$ (step 3 in Algorithm 4): $\mathbf{d}^T \mathbf{x}_i = d_j x_{ij}$ in CDN because only the jth feature is updated, while $\mathbf{d}^T \mathbf{x}_i = \sum_{j=1}^{P} d_j x_{ij}$ in BCDN. However, $\mathbf{d}^T \mathbf{x}_i$ in BCDN could be computed in parallel with P threads and a reduction-sum operation, so the time cost is equivalent for CDN and BCDN.

Summarizing the above analysis, the ideal speedup of BCDN is the bundle size P. It is worth noting that P can be very large in practice. In our experiments, P can be at least 1250 for the dataset real-sim. See Table 2 for details.

4   Global Convergence of BCDN

Our BCDN conducts P-dimensional line search for all features of a bundle $\mathcal{B}^q$ to ensure its global convergence, under high DOP. In this section, we will theoretically analyze the convergence of BCDN on two aspects: the convergence of P-dimensional line search and the global convergence.

Lemma 1. *BCDN (Algorithm 3) is a special case of CGD [16] with the specification $H \equiv \text{diag}(\nabla^2 L(\mathbf{w}))$.*

Proof. Note that, the selection of bundle set in Eq. (5) is consistent with that used in CGD (Eq. (12) in [16]). Then, for descent direction computing for a bundle in Algorithm 3, we have,

$$\mathbf{d}^{k,\mathcal{B}^q} = \sum_{j \in \mathcal{B}^q} d(\mathbf{w}; j)\mathbf{e}_j$$

$$= \sum_{j \in \mathcal{B}^q} \arg\min_{d}\{\nabla_j L(\mathbf{w})^T d + \frac{1}{2}\nabla_{jj}^2 L(\mathbf{w})d^2 + |w_j + d|\}\mathbf{e}_j \qquad (7)$$

$$= \arg\min_{d}\{\nabla L(\mathbf{w})^T \mathbf{d} + \frac{1}{2}\mathbf{d}^T H \mathbf{d} + \|\mathbf{w} + \mathbf{d}\|_1 \mid d_t = 0, \forall t \notin \mathcal{B}^q\} \qquad (8)$$

$$\equiv d_H(\mathbf{w}; \mathcal{B}^q) \qquad (9)$$

where Eq. (7) is derived by considering the definition of $d(\mathbf{w}; j)$ in Eq. (3); Eq. (8) is obtained by applying the setting of $H \equiv \text{diag}(\nabla^2 L(\mathbf{w}))$; Eq. (9) follows the definition of the descent direction by Tseng et al (Eq. (6) in [16]). Therefore the definition of direction computing is in a CGD manner.

Moreover, since BCDN conducts line Armijo search for $\mathbf{d}^{k,\mathcal{B}^q}$, it is clear that BCDN is a special case of CGD by setting $H = \text{diag}(\nabla^2 L(\mathbf{w}))$. ∎

By means of Lemma 1, we can use conclusions of Lemma 5 and Theorem 1(e) in [16] to prove the following theorems.

Theorem 1 (Convergence of P-dimensional line search). *For ℓ_1-regularized logistic regression, the P-dimensional line search in Algorithm 3 will converge within finite steps. That is, the descent condition in Eq. (4) $F_c(\mathbf{w} + \alpha \mathbf{d}) - F_c(\mathbf{w}) \leq \sigma \alpha \Delta$ is satisfied for any $\sigma \in (0, 1)$ within finite steps.*

Proof. According to Lemma 5 of [16], to have finite steps of line search, it needs two requirements. The first requirement is,

$$\nabla_{jj}^2 L(\mathbf{w}) > 0, \ \forall \, j \in \mathcal{B}^q \qquad (10)$$

and the second requirement is that there exists $E > 0$ such that,

$$\|\nabla L(\mathbf{w}_1) - \nabla L(\mathbf{w}_2)\| \leq E\|\mathbf{w}_1 - \mathbf{w}_2\| \qquad (11)$$

First, we prove that the first requirement in Eq. (10) can be satisfied. Note that, we can easily obtain the closed form solution of Eq. (3):

$$d(\mathbf{w}; j) = \begin{cases} -\frac{\nabla_j L(\mathbf{w})+1}{\nabla_{jj}^2 L(\mathbf{w})} & \text{if } \nabla_j L(\mathbf{w}) + 1 \leq \nabla_{jj}^2 L(\mathbf{w})w_j, \\ -\frac{\nabla_j L(\mathbf{w})-1}{\nabla_{jj}^2 L(\mathbf{w})} & \text{if } \nabla_j L(\mathbf{w}) - 1 \geq \nabla_{jj}^2 L(\mathbf{w})w_j, \\ -w_j & \text{otherwise} \end{cases} \qquad (12)$$

where for logistic regression we have,

$$\nabla_j L(\mathbf{w}) = c \sum_{i=1}^{s} (\tau(y_i \mathbf{w}^T \mathbf{x}_i) - 1)y_i x_{ij}$$

$$\nabla_{jj}^2 L(\mathbf{w}) = c \sum_{i=1}^{s} \tau(y_i \mathbf{w}^T \mathbf{x}_i)(1 - \tau(y_i \mathbf{w}^T \mathbf{x}_i))x_{ij}^2 \qquad (13)$$

where $\tau(s) \equiv \frac{1}{1+e^{-s}}$ is the derivative of the logistic loss function. Eq. (13) shows that $\nabla_{jj}^2 L(\mathbf{w}) > 0$ except for $x_{ij} = 0, \forall i = 1, \cdots, s$. In this exception, $\nabla_j L(\mathbf{w})$ and $\nabla_{jj}^2 L(\mathbf{w})$ always remain zero. There are two situations: $w_j = 0$ and $w_j \neq 0$.

- $w_j \neq 0$: according to Eq. (12), we have $d(\mathbf{w}; j) = -w_j$. Note that $d(\mathbf{w}; j) = -w_j$ also satisfies the sufficient decrease condition in Eq. (4), so w_j becomes zero in the first iteration. Then, in the following iterations, w_j will always satisfy the shrinking condition[4] in BCDN and will always be removed from the working set and remains zero.
- $w_j = 0$: w_j will always satisfy the shrinking condition and will be removed from the working set from the first iteration to the end.

Under the above analysis, in the exception of $\nabla_{jj}^2 L(\mathbf{w}) = 0$, w_j will becomes zero and have no effect on the optimization procedure at least from the second iteration to the end. Therefore Eq. (10) always holds for the working set.

Second, we prove the second requirement in Eq. (11). We follow the analysis in Appendix D of [17]. For logistic regression, we have,

$$\|\nabla L(\mathbf{w}_1) - \nabla L(\mathbf{w}_2)\| \le \|\nabla^2 L(\bar{\mathbf{w}})\| \|\mathbf{w}_1 - \mathbf{w}_2\|$$

where $\bar{\mathbf{w}} = t\mathbf{w}_1 + (1-t)\mathbf{w}_2, 0 \le t \le 1$. Note that, Hessian of the logistic loss can be expressed as,

$$\nabla^2 L(\mathbf{w}) = cX^T DX \tag{14}$$

where $D = \mathrm{diag}(D_{11}, D_{22}, \cdots, D_{ss})$ with $D_{ii} = \tau(y_i \mathbf{w}^T \mathbf{x}_i)(1 - \tau(y_i \mathbf{w}^T \mathbf{x}_i))$. Considering Eq. (14) and the fact that $D_{ii} < 1$, we have,

$$\|\nabla^2 L(\bar{\mathbf{w}})\| < c\|X^T\| \|X\|$$

Therefore, $E = c\|X^T\| \|X\|$ will fulfill the second requirement of Eq. (11). ■

The experimental results in Section 6.1 support the analysis in Theorem 1.

Theorem 2 (Global Convergence of BCDN). *Let $\{\mathbf{w}^k\}$, $\{\alpha^k\}$ be the sequences generated by Algorithm 3. If $\sup_k \alpha^k < \infty$, then every cluster point of $\{\mathbf{w}^k\}$ is a stationary point of $F_c(\mathbf{w})$.*

Proof. In Algorithm 4, $\alpha^k \le 1, k = 1, 2, \cdots$, which satisfies $\sup_k \alpha^k < \infty$. To ensure the global convergence, Tseng et al made the following assumption,

$$0 < \nabla_{jj}^2 L(\mathbf{w}^k) \le \bar{\lambda}, \; \forall j = 1, \cdots, n, \; k = 1, 2, \cdots \tag{15}$$

Considering Eq. (14), we have $\nabla_{jj}^2 L(\mathbf{w}^k) < c\|X^T\| \|X\|$. Setting $\bar{\lambda} = c\|X^T\| \|X\|$ and following the same analysis in Theorem 1, we have $\nabla_{jj}^2 L(\mathbf{w}^k) > 0$ for all features in the working set. Therefore Eq. (15) is fulfilled. According to Theorem 1(e) in [16], any cluster point of $\{\mathbf{w}^k\}$ is a stationary point of $F_c(\mathbf{w})$. ■

---

[4] BCDN uses the same shrinking strategy as that in CDN (Eq. (32) in [17]).

Theorem 2 guarantees that our proposed BCDN will converge globally for any bundle size $P \in [1, n]$.

5 Datasets and Implementation

In this section, various aspects of the performances of CDN, SCDN and BCDN will be investigated by extensive experiments on five public datasets. For fair comparison, in these experiments, we use these methods to solve logistic regression with a bias term b,

$$\min_{\mathbf{w} \in \mathbb{R}^n, b} F_c(\mathbf{w}, b) \equiv c \sum_{i=1}^{s} \log(1 + e^{-y_i(\mathbf{w}^T \mathbf{x}_i + b)}) + \|\mathbf{w}\|_1. \tag{16}$$

5.1 Datasets

The five datasets used in our experiments are summarized in Table 1[5]. news20, rcv1, a9a and real-sim are document datasets, whose instances are normalized to unit vectors. gisette is a handwriting digit problem from NIPS 2003 feature selection challenge, whose features are linearly scaled to the [-1,1] interval. The bundle size P for BCDN in Algorithm 3 is set according to Table 2. Note that P^* is only BCDN's conservative setting, under which BCDN can quickly converge with the most strict stopping criteria $\epsilon = 10^{-8}$ (defined in Eq. (17)). Moreover, P can be larger for a common setting such as $\epsilon = 10^{-4}$ (see Section 6.4).

Table 1. Summary of data sets. #test is the number of test samples. The best regularization parameter c is set according to Yuan et al [17]. Spa. means optimal model sparsity ($\frac{n-\|\mathbf{w}^*\|_0}{n}$), Acc. is the test accuracy for optimal model.

Dataset	s	#test	n	best c	Spa./%	Acc./%
a9a	26,049	6,512	123	2	17.89	84.97
real-sim	57,848	14,461	20,958	4	83.36	97.16
news20	15,997	3,999	1,355,191	64	99.80	95.62
gisette	6,000	1,000	5,000	0.25	90.7	98.10
rcv1	541,920	135,479	47,236	4	76.77	97.83

Table 2. Conservative bundle size P^* for each dataset

Dataset	a9a	real-sim	news20	gisette	rcv1
Bundle size P^*	25	1,250	150	15	200

---

[5] All these datasets can be downloaded at
http://www.csie.ntu.edu.tw/~cjlin/libsvmtools/datasets/

5.2 Implementation

All the three algorithms, CDN, SCDN and BCDN, are implemented in C/C++ language. Since the shrinking procedure cannot be performed inside the parallel loop in SCDN and BCDN, to enable fair comparison, we use equivalent implementation where the shrinking is conducted outside the parallel loop in CDN. Further, we set $\sigma = 0.01$ and $\beta = 0.5$ for the line search procedure in the three algorithms. OpenMP is used as the parallel programming model. Work in parallel is distributed among a team of threads using OpenMP `parallel for` construct and the `static` scheduling of threads is used because it proves to be very efficient in our experiments.

The stopping condition similar to the outer stopping condition in [18] is used in our implementation.

$$\|\nabla^S F_c(\mathbf{w}^k)\|_1 \leq \epsilon_{end} \equiv \epsilon \cdot \frac{\min(\#pos, \#neg)}{s} \cdot \|\nabla^S F_c(\mathbf{w}^1)\|_1, \qquad (17)$$

where ϵ is user-defined stopping tolerance; $\#pos$ and $\#neg$ respectively denote the numbers of positive and negative labels in the training set; $\nabla^S F_c(\mathbf{w}^k)$ is the minimum-norm sub-gradient,

$$\nabla^S_j F_c(\mathbf{w}) \equiv \begin{cases} \nabla_j L(\mathbf{w}) + 1 & \text{if } w_j > 0, \\ \nabla_j L(\mathbf{w}) - 1 & \text{if } w_j < 0, \\ \text{sgn}(\nabla_j L(\mathbf{w})) \max(|\nabla_j L(\mathbf{w})| - 1, 0) & \text{otherwise} \end{cases}$$

We run CDN with an extremely small stopping criteria $\epsilon = 10^{-8}$ to get the optimal value F_c^*, which is used to compute the relative difference to the optimal function value (*relative error*),

$$(F_c(\mathbf{w}, b) - F_c^*)/F_c^* \qquad (18)$$

Some private implementation details are listed as follows:

- CDN: we use the source code included in LIBLINEAR[6]. Shrinking procedure is modified to be consistent with the parallel algorithms BCDN and SCDN.
- SCDN: though Bradley et al [2] released the source code for SCDN, for fair comparison, we reimplement it in C language based on CDN implementation.
- BCDN: we implement BCDN carefully, including the data type and the atomic operation. For atomic operation, we use a compare-and-swap implementation using inline assembly.

6 Experimental Results

In this section we provide several groups of experimental results, including line search step number, scalability and timing results about relative error, testing accuracy and number of nonzeros (NNZ) in the model. To estimate the testing

---

[6] Version 1.7, http://www.csie.ntu.edu.tw/~cjlin/liblinear/oldfiles/

accuracy, for each dataset, 25% samples are used for testing and the rest samples are used for training.

All experiments are conducted on a 64-bit machine with Intel(R) Core(TM) i7 CPU (8 cores) and 12GB main memory. We set $\bar{P} = 8$ for SCDN in Algorithm 2 and use 8 threads to run the parallel updates with DOP of $\bar{P}=8$.

For BCDN, the descent direction computing (step 7 in Algorithm 3) can have the DOP of bundle size P, which is several hundreds even to thousand according to Table 2, while our 8-core machine is unable to fully exhibit its parallelism potential. To justify time performances of three algorithms impartially, we need to estimate time cost for BCDN by running with at least P cores. Assuming we have a machine with P cores, to distribute step 7 in Algorithm 3 to P cores, the extra data transfer (if needed) is little: the training data $(\mathbf{X}, \mathbf{y})$ only needs to be transferred once before all iterations, so its time cost can be omitted. Arrays with the size of $s \times sizeof(double)$ bytes containing values of $e^{\mathbf{w}^T \mathbf{x}_i}, \mathbf{d}^T \mathbf{x}_i, i = 1, \cdots, s$ need to be transferred each time, which costs very little extra transfer time. Taking into count the scheduling time, we estimate the fully parallelized computing time of descent direction computing t_p by multiply the ideal parallel time cost with a reasonable factor 2,

$$t_p = (2 \cdot t_{\text{serial}})/P,$$

where t_{serial} is the serial time cost of step 7 in Algorithm 3.

6.1   Empirical Performance of Line Search

Table 3 reports the average number of line search steps per outer iteration. These statistics support the analysis in Theorem 1: for all datasets, line search of BCDN terminates in finite steps, which is far less than that of CDN and SCDN. It is also in line with the analysis in Section 3.1: BCDN conducts about $1/P$ times of line search compared to CDN, while the time cost of each line search is about the same for both. In Table 3, BCDN's number of line search steps is a little larger than $1/P$ times of that of CDN. This is because the parallel direction computing procedure in BCDN slows its convergence rate, which increases the number of line search steps. SCDN conducts more line search steps than BCDN and CDN, which indicates that its parallel strategy cannot well deal with interference among features and tends to diverge, thus needing more line search steps to fulfill the descent condition in Eq. (4).

Table 3. Average number of line search steps per outer iteration with $\epsilon = 10^{-4}$

Datasets	a9a	real-sim	news20	gisette	rcv1
CDN	96.2	3,455.3	3,022.3	472.2	10,272.4
SCDN	149.8	3,560.4	2,704.0	679.4	10,504.5
BCDN	6.0	5.7	62.4	97.6	57.3

6.2   Empirical Performance of Global Convergence

We verify the global convergence of all compared algorithms by setting the most strict stopping criteria $\epsilon = 10^{-8}$. Fig. 1 of the relative error (see Eq. (18)) shows that in all the cases, BCDN could converge to the final value at the fastest speed. Meanwhile, though SCDN behaves faster compared to CDN in the beginning, it cannot converge in a limited time (Fig. 1(b),(c)) or cannot converge faster than CDN (Fig. 1(a)). Table 4 with the runtime and iteration number reaches the same conclusion. The iteration number of SCDN is more than that of CDN, while the iteration number of BCDN is less than SCDN except for real-sim. This again indicates that SCDN tends to diverge with $\bar{P} = 8$ while BCDN can quickly converge under strict stopping criteria with extremely high DOP (large bundle size P, even to thousand).

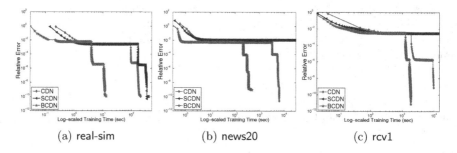

(a) real-sim   (b) news20   (c) rcv1

Fig. 1. Relative error under most strict stopping condition of $\epsilon = 10^{-8}$

Table 4. Runtime (sec) and iteration number. The number marked a "$*$" means less than that of CDN

Methods	real-sim		news20		rcv1	
	time	#iter	time	#iter	time	#iter
CDN	210.2	1,311	6,426.4	50,713	11,583.9	2,970
SCDN	399.9	$*1,158$	$> 54,672.9$	$> 82,520$	$> 13,900.6$	$> 3,029$
BCDN	$*13.2$	1,838	$*489.5$	73,986	$*2,298.4$	$*2,774$

6.3   Time Performance under Common Setting

Fig. 2 plots relative error (see Eq. 18), testing accuracy and model NNZ, with a common setting of $\epsilon = 10^{-4}$. We use the conservative setting P^* in Table 2 for BCDN. For all datasets, BCDN is much faster than CDN and SCDN, which highlights its higher DOP and strong convergence guarantee. Note that for gisette SCDN is even slower than CDN, which comes from its tend to diverge at a DOP of 8 ($\bar{P} = 8$).

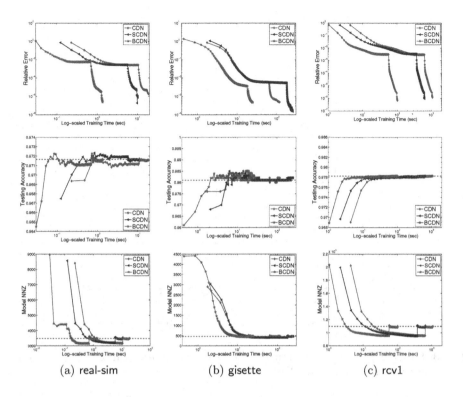

Fig. 2. Time performance under common setting on 3 datasets. Top, middle and bottom plot traces of relative error, testing accuracy and model NNZ over training time, respectively. The dotted horizontal line is the value obtained by running CDN with $\epsilon = 10^{-8}$.

6.4   Scalability of BCDN

This section evaluates the scalability of BCDN (runtime and number of outer iterations w.r.t varying bundle size P) with the common setting $\epsilon = 10^{-4}$. From Fig. 3 one can see that the runtime (blue lines) becomes shorter as the increase of bundle size P, which is in line with the analysis of BCDN in Section 3.1: BCDN has an ideal speedup of P. At the same time, the increase of P brings about more outer iterations (see the green lines in Fig. 3), which is because more parallelism causes slower convergence rate. However, it will not introduce extra runtime because of the more parallelism with larger P. With the feature number of 20,958 (Table 1), real-sim could acquire an amazing high DOP of 1,300 or higher in Fig. 3 (b), which comes from the strong convergence guarantee of BCDN.

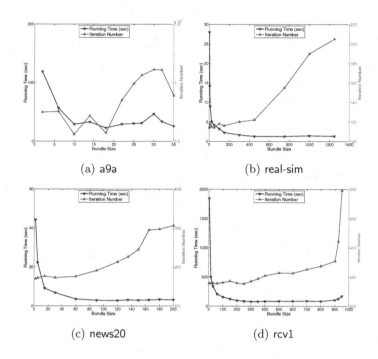

(a) a9a   (b) real-sim

(c) news20   (d) rcv1

Fig. 3. Scalability of BCDN. Runtime (blue lines) and outer iteration number (green lines) w.r.t varying bundle size P.

7 Conclusions

This paper introduces the Bundle CDN, a highly parallelized approach with DOP of the bundle size P, for training high dimensional ℓ_1-regularized logistic regression models. It has a strong convergence guarantee under theoretical analysis, which is consistent with empirical experiments. A set of experimental comparisons on 5 public large-scale datasets demonstrate that the proposed BCDN is superior among state-of-the-art ℓ_1 solvers for speed, which comes from its high DOP to better exploit parallelism among features.

High DOP of BCDN makes it possible to develop highly parallelized algorithm on clusters with many more cores or heterogeneous computing frameworks [6] such as GPU and FPGA. BCDN can also be used to solve other ℓ_1-regularized problems with a higher speed, such as Lasso, compressed sensing, Support Vector Machine and other related discriminative models [9,8].

Acknowledgments. Thanks China 973 Program 2011CB302203, STCSM Project 11JC1405200 and NSFC CB0300108 for support, and to Jiuliang Guo and Hong Wei for helpful discussions.

References

1. Bian, Y., Zhao, X., Song, J., Liu, Y.: Parallelized annealed particle filter for real-time marker-less motion tracking via heterogeneous computing. In: ICPR, pp. 2444–2447 (2012)
2. Bradley, J.K., Kyrola, A., Bickson, D., Guestrin, C.: Parallel coordinate descent for l1-regularized loss minimization. In: ICML, pp. 321–328 (2011)
3. Burke, J.: Descent methods for composite nondifferentiable optimization problems. Math. Program. 33(3), 260–279 (1985)
4. Chang, K.W., Hsieh, C.J., Lin, C.J.: Coordinate descent method for large-scale l2-loss linear support vector machines. Journal of Machine Learning Research 9, 1369–1398 (2008)
5. Fan, R.E., Chang, K.W., Hsieh, C.J., Wang, X.R., Lin, C.J.: Liblinear: A library for large linear classification. Journal of Machine Learning Research 9, 1871–1874 (2008)
6. Gaster, B., Howes, L., Kaeli, D., Mistry, P., Schaa, D.: Heterogeneous Computing with OpenCL. Elsevier Science & Technology Books (2012)
7. Langford, J., Li, L., Zhang, T.: Sparse online learning via truncated gradient. Journal of Machine Learning Research 10, 777–801 (2009)
8. Li, X., Lee, T., Liu, Y.: Stochastic feature mapping for pac-bayes classification. arXiv:1204.2609 (2012)
9. Li, X., Lee, T.S., Liu, Y.: Hybrid generative-discriminative classification using posterior divergence. In: CVPR, pp. 2713–2720 (2011)
10. Li, Y., Osher, S.: Coordinate descent optimization for l1 minimization with application to compressed sensing; a greedy algorithm. Inverse Probl. Imaging 3(3), 487–503 (2009)
11. Lin, C.J., Moré, J.J.: Newton's method for large bound-constrained optimization problems. SIAM J. on Optimization 9(4), 1100–1127 (1999)
12. Ng, A.Y.: Feature selection, l1 vs. l2 regularization, and rotational invariance. In: ICML, pp. 78–85 (2004)
13. Niu, F., Recht, B., Re, C., Wright, S.J.: Hogwild: A lock-free approach to parallelizing stochastic gradient descent. In: NIPS, pp. 693–701 (2011)
14. Scherrer, C., Tewari, A., Halappanavar, M., Haglin, D.: Feature clustering for accelerating parallel coordinate descent. In: NIPS, pp. 28–36 (2012)
15. Shalev-Shwartz, S., Tewari, A.: Stochastic methods for l_1 regularized loss minimization. In: ICML, p. 117 (2009)
16. Tseng, P., Yun, S.: A coordinate gradient descent method for nonsmooth separable minimization. Math. Program. 117(1-2), 387–423 (2009)
17. Yuan, G.X., Chang, K.W., Hsieh, C.J., Lin, C.J.: A comparison of optimization methods and software for large-scale l1-regularized linear classification. Journal of Machine Learning Research 11, 3183–3234 (2010)
18. Yuan, G.X., Ho, C.H., Lin, C.J.: An improved glmnet for l1-regularized logistic regression. In: KDD, pp. 33–41 (2011)
19. Zinkevich, M., Smola, A., Langford, J.: Slow learners are fast. In: NIPS, pp. 2331–2339 (2009)

MORD: Multi-class Classifier for Ordinal Regression

Kostiantyn Antoniuk, Vojtěch Franc, and Václav Hlaváč

Czech Technical University in Prague,
Faculty of Electrical Engineering,
Technicka 2, 166 27, Praha 6
{antonkos,xfrancv,hlavac}@cmp.felk.cvut.cz

Abstract. We show that classification rules used in ordinal regression
are equivalent to a certain class of linear multi-class classifiers. This observation not only allows to design new learning algorithms for ordinal regression using existing methods for multi-class classification but
it also allows to derive new models for ordinal regression. For example,
one can convert learning of ordinal classifier with (almost) arbitrary loss
function to a convex unconstrained risk minimization problem for which
many efficient solvers exist. The established equivalence also allows to increase discriminative power of the ordinal classifier without need to use
kernels by introducing a piece-wise ordinal classifier. We demonstrate
advantages of the proposed models on standard benchmarks as well as
in solving a real-life problem. In particular, we show that the proposed
piece-wise ordinal classifier applied to visual age estimation outperforms
other standard prediction models.

Keywords: Ordinal regression, linear multi-class classification.

1 Introduction

The classification problem consists of predicting a hidden class label $y \in \mathcal{Y}$
based on observations $x \in \mathcal{X}$ using a classifier $h\colon \mathcal{X} \to \mathcal{Y}$. In the statistical
classification, the pairs of (x, y) are assumed to be a realization of some ran-
dom variables distributed according to $P(x, y)$. This paper analyses a class of
classification problems fitting under the ordinal regression setting which imposes
additional assumptions on the distribution $P(x, y)$. In particular, the labels in
$\mathcal{Y}$ are assumed to be ordered, w.l.o.g. we use $\mathcal{Y} = \{1, \ldots, Y\}$ equipped with
a natural order, and they are modeled as a result of a course measurement of
some continuous random variable $\chi(x)$. More precisely, let as define a set of Y
intervals

$$U(1) = (-\infty, \theta_1], \quad U(2) = (\theta_1, \theta_2], \quad \ldots, \quad U(Y) = (\theta_{Y-1}, \infty),$$

determined by a sequence of non-decreasing thresholds $\theta_1, \theta_2, \ldots, \theta_{Y-1}$. The
standard model of [1] assumes that we observe label $y \in \mathcal{Y}$ if a realization of
the random variable $\chi(x)$ is in the interval $U(y)$. Thus the classes correspond to

H. Blockeel et al. (Eds.): ECML PKDD 2013, Part III, LNAI 8190, pp. 96–111, 2013.

contiguous ordered intervals on some continuous scale. Based on this assumption various ordinal regression models have been proposed and they are routinely applied in fields like social sciences, epidemiology, information retrieval or, recently in computer vision.

A typical problem is how to learn the classifier given a set of training examples $\{(x^1, y^1), \ldots, (x^m, y^m)\} \in (\mathcal{X} \times \mathcal{Y})^m$ drawn from i.i.d. random variables distributed according to some unknown distribution $P(x, y)$ which satisfies the "ordering" assumption mentioned above. In this paper, we consider the formulation which defines the target classifier to be the one with minimal expected risk (also called Bayes classifier)

$$R(h) = E_{p(x,y)}\big(\Delta(y, h(x))\big)$$

where $\Delta \colon \mathcal{Y} \times \mathcal{Y} \to \mathbb{R}_+$ is a given application specific loss function penalizing responses of the classifier.

In statistics, the learning problem is typically solved by constructing a plug-in Bayes classifier which replaces the true distribution $P(x, y)$ by its Maximum-Likelihood estimate. This approach requires to guess the shape of the underlying distribution $P(x, y)$ which can be difficult in practice. A different approach based on the risk minimization paradigm has been put forward in the machine learning literature. The idea is to learn the classifier directly from the examples without the need to estimate the generating distribution [2]. This approach selects the best classifier from a prescribed class of classifiers by minimizing a surrogate of the expected risk $R(h)$. The typical class of classifiers considered in the context of ordinal regression is the linear thresholded rule

$$h(x; w, \theta) = 1 + \sum_{k=1}^{Y-1} [\![\langle x, w \rangle > \theta_k]\!], \tag{1}$$

where $x \in \mathcal{X} = \mathbb{R}^n$ is a vector of real-valued features, $w \in \mathbb{R}^n$ is a parameter vector and $\theta = (\theta_1, \ldots, \theta_{Y-1}) \in \mathbb{R}^{Y-1}$ a vector of thresholds. In the sequel we refer to (1) as the ordinal (ORD) classifier. We call the vector θ admissible iff its components are non-decreasing i.e. $\theta \in \Theta = \{\theta' \in \mathbb{R}^{Y-1} \mid \theta'_k \leq \theta'_{k+1}, k = 1, \ldots, Y-1\}$. The form of the ORD classifier reflects the assumption that the classes correspond to intervals on $\mathbb{R}$. It is seen that ORD classifier predicts y iff the value $\langle x, w \rangle$ is in the interval $U(y)$.

A Perceptron-like algorithm called PRank learning the ORD classifier in an on-line fashion has been proposed in [3]. They formulate learning as minimization of the empirical risk with the Mean Absolute Error (MAE) loss function $\Delta(y, y') = |y - y'|$ (also called ranking loss) and provide mistake bounds for the case of separable examples. The authors of [4] proposed to learn the ORD classifier by a modified Support Vector Machine algorithm originally designed for two-class classification. The paper [5] improves the algorithms of [4] by enforcing the learned thresholds to be admissible. A generic framework which allows to convert learning of the ORD classifier to the problem of learning a two-class linear SVM classifier (with modified example weights) have been proposed in [6].

They show that appropriately weighted SVM hinge-loss is an upper bound of so called V-shaped loss (e.g. MAE and the 0/1-loss are V-shaped) evaluated on the ORD classifier. The paper [7] analyses a relation between the ordinal regression and the multi-class classifiers, however, by their definition the ordinal classifier as any Bayesian classifier with the V-shaped loss function, i.e. they do not considered ordering of the labels at all.

The previous works in their core convert learning of the ORD classifier into learning a set of two-class classifiers. The resulting two-class classifiers are trained by a modified SVM algorithm [4][5][6] or Perceptron [3]. In this paper we show that such conversions are not necessary. We prove that the ORD classifier is equivalent to a linear multi-class classifier whose class parameter vectors are collinear and their magnitude is linearly increasing with the label. We call the new representation the Multi-class ORDinal (MORD) classifier. Our equivalence proof is constructive so that we can convert any ORD classifier to the MORD classifier and vice-versa. We show that the new representation can be beneficial for learning. In particular, the well understood methods for learning multi-class linear classifiers can be readily applied. We experimentally show that a generic multi-class SVM algorithm used to learn MORD delivers the same (or slightly better) results when compared to the specialized learning algorithms derived for the ORD classifier. The proposed approach works for (almost) arbitrary loss function unlike the existing methods which require V-shaped losses. In addition, we show that the new representation allows to increase discriminative power of the ordinal classifier without need to use kernels by introducing a piece-wise ordinal classifier. We demonstrate advantages of the proposed models on standard benchmarks as well as in solving a real-life problem. We show that the proposed piece-wise ordinal classifier applied to visual age estimation outperforms other prediction models and is also comparable to commercial solutions.

The paper is organized as follows. The equivalence between the ORD classifier and the linear multi-class classifiers is described in Section 2. In Section 3, we define a new model for ordinal regression. In Section 4, we compare several classification models for ordinal regression in an unified view. In Section 5, we described a generic algorithm for learning the proposed models. Experiments are presented in Section 6 and Section 7 concludes the paper.

2   Ordinal Regression as Linear Multi-class Classification

Let us consider one-dimensional observations $x \in \mathcal{X} = \mathbb{R}$ in which case the ORD classifier $h(x) = 1 + \sum_{k=1}^{Y-1} [\![x > \theta_k]\!]$ splits the real axis into Y intervals defined by thresholds $\theta_1 \leq \theta_2 \leq \cdots \leq \theta_{Y-1}$. One may think of representing the ORD classifier in the form

$$h'(x) = \underset{y \in \mathcal{Y}}{\operatorname{argmax}} f(x, y), \qquad (2)$$

where $f \colon \mathbb{R} \times \mathcal{Y} \to \mathbb{R}$ is a discriminant function. If we manage to construct the discriminant functions such that $f(x, y) \geq f(x, y')$, $y' \in \mathcal{Y} \setminus \{y\}$ iff $h(x) = y$ then both representations will be equivalent i.e. $h'(x) = h(x)$, $x \in \mathbb{R}$. Let us consider a linear discriminant function with the slope equal to y, i.e. $f(x, y) = x \cdot y + b_y$, in which case (2) becomes a linear multi-class classifier. It is not difficult to see that such linear classifier also splits the real axis into intervals. Fig 1 shows an example of the ORD classifier and its equivalent linear classifier $h'(x)$.

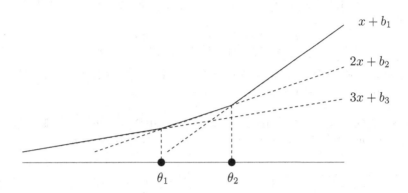

Fig. 1. The figure illustrates relation between the ORD classifier $h(x) = 1 + \sum_{k=1}^{Y-1} [\![x > \theta_k]\!]$ and its alternative representation $h'(x) = \mathrm{argmax}_{y \in \mathcal{Y}}(x \cdot y + b_y)$ for the $(Y = 3)$-class problem. Note, that x and y-axes have different scale in order to save space.

The same idea can be applied for n-dimensional observations $\boldsymbol{x} \in \mathcal{X} = \mathbb{R}^n$. The multi-class linear classifier which can represent the ORD classifier (1) reads

$$h'(\boldsymbol{x}; \boldsymbol{w}, \boldsymbol{b}) = \underset{y \in \mathcal{Y}}{\mathrm{argmax}} \left(\langle \boldsymbol{x}, \boldsymbol{w} \rangle \cdot y + b_y \right), \qquad (3)$$

where $\boldsymbol{w} \in \mathbb{R}^n$ is parameter vector and $\boldsymbol{b} = (b_1, \ldots, b_Y) \in \mathbb{R}^Y$ is a vector of intercepts. We denote (3) as the Multi-class ORDinal (MORD) classifier. Inside the paper we assume that the "argmax" operator returns the minimal label in the case of more than one maximizer.

A natural question is whether both representations are equivalent in the sense that any ORD classifier can be represented by some MORD classifier and vice-versa. The following theorem gives a positive answer to the question.

Theorem 1. *The ORD classifier (1) and the MORD classifier (3) are equivalent in the following sense. For any $\boldsymbol{w} \in \mathbb{R}^n$ and admissible $\boldsymbol{\theta} \in \Theta$ there exists $\boldsymbol{b} \in \mathbb{R}^Y$ such that $h(\boldsymbol{x}, \boldsymbol{w}, \boldsymbol{\theta}) = h'(\boldsymbol{x}, \boldsymbol{w}, \boldsymbol{b}), \forall \boldsymbol{x} \in \mathbb{R}^n$. For any $\boldsymbol{w} \in \mathbb{R}^n$ and $\boldsymbol{b} \in \mathbb{R}^n$ there exists admissible $\boldsymbol{\theta} \in \Theta$ such that $h(\boldsymbol{x}, \boldsymbol{w}, \boldsymbol{\theta}) = h'(\boldsymbol{x}, \boldsymbol{w}, \boldsymbol{b}), \forall \boldsymbol{x} \in \mathbb{R}^n$.*

A proof is given in Appendix A.

Our proof is constructive in the sense that we can provide a conversion from the ORD classifier to the MORD classifier and vice-versa. In exotic cases, which however may appear in practice, some classes can collapse to a single point and effectively disappear. To cover all such situations, we first define the concept of non-degenerated classifier and then we give formulas for the conversions.

Definition 1 (Degenerated and non-degenerated classifier). *We call class* $y \in \mathcal{Y}$ *non-degenerated for classifier* $h'(\boldsymbol{x})$ *iff* $\mathcal{X}_y = \mathrm{interior}(\{\boldsymbol{x} \in \mathcal{X} : h'(\boldsymbol{x}) = y\}) \neq \emptyset$. *Classifier* $h'(\boldsymbol{x})$ *is non-degenerated iff all classes are non-degenerated. In opposite case the classifier is called degenerated.*

Given a MORD classifier, the class $\hat{y} \in \mathcal{Y}$ is non-degenerated iff the linear inequalities

$$
\begin{aligned}
z\hat{y} + b_{\hat{y}} &> z(\hat{y} - k) + b_{\hat{y}-k}, \; 1 \le k < \hat{y}, \\
z\hat{y} + b_{\hat{y}} &\ge z(\hat{y} + t) + b_{\hat{y}+k}, \; 1 < t \le Y - \hat{y},
\end{aligned}
\tag{4}
$$

are solvable w.r.t. $z \in \mathbb{R}$. It is seen that we can check it in $\mathcal{O}(Y)$ time. We refer to the proof for more details.

Conversion formulas. Given parameters of the ORD classifier $\boldsymbol{w} \in \mathbb{R}^n$, $\boldsymbol{\theta} \in \Theta$, the equivalent MORD classifier has parameters $\boldsymbol{w}$ and $\boldsymbol{b}$ given by

$$
b_1 = 0 \qquad \text{and} \qquad b_y = -\sum_{i=1}^{y-1} \theta_i, \; y = 2, \dots, Y.
\tag{5}
$$

The conversion from the MORD classifier to the ORD classifier is done differently for the non-generated and the degenerated classifier. Given parameters of a non-degenerated MORD classifier $\boldsymbol{w} \in \mathbb{R}^n$ and $\boldsymbol{b} \in \mathbb{R}^Y$, we can compute thresholds $\boldsymbol{\theta} \in \Theta$ of the equivalent ORD classifier by

$$
\theta_y = b_y - b_{y+1}, \qquad y = 1, \dots, Y - 1.
\tag{6}
$$

Given parameters of a degenerated MORD classifier $\boldsymbol{w} \in \mathbb{R}^n$ and $\boldsymbol{b} \in \mathbb{R}^Y$, we compute thresholds $\boldsymbol{\theta} \in \Theta$ of the equivalent ORD classifier by

$$
\theta_{y_i} = \dots = \theta_{y_{i+1}-1} = \frac{b_{y_i} - b_{y_{i+1}}}{(y_{i+1} - y_i)}, \; i = 1, \dots, p,
\tag{7}
$$

where $y_i \in \mathcal{Y}$, $i = 1, \dots, p$ is an increasing subsequence of non-degenerated classes.

Finally, let us note that the MORD classifier is represented by $n + Y$ parameters instead of $n + Y - 1$ parameters of the ORD classifier. However, the parameters of the MORD classifier are unconstrained which makes the MORD representation attractive for learning because no additional constraints on the intercepts $\boldsymbol{\theta} \in \Theta$ are not needed.

3   Piece-Wise Ordinal Regression Classifier

The discriminative power of the ORD classifier can be limiting in some cases. Mapping the observations into higher dimensional space via usage of kernel functions is one way to make the linear ORD classifier more discriminative. Though the "kernalization" of the ORD classifier is straightforward it is not suitable in all cases. For example, the kernels are prohibitive in applications which require processing of large amounts of training examples and/or if a real-time response of the classifier is the must. Instead, we proposed to stay in the original feature space where we construct a combined classifier from a set of simpler component classifiers. In our case, the component classifiers will be the MORD classifiers, each responsible for a subset of labels.

Let $Z > 1$ be a number of cut labels $(\hat{y}_1, \hat{y}_2, \ldots, \hat{y}_Z) \in \mathcal{Y}^Z$ such that $\hat{y}_1 = 1$, $\hat{y}_Z = Y$ and $\hat{y}_z \leq \hat{y}_{z+1}$, $z \in \mathcal{Z} = \{1, \ldots, Z - 1\}$. The cut labels define a partitioning of $\mathcal{Y}$ into Z subsets $\mathcal{Y}_z = \{y \in \mathcal{Y} \mid \hat{y}_z \leq y \leq \hat{y}_{z+1}\}$, $z \in \mathcal{Z}$. We will model dependence between the observation $\boldsymbol{x}$ and a subset of labels $\mathcal{Y}_z$ by a component classifier

$$h_z(\boldsymbol{x}) = \operatorname*{argmax}_{y \in \mathcal{Y}_z} f_z(\boldsymbol{x}, y) \tag{8}$$

where $f_z \colon \mathbb{R}^n \times \mathcal{Y}_z \to \mathbb{R}$ is a discriminant function. We define a combined classifier whose discriminant function is composed of discriminant functions of the component classifiers as follows

$$h''(\boldsymbol{x}) = \operatorname*{argmax}_{z \in \mathcal{Z}} \max_{y \in \mathcal{Y}_z} f_z(\boldsymbol{x}, y) . \tag{9}$$

We set the discriminant functions to be

$$f_z(\boldsymbol{x}, y) = \langle \boldsymbol{x}, \boldsymbol{w}_z(1 - \alpha(y, z)) + \boldsymbol{w}_{z+1}\alpha(y, z) \rangle + b_y \tag{10}$$

where

$$\alpha(y, z) = \frac{y - \hat{y}_z}{\hat{y}_{z+1} - \hat{y}_z}$$

and $\boldsymbol{W} = [\boldsymbol{w}_1, \ldots, \boldsymbol{w}_Z] \in \mathbb{R}^{n \times Z}$, $\boldsymbol{b} \in \mathbb{R}^Y$ are parameters. With these definitions it can be shown that: i) the component classifiers (8) are the ORD classifiers and ii) the combined classifier (9) is well defined because all its neighboring discriminant functions are consistent at the cut labels, i.e. $f_z(\boldsymbol{x}, \hat{y}_{z+1}) = f_{z+1}(\boldsymbol{x}, \hat{y}_{z+1})$, $z \in \mathcal{Z}$, holds. The claim i) is seen after substituting (10) into (8) which after some algebra yields

$$h_z(\boldsymbol{x}) = \operatorname*{argmax}_{y \in \mathcal{Y}_z} \left(\langle \boldsymbol{x}, \boldsymbol{w}_{z+1} - \boldsymbol{w}_z \rangle \alpha(y, z) + b_y \right)$$

and since $\alpha(y, z)$ is linearly increasing with y, Theorem 1 guarantees that $h_z(\boldsymbol{x})$ is the MORD classifier equivalent to the ORD classifier. The claim ii) follows from the fact that $\alpha(\hat{y}_{z+1}, z) = 1$ and $\alpha(\hat{y}_{z+1}, z + 1) = 0$, and thus $f_z(\boldsymbol{x}, \hat{y}_{z+1}) = \langle \boldsymbol{x}, \boldsymbol{w}_{z+1} \rangle + b_{\hat{y}_{z+1}} = f_{z+1}(\boldsymbol{x}, \hat{y}_{z+1})$.

We can explicitly write the component classifier, which we call the Piece-Wise Multi-class ORDinal (PW-MORD) classifier, as follows

$$h''(\boldsymbol{x}, \boldsymbol{W}, \boldsymbol{b}) = \operatorname*{argmax}_{z \in \mathcal{Z}} \operatorname*{argmax}_{y \in \mathcal{Y}_z} \left(\langle \boldsymbol{x}, \boldsymbol{w}_z(1 - \alpha(y, z)) + \boldsymbol{w}_{z+1}\alpha(y, z) \rangle + b_y \right) . \tag{11}$$

Figure 2 visualizes the ORD (=MORD) and the PW-MORD classifier on a toy data. It is seen that the distribution of the data cannot be well described by the ORD classifier while the PW-MORD composed of 3 ORD classifiers provides much better model in this case.

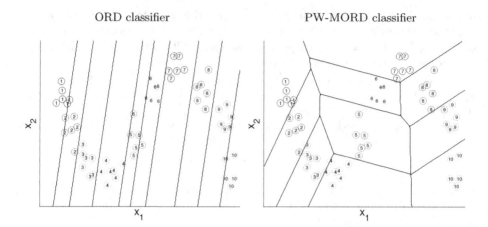

Fig. 2. The figure shows the partitioning of 2-dimensional feature space realized by the ORD classifier and the PW-MORD classifier with $Z = 3$ components. The cut labels for the PW-MORD classifier where set to $\{1, 4, 7, 10\}$.

4   Unified View of Classifiers for Ordinal Regression

Let us consider the linear multi-class classifier

$$h(\boldsymbol{x}, \boldsymbol{W}, \boldsymbol{b}) = \operatorname*{argmax}_{y \in \mathcal{Y}} \left(\langle \boldsymbol{x}, \sum_{z=1}^{Z} \beta(y, z) \boldsymbol{w}_z \rangle + b_y \right) \tag{12}$$

where $\boldsymbol{W} = [\boldsymbol{w}_1, \dots, \boldsymbol{w}_Z] \in \mathbb{R}^{n \times Z}$, $\boldsymbol{b} = [b_1; \dots; b_Y] \in \mathbb{R}^Y$ are parameters and $\beta \colon \mathcal{Y} \times \{1, \dots, Z\} \to \mathbb{R}$ are fixed numbers. We are going to describe several instances of the classifier (12) which can be useful models for ordinal regression.

1. Rounded linear-regression rule

$$h(\boldsymbol{x}, \boldsymbol{w}, b) = \max(1, \min(Y, \operatorname{round}(\langle \boldsymbol{w}, \boldsymbol{x} \rangle + b))) \tag{13}$$

is the most simplest model for the ordinal regression obtained by clipping a rounded response of the standard linear regression rule to the interval $[1, Y]$. It is easy to show that (13) is an instance of (12) recovered after setting $Z = 1$, $\beta(1, y) = 2y$, $y \in \mathcal{Y}$, and fixing the components of the intercept vector $\boldsymbol{b}$ to $b_y = 2by - y^2$. Using the conversion formula (6) we can show that the rounded linear-regression rule is equivalent to the ORD classifier with equal width of the decision intervals, namely, with $\theta_{k+1} - \theta_k = 2$, $k = 1, \dots, Y - 2$.

2. Multi-class linear classifier

$$h(\boldsymbol{x}, \boldsymbol{W}, \boldsymbol{b}) = \operatorname*{argmax}_{y \in \mathcal{Y}} \left(\langle \boldsymbol{w}_y, \boldsymbol{x} \rangle + b_y \right) \tag{14}$$

is recovered after setting $Z = Y$ and $\beta(y, z) = [\![y = z]\!]$, $y \in \mathcal{Y}$, $z \in \{1, \ldots, Z\}$. It is the most generic (and also most discriminative) form of (12) which completely ignores ordering of the labels.

3. *The proposed MORD* classifier (3) is recovered after setting $Z = 1$, $\boldsymbol{W} = \boldsymbol{w}_1$, and $\beta(y, 1) = y$, $y \in \mathcal{Y}$. We showed that the MORD classifier is equivalent to the standard ORD classifier (1) most frequently used in the ordinal regression.

4. *The proposed PW-MORD* classifier (11) is recovered after setting $\beta(y, z)$ according to

$$
\begin{array}{lll}
\beta(y, z) = 1 - \alpha(y, z) & \text{for } z = 1, \ldots, Z - 1, y \in \mathcal{Y}_z, & \\
\beta(y, z) = \alpha(y, z - 1) & \text{for } z = 2, \ldots, Z, y \in \mathcal{Y}_z, & (15) \\
\beta(y, z) = 0 & \text{otherwise.} &
\end{array}
$$

The PW-MORD is composed from $Z - 1$ MORD classifiers each modeling a subset of labels (see Section 3). The PW-MORD is most flexible as it allows to smoothly control its the complexity by a single parameter Z. It is easy to see that for $Z = 2$ the PW-MORD is equivalent to the MORD (=ORD) classifier while for $Z = Y$ it becomes the Multi-class linear classifier.

5 Generic Learning Algorithm for Ordinal Regression

In this section we consider problem of learning parameters of the generic linear multi-class classifier (12) from given example set $\{(\boldsymbol{x}^1, y^1), \ldots, (\boldsymbol{x}^m, y^m)\} \in (\mathbb{R}^n \times \mathcal{Y})^m$. We propose to use a generic and well understood framework originally developed for the structured output learning [8]. Following [8], we define an approximate empirical risk

$$
R(\boldsymbol{W}, \boldsymbol{b}) = \frac{1}{m} \sum_{i=1}^{m} \max_{y \in \mathcal{Y}} \left[\Delta(y, y^i) + \left\langle \boldsymbol{x}^i, \sum_{z \in \mathcal{Z}} \beta(y, z) \boldsymbol{w}_z \right\rangle (y - y^i) + b_y - b_{y^i} \right], \quad (16)
$$

where $\Delta \colon \mathcal{Y} \times \mathcal{Y} \to \mathbb{R}$ is any loss function satisfying

$$
\Delta(y, y) = 0, \forall y \in \mathcal{Y} \quad \text{and} \quad \Delta(y, y') > 0, \forall (y, y') \in \mathcal{Y}^2 \quad \text{such that} \quad y \neq y'. \tag{17}
$$

This risk approximation uses the idea of the margin-rescaling loss functions [8] applied on the classifier (12). It is easy to prove that $R(\boldsymbol{w}, \boldsymbol{b})$ is a convex upper bound on the true empirical risk

$$
R_{\text{emp}}(\boldsymbol{W}, \boldsymbol{b}) = \frac{1}{m} \sum_{i=1}^{m} \Delta(y^i, h(\boldsymbol{x}^i, \boldsymbol{W}, \boldsymbol{b})).
$$

We can formulate learning of the classifier (12) as the following convex unconstrained minimization problem

$$
(\boldsymbol{W}^*, \boldsymbol{b}^*) = \operatorname*{argmin}_{\boldsymbol{W} \in \mathbb{R}^n, \boldsymbol{b} \in \mathbb{R}^Y} \left[\frac{\lambda}{2} \left(\|\boldsymbol{W}\|^2 + \|\boldsymbol{b}\|^2 \right) + R(\boldsymbol{W}, \boldsymbol{b}) \right]. \tag{18}
$$

where $\| \cdot \|$ denotes the Frobenius norm and $\lambda > 0$ is a prescribed (regularization) constant used to control over-fitting. The setting with $\lambda = 0$, referred to as the empirical risk minimization learning, means that we simply try to find the classifier with minimal upper bound $R(\boldsymbol{W}, \boldsymbol{b})$ on the empirical risk $R_{\mathrm{emp}}(\boldsymbol{W}, \boldsymbol{b})$, in other words the one which performs best on training examples measured in terms of the prescribed loss $\Delta(y, y')$. The setting $\lambda > 0$, referred to as the regularized risk minimization learning, is equivalent to minimizing the risk $R(\boldsymbol{W}, \boldsymbol{b})$ w.r.t. parameters constrained to be inside a ball with radius inversely proportional to λ. The latter setting can be also interpreted as trying to maximize a generalized margin between the training examples and the classifier.

A big effort has been put by the ML community into development of efficient solvers for the problem (18). For example, a generic bundle method for risk minimization [9] or its accelerated variant [10] can be readily applied to solve (18).

Let us compare our framework with the existing algorithms for learning the ORD classifier. The existing algorithms consider a limited set of loss functions $\Delta(y, y')$. The most generic approach of [6] derives an upper bound for so called V-shaped losses: a loss is called V-shaped if it satisfies (17) and in addition

$$\Delta(y, y') \geq \Delta(y, y' + 1) \quad \text{if} \quad y' \leq y \quad \text{and} \quad \Delta(y, y') \leq \Delta(y, y' + 1) \quad \text{if} \quad y' \geq y.$$
(19)

The V-shaped loss (19) subsumes the most frequently used losses, i.e. the MAE loss $\Delta(y, y') = |y - y'|$ and the 0/1-loss $\Delta(y, y') = [\![y \neq y]\!]$, yet it is not as generic as the loss (17) applicable in our framework. Next limitation of the existing algorithms is that they have to care about feasibility of the thresholds $\boldsymbol{\theta} \in \Theta$ because they work directly on the parameters of the ORD classifier. This requires to either introduce additional constraints on the thresholds $\boldsymbol{\theta} \in \Theta$ or to impose additional constraints on the loss function, namely, that the loss must be convex [6]. For instance, the 0/1-loss is not convex hence the learning algorithms require extra inequality constraints (like the SVOR-EXP algorithm of [5]) which may complicate the optimization. Note that in our approach the problem (18) remains unconstrained irrespectively to the selected loss.

The generality of our framework, however, does not automatically imply that the risk approximation (16) is better (tighter) than those used in existing methods. We experimentally show that in the case of the MAE loss, i.e. the most frequently used one, the proposed approximation (16) provides slightly but consistently better approximation than the existing ones.

6   Experiments

In this section we empirically compare the proposed methods with existing algorithms. In Section 6.1 we present experiments on standard benchmarks. Experiments on real-life problem of visual age estimation are described in Section 6.2.

In our experiments we compare the following methods:

1. MORD. Proposed classifier (3) trained by (18) using the MAE loss.

2. PW-MORD. Proposed classifier (11) trained by (18) using the MAE loss.
3. LinReg. Rounded linear regressor (13) trained by (18) using the MAE loss.
4. LinClass. Standard multi-class linear classifier (14) trained by (18) using the MAE loss. It is an instance of the Structured Output Classifier [8]. Note that LinClass is up to the loss very similar to the standard multi-class SVM.
5. SVOR-EXP. Support Vector Ordinal Regression with explicit constraints [5].
6. SVOR-IMC. Support Vector Ordinal Regression with implicit constraints [5].

The SVOR-IMC and SVOR-EXP are instances of a generic framework of [6] developed for learning the ORD classifier (1). It was shown that the algorithms minimize a convex upper bound on the MAE-loss (SVOR-IMC) and the 0/1-loss (SVOR-EXP), respectively. Other methods for learning the ORD classifier have been proposed like SVM-based algorithms of [4] or the Support Vector Regression [2]. However, they are consistently outperformed by the SVOR-EXP and SVOR-IMC hence we compare only against the latter two.

We consider the MAE $\Delta(y, y') = |y - y'|$ as the desired metric because it is by far the most frequently used loss in the ordinal regression context as well as it is suitable for the real-life problem we consider.

All tested algorithms are instances of the regularized risk minimization framework (18). Note that SVOR-IMC and SVOR-EXP were originally formulated as quadratic programs but can be easily converted to (18). In the case of SVOR-IMC the problem (18) uses additional constraints $\theta \in \Theta$. The learning problem (18) is specified up to the regularization constant λ tuned on validation data from a fixed set of values Λ. In particular, we set $\Lambda = \{1, 0.1, 0.01, 0.001, 0\}$. We used two optimization algorithms to solve (18). For $\lambda > 0$ we used the Bundle Methods for Risk Minimization (BMRM) [9]. For $\lambda = 0$ we used the Analytic Center Cutting Plane Method (ACCPM) proposed in [11]. To avoid implementation bias, we wrote all algorithms by ourselves using mainly Matlab and C only to program a QP solver called inside the BMRM algorithm. In both cases we set the solvers to find the ε-optimal solution of the learning objective, in particular, we stopped the solver if the objective was below factor of 1.01 of the optimal value (we use the Lagrange dual to get the optimality certificate).

6.1 Standard Benchmarks

We performed experiments with seven data sets[1] used in [5][6]. We followed exactly the same evaluation protocol. The data were produced by discretising metric regression problems into $Y = 10$ bins. Data are randomly partitioned to train/test part. The partitioning was repeated 20 times. The features are normalized to have zero mean and unit variance coordinate wise. The reported results are averages and standard deviations computed over the 20 partitions. The feature dimension and train/test ratios are listed in Table 1.

We performed two experiments. The goal of the first experiment is to assess the ability of the proposed training algorithm (18) to minimize the empirical

---

[1] The link http://www.dcc.fc.up.pt/~ltorgo/Regression/census.tar.gz to the eight dataset "Census" was broken hence we could not include it.

Table 1. Comparison of the MORD, SVOR-IMC and SVOR-EXC in terms of the ability to minimize the empirical risk measured in terms of the MAE and 0/1-loss. The columns 2 and 3 show data dimension and training/testing split ratio, respectively.

	n	train/test	TrnRisk	MORD	SVOR-IMC	SVOR-EXC
Pyrimidines	27	50/24	MAE	**0.433 (0.093)**	0.482 (0.104)	0.491 (0.125)
			0/1	0.343 (0.064)	0.391 (0.069)	**0.329 (0.078)**
MachineCPU	6	150/59	MAE	**0.914 (0.052)**	0.920 (0.046)	0.972 (0.068)
			0/1	0.602 (0.035)	0.611 (0.027)	**0.594 (0.029)**
Boston	13	300/206	MAE	**0.812 (0.043)**	0.823 (0.047)	0.869 (0.050)
			0/1	0.558 (0.026)	0.573 (0.027)	**0.551 (0.028)**
Abalone	8	1000/3177	MAE	**1.412 (0.038)**	1.422 (0.041)	1.632 (0.063)
			0/1	0.734 (0.015)	0.748 (0.017)	**0.715 (0.016)**
Bank	32	3000/5192	MAE	**1.421 (0.021)**	1.429 (0.021)	1.913 (0.051)
			0/1	0.700 (0.006)	0.716 (0.007)	**0.690 (0.005)**
Computer	21	4000/4192	MAE	0.632 (0.010)	**0.632 (0.010)**	0.653 (0.012)
			0/1	**0.477 (0.006)**	0.480 (0.006)	0.477 (0.008)
California	8	5000/15640	MAE	**1.178 (0.013)**	1.182 (0.014)	1.233 (0.014)
			0/1	0.692 (0.008)	0.697 (0.007)	**0.681 (0.008)**

risk if compared to the existing algorithms SVOR-IMC and SVOR-EXC. Note that all the tested methods learn exactly the same ORD classifier by minimizing a convex approximation of the empirical risk whose direct optimization is not tractable. SVOR-IMC and SVOR-EXC use a specific risk approximation tailored for the ORD classifier. Our method makes it possible to train the ORD classifier using the standard margin-rescaling. In this experiment we set $\lambda = 0$, i.e. we minimized just the risk approximation which we want to assess. Table 1 summarizes the results. It is seen that our method slightly but consistently (up to one near draw for "Computer" data) outperforms the SVOR-IMC approximation in terms of the MAE loss for which both methods were intended. The results of SVOR-EXC optimizing the 0/1-loss are included just for completeness.

The goal of the second experiment is to assess the ability to minimize the test risk (generalization error). We compare the proposed methods PW-MORD against the standard models. We considered the PW-MORD with $Z = 2, 3, 4$ and the cut labels set symmetrically, i.e. $\{1, 10\}$, $\{1, 5, 10\}$ and $\{1, 4, 7, 10\}$. Note that PW-MORD with Z=2 corresponds to the MORD classifier hence not included in testing. The optimal complexity of the PW-MORD classifier, i.e. the number Z, as well as the regularization constant $\lambda \in \{1, 0.1, 0.01, 0.001, 0\}$ for all methods were selected based on 5-fold cross-validation estimate of the MAE (on training split). Table 2 summarizes the results. In most cases the PW-MORD classifier outperformed the competitors in terms of the target MAE metric. We attribute this fact to its flexible complexity and the ability of the proposed training algorithm to well approximate the loss function (see previous experiment). The PW-MORD was outperformed only by the LinReg on the "Pyrimids" data and by the LinCls on the "California" data. This is result is not surprising because the "Pyrimids" data have very few training examples hence the simplest regression model best avoids over-fitting. On the other hand, the "California" data are low dimensional with high number of training examples and thus LinCls, the most flexible model, can best describe the data without overfitting, i.e. the ordering prior imposed by the other models is not needed here. A surprising result is that the winner in terms of the MAE loss is in most cases the best method for the 0/1-loss, i.e. it is better than the SVOR-EXC algorithm directly optimizing the 0/1-loss. Currently we do not have a good explanation of this observation.

Table 2. Comparison of various classification models in terms of the test risk measured in terms of the MAE and the 0/1-loss. The column (Z) shows the best complexity of the PW-MORD classifier selected in the cross-validation stage.

	TstRisk	LinCls	LinReg	PW-MORD	(Z)	SVOR-IMC	SVOR-EXC
Pyrimidines	MAE	1.59 (0.25)	**1.37 (0.27)**	1.50 (0.38)	4	1.52 (0.29)	1.63 (0.28)
	0/1	0.76 (0.10)	0.76 (0.10)	**0.74 (0.09)**		0.79 (0.07)	0.80 (0.08)
MachineCPU	MAE	1.00 (0.15)	1.03 (0.10)	**0.95 (0.12)**	2	0.95 (0.11)	1.01 (0.13)
	0/1	0.65 (0.06)	0.70 (0.06)	**0.62 (0.06)**		0.63 (0.06)	0.65 (0.05)
Boston	MAE	0.94 (0.07)	0.95 (0.06)	**0.86 (0.05)**	3	0.91 (0.06)	0.97 (0.08)
	0/1	0.62 (0.03)	0.64 (0.03)	**0.58 (0.03)**		0.61 (0.03)	0.62 (0.04)
Abalone	MAE	1.42 (0.02)	1.51 (0.01)	**1.41 (0.02)**	4	1.47 (0.01)	1.68 (0.04)
	0/1	0.73 (0.01)	0.79 (0.01)	**0.73 (0.01)**		0.76 (0.01)	0.73 (0.01)
Bank	MAE	1.45 (0.01)	1.51 (0.01)	**1.45 (0.01)**	4	1.45 (0.01)	1.94 (0.05)
	0/1	0.70 (0.01)	0.77 (0.01)	0.70 (0.01)		0.72 (0.01)	**0.69 (0.00)**
Computer	MAE	0.62 (0.01)	0.72 (0.01)	**0.61 (0.01)**	4	0.63 (0.01)	0.65 (0.01)
	0/1	0.47 (0.00)	0.56 (0.01)	**0.47 (0.01)**		0.48 (0.01)	0.48 (0.00)
California	MAE	**1.12 (0.00)**	1.21 (0.01)	1.14 (0.00)	4	1.18 (0.01)	1.23 (0.01)
	0/1	**0.67 (0.00)**	0.71 (0.00)	0.68 (0.00)		0.70 (0.00)	0.68 (0.00)

6.2 Visual Age Prediction

We consider problem of predicting an apparent age of a person from an image of his/her face. We experimented on a dataset containing 37,668 face images obtained by putting together standard face-recognition benchmarks (Feret, PAL, LFW, BioID, FaceTracer, xm2vts) and completing the rest by images downloaded from the Internet. Images were manually annotated by age which was in range from 0 to 100 years. The database has equal ratio of male and females. Each face was registered by a landmark detector [12], normalized to a canonical image 30×20 by affine transform and described by pyramid-of-LBP descriptor [13]. Each face is represented by $n = 159,488$-dimensional sparse binary vector. We randomly split the data into training/validation/test part in ration 60/20/20. The validation part is used to tune the regularization constant. The reported results are averages and standard deviations of test errors computed over 3 splits.

We compared the linear multi-class SVM classifier (LinCls), the standard ordinal regression model implemented via the MORD representation and the PW-MORD classifier. In the case of LinCls we had to discretize the age into 10 equal bins because modeling all 101 classes would not be feasible (only representation of the classifier would require 120MB). We used PW-MORD with Z=11 and set the cut labels to equally cover the range of 101 years. Thus the PW-MORD classifier models each decade by a single linear ordinal regression classifier. We also compared against a commercial face recognition system developed by FACE.COM[2]. Results are summarized in Table 3 reporting the target MAE loss as well as the error levels. The proposed PW-MORD significantly outperformed all competing ordinal regression models by significant margin. The MORD classifier (=standard ORD model) is apparently not sufficiently discriminative. On the other hand, training full multi-class classier for all 101 classes is not feasible. The PW-MORD model also compares favorably with the FACE.COM system. Namely, in terms of MAE metric the FACE.COM is slightly better, however, the PW-MORD provides substantially better results for lower error levels what is typically preferred in practice.

Table 3. Comparison of various classifiers on the visual age estimation problem. The upper table shows the test MAE, i.e. average prediction error in years. The bottom table shows error levels for the tested classifiers, e.g. the first row tells the percentage of examples with MAE not greater than 5 years.

TstRisk	LinCls	PW-MORD	MORD	FaceCom
MAE	11.19 (0.16)	7.92 (0.06)	14.53 (0.13)	7.89 (NA)

	Occurrence in [%]			
Error level	LinCls	PW-MORD	MORD	FaceCom
5	44.8	52.9	28.3	47.4
10	65.2	74.5	47.8	73.7
20	84.6	91.1	74.4	93.3
30	91.7	97.2	88.9	98.2
40	94.8	99.2	95.6	99.5
50	96.4	99.8	98.4	99.9
60	97.8	99.9	99.6	100.0
70	98.9	100.0	99.9	100.0

7   Conclusions

We have shown equivalence between the classification rule used in ordinal regression and a class of linear multi-class classifiers. The established equivalence has the following benefits. First, it allows to better understand various classification models. Second, it provides a path to develop new learning algorithms for ordinal regression borrowing from well understand multi-class classification. Third, it allows to design new models for ordinal regression with higher discriminative power. Experiments on standard benchmarks as well as a real-life problem of visual age estimation demonstrate usefulness of the proposed method.

[2] FACE.COM (www.face.com) provided a free access server with face recognition technology. We passed our data though the server between July 15 and August 15, 2012. Recently, the company has been acquired by Facebook and the server closed.

Acknowledgments. KA was supported by the Grant agency of CTU project SGS12187/OHK3/3T/13 and the Visegrad Scholarship contract No. 51200430, VF by the Grant Agency of the Czech Republic under Project P202/12/2071, VH by EC project FP7-288533 CLOPEMA and the Technology Agency of the Czech Republic under Project TE01020197.

References

1. McCullagh, P.: Regression models for ordinal data. Journal of the Royal Statical Society 42(2), 109–142 (1980)
2. Vapnik, V.N.: Statistical learning theory. In: Adaptive and Learning Systems. Wiley, New York (1998)
3. Crammer, K., Singer, Y.: Pranking with ranking. In: Advances in Neural Information Processing Systems (NIPS), pp. 641–647 (2001)
4. Shashua, A., Levin, A.: Ranking with large margin principle: Two approaches. In: Proceedings of Advances in Neural Information Processing Systems, NIPS (2002)
5. Chu, W., Keerthi, S.S.: New approaches to support vector ordinal regression. In: Proc. of the International Conference on Machine Learning (ICML), pp. 145–152 (2005)
6. Li, L., Lin, H.T.: Ordinal regression by extended binary classification. In: Advances in Neural Information Processing Systems, NIPS (2006)
7. Fen, X., Liang, Z., Yang, Y., Zhang, W.: Ordinal regression as multiclass classification. Intern. Journal of Intelligent Control and Systems 12(3), 230–236 (2007)
8. Tsochantaridis, I., Joachims, T., Hofmann, T., Altun, Y., Singer, Y.: Large margin methods for structured and interdependent output variables. Journal of Machine Learning Research 6, 1453–1484 (2005)
9. Teo, C.H., Vishwanthan, S., Smola, A.J., Le, Q.V.: Bundle methods for regularized risk minimization. Journal of Machine Learning Research 11, 311–365 (2010)
10. Franc, V., Sonneburg, S.: Optimized cutting plane algorithm for large-scale risk minimization. Journal of Machine Learning Research 10, 2157–2232 (2009)
11. Antoniuk, K., Franc, V., Hlaváč, V.: Learning markov networks by analytic center cutting plane method. In: Proceedings of International Conference on Pattern Recognition (ICPR), pp. 2250–2253 (2012)
12. Uřičář, M., Franc, V., Hlaváč, V.: Detector of facial landmarks learned by the structured output SVM. In: Proceedings of the International Conference on Computer Vision Theory and Applications (VISAPP), pp. 547–556 (2012)
13. Sonnenburg, S., Franc, V.: Coffin: A computational framework for linear svms. In: Proceedings of the 27th Annual International Conference on Machine Learning (ICML 2010), Madison, USA, Omnipress (June 2010)

Appendix: Proof of Theorem 1

Let us prove the first part of theorem stating that for any $w \in \mathbb{R}^n$ and admissible $\theta \in \Theta$ there exists $b \in \mathbb{R}^Y$ such that $h(x, w, \theta) = h'(x, w, b)$, $\forall x \in \mathbb{R}^n$. In particular we show that $b \in \mathbb{R}^Y$ given by the formula (5) satisfies theorem.

First, suppose the ORD classifier $h(x; w, \theta)$ outputs $y \in \mathcal{Y}$ for some $x \in \mathcal{X}$, i.e. $\theta_y \geq \langle w, x \rangle > \theta_{y-1}$ holds[1]. The MORD classifier $h'(x, w, b)$ outputs the same y iff the system of inequalities

$$\langle w, x \rangle y + b_y > \langle w, x \rangle (y - k) + b_{y-k}, \ 1 \leq k < y,$$
$$\langle w, x \rangle y + b_y \geq \langle w, x \rangle (y + t) + b_{y+t}, \ 1 \leq t \leq Y - y \tag{20}$$

holds. The system (20) can be rewitten as[2]

$$\langle w, x \rangle k > \sum_{i=y-k}^{y-1} \theta_i, \ 1 \leq k < y,$$
$$\langle w, x \rangle t \leq \sum_{i=y}^{y+t-1} \theta_i, \ 1 \leq t \leq Y - y. \tag{21}$$

The validity of (21) follows from

$$\langle w, x \rangle k > \theta_{y-1} k \geq \sum_{i=y-k}^{y-1} \theta_i, \ 1 \leq k < y,$$
$$\langle w, x \rangle t \leq \theta_y t \leq \sum_{i=y}^{y+t-1} \theta_i, \ 1 \leq t \leq Y - y, \tag{22}$$

where the first inequality (on both lines) is induced by $\theta_y \geq \langle w, x \rangle > \theta_{y-1}$ and the second inequality (also on both lines) is due to $\theta_1 \leq \theta_2 \leq \cdots \leq \theta_{Y-1}$.

Second, suppose the MORD classifier $h'(x, w, b)$ outputs $y \in \mathcal{Y}$ for some $x \in \mathcal{X}$, which means that

$$\langle w, x \rangle y + b_y > \langle w, x \rangle (y - 1) + b_{y-1},$$
$$\langle w, x \rangle y + b_y \geq \langle w, x \rangle (y + 1) + b_{y+1}, \tag{23}$$

which is equivalent to

$$b_y - b_{y+1} \geq \langle w, x \rangle > b_{y-1} - b_y. \tag{24}$$

Finally, after combining (24) with (5) we obtain $\theta_y \geq \langle w, x \rangle > \theta_{y-1}$, which implies that the ORD classifier $h(x, w, \theta)$ outputs the same y.

Let us make an observation before proving the second part of the theorem. Let $y_1, \ldots, y_p$, denote an increasing subsequence of the non-degenerated classes of the MORD classifier $h'(x, w, b)$. For arbitrary $x_{y_i} \in \mathcal{X}_{y_i} = \{x \in \mathbb{R}^n \mid h'(x, w, b) = y_i\}$, $i = 1, \ldots, p$, it holds that

$$\langle w, x_{y_i} \rangle y_i + b_{y_i} > \langle w, x_{y_{i-1}} \rangle y_{i-1} + b_{y_i-1},$$
$$\langle w, x_{y_i} \rangle y_i + b_{y_i} \geq \langle w, x_{y_{i+1}} \rangle y_{i-1} + b_{y_i+1}, \tag{25}$$

It follows that

$$\frac{b_{y_i} - b_{y_{i+1}}}{y_{i+1} - y_i} \geq \langle w, x_{y_i} \rangle > \frac{b_{y_{i-1}} - b_{y_i}}{y_i - y_{i-1}}, \quad i = 1, \ldots, p - 1.$$

---

[1] The inequalities are different in the case of $y \in \{1, Y\}$, however, the analysis remains similar thus it is omited here.

[2] We use convention that a sum is zero if its upper index is less than the lower one.

Thus, for any MORD classifier $h'(\boldsymbol{x}, \boldsymbol{w}, \boldsymbol{b})$ with non-degenerated classes $y_1, \ldots, y_p$, it holds that

$$\frac{b_{y_{p-1}} - b_{y_p}}{y_p - y_{p-1}} > \cdots > \frac{b_{y_{i-1}} - b_{y_i}}{y_i - y_{i-1}} > \cdots > \frac{b_{y_1} - b_{y_2}}{y_2 - y_1}. \tag{26}$$

We are now ready to proof the second part of the theorem stating that for any $\boldsymbol{w} \in \mathbb{R}^n$, $\boldsymbol{b} \in \mathbb{R}^Y$ and the admissible vector $\boldsymbol{\theta} \in \Theta$ computed by the formula (7) the equality $h(\boldsymbol{x}, \boldsymbol{w}, \boldsymbol{\theta}) = h'(\boldsymbol{x}, \boldsymbol{w}, \boldsymbol{b})$ holds $\forall \boldsymbol{x} \in \mathbb{R}^n$. It is enough to show that for arbitrary $\boldsymbol{x} \in \mathcal{X}$ the ORD classifier $h(\boldsymbol{x}, \boldsymbol{w}, \boldsymbol{\theta})$ outups y_i iff the MORD classifier $h'(\boldsymbol{x}, \boldsymbol{w}, \boldsymbol{b})$ outputs the same output y_i.

First, suppose the MORD classifier $h'(\boldsymbol{x}; \boldsymbol{w}, \boldsymbol{b})$ outputs $y_i \in \mathcal{Y}$ for some $\boldsymbol{x} \in \mathcal{X}$. We want to show that the ORD classifier $h(\boldsymbol{x}; \boldsymbol{w}, \boldsymbol{\theta})$ outputs the same label y_i. We shall analyse only the cases $1 < i < p$, however, the prove for $i \in \{1, p\}$ is similar and hence omitted. The equality $h'(\boldsymbol{x}, \boldsymbol{b}) = y_i$ implies that

$$\begin{aligned}
\langle \boldsymbol{w}, \boldsymbol{x} \rangle y_i + b_{y_i} &> \langle \boldsymbol{w}, \boldsymbol{x} \rangle y_{i-1} + b_{y_{i-1}}, \\
\langle \boldsymbol{w}, \boldsymbol{x} \rangle y_i + b_{y_i} &\geq \langle \boldsymbol{w}, \boldsymbol{x} \rangle y_{i+1} + b_{y_{i+1}},
\end{aligned} \tag{27}$$

which is equivalent to $\frac{b_{y_i} - b_{y_{i+1}}}{y_{i+1} - y_i} \geq \langle \boldsymbol{w}, \boldsymbol{x} \rangle > \frac{b_{y_{i-1}} - b_{y_i}}{y_i - y_{i-1}}$ and after combining with (7) we see that the ORD classifier $h(\boldsymbol{x}, \boldsymbol{w}, \boldsymbol{\theta})$ outputs the same y_i.

Second, suppose the ORD classifier $h(\boldsymbol{x}, \boldsymbol{w}, \boldsymbol{\theta})$ outputs y_i for some arbitrary $\boldsymbol{x} \in \mathcal{X}$, i.e. $\frac{b_{y_i} - b_{y_{i+1}}}{y_{i+1} - y_i} \geq \langle \boldsymbol{w}, \boldsymbol{x} \rangle > \frac{b_{y_{i-1}} - b_{y_i}}{y_i - y_{i-1}}$ holds. To show that MORD classifier $h'(\boldsymbol{x}; \boldsymbol{w}, \boldsymbol{\theta})$ outputs the same y_i it is enough to prove that the system

$$\langle \boldsymbol{w}, \boldsymbol{x} \rangle y_i + b_{y_i} > \langle \boldsymbol{w}, \boldsymbol{x} \rangle y_j + b_{y_j}, \ \forall y_j < y_i, \tag{28}$$

$$\langle \boldsymbol{w}, \boldsymbol{x} \rangle y_i + b_{y_i} \geq \langle \boldsymbol{w}, \boldsymbol{x} \rangle y_j + b_{y_j}, \ \forall y_j > y_i \tag{29}$$

holds. Indeed, from inequality $\langle \boldsymbol{w}, \boldsymbol{x} \rangle > \frac{b_{y_{i-1}} - b_{y_i}}{y_i - y_{i-1}}$ after some algebra and applying (26) (after third line) we have

$$\begin{aligned}
\langle \boldsymbol{w}, \boldsymbol{x} \rangle (y_i - y_j) &> (y_i - y_j) \frac{b_{y_{i-1}} - b_{y_i}}{y_i - y_{i-1}} \\
&= (-y_j + y_{j+1} - y_{j+1} + \cdots + y_{i-1} - y_{i-1} + y_i) \frac{b_{y_{i-1}} - b_{y_i}}{y_i - y_{i-1}} \\
&= (y_{j+1} - y_j) \frac{b_{y_{i-1}} - b_{y_i}}{y_i - y_{i-1}} + \cdots + (y_i - y_{i-1}) \frac{b_{y_{i-1}} - b_{y_i}}{y_i - y_{i-1}} \\
&> (y_{j+1} - y_j) \frac{b_{y_j} - b_{y_{j+1}}}{y_{j+1} - y_j} + \cdots + (y_i - y_{i-1}) \frac{b_{y_{i-1}} - b_{y_i}}{y_i - y_{i-1}} \\
&= b_{y_j} - b_{y_{j+1}} + b_{y_{j+1}} - \cdots - b_{y_{i-1}} + b_{y_{i-1}} - b_{y_i} = b_{y_j} - b_{y_i},
\end{aligned}$$

from which the inequalities (28) follow for $\forall y_j < y_i$. The proof of the inequalities (29) is analogical. ∎

Identifiability of Model Properties
in Over-Parameterized Model Classes

Manfred Jaeger

Aalborg University, Denmark
jaeger@cs.aau.dk

Abstract. Classical learning theory is based on a tight linkage between hypothesis space (a class of function on a domain X), data space (function-value examples $(x, f(x))$), and the space of queries for the learned model (predicting function values for new examples x). However, in many learning scenarios the 3-way association between hypotheses, data, and queries can really be much looser. Model classes can be over-parameterized, i.e., different hypotheses may be equivalent with respect to the data observations. Queries may relate to model properties that do not directly correspond to the observations in the data. In this paper we make some initial steps to extend and adapt basic concepts of computational learnability and statistical identifiability to provide a foundation for investigating learnability in such broader contexts. We exemplify the use of the framework in three different applications: the identification of temporal logic properties of probabilistic automata learned from sequence data, the identification of causal dependencies in probabilistic graphical models, and the transfer of probabilistic relational models to new domains.

1 Introduction

This paper is originally motivated by ongoing research in learning probabilistic automata models for applications in model-based design in verification [16,10]. In model-based design, various forms of finite probabilistic automata models are used to model hard- or software systems. Relevant properties of the system are expressed using a formal, logical representation language, such as probabilistic linear time logic (PLTL), or probabilistic computation tree logic (PCTL) [2]. Efficient algorithms exist to check whether such a property is satisfied by a given automaton model, i.e., whether the design or actual system represented by the model satisfies a certain specification. Traditionally, the formal models used in this process are constructed manually. This, however, can be a very time-consuming and error-prone process. We are therefore interested in the possibility of automatically learning finite automata models from data consisting of observations of visible system behaviors. Adapting standard automata learning algorithms [3,4] we obtain learning methods for our application that come with certain convergence guarantees for the large-sample limit. However, these convergence guarantees do not directly imply what is of ultimate interest, namely,

H. Blockeel et al. (Eds.): ECML PKDD 2013, Part III, LNAI 8190, pp. 112–127, 2013.

that in the large-sample limit the learned model will agree with the actual observed system on properties representable in the formal representation language we use in model-checking. Concretely, building on the given convergence guarantees, one can show that the learned model will in the limit agree with the actual system on PLTL properties [10]. However, the same does not appear to be true for PCTL properties.

Abstracting from this specific learning problem, we are faced with the more general question: what classes of properties that we will want to query our learned model for, can, in principle, be learned from the observations that are represented by our data? This question is closely connected to *learnability* in the sense of computational learning theory, as well as to *identifiability* in the sense of statistics. However, it seems that neither these two existing theoretical frameworks are quite sufficient to analyze the scenario we are here considering.

Computational learning theory uses at its conceptual foundation a very close linkage between *hypothesis space*, *data space*, and what one may call *query space*: the hypothesis space is taken to consist of a set $\mathcal{F}$ of functions, data consists of a set of observed pairs $(x, f(x))$ of arguments and function values, and a learned function $f \in \mathcal{F}$ will be queried for its function values $f(x)$. This setup does not incorporate the possibility of an over-parameterized hypothesis space, i.e., the existence of two distinct hypotheses h, h' that define the same function, and that therefore would lead to equivalent data observations. Under the assumption that a learned hypothesis will be queried for its function values, this possibility may also be safely neglected, since it would make no difference whether hypothesis h or h' is learned. This radically changes, however, when also the close linkage between data space and query space is lost, and the learned hypothesis will be queried for properties that may not exactly match the type of observations found in the data. This is precisely the situation we find ourselves in when learning probabilistic automata for model-checking purposes: two distinct automata can induce the same data-distribution of observable behaviors, but differ with respect to some properties in our formal query languages. On the other hand, some relaxation of the three-way linkage between data, hypotheses and queries does not necessarily preclude learnability: in our positive results about learning PLTL properties we have data consisting of finite strings, hypotheses consisting of finite automata, and queries consisting of logical formulas.

The issue of over-parameterized model spaces containing distinct hypotheses that generate indistinguishable data is exactly the subject of the statistical notion of identifiability. However, statistical identifiability theory does not relate hypothesis and data space to classes of queries over the learned model. It is implicitly assumed that the purpose of learning, which here comes down to parameter estimation, is to identify the true parameter. In contrast, we may be satisfied with learning a hypothesis that is distinct from the "true" one, as long as it is equivalent with regard to a certain class of query properties.

The problem of learnability of certain query properties in an over-parameterized model class is a quite common one, and certainly not limited to, or first encountered in, our problem of identifying logical properties of a finite automaton. Another motivating example of the same problem which we shall consider in this paper is *causal discovery* from observational data: a directed graphical model (Bayesian network) is sometimes regarded as a causal model, where directed connections between random variables represent causal dependencies. However, it is well known that a Bayesian network learned from statistical data only allows a limited interpretation as a causal model, since networks with different directed edge structures can induce the same data distribution, and hence be indistinguishable based on observational data. The possibilities and inherent limitations of using Bayesian network learning for causal discovery are now well understood [9]. However, the sometimes controversial debate into this issue has not been fully phrased within a formal theory of learnability or identifiability, which, one could imagine, sometimes might have helped to elucidate matters more clearly [15,6].

With an increasing ambition of learning models in more and more expressive model classes, for example probabilistic programming languages [12,7,5], one also encounters more and more complex relationships between the wide range of model properties that could be queried, and the empirical content of the original data. Broadening existing theories of learnability to enable a systematic and principled analysis of these relationships and dependencies, thus, could be useful in a variety of contexts.

In this paper we are going to propose a formal framework combining elements of computational learnability and statistical identifiability that enables a systematic study of what model properties can, or can not, be identified in a certain model class, given a particular type of training data. To this end we first introduce a very general setup of learning as maximization of a score function (Section 2). We then propose a definition of *identifiability* that makes no assumptions on structural correspondences between data-, model-, and query space. Based on these definitions, we obtain a first theorem about non-identifiability. We demonstrate the applicability of our framework by deriving non-identifiability results for PCTL properties of probabilistic automata, directed edge relationships in Bayesian networks, and probabilistic queries on varying domains in statistical relational models. Finally, in Section 4 we adapt the initial very general framework for the special case of statistical learning, and discuss its relationship with PAC learnability [18].

2   Learning as Score Maximization

We characterize learning as the maximization of a score function σ that defines for each M in a model (or hypothesis) space $\mathcal{M}$, and each dataset D in a dataspace $\mathcal{D}$ a score $\sigma(M, D)$. This, on the one hand, is the most natural description of a wide range of learning methods that in fact operate by using heuristic or stochastic search to maximize a given score function. On the other hand, it also accommodates in a trivial way any other algorithmic learning procedure not based on an explicit score function by representing an algorithm l that on input D outputs a hypothesis $l(D) \in \mathcal{M}$ via a score function σ_l with $\sigma_l(M, D) = 1$ iff $M = l(D)$, and $\sigma(M, D) = 0$ otherwise.

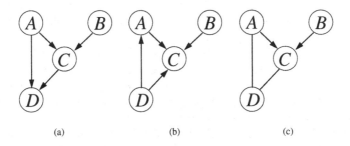

Fig. 1. Two Bayesian Networks and their Essential Graph

One restriction we impose on the scoring function σ is that for all $D \in \mathcal{D}$ the supremum of $\sigma(\cdot, D)$ is attained for some $M \in \mathcal{M}$. We write

$$\mathcal{M}(\sigma, D) := \{M \in \mathcal{M} \mid \sigma(M, D) = \max_{\tilde{M} \in \mathcal{M}} \sigma(\tilde{M}, D)\}$$

for the set of all M maximizing $\sigma(\cdot, D)$. Thus, $\mathcal{M}(\sigma, D)$ is the set of models learned from D using σ. $\mathcal{M}(\sigma, D)$ may contain more than one element, but is assumed to be nonempty.

In the following we give two examples of learning algorithms not usually seen from the perspective of score-based learning. We here develop a more meaningful characterization of these learning methods in terms of score maximization than simply by the trivial σ_l representation mentioned above.

Example 1. Bayesian Networks are probabilistic models for the joint distribution of a set of random variables [13]. They consist of a directed acyclic graph over the random variables, and, for each variable, the specification of the conditional probability distribution of that variable given its parents in the graph. Figure 1 (a),(b) shows two different Bayesian Networks for random variables A, B, C, D. The graph structure of a Bayesian network encodes conditional independence relations among the variables.

Probabilistic automata represent probability distributions over infinite sequences of symbols from a given alphabet Σ. Figure 2 shows three different probabilistic automata defining probability distributions on $\{a, b\}^\infty$. In the figure, states marked with * are the initial states of the automaton. Solid edges denote state transitions taken with probability 1, and dashed edges transitions with probability $1/2$. All automata in Figure 2 define the same probability distribution P_M, which is the uniform distribution on $aa\{a, b\}^\infty$.

M_a and M_b are deterministic automata: for every state in the automaton, and every $s \in \Sigma$ there exists at most one possible successor state labeled with s. M_c, on the other hand, is non-deterministic, since the initial state has two different a-successors.

A classic algorithm for learning Bayesian networks is the PC-algorithm [17], and a standard algorithm for learning deterministic probabilistic automata is the Alergia algorithm and its variants [3,4,8,10].

Even though quite different with respect to the learning tasks they solve, and the algorithmic details, the PC algorithm and Alergia share some common features which can be expressed in a common structure of a high-level score function representation of these approaches. Both PC and Alergia identify the graphical structure of the model based on statistical tests performed on the data. In the PC algorithm, these are conditional independence tests for subsets of the random variables. In Alergia, one performs tests for the identity of the conditional distributions on (infinite) suffixes $s \in \Sigma^\infty$ given different (finite) prefixes from Σ^*.

In Alergia, the outcome of the statistical tests determines the structure of the model uniquely. In the PC-algorithm, the tests determine the structure only up to membership in a class of network structures encoding the same conditional independence constraints. This equivalence class can be represented by an *essential graph*, which is a mixed graph with directed and undirected edges. Figure 1 (c) shows the essential graph representing the equivalence class consisting of the networks (a) and (b). In order to obtain a final directed graph, one may employ any method to select a representative directed graph from the equivalence class.

Once the model structure is fixed, maximum likelihood parameters are fitted. The overall learning process can thus be described as maximizing a score function of the form

$$\sigma(M, D) = R(M) \cdot \text{Tests}(M, D) \cdot \text{Lhood}(M, D) \qquad (1)$$

where $\text{Tests}(M, D)$ is a 0/1-valued function that evaluates to 1 iff the structure of M is consistent with the outcome of all relevant statistical tests performed on D, and $\text{Lhood}(M, D) = P_M(D)$ is the likelihood function (also depending on the parameters of M). $R(M)$ is a factor only needed for capturing the PC algorithm. It is a another 0/1-valued function that evaluates to 1 iff the structure of M is the representative directed graph for its underlying essential graph. For Alergia, we may just assume that $R(M)$ is constant equal to 1.

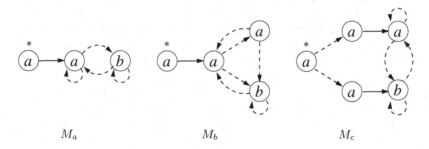

$$M_a \qquad\qquad M_b \qquad\qquad M_c$$

Fig. 2. Probabilistic Automata

A second example looks at a more standard form of score-based learning.

Example 2. (Penalized likelihood score (PLS)) Standard model scores like Akaike Information Criterion (AIC), Bayesian Information Criterion (BIC), or Minimum Description Length (MDL) are used for a wide range of probabilistic model classes. They are penalized likelihood scores of the form

$$\sigma_{PLS}(M, D) := c \cdot \log(Lhood(M, D)) - g(|D|)h(|M|), \tag{2}$$

where c is a constant, $g(|D|)$ is a function of the size of the data D, and $h(|M|)$ is a function of the size of M, where g and h are non-negative, and h is monotone in $|M|$. For Bayesian networks, as well as for many other probabilistic types of models, the size $|M|$ typically is defined as the number of free parameters in M.

Assuming that the two network structures (a) and (b) in Figure 1 are equipped with their maximum likelihood parameters, they will both define the same distribution P_M, and, thus, obtain the same log-likelihood score $\log(P_M(D))$.

3   Identifiability

We are interested in learnability in the limit of large datasets. Apart from datasets consisting of independent samples, we also want to consider datasets that may consist, e.g., of a single long sequential observation of a temporal process, or a (large) graph. We therefore only assume that the space of all datasets is structured as follows.

Definition 1. *A stratified dataspace is a set*

$$\mathcal{D} = \cup_n \mathcal{D}^{(n)}$$

which is equipped with a partial order relation $\prec$, so that $D \prec D'$ implies that $D \in \mathcal{D}^{(j)}$, $D' \in \mathcal{D}^{(k)}$ with $j < k$. Furthermore, it is required that for every n and $D \in \mathcal{D}^{(n)}$ there exists $D' \in \mathcal{D}^{(n+1)}$ with $D \prec D'$.

The intention is that $D \prec D'$ means that D' is an extension of the dataset D.

Example 3. (a) Suppose the data consists of independent samples $s = s_1, \ldots, s_n$ from a sample space S. Then $\mathcal{D}^{(n)} = S^n$, and $s \prec s'$ iff s' extends s, i.e., $s \in S^n$, $s' \in S^{n'}$ with $n' > n$, and $s'_i = s_i$ for $i = 1, \ldots, n$.

(b) In learning from a single sequence of symbols from an alphabet Σ, we have $\mathcal{D}^{(n)} = \Sigma^n$, and $D \prec D'$ iff D is a prefix of D'.

(c) When learning from graphs (or, more generally, colored graphs, or relational data), we can have that $\mathcal{D}^{(n)}$ is the set $\mathcal{G}^n$ of all graphs with n nodes, and $G \prec G'$ iff G is embedded as a sub-graph in G'.

Learning in the limit of large datasets now is analyzed in terms of increasing sequences $D_1 \prec D_2 \prec \cdots$. We write $\uparrow D_n$ for such a sequence.

As in (1) and (2), most score functions will be composed from a number of simpler scores or measurements:

Definition 2. *A score feature is any function*

$$F : \; \mathcal{M} \times \mathcal{D} \to \mathcal{F}$$

with $\mathcal{F} \subseteq \mathbb{R}$. When $\mathcal{F} = \{0,1\}$ then F is called a Boolean feature. A score feature $F : \mathcal{M} \to \mathcal{F}$ that only depends on $\mathcal{M}$ also is called a model feature, *and a feature $F : \mathcal{D} \to \mathcal{F}$ that is only a function of $\mathcal{D}$ is called a* data feature. *A* query space *is a set Φ of Boolean model features.*

The score function $\sigma(\cdot, \cdot)$ itself is a score feature. For probabilistic models, a key score feature is the likelihood $Lhood : (M, D) \mapsto P_M(D)$.

For a query space Φ we write $M \equiv_\Phi M'$ if $\phi(M) = \phi(M')$ for all $\phi \in \Phi$. $[M]_{\equiv_\Phi}$ denotes the equivalence class of M in $\mathcal{M}$ with regard to the equivalence relation $\equiv_\Phi$. We are now ready to introduce our central definition.

Definition 3. *Let $\mathcal{M}$ be a model space, $\mathcal{D}$ a stratified dataspace, and Φ a query space for $\mathcal{M}$. Let σ be a score function on $\mathcal{M} \times \mathcal{D}$. We say that Φ is $(\mathcal{D}, \sigma)$-identifiable in $\mathcal{M}$, if for all $M \in \mathcal{M}$ there exists $\uparrow D_n$ so that for all $\phi \in \Phi$ there exists $n_\phi \in \mathbb{N}$, so that for all $n \geq n_\phi$:*

$$\mathcal{M}(\sigma, D_n) \subseteq [M]_{\equiv_\phi} \qquad (3)$$

Definition 3, thus, demands that for every possible M there exists some ideal data sequence $\uparrow D_n$ which enables us to obtain from the learned models in the large sample limit the same answers for all queries ϕ that we would obtain from the "true" model M.

Definition 3 only refers to Boolean queries. In many cases, however, we will also be interested in querying other kinds of model features. For probabilistic models M, for instance, a typical query can be the probability $P_M(E)$ assigned by the model to some event E, or the most probable configuration of a set of unobserved random variables. Identifiability of queries of this kind can be approximated by identifiability of suitable Boolean query spaces. For probability queries $P_M(E)$, for instance, one should only require that the answers obtained from the learned models converge to the true value in the large sample limit. Such a convergence is exactly captured by Definition 3 as identifiability of the Boolean features $P_M(E) \in I$, for all open intervals I.

Definition 3 only provides a basis for analyzing identifiability questions. Neither positive nor negative identifiability results purely in the sense of this definition are very interesting per se. A positive identifiability result in the sense of Definition 3 is not particularly useful in itself, since it would only say that based on some ideal data sequence $\uparrow D_n$ we would be able to identify Φ. To be practically relevant, we would need the sharper result that this will be true for all data sequences that in some sense are sufficiently informative, or representative of the true model, and that are the datasets we are likely to have for learning in practice. We will extend Definition 3 in this direction in Section 4.

The main focus of this paper, however, is to establish negative identifiability results, i.e., impossibility results for learning. A negative result just in the sense of Definition 3 would not be very strong either, since it would only tell us that learning with a particular score function σ does not enable us to identify Φ. For non-identifiability results, however, we are interested in whether it is impossible to identify Φ using any member in a certain class of score functions, which represents all in some sense applicable learning algorithms. In fact, one may wonder why Definition 3 refers to a score function at all. Based on our original motivation described in Section 1, one may rather want to describe identifiability as an inherent relationship between the model, query, and data spaces, so that identifiability just means that the data contains a sufficient amount of query-relevant information about the model – which should be independent of any learning procedures. However, as the following example illustrates, it would likely be futile to try to address the identifiability problem without explicit consideration of admissible score functions, since otherwise one can always obtain trivial identifiability results via artificial score functions that do not capture realistic learning methods, but rather describe querying a perfect oracle:

Example 4. Assume that the cardinality of $\mathcal{D}$ is at least as big as the cardinality of $\mathcal{M}$, so that there is a one-to-one function $f : \mathcal{M} \to \mathcal{D}$. Then we can define

$$\sigma(M, D) := \begin{cases} 1 \text{ if } D = f(M) \\ 0 \text{ otherwise} \end{cases} \tag{4}$$

For every $M \in \mathcal{M}$ then $\mathcal{M}(\sigma, f(M)) = \{M\}$, and (3) is satisfied by the constant sequence $D_n = f(M)$.

In the following, thus, we will derive non-identifiability results in the sense of Definition 3 for all σ in certain classes of score functions. We specify classes of score functions that correspond to realistic classes of learning algorithms in terms of the score features that they are based on. We write $\sigma(F_0(M, D), \ldots, F_k(M, D))$ for a score function σ that depends on M, D only through the features $F_i(M, D)$ (which may also include pure model or data features).

Definition 4. *A score function* $\sigma(F_0(M, D), \ldots, F_k(M, D))$ *is* monotone in the *score feature* F_0, *if for all* $f_i \in \mathcal{F}_i$ *($i = 1, \ldots, k$), and $r, s \in \mathcal{F}_0$: $r \leq s \Rightarrow$* $\sigma(r, f_1, \ldots, f_k) \leq \sigma(s, f_1, \ldots, f_k)$.

Example 5. The PC/Alergia-learning score function (1) is monotone in all its features $R(M)$, $Tests(M, D)$, and $Lhood(M, D)$. The PLS score function (2) is monotone in $Lhood(M, D)$, as well as $-h(|D|)$ and $-h(|M|)$.

Both (1) and (2) are monotone in a specific model feature ($R(M)$, respectively $-h(|M|)$) that represents a preference or bias function on $\mathcal{M}$. Such a feature can express a bias towards small or canonical models, or, in a Bayesian framework, a prior probability of M.

In the following we pay particular attention to such bias features expressing prior knowledge or preferences, and we write $\sigma(B, \boldsymbol{F})$ for a score function that is monotone in a designated numerical bias model feature B, and that further depends in an arbitrary manner on other score features $\boldsymbol{F} = F_1, \ldots, F_k$.

For any subsets $M \subseteq \mathcal{M}$, $D \subseteq \mathcal{D}$ and feature vector $(F_0, \ldots, F_k)$ we write $(F_0, \ldots, F_k)(M, D)$ for the set $\{(F_0(M, D), \ldots, F_k(M, D)) \mid M \in M, D \in D\}$. In the following we will be concerned with sets M that are equivalence classes $[]_\Phi$ for some set Φ of query properties.

Definition 5. *Let $M \subseteq \mathcal{M}$, $D \subseteq \mathcal{D}$. We say that B and F are orthogonal on $M \times D$, denoted $B \bot F(M, D)$, if $(B, F)(M, D) = B(M) \times F(M, D)$.*

In other words, $B \bot F(M, D)$ means that on the set $M \times D$ a given value of $F()$ does not constrain the possible values of $B()$, and vice-versa.

Proposition 1. *Let $\sigma(B, F)$ be a score function that is monotone in B. Let Φ be a query space. If there exist distinct, nonempty equivalence classes $[M]_\Phi, [M']_\Phi$, such that for all sufficiently large n, and all $D \in \mathcal{D}^{(n)}$:*

(i) $F([M]_\Phi, D) = F([M']_\Phi, D)$
(ii) $B \bot F([M]_\Phi, D)$ *and* $B \bot F([M']_\Phi, D)$,

then Φ is not $(\mathcal{D}, \sigma)$-identifiable.

Proof. Without loss of generality assume that

$$\sup B([M]_\Phi) \geq \sup B([M']_\Phi). \tag{5}$$

Then, for all sufficiently large n, all $D \in \mathcal{D}^{(n)}$, and all $\tilde{M}' \in [M']_\Phi$ there exists $\tilde{M} \in [M]_\Phi$ with $\sigma(\tilde{M}, D) \geq \sigma(\tilde{M}', D)$, and (3) does not hold for $[M']_\Phi$.

Example 6. (Non-identifiability of BN structure) Let $\mathcal{M}$ be the set of Bayesian networks over the variables A, B, C, D, and $\mathcal{D}^{(n)} = S^n$, where S is the sample space of joint observations of the variables. Let $\Phi = \{\phi_{X \to Y} \mid X, Y \in A, B, C, D\}$, where $\phi_{X \to Y}(M)$ is true iff the edge $X \to Y$ is included in M. Φ here is finite. The score functions (1) and (2) are based on the score features *Test()*, *Lhood()*, the data feature $g()$, as well as the bias features $R()$ (for PC) and $-h()$ (for PLS). We use Proposition 1 to show that Φ is not identifiable from data consisting of joint observations of A, B, C, D by any score function that is based on the score features *Test* and *Lhood*, together with any data features $F(D)$, and bias features such as $R()$ or $-h()$.

We consider the Φ-equivalence classes of M and M' given by Figure 1 (a) and (b). We first verify that (i) and (ii) hold for $[M]_\Phi, [M']_\Phi$, and $F = (Test, Lhood)$. We need not consider any pure data features $F(D)$ here: since D is fixed in (i) and (ii), any addition of pure data features $F(D)$ to F has no impact on the validity of (i) and (ii).

$[M]_\Phi$ and $[M']_\Phi$ contain exactly the Bayesian networks with the same structure as M and M', respectively. Since both structures represent the same conditional independencies, we have for all D that $Test([M]_\Phi, D) = Test([M']_\Phi, D) = \{0\}$ (the outcome of the independence tests on D are not compatible with the structures of M and M'), or $Test([M]_\Phi, D) = Test([M']_\Phi, D) = \{1\}$ (otherwise). Also, since exactly the same class of distributions can be represented by models in $[M]_\Phi$ and in $[M']_\Phi$, one has $Lhood([M]_\Phi, D) = Lhood([M']_\Phi, D)$. This together shows (i) (note that in general it is not enough to show these identities separately for the value sets of each feature $F \in \boldsymbol{F}$; here it is sufficient because for the *Test* feature the value sets turned out to be just singletons).

Both the bias features $R()$ and $-h()$ have a unique value on the equivalence classes $[M]_\Phi$ and $[M']_\Phi$. This makes condition (ii) trivially satisfied for these and similar bias features.

Example 7. (Non-identifiability of PCTL) Let $\mathcal{M}^{nd}$ be the class of non-deterministic probabilistic finite automata, and let $\mathcal{D}^{(n)} = (\Sigma^*)^n$. Thus, even though the automata define a distribution over Σ^∞, we assume that data $D = (s_1, \ldots, s_n)$ consists of finite strings $s_i \in \Sigma^*$ only, where the s_i are obtained by sampling traces of the automaton up to some random length $l_i = |s_i|$. As mentioned in Section 1 we are interested in queries from the formal languages PLTL and PCTL. Full syntax and semantics definitions for these languages can be found in [2]. In the following we will only somewhat informally introduce specific properties they can represent. For this we will be needing formulas built using only the temporal operator "next (time point)", written $\bigcirc$. With PLTL one can specify probability bounds on the distribution over sequences defined by the automaton. For example, a PLTL-expressible property would be that the probability that the sequence starts with aaa is > 0.49, formally expressed by $P_{>0.49}(a \wedge \bigcirc a \wedge \bigcirc \bigcirc a)$. This property is satisfied by all three automata in Figure 2.

PCTL syntactically allows formulas in which probability quantifiers $P_{>\ldots}$ and temporal operators are nested. Semantically they represent model properties that refer to internal states of the automaton. One such property that we can formulate reads in formal PCTL syntax

$$\phi_1 \equiv P_{>0.4} \bigcirc P_{>0.9} \bigcirc a. \tag{6}$$

The meaning of this is that (from the initial state) there is a probability > 0.4 of reaching a state from which the probability then is > 0.9 of reaching by the next transition a state labeled with a. For the automata of Figure 2, ϕ_1 is false for M_a and M_b, and true for M_c. All automata of Figure 2 satisfy the following two formulas:

$$\phi_2 \equiv P_{=1}(a \wedge \bigcirc a) \tag{7}$$

$$\phi_3 \equiv P_{=0.5} \bigcirc \bigcirc a \tag{8}$$

Given that an automaton M satisfies ϕ_2 and ϕ_3, one has that $P_M(aaa\Sigma^\infty) = P_M(aab\Sigma^\infty) = 1/2$, and the distribution P_M then is fully specified by the two conditional distributions $P_M(\cdot \mid aaa)$ and $P_M(\cdot \mid aab)$. For the automata of Figure 2, both these conditionals are the uniform distribution on Σ^∞. One can show that also any other pair of conditional distributions that can be defined by a finite probabilistic automaton can be implemented both by automata for which ϕ_1 is true, and by automata for which ϕ_1 is false. For the first case, one can follow the basic structure of M_c, where one branches from the initial state in the first step, followed by a deterministic transition in the second. For the second case, one builds on the structure of M_a, and constructs a model in which from every state reachable by the first transition there is a probability of 0.5 each for reaching an a, respectively b, state by the second transition.

Letting $\Phi = \{\phi_1, \phi_2, \phi_3\}$, the above considerations mean that the models in the two equivalence classes $[M_a]_\Phi$ and $[M_c]_\Phi$ can represent exactly the same distributions. This implies that for any vector of score features F that only contain features such as *Tests* and *Lhood*, which depend on M only through the distribution P_M, one has $F([M_a]_\Phi, D) = F([M_c]_\Phi, D)$ for all D. Thus Proposition 1 (i) is satisfied for F of this form, which includes Alergia-style learning (cf. (1)).

We here do not consider condition (ii) for a bias feature B. Alergia-style algorithms do not use a bias feature, and so there are no immediate canonical candidates. However, Alergia-style algorithms also learn deterministic automata only, whereas here we considered non-deterministic ones. Within the class $\mathcal{M}^{nd}$ a measure of the complexity of the model may be a reasonable bias to include in a learning process.

Proposition 1 may appear rather specific, and possibly narrow, due to the assumption that we are dealing with score functions that can be written as $\sigma(B, F)$. Furthermore, the structural conditions (i) and (ii) appear quite strong, and especially (ii) will probably not always hold, even in the case of non-identifiability. Thus, (i) and (ii) are simple sufficient, but certainly not necessary conditions for non-identifiability. On the other hand, Examples 6 and 7 show that Proposition 1 already does cover a certain range of different identifiability problems. The following example further indicates that the structural form of σ that we here assumed is in some sense natural and most general: already allowing a slight generalization in the form of σ, where one can have two different bias features B_1, B_2, again leads to trivial identifiability results similar to Example 4:

Example 8. Assume that both $\mathcal{M}$ and $\mathcal{D}$ are countably infinite. Let $f : \mathcal{M} \to \mathbb{N}$ and $g : \mathcal{D} \to 2\mathbb{N}$, where $2\mathbb{N}$ stands for the set of even natural numbers, such that both f and g are one-to-one and onto. Define the two model features $B_1(M) := f(M)$, $B_2(M) := -f(M)$, and consider $g(D)$ as a data feature. We can then define the score function

$$\sigma(M, D) := \begin{cases} B_1(M) & \text{if } B_1(M) \leq B_2(M) + g(D) \\ B_2(M) + g(D) & \text{otherwise} \end{cases} \tag{9}$$

It is straightforward to verify that σ is monotone in both B_1 and B_2. Also, Proposition 1 (i) and (ii) are trivially satisfied for $F = g$ and $B = B_i$ $(i = 1, 2)$. For any given D, $\sigma(M, D)$ is maximized when $B_1(M) = B_2(M) + g(D)$, i.e., $f(M) = -f(M) + g(D)$, or $f(M) = g(D)/2$. By the assumptions on f and g there is a unique M that satisfies this condition. Thus, similarly as in Example 4 we can choose for a given M the constant sequence $D_n = D$, for the D with $g(D) = 2f(M)$.

We end this section by considering an example in relational learning. Probabilistic relational (or probabilistic logical) models define probability distributions over relational structures, i.e., over the interpretations I over finite domains

$C = \{c_1, \ldots, c_m\}$ of the relation symbols in a signature Σ. Most types of probabilistic relational models only define the conditional distributions $P(I \mid C)$ of interpretations for given domains. Only in some expressive representation languages such as BLOG [12] one can also specify distributions $P(C)$ over domains. For relational learning one can distinguish several types of stratified dataspaces:

$\mathcal{D}_1$: $\mathcal{D}^{(n)}$ consists of n independent samples $I_1, \ldots, I_n$ of interpretations over a fixed domain C.

$\mathcal{D}_2$: $\mathcal{D}^{(n)}$ consists of n independent samples $(I_1, D_1), \ldots, (I_n, D_n)$ of interpretations over different domains C_i. Example: molecular data, where each I_i represents a molecule described by attributes and bond-relations over a domain D_i of atoms.

$\mathcal{D}_3$: (cf. Example 3 (c)) $\mathcal{D}^{(n)}$ consists of one observation of an interpretation I over the domain $C_n = \{c_1, \ldots, c_n\}$, and $I \prec I'$ iff I is an interpretation over C_n, I' an interpretation over $C_{n'}$, $n < n'$, and I is the substructure induced by I' on C_n. Example: learning from a single database, such as the IMDB movie database, or DBLP bibliographic database. Increasing data here means that the database grows by adding more objects (movies, bibliographic entries...).

Probabilistic relational models learned from data for domains C_i may be used to perform inference for new domains C not represented in the data. This can be seen as the weakest form of model transfer, which in a much more ambitious form (also including a transformation of the relational signatures) becomes transfer learning in the sense of [11]. We now derive within our framework an impossibility result for relational model transfer. Again, the interest of the analysis does not so much lie in the concrete impossibility result we obtain, but in the demonstration that our general framework allows us to express in a rigorous manner what appears to be intuitively rather obvious.

Example 9. For concreteness' sake, let $\mathcal{M}$ be the class of Markov Logic Networks (MLNs) [14] for a signature of just a single attribute (unary relation) symbol $a(X)$. We assume that we learn from data of type $\mathcal{D}_3$, i.e., a dataset of size n consists of the domain C_n, and for each c_i the information whether $a(c_i)$ is true or false. We consider the query $\phi = P(a(c_1) \mid C = \{c_1\}) > 0.5$, i.e. we ask whether $P(a(c_1)) > 0.5$, for the single object c_1 in a domain of size one.

We now apply Proposition 1 to show that $\Phi = \{\phi\}$ is not $(\mathcal{D}_3, \sigma)$-identifiable by likelihood-based learning, i.e., for any σ that only uses the likelihood feature $Lhood(M, (I, C_n)) = P_M(I \mid C_n)$. Thus, we have to show (i) for $\mathbf{F} = Lhood$. Since Φ consists of a single Boolean query, there are only two equivalence classes $[]_\Phi$. Let $n \geq 2$, and $D = (I, C_n)$ a dataset of size n. Assume that in I $a(c_i)$ is true for $i = 1, \ldots, k$, and false for $i = k+1, \ldots, n$. Consider the two formulas

$$p(X_1) \wedge \ldots \wedge p(X_k) \wedge \neg p(X_{k+1}) \wedge \ldots \wedge \neg p(X_n) \wedge \bigwedge_{i \neq j} X_i \neq X_j \quad (10)$$

$$p(X) \quad (11)$$

Consider an MLN M consisting of formula (10) with a weight w. As $w \to \infty$, the MLN defines over the interpretations on C_n a distribution that is concentrated on the structures where $a()$ is true for exactly k objects, and false for $n - k$ objects. Since there are $\binom{n}{k}$ such structures, one obtains for these models likelihood values $P_M(I \mid C_n)$ converging to $1/\binom{n}{k}$. Since all possible MLNs for the given signature must assign equal probabilities to isomorphic structures, no higher likelihoods are obtainable by any other model. On the other hand, if $w \to -\infty$, then $P_M(I \mid C_n) \to 0$. For all settings of w one has $P_M(a(c_1) \mid C = \{c_1\}) = 0.5$, i.e., the query ϕ evaluates to false.

Now consider an MLN M' consisting of (10) with a weight w, and (11) with a weight $u = 1$. As w ranges in $]-\infty, \infty[$ one again has that the likelihood $P_{M'}(I \mid C_n)$ ranges in $]0, 1/\binom{n}{k}[$. Now, however, for all such M' $P_{M'}(a(c_1) \mid C = \{c_1\}) > 0.5$. Thus, one has that M and M' define two different Φ-equivalence classes, and $\boldsymbol{F}([M]_\Phi, D) = \boldsymbol{F}([M]_\Phi, D) =]0, 1/\binom{n}{k}[$.

For pure likelihood-based learning, i.e., in the absence of a bias feature $B(M)$, we thus obtain that ϕ is not identifiable. What happens if we were to add a bias feature that expresses a preference for syntactically simpler MLNs, as measured, for example, in terms of the number and length of formulas? MLNs containing formulas such as (10) would then obtain a low bias value $B(M)$. Simple MLNs with high $B(M)$ values, on the other hand, would most likely be unable to express the distribution that leads to the maximal likelihood value of $1/\binom{n}{k}$ for the data. Thus, condition (ii) would not be satisfied, and Proposition 1 does not establish non-identifiability for these scenarios.

Our examples show that Proposition 1 can be used to establish non-identifiability in some relevant cases, but it is far from being applicable in all cases. However, it appears that Proposition 1 is about as far as one can go without imposing some further restrictions on admissible score functions σ, or on the available data sequences $\uparrow D_n$.

4   Consistent Identifiability and PAC Learning

We now take a closer look at how Definition 3 can be strengthened by replacing the mere existence condition for a data sequence $\uparrow D_n$ with a condition only for the data-sequences that we are likely to see in practice. For this we now only consider spaces of probabilistic models that induce a distribution on the sample data.

Definition 6. *A model class $\mathcal{M}$ is probabilistic with associated stratified data space $\mathcal{D}$, if each $M \in \mathcal{M}$ defines*

- *a probability distribution $P_M^{(1)}$ on $\mathcal{D}^{(1)}$, and*
- *for each $n > 1$ a conditional distribution $P_M^{(n|n-1)}$ on $\mathcal{D}^{(n)}$ given $\mathcal{D}^{(n-1)}$, so that for $D \in \mathcal{D}^{(n-1)}$*

$$P_M^{(n|n-1)}(\{D' \mid D \prec D'\} \mid D) = 1.$$

The distributions $P_M^{(1)}, P_M^{(n|n-1)}$ jointly define probability distributions $P_M^{(n)}$ on $\mathcal{D}^{(n)}$ for all n.

We note that for continuous spaces $\mathcal{D}$ the above definition implicitly assumes some measurability conditions that we have not spelled out.

For probabilistic models we can now introduce *consistent identifiability*, which is an adaptation of the statistical concept of consistency for our type of learning scenario.

Definition 7. *Let $\mathcal{M}$ be a probabilistic model space with associated data space $\mathcal{D}$, σ a score function. Φ is consistently σ-identifiable, if for all $M \in \mathcal{M}$, all $\epsilon > 0$, and for all $\phi \in \Phi$ there exists $n_0 \geq 1$, such that for all $n \geq n_0$:*

$$P_M^{(n)}\{D_n \mid \mathcal{M}(\sigma, D_n) \subseteq [M]_{\equiv_\phi}\} \geq 1 - \epsilon \tag{12}$$

We can now formulate a positive identifiability result reported in [10]: for $\mathcal{M}^d$ the set of deterministic probabilistic finite automata, and Φ the class of PLTL queries: Φ is consistently σ-identifiable, where σ is an Alergia-style score function based on the model features *Tests* and *Lhood*. Comparing this result with Example 7 we find that PCTL is not identifiable in $\mathcal{M}^{nd}$ (under certain assumptions on the data and model features available for learning), whereas PLTL is identifiable in $\mathcal{M}^d$. This leaves as two open and important questions whether PCTL is identifiable in $\mathcal{M}^d$, or PLTL in $\mathcal{M}^{nd}$.

We close this section with some remarks on the relationship between condition (12) and *PAC*-learnability. For this assume that each $M \in \mathcal{M}$ defines a Boolean function f_M defined on a countable domain X, that $\mathcal{D}^{(n)}$ consists of sets of size n of pairs $(x, f_M(x))$ ($x \in X$), and that $\Phi = X$. In addition, let M also define a probability distribution P_M over X, which then induces a distribution on $\mathcal{D}^{(n)}$ for each n. Finally, assume that here $\mathcal{M}(\sigma, D)$ is a singleton, which we denote $M(D)$. From consistent identifiability in the sense of Definition 7 we then obtain that for all $\epsilon, \delta > 0$ there exists n_0, so that for all $n > n_0$

$$P_M^{(n)}\{D_n \mid P_M\{x \mid f_{M(D)}(x) \neq f_M(x)\} \leq \delta\} \geq 1 - \epsilon \tag{13}$$

To obtain (13) from (12) we first observe that there exists a finite subset $X' \subseteq X$ with $P_M(X') > 1 - \delta$. Assume $|X'| = N$, and set ϵ to ϵ/N in (12). Then, for sufficiently large n_0, (12) holds simultaneously for all $\phi = x \in X'$, i.e., for all $n \geq n_0$

$$P_M^{(n)}\{D_n \mid M(D) \in [M]_{X'}\} \geq 1 - \epsilon$$

Since $M(D) \in [M]_{X'}$ implies $P_M\{x \mid f_{M(D)}(x) \neq f_M(x)\} \leq \delta$, we obtain (13).

Property (13) is structurally similar to PAC-learnability [18,1]. One superficial difference between (13) and PAC-learnability is that the latter does not assume that there is an association between the (true) model (or hypothesis) and the distribution over X. Thus, where our definition says "for all M...", PAC is expressed in terms of "...for all h (hypotheses) and all μ (distributions on X) ...". This, however, is no real difference, since if our model space contains models for all possible combinations of functions f_M and distributions P_M, then the quantification over all M is the same as a quantification over all functions and all distributions. A second difference is much more significant: PAC is a uniform condition where the ϵ, δ-bounds are independent of h and μ, i.e., it is defined by the quantifier string $\forall \epsilon, \delta \exists n_0 \forall h, \mu \forall n > n_0 \ldots$, whereas our condition (13) is $\forall M, \epsilon, \delta \exists n_0 \forall n > n_0 \ldots$. Requiring in our Definition 7 a threshold n_0 that is uniform for M and ϕ would be too strong for most intended applications, since models M and queries ϕ may differ widely with respect to their complexity, and so it will be unrealistic to ask for uniform bounds on the necessary sample size for their identification.

5   Conclusion

We have developed a formal framework for analyzing learnability, or identifiability questions in learning scenarios where there may only be a loose association between model, data, and query space. The main contribution of this paper is to provide the conceptual tools for a rigorous and uniform analysis of a wide spectrum of such identifiability problems.

A key element of the proposed approach is to conceptualize learning as maximization of a score function with a dependence on a single distinguished model bias feature. Based on such score functions, we can formulate a first general sufficient condition for non-identifiability. This result is still somewhat limited in two ways: first, while easy to prove, the result is not necessarily easy to apply, since verifying conditions (i) and (ii) may require some non-trivial analysis in different applications. Second, Proposition 1 is a rather strong sufficient condition that may not actually be satisfied in many cases of non-identifiability. However, as Examples 4 and 8 indicate, it may prove difficult to obtain stronger results at the same high level of generality, and without restrictions to specific types of model spaces or learning approaches. A particularly relevant more specialized setting is that of consistent identifiability for probabilistic models. Future work will be focussed on obtaining more powerful analysis tools than Proposition 1 for this setting.

References

1. Angluin, D.: Queries and concept learning. Machine Learning 2, 319–342 (1988)
2. Baier, C., Katoen, J.P.: Principles of Model Checking. MIT Press (2008)
3. Carrasco, R., Oncina, J.: Learning stochastic regular grammars by means of a state merging method. In: Carrasco, R.C., Oncina, J. (eds.) ICGI 1994. LNCS, vol. 862, pp. 139–152. Springer, Heidelberg (1994)
4. Carrasco, R.C., Oncina, J.: Learning deterministic regular grammars from stochastic samples in polynomial time. In: ITA, pp. 1–20 (1999)
5. De Raedt, L., Frasconi, P., Kersting, K., Muggleton, S.H. (eds.): Probabilistic Inductive Logic Programming. LNCS (LNAI), vol. 4911. Springer, Heidelberg (2008)
6. Glymour, C., Spirtes, P., Richardson, T.: On the possibility of inferring causation from association without background knowledge. In: Glymour, C., Cooper, G.F. (eds.) Computation, Causation & Discovery, ch. 9, pp. 323–331. AAAI Press, MIT Press (1999)

7. Goodman, N.D., Mansinghka, V.K., Roy, D., Bonawitz, K., Tenenbaum, J.B.: Church: a language for generative models. In: Proceedings of the 24th Conference on Uncertainty in Artificial Intelligence, UAI 2008 (2008)
8. de la Higuera, C.: Grammatical Inference: Learning Automata and Grammars. Cambridge University Press (2010)
9. Korb, K., Nicholson, A.: The causal interpretation of Bayesian networks. In: Holmes, D., Jain, L. (eds.) Innovations in Bayesian Networks. SCI, vol. 156, pp. 83–116. Springer, Berlin (2008)
10. Mao, H., Chen, Y., Jaeger, M., Nielsen, T.D., Larsen, K.G., Nielsen, B.: Learning probabilistic automata for model checking. In: Proceedings of the 8th International Conference on Quantitative Evaluation of SysTems, QEST (2011)
11. Mihalkova, L., Huynh, T., Mooney, R.J.: Mapping and revising markov logic networks for transfer learning. In: Proc. of AAAI 2007 (2007)
12. Milch, B., Marthi, B., Russell, S., Sontag, D., Ong, D., Kolobov, A.: Blog: Probabilistic logic with unknown objects. In: Proceedings of the 19th International Joint Conference on Artificial Intelligence (IJCAI 2005), pp. 1352–1359 (2005)
13. Pearl, J.: Probabilistic Reasoning in Intelligent Systems: Networks of Plausible Inference, 2nd pr. edn. The Morgan Kaufmann series in representation and reasoning. Morgan Kaufmann, San Mateo (1988)
14. Richardson, M., Domingos, P.: Markov logic networks. Machine Learning 62(1-2), 107–136 (2006)
15. Robins, J.M., Wasserman, L.: On the impossibility of inferring causation from association without background knowledge. In: Glymour, C., Cooper, G.F. (eds.) Computation, Causation & Discovery, ch. 8, pp. 305–321. AAAI Press, MIT Press (1999)
16. Sen, K., Viswanathan, M., Agha, G.: Learning continuous time Markov chains from sample executions. In: Proceedings of the 1st International Conference on Quantitative Evaluation of SysTems (QEST), pp. 146–155 (2004)
17. Spirtes, P., Glymour, C., Scheines, R.: Causation, Prediction and Search. Springer (1993)
18. Valiant, L.G.: A theory of the learnable. Communications of the ACM 27(11), 1134–1142 (1984)

Exploratory Learning

Bhavana Dalvi, William W. Cohen, and Jamie Callan

School of Computer Science,
Carnegie Mellon University,
Pittsburgh, PA 15213
{bbd,wcohen,callan}@cs.cmu.edu

Abstract. In multiclass semi-supervised learning (SSL), it is sometimes the case that the number of classes present in the data is not known, and hence no labeled examples are provided for some classes. In this paper we present variants of well-known semi-supervised multiclass learning methods that are robust when the data contains an unknown number of classes. In particular, we present an "exploratory" extension of expectation-maximization (EM) that explores different numbers of classes while learning. "Exploratory" SSL greatly improves performance on three datasets in terms of F1 on the classes *with* seed examples—i.e., the classes which are expected to be in the data. Our Exploratory EM algorithm also outperforms a SSL method based non-parametric Bayesian clustering.

1 Introduction

In multiclass semi-supervised learning (SSL), it is sometimes the case that the number of classes present in the data is not known. For example, consider the task of classifying noun phrases into a large hierarchical set of categories such as "person", "organization", "sports team", etc., as is done in broad-domain information extraction systems (e.g., [5]). A sufficiently large corpus will certainly contain some unanticipated natural clusters—e.g., kinds of musical scales, or types of dental procedures. Hence, it is unrealistic to assume some examples have been provided for each class: a more plausible assumption is that an unknown number of classes exist in the data, and that labeled examples have been provided for some subset of these classes.

This raises the natural question: how robust are existing SSL methods to unanticipated classes? As we will show experimentally below, SSL methods can perform quite poorly in this setting: the instances of the unanticipated classes might be forced into one or more of the expected classes, leading to a cascade of errors in class parameters, and then to class assignments to other unlabeled examples. To address this problem, we present an "exploratory" extension of expectation-maximization (EM) which explores different numbers of classes while learning.

More precisely, in a traditional SSL task, the learner assumes a fixed set of classes $C_1, C_2, \ldots C_k$, and the task is to construct a k-class classifier using labeled datapoints X^l and unlabeled datapoints X^u, where X^l contains a (usually small) set of "seed" examples of each class. In exploratory SSL, we assume the same inputs, but allow the classifier to predict labels from the set $C_1, \ldots, C_m$, where $m \geq k$: in other words, every example x may be predicted to be in either a known class $C_i \in C_1 \ldots C_k$, or an unknown class $C_i \in C_{k+1} \ldots C_m$.

H. Blockeel et al. (Eds.): ECML PKDD 2013, Part III, LNAI 8190, pp. 128–143, 2013.
© Springer-Verlag Berlin Heidelberg 2013

We will show that exploratory SSL can greatly improve performance on noun-phrase classification tasks and document classification tasks, for several well-known SSL methods. E.g. Figure 1 (b) top row shows, the confusion matrices for a traditional SSL method on a 20-class problem at the end of iteration 1 and 15, when the algorithm is presented with seeds for 6 of the classes. Here, red indicates overlap between classes, and dark blue indicates no overlap. So we see that many of the seed classes are getting confused with the unknown classes at the end of 15 iterations of SSL showing semantic drift. With the same inputs, our novel "exploratory" EM algorithm performs quite well (Figure 1 (b) bottom row); i.e. it introduces additional clusters and at the end of 15 iterations improves F1 on classes for which seed examples were provided.

Contributions. We focus on the novel problem of dealing with learning when only fraction of classes are known upfront, and there are unknown classes hidden in the data. We propose a variant of the EM algorithm where new classes can be introduced in each EM iteration. We discuss the connections of this algorithm to the structural EM algorithm. Next we propose two heuristic criteria for predicting when to create new class during an EM iteration, and show that these two criteria work well on three publicly available datasets. Further we evaluate third criterion, that introduces classes uniformly at random and show that our proposed heuristics are more effective than this baseline. Experimentally, Exploratory EM outperforms a semi-supervised variant of non-parametric Bayesian clustering (Gibbs sampling with Chinese Restaurant Process)—a technique which also "explores" different numbers of classes while learning. We also compare our method against a semi-supervised EM method with m extra classes (trying different values of m).

In this paper, Exploratory EM is instantiated to produce exploratory versions of three well-known SSL methods: semi-supervised Naive Bayes, seeded K-Means, and a seeded version of EM using a von Mises-Fisher distribution [1]. Our experiments focus on improving accuracy on the classes that do have seed examples—i.e., the classes which are expected to be in the data.

Outline. In Section 2, we first introduce an exploratory version of EM, and then discuss several instantiations of it, based on different models for the classifiers (mixtures of multinomials, K-Means, and mixtures of von Mises-Fisher distributions) and different approaches to introducing new classes. We then compare against an alternative exploratory SSL approach, namely Gibbs sampling with Chinese restaurant process [14]. Section 3 presents experimental results, followed by related work and conclusions.

2   Exploratory SSL Methods

2.1   A Generic Exploratory Learner

Many common approaches to SSL are based on EM. In a typical EM setting, the M-step finds the best parameters θ to fit the data, $X^l \cup X^u$, and the E-step probabilistically labels the unknown points with a distribution over the known classes $C_1, C_2, \ldots C_k$. In some variants of EM, including the ones we consider here, a "hard" assignment is made to classes instead, an approach named *classification EM* [6]. Our exploratory version of EM differs in that it can introduce new classes $C_{k+1} \ldots C_m$ during the E-step.

Algorithm 1. EM algorithm for exploratory learning with model selection

---

1: **function Exploratory EM** $(X^l, Y^l, X^u, \{C_1 \ldots C_k\})$: $\{C_{k+1} \ldots C_m\}, \theta^m, Y^u$
2: **Input**: X^l labeled data points; Y^l labels for datapoints X^l; X^u unlabeled datapoints (same feature space as X^l); $\{C_1 \ldots C_k\}$ set of known classes to which x's belong.
3: **Output**: $\{C_{k+1} \ldots C_m\}$ newly-discovered classes; $\{\theta^1, \ldots, \theta^m\}$ parameters for all m classes; Y^u labels for unlabeled data points X^u
   {Initialize model parameters using labeled data}
4: $\theta_0^1, \ldots, \theta_0^k = argmax_\theta L(X^l, Y^l | \theta^k)$
5: i is # new classes ; $i = 0$; *CanAddClasses* = *true*
6: **while** data likelihood not converged AND #classes not converged **do**
   {E step: (Iteration t) Make predictions for the unlabeled data-points}
7:   $i_{old} = i$; Compute baseline log-likelihood *BaselineLL* $= log P(X | \theta_t^1, \ldots, \theta_t^{k+i_{old}})$
8:   **for** $x \in X^u$ **do**
9:     Predict $P(C_j | x, \theta_t^1, \ldots, \theta_t^{k+i})$ for all labels $1 \leq j \leq k+i$
10:     **if** nearlyUniform$(P(C_1|x), \ldots, P(C_{k+i}|x))$ AND *CanAddClasses* **then**
11:       Increment i; Let C_{k+i} be the new class.
12:       Label x with C_{k+i} in Y^u, and compute parameters θ_t^{k+i} for the new class.
13:     **else**
14:       Assign x to $(argmax_{C_j} P(C_j|x))$ in Y^u where $1 \leq j \leq k+i$
15:     **end if**
16:   **end for**
17:   $i_{new} = i$; Compute ExploreEM loglikelihood *ExploreLL* $= log P(X | \theta_t^1, \ldots, \theta_t^{k+i_{new}})$
   {M step : Recompute model parameters using current assignments for X^u}
18:   **if** Penalized data likelihood is better for exploratory model than baseline model **then**
     {Adopt the new model with $k + i_{new}$ classes}
19:     $\theta_{t+1}^{k+i_{new}} = argmax_\theta L(X^l, Y^l, X^u, Y_t^u | \theta^{k+i_{new}})$
20:   **else**
     {Keep the old model with $k + i_{old}$ classes}
21:     $\theta_{t+1}^{k+i_{old}} = argmax_\theta L(X^l, Y^l, X^u, Y_t^u | \theta^{k+i_{old}})$
22:     *CanAddClasses* = $false$
23:   **end if**
24: **end while**
25: **end function**

Algorithm 1 presents a generic Exploratory EM algorithm (without specifying the model being used). There are two main differences between the algorithm and standard classification-EM approaches to SSL. First, in the E step, each of the unlabeled datapoint x is either assigned to one of the existing classes, or to a newly-created class. We will discuss the "nearUniform" routine below, but the intuition we use is that a new class should be introduced to hold x when the probability of x belonging to existing classes is close to uniform. This suggests that x is not a good fit to any existing classes, and that adding x to any existing class will lower the total data likelihood substantially. Second, in the M-step of iteration t, we choose either the model proposed by Exploratory EM method that might have more number of classes than previous iteration

$t-1$ or the baseline version with same number of classes as iteration $t-1$. This choice is based on whether exploratory model satisfies a model selection criterion in terms of increased data likelihood and model complexity. If the algorithm decides that baseline model is better than exploratory model in iteration t, then from iteration $t+1$ onwards the algorithm won't introduce any new classes.

2.2 Discussion

Friedman [13] proposed the Structural EM algorithm that combines the standard EM algorithm, which optimizes parameters, with structure search for model selection. This algorithm learns networks based on penalized likelihood scores, in the presence of missing data. In each iteration it evaluates multiple models based on the expected scores of models with missing data, and selects the model with best expected score. This algorithm converges at local maxima for penalized log likelihood (the score includes penalty for increased model complexity).

Similar to Structural EM, in each iteration of Algorithm 1, we evaluate two models, one with and one without adding extra classes. These two models are scored using a model selection criterion like AIC or BIC, and the model with best penalized data likelihood score is selected in each iteration. Further when the model selection criterion fails, the algorithm reverts to standard semi-supervised EM algorithm. Say this model switch happens at iteration t_{switch}, then from iteration 1 to t_{switch}, Algorithm 1 acts like the structural EM algorithm [13]. From iteration $t_{switch}+1$ till the data likelihood converges, the algorithm acts as semi-supervised EM algorithm.

Next let us discuss the applicability of this algorithm for clustering as well as classification tasks. Notice that Algorithm 1 reverts to an unsupervised clustering method if X_l is empty, and reverts to a supervised generative learner if X_u is empty. Likewise, if no new classes are generated, then it behaves as a multiclass SSL method; for instance, if the classes are well-separated and X_l contains enough labels for every class to approximate these classes, then it is unlikely that the criterion of nearly-uniform class probabilities will be met, and the algorithm reverts to SSL. Henceforth we will use the terms "class" and "cluster" interchangeably.

2.3 Model Selection

For model penalties we tried multiple well known criteria like BIC, AIC and AICc. Burnham and Anderson [4] have experimented with AIC criteria and proposed AICc for datasets where, the number of datapoints is less than 40 times number of features. The formulae for scoring a model using each of the three criteria that we tried are:

$$BIC(g) = -2 * L(g) + v * ln(n) \tag{1}$$
$$AIC(g) = -2 * L(g) + 2 * v \tag{2}$$
$$AICc(g) = AIC(g) + 2 * v * (v+1)/(n-v-1) \tag{3}$$

where g is the model being evaluated, $L(g)$ is the log-likelihood of the data given g, v is the number of free parameters of the model and n is the number of data-points. While comparing two models, a lower value is preferred. The extended Akaike information criterion (AICc) suited best for our experiments since our datasets have large number of features and small number of data points. With AICc criterion, the objective function that Algorithm 1 optimizes is:

$$\max_{m,\{\theta^1...\theta^m\},m \geq k} \{\text{Log Data Likelihood} - \text{Model penalty}\}$$

i.e.,
$$\max_{m,\{\theta^1...\theta^m\},m \geq k} \{\log P(X|\theta^1, \ldots, \theta^m)\} - \{v + (v * (v + 1)/(n - v - 1))\} \quad (4)$$

Here, k is the number of seed classes given as input to the algorithm and m is the number of classes in the resultant model ($m \geq k$).

2.4   Exploratory Versions of Well-Known SSL Methods

In this section we will consider various SSL techniques, and propose exploratory extensions of these algorithms.

Semi-Supervised Naive Bayes. Nigam et al. [21] proposed an EM-based semi-supervised version of multinomial Naive Bayes. In this model $P(C_j|x) \propto P(x|C_j) * P(C_j)$, for each unlabeled point x. The probability $P(x|C_j)$ is estimated by treating each feature in x as an independent draw from a class-specific multinomial. In document classification, the features are word occurrences, and the number of outcomes of the multinomial is the vocabulary size. This method can be naturally used as an instance of Exploratory EM, using the multinomial model to compute $P(C_j|x)$ in Line 1. The M step is also trivial, requiring only estimates of $P(w|C_j)$ for each word/feature w.

Seeded K-Means. It has often been observed that K-Means and EM are algorithmically similar. Basu and Mooney [2] proposed a seeded version of K-Means, which is very analogous to Nigam et al's semi-supervised Naive Bayes, as another technique for semi-supervised learning. Seeded K-Means takes as input a number of clusters, and seed examples for each cluster. The seeds are used to define an initial set of cluster centroids, and then the algorithm iterates between an "E step" (assigning unlabeled points to the closest centroid) and an "M step" (recomputing the centroids).

In the seeded K-Means instance of Exploratory EM, we again define $P(C_j|x) \propto P(x|C_j) * P(C_j)$, but define $P(x|C_j) = x \cdot C_j$, i.e., the inner product of a vector representing x and a vector representing the centroid of cluster j. Specifically, x and C_j both are represented as L_1 normalized TFIDF feature vectors. The centroid of a new cluster is initialized with smoothed counts from x. In the "M step", we recompute the centroids of clusters in the usual way.

Seeded Von-Mises Fisher. The connection between K-Means and EM is explicated by Banerjee et al. [1], who described an EM algorithm that is directly inspired by K-Means and TFIDF-based representations. In particular, they describe generative cluster models based on the von Mises-Fisher (vMF) distribution, which describes data distributed on the unit hypersphere. Here we consider the "hard-EM" algorithm proposed by Banerjee et al, and use it in the seeded (semi-supervised) setting proposed by Basu et al. [2]. This natural extension of Banerjee et al[1]'s work can be extended to our exploratory setting.

As in seeded K-Means, the parameters of vMF distribution are initialized using the seed examples for each known cluster. In each iteration, we compute the probability

---

Algorithm 2. JS criterion for new class creation

1: **function JSCriterion**($[P(C_1|x) \ldots P(C_k|x)]$):
2: **Input**: $[P(C_1|x) \ldots P(C_k|x)]$ probability distribution of existing classes for a data point x
3: **Output:** *decision* : **true** iff new class needs to be created
4: $u = [1/k \ldots 1/k]$ {i.e., the uniform distribution with current number of classes = k}
5: *decision* = false
6: **if** Jensen-Shannon-Divergence(u, $P(C_j|x)$) $< \frac{1}{k}$ **then**
7:    *decision* = true
8: **end if**
9: **end function**

Algorithm 3. MinMax criterion for new class creation

1: **function MinMaxCriterion**($[P(C_1|x) \ldots P(C_k|x)]$):
2: **Input**: $[P(C_1|x) \ldots P(C_k|x)]$ probability distribution of existing classes for a data point x
3: **Output:** *decision* : **true** iff new class needs to be created
4: k is the current number of classes
5: $maxProb = \max(P(C_j|x))$; $minProb = \min(P(C_j|x))$
6: **if** $\frac{maxProb}{minProb} < 2$ **then**
7:    *decision* = true
8: **end if**
9: **end function**

---

of C_j given data point x, using vMF distribution, and then assign x to the cluster for which this probability is maximized. The parameters of the vMF distribution for each cluster are then recomputed in the M step. For this method, we use a TFIDF-based L_2 normalized vectors, which lie on the unit hypersphere.

Seeded vMF and seeded K-Means are closely related—in particular, seeded vMF can be viewed as a more probabilistically principled version of seeded K-Means. Both methods allow use of TFIDF-based representations, which are often preferable to unigram representations for text: for instance, it is well-known that unigram representations often produce very inaccurate probability estimates.

2.5 Strategies for Inducing New Clusters/Classes

In this section we will formally describe some possible strategies for introducing new classes in the E step of the algorithm. They are presented in detail in Algorithms 2 and 3, and each of these is a possible implementation of the "nearUniform" subroutine of Algorithm 1. As noted above, the intuition is that new classes should be introduced to hold x when the probabilities of x belonging to existing classes are close to uniform. In the JS criterion, we require that Jensen-Shanon divergence[1] between the posterior class distribution for x to the uniform distribution be less than $\frac{1}{k}$. The MinMax criterion is a somewhat simpler approximation to this intuition: a new cluster is introduced if the maximum probability is no more than twice the minimum probability.

---

[1] The Jensen-Shannon divergence between p and q is the average Kullback-Leiber divergence of p and q to a, the average of p and q, i.e., $\frac{1}{2}(KL(p||a + KL(q||a))$.

Algorithm 4. Exploratory Gibbs Sampling with Chinese Restaurant Process

---
 1: **function GibbsCRP** $(X^l, Y^l, X^u, \{C_1 \ldots C_k\}) : C_{k+1} \ldots C_m, Y^u$
 2: **Input**: X^l labeled data points; Y^l labels of X^l; X^u unlabeled data points;
    $\{C_1 \ldots C_k\}$ set of known classes x's belong to; P_{new} probability of creating a new class.
 3: **Output:** $C_{k+1} \ldots C_m$ newly-discovered classes; Y^u labels for X^u
 4: **for** x in X^u **do**
 5:     Save a random class from $\{C_1 \ldots C_k\}$ for x in Y^u
 6: **end for**
 7: Set $m = k$
 8: **for** $t : 1$ to $numEpochs$ **do**
 9:     **for** x_i in X^u **do**
10:         Let y_i's be x_i's label in epoch $t - 1$
11:         predict $P(C_j|x_i, Y^l \cup Y^u - \{y_i\})$ for all labels $1 \leq j \leq m$
12:         y_i', m' = CRPPick$(P_{new}, P(C_1|x_i), \ldots, P(C_{m+1}|x_i))$
13:         Save y_i' as x_i's label in epoch t
14:         $m = m'$
15:     **end for**
16: **end for**
17: **end function**

2.6   Baseline Methods

Next, we will take a look at various baseline methods that we implemented to measure the effectiveness of our proposed approach.

Random New Class Creation Criterion: To measure the effectiveness of criteria proposed in Algorithms 2 and 3, we experimented with a random baseline criterion, that returns "true" uniformly at random with probability equal to that of MinMax or JS criterion returning true for the same dataset. This is referred to as Random criterion below.

Semi-supervised EM with m Extra Classes: One might argue that the goal of the Exploratory EM algorithm can also be achieved by adding a random number of empty classes to the semi-supervised EM algorithm. We compare our method against the best possible value of this baseline, i.e. by choosing the number of classes that maximizes F1 on the seed classes. Note that in practice, the test labels are not available, so this is the upper bound on performance of this baseline. We compare our method with this upper bound in Section 3. Our method is different from this baseline in two ways. First, it does not need the number of extra clusters as input. Second, it seeds the extra clusters with those datapoints that are unlikely to belong to existing classes, as compared to initializing them randomly.

A Seeded Gibbs Sampler with CRP: The Exploratory EM method is broadly similar to non-parametric Bayesian methods, such as the Chinese Restaurant process (CRP) [14]. CRP is often used in non-parametric models (e.g., topic models) that are based on Gibbs sampling, and indeed, since it is straightforward to replace EM with Gibbs-sampling, one can use this approach to estimate the parameters of any of the models considered here (i.e., multinomial Naive Bayes, K-Means, and the von Mises-Fisher

---

Algorithm 5. Modified CRP criterion for new class creation

1: **function ModCRPPick** $(P_{new}, P(C_1|x), \ldots, P(C_{k+i}|x)) : y, i'$
2: **Input**: P_{new} probability of creating new class;
   $P(C_1|x), \ldots, P(C_{k+i}|x)$ probability of existing classes given x
3: **Output**: y class for x; i' new number of classes
4: $u = [1/k + i \ldots 1/k + i]$ {uniform distribution with $k + i$ classes}
5: $d = $ Jensen-Shannon-Divergence$(u, P(C_j|x))$
6: $q = \frac{P_{new}}{((k+i)*d)}$
7: **if** a coin with bias q is heads **then**
      {create a new class and assign to that}
8:    $y = k + i + 1$ and $i' = i + 1$
9: **else**
      {assign to an existing class}
10:   $i' = i$ and $y = $ sample from distribution $[P(C_1|x) \ldots P(C_k|x)]$
11: **end if**
12: **end function**

---

distribution). Algorithm 4 presents a seeded version of a Gibbs sampler based on this idea. In brief, Algorithm 4, starts with a classifier trained on the labeled data. Collapsed Gibbs sampling is then performed over the latent labels of unlabeled data, incorporating the CRP into the Gibbs sampling to introduce new classes. (In fact, we use block sampling for these variables, to make the method more similar to the EM variants.)

Note that this algorithm is naturally "exploratory", in our sense, as it can produce a number of classes larger than the number of classes for which seed labels exist. However, unlike our exploratory EM variants, the introduction of new classes is not driven by examples that are "hard to classify"—i.e., have nearly-uniform posterior probability of membership in existing classes. In CRP method, the probability of creating a new class depends on the data point, but it does not explicitly favor cases where the posterior over existing classes is nearly uniform.

To address this issue, we also implemented a variant of the seeded Gibbs sampler with CRP, in which the examples with nearly-uniform distributions are more likely to be assigned to new classes. This variant is shown in Algorithm 5, which replaces the routine CRPPick in the Gibbs sampler—in brief, we simply scale down the probability of creating a new class by the Jensen-Shannon divergence of the posterior class distribution for x to the uniform distribution. Hence the probability of creating new class explicitly depends on how well the given data point fits in one of the existing classes. An experimental comparison of our proposed method with Gibbs sampling and CRP based baselines is shown in Section 3.2.

3   Experimental Results

We now seek to experimentally answer the questions raised in the introduction. How robust are existing SSL methods, if they are given incorrect information about the number of classes present in the data, and seeds for only some of these classes? Do the exploratory versions of the SSL methods perform better? How does Exploratory EM compare with the existing "exploratory" method of Gibbs sampling with CRP?

Table 1. Comparison of Exploratory EM w.r.t. SemisupEM for different datasets and class creation criteria. For each exploratory method we report the macro avg. F1 over seed classes followed by avg number of clusters generated. e.g. For 20-Newsgroups dataset, Exploratory EM with K-Means and MinMax results in 57.4 F1 and generates 22 clusters on avg. ▲ (and △) indicates that improvements are statistically significant w.r.t SemisupEM with 0.05 (and 0.1) significance level.

Dataset (#seed / #total classes)	Algorithm	SemisupEM	Exploratory EM			Best m extra classes SemisupEM
			MinMax	JS	Random	
Delicious_Sports	KM	60.9	89.5 (30) ▲	**90.6** (46) ▲	84.8 (55) ▲	69.4 (10) ▲
(5/26)	NB	46.3	45.4 (06)	**88.4** (51) ▲	67.8 (38) ▲	65.8 (10) ▲
	VMF	64.3	72.8 (06) △	63.0 (06)	66.7 (06)	**78.2** (09) ▲
20-Newsgroups	KM	44.9	**57.4** (22) ▲	39.4 (99) ▼	53.0 (22) ▲	49.8 (11) ▲
(6/20)	NB	34.0	34.6 (07)	34.0 (06)	34.0 (06)	**35.0** (07)
	VMF	18.2	09.5 (09) ▼	19.8 (06)	18.2 (06)	**20.3** (10) ▲
Reuters	KM	8.9	12.0 (16) △	**27.4** (100) ▲	13.7 (19) ▲	16.3 (14) ▲
(10/65)	NB	6.4	10.4 (10)	**18.5** (77) ▲	10.6 (10)	16.1 (15)
	VMF	10.5	20.7 (11) ▲	**30.4** (62) ▲	10.5 (10)	20.6 (16) △

We used three publicly available datasets for our experiments. The first is the widely-used 20-Newsgroups dataset [23]. We used the "bydate" dataset, which contains total of 18,774 text documents, with vocabulary size of 61,188. There are 20 non-overlapping classes and the entire dataset is labeled. The second dataset is the Delicious_Sports dataset, published by [9]. This is an entity classification dataset, which contains items extracted from 57K HTML tables in the sports domain (from pages that had been tagged by the social bookmarking system del.icio.us). The features of an entity are ids for the HTML table columns in which it appears. This dataset contains 282 labeled entities described by 721 features and 26 non-overlapping classes (e.g., "NFL teams", "Cricket teams"). The third dataset is the Reuters-21578 dataset published by Cai et al. [10]. This corpus originally contained 21,578 documents from 135 overlapping categories. Cai et al. discarded documents with multiple category labels, resulting in 8,293 documents (vocabulary size=18,933) in 65 non-overlapping categories.

3.1 Exploratory EM vs. SemisupEM with Few Seed Classes

Table 1 shows the performance of seeded K-Means, seeded Naive Bayes, and seeded vMF using 5 different algorithms. For each dataset only a few of the classes present in the data (5 for Delicious_Sports, and 6 for 20-Newsgroups and 10 for Reuters), are given as seed classes to all the algorithms. Five percent datapoints were given as training data for each "seeded" class. The first method, shown in the column labeled SemisupEM, uses these methods as conventional SSL learners. The second method is Exploratory EM with the simple MinMax new-class introduction criterion, and the third is Exploratory EM with the JS criterion. Forth method is Exploratory EM with the Random criterion. The last one is upper bound on SemisupEM with m extra classes.

ExploreEM performs hard clustering of the dataset i.e. each datapoint belongs to only one cluster. For all methods, for each cluster we assign a label that maximizes accuracy (i.e. majority label for the cluster). Thus using complete set of labels we can generate a single label per datapoint. Reported Avg. F1 value is computed by macro

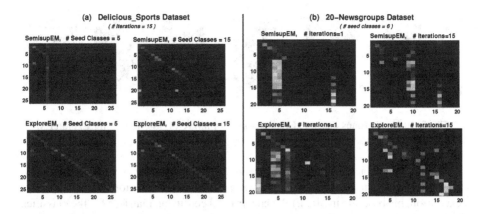

Fig. 1. (a) Confusion matrices, varying number of seed classes, for the Delicious_Sports dataset. (b) Confusion matrices, varying the number of EM iterations for the 20-Newsgroups dataset. Each is using Explore-KMeans with the MinMax criterion.

averaging F1 values of seed classes only. Note that, for a given dataset, number of seed classes and training percentage per seed class there are many ways to generate a train-test partition. We report results using 10 random train-test partitions of each dataset. The same partitions are used to run all the algorithms being compared and to compute the statistical significance of results.

We first consider the value of exploratory learning. With the JS criterion, the exploratory extension gives comparable or improved performance on 8 out of 9 cases. In 5 out of 8 cases the gains are statistically significant. With the simpler MinMax criterion, the exploratory extension results in performance improvements in 6 out of 8 cases, and significantly reduces performance only in one case. The number of classes finally introduced by the MinMax criterion is generally smaller than those introduced by JS criterion.

For both SSL and exploratory systems, the seeded K-Means method gives good results on all 3 datasets. In our MATLAB implementation, the running time of Exploratory EM is longer, but not unreasonably so: on average for 20-Newsgroups dataset Semisup-KMeans took 95 sec. while Explore-KMeans took 195 sec. and for Reuters dataset, Semisup-KMeans took 7 sec. while Explore-KMeans took 28 sec.

We can also see that Random criterion shows significant improvements over the baseline SemisupEM method in 4 out of 9 cases. While Exploratory EM method with MinMax and JS criterion shows significant improvements in 5 out of 9 cases. In terms of magnitude of improvements, JS is superior to Random criterion.

Next we compare Exploratory EM with baseline named "SemisupEM with m extra classes". The last column of Table 1 shows the best performance of this baseline by varying $m = \{0, 1, 2, 5, 10, 20, 40\}$, and choosing that value of m for which seed class F1 is maximum. Since the "best m extra classes" baseline is making use of the test labels to pick right number of classes, it cannot be used in practice; however Exploratory EM methods produce comparable or better performance with this strong baseline.

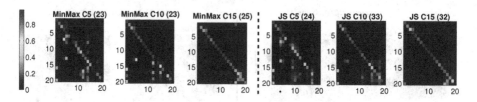

Fig. 2. 20-Newsgroups dataset : Comparison of MinMax vs. JS criterion for ExploreEM

To better understand the qualitative behavior of our methods, we conducted some further experiments with Semisup-KMeans with the MinMax criterion (which appears to be a reasonable baseline method.) We constructed confusion matrices for the classification task, to check how different methods perform on each dataset.[2] Figure 1 (a) shows the confusion matrices for SemisupEM (top row) and Exploratory EM (bottom row) with five and fifteen seeded classes. We can see that SemisupEM with only five seed classes clearly confuses the unexpected classes with the seed classes, while Exploratory EM gives better quality results. Having seeds for more classes helps both SemisupEM and Exploratory EM, but SemisupEM still tends to confuse the unexpected classes with the seed classes. Figure 1 (b) shows similar results on the 20-Newsgroups dataset, but shows the confusion matrix after 1 iteration and after 15 iterations of EM. It shows that SemisupEM after 15 iterations has made limited progress in improving its classifier when compared to Exploratory EM.

Finally, we compare the two class creation criteria, and show a somewhat larger range of seeded classes, ranging from 5 to 15 (out of 20 actual classes). In Figure 2 each of the confusion-matrices is annotated with the strategy, the number of seed classes and the number of classes produced. (E.g., plot "MinMax-C5(23)" describes Explore-KMeans with MinMax criterion and 5 seed classes which produces 23 clusters.) We can see that MinMax criterion usually produces a more reasonable number of clusters, closer to the ideal value of 20; however, performance of the JS method in terms of seed class accuracy is comparable to the MinMax method.

These trends are also shown quantitatively in Figure 3, which shows the result of varying the number of seeded classes (with five seeds per class) for Explore-KMeans and Semisup-KMeans; the top shows the effect on F1, and the bottom shows the effect on the number of classes produced (for Explore-KMeans only). Figure 4 shows a similar effect on the Delicious_Sports dataset: here we systematically vary the number of seeded classes (using 5 seeds per seeded class, on the top), and also vary the number of seeds per class (using 10 seeded classes, on the bottom.) The left-hand side compares the F1 for Semisup-KMeans and Explore-KMeans, and the right-hand side shows the number of classes produced by Explore-KMeans. For all parameter settings, Explore-KMeans is better than or comparable to Semisup-KMeans in terms of F1 on seed classes.

---

[2] For purposes of visualization, introduced classes were aligned optimally with the true classes.

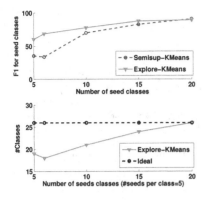 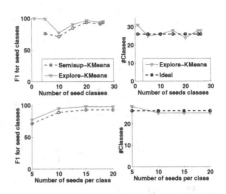

Fig. 3. 20-Newsgroups dataset: varying the number of seed classes (using the MinMax criterion)

Fig. 4. Delicious Sports dataset: Top, varying the number of seed classes (with five seeds per class). Bottom, varying the number of seeds per class (with 10 seed classes).

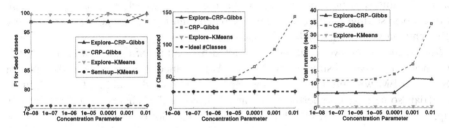

Fig. 5. Delicious_Sports dataset: Varying the concentration parameter, with five seed classes

3.2 Comparison with the Chinese Restaurant Process

As discussed in Section 2.6, a seeded version of the Chinese Restaurant Process with Gibbs sampling (CRPGibbs) is an alternative exploratory learning algorithm. In this section we compare the performance of CRPGibbs with Explore-KMeans and Semisup-KMeans. We consider two versions of CRP-Gibbs, one using the standard CRP and one using our proposed modified CRP criterion for new class creation that is sensitive to the near-uniformity of instance's posterior class distribution. CRP-Gibbs uses the same instance representation as our K-Means variants i.e. L_1 normalized TFIDF features.

It is well-known that CRP is sensitive to the concentration parameter P_{new}. Figures 5 and 6 show the performance of all the exploratory methods, as well as Semisup-KMeans, as the concentration parameter is varied from 10^{-8} to 10^{-2}. (For Explore-KMeans and Semisup-KMeans methods, this parameter is irrelevant). We show F1, the number of classes produced, and run-time (which is closely related to the number of classes produced.) The results show that a well-tuned seeded CRP-Gibbs can obtain good F1-performance, but at the cost of introducing many unnecessary clusters. The modified Explore-CRP-Gibbs performs consistently better, but not better than Explore-KMeans, and Semisup-KMeans performs the worst.

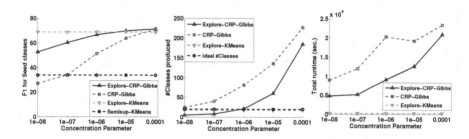

Fig. 6. 20-Newsgroups dataset: Varying concentration parameter, with six seed classes

4   Related Work

In this paper we describe and evaluate a novel multiclass SSL method that is more robust when there are unanticipated classes in the data—or equivalently, when the algorithm is given seeds from only some of the classes present in the data. To the best of our knowledge this specific problem has not been explored in detail before, even though in real-world settings, there can be unanticipated (and hence unseeded) classes in any sufficiently large-scale multiclass SSL task.

More generally, however, it has been noted before that SSL may suffer due to the presence of unexpected structure in the data. For instance, Nigam et al's early work on SSL based EM with multinomial Naive Bayes [21] noted that adding too much unlabeled data sometimes hurt performance on SSL tasks, and discusses several reasons this might occur, including the possibility that there might not be a one-to-one correspondence between the natural mixture components (clusters) and the classes. To address this problem, they considered modeling the positive class with one component, and the negative class with a mixture of components. They propose to choose the number of such components by cross-validation; however, this approach is relatively expensive, and inappropriate when there are only a small number of labeled examples (which is a typical case in SSL). More recently, McIntosh [18] described heuristics for introducing new "negative categories" in lexicon bootstrapping, based on a domain-specific heuristic for detecting semantic drift with distributional similarity metrics. Our setting is broadly similar to these works, except that we consider this task in a general multiclass-learning setting, and do not assume seeds from an explicitly-labeled "negative" class, which is a mixture; instead, we assume seeds from known classes only. Thus we assume that data fits a mixture model with a one-to-one correspondence with the classes, but only after the learner introduces new classes hidden in the data. We also explore this issue in much more depth experimentally, by systematically considering the impact of having too few seed classes, and propose and evaluate a solution to the problem. There has also been substantial work in the past to automatically decide the right "number of clusters" in unsupervised learning [11,22,15,7,19,27]. Many of these techniques are built around K-Means and involve running it multiple times for different values of K. Exploratory learning differs in that we focus on a SSL setting, and evaluate specifically the performance difference on the seeded classes, rather than overall performance differences.

There is also a substantial body of work on constrained clustering; for instance, Wagstaff et al [26] describe a constrained clustering variant of K-Means "must-link" and "cannot-link" constraints between pairs. This technique changes the cluster assignment phase of K-Means algorithm by assigning each example to the closest cluster such that none of the constraints are violated. SSL in general can be viewed as a special case of constrained clustering, as seed labels can be viewed as constraints on the clusters; hence exploratory learning can be viewed as a subtype of constrained clustering, as well as a generalization of SSL. However, our approach is different in the sense that there are more efficient methods for dealing with seeds than arbitrary constraints.

In this paper we focused on EM-like SSL methods. Another widely-used approach to SSL is label propagation. In the modified adsorption algorithm [25], one such graph-based label propagation method, each datapoint can be marked with one or more known labels, or a special dummy label meaning "none of the above". Exploratory learning is an extension that applies to a different class of SSL methods, and has some advantages over label propagation: for instance, it can be used for inductive tasks, not only transductive tasks. Exploratory EM also provides more information by introducing multiple "dummy labels" which describe multiple new classes in the data.

A third approach to SSL involves unsupervised dimensionality reduction followed by supervised learning (e.g., [8]). Although we have not explored their combination, these techniques are potentially complementary with exploratory learning, as one could also apply EM-like methods, in a lower-dimensional space (as is typically done in spectral clustering). If this approach were followed then an exploratory learning method like Exploratory EM could be used to introduce new classes, and potentially gain better performance, in a semi-supervised setting.

One of our benchmark tasks, entity classification, is inspired by the NELL (Never Ending Language Learning) system [5]. NELL performs broad-scale multiclass SSL. One subproject within NELL [20] uses a clustering technique for discovering new relations between existing noun categories—relations not defined by the existing hand-defined ontology. Exploratory learning addresses the same problem, but integrates the introduction of new classes into the SSL process. Another line of research considers the problem of "open information extraction", in which no classes or seeds are used at all [28,12,9]. Exploratory learning, in contrast, can exploit existing information about classes of interest and seed labels to improve performance.

Another related area of research is novelty detection. Topic detection and tracking task aims to detect novel documents at time t by comparing them to all documents till time $t - 1$ and detects novel topics. Kasiviswanathan et al. [16] assumes the number of novel topics is given as input to the algorithm. Masud et al. [17] develop techniques on streaming data to predict whether next data chunk is novel or not. Our focus is on improving performance of semi-supervised learning when the number of new classes is unknown. Bouveyron [3] worked on the EM approach to model unknown classes, but the entire EM algorithm is run for multiple numbers of classes. Our algorithm jointly learns labels as well as new classes. Schölkopf et al. [24] defines a problem of learning a function over the data space that isolates outliers from class instances. Our approach is different in the sense we do not focus on detecting outliers for each class.

5 Conclusion

In this paper, we investigate and improve the robustness of SSL methods in a setting in which seeds are available for only a subset of the classes—the subset of most interest to the end user. We performed systematic experiments on fully-labeled multiclass problems, in which the number of classes is known. We showed that if a user provides seeds for only some, but not all, classes, then SSL performance is degraded for several popular EM-like SSL methods (semi-supervised multinomial Naive Bayes, seeded K-Means, and a seeded version of mixtures of von Mises-Fisher distributions). We then described a novel extension of the EM framework called Exploratory EM, which makes these methods much more robust to unseeded classes. Exploratory EM introduces new classes on-the-fly during learning based on the intuition that hard-to-classify examples—specifically, examples with a nearly-uniform posterior class distribution—should be assigned to new classes. The exploratory versions of these SSL methods often obtained dramatically better performance—e.g., on Delicious_Sports dataset up to 90% improvements in F1, on 20-Newsgroups dataset up to 27% improvements in F1, and on Reuters dataset up to 200% improvements in F1. In comparative experiments, one exploratory SSL method, Explore-KMeans, emerged as a strong baseline approach.

Because Exploratory EM is broadly similar to non-parametric Bayesian approaches, we also compared Explore-KMeans to a seeded version of an unsupervised mixture learner that explores differing numbers of mixture components with the Chinese Restaurant process (CRP). Explore-KMeans is faster than this approach, and more accurate as well, unless the parameters of the CRP are very carefully tuned. Explore-KMeans also generates a model that is more compact, having close to the true number of clusters. The seeded CRP process can be improved, moreover, by adapting some of the intuitions of Explore-KMeans, in particular by introducing new clusters most frequently for hard-to-classify examples (those with nearly-uniform posteriors).

The exploratory learning techniques we described here are limited to problems where each data point belongs to only one class. An interesting direction for future research can be to develop such techniques for multi-label classification, and hierarchical classification. Another direction can be create more scalable parallel versions of Explore-KMeans for much larger datasets, e.g., large-scale entity-clustering task.

Acknowledgments. This work is supported in part by the Intelligence Advanced Research Projects Activity (IARPA) via Air Force Research Laboratory (AFRL) contract number FA8650-10-C-7058. The U.S. Government is authorized to reproduce and distribute reprints for Governmental purposes notwithstanding any copyright annotation thereon. This work is also partially supported by the Google Research Grant. The views and conclusions contained herein are those of the authors and should not be interpreted as necessarily representing the official policies or endorsements, either expressed or implied, of Google, IARPA, AFRL, or the U.S. Government.

References

1. Banerjee, A., Dhillon, I.S., Ghosh, J., Sra, S.: Clustering on the unit hypersphere using von mises-fisher distributions. In: JMLR (2005)
2. Basu, S., Banerjee, A., Mooney, R.: Semi-supervised clustering by seeding. In: ICML (2002)

3. Bouveyron, C.: Adaptive mixture discriminant analysis for supervised learning with unobserved classes (2010)
4. Burnham, K.P., Anderson, D.R.: Multimodel inference understanding aic and bic in model selection. Sociological Methods & Research (2004)
5. Carlson, A., Betteridge, J., Wang, R.C., Hruschka Jr., E.R., Mitchell, T.M.: Coupled semi-supervised learning for information extraction. In: WSDM (2010)
6. Celeux, G., Govaert, G.: A classification em algorithm for clustering and two stochastic versions. Computational Statistics & Data Analysis (1992)
7. Chiang, M.M.-T., Mirkin, B.: Intelligent choice of the number of clusters in k-means clustering: An experimental study with different cluster spreads. J. Classification (2010)
8. Dalvi, B., Cohen, W.: Very fast similarity queries on semi-structured data from the web. In: SDM (2013)
9. Dalvi, B., Cohen, W., Callan, J.: Websets: Extracting sets of entities from the web using unsupervised information extraction. In: WSDM (2012)
10. Deng Cai, X.W., He, X.: Probabilistic dyadic data analysis with local and global consistency. In: ICML (2009)
11. Dutta, H., Passonneau, R., Lee, A., Radeva, A., Xie, B., Waltz, D., Taranto, B.: Learning parameters of the k-means algorithm from subjective human annotation. In: FLAIRS (2011)
12. Etzioni, O., Cafarella, M., Downey, D., Kok, S., Popescu, A.-M., Shaked, T., Soderland, S., Weld, D.S., Yates, A.: Web-scale information extraction in knowitall. In: WWW (2004)
13. Friedman, N., Ninio, M., Pe'er, I., Pupko, T.: A structural em algorithm for phylogenetic inference. Journal of Computational Biology (2002)
14. Griffiths, D., Tenenbaum, M.: Hierarchical topic models and the nested chinese restaurant process. In: NIPS (2004)
15. Hamerly, G., Elkan, C.: Learning the k in k-means. In: NIPS (2003)
16. Kasiviswanathan, S.P., Melville, P., Banerjee, A., Sindhwani, V.: Emerging topic detection using dictionary learning. In: CIKM (2011)
17. Masud, M.M., Gao, J., Khan, L., Han, J., Thuraisingham, B.: Integrating novel class detection with classification for concept-drifting data streams. In: Buntine, W., Grobelnik, M., Mladenić, D., Shawe-Taylor, J. (eds.) ECML PKDD 2009, Part II. LNCS, vol. 5782, pp. 79–94. Springer, Heidelberg (2009)
18. McIntosh, T.: Unsupervised discovery of negative categories in lexicon bootstrapping. In: EMNLP (2010)
19. Menasce, D.A., Almeida, V.A.F., Fonseca, R., Mendes, M.A.: A methodology for workload characterization of e-commerce sites. In: EC (1999)
20. Mohamed, T., Hruschka Jr., E., Mitchell, T.: Discovering relations between noun categories. In: EMNLP (2011)
21. Nigam, K., McCallum, A., Thrun, S., Mitchell, T.: Text classification from labeled and unlabeled documents using em. Machine Learning (2000)
22. Pelleg, D., Moore, A., et al.: X-means: Extending k-means with efficient estimation of the number of clusters. In: ICML (2000)
23. Rennie, J.: 20-newsgroup dataset (2008)
24. Schölkopf, B., Williamson, R.C., Smola, A.J., Shawe-Taylor, J., Platt, J.: Support vector method for novelty detection. In: NIPS (2000)
25. Talukdar, P.P., Crammer, K.: New regularized algorithms for transductive learning. In: Buntine, W., Grobelnik, M., Mladenić, D., Shawe-Taylor, J. (eds.) ECML PKDD 2009, Part II. LNCS, vol. 5782, pp. 442–457. Springer, Heidelberg (2009)
26. Wagstaff, K., Cardie, C., Rogers, S., Schrodl, S.: Constrained k-means clustering with background knowledge. In: ICML (2001)
27. Welling, M., Kurihara, K.: Bayesian k-means as a maximization-expectation algorithm. In: ICDM (2006)
28. Yates, A., Cafarella, M., Banko, M., Etzioni, O., Broadhead, M., Soderland, S.: Textrunner: Open information extraction on the web. In: NAACL (2007)

Semi-supervised Gaussian Process Ordinal Regression

P.K. Srijith[1], Shirish Shevade[1], and S. Sundararajan[2]

[1] Computer Science and Automation, Indian Institute of Science, India
{srijith,shirish}@csa.iisc.ernet.in
[2] Microsoft Research, Bangalore, India
ssrajan@microsoft.com

Abstract. Ordinal regression problem arises in situations where examples are rated in an ordinal scale. In practice, labeled ordinal data are difficult to obtain while unlabeled ordinal data are available in abundance. Designing a probabilistic semi-supervised classifier to perform ordinal regression is challenging. In this work, we propose a novel approach for semi-supervised ordinal regression using Gaussian Processes (GP). It uses the expectation-propagation approximation idea, widely used for GP ordinal regression problem. The proposed approach makes use of unlabeled data in addition to the labeled data to learn a model by matching ordinal label distributions approximately between labeled and unlabeled data. The resulting mixed integer programming problem, involving model parameters (real-valued) and ordinal labels (integers) as variables, is solved efficiently using a sequence of alternating optimization steps. Experimental results on synthetic, bench-mark and real-world data sets demonstrate that the proposed GP based approach makes effective use of the unlabeled data to give better generalization performance (on the absolute error metric, in particular) than the supervised approach. Thus, it is a useful approach for probabilistic semi-supervised ordinal regression problem.

Keywords: Gaussian processes, ordinal regression, semi-supervised learning, annealing.

1 Introduction

We consider the problem of predicting variables of ordinal scale, a setting referred to as *ordinal regression*. These problems arise in many different domains like Social Sciences, Bioinformatics and Information Retrieval. For example, a user can label a retrieved document using one of the following categories: *highly relevant, relevant, average, irrelevant* and *highly irrelevant*. There exists an order among the labels, which makes the ordinal regression problems different from classification problems. Further, the labels are discrete and not continuous, unlike in the regression problems.

Although the problem of ordinal regression is well studied in Statistics [1,2,3], there has been a surge of interest, in recent years, in solving this problem in a

H. Blockeel et al. (Eds.): ECML PKDD 2013, Part III, LNAI 8190, pp. 144–159, 2013.

learning framework. The ordinal regression problem can be solved by treating it as a regression problem after transforming the ordinal scales into numeric values [4], or by converting it into nested binary classification problems that encode the ordering of the original ranks [5]. This solution strategy can be referred to as a reduction framework. Alternatively, the problem can be solved directly using machine learning algorithms like support vector machines (SVM) [6] or Gaussian Processes (GP) [7].

In many practical applications, labeled data are scarce to obtain. For example, in the domain of Bioinformatics, time consuming experiments and domain knowledge (biological experts) are required to label the data. Thus, obtaining the label information is expensive and time consuming. However, unlabeled data are easily available and are present in abundance. *Semi-supervised learning* [8] uses the unlabeled data along with the labeled data to learn better predictive models. Many approaches have been developed for the semi-supervised learning of regression and classification tasks. These approaches are based on various assumptions on the unlabeled data like clustering, smoothness or manifold [8]. They can be broadly classified as generative approaches, graph based approaches and approaches implementing low-density separation [8]. There exists a rich literature on semi-supervised regression and classification. See [8] and the references therein for more details. However, there is not much work reported in the literature to solve semi-supervised ordinal regression problem.

Semi-supervised ordinal regression problems arise quite naturally in several contexts. For instance, in recommendation systems, every user may rate only a few items. Often, the labeled ordinal data are insufficient to learn a good ordinal regression model. Most of the literature on ordinal regression [6,7,9,10,11,12] focused on the supervised learning setting. Recently, transductive ordinal regression (TOR) [13] approach was proposed to perform ordinal regression in a semi-supervised setting. The approach uses the reduction framework to solve the ordinal regression problem and learns the labels of the unlabeled examples and the decision function iteratively. The approach can be used for a general class of loss functions and was shown to give better performance than the approach which used only labeled examples. Semi-supervised manifold ordinal regression [14] is a new approach for semi-supervised ordinal regression for image ranking. This approach uses the assumption that is most appropriate for image analysis: the high dimensional observations lie on or close to a low-dimensional manifold. However, none of these approaches offer a solution to the semi-supervised ordinal regression problem in the Bayesian setting.

In the Bayesian setting, Bayesian committee machine [15] is one of the early attempts to solve a transductive regression problem using Gaussian processes. Though computationally expensive, it performs well on low noise data sets. Null category noise model [16] provides a semi-supervised approach to Gaussian process classification. A disadvantage of this approach is that the Gaussian approximation to the noise model can have negative variance. Semi-supervised Gaussian process classifiers [17] use a graph based approach to learn semi-supervised GP classifiers. It is based on using geometric properties of unlabeled data within globally defined kernel functions. It is extended to regression problems in [18]. They also propose a feedback mechanism in which the model is retrained by considering some unlabeled data and its predictions as labeled data. The Archipelago model [19] presents a generative approach for semi-supervised GP classification. It uses a GP to specify priors over label distribution and uses it along with a base distribution to model data distribution. More closely related to our work is the "Distribution Matching" approach for transductive regression and classification [20]. This approach is designed for a large margin setting. In a GP setting, similar ideas are used in [21] and [22] for transductive GP regression and multi-category classification, respectively. However, none of these transductive or semi-supervised GP based approaches are extended to semi-supervised ordinal regression problem.

Contributions: We propose a novel approach for semi-supervised ordinal regression using Gaussian Processes. GPs are non-parametric Bayesian models and provide a probabilistic kernel based approach for learning. Our method, hereafter abbreviated as SSGPOR, learns decision boundaries which pass through a low density region. The proposed approach is based on the assumption that the output distributions corresponding to labeled and unlabeled data are similar, a well founded assumption explored in the transductive classification and regression settings [20]. The proposed approach models the similarity by minimizing the Kullback-Leibler (KL) divergence between the predictive distribution over the unlabeled data outputs and an approximate distribution . The approximate distribution has properties similar to the labeled data output distribution. Obtaining the approximate distribution satisfying these properties is challenging. Our approach involves solving two sub-problems iteratively: (1) We learn the model by minimizing an upper bound on the negative logarithm of the evidence and the KL divergence, (2) we estimate the approximate distribution efficiently using the label switching method [23] that solves an underlying integer programming problem. To avoid bad local minima that typically arise with the unlabeled data in the semi-supervised setting, we use an annealing technique where the contribution of the unlabeled loss term is gradually increased [24].

Our method can be seen as an extension of the supervised Gaussian process ordinal regression approach using expectation propagation (EPGPOR) [7], to the semi-supervised setting. The EPGPOR approach is among the state-of-the-art approaches for ordinal regression. We compare the performance of the proposed SSGPOR approach with the EPGPOR approach. The experiments on synthetic, benchmark and real-world data sets show that, the performance of the EPGPOR approach could be significantly improved using our method when unlabeled data are available. It is also observed that the SSGPOR approach performs better than the TOR approach [13] in the transductive setting. Large improvements are observed on the absolute error metric than zero-one error metric. Note that unlike classification problems where zero-one error is important, absolute error metric is more meaningful in ordinal regression problems.

The rest of the paper is organized as follows. In Sect. 2, we introduce the Gaussian process and discuss the Gaussian process ordinal regression approach using expectation propagation (EPGPOR). Section 3 discusses the proposed approach, semi-supervised Gaussian process ordinal regression (SSGPOR), in detail. Comparisons of the SSGPOR, EPGPOR and TOR approaches on synthetic, benchmark and real-world data sets are presented in Sect. 4. Finally, some conclusions are drawn in Sect. 5.

We use the following notations for the discussion ahead. Given a sample of n_l labeled independent examples $\mathcal{D}_l = (X_l, \mathbf{y}_l) = \{(\mathbf{x}_i, y_i)\}_{i=1}^{n_l}$ and n_u unlabeled independent examples $\mathcal{D}_u = (X_u) = \{(\mathbf{x}_i)\}_{i=1}^{n_u}$. Let $\mathcal{D} = \mathcal{D}_l \cup \mathcal{D}_u$ denote the set of all training examples of size n ($n = n_l + n_u$). Let $\mathcal{D}_*$ be the set consisting of n_* test data points X_*. We assume $\mathbf{x}_i \in \mathbb{X} \subseteq \mathbb{R}^d$ and $y_i \in \mathbb{Y} = \{c_1, c_2, \dots, c_r\}$, where $c_1 < c_2 < \dots < c_r$. We consider an ordinal regression problem with r ordered categories and without loss of generality, we denote them by r consecutive integers $\{1, 2, \dots, r\}$. Our goal is to learn a decision function $h : \mathbb{X} \to \mathbb{Y}$ from both labeled and unlabeled data, such that it generalizes well on test data.

2 Background

A Gaussian process (GP) is a collection of random variables with the property that the joint distribution of any finite subset of the variables is a Gaussian [25]. It generalizes the Gaussian distribution to infinitely many random variables. The GP is used to define a prior distribution over latent functions underlying a model. It is completely specified by a mean function and a covariance function. The covariance function is defined over latent function values of a pair of input examples and is typically evaluated using the Mercer kernel function over the pair of input examples. The covariance function expresses some general properties of functions such as their smoothness, and length-scale. A commonly used covariance function is the squared exponential (SE) or the Gaussian kernel

$$cov(t_i, t_j) = k(\mathbf{x}_i, \mathbf{x}_j) = \exp(-\frac{\kappa}{2}\|\mathbf{x}_i - \mathbf{x}_j\|^2). \qquad (1)$$

Here $t_i = t(\mathbf{x}_i)$ and $t_j = t(\mathbf{x}_j)$ are latent function values associated with the inputs $\mathbf{x}_i$ and $\mathbf{x}_j$ respectively. $\kappa > 0$ is the hyper-parameter associated with the covariance function and $\|\cdot\|$ is the L_2 norm. The latent function sampled from a GP is denoted by t and in particular we denote the latent functions associated with labeled data as $\mathbf{t_l}$, unlabeled data as $\mathbf{t_u}$ and test data as $\mathbf{t_*}$. Let $K_{ll} = k(X_l, X_l)$, $K_{l*} = k(X_l, X_*)$ and $K_{**} = k(X_*, X_*)$. Here $k(X_l, X_*)$ is an $n_l \times n_*$ matrix of covariances evaluated at all pairs of labeled training and test input data. The matrices $k(X_l, X_l)$, $K(X_*, X_l)$ and $K(X_*, X_*)$ are defined similarly.

Gaussian Process Ordinal Regression: The Gaussian process ordinal regression (GPOR) [7] approach uses a non Gaussian likelihood function for modeling the ordinal labels. It uses a zero mean Gaussian process prior on the latent function values $t(\mathbf{x})$. Under noisy observations, for an input $\mathbf{x}$, the likelihood function for an ordinal output y is defined as

$$p(y|t(\mathbf{x})) = \Phi\left(\frac{b_y - t(\mathbf{x})}{\sigma}\right) - \Phi\left(\frac{b_{y-1} - t(\mathbf{x})}{\sigma}\right), \qquad (2)$$

where σ is the standard deviation of the Gaussian noise and Φ is the Gaussian cumulative distribution function i.e. $\Phi(z) = \int_{-\infty}^{z} \mathcal{N}(\delta; 0, 1)d\delta$. The thresholds $b_0, b_1, \ldots, b_r \in \mathcal{R}$ ($b_0 \le b_1 \le \ldots \le b_r$ where $b_0 = -\infty$ and $b_r = \infty$) are fixed so that the likelihood function represents a valid probability distribution over the ordinal outputs. The thresholds $b_1 \le b_2 \le \ldots \le b_{r-1}$ divide a real line into r contiguous intervals. A real latent function value is mapped to a discrete ordinal output based on the interval in which it lies. The likelihood (2) is not a Gaussian and therefore the posterior, $p(\mathbf{t_l}|\mathcal{D}_l)$, could not be obtained in closed form. The GPOR approach works by approximating the posterior as a Gaussian distribution using either Laplace approximation (MAPGPOR) or using expectation propagation (EPGPOR).

Learning: The Expectation propagation (EP) [26] approach approximates the posterior $p(\mathbf{t_l}|\mathcal{D}_l) \propto \prod_{i=1}^{n_l} p(y_i|t_i)p(\mathbf{t_l})$ as a product of Gaussian distributions $r(\mathbf{t_l}; \mathbf{h}, A) = \prod_{i=1}^{n_l} \hat{p}(t_i)p(\mathbf{t_l})$, where $\hat{p}(t_i) = s_i \exp(-\frac{1}{2}p_i(t_i - m_i)^2)$, $A = (K_{ll}^{-1} + \Pi)^{-1}$, and $\mathbf{h} = A\Pi\mathbf{m}$. Here, Π is a $n_l \times n_l$ diagonal matrix with elements in the diagonal given by $\{p_i\}_{i=1}^{n_l}$ and $\mathbf{m}$ is a n_l dimensional column vector with elements given by $\{m_i\}_{i=1}^{n_l}$. The parameters $\{s_i, m_i, p_i\}_{i=1}^{n_l}$ are called the site parameters of the EP approximation. The site parameters are obtained iteratively where in each iteration i, $\{s_i, m_i, p_i\}$ are obtained by minimizing the Kullback-Leibler (KL) divergence [8], $KL(r_{-i}(t_i)p(y_i|t_i) \| r_{-i}(t_i)\hat{p}(t_i))$. Here $r_{-i}(t_i)$ is the marginal cavity distribution over t_i obtained after leaving out the ith likelihood term $\hat{p}(y_i|t_i)$ from the approximated posterior $r(\mathbf{t_l})$ and then marginalizing over the remaining variables.

The EPGPOR approach performs model selection by minimizing an upper bound ($\mathcal{F}(\theta)$) on the negative logarithm of evidence ($p(\mathcal{D}_l|\theta)$) ,

$$\operatorname*{argmin}_{\theta} \mathcal{F}(\theta) = \operatorname*{argmin}_{\theta} - \sum_{i=1}^{n_l} \int r(t_i; h_i, A_{ii}) \log(\phi(\frac{b_{y_i} - t_i}{\sigma}) - \phi(\frac{b_{y_i-1} - t_i}{\sigma}))dt_i$$

$$+ \frac{1}{2}\log|I + K_{ll}\Pi| + \frac{1}{2}tr(I + K_{ll}\Pi)^{-1}$$

$$+ \frac{1}{2}\mathbf{m}^{\top}(K_{ll} + \Pi^{-1})^{-1}K_{ll}(K_{ll} + \Pi^{-1})^{-1}\mathbf{m} \quad (3)$$

where θ is the model parameter vector which includes the kernel parameter κ in the covariance function, the threshold parameters $(b_1, b_2, \ldots, b_{r-1})$ and the noise parameter σ in the likelihood function. Here, $tr(B)$ denotes the trace of the matrix B. The optimization can be done using any standard gradient based techniques like conjugate gradient. During optimization, for every new model parameter values, the site parameters and the approximated posterior $r(\mathbf{t_l})$ are re-estimated using the EP approach.

Prediction: The learnt model parameters and the EP approximated posterior are used to make predictions on test data. The predictive distribution of the latent function t_* for a test data $\mathbf{x}_*$ is $p(t_*|\mathbf{x}_*, \mathcal{D}_l) \sim N(t_*; \mu_*, \sigma_*^2)$, where $\mu_* = K_{l*}^{\top}(K_{ll} + \Pi^{-1})^{-1}\mathbf{m}$ and $\sigma_*^2 = K_{**} - K_{l*}^{\top}(K_{ll} + \Pi^{-1})^{-1}K_{l*}$. The predictive distribution for test output is $p(y_*|\mathbf{x}_*, \mathcal{D}_l) = \phi\left(\frac{b_{y_*}-\mu_*}{\sqrt{\sigma^2+\sigma_*^2}}\right) - \phi\left(\frac{b_{y_*-1}-\mu_*}{\sqrt{\sigma^2+\sigma_*^2}}\right)$.

The EPGPOR approach is a supervised approach. It does not perform well when the size of the labeled data are small. In most of the practical scenarios, labeled data are limited while unlabeled data are available in abundance. We propose a semi-supervised approach which extends the EPGPOR approach to a semi-supervised setting. The proposed approach make use of the unlabeled data along with the labeled data to learn a better decision function than the EPGPOR approach.

3 Semi-supervised Gaussian Process Ordinal Regression

The proposed approach, semi-supervised Gaussian process ordinal regression (SSGPOR), is based on the idea of "Distribution Matching" [20,21,22] and is derived by extending the transductive GP regression (TGPR) [21] approach to the ordinal regression setting. The basic assumption is that the predictive distribution on unlabeled data should have properties similar to the output distribution on labeled data. In particular, it requires the average number of examples for an ordinal category in unlabeled data should match approximately with the average number of examples for that category in labeled data. The assumption is justified by the independent and identically distributed (i.i.d.) nature of the data and is true for many real-world data sets [21]. The model parameters are estimated subject to these assumptions. It results in distributions which are consistent across labeled and unlabeled data. We now briefly describe the TGPR approach and then explain the proposed approach in detail.

The TGPR approach [21] models the regression problem where the output is real valued and the likelihood is a Gaussian. It considers a transductive setting where the training data set is $\mathcal{D}_l \cup \mathcal{D}_u$ and the designed GP model is used to predict the labels of the examples in $\mathcal{D}_u$. The TGPR approach requires the predictive Gaussian distribution over unlabeled data to be close to a family of Gaussian distributions $\hat{Q}$. The family $\hat{Q}$ is such that the first and second moments of its members on unlabeled data are close to the corresponding moments obtained using labeled data. The model parameters $(\hat{\theta})$ are obtained by minimizing the negative logarithm of evidence $(p(\mathcal{D}_l|\hat{\theta}))$, subject to the constraint that the predictive distribution over unlabeled data $p(\mathbf{y}_u|\mathcal{D}_l, \mathcal{D}_u, \hat{\theta})$, belongs to the approximating family $\hat{Q}$. The constraint could be enforced by minimizing the Kullback-Leibler (KL) divergence between $p(\mathbf{y}_u|\mathcal{D}_l, \mathcal{D}_u, \hat{\theta})$ and some $\hat{q} \in \hat{Q}$ [21]. The model parameters $(\hat{\theta})$ and $\hat{q} \in \hat{Q}$ are estimated by solving the joint optimization problem ;

$$\underset{\hat{q} \in \hat{Q}, \hat{\theta}}{\text{argmin}} \; -\log p(\mathcal{D}_l|\hat{\theta}) + \lambda \, KL(\hat{q}(\mathbf{y}_u)||p(\mathbf{y}_u|\mathcal{D}_l, \mathcal{D}_u, \hat{\theta})). \tag{4}$$

Here, λ is a regularization parameter and for two distributions q and p, $KL(q||p)$ $= \int q(y) \log \frac{q(y)}{p(y)} dy$. The parameters are obtained using an alternating optimization approach [21].

It is not easy to extend the TGPR approach to the ordinal regression setting. This is due to the nature of the labels and the likelihood. In ordinal regression, the labels are discrete and ordered. Further, the likelihood is non-Gaussian. Since labels are discrete and ordered, we have to consider a discrete approximating distribution. Because of the non-Gaussian nature of the likelihood, we have to use approximation techniques like expectation propagation to obtain a Gaussian approximated posterior [7]. The discrete nature of the labels results in an integer programming problem which needs to be solved efficiently. We now give the details of the proposed approach.

Proposed Approach: The SSGPOR approach considers the setting where the training data set is $\mathcal{D}_l \cup \mathcal{D}_u$ and the designed GP model is tested on $\mathcal{D}_*$. It uses the likelihood (2) and the expectation propagation approach [7], to obtain a Gaussian approximation of the posterior distribution. The resulting predictive distribution on an ordinal output y_u of an unlabeled example $\mathbf{x}_u \in \mathcal{D}_u$ is given as

$$p(y_u|\mathbf{x}_u, \mathcal{D}_l) = \phi\left(\frac{b_{y_u} - \mu_u}{\sqrt{\sigma^2 + \sigma_u^2}}\right) - \phi\left(\frac{b_{y_u-1} - \mu_u}{\sqrt{\sigma^2 + \sigma_u^2}}\right), \quad y_u = 1, \dots, r \qquad (5)$$

where $\mu_u = K_{lu}^\top (K_{ll} + \Pi^{-1})^{-1}\mathbf{m}$ and $\sigma_u^2 = K_{uu} - K_{lu}^\top (K_{ll} + \Pi^{-1})^{-1}K_{lu}$.

The SSGPOR approach requires the predictive distribution (5) over the unlabeled data to have some properties similar to the output distribution over the labeled data. We achieve this by considering an approximate distribution over the unlabeled data output with properties similar to the labeled data output distribution, and constrain the predictive distribution to be close to the approximate distribution. Since outputs are discrete in the ordinal regression setting, the approximate distribution takes the form of a multinomial distribution. In particular, we consider a multinomial distribution with r categories such that probability of success, p_j, for each category is defined by the average number of examples of that category in labeled data, i.e. $p_j = \gamma_j$, where $\gamma_j = \frac{1}{n_l}\sum_{i=1}^{n_l} \mathbb{I}(y_i = j)$ ($\mathbb{I}(\cdot)$ is an Indicator function). We define a label matrix q of size $n_u \times r$, where each row q_i is an i.i.d. random vector following the multinomial distribution for a single trial and provides a label for the ith unlabeled example. The ith unlabeled example is assigned a label j, if $q_{ij} = 1$. We have $q_{ij} \in \{0,1\}$ and $\sum_{j=1}^r q_{ij} = 1 \ \forall i = 1, \dots, n_u$. Also, q satisfies the label constraints $\frac{1}{n_u}\sum_{i=1}^{n_u} q_{ij} = \gamma_j \ \forall j = 1, \dots, r$, which ensures that the distribution over the unlabeled data are similar to the labeled data distribution. The label constraints are important in a semi-supervised setting as they avoid trivial solutions like assigning all unlabeled data to a single category [8]. Let Q be the set of all q satisfying all these constraints, i.e. $Q = \{q : q \in \{0,1\}^{n_u \times r}, \sum_{j=1}^r q_{ij} = 1 \ \forall i, \frac{1}{n_u}\sum_{i=1}^{n_u} q_{ij} = \gamma_j \ \forall j\}$. The SSGPOR approach requires the predictive distribution over all the unlabeled data $p(\mathbf{y}_u|\mathcal{D}_l, \mathcal{D}_u)$ to be close enough to some $q \in Q$. This can be achieved by minimizing the KL-divergence between q and $p(\mathbf{y}_u|\mathcal{D}_l, \mathcal{D}_u)$. Since obtaining the joint distribution $p(\mathbf{y}_u|\mathcal{D}_l, \mathcal{D}_u)$ is difficult, we instead minimize the sum of the KL divergence between q_u and $p(y_u|\mathbf{x}_u, \mathcal{D}_l)$ over all unlabeled examples.

Objective Function: The SSGPOR approach estimates the model parameters $\theta = (b_1, b_2, \dots, b_{r-1}, \kappa, \sigma)$ and $q \in Q$, by minimizing the upper bound on the negative logarithm of evidence (3) and the sum of the KL-divergences over all unlabeled data. It results in the following joint optimization problem;

$$\underset{q \in Q, \theta}{\operatorname{argmin}} \ \mathcal{F}(\theta) + \lambda \sum_{i=1}^{n_u} KL(q_i \| p(y_i | \mathbf{x}_i, \mathcal{D}_l, \theta)). \tag{6}$$

Here, the variable λ serves as a regularization parameter determining the importance that should be given to the unlabeled data term. The model parameters θ and q are obtained by an alternating optimization approach. It is an iterative approach, where in each iteration, we first solve the model parameters keeping q fixed. Then, we estimate $q \in Q$ keeping the model parameters fixed.

Alternating Optimization

(i) **Estimating θ** For a fixed q, the model parameters (θ) are obtained as

$$
\begin{aligned}
\underset{\theta}{\operatorname{argmin}} \quad & \frac{1}{n_l}\mathcal{F}(\theta) - \lambda \frac{1}{n_u} \sum_{i=1}^{n_u} KL(q_i \| p(y_i | \mathbf{x}_i, \mathcal{D}_l, \theta)) \\
= \underset{b_1, \ldots, b_r, \sigma^2, \kappa}{\operatorname{argmin}} \quad & -\frac{1}{n_l} \sum_{i=1}^{n_l} \int r(t_i; h_i, A_{ii}) \log(\phi(\frac{b_{y_i} - t_i}{\sigma}) - \phi(\frac{b_{y_i-1} - t_i}{\sigma})) dt_i \\
& -\lambda \frac{1}{n_u} \sum_{i=1}^{n_u} \sum_{j=1}^{r} q_{ij} \log(\phi(\frac{b_j - \mu_i}{\sqrt{\sigma^2 + \sigma_i^2}}) - \phi(\frac{b_{j-1} - \mu_i}{\sqrt{\sigma^2 + \sigma_i^2}})) + \frac{1}{2} \log|I + K_{ll}\Pi| \\
& +\frac{1}{2} tr((I + K_{ll}\Pi)^{-1}) + \frac{1}{2} \mathbf{m}^\top (K_{ll} + \Pi^{-1})^{-1} K_{ll} (K_{ll} + \Pi^{-1})^{-1} \mathbf{m} \\
& \text{s.t. } b_1 \leq \ldots \leq b_r
\end{aligned} \tag{7}
$$

This problem can be converted to an unconstrained optimization problem and can be solved using any standard optimization technique like conjugate gradient. During optimization, the site parameters and the approximated posterior $r(\mathbf{t}_1)$ are re-estimated using the EP approach.

(ii) **Estimating q** For fixed model parameters, q is estimated by minimizing the sum of the KL-divergences over all the unlabeled data subject to the constraint that $q \in Q$. It results in the following optimization problem.

$$
\begin{aligned}
\underset{q \in \{0,1\}^{n_u \times r}}{\operatorname{argmin}} \quad & -\sum_{i=1}^{n_u} \sum_{j=1}^{r} q_{ij} \log(\phi(\frac{b_j - \mu_i}{\sqrt{(\sigma^2 + \sigma_i^2)}}) - \phi(\frac{b_{j-1} - \mu_i}{\sqrt{(\sigma^2 + \sigma_i^2)}})) \\
\text{s.t. } & \frac{1}{n_u} \sum_{i=1}^{n_u} q_{ij} = \gamma_j \ \forall j = 1, \ldots, r \ , \ \sum_{j=1}^{r} q_{ij} = 1 \ \forall i = 1, \ldots, n_u
\end{aligned} \tag{8}
$$

Estimation of q is a binary integer programming problem and is done efficiently using the label switching algorithm [23].

We now discuss the proposed SSGPOR algorithm to solve (6) in detail.

Algorithm: The SSGPOR algorithm (Algorithm 1) consists of two parts: (i) initialization part (steps 2 and 3) and (ii) iterative part (steps 4–9).

The initialization of model parameters θ (step 2) is done by solving the supervised learning problem using the EPGPOR approach on labeled data, $\mathcal{D}_l$. It is then used to initialize the label matrix q (step 3) so that constraints are satisfied. This is done as follows. The initialized model parameters are used to find the prediction probability (5) for every category of unlabeled data. For every category, the unlabeled data examples are ranked based on the descending order of their prediction probability for that category. Starting from category 1 to r, the top ranked unlabeled data examples are assigned to the respective categories such that the number of examples assigned to each category does not exceed the expected number $(n_u \times \gamma_j)$. Care should be taken to remove examples from the sorted list corresponding to other categories, once they have been assigned to a particular category.

The iterative part of the algorithm corresponds to solving the problem (6) for different values of the regularization parameter λ. To avoid drastic switching of the labels in q, λ is varied from a small value to a final value 1 in annealing steps. That is, little importance is given to the unlabeled examples in the beginning ($\lambda = 10^{-3}$) and the importance of the unlabeled examples is increased gradually as λ is increased. This helps the algorithm to avoid poor local minima and achieve better performance. Step 4 of Algorithm 1 corresponds to this outer loop.

The inner loop (steps 5–8) does alternating minimization of θ and q in (6), for a given λ. In particular, optimization of θ (or q) for a fixed q (or θ) corresponds to solving (7) (or (8)). This alternating minimization procedure is repeated until no label switching happens. Algorithm 1 can be made more efficient by ensuring that steps 6 and 7 use the most recent θ and q as the starting points. For step 6, we employed the standard conjugate gradient method to solve (7), by converting it to an unconstrained optimization problem. For step 7, the label switching algorithm [23] was used.

The label switching algorithm assumes that the constraints are satisfied initially. It then proceeds by switching the labels of a pair of examples from two consecutive categories if the objective function decreases after such switching. The algorithm greedily performs as many such switches as possible for every consecutive categories. The pairwise switching of labels ensures that the constraints are satisfied throughout the label switching algorithm. The algorithm converges after a few iterations and the overall cost is proportional to $\mathcal{O}(n_u r)$.

Algorithm 1. SSGPOR Algorithm

1: **procedure** SSGPOR(D_l, D_u)
2:     Initialize θ by solving (3).
3:     Initialize the label matrix q.
4:     **for** $\lambda = \{10^{-3}, 3 \times 10^{-3}, 10^{-2}, 3 \times 10^{-2}, 10^{-1}, 3 \times 10^{-1}, 1\}$ **do**
5:         **repeat**
6:             Estimate θ by solving the optimization problem (7) for fixed q.
7:             Estimate q by solving the optimization problem (8) for fixed θ.
8:         **until** q is unchanged during step 7
9:     **end for**
10: **return** θ
11: **end procedure**

---

4   Experimental Results

We perform experiments on synthetic, benchmark and real-world data sets to compare the performance of the proposed SSGPOR approach (in the semi-supervised setting) with the EPGPOR approach. The EPGPOR approach is a supervised approach and does not use unlabeled data. We also compare the SS-GPOR approach with the transductive ordinal regression (TOR) [13] approach. For brevity, we refer to these approaches as EPGPOR, SSGPOR and TOR. TOR used a transductive setting and therefore, for fair comparison, we also used SS-GPOR in the transductive setting. The SSGPOR and EPGPOR approaches use the Gaussian kernel (1) in all the experiments. First, we conduct experiments on a synthetic data set to visualize the decision boundaries obtained using EPG-POR and SSGPOR. The generalization performance of the models is studied on several benchmark data sets. Finally, the effectiveness of SSGPOR is demonstrated on a real-world sentiment data set.

The generalization performance is compared using two metrics, *zero-one error* and *absolute error* [7]. Let the actual test outputs be $\{y_1, \ldots, y_{n_*}\}$ and the predicted test outputs be $\{\hat{y}_1, \ldots, \hat{y}_{n_*}\}$. Then the *zero-one error* and *absolute error* are defined as follows.

zero-one error gives the fraction of incorrect predictions on test data *i.e.* $\frac{1}{n_*} \sum_{i=1}^{n_*} \mathbb{I}(\hat{y}_i \neq y_i)$, where $\mathbb{I}(\cdot)$ is an indicator function.
absolute error gives the average deviation of predicted outputs from the actual outputs *i.e.* $\frac{1}{n_*} \sum_{i=1}^{n_*} |\hat{y}_i - y_i|$, where $|\cdot|$ denotes the absolute function.

Ordinal regression problems require the predicted category to be close enough to the actual category. The *absolute error* captures this and hence, it is more meaningful than the *zero-one error* for ordinal regression problems. One prefers approaches with low *zero-one* and *absolute* errors.

Synthetic Data: We conduct experiments on a two dimensional synthetic data set to visualize the decision boundaries obtained using EPGPOR and SSGPOR. The data set consists of three ordinal categories with 10 labeled examples and 100 unlabeled examples in each category. The labeled and unlabeled data for each category were generated from a Gaussian distribution with different mean and covariance. We consider two synthetic data sets. In the first, the labeled data distribution is similar to the unlabeled data distribution while in the second, they are different. The decision boundaries obtained using SSGPOR and EPGPOR for the two data sets are depicted in Fig. 1a and Fig. 1b. The decision boundary is the predictive mean value indexed by the thresholds. Table 1 provides the zero-one and absolute errors on the unlabeled data using EPGOR and SSGPOR for both the synthetic data sets. The zero-one and absolute errors are the same in this experiment because error occurred only between the neighboring classes.

In Fig. 1a, where labeled and unlabeled data distributions are similar, both SSGPOR and EPGPOR are able to learn decision boundaries passing through a low density region. In Fig. 1b, where the labeled data distribution differs from the unlabeled data distribution, SSGPOR learns a better decision boundary

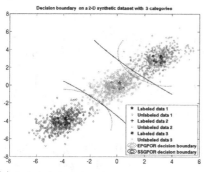
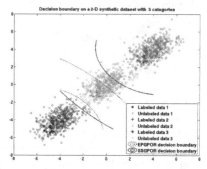

(a) Labeled data distribution similar to unlabeled data distribution

(b) Labeled data distribution not similar to unlabeled data distribution

Fig. 1. The decision boundaries obtained with SSGPOR and EPGPOR on a 2-dimensional synthetic data set with 3 ordinal categories

Table 1. Zero-one and absolute errors on the synthetic dataset using EPGPOR and SSGPOR. The numbers in bold face style indicate the best results.

Method	distributions similar		distributions different	
	zero-one	absolute	zero-one	absolute
EPGPOR	0.0456	0.0456	0.1489	0.1489
SSGPOR	**0.0267**	**0.0267**	**0.0733**	**0.0733**

than EPGPOR. The unlabeled data help SSGPOR to shift its decision boundary towards a region of low data density. From Table 1, we observe that in either cases, SSGPOR gives lower errors than EPGPOR. It is important to note that the increase in the error is significantly higher ($\sim 10\%$) for EPGPOR compared to SSGPOR ($\sim 5\%$). This corroborates well with the observation that effective decision boundary is learnt by SSGPOR using unlabeled data.

Benchmark Data: We conduct experiments on benchmark data sets to study the generalization performance of the proposed SSGPOR approach. The experiments are conducted on six benchmark data sets [7] with varying sizes. The properties of these benchmark data sets are summarized in Table 2. These are regression data sets. The continuous target values are discretized into ordinal values using equal frequency binning. For each data set, we discretize the target values in the original data set into 5 ordinal categories. Each data set is randomly partitioned into training and test data sets as mentioned in Table 2. We generate 10 such training and test data set instances by repeated independent partitioning. For each data set, zero-one and absolute errors are obtained on all the 10 instances of training and test data sets. The mean of the zero-one and absolute errors, along with their standard deviation, are used to compare the performance of the approaches.

Semi-supervised Setting: Figures 2 and 3 provide a comparison of SSGPOR and EPGPOR on the benchmark data sets using mean zero-one error and mean

Table 2. Benchmark data sets and their properties

Data sets	Boston	Stocks	Abalone	Bank	California	Census
Attributes	13	9	8	32	8	16
Training Instances	300	600	1000	2000	3000	4000
Test Instances	206	350	3177	6192	17,640	18,784

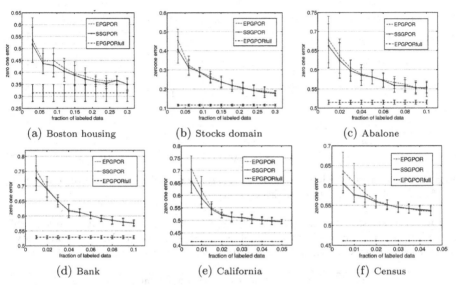

(a) Boston housing   (b) Stocks domain   (c) Abalone

(d) Bank   (e) California   (f) Census

Fig. 2. Comparison of SSGPOR and EPGPOR using mean zero-one error on varying the fraction of labeled examples. Error bars denote the standard deviation.

absolute error, respectively. Here, a fraction of the training data acts as labeled data and the rest as unlabeled data. For each benchmark data set, we plot the performance of the approaches as we vary the fraction of labeled data. We also plot the performance that can be obtained using EPGOR when the entire training set is used as the labeled data, and is denoted as EPGPORfull.

We observe from Fig. 2 and Fig. 3 that SSGPOR performs better than EPGPOR for both zero-one and absolute errors. The improvement in performance is higher when the fraction of labeled data are small. As we increase the fraction of labeled data, the improvement in performance decreases, and both the approaches start giving similar results. Eventually, the performance of both the approaches converges to the case of using full training data as the labeled data set. We observe that the improvement in performance is greater for the absolute error than for the zero-one error. That is, the labels predicted by SSGPOR are more closer to the true labels, as one would desire in an ordinal regression problem. SSGPOR gives better results on large data sets like California and Census, than on small data sets. This is due to the availability of more unlabeled data in large data sets. SSGPOR is thus able to make effective use of unlabeled data to improve the generalization performance on benchmark data sets.

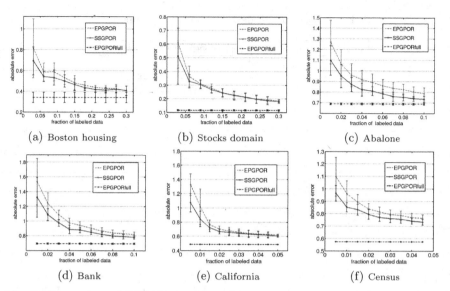

Fig. 3. Comparison of SSGPOR and EPGPOR using mean absolute error on varying the fraction of labeled examples. Error bars denote the standard deviation.

Table 3. T-test statistic computed with respect zero-one and absolute errors for different datasets for the smallest fraction of labeled examples. We use the bold face style to indicate the cases for which the t-test statistic is greater than the critical value.

Error	Boston	Stocks	Abalone	Bank	California	Census
Zero-one	**1.8331**	**2.9394**	1.0179	**2.2251**	**4.4971**	**2.4149**
Absolute	**2.3141**	**4.0553**	**3.4269**	**3.2525**	**4.8434**	**2.9454**

Statistical Significance Test: We use the paired t-test [27] to check if the proposed SSGPOR performs significantly better than EPGPOR. For each data set, we compute the t-test statistic with respect to zero-one and absolute errors for the smallest fraction of labeled data. The errors are obtained on 10 instances of training and test data sets. The null hypothesis is that both SSGPOR and EPGPOR have similar performance. Under the null hypothesis, the t-test statistic follows the Student's t-distribution with 9 degrees of freedom[1]. For the confidence level of 95% and degrees of freedom 9, critical value for the one-sided t-test is 1.833. We reject the null hypothesis if the computed t-test statistic is greater than the critical value. Table 3 reports the t-test statistic computed for each dataset. From Table 3, we observe that the computed t-statistic with respect to zero-one error is greater than the critical value for all datasets except for the Abalone data set. With respect to absolute error, it is greater than the critical value for all the data sets. Therefore, the performance of SSGPOR is significantly better than that of EPGPOR and is a better approach than EPGPOR to perform semi-supervised ordinal regression.

---

[1] Under null hypothesis t-statistic follows the Student's t-distribution with $s-1$ degrees of freedom, where s is the sample size.

Table 4. Comparison of SSGPOR and EPGPOR in the transductive setting for different labeled data sizes. The numbers in bold face style indicate the best results.

	50 labeled examples				100 labeled examples			
	zero-one error		absolute error		zero-one error		absolute error	
Data	EPGPOR	SSGPOR	EPGPOR	SSGPOR	EPGPOR	SSGPOR	EPGPOR	SSGPOR
Boston	0.3860	**0.3816**	0.4656	**0.4498**	0.3590	**0.3538**	0.4192	**0.4039**
Stocks	0.2732	**0.2503**	0.2894	**0.2669**	0.2079	**0.1977**	0.2165	**0.2059**
Abalone	0.5764	**0.5643**	0.8834	**0.7947**	0.5453	**0.5407**	0.7781	**0.7378**
Bank	0.6626	**0.6571**	1.1657	**1.0287**	0.6130	**0.6091**	0.9358	**0.8756**
California	0.5253	**0.5141**	0.6998	**0.6649**	0.4976	**0.4934**	0.6331	**0.6201**
Census	0.5837	**0.5823**	0.9028	**0.8566**	0.5553	**0.5540**	0.8215	**0.7822**

Table 5. Comparison of EPGPOR, SSGPOR and TOR when labeled data size is 100. The numbers in bold face style indicate the best results.

	zero-one error			absolute error		
Data	EPGPOR	SSGPOR	TOR	EPGPOR	SSGPOR	TOR
Abalone	0.5453	**0.5407**	0.5420	0.7781	**0.7378**	0.7700
Bank	0.6130	**0.6091**	0.6220	0.9358	**0.8756**	0.9200
California	0.4976	**0.4934**	0.5200	0.6331	**0.6201**	0.6750
Census	0.5553	**0.5540**	0.5700	0.8215	**0.7822**	0.7900

Transductive Setting: We conduct experiments to study the performance of the proposed approach in a transductive setting. Here, we assume the unlabeled test examples are available at the time of training. The experiments are conducted on all the data sets. The mean zero-one and absolute errors (over 20 independent partitions of training and test data), when labeled data sizes are 50 and 100, are given in Table 4. Transductive setting experiments show a similar behavior as that of the semi-supervised setting. Comparison with EPGPOR shows that the improvement in performance is higher when the fraction of labeled data are small and the improvement decreases with more labeled data. Again, we observe that the improvements are larger for the absolute error than for the zero-one error.

Comparison with TOR [13]: The transductive setting experiments provide us an opportunity to compare EPGPOR and SSGPOR with TOR. We note that TOR uses a Perceptron kernel [13]. Table 5 compares the mean zero-one and absolute errors obtained for EPGPOR and SSGPOR with the reported TOR results [13] on Abalone, Bank, California and Census data sets, when the labeled data size is fixed to 100. We observe that the performance of EPGPOR is comparable with that of TOR whereas, SSGPOR performs better than TOR. Also, we get the predictive probability information using SSGPOR unlike TOR.

Sentiment Data: We conduct experiments on real-world sentiment data sets[2]. The data sets consist of reviews and ratings of users on products at

---

[2] http://www.cs.jhu.edu/~mdredze/datasets/sentiment/

Table 6. Mean zero one and absolute errors on the sentiment data when labeled data size is 100. The numbers in bold face style indicate the best results.

Data	zero-one error		absolute error	
	EPGPOR	SSGPOR	EPGPOR	SSGPOR
Book	0.7385	**0.6546**	1.3022	**0.9424**
Kitchen	0.7266	**0.6547**	1.2370	**0.9642**
Dvd	0.7276	**0.6476**	1.1558	**0.9288**
Electronics	0.7327	**0.6613**	1.3714	**0.9696**

Amazon.com [13]. The task is to predict the rating of a user review on a scale of 1 to 5. We consider four categories of products, Book, Kitchen, Dvd and Electronics. The data sets are preprocessed and the best 1000 words are selected based on the *tf-idf* value to form the feature vector. The data sets consist of around 5000 samples. We conduct the transductive setting experiments on the data sets with the labeled data size as 100. Table 6 reports the mean zero one and mean absolute errors obtained using SSGPOR and EPGPOR for the data sets. We observe that SSGPOR significantly boosts the performance with the additional unlabeled data, on the sentiment data sets.

5   Conclusion

In this work, we proposed an approach to perform ordinal regression using Gaussian processes in a semi-supervised setting. A semi-supervised approach to ordinal regression is important as it is expensive to obtain labeled data, whereas unlabeled data are easily available. The proposed approach, semi-supervised Gaussian process ordinal regression (SSGPOR), was based on the assumption that the distribution on unlabeled data is similar to that on labeled data. The approach used an alternating optimization method to learn the model parameters and the label matrix. The label matrix was learnt efficiently using the label switching algorithm. Experimental results on synthetic, benchmark and real-world data sets showed that the SSGPOR approach performed better than the supervised EPGPOR approach and the TOR approach. Thus, it is a useful approach for semi-supervised ordinal regression.

References

1. McCullagh, P.: Regression Models for Ordinal Data. Journal of the Royal Statistical Society 42, 109–142 (1980)
2. McCullagh, P., Nelder, J.A.: Generalized Linear Models, 2nd edn. Chapman & Hall, London (1989)
3. Johnson, V.E., Albert, J.H.: Ordinal Data Modeling (Statistics for Social and Behavioral Sciences). Springer (2001)
4. Kramer, S., Widmer, G., Pfahringer, B., De Groeve, M.: Prediction of Ordinal Classes Using Regression Trees. Fundam. Inform. 47, 1–13 (2001)

5. Frank, E., Hall, M.: A simple approach to ordinal classification. In: Flach, P.A., De Raedt, L. (eds.) ECML 2001. LNCS (LNAI), vol. 2167, pp. 145–156. Springer, Heidelberg (2001)
6. Chu, W., Keerthi, S.S.: New Approaches to Support Vector Ordinal Regression. In: International Conference on Machine Learning, pp. 145–152 (2005)
7. Chu, W., Ghahramani, Z.: Gaussian Processes for Ordinal Regression. J. Mach. Learn. Res. 6, 1019–1041 (2005)
8. Chapelle, O., Schölkopf, B., Zien, A. (eds.): Semi-Supervised Learning. MIT Press, Cambridge (2006)
9. Herbrich, R., Graepel, T., Obermayer, K.: Large Margin Rank Boundaries for Ordinal Regression. In: Advances in Large Margin Classifiers. MIT Press (2000)
10. Shashua, A., Levin, A.: Ranking with Large Margin Principle: Two Approaches. In: Advances in Neural Information Processing Systems, pp. 937–944 (2003)
11. Li, L., Lin, H.T.: Ordinal Regression by Extended Binary Classification. In: Advances in Neural Information Processing Systems, pp. 865–872 (2006)
12. Sun, B.Y., Li, J., Wu, D.D., Zhang, X.M., Li, W.B.: Kernel Discriminant Learning for Ordinal Regression. IEEE Trans. on Knowl. and Data Eng. 22, 906–910 (2010)
13. Seah, C.W., Tsang, I., Ong, Y.S.: Transductive Ordinal Regression. IEEE Transactions on Neural Networks and Learning Systems 23(7), 1074–1086 (2012)
14. Liu, Y., Liu, Y., Zhong, S., Chan, K.C.: Semi-Supervised Manifold Ordinal Regression for Image Ranking. In: ACM Multimedia, pp. 1393–1396 (2011)
15. Tresp, V.: A Bayesian Committee Machine. Neural Computation 12(11) (2000)
16. Lawrence, N.D., Jordan, M.I.: Semi-supervised Learning via Gaussian Processes. In: Advances in Neural Information Processing Systems, pp. 753–760 (2004)
17. Sindhwani, V., Chu, W., Keerthi, S.S.: Semi-supervised Gaussian process classifiers. In: International Joint Conference on Artificial Intelligence, pp. 1059–1064 (2007)
18. Guo, X., Yasumura, Y., Uehara, K.: Semi-supervised gaussian process regression and its feedback design. In: Zhou, S., Zhang, S., Karypis, G. (eds.) ADMA 2012. LNCS, vol. 7713, pp. 353–366. Springer, Heidelberg (2012)
19. Adams, R.P., Ghahramani, Z.: Archipelago: Nonparametric Bayesian Semi-Supervised Learning. In: International Conference on Machine Learning (2009)
20. Quadrianto, N., Petterson, J., Smola, A.: Distribution Matching for Transduction. In: Advances in Neural Information Processing Systems, pp. 1500–1508 (2009)
21. Le, Q.V., Smola, A.J., Gärtner, T., Altun, Y.: Transductive Gaussian Process Regression with Automatic Model Selection. In: Fürnkranz, J., Scheffer, T., Spiliopoulou, M. (eds.) ECML 2006. LNCS (LNAI), vol. 4212, pp. 306–317. Springer, Heidelberg (2006)
22. Gärtner, T., Le, Q.V., Burton, S., Smola, A.J., Vishwanathan, S.V.N.: Large-Scale Multiclass Transduction. In: Advances in Neural Information Processing Systems, pp. 411–418 (2006)
23. Keerthi, S.S., Sellamanickam, S., Shevade, S.K.: Extension of TSVM to Multi-Class and Hierarchical Text Classification Problems With General Losses. In: International Conference on Computational Linguistics, pp. 1091–1100 (2012)
24. Sindhwani, V., Keerthi, S.S., Chapelle, O.: Deterministic Annealing for Semi-Supervised Kernel Machines. In: International Conference on Machine Learning, pp. 841–848 (2006)
25. Rasmussen, C.E., Williams, C.K.I.: Gaussian Processes for Machine Learning (Adaptive Computation and Machine Learning). The MIT Press (2005)
26. Minka, T.: A Family of Algorithms for Approximate Bayesian Inference. PhD thesis, Massachusetts Institute of Technology (2001)
27. Dietterich, T.G.: Approximate Statistical Tests for Comparing Supervised Classification Learning Algorithms. Neural Computation 10, 1895–1923 (1998)

Influence of Graph Construction
on Semi-supervised Learning

Celso André R. de Sousa, Solange O. Rezende,
and Gustavo E.A.P.A. Batista

Instituto de Ciências Matemáticas e de Computação, Universidade de São Paulo,
Campus de São Carlos, Brazil
{sousa,solange,gbatista}@icmc.usp.br

Abstract. A variety of graph-based semi-supervised learning (SSL) algorithms and graph construction methods have been proposed in the last few years. Despite their apparent empirical success, the field of SSL lacks a detailed study that empirically evaluates the influence of graph construction on SSL. In this paper we provide such an experimental study. We combine a variety of graph construction methods as well as a variety of graph-based SSL algorithms and empirically compare them on a number of benchmark data sets widely used in the SSL literature. The empirical evaluation proposed in this paper is subdivided into four parts: (1) best case analysis; (2) classifiers' stability evaluation; (3) influence of graph construction; and (4) influence of regularization parameters. The purpose of our experiments is to evaluate the trade-off between classification performance and stability of the SSL algorithms on a variety of graph construction methods and parameter values. The obtained results show that the mutual k-nearest neighbors (mutKNN) graph may be the best choice for adjacency graph construction while the RBF kernel may be the best choice for weighted matrix generation. In addition, mutKNN tends to generate smoother error surfaces than other adjacency graph construction methods. However, mutKNN is unstable for a relatively small value of k. Our results indicate that the classification performance of the graph-based SSL algorithms are heavily influenced by the parameters setting and we found no evident explorable pattern to relay to future practitioners. We discuss the consequences of such instability in research and practice.

Keywords: Semi-supervised learning, graph-based methods, experimental study, classification.

1 Introduction

Semi-supervised learning (SSL) has gained increased attention in the last few years [3,15]. Among all SSL algorithms, graph-based methods are widely used because the weighted graph may approximate the low dimensional manifold in which the data should lie. The research community has proposed a variety of graph-based SSL algorithms [1,8,14,16] as well as a variety of graph construction

H. Blockeel et al. (Eds.): ECML PKDD 2013, Part III, LNAI 8190, pp. 160–175, 2013.

methods [5,8,13]. Despite its increasing popularity, the SSL literature lacks a comprehensive and unbiased empirical study that shows the influence that graph construction methods have in both classification performance and stability of the graph-based SSL algorithms.

1.1 Contributions

In this paper, we provide a detailed empirical comparison of the state-of-the-art, graph-based SSL algorithms combined with a variety of graph construction methods. The empirical analysis proposed in this paper is subdivided into four parts as follows:

Best case analysis. We evaluate the best error rates of each combination of SSL algorithm and graph construction method for a number of sparsification parameter values. Although this is the most common approach to evaluate SSL algorithms in the literature [3], this empirical setting alone may not provide all the necessary information to choose the best classifiers for real applications. For instance, stable classifiers may be preferable over classifiers which are able to provide excellent performance for a very narrow range of parameter values and mediocre performance for the remaining values;

Classifiers' stability evaluation. We evaluate the stability of the SSL algorithms combined with the graph construction methods as we vary the value of the sparsification parameter. As we mentioned before, this analysis is important because a classifier may achieve the best overall classification performance for a very narrow range of the parameter values. Then, this analysis is an invaluable tool to identify which classifiers provide a good trade-off between classification performance and stability;

Influence of graph construction. We also evaluate the graph construction methods combined with the SSL algorithms over a wide range of sparsification parameter values. We want to verify: (1) how the graph construction methods affect the classification performance of each SSL algorithm and (2) the stability of the graph construction methods as we vary the sparsification parameter values. For the classifiers that have at least one regularization parameter, we fixed the regularization parameter(s) with the value that achieved the best average error rate and then varied the sparsification parameter value;

Influence of regularization parameters. We evaluate the error surfaces generated by the SSL algorithms that have regularization parameters. We first chose the sparsification parameter that achieved the best average error rate and then we varied the regularization parameters of the SSL algorithms.

The obtained results show that the mutual k-nearest neighbors (mutKNN) graph may be the best choice for adjacency graph construction while the RBF kernel may be the best choice for weighted matrix generation. In addition, mutKNN tends to generate smoother error surfaces than other adjacency graph construction methods. However, mutKNN is unstable for a relatively small value of k.

Our results indicate that the classification performance of the graph-based SSL algorithms are heavily influenced by *internal* parameters (such as regularization parameters) and *external* parameters (such as the number of neighbors in a k-nearest neighbor graph). Such variability showed no evident explorable pattern to relay to future practitioners. In addition, the SSL assumption that only a very restricted set of labeled examples exists may make parameter estimation techniques commonly used in classification unfeasible.

We believe that our results have two major consequences:

For practitioners. Given a data set, it is difficult to recommend an SSL algorithm, a graph sparsification parameter value or a regularization parameter value that is expected to provide good classification performance. As the number of labeled examples is usually very restricted in SSL applications, the practitioner has no tools to make an informed choice of these parameter values. As we will show, an incorrect choice of the parameter values may seriously affect the classification results;

For researchers. Changes in the parameter values also cause changes in the relative ranking among the classifiers. It means that for a specific data set several methods may figure as the best classifier for a certain range of parameter values. This is a serious issue since the empirical evidence that one method outperforms the competitors might be confirmed only for a restricted set of the parameter values. In addition, this performance variability may hinder the reproduction of the experimental results for papers that do not clearly report every parameter value used in the empirical evaluation.

1.2   Outline

The remainder of this paper is organized as follows. Section 2 describes the notation used throughout the paper and revises the graph construction methods. Section 3 revises the state-of-the-art, graph-based SSL algorithms. Section 4 empirically evaluates the graph construction methods combined with the graph-based SSL algorithms. Finally, Section 5 concludes the paper and suggests directions for future research.

2   Graph Construction

In this section we revise widely used methods to generate sparse weighted graphs, which are frequently considered the heart of graph-based SSL [15]. Section 2.1 describes the notation used throughout the paper. Section 2.2 revises approaches used to generate a sparse undirected[1] graph (or adjacency matrix) from the training sample. Section 2.3 revises approaches used to generate a weighted matrix from the sparse graph.

---

[1] This paper focus on undirected graphs, which are commonly used in SSL [15].

2.1   Notation and Preliminaries

Consider a training sample $\mathcal{X} := \{x_i\}_{i=1}^n \subset \mathbb{R}^d$ in which the first l examples are labeled, i.e., x_i has label $y_i \in \mathbb{N}_c$ where $\mathbb{N}_p := \{i \in \mathbb{N}^* | 1 \le i \le p\}$ with $p \in \mathbb{N}^*$ and c being the number of classes. Let $u := n - l$ be the amount of unlabeled examples and $Y \in \mathbb{B}^{n \times c}$ be a label matrix in which $Y_{ij} = 1$ if and only if x_i has label $y_i = j$. Consider an undirected graph $\mathcal{G} := (\mathcal{X}, \mathcal{E})$ in which each x_i is a node of $\mathcal{G}$. Let $\mathcal{N}_i \subset \mathcal{X}$ be the set of neighbors of x_i and x_{i_k} the k-th nearest neighbor of x_i. In order to generate a sparse weighted matrix $W \in \mathbb{R}^{n \times n}$ from $\mathcal{G}$ one uses a similarity function $\mathcal{K} : \mathbb{R}^d \times \mathbb{R}^d \mapsto \mathbb{R}$ to compute the weights W_{ij}.

The graph Laplacians are important tools for machine learning. The *combinatorial* Laplacian is defined by $\Delta := D - W$ where $D := \mathrm{diag}(W 1_n)$ such that 1_n is an n-dimensional 1-entry vector. The *normalized* Laplacian is defined by $L := I_n - D^{-1/2} W D^{-1/2}$ where I_n is the n-by-n identity matrix.

All matrices can be subdivided into labeled and unlabeled submatrices. Let $F \in \mathbb{R}^{n \times c}$ be the output of a given graph-based SSL algorithm. The F and Y matrices are subdivided into two submatrices while all others are subdivided into four submatrices. For instance:

$$W := \begin{bmatrix} W_{\mathcal{L}\mathcal{L}} & W_{\mathcal{L}\mathcal{U}} \\ W_{\mathcal{U}\mathcal{L}} & W_{\mathcal{U}\mathcal{U}} \end{bmatrix} \qquad Y := \begin{bmatrix} Y_{\mathcal{L}} \\ Y_{\mathcal{U}} \end{bmatrix}$$

where $W_{\mathcal{L}\mathcal{L}} \in \mathbb{R}^{l \times l}$ and $Y_{\mathcal{L}} \in \mathbb{B}^{l \times c}$ are the submatrices of W and Y, respectively, on labeled examples, and so on. By definition, $Y_{\mathcal{U}}$ is an $u \times c$ null matrix. This paper focus on the multi-class problem; hence, $Y_{\mathcal{L}} 1_c = 1_l$.

2.2   Adjacency Graph Construction

The adjacency graph construction process generates a graph $\mathcal{G}$ (or adjacency matrix A) from $\mathcal{X}$ using a distance function $\Psi : \mathbb{R}^d \times \mathbb{R}^d \mapsto \mathbb{R}$. Let $\Psi \in \mathbb{R}^{n \times n}$ be a distance matrix in which $\Psi_{ij} := \Psi(x_i, x_j)$ and $A \in \mathbb{B}^{n \times n}$ be an adjacency matrix[2] in which $A_{ij} = 1$ if and only if $x_j \in \mathcal{N}_i$. We now describe the two most used adjacency graph construction methods for graph-based learning.

ϵ-**neighborhood (ϵN).** There exists an undirected edge between x_i and x_j in an ϵN graph if and only if $\Psi(x_i, x_j) \le \epsilon$ where $\epsilon \in \mathbb{R}_+^*$ is a free parameter. In general, ϵN graphs are not widely used in practical situations because they can generate graphs with many disconnected components for an improper value of ϵ. Due to this fact, we did not use the ϵN graph in our experiments.

k-**nearest neighbors (kNN).** There exists an edge from x_i to x_j if and only if x_j is one of the k closest examples of x_i. Because the adjacency matrix of a kNN graph may not be symmetric, three strategies are commonly used to symmetrize it: *mutual kNN* (mutKNN), which generates $\widehat{A} = \min(A, A^\top)$; *symmetric kNN* (symKNN), which generates $\widehat{A} = \max(A, A^\top)$; and *symmetry-favored kNN* (symFKNN) [8], which generates $\widehat{A} = A + A^\top$ (a non-binary adjacency matrix).

---

[2] Non-binary adjacency matrices may also be applied.

2.3   Weighted Matrix Generation

Given an adjacency matrix $\mathbf{A}$, we generate a sparse weighted matrix $\mathbf{W}$ using a similarity function $\mathcal{K} : \mathbb{R}^d \times \mathbb{R}^d \mapsto \mathbb{R}$. We describe three widely used approaches to generate $\mathbf{W}$. Two of them, RBF kernel and similarity function of Hein & Maier [5], define the $\mathbf{W}$ matrix using the relation $\mathbf{W}_{ij} = \mathbf{A}_{ij}\mathcal{K}(\boldsymbol{x}_i, \boldsymbol{x}_j)$. The third approach, based on local reconstruction minimization [13], generates a sparse weighted matrix $\mathbf{W}$, not necessarily symmetric, without an explicit $\mathcal{K}$.

RBF kernel. The RBF (or Gaussian) kernel computes the similarity between $\boldsymbol{x}_i$ and $\boldsymbol{x}_j$ by $\mathcal{K}(\boldsymbol{x}_i, \boldsymbol{x}_j) := \exp\left(-\Psi^2(\boldsymbol{x}_i, \boldsymbol{x}_j)/\left(2\sigma^2\right)\right)$ in which $\sigma \in \mathbb{R}_+^*$ is the kernel bandwidth parameter.

Similarity function of Hein & Maier [5] (HM). Given a function $\psi(\cdot, \cdot)$ in which $\psi(\boldsymbol{x}_i, k) := \Psi(\boldsymbol{x}_i, \boldsymbol{x}_{i_k})$ with $k \in \mathbb{N}^*$, the HM similarity function is defined by $\mathcal{K}(\boldsymbol{x}_i, \boldsymbol{x}_j) := \exp\left(-\Psi^2(\boldsymbol{x}_i, \boldsymbol{x}_j)/\left(\max\left\{\psi(\boldsymbol{x}_i, k), \psi(\boldsymbol{x}_j, k)\right\}\right)^2\right)$. This is an RBF kernel with an adaptive kernel size.

Local Linear Embedding (LLE). The LLE approach [13] generates the $\mathbf{W}$ matrix by solving the following optimization problem:

$$\min_{\mathbf{W} \in \mathbb{R}^{n \times n}} \sum_{i=1}^{n} \left\| \boldsymbol{x}_i - \sum_{\boldsymbol{x}_j \in \mathcal{N}_i} \mathbf{W}_{ij}\boldsymbol{x}_j \right\|_2^2 \quad \text{s.t.} \quad \mathbf{W}\mathbf{1}_n = \mathbf{1}_n, \ \ \mathbf{W} \geq 0 \quad (1)$$

The symbol $\|\cdot\|_2$ represents the l_2-norm.

3   Label Diffusion

Given a weighted matrix $\mathbf{W}$, a graph-based SSL algorithm uses $\mathbf{W}$ and the label matrix $\mathbf{Y}$ to generate the output matrix $\mathbf{F}$ by label diffusion in the weighted graph. We now revise the state-of-the-art graph-based SSL algorithms used in our empirical comparison. We should note that these algorithms have an intrinsic condition to classify all unlabeled examples in $\mathcal{X}$, which frequently is not explicit in the literature. Assumption 1 describes this condition.

Assumption 1. *Each unlabeled example is on a connected subgraph in which there exists at least one labeled example.*

Gaussian Random Fields (GRF). The GRF algorithm [16] solves the optimization problem $\mathbf{F} = \arg\min_{\mathbf{F} \in \mathbb{R}^{n \times c}} \text{tr}\left(\mathbf{F}^\top \mathbf{\Delta} \mathbf{F}\right)$ s.t. $\mathbf{F}_{\mathcal{L}} = \mathbf{Y}_{\mathcal{L}}$, which gives the closed-form solution $\mathbf{F}_{\mathcal{U}} = -\mathbf{\Delta}_{\mathcal{U}\mathcal{U}}^{-1}\mathbf{\Delta}_{\mathcal{U}\mathcal{L}}\mathbf{Y}_{\mathcal{L}}$.

Local and Global Consistency (LGC). The LGC algorithm [14] solves the optimization problem $\mathbf{F} = \arg\min_{\mathbf{F} \in \mathbb{R}^{n \times c}} \text{tr}\left(\mathbf{F}^\top \mathbf{L}\mathbf{F} + \mu(\mathbf{F} - \mathbf{Y})^\top(\mathbf{F} - \mathbf{Y})\right)$, which gives the closed-form solution $\mathbf{F} = \left(\mathbf{I}_n + \mathbf{L}/\mu\right)^{-1}\mathbf{Y}$.

Laplacian Regularized Least Squares (LapRLS). The LapRLS algorithm [1] minimizes the following regularization framework:

$$\min_{f \in \mathcal{H}_\mathcal{K}} \frac{1}{l} \sum_{i=1}^{l} \mathcal{V}(\boldsymbol{x}_i, y_i, f) + \gamma_A \|f\|_{\mathcal{H}_\mathcal{K}} + \gamma_I \mathbf{f}^\top \mathbf{\Delta}\mathbf{f} \quad (2)$$

where $\mathcal{V}(\boldsymbol{x}_i, y_i, f) = (y_i - f(\boldsymbol{x}_i))^2$, $\mathcal{H}_{\mathcal{K}}$ is the *Reproducing Kernel Hilbert Space (RKHS)* for the kernel $\mathcal{K}$, $\mathbf{f} := [f(\boldsymbol{x}_1), \cdots, f(\boldsymbol{x}_n)]^{\top} \in \mathbb{R}^n$, $\|\cdot\|_{\mathcal{H}_{\mathcal{K}}}$ is the norm in $\mathcal{H}_{\mathcal{K}}$, and γ_A and γ_I are the regularization parameters. Let $\boldsymbol{y} := [y_1, \cdots, y_l, 0, \cdots, 0] \in \mathbb{R}^n$ be the label vector in which $y_i \in \{-1, +1\}$ and $\mathbf{K} \in \mathbb{R}^{n \times n}$ a gram matrix such that $\mathbf{K}_{ij} := \mathcal{K}(\boldsymbol{x}_i, \boldsymbol{x}_j)$. Due to the *Representer Theorem* in [1], the solution of (2) can be written as an expansion of kernel functions over both labeled and unlabeled examples, i.e., $f(\boldsymbol{x}) = \sum_{i=1}^n \mathcal{K}(\boldsymbol{x}, \boldsymbol{x}_i)\alpha_i$ with $\boldsymbol{\alpha} \in \mathbb{R}^n$. Solving (2) using this expansion, we get $\boldsymbol{\alpha} = (\mathbf{J}\mathbf{K} + \gamma_A l \mathbf{I}_n + \gamma_I l \boldsymbol{\Delta}\mathbf{K})^{-1} \boldsymbol{y}$ where $\mathbf{J} := \mathrm{diag}\left([1, \cdots, 1, 0, \cdots, 0]^{\top}\right)$ whose first l diagonal entries are 1 and the rest 0.

Laplacian Support Vector Machine (LapSVM). The LapSVM algorithm [1] minimizes the problem in (2) with $\mathcal{V}(\boldsymbol{x}_i, y_i, f) = \max(0, 1 - y_i f(\boldsymbol{x}_i))$. Solving (2) using the expansion $f(\boldsymbol{x}) = \sum_{i=1}^n \mathcal{K}(\boldsymbol{x}, \boldsymbol{x}_i)\alpha_i$, we get the solution $\boldsymbol{\alpha} = \frac{1}{2}(\gamma_A \mathbf{I}_n + \gamma_I \boldsymbol{\Delta}\mathbf{K})^{-1} \overline{\mathbf{J}}^{\top} \overline{\mathbf{Y}} \boldsymbol{\beta}^*$ where $\overline{\mathbf{J}} := [\mathbf{I}_l \ \mathbf{O}_{l \times u}]$ such that $\mathbf{O}_{l \times u}$ is an $l \times u$ null matrix, $\overline{\mathbf{Y}} := \mathrm{diag}\left([y_1, \cdots, y_l]^{\top}\right)$, and $\boldsymbol{\beta}^* \in \mathbb{R}^l$ is given by

$$\boldsymbol{\beta}^* = \underset{\boldsymbol{\beta} \in \mathbb{R}^l}{\arg\min} \quad \mathbf{1}_l^{\top}\boldsymbol{\beta} - \frac{1}{2}\boldsymbol{\beta}^{\top}\mathbf{Q}\boldsymbol{\beta} \quad \text{s.t.} \quad \boldsymbol{y}^{\top}\boldsymbol{\beta} = 0, \ 0 \le \boldsymbol{\beta} \le \frac{1}{l}$$

such that $\mathbf{Q} = \frac{1}{2}\overline{\mathbf{Y}} \ \overline{\mathbf{J}}\mathbf{K} \left(\gamma_A \mathbf{I}_n + \gamma_I \boldsymbol{\Delta}\mathbf{K}\right)^{-1} \overline{\mathbf{J}}^{\top} \overline{\mathbf{Y}}$.

Robust Multi-class Graph Transduction (RMGT). The RMGT algorithm [8] solves the convex optimization problem $\mathbf{F} = \arg\min_{\mathbf{F} \in \mathbb{R}^{n \times c}} \mathrm{tr}\left(\mathbf{F}^{\top}\boldsymbol{\Delta}\mathbf{F}\right)$ s.t. $\mathbf{F}_{\mathcal{L}} = \mathbf{Y}_{\mathcal{L}}$, $\mathbf{F}\mathbf{1}_c = \mathbf{1}_n$, $\mathbf{F}^{\top}\mathbf{1}_n = n\boldsymbol{\omega}$ where $\boldsymbol{\omega} \in \mathbb{R}^c$ is the class prior probabilities. The solution of this optimization problem is given by:

$$\mathbf{F}_{\mathcal{U}} = -\boldsymbol{\Delta}_{\mathcal{U}\mathcal{U}}^{-1}\boldsymbol{\Delta}_{\mathcal{U}\mathcal{L}}\mathbf{Y}_{\mathcal{L}} + \frac{\boldsymbol{\Delta}_{\mathcal{U}\mathcal{U}}^{-1}\mathbf{1}_u}{\mathbf{1}_u^{\top}\boldsymbol{\Delta}_{\mathcal{U}\mathcal{U}}^{-1}\mathbf{1}_u}\left(n\boldsymbol{\omega}^{\top} - \mathbf{1}_l^{\top}\mathbf{Y}_{\mathcal{L}} + \mathbf{1}_u^{\top}\boldsymbol{\Delta}_{\mathcal{U}\mathcal{U}}^{-1}\boldsymbol{\Delta}_{\mathcal{U}\mathcal{L}}\mathbf{Y}_{\mathcal{L}}\right)$$

4   Experimental Evaluation

In this section we provide a detailed empirical comparison of the graph-based SSL algorithms described in Section 3 combined with the graph construction methods described in Section 2 on a number of benchmark data sets. The objective of these experiments is to evaluate the influence that graph construction methods have in the classifiers' performance. We performed experiments in a transductive setting using different sets of labeled and unlabeled examples in each execution.

For a fair comparison and ease of reproducibility, we used the source code of the authors of the algorithms when possible. As some authors implemented their methods in Matlab, we used the matlabcontrol[3] library to link the Matlab code and Java. Due to reasons concerning reproducibility, all source codes and data sets used in our experiments are freely available[4].

---

[3] https://code.google.com/p/matlabcontrol/downloads/list
[4] http://www.icmc.usp.br/~gbatista/ECML2013

4.1   Data Sets

We used in our experiments the USPS, $COIL_2$, DIGIT-1, G-241C, G-241N, and TEXT data sets. These data sets are freely available[5] and very popular in the SSL literature [3]. USPS and DIGIT-1 are data sets for digit recognition, TEXT is a data set for text classification, G-241N and G-241C are data sets for classification of Gaussian distributions, and $COIL_2$ is a data set for image classification. We used the data splits of 10 labeled examples suggested in [3].

We run *principal component analysis* (PCA) to reduce the dimensionality of the data sets. In high-dimensional data, the distance to the nearest neighbor approaches the distance of the farthest neighbor [2]. It degenerates the quality of the graph and possibly decreases the classification performance of the SSL algorithms. After some preliminary experimental evaluation, we decided to reduce the dimensionality of the data to 50 features using the *Matlab Toolbox for Dimensionality Reduction*[6] library. We did not run PCA only on the TEXT data set to maintain the sparseness property of these data.

4.2   Empirical Setup

In this section, we describe the experimental design decisions that we have taken in our experiments in order to facilitate the reproduction of our results.

Distance functions. Due to its high popularity in the text classification literature, we used the cosine distance in the experiments using the TEXT data set. The cosine distance is defined as $\Psi(\boldsymbol{x}_i, \boldsymbol{x}_j) = 1 - \langle \boldsymbol{x}_i, \boldsymbol{x}_j \rangle_d / (\|\boldsymbol{x}_i\|_2 \|\boldsymbol{x}_j\|_2)$ where $\langle \cdot, \cdot \rangle_d$ is the inner product of vectors in $\mathbb{R}^d$. For all other data sets we used the l_2 norm as a distance function.

Graph Laplacians. Since the normalized Laplacian $\mathbf{L}$ may lead to better empirical results in comparison with the combinatorial Laplacian $\boldsymbol{\Delta}$ [7], we used $\mathbf{L}$ instead of $\boldsymbol{\Delta}$ in the formulation of the graph-based SSL algorithms. We obtained poor results using $\mathbf{L}$ in the RMGT algorithm during preliminary experiments; therefore, we report the results of RMGT using $\boldsymbol{\Delta}$. In preliminary experiments, we observed some errors using RMGT in the $COIL_2$ data set. These errors occurred because at least one of the eigenvalues of the graph Laplacian was equal to (or approximately) zero. In an attempt to avoid numerical instabilities while solving linear systems using the graph Laplacians, we generated the combinatorial Laplacian as $\boldsymbol{\Delta} = \gamma \mathbf{D} - \mathbf{W}$ and the normalized Laplacian as $\mathbf{L} = \gamma \mathbf{I}_n - \mathbf{D}^{-1/2} \mathbf{W} \mathbf{D}^{-1/2}$ where a small $\gamma > 1$ is used to increase the eigenvalues of the graph Laplacians. In our experiments, we set $\gamma = 1.01$.

Mutual kNN. The procedure $\widehat{\mathbf{A}} = \min(\mathbf{A}, \mathbf{A}^\top)$ may generate a graph with isolated vertices. It may degenerate the output of the SSL algorithms because the label diffusion process could not be effective. In an attempt to avoid this

---

[5] http://olivier.chapelle.cc/ssl-book/benchmarks.html.

[6] http://homepage.tudelft.nl/19j49/
Matlab_Toolbox_for_Dimensionality_Reduction.html.

problem, we created an undirected edge between each isolated vertex and its nearest neighbor. Other strategies may also be applied as well [12].

LLE. We used the *Local Anchor Embedding* (LAE) method [9][7] to solve the optimization problem in (1). LLE is an example of LAE if we generate a bipartite graph whose "anchor" points are exactly the training examples. Since LLE may not generate a symmetric weighted matrix, we symmetrize the output matrix of LLE, $\mathbf{W}_{LLE}$, as $\mathbf{W} = \frac{1}{2}\left(\mathbf{W}_{LLE} + \mathbf{W}_{LLE}^{\top}\right)$.

SymFKNN + LLE. Because the adjacency matrix of the symFKNN graph is non-binary, we compute $\widehat{\mathbf{W}} = \mathbf{W}_{LLE} \odot \mathbf{A}$ where $\odot$ is the Hadamard product. Then, we generate $\mathbf{W} = \frac{1}{2}\left(\widehat{\mathbf{W}} + \widehat{\mathbf{W}}^{\top}\right)$.

LapSVM. We run LapSVM using the source code in [11][8]. We trained LapSVM using Newton's method, which gave better results than the preconditioned conjugate gradient method during preliminary experiments.

LapRLS. We used the multi-class version of LapRLS; hence, we compute $\boldsymbol{\alpha}$ as $\boldsymbol{\alpha} = (\mathbf{J}\mathbf{K} + \gamma_A l \mathbf{I}_n + \gamma_I l \mathbf{\Delta} \mathbf{K})^{-1}\mathbf{Y}$ and get the output matrix $\mathbf{F} = \mathbf{K}\boldsymbol{\alpha}$.

Classification. In order to classify the unlabeled examples, we used the *class mass normalization* (CMN) procedure [16]. This is an useful procedure when we are dealing with data sets with imbalanced labels. We obtained poor results using CMN in RMGT; therefore, we report the results for RMGT using the *argmax* operator. We report the results for GRF, LGC, and LapRLS using CMN while the results for LapSVM are reported using the *sign* function. For GRF, we computed CMN using $\mathbf{F}_{\mathcal{U}}$ instead of $\mathbf{F}$, as suggested in [16].

4.3   Parameter Setting

We now describe the parameter setting used in our experimental evaluation.

SymKNN, mutKNN, and symFKNN. The sparsification parameter k was chosen at the range $\{4, 6, 8, \cdots, 40\}$.

RBF kernel. Because it is not straightforward to find an adequate value for the kernel bandwidth σ when labeled examples are scarce, we estimate its value by $\sigma = \sum_{i=1}^{n} \Psi(\boldsymbol{x}_i, \boldsymbol{x}_{i_k})/(3n)$, as suggested in [6].

Gram matrix. We generated the gram matrix $\mathbf{K}$ using the RBF kernel. We used the same distance function $\Psi(\cdot, \cdot)$, the sparsification parameter k, and the kernel bandwidth σ used during graph construction to compute $\mathbf{K}$.

LGC. The regularization parameter μ in the LGC framework was chosen at range $\{0.01, 0.05, 0.1, 0.5, 1, 2, 5, 10, 50, 100\}$.

LapRLS and LapSVM. The regularization parameters γ_A and γ_I were chosen at range $\{10^{-6}, 10^{-4}, 10^{-2}, 10^{-1}, 1, 10, 100\}$, as suggested in [11]. All other parameters were set to their default values.

RMGT. For the RMGT algorithm, we assumed a uniform class distribution, i.e., we set $\boldsymbol{\omega} = \mathbf{1}_c/c$ instead of using the class prior probabilities, as suggested in [8]. We achieved better results in preliminary experiments using

---

[7] http://www.ee.columbia.edu/ln/dvmm/downloads/WeiGraphConstructCode/dlform.htm.

[8] http://www.dii.unisi.it/~melacci/lapsvmp/index.html.

the uniform class distribution in most data sets; therefore, we report the results for RMGT using this setting for all data sets, excluding USPS. For the USPS data set, we used the class prior probabilities, which achieved the best results.

4.4   Analysis of the Results

In this section we analyze the obtained results. Our empirical analysis is subdivided into four parts: (1) best cases analysis; (2) graph-based SSL algorithm comparison; (3) influence of graph construction on SSL; and (4) influence of regularization parameters on the classifiers' performance.

Best case analysis. Table 1 shows the obtained results for the best case analysis. Each numerical result in this table is the lowest average error rate obtained by a combination of an SSL algorithm, a graph construction method and a data set for all parameter values (sparsification and regularization, if applicable), as described in Section 4.3. The four worst results obtained by an SSL algorithm in each data set have a grey background while the best one is in bold. The best overall result for each data set is boxed.

We can see in Table 1 that the symKNN-LLE and symFKNN-LLE graphs may not be adequate for GRF, LGC, and LapRLS because they achieved unsatisfactory results in all data sets. We also see that mutKNN outperformed the symKNN and symFKNN graphs in most situations, independent of the weighted matrix generation method or the SSL algorithm used. Therefore, for the data sets considered in this study, mutKNN presented the best performance among all adjacency graph construction methods.

We ran the Friedman's test[9] with Nemenyi's post test using a confidence level of 0.05 to statistically compare the performance of the graph construction methods. Table 2 shows the average rankings. The best rankings are marked in bold face and the results that were outperformed by the best ranked method are marked with grey background. We can see that symFKNN-RBF and mutKNN-RBF obtained the best rankings for most SSL algorithms. However, the statistical test found significant differences for only 7 cases.

After analyzing the classifiers, we see that RMGT achieved the best overall classification performance in 4 out of 6 data sets. Although RMGT achieved satisfactory results on most data sets, it did not perform well on the USPS data set.

Classifiers' stability evaluation. As we mentioned earlier, the best case analysis does not allow us to investigate the stability of the classifiers. In this analysis, we investigate the stability of the SSL algorithms as we vary the graph sparsification parameter value. Due to space restrictions and because the mutKNN-RBF graph achieved the best overall classification performance in the best case analysis, we show here only the results obtained with the mutKNN-RBF graph. The interested reader will find the results for other graph construction methods on the paper's website.

---

[9] See [4] and references therein for a review on statistical tests for machine learning.

Table 1. Average error rates and standard deviations of the SSL algorithms for each graph construction method and data set

Data sets	USPS	COIL$_2$	DIGIT-1	G-241N	G-241C	TEXT
GRF-symKNN-RBF	11.07 (3.33)	35.13 (6.92)	10.19 (4.27)	**46.12 (7.61)**	46.28 (6.98)	39.15 (5.69)
GRF-mutKNN-RBF	**9.75 (4.50)**	**35.07 (3.82)**	**9.35 (4.51)**	46.94 (4.81)	46.72 (4.85)	**37.51 (6.85)**
GRF-symFKNN-RBF	10.75 (3.77)	35.22 (6.92)	10.01 (3.93)	**46.12 (7.65)**	46.34 (6.83)	38.50 (5.87)
GRF-symKNN-HM	15.53 (2.76)	38.55 (6.06)	10.73 (4.27)	46.86 (5.28)	**46.19 (7.25)**	42.32 (8.54)
GRF-mutKNN-HM	11.01 (3.59)	35.30 (3.87)	10.02 (6.36)	46.66 (6.18)	46.58 (5.06)	41.18 (9.87)
GRF-symFKNN-HM	15.17 (3.09)	37.77 (6.25)	10.24 (4.36)	46.77 (5.38)	46.27 (7.05)	42.20 (8.62)
GRF-symKNN-LLE	16.03 (2.47)	36.04 (5.60)	10.94 (4.69)	47.54 (3.77)	47.33 (4.77)	43.56 (6.96)
GRF-mutKNN-LLE	11.64 (3.39)	35.20 (3.79)	10.30 (5.89)	47.14 (2.87)	46.98 (3.64)	42.34 (6.48)
GRF-symFKNN-LLE	15.55 (2.74)	36.10 (5.88)	10.31 (4.68)	47.25 (4.14)	47.46 (4.36)	43.54 (6.95)
LGC-symKNN-RBF	11.22 (3.07)	34.96 (6.69)	10.68 (4.91)	38.06 (6.91)	40.24 (5.13)	35.42 (5.58)
LGC-mutKNN-RBF	9.93 (4.34)	35.07 (3.82)	10.54 (5.21)	39.82 (5.36)	41.85 (4.32)	**34.78 (6.55)**
LGC-symFKNN-RBF	10.97 (3.00)	**34.81 (6.22)**	**10.47 (4.66)**	**37.95 (6.66)**	40.10 (5.46)	35.51 (5.64)
LGC-symKNN-HM	14.49 (5.25)	37.20 (7.32)	11.53 (5.00)	38.36 (6.83)	40.27 (4.48)	37.51 (4.48)
LGC-mutKNN-HM	10.79 (3.75)	35.19 (4.90)	10.96 (5.34)	39.51 (5.80)	41.94 (4.20)	36.01 (5.63)
LGC-symFKNN-HM	14.63 (3.33)	36.36 (8.25)	11.07 (4.76)	38.13 (7.11)	40.17 (4.75)	37.49 (4.35)
LGC-symKNN-LLE	15.05 (4.33)	35.95 (6.09)	11.49 (5.41)	41.22 (4.12)	43.33 (3.03)	39.18 (4.02)
LGC-mutKNN-LLE	11.04 (3.82)	35.18 (3.77)	10.96 (6.34)	42.12 (3.90)	42.94 (3.09)	35.89 (9.20)
LGC-symFKNN-LLE	14.51 (2.81)	35.98 (6.07)	10.97 (5.01)	41.24 (4.37)	43.06 (3.34)	39.03 (3.82)
LapRLS-symKNN-RBF	10.99 (3.05)	34.92 (5.98)	10.22 (4.25)	38.09 (6.76)	40.35 (6.23)	35.12 (5.68)
LapRLS-mutKNN-RBF	**9.75 (4.53)**	33.56 (7.32)	**9.33 (4.48)**	38.36 (5.96)	40.66 (5.45)	**34.58 (6.14)**
LapRLS-symFKNN-RBF	10.57 (2.90)	35.50 (5.84)	10.02 (3.92)	38.08 (6.64)	40.36 (6.02)	35.34 (5.73)
LapRLS-symKNN-HM	14.56 (3.89)	37.58 (5.91)	10.76 (4.24)	38.18 (6.70)	40.24 (6.07)	37.12 (4.52)
LapRLS-mutKNN-HM	10.57 (4.66)	32.80 (7.67)	9.92 (5.50)	38.29 (6.00)	40.67 (5.50)	35.90 (5.61)
LapRLS-symFKNN-HM	14.38 (4.14)	36.93 (4.95)	10.28 (4.32)	**38.06 (6.52)**	**40.11 (6.06)**	37.32 (4.38)
LapRLS-symKNN-LLE	14.73 (3.24)	36.85 (5.25)	10.93 (4.66)	38.68 (5.60)	40.61 (5.51)	38.49 (4.00)
LapRLS-mutKNN-LLE	11.28 (4.09)	**31.78 (7.81)**	10.19 (5.92)	38.66 (5.72)	40.59 (5.61)	37.28 (5.26)
LapRLS-symFKNN-LLE	14.55 (3.37)	36.17 (4.81)	10.31 (4.63)	38.69 (5.56)	40.61 (5.52)	38.62 (4.13)
LapSVM-symKNN-RBF	11.42 (4.03)	34.96 (6.81)	9.42 (3.97)	39.16 (6.07)	40.91 (6.08)	39.88 (6.02)
LapSVM-mutKNN-RBF	9.91 (2.51)	34.37 (6.47)	8.67 (3.89)	**38.90 (6.50)**	40.90 (6.08)	37.49 (7.07)
LapSVM-symFKNN-RBF	11.04 (3.43)	34.04 (6.92)	9.47 (4.19)	39.16 (6.07)	40.91 (6.08)	39.45 (6.30)
LapSVM-symKNN-HM	14.63 (5.47)	36.40 (4.07)	10.13 (3.65)	39.15 (6.04)	40.91 (6.08)	43.06 (4.99)
LapSVM-mutKNN-HM	10.04 (2.83)	33.08 (6.35)	9.58 (4.73)	39.00 (6.43)	40.90 (6.08)	42.10 (6.03)
LapSVM-symFKNN-HM	14.35 (4.29)	36.57 (3.57)	9.93 (3.95)	39.14 (6.05)	40.91 (6.08)	42.71 (5.15)
LapSVM-symKNN-LLE	14.82 (3.38)	35.39 (4.80)	10.31 (4.11)	39.12 (6.43)	40.90 (6.07)	42.77 (6.20)
LapSVM-mutKNN-LLE	10.61 (2.49)	**31.54 (6.24)**	10.22 (5.52)	38.95 (6.46)	**40.82 (6.66)**	41.80 (7.65)
LapSVM-symFKNN-LLE	14.41 (3.23)	35.21 (4.58)	9.83 (3.99)	39.00 (6.40)	40.84 (6.40)	42.62 (4.92)
RMGT-symKNN-RBF	16.62 (2.90)	31.05 (4.81)	8.63 (3.35)	44.99 (6.97)	38.44 (6.22)	30.43 (6.26)
RMGT-mutKNN-RBF	**13.08 (3.41)**	28.95 (3.88)	8.13 (3.14)	46.11 (4.50)	42.76 (6.11)	**27.77 (5.95)**
RMGT-symFKNN-RBF	16.02 (2.85)	32.94 (4.20)	8.55 (3.84)	45.25 (6.07)	**38.31 (6.02)**	29.65 (6.46)
RMGT-symKNN-HM	19.08 (1.22)	31.20 (6.14)	8.07 (2.69)	44.31 (9.03)	38.48 (6.91)	34.86 (6.04)
RMGT-mutKNN-HM	16.99 (2.45)	**28.00 (4.67)**	**7.50 (2.43)**	44.73 (5.48)	40.53 (4.37)	31.12 (6.35)
RMGT-symFKNN-HM	18.88 (2.26)	30.56 (5.52)	7.92 (2.58)	44.68 (7.89)	38.48 (6.67)	34.61 (6.25)
RMGT-symKNN-LLE	19.04 (1.19)	30.63 (3.94)	7.91 (2.49)	42.83 (6.00)	42.25 (3.32)	36.61 (4.79)
RMGT-mutKNN-LLE	17.85 (1.95)	29.49 (4.16)	7.53 (2.11)	43.75 (6.40)	42.12 (4.08)	33.89 (5.32)
RMGT-symFKNN-LLE	18.97 (1.18)	30.41 (3.71)	7.73 (2.43)	**42.75 (7.33)**	41.77 (3.33)	36.25 (4.78)

Table 2. Average rankings of the graph construction methods for each SSL algorithm

	GRF	LGC	LapRLS	LapSVM	RMGT	mean
symKNN-RBF	2.9167	2.8333	3.5	5.1667	5.1667	3.9167
mutKNN-RBF	2.6667	3	**3.1667**	2	4.8333	3.1333
symFKNN-RBF	**2.5833**	**1.6667**	3.25	4.6667	5	3.4333
symKNN-HM	6	6.5	6	7.75	6.25	6.5
mutKNN-HM	3.8333	4.5833	4.25	3.4167	**3.5**	3.9167
symFKNN-HM	5	5.6667	4.8333	6.9167	5.0833	5.5
symKNN-LLE	8.3333	8	7.9167	7	6.1667	7.4833
mutKNN-LLE	5.8333	5.4167	4.6667	3.1667	4.1667	4.65
symFKNN-LLE	7.8333	7.3333	7.4167	4.9167	4.8333	6.4667

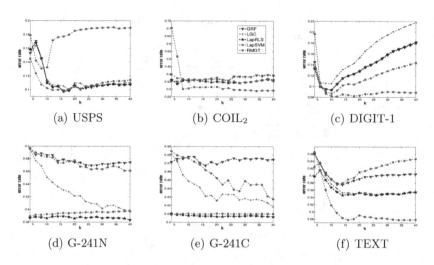

(a) USPS   (b) COIL$_2$ (c) DIGIT-1

(d) G-241N   (e) G-241C   (f) TEXT

Fig. 1. Average error rates of the SSL algorithms using the mutKNN-RBF graph

Fig. 1 shows the results for this empirical analysis using the mutKNN-RBF graph as we vary the sparsification parameter value. Notice that the legend for all graphics in Fig. 1 can be found in Fig. 1(b). The RMGT algorithm achieved good classification performance and stability on the COIL$_2$, TEXT, and DIGIT-1 data sets when $k \geq 14$. However, RMGT was generally the worst classifier for the USPS data set and the second worst for the G-241N and G-241C data sets. Moreover, RMGT appears to be unstable for relatively small values of k. For instance, the instability of RMGT is evidenced in the COIL$_2$ data set for $k \leq 6$ while all other classifiers achieved satisfactory results with this setting.

LapRLS and LapSVM achieved exceptional stability on the G-241C and G-241N data sets. Due to this high stability, we suppose that LapRLS and LapSVM may be the best SSL algorithms for classification of Gaussian distributions. We also note in Fig. 1 that the assumption that sparse graphs give better results than dense graphs may not necessarily be true. For instance, the results for the GRF, LGC, and RMGT algorithms on the G-241C and G-241N data sets using dense graphs are better than those for sparse graphs. In addition, the results for all SSL algorithms on the TEXT data set for relatively small values of k are not satisfactory while the results for the LGC, LapRLS, and RMGT algorithms with dense graphs are.

Influence of graph construction. We now evaluate how different graphs can influence the classification performance of the SSL algorithms. Once again, we perform this analysis as we vary the sparsification parameter value in order to analyze the stability of the graph construction methods combined with the SSL algorithms. Due to lack of space, we only present the plots for the USPS data set in Fig. 2. Once again, we invite the interested reader to check the paper's website. It is clear from Fig. 2 that the results show a lot of variability. For a given classifier, we can observe that several graph construction methods figure among the best and the worst method as we vary the value of k. The variability problem is more intense for small values of k, specially $k \leq 14$. This seems to be a permissive problem since small values of k performed better for this specific data set, but too small values might greatly degrade the classifiers' performance.

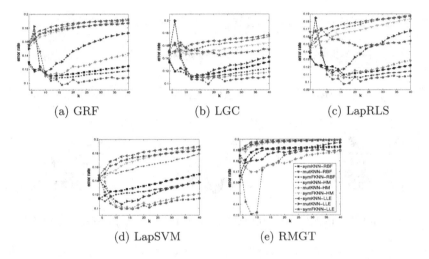

(a) GRF   (b) LGC   (c) LapRLS

(d) LapSVM   (e) RMGT

Fig. 2. Average error rates of the graph construction methods on the USPS data set

The USPS data set is an excellent example of the high influence of the graph construction methods and the sparsification parameter values over the SSL algorithms. As we vary the k parameter, the classifiers' performance vary significantly; some of them in a range of almost 10%. Such performance variation is certainly a concern for the practitioner, who would have difficulties in finding a parameter setting that guarantees a good classification performance. Moreover, such high variability causes several changes in the relative rankings of the classifiers. In some cases, the same classifier might figure among the best and the worst methods as we vary the k parameter in the narrow range of $[4, 14]$. These changes of relative order may cause some serious concerns for the research community. Without an extensive analysis of the influence of parameter values, some studies may experimentally show that a proposed algorithm outperforms the state-of-the-art algorithms, being that this conclusion only holds for certain parameter values. We are not claiming here that such an incident has ever happened, and we have not observed any such evidence, however; it is certainly undesirable for the research community to be affected of such a situation.

We suggest that every research paper that proposes a new SSL algorithm or graph construction method to fully analyze the influence of its parameters. The experimental setup used in this paper is a proposal of how newly proposed methods should be evaluated. It is important to evaluate the algorithms' performance for a wide range of *external* parameters, such as k, and graph construction methods. Some algorithms also have *internal* parameters, such as regularization parameters, that also need to be evaluated, as we show next.

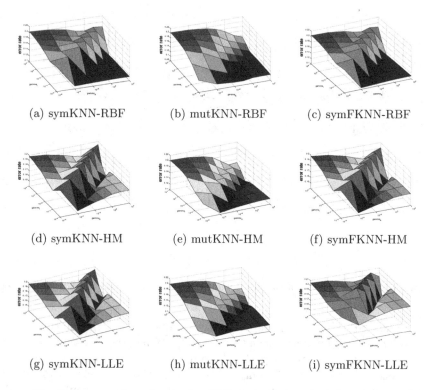

(a) symKNN-RBF   (b) mutKNN-RBF   (c) symFKNN-RBF

(d) symKNN-HM   (e) mutKNN-HM   (f) symFKNN-HM

(g) symKNN-LLE   (h) mutKNN-LLE   (i) symFKNN-LLE

Fig. 3. Error surfaces for the LapRLS algorithm on the USPS data set

Influence of regularization parameters. We evaluate the influence of regularization parameters on the classification performance of the graph-based SSL algorithms. We evaluate the error surfaces generated by the SSL algorithms for each graph construction method and data set. Due to lack of space, we only show the most relevant results for LapRLS and LapSVM. We fixed the value of k that achieved the best error rate for each combination of SSL algorithm and graph construction method. In the sequence, we varied the values of γ_A and γ_I, as described in Section 4.3.

Fig. 3 shows the results for LapRLS on the USPS data set for different graph construction methods. We see that mutKNN generated smoother error surfaces than symKNN and symFKNN graphs, independent of the weighted matrix generation method used.

Many of the obtained results for this analysis are qualitatively equivalent fixing an SSL algorithm and a data set. However, we found some specific results that have an explicit pattern for parameters choice and others which may not have any evident pattern. We discuss them in the following.

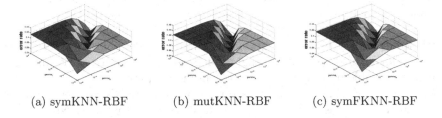

(a) symKNN-RBF   (b) mutKNN-RBF   (c) symFKNN-RBF

Fig. 4. Error surfaces for LapRLS on the TEXT data set using the RBF kernel

Fig. 4 shows the obtained results for LapRLS on the TEXT data set using the RBF kernel combined with the adjacency graph construction methods. We see that the "optimal region" occurs only when $\gamma_A = \gamma_I$.

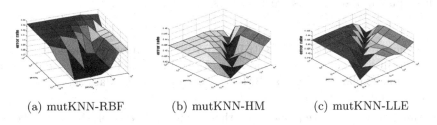

(a) mutKNN-RBF   (b) mutKNN-HM   (c) mutKNN-LLE

Fig. 5. Error surfaces for LapSVM on the TEXT data set using the mutKNN graph

Fig. 5 shows the obtained results for LapSVM on the TEXT data set using the mutKNN graph combined with the weighted matrix generation methods. We can not see any evident pattern that could help parameter choice. This may be an obstacle to apply LapSVM on real applications on text classification. For instance, the "optimal region" for the mutKNN-LLE graph occurs when $\gamma_A = \gamma_I$, which is not a good setting for the other graphs.

5   Conclusions and Further Research

In this paper, we provided a detailed empirical comparison of five state-of-the-art, graph-based SSL algorithms combined with three adjacency graph construction methods and three weighted matrix generation methods. Our experimental evaluation indicated that the SSL algorithms are strongly affected by the graph sparsification parameter value and the choice of the adjacency graph construction and weighted matrix generation methods. The algorithms that have regularization parameters were also very dependent on a good setting of these parameters.

Consequently, we proposed an experimental setup that should be used in empirical comparisons in future work in SSL. Our idea is that a newly proposed algorithm should not be compared to other state-of-the-art algorithms using only the best case analysis. We believe that a detailed evaluation of all parameters is necessary. Due to the nature of SSL, in which there exists only a limited number of labeled examples, tuning all parameters might be unfeasible. Therefore, there is a need for algorithms that are slightly dependable on parameter tuning, i.e., that have a stable performance over the parameter space.

Our experimental results showed a superiority of mutKNN over the symKNN and symFKNN graphs. However, our results also showed that mutKNN is unstable for a relatively small value of k. In addition, we showed that mutKNN tends to generate smoother error surfaces than symKNN and symFKNN graphs. Our experiments also indicated a superiority of the RBF kernel in comparison to the HM and LLE methods.

As we analyzed our experimental results, we noticed other interesting patterns that we could not verify given the lack of experimental evidence. We propose an investigation concerning the validity of these observations as future research. Our empirical observations are as follows:

- Although RMGT achieved satisfactory results on most data sets, it did not perform well on the USPS data set. As USPS is an imbalanced dataset, a possible explanation is that RMGT is not effective on data sets with imbalanced labels;
- Maier et al. [10] have pointed out that the mutKNN graph should be chosen if one is only interested in identifying the "most significant" cluster. Based on this statement, we suppose that mutKNN is the best graph when we are dealing with data sets with imbalanced labels because it may identify the "most significant" class (the minority class in this case). This hypothesis is supported by the fact that, in Table 1, mutKNN achieved better classification performance than symKNN and symFKNN for all combinations of SSL algorithm and weighted matrix generation method on the USPS data set;
- Table 1 shows that RMGT achieved the best overall classification performance in 4 out of 6 data sets. This surprising classification performance may be due to the addition of the normalization constraints $\mathbf{F}\mathbf{1}_c = \mathbf{1}_n$ and $\mathbf{F}^\top \mathbf{1}_n = n\boldsymbol{\omega}$ in the optimization framework. It would be interesting to investigate if other SSL algorithms' classification performances could be improved if these constraints were included in their optimization framework;
- In Fig. 4, we observed that the "optimal region" occurs only when $\gamma_A = \gamma_I$. Since this behavior occurred for all graphs (the other results are not shown here due to lack of space), we ask if this setting should be chosen for text classification tasks when using LapRLS.

Acknowledgments. This research was supported by the Brazilian agencies CAPES and FAPESP. Thanks to Diego F. Silva, Vinícius M. A. de Souza, Rafael Giusti, Ricardo M. Marcacini, and Rafael G. Rossi for their help in the experiments.

References

1. Belkin, M., Niyogi, P., Sindhwani, V.: Manifold regularization: A geometric framework for learning from labeled and unlabeled examples. JMLR 7, 2399–2434 (2006)
2. Beyer, K., Goldstein, J., Ramakrishnan, R., Shaft, U.: When is "nearest neighbor" meaningful? In: Beeri, C., Bruneman, P. (eds.) ICDT 1999. LNCS, vol. 1540, pp. 217–235. Springer, Heidelberg (1998)
3. Chapelle, O., Schölkopf, B., Zien, A.: Semi-supervised learning. The MIT Press (2006)
4. Demšar, J.: Statistical comparisons of classifiers over multiple data sets. JMLR 7, 1–30 (2006)
5. Hein, M., Maier, M.: Manifold denoising. In: NIPS 19, pp. 561–568 (2007)
6. Jebara, T., Wang, J., Chang, S.F.: Graph construction and b-matching for semi-supervised learning. In: ICML, pp. 441–448 (2009)
7. Johnson, R., Zhang, T.: On the effectiveness of laplacian normalization for graph semi-supervised learning. JMLR 8, 1489–1517 (2007)
8. Liu, W., Chang, S.F.: Robust multi-class transductive learning with graphs. In: CVPR, pp. 381–388 (2009)
9. Liu, W., He, J., Chang, S.F.: Large graph construction for scalable semi-supervised learning. In: ICML, pp. 679–686 (2010)
10. Maier, M., Hein, M., von Luxburg, U.: Optimal construction of k-nearest-neighbor graphs for identifying noisy clusters. Theoretical Computer Science 410(19), 1749–1764 (2009)
11. Melacci, S., Belkin, M.: Laplacian support vector machines trained in the primal. JMLR 12, 1149–1184 (2011)
12. Ozaki, K., Shimbo, M., Komachi, M., Matsumoto, Y.: Using the mutual k-nearest neighbor graphs for semi-supervised classification of natural language data. In: CoNLL, pp. 154–162 (2011)
13. Roweis, S., Saul, L.: Nonlinear dimensionality reduction by locally linear embedding. Science 290(5500), 2323–2326 (2000)
14. Zhou, D., Bousquet, O., Lal, T., Weston, J., Schölkopf, B.: Learning with local and global consistency. In: NIPS 16, pp. 321–328 (2004)
15. Zhu, X.: Semi-supervised learning literature survey. Tech. Rep. 1530, Computer Sciences, University of Wisconsin-Madison (2005)
16. Zhu, X., Ghahramani, Z., Lafferty, J.: Semi-supervised learning using gaussian fields and harmonic functions. In: ICML, pp. 912–919 (2003)

Tractable Semi-supervised Learning of Complex Structured Prediction Models

Kai-Wei Chang[1], S. Sundararajan[2], and S. Sathiya Keerthi[3]

[1] Dept. of Computer Science, University of Illinois at Urbana-Champaign, IL USA
[2] Microsoft Research India, Bangalore, India
[3] Cloud and Information Services Lab, Microsoft, Mountain View, CA
kchang10@illinois.edu, {ssrajan,keerthi}@microsoft.com

Abstract. Semi-supervised learning has been widely studied in the literature. However, most previous works assume that the output structure is simple enough to allow the direct use of tractable inference/learning algorithms (e.g., binary label or linear chain). Therefore, these methods cannot be applied to problems with complex structure. In this paper, we propose an approximate semi-supervised learning method that uses piecewise training for estimating the model weights and a dual decomposition approach for solving the inference problem of finding the labels of unlabeled data subject to domain specific constraints. This allows us to extend semi-supervised learning to general structured prediction problems. As an example, we apply this approach to the problem of multi-label classification (a fully connected pairwise Markov random field). Experimental results on benchmark data show that, in spite of using approximations, the approach is effective and yields good improvements in generalization performance over the plain supervised method. In addition, we demonstrate that our inference engine can be applied to other semi-supervised learning frameworks, and extends them to solve problems with complex structure.

1 Introduction

Over the past decade, a variety of semi-supervised learning methods have been suggested in the literature; these methods use unlabeled data to yield a good lift in generalization performance when labeled data is sparse. One particular model, initially proposed by Joachims [12] for binary support vector machines, has given impressive results on problems involving large feature spaces, such as those encountered in text classification and natural language processing. This model has been nicely extended to multi-class classification, ordinal regression and structured output problems [31,5,2]. The key ideas behind these methods are: (a) using the labels of the unlabeled data ($\mathbf{Y}_U$) as extra variables and the associated loss function in training; (b) optimizing the model weight vector (θ) and $\mathbf{Y}_U$ via alternating optimization steps; (c) using constraints on $\mathbf{Y}_U$ that come from domain knowledge to effectively guide the training towards good solutions; and (d) employing annealing to avoid getting caught in local minima.

The probabilistic structured output model in [5] is useful only when the output structure is simple, e.g., linear chains, that allows inference and learning computations to be done in a tractable fashion. In this paper we go beyond and focus on problems with more complex output structure. For such problems, making approximations in inference and

H. Blockeel et al. (Eds.): ECML PKDD 2013, Part III, LNAI 8190, pp. 176–191, 2013.
© Springer-Verlag Berlin Heidelberg 2013

learning is inevitable. In many practical situations, the complex output structure is such that: (a) the output has several components with sparse inter-component intersections; and (b) learning computations involving each component viewed as a separate piece is tractable (known as piecewise training [32]). Thus, if we replace the likelihood terms in the semi-supervised training objective function with composite likelihood [20] terms involving the components, then the θ determination step becomes tractable; we do not obtain the component marginals from the underlying intractable joint distribution. The components-based structure also lets us to use dual decomposition based inference techniques [16,14] which have matured over recent years. The use of these approximations in the semi-supervised learning method is neat since alternating optimization allows the two approximations to be used independently in an iterative loop without crossing each other.

We illustrate our approach by applying it to the problem of multi-label classification and developing all the details for it. Detailed empirical analysis on several benchmark datasets shows the effectiveness of the approach. Despite the approximations, the semi-supervised method gives good lift in performance over the plain supervised method. Also, when tested on small datasets where exact inference and learning are feasible, the generalization performance of our method is competitive with that obtained by the semi-supervised learning method using exact inference and exact learning. Furthermore, our inference engine could be used with other semi-supervised learning methods; experimental comparison with such methods show that the proposed method performs significantly better.

The rest of the paper is organized as follows. We review the related work in Section 2. In Section 3 we provide a background on semi-supervised learning, label assignment problem, and composite likelihood. Section 4 gives the details behind the determination of $\mathbf{Y}_U$ subject to domain constraints. Section 5 develops the details of the proposed method for the multi-label classification problem. Experimental results are given in Section 6. Section 7 concludes the paper.

2   Related Work

Some related works have a resemblance to the method proposed in this paper. Composite likelihood methods [20], Piecewise training [32], message-passing algorithm [10], and large-margin methods with approximate inference [30,6,18] have been proposed for approximate learning. However, they have been used only in the supervised learning. Dual decomposition methods have been combined with training in the supervised learning of large margin models [26,15], but the ideas do not transfer to semi-supervised learning and probabilistic models.

Approximate inference algorithms for general structure (e.g., [9,14,16,28,25]) have been widely studied in the literature. However, most of them do not consider solving the inference problem with constraints from prior knowledge. On the other hand, many studies have been conducted to solve specific inference problems with constraints (e.g., [17,3]). Recently, Martins et al. [24] proposed a decomposition method to solve general label assignment problems with first-order logic constraints. However, it is not clear how to inject the corpus-level constraints into their framework. In this paper, we extend a dual decomposition method [16] to solve general constrained inference problem involved in our semi-supervised learning framework.

Several semi-supervised learning algorithms [2,23,36,40,1,34,7] have been proposed in the literature. However, with a few exceptions, they all focus on the cases that exact inference and learning are tractable. In such cases, there is no need to use an approximate algorithm. Note that for some problems with a special structure, efficient exact algorithms have been developed (e.g., Viterbi algorithm for linear chain structure). They might be faster than a general approximate inference/learning tool. We refer the readers to an example in [3, Section 6.1], where they showed that by using a dynamic programming algorithm to leverage the problem structure, a Lagrangian relaxation method is more efficient than a general approximate inference solver. In addition, some works have been conducted to study semi-supervised learning approach for multi-label classification problems [4,8,22,38]. However, they do not use constraints from domain knowledge and the settings are different from ours.

In the following, we briefly discuss the connection of our method to other semi-supervised learning frameworks. Our work is closely related to [5] and [29]. These probability models have been shown to generate the state-of-the-art results on problems with a tractable structure, but they are not directly applicable to problems with complex structured outputs. When exact inference and learning are tractable, our model is reduced to a probability model similar to them. Without corpus level constraints, our model is also related to CoDL [2]. However, CoDL only demonstrate the results using exact inference algorithm and Perceptron style learning steps, while our probabilistic method is more general. The posterior regularization [7] and Generalized Exception [23] are probabilistic methods that has several key differences from our method: (a) they enforce domain constraints only in an expectation sense; and (b) it is unclear if they applied to problems with complex structured outputs. Transductive SVM proposed in [40] considers extending structured SVM to a semi-supervised setting. The method is complicated and the study is only conducted on problems with simple structures (linear chain and multi-class). Moreover, in their experiments they do not incorporate prior knowledge via constraints, which are crucial to get good performance in semi-supervised learning. Yu [36] also considers an extension of a large-margin method and regularizes the model with the labels assigned to the unlabeled data. Their framework is different from us, and it is unclear how to train their model on problems with complex structure. Lee et al. [19] proposed a semi-supervised discriminative random field algorithm with approximate inference. However, their method doesn't incorporate constraints. We will show the value of using constraints in Section 6.2.

Our primary focus is on semi-supervised structured prediction problems whose output structure is too complex to allow exact inference and learning. Therefore, comparing different semi-supervised learning framework on the problems with tractable structure is not the focus of this paper. Nevertheless, we show that our inference engine can be plugged in other semi-supervised learning frameworks such as transductive SVM [40] and CoDL [2] in Section 6.5.

3 Semi-supervised Learning

3.1 Learning Problem

Consider a structured output problem in which a data instance consists of an input vector $\mathbf{x} \in \mathcal{X}$ and a label vector $\mathbf{y} \in \mathcal{Y}$. For example, in sequence labeling, $\mathbf{x}$ is a sequence of tokens $\{x^1, \ldots, x^l\}$ and $\mathbf{y}$ is a sequence of scalar labels $\{y^1, \ldots, y^l\}$. We are interested in discriminative models to determine $\mathbf{y}$ for given $\mathbf{x}$. This is done by using a feature vector $f(\mathbf{x}, \mathbf{y})$ and a parameter vector θ which define a scoring function $s(\mathbf{x}, \mathbf{y}, \theta) = \theta \cdot f(\mathbf{x}, \mathbf{y})$. Then inference is done as $\mathbf{y}^* = \arg\max_{\mathbf{y}} s(\mathbf{x}, \mathbf{y}, \theta)$. We assume that the scoring function $s(\mathbf{x}, \mathbf{y}, \theta)$ can be written as:

$$s(\mathbf{x}, \mathbf{y}, \theta) = \sum_c \phi_c(\mathbf{y}_{\pi_c}), \tag{1}$$

where c is some break-up of $s(\cdot)$ into sub-scores, $\pi_c \subset \{1, \ldots, N\}$ is an index set associated with c, and $\mathbf{y}_{\pi_c} = \{y_j : j \in \pi_c\}$ is the label assignment on c. For this paper c can be taken as a break up into sub-problems such that each sub-problem, $\arg\max_{\mathbf{y}_{\pi_c}} \phi_c(\mathbf{y}_{\pi_c})$ is easy to solve, and the dependency among the variables $\mathbf{y}_{\pi_c}$ is considered only within the component c. For example, when single variables are considered, we are ignoring the label dependency and using a simple model for marginal probabilities instead of computing them using the entire graph. More examples can be found in, for example, [39]. Note that we have suppressed the dependency of the potentials on $\mathbf{x}$ and θ in (1) for ease of notation.

For probabilistic models, we can define conditional probability using the scoring function: $p(\mathbf{y}|\mathbf{x}, \theta) \propto \exp(s(\mathbf{x}, \mathbf{y}, \theta))$. If $(\mathbf{X}, \mathbf{Y})$ is a set of data instances $\{(\mathbf{x}, \mathbf{y})\}$, then $p(\mathbf{Y}|\mathbf{X}, \theta)$ can be re-written as the product of $p(\mathbf{y}|\mathbf{x}, \theta)$ assuming samples are drawn i.i.d from a fixed distribution. For ease of notation we will simply refer to these quantities as $p_\theta(\mathbf{Y})$ and $p_\theta(\mathbf{y})$.

Let $(\mathbf{X}_L, \mathbf{Y}_L) = \{(\mathbf{x}_L, \mathbf{y}_L)\}$ denote the set of all labeled instances. Consider the supervised learning problem of determining θ by maximizing the regularized log-likelihood:

$$\max_\theta S(\theta) = R(\theta) + \mathcal{L}(\mathbf{Y}_L; \mathbf{X}_L, \theta),$$

where $R(\theta) = -\|\theta\|^2/2\sigma^2$ is a regularizer and $\mathcal{L}(\mathbf{Y}_L; \mathbf{X}_L, \theta) = \frac{1}{n_L} \sum_{\mathbf{x}_L} \mathcal{L}_{\mathbf{x}_L}(\mathbf{y}_L; \mathbf{x}_L, \theta)$ is the log likelihood term and n_L is the number of labeled instances. $\mathcal{L}_{\mathbf{x}_L}$ is the instance level log likelihood; in the probabilistic model

$$\mathcal{L}_{\mathbf{x}_L}(\mathbf{y}_L; \mathbf{x}_L, \theta) = \log \frac{\exp(s(\mathbf{x}_L, \mathbf{y}_L, \theta))}{\sum_{\mathbf{y} \in \mathcal{Y}} \exp(s(\mathbf{x}_L, \mathbf{y}, \theta))}.$$

In semi-supervised learning, θ is learned from both labeled data $(\mathbf{X}_L, \mathbf{Y}_L)$ and unlabeled data $\mathbf{X}_U$, and we consider the following optimization problem instead:

$$\max_{\theta, \mathbf{Y}_U} S(\theta) + \mathcal{L}(\mathbf{Y}_U; \mathbf{X}_U, \theta) \quad \text{s.t.} \quad \mu(\mathbf{X}_U, \mathbf{Y}_U) \geq \mathbf{c}, \tag{2}$$

where $\mathbf{Y}_U$ is the labels assigned to the unlabeled data during the learning. The vector valued domain constraints $\mu(\mathbf{X}_U, \mathbf{Y}_U) \geq \mathbf{c}$ (discussed below) are included to guide the semi-supervised learning algorithm towards good solutions.

We can perform alternating optimization over θ and $\mathbf{Y}_U$ to solve (2). While the optimization of θ can be done using a standard gradient based optimization routine such as LBFGS, optimization of $\mathbf{Y}_U$ can be done by solving a label assignment problem with constraints. If the structure is complex, both, optimizing θ and optimizing $\mathbf{Y}_U$ will be difficult. In the following, we will discuss how to use approximate inference and learning algorithms to learn θ and $\mathbf{Y}_U$.

3.2   Label Assignment Problem (Inference)

To simplify notations let us use $\mathbf{Y}$, $\mathbf{X}$, $\mathbf{y}$ and $\mathbf{x}$ to represent $\mathbf{Y}_U$, $\mathbf{X}_U$, $\mathbf{y}_U$ and $\mathbf{x}_U$. The label assignment problem subject to inequality constraints is given by:

$$\max_{\mathbf{Y}} \mathcal{L}(\mathbf{Y}; \mathbf{X}, \theta) \quad \text{s.t.} \quad \mu(\mathbf{X}, \mathbf{Y}) \geq \mathbf{c}. \tag{3}$$

We can use (1) to rewrite (3) as:

$$\max_{\mathbf{Y}} \frac{1}{n_u} \sum_i \sum_c \phi_{i,c}(\mathbf{y}_{i,\pi_c}) \quad \text{s.t.} \quad \mu(\mathbf{X}, \mathbf{Y}) \geq \mathbf{c}, \tag{4}$$

where the index i refers to i^{th} unlabeled example, c refers to a clique and n_u denotes the number of unlabeled examples. We assume that the constraint function μ can be written as: $\mu(\mathbf{X}, \mathbf{Y}) = \frac{1}{n_u} \sum_i \mu_i(\mathbf{x}_i, \mathbf{y}_i)$. We will also assume that, for each example, say the i-th, μ_i decomposes clique-wise, in the same way as (1). Thus, $\mu_i(\mathbf{x}, \mathbf{y})$ can be written as: $\sum_c \gamma_{ic}\mu_{ic}(\mathbf{x}, \mathbf{y}_c)$ where some γ_cs can be zero. These assumptions hold in most practical structured output scenarios; see [2,7,5] for many examples of constraints arising in different problems. Later we will describe constraints arising in multi-label classification. A constraint could be *instance level* (e.g., a particular label has to occur only once in an example) or *corpus level* (e.g., the number of occurrences of a particular label in all the examples is some number). Both these types of constraints fall in the general constraint function format described above. Note that an instance level constraint is a special case of a corpus level constraint and is obtained by setting $\mu_i(\mathbf{x}_i, \mathbf{y}_i) = 0$ for all i except one of them.

If all the constraints are *instance level*, the solution to (4) can be obtained by solving the inference problem on each example independently. Otherwise, joint inference is required. We will discuss the joint inference approach later. The inference problem can be solved exactly and efficiently only for restricted structured output types (e.g., linear chain, tree with low-width). For general structured output problems, *Master-Slave* type methods have been proposed to solve the inference problem.

3.3   Composite Likelihood Maximization (Learning)

In general structured prediction problems with graphs having cycles, learning algorithms for probabilistic models are intractable due to the need to handle the partition function. To alleviate the computational intractability of parameter estimation in supervised learning, *composite marginal or conditional likelihood* maximization [20] and *piecewise training* methods [32] have been proposed. Such models are useful not only to reduce computational complexity but also to provide robustness to model misspecification via using the simpler interactions. We make use of the piecewise training approach to learn the model parameter vector θ in our semi-supervised learning setting. In this approach, the likelihood is approximated by the composition of likelihoods of the components. That is, $p_\theta(\mathbf{y}) = \exp\{\mathcal{L}(\mathbf{y}; \theta, \mathbf{x})\} \approx \prod_c p(\mathbf{y}_{\pi_c}; \theta_c)$, where $p(\mathbf{y}_{\pi_c}; \theta_c) = \frac{\exp(\phi_c(\mathbf{y}_{\pi_c}; \theta_c))}{\mathcal{Z}_c}$ and $\mathcal{Z}_c = \sum_{\mathbf{y}_{\pi_c}} \exp(\phi_c(\mathbf{y}_{\pi_c}; \theta_c))$. We assume that each component c is tractable (e.g., $\mathbf{y}_{\pi_c}$ is a small subset of variables or c is a tree). Note that we are not marginalizing any underlying intractable joint distribution $p(\mathbf{y}; \theta)$.

Thus we replace the log-likelihood term $\mathcal{L}(\mathbf{Y}; \mathbf{X}, \theta)$ with $\mathcal{L}_C(\mathbf{Y}; \mathbf{X}, \theta) = \sum_c \mathcal{L}_C^c$, $\mathcal{L}_C^c = \sum_i \phi_c(\mathbf{y}_{i,c}) - \sum_i \log(\mathcal{Z}_{i,c})$. Note that the first term (which is critical for inference) remains the same as in the *full* likelihood maximization of general structured prediction models. It is in the second term (involving the partition function) where we make a tractable simplification. It is possible to learn the model parameters of components independently in situations where there is no overlap of parameters between components. In general, we allow overlaps (i.e., share parameters across the components) keeping tractability in mind so that the parameters of components are optimized together. However, unlike [32] where the components considered are factors of the graphical model, we allow general user specified components involving more than one factor, as long as inference on them is tractable. (In section 5, we illustrate this approach on the multi-label classification problem, using trees as components sharing model parameters in a fully connected graph.)

With this approach, the semi-supervised learning problem (2) can be written as:

$$\max_{\theta, \mathbf{Y}_U} R(\theta) + \mathcal{L}_C(\mathbf{Y}_L; \mathbf{X}_L, \theta) + C_m \mathcal{L}_C(\mathbf{Y}_U; \mathbf{X}_U, \theta) \quad \text{s.t.} \quad \mu(\mathbf{X}_U, \mathbf{Y}_U) \geq \mathbf{c}. \quad (5)$$

where C_m is a regularization parameter introduced to provide annealing capability, which we discuss next.

3.4 Annealing Steps for Solving (5)

The objective function in (5) (and in (2)) is a non-concave function. In practice, we can apply annealing steps using the parameter C_m to avoid being trapped in a bad local minimum. The optimization procedure has a double loop. In the outer loop, we gradually increase C_m from a small positive value (e.g., tripling the value in every iteration starting from 10^{-4}) to one. This allows the unlabeled data to gradually influence the modeling process in achieving a better optimum. In the inner loop, we alternatively update the label assignment $\mathbf{Y}_U$ and the model θ, where θ is updated using the LBFGS routine [21]. For efficiency sake we set the maximal number of LBFGS iterations to be 25 and stop the inner loop after five rounds; we found this sufficient to get good solutions.

4 Solving Label Assignment Problems with Constraints

We begin by discussing a master-slave approach [16] for the inference problem without constraints. Then, we describe a joint approximate inference algorithm for the label assignment problem with corpus level constraints.

4.1 Master-Slave Approach

Let $\mathbf{y} = \{y_1, \ldots, y_N\}$ be a vector of random variables associated with an example; $y_j \in \mathcal{Y}_j$ where $\mathcal{Y}_j$ is a discrete set. Consider the inference problem for one example without any constraints.

$$\mathbf{y}^* = \arg\max_{\mathbf{y}} \sum_c \phi_c(\mathbf{y}_{\pi_c}). \tag{6}$$

Note that to avoid notational complexity, we use the same notations as in (1) and assume that the score is re-written as a composition of scores on sub-problems such that each sub-problem c is tractable. Let $\pi_j^{-1} = \{c : j \in \pi_c\}$. For each $a \in \mathcal{Y}_j$, introduce a binary integer variable $z_{j,a}$ indicating whether y_j assumes the value a. Let $\mathbf{z}_j = (z_{j,1}, \ldots, z_{j,|\mathcal{Y}_j|})$, a binary vector. Let Z_j be the set of vector values taken by $\mathbf{z}_j$ and $\mathbf{z}$ be a single vector that collects all $\mathbf{z}_j$. So, there exists an invertible mapping between $\mathbf{z}$ and $\mathbf{y}$. Denote it by $\mathbf{y} = \mathrm{bip}(\mathbf{z})$ and $\mathbf{z} = \mathrm{bip}^{-1}(\mathbf{y})$. Let $\mathbf{z}_{\pi_c} = \{z_j : j \in \pi_c\}$. We slightly abuse notations and write $\phi_c(\mathbf{z}_{\pi_c}) = \phi_c(\mathrm{bip}(\mathbf{z}_{\pi_c}))$. Let $Z_{\pi_c} = \prod_{j \in \pi_c} Z_j$. With these notations, we can rewrite (6) as

$$\max_{\mathbf{z}} \sum_c \phi_c(\mathbf{z}_{\pi_c}) \quad \text{s.t.} \quad z_j \in Z_j, \; j = 1, \ldots, N. \tag{7}$$

In the master-slave approach, new variable vectors $\{\mathbf{z}_{\pi_c}^c\}$ are introduced for each sub-problem and constraints connecting them are introduced via a variable vector $\bar{\mathbf{z}}$ controlled by the master that coordinates an iterative optimization process. Using these variables we can rewrite (7) as

$$\max_{\bar{\mathbf{z}}, \{\mathbf{z}_{\pi_c}^c\}} \sum_c \phi_c(\mathbf{z}_{\pi_c}^c) \quad \text{s.t.} \quad \mathbf{z}_{\pi_c}^c \in Z_{\pi_c} \; and \; \mathbf{z}_{\pi_c}^c = \bar{\mathbf{z}}_{\pi_c} \; \forall c.$$

Then, the Lagrangian min-max dual problem can be written as [16]

$$\min_{\nu} \sum_c \max_{\mathbf{z}_{\pi_c}^c} \left(\phi_c(\mathbf{z}_{\pi_c}^c) + \sum_{j \in \pi_c} \langle \nu_{c,j}, (\mathbf{z}_{\pi_c}^c)_j \rangle \right) \quad \text{s.t.} \sum_{c \in \pi_j^{-1}} \nu_{c,j} = 0 \; \forall j.$$

This problem can be solved using methods such as the projected sub-gradient method [16] or the accelerated dual method [14]. Since we are discussing the case without constraints, the inference problem for each example is independent and so we simply have to repeat the above described method to all examples.

4.2 Joint Inference with Constraints

For corpus level constraints joint inference over all examples is needed. Notations become a bit clumsy: when dealing with all examples we need to use $\mathbf{y}_{i,\pi_c}, \bar{\mathbf{z}}_{i,\pi_c}$ etc., instead of $\mathbf{y}_{\pi_c}, \bar{\mathbf{z}}_{\pi_c}$ etc. We also use $\mathbf{z}_{i,\pi_c}$ as a shorthand for $\mathbf{z}_{i,\pi_c}^c$. Let C_i be the set of components associated with example i. The m-th constraint can be written as: $\sum_i \sum_{c \in C_i} \gamma_{m,i,c} \mu_{m,i,c}(\mathbf{y}_{i,\pi_c}) \geq c_m$. Then the joint optimization problem is given by:

$$\max_{\{\bar{\mathbf{z}}_i\}, \{\mathbf{z}_{i,\pi_c}\}} \sum_i \sum_{c \in C_i} \phi_{i,c}(\mathbf{z}_{i,\pi_c})$$

$$\text{s.t.} \quad \mathbf{z}_{i,\pi_c} \in Z_{\pi_c} \; and \; \mathbf{z}_{i,\pi_c} = \bar{\mathbf{z}}_{i,\pi_c} \; \forall i,c$$

$$\sum_i \sum_{c \in C_i} \gamma_{m,i,c} \mu_{m,i,c}(\mathbf{z}_{i,\pi_c}) \geq c_m, \forall m.$$

Let us define the following functions: $\tilde{\phi}_{i,c}(\mathbf{z}_{i,\pi_c}) = \phi_{i,c}(\mathbf{z}_{i,\pi_c}) + g(\mathbf{z}_{i,\pi_c})$, $g(\mathbf{z}_{i,\pi_c}) = \sum_m \eta_m(\gamma_{m,i,c}\mu_{m,i,c}(\mathbf{z}_{i,\pi_c}) - c_m)$, $h(\nu_c^{(i)}; \mathbf{z}_{i,\pi_c}) = \sum_{j \in \pi_{i,c}} \langle \nu_{c,j}^{(i)}, (\mathbf{z}_{i,\pi_c})_j \rangle$. Then, the corresponding Lagrangian min-max problem is given by:

$$\min_{\{\{\nu^{(i)}\},\{\eta_m\}\}} \max_{\{\mathbf{z}_{i,\pi_c}\}} \sum_i \sum_{c \in \mathcal{C}_i} \left(\tilde{\phi}_{i,c}(\mathbf{z}_{i,\pi_c}) + h(\nu_c^{(i)}; \mathbf{z}_{i,\pi_c}) \right)$$

$$\text{s.t.} \quad \mathbf{z}_{i,\pi_c} \in Z_{\pi_c} \quad \sum_{c \in \pi_{i,j}^{-1}} \nu_{c,j}^{(i)} = 0 \ \forall i, j, \eta_m \geq 0, \ \forall m.$$

We use the projected sub-gradient method to solve this problem since the inner *maximization* is not differentiable. Note that for fixed $\{\eta_m\}$, the dual variables $\{\nu^{(i)}\}$ can be solved independently for each i; this is possible due to the decomposable nature of the constraint functions across the examples. The examples get coupled only via the domain constraint dual variables. Therefore, we use an alternate optimization strategy of optimizing $\{\eta_m\}$ and $\{\nu^{(i)}\}$. Essentially, we run a projected sub-gradient based algorithm over several iterations in an inner loop, and run a similar algorithm in an outer loop.

Optimizing $\{\nu^{(i)}\}$ with Fixed $\{\eta_m\}$. Assume that $\{\eta_m\}$ is fixed and consider the sub-problem involving the *i-th* example given below:

$$\min_{\{\{\nu^{(i)}\}\}} \max_{\{\mathbf{z}_{i,\pi_c}\}} \sum_{c \in \mathcal{C}_i} \mathcal{U}(\nu_c^{(i)}, \mathbf{z}_{i,\pi_c}; \{\eta_m\}) \quad \text{s.t.} \quad \mathbf{z}_{i,\pi_c} \in Z_{\pi_c}, \quad \sum_{c \in \pi_{i,j}^{-1}} \nu_{c,j}^{(i)} = 0 \ \forall j$$

where $\mathcal{U}(\nu_c^{(i)}, \mathbf{z}_{i,\pi_c}; \{\eta_m\}) = \tilde{\phi}_{i,c}(\mathbf{z}_{i,\pi_c}) + h(\nu_c^{(i)}; \mathbf{z}_{i,\pi_c})$. We solve the inner maximization problem (i.e., $\hat{\mathbf{z}}_{i,\pi_c} = \arg\max_{\mathbf{z}_{i,\pi_c}} \mathcal{U}(\nu_c^{(i)}, \mathbf{z}_{i,\pi_c}; \{\eta_m\})$ for fixed $\nu_c^{(i)}$. Then, the sub-gradient of $\mathcal{U}(\cdot)$ with respect to $\nu_{c,j}^{(i)}$ is $(\hat{\mathbf{z}}_{i,\pi_c})_j$. (Note that $h(\nu_c^{(i)}; \mathbf{z}_{i,\pi_c})$ is linear in $\nu_c^{(i)}$.) Using the prediction, we make the update: $\nu_{c,j}^{(i)}(t) \leftarrow \nu_{c,j}^{(i)}(t-1) - \gamma_t \Delta \nu_{c,j}^{(i)}$ where t denotes the t-th step and γ_t is the learning rate. Assuming that we start with $\nu^{(i)}$ satisfying the equality constraint, the constraint will be satisfied if $\sum_{c \in \bar{C}_j} \Delta \nu_{c,j}^{(i)} = 0$ where $\bar{C}_j = \{c : c \in \pi_{i,j}^{-1}\}$. This can be ensured by setting $\Delta \nu_{c,j}^{(i)} = (\hat{\mathbf{z}}_{i,\pi_c})_j - \frac{1}{|\bar{C}_j|} \sum_{c \in \bar{C}_j} (\hat{\mathbf{z}}_{i,\pi_c})_j$ (i.e., removing the *mean* from each component's optimal assignment). This update for ν is indeed the Euclidean projection on the feasible set. By assumption, each component c is tractable; therefore, the optimal assignment $\hat{\mathbf{z}}_{i,\pi_c}$ can be easily found. For example, in simple cases involving only a single node or a pair nodes in the graph, the optimal assignment can be found by enumeration. For more complex component such as trees, the *max-product* algorithm [27] can be used to find the optimal assignment.

Optimizing $\{\eta_m\}$ with Fixed $\{\nu^{(i)}\}$. Consider the sub-problem of optimizing $\{\eta_m\}$ given by:

$$\min_{\{\eta_m\}} \max_{\{\mathbf{z}_{i,\pi_c}\}} \sum_i \sum_{c \in \mathcal{C}_i} \mathcal{U}(\{\eta_m\}, \mathbf{z}_{i,\pi_c}; \nu_c^{(i)}) \quad \text{s.t.} \quad \mathbf{z}_{i,\pi_c} \in Z_{\pi_c} \quad \eta_m \geq 0 \ \forall m.$$

We solve this problem using the projected sub-gradient method; the parameters $\eta_m^{(t)}$ are updated as: $\eta_m^{(t)} \leftarrow [\eta_m^{(t-1)} - \tilde{\gamma}_t \Delta \eta_m]^+$ where $+$ indicates projection on the non-negative orthant, and $\Delta \eta_m = \sum_{i,c \in \mathcal{C}_i} (\gamma_{m,i,c}\mu_{m,i,c}(\hat{\mathbf{z}}_{i,\pi_c}) - c_m)$. For fixed $\eta_m^{(t)}$, the optimal assignments $\hat{\mathbf{z}}_{i,\pi_c}, \forall c, i$ can be found as earlier. Once the optimal assignments are obtained, we follow [16, Section IV.B] to obtain the final primal solution.

5 Multi-label Classification Example

The method proposed in Sections 3-4 is applicable to any general structured prediction problem. To demonstrate the usefulness of this method, we show how our method can be applied to the multi-label classification problem and conduct related experiments. Here we use one formulation, given in [6]. We consider a pair-wise fully connected graph (to take care of all label correlations). Then, the scoring function $s(\mathbf{x}, \mathbf{y}; \theta)$ can be written as:

$$s(\mathbf{x}, \mathbf{y}; \theta) = \sum_p s_p(\mathbf{z}_p; \theta_p(\mathbf{z}_p)) + \sum_{p, q \neq p} s_{pq}((\mathbf{z}_{pq}); \theta_{pq}(\mathbf{z}_{pq})) \qquad (8)$$

where the indices p and q run over the nodes (classes), $s_p(\cdot)$, $s_{pq}(\cdot)$ denote the node and edge scores computed using the class and label dependent model parameters $\theta_p(\mathbf{z}_p)$ and $\theta_{pq}(\mathbf{z}_{pq})$; $\mathbf{y} = \{-1, 1\}^K$ is a K-dimensional vector, where K is the number of classes. With binary label assignment for each class, $\mathbf{z}_p \in \mathcal{Z}_p$ is a 2-dimensional vector, and $\mathcal{Z}_p = \{(1, 0), (0, 1)\}$. $\mathbf{z}_{pq} \in \mathcal{Z}_{pq}$ is a 4-dimensional binary vector with only one *unit* element. There exists a mapping between $\mathbf{y}$ and $\mathbf{z}$ as described in Section 4.1: $\mathbf{z}_p = \text{bip}^{-1}(y_p)$ and $\mathbf{z}_{pq} = \text{bip}^{-1}(y_p, y_q)$. For linear models, the scores are computed as: $s_p(\mathbf{z}_p; \theta_p(\mathbf{z}_p)) = \theta_p(\mathbf{z}_p) \cdot \mathbf{x}$ and $s_{pq}(\mathbf{z}_{pq}; \theta_{pq}(\mathbf{z}_{pq})) = \theta_{pq}(\mathbf{z}_{pq}) \cdot \mathbf{x}$, where $\mathbf{x}$ is the feature vector. This setting has been used to study approximate learning or inference in the supervised learning setting [6,26,39]. However, we are not aware of any existing work studying this formulation in the semi-supervised setting.

Composite Likelihood. Given the score having the form given in (8), composite likelihood can be defined in many different ways (e.g., [39]). In this paper, we define a composite likelihood function composed of K *spanning* tree models where each tree has one class at its *root* and the remaining classes as leaf nodes. Then, we can write the score for each tree as:

$$s_k(\mathbf{x}, \mathbf{y}; \bar{\theta}_k) = \frac{1}{K} \sum_p \theta_p(\mathbf{z}_p) \cdot \mathbf{x} + \frac{1}{2} \sum_{q \neq k} \theta_{pq}(\mathbf{z}_{pq}) \cdot \mathbf{x}$$

and the likelihood function as $p_\theta(\mathbf{Y}) = \prod_k p_k(\mathbf{Y}; \bar{\theta}_k)$ where $\bar{\theta}_k = \{\theta_p, \theta_{p,q} : q \neq k, p = 1, \ldots, K\}$, $p_k(\mathbf{Y}; \bar{\theta}_k) = \frac{\exp(s_k(\mathbf{x}, \mathbf{y}; \bar{\theta}_k))}{\sum_\mathbf{y} \exp(s_k(\mathbf{x}, \mathbf{y}; \bar{\theta}_k))}$. Note that $s(\mathbf{x}, \mathbf{y}; \theta) = \sum_k s_k(\mathbf{x}, \mathbf{y}; \theta)$ (scaling factors ensure the equality), and potentials are shared across the models. Therefore, all the models are learned jointly. [1] Since each sub-problem is a tree, the partition function and its gradient can be easily computed[2]; also, inference can be efficiently done. [3] See Komodakis et al, [16] for a discussion on the choice of sub-problems used in the decomposition.

Constraints. Following the discussions in Section 3.2, we consider two types of corpus-level constraints: label distribution constraint (LDC) and label correlation constraint

---

[1] For other complex structured outputs, if the potentials are not shared then components could be learned independently.

[2] Evaluating the partition function of each sub-problem can be done in $O(Kln)$, where K, l, n are the numbers of classes, features, instances, respectively. Therefore, compute the composite likelihood requires $O(K^2 ln)$.

[3] Our inference algorithm is an iterative process. If the number of iterations is fixed, labeling l instances cost $O(K^2 ln)$.

Table 1. Data statistics (l: total number of samples, l_{tri}: number of train samples, l_{tst}: number of test samples, n: number of features, and K: number of classes). The train set is further split into two parts: labeled train and unlabeled train (see text for details).

Data set	l	l_{tri}	l_{tst}	n	K
scene	2,407	1,684	723	294	6
yeast	2,417	1,691	726	103	14
emotions	593	415	178	72	6
rcv	6,000	4,200	1,800	47,236	30
tmc2007	28,596	20,018	8,578	30,438	22

(LCC). LDC constrains the number of times a given label occurs throughout the entire data set and LCC constrains the number of times one label co-occurs with another throughout the data set. In the context of multi-label classification, LDC can be written in the following forms $\sum_{\mathbf{y} \in \mathbf{Y}} \delta(\text{bip}^{-1}(y_p), \mathbf{z}'_p) = n(\mathbf{z}'_p), \forall p = 1, ...K, \mathbf{z}'_p \in \mathcal{Z}_p$, where δ is the Kronecker delta function. LCC is $\sum_{\mathbf{y} \in \mathbf{Y}} \delta(\text{bip}^{-1}(y_p, y_q), \mathbf{z}'_{pq}) = n(\mathbf{z}'_{pq}), \forall p, q \neq p, \mathbf{z}'_{pq} \in \mathcal{Z}_{pq}$. $n(\mathbf{z}'_p)$ and $n(\mathbf{z}'_{pq})$ are given and estimated by counting the occurrence/co-occurrence of labels in the data. There are many other possible constraints that can be used in multi-label problems. For example, we can restrict the number of assigned labels for each sample. Local correlations [11] can be also incorporated in our model via instance level constraints. In the experiments we use LDC and LCC only. We will show the value of these constraints in Section 6.2.

6   Experiments

In this section, we do experiments on multi-label classification to understand the following: (a) how well the proposed semi-supervised learning framework improves over the supervised classifier; (b) the role of different constraints on the performance; (c) the impact of the inference and learning approximations on the performance; and (d) differences between transductive and semi-supervised solutions.

Datasets and Setting. We considered five multi-label datasets from various applications[4]. Table 2 lists the data statistics. For rcv, we only used the first set of data. Because several labels in rcv have only a few positive examples, we removed such labels and only considered the 30 most frequent labels. For each data set, we constructed 10 random train and test splits of 70% and 30%. Then, we performed experiments on different degrees of labeling (d) by further splitting the train set into two parts: labeled train data ($d\%$) and unlabeled train data (($100 - d)\%$). In the training phase, only the labels of labeled train data are used. Because there are only few labels in the training set when the degree of labeling is low, we fixed the regularization parameter to a default value ($\sigma^2 = 1$) instead of tuning it. We checked in our experiments that the conclusions are still valid when σ is tuned using a validation set. We evaluate performance in terms of the Micro-F1 score [35] and conduct Wilcoxon sign-rank test with the significance level of 5%. All results in the tables are reported in terms of the mean and standard deviation over the 10 splits. Unless stated otherwise all results correspond to the case where both, label distribution and

---

[4] Data is available at http://mulan.sourceforge.net.

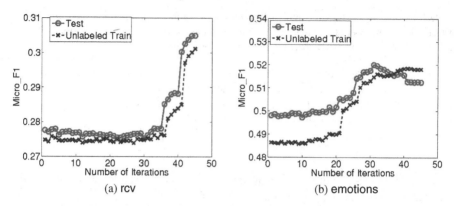

(a) rcv   (b) emotions

Fig. 1. Performance on test set (red) and unlabeled train set (blue) along iterations for $d = 1\%$

label correlation constraints are used. Test set inference is done collectively by treating it as a sample and applying collective inference with the constraints.

6.1   Performance of the Proposed Method

We begin by showing the performance (as a function of optimization iterations) of the proposed semi-supervised learning algorithm with approximate inference and learning. As we mentioned in Section 3.4, we increase the weight of unlabeled data C_m from 10^{-4} to 1 by a factor of 3 and for each C_m we conduct five inner iterations. Therefore, there are 45 total iterations for each trial. Figure 1 shows Micro-F1 scores evaluated on unlabeled data and test data along iterations when $d = 1\%$. We omit the corresponding plots for scene, yeast, and tmc2007 because they are somewhat similar to that on rcv. In most cases, the model achieves better performance as the number of iterations grows. This result is consistent with the observations made in the semi-supervised learning literature, e.g., [2,5]. However, a key difference is that, our results are demonstrated on problems with complex structured outputs. The test set curve of emotions dips a bit at the end in Figure 1(b). This because, in the annealing step, we increase C_m along iterations. As the amount of unlabeled data in emotions is small, large C_m may overemphasize the unlabeled data, resulting in slightly inferior performance. When degree of labeling is 2% or higher, the performance on emotions does not drop in the end. In general, when the amount of unlabeled data is small, over-emphasizing it via large C_m leads to over-fitting.

Next, we show that our semi-supervised classifier (Semi-Sup) is better than the supervised classifier (Sup) that is trained only on the labeled data. We compare the results with various degrees of labeling $d \in \{1, 2, 4, 8\}$. For a fair comparison, when evaluating Sup on test set, the constraints are used. Table 2 shows the results. Overall, the semi-supervised learning algorithm outperforms the baseline classifier. The improvement is statistically significant. As d increases, the Micro-F1 score of the methods increases while the standard deviation decreases, as expected.

In section 3.4, we mentioned the use of annealing to deal with the non-concavity of the objective function in (5). Experiments show the importance of annealing. For

Table 2. Comparison of semi-supervised (Semi-Sup) and supervised (Sup) learning algorithms. Boldface indicates significant improvement of Semi-Sup over Sup.

Dataset	Degree of labeling (d)							
	1%		2%		4%		8%	
	Semi-Sup	Sup	Semi-Sup	Sup	Semi-Sup	Sup	Semi-Sup	Sup
scene	**55.2±3.7**	51.5±2.7	**59.3±2.1**	56.9±2.4	**63.8±1.4**	61.8±2.5	**66.9±1.4**	66.0±1.6
yeast	**42.9±2.1**	42.5±1.8	**43.4±1.0**	43.2±1.2	**45.2±1.2**	44.5±1.0	**45.2±0.9**	44.8±0.9
emotions	**51.3±5.9**	49.9±4.5	**55.5±4.3**	53.8±3.9	**58.8±3.8**	58.3±4.3	**62.2±2.5**	60.5±2.7
rcv	**30.5±2.1**	27.8±2.0	**33.6±1.8**	32.2±1.8	**36.4±0.9**	34.2±1.4	**36.3±1.3**	34.0±1.4
tmc2007	**42.0±1.2**	41.3±1.2	**43.0±1.1**	42.4±1.2	**44.1±0.6**	41.4±1.7	**43.8±0.7**	41.4±1.6

Table 3. Columns 1-4 compare various situations with constraints: without incorporating any constraint (No), using label distribution constraints (LDC), using label correlation constraints (LCC) and using both constraints (Both). Columns 4-7 investigate approximate/exact learning/inference. We use abbreviations to represent the combinations (e.g., ALEI stands for (A)pproximate (L)earning with (E)xact (I)nference). Results are in Micro-F1 (%). The best result (mean) of each column is boldfaced.

Data	d	ALAI			ALAI	ALEI	ELAI	ELEI
		No	LDC	LCC	Both		Both	
scene	1	51.6±2.4	54.6±3.5	53.5±3.6	55.2±3.7	**56.8±5.9**	53.7±5.4	56.6±5.5
	2	58.0±1.5	59.1±2.1	59.1±1.9	59.3±2.1	**62.6±2.0**	60.1±2.6	61.8±2.4
	4	61.2±3.2	63.5±1.6	63.2±1.8	63.8±1.4	**65.2±1.8**	64.2±1.6	64.7±1.9
yeast	1	42.3±2.7	42.8±2.1	42.8±2.3	42.9±2.1	42.1±2.0	**42.9±1.7**	42.2±1.8
	2	42.8±1.5	43.2±1.1	43.0±0.8	**43.4±1.0**	42.5±1.4	43.1±1.3	43.0±1.5
	4	45.0±1.8	**45.2±1.2**	45.2±1.4	45.1±1.2	44.5±1.3	44.5±1.6	44.0±1.4
emotions	1	48.3±6.2	51.0±5.9	51.1±5.7	51.2±5.9	**54.0±5.0**	51.7±6.9	52.9±5.6
	2	53.2±4.5	55.2±3.9	54.7±4.4	**55.5±4.3**	55.2±4.8	54.0±3.7	54.6±5.7
	4	58.0±3.9	**59.0±4.0**	59.0±3.9	58.8±3.8	58.4±3.1	57.0±3.2	58.1±3.1

example, without using annealing, the mean performance of Semi-Sup on scene ($d = 1$) drops from 55.2% to 54.2%. In addition, incorporating constraints during testing improves the performance of the classifier. For example, without using constraints, the mean Micro-F1 scores of Semi-Sup and Sup on scene ($d = 1$) dropped to 54.9% and 49.6% from 55.2% and 51.5% respectively (see Table 2).

6.2 Using Different Sets of Constraints

The first four columns in Table 3 show the results for different uses of constraints. Without using constraints, the performance of Semi-Sup is suboptimal. With label distribution constraint (LDC) the performance is significantly better. In most cases, using label correlation constraints (LCC) obtains similar results as using LDC. However, when both constraints are used (Both), the performance gain is enhanced even further. These results show the importance of using constraints to improve the model.

6.3   Exact versus Approximation

Next, we investigate the performance difference between using an exact algorithm and an approximate algorithm during training and inference. We show the results on three small data sets, scene, yeast and emotions. The number of labels in these data sets is less than 15. For such cases, exact learning and inference algorithms are tractable. For exact inference, we explicitly enumerated all possible assignments of labels and chose the one with the highest objective function value in (6) as the solution. For exact learning, instead of computing composite likelihood, we compute the full likelihood. Computation of the partition function is the main bottleneck associated with exact learning. However, when the size of labels is small, the partition function can be computed exactly using scores for all enumerated label assignments. Columns 4-7 of in Table 3 show the results. In some cases, the approximate algorithm even achieves better performance than the exact one (e.g., ALAI achieves the best result on yeast). However, except on yeast ($d = 1\%, 4\%$), the difference between ELEI and the best result are not statistically significant. In general, the performance of approximate inference/learning is competitive with that of exact inference/learning. Regarding the running time, ALAI takes 4,167 seconds to train a model using yeast data set, while ELEI takes 9,175 seconds. Therefore, the approximation is faster than the exact method, especially when the number of labels is large. This shows the effectiveness of our semi-supervised learning framework with approximate inference and training.

6.4   Transductive Setting Experiments

If the test data is known in advance, it can be used in the semi-supervised learning process as additional unlabeled data. This has the potential to yield better performance on the test data. We refer this as the transductive setting, as opposed to the original inductive setting. Comparing the results of the transductive setting with those in Table 2 for the inductive setting, we found that the transductive setting is statistically significantly better on emotions data but achieves similar performance as the inductive setting on the other two data sets. For example, when degree of labeling is 1%, the mean performances of the transductive setting are 55.1%, 42.9% and 51.7% on scene, yeast, and emotions, respectively; compare this with 55.2%, 42.9% and 51.3% for the inductive setting. The key reason for this is that the original unlabeled train set of emotions is small, and therefore, including test data in training helps it do better.

Moreover, the transductive setting improves the supervised learning setting. For example, using the full train set as labeled data and the test set as the unlabled data to train the model improves the plain supervised learning setting from 74.3% to 74.8% in Micro-F1 on scene data set.

6.5   Incorporate the Inference Engine Proposed in Sec. 4 with Existing Methods

As we mentioned in Section 2, the proposed method is related to several methods in semi-supervised learning literature. However, most existing papers focus on problems with tractable structure outputs (e.g., linear chains), for which there is no need to use an approximate algorithm. Therefore, we are not aware of any existing method that we

Table 4. Comparison of semi-supervised (Semi-Sup) and supervised (Sup) learning algorithms of TSVM+ and CoDL+. Boldface indicates significant improvement of Semi-Sup over Sup or vice versa. We reproduce the results of our model from Table 2 for ease of reference.

Dataset	Degree of labeling (d)							
	1%		2%		4%		8%	
	Semi-Sup	Sup	Semi-Sup	Sup	Semi-Sup	Sup	Semi-Sup	Sup
	TSVM+							
scene	**45.7±4.0**	42.1±2.6	**50.7±2.4**	45.5±3.2	**60.4±1.5**	55.9±1.9	**64.9±1.2**	63.0±1.2
yeast	40.1±2.1	**41.1±2.2**	40.3±1.3	**41.8±1.6**	41.2±1.5	**43.3±1.2**	40.4±1.7	**43.2±1.6**
emotions	46.8±6.4	50.0±5.7	51.9±4.3	51.5±4.2	54.2±3.5	55.3±5.1	56.7±2.9	57.6±1.8
	CoDL+							
scene	**50.2±9.1**	33.8±6.8	**56.0±3.5**	37.8±3.8	**60.8±1.9**	45.4±4.7	**65.3±1.8**	55.4±3.8
yeast	40.9±3.3	39.6±3.0	39.8±2.4	**41.5±1.5**	40.9±1.7	**43.2±1.9**	40.5±1.3	**44.3±1.9**
emotions	**52.6±4.3**	43.0±8.2	**55.9±6.5**	44.8±6.3	**58.6±3.8**	52.5±4.7	**62.0±4.0**	55.8±4.2
	The proposed method							
scene	**55.2±3.7**	51.5±2.7	**59.3±2.1**	56.9±2.4	**63.8±1.4**	61.8±2.5	**66.9±1.4**	66.0±1.6
yeast	**43.9±2.1**	42.5±1.8	**43.4±1.0**	43.2±1.2	**45.2±1.2**	44.5±1.0	**45.2±0.9**	44.8±0.9
emotions	**51.3±5.9**	49.9±4.5	**55.5±4.3**	53.8±3.9	**58.8±3.8**	58.3±4.3	**62.2±2.5**	60.5±2.7

can directly compare with. However, our inference method introduced in Section 4 can be applied to other semi-supervised models. In the following, we show two examples, where we combine our inference engine with CoDL [2] and transductive structured SVM [40,6]. We refer the combined methods as CoDL+ and TSVM+, respectively.

TSVM+. TSVM [40] extends a binary transductive SVM model [12] to deal with structured outputs. However, it cannot deal with complex structured outputs, because it relies on an exact inference solver. In addition, the model does not use constraints from domain knowledge to guide learning. To extend TSVM, we modify the learning algorithm with an approximate structured SVM approach [6]. The augmented inference problems involved during the learning are solved approximately using our inference algorithm. Then, we add prior constraints over y in [40, Eq. OP3] to incorporate with corpus-level constraints. The resulted optimization problem is solved by a CCCP procedure [37]. We implemented TSVM+ based upon the Matlab version of structured SVM [33,13].

CoDL+. CoDL [2] proposed a general framework to incorporate declarative constraints for structured learning. However, they did not show results on problems with complex structured output. We implement CoDL using an Averaged Structured Perceptron algorithm and our inference engine. Specifically, Steps 4-7 in [2, Algorithm 2] and Step 4 in [2, Algorithm 3] are replaced by our inference algorithm. We use an L+I setting with $\rho_k = \infty$. CoDL uses a smoothing parameter to combine the models learned from labeled and unlabeled data instead of using annealing steps. For a better comparison, we implement the same annealing steps introduced in 3.4 for TSVM+ and CODL+.

Comparison and Discussion. Table 4 shows the performance in micro-F1. Results show that TSVM+ significantly improves the plain supervised setting on scene, but achieves a sub-optimal solution on yeast. We suspect that the performance dip is due to over-fitting (a similar situation is shown in Figure 1(b)). In fact, TSVM+ achieves better generative performance at early iterations on yeast. For example, it achieves

42.4% F1 at iteration 17 when $C_m = 0.005$ and the performance goes down afterwards. Regarding CODL+, it significantly improves its supervised counterpart on both scene and emotions. However, it also suffers from over-fitting on yeast data set.

The results in this section are mainly to demonstrate that our inference method can be incorporated with other semi-supervised models. We noted that extending existing semi-supervised learner for complex structured output problems is nontrivial. Therefore, a careful study might further improve the performance of CoDL+ or TSVM+. Nevertheless, results in Tables 4 show that our method achieves better or competitive performance with CODL+ and TSVM+ in all cases.

7   Conclusion and Discussion

In summary, we presented an effective semi-supervised learning framework for structured prediction problems with complex output structure. The proposed framework is general and can be easily applied to problems other than multi-label classification. Evaluating the framework on more complex problems is an important direction. A detailed and thoughtful comparison with other state-of-the-art semi-supervised multi-label classification methods is an interesting topic for future study.

References

1. Brefeld, U., Scheffer, T.: Semi-supervised learning for structured output variables. In: ICML (2006)
2. Chang, M.W., Ratinov, L.A., Roth, D.: Structured learning with constrained conditional models. Machine Learning 88(3), 399–431 (2012)
3. Chang, Y.W., Collins, M.: Exact decoding of phrase-based translation models through lagrangian relaxation. In: EMNLP (2011)
4. Chen, G., Song, Y., Wang, F., Zhang, C.: Semi-supervised multi-label learning by solving a sylvester equation. In: SDM, pp. 410–419 (2008)
5. Dhillon, P.S., Keerthi, S.S., Bellare, K., Chapelle, O., Sellamanickam, S.: Deterministic annealing for semi-supervised structured output learning. In: AISTATS (2012)
6. Finley, T., Joachims, T.: Training structural SVMs when exact inference is intractable. In: ICML, pp. 304–311 (2008)
7. Ganchev, K., Graca, J., Gillenwater, J., Taskar, B.: Posterior regularization for structured latent variable models. JMLR 11 (2010)
8. Guo, Y., Schuurmans, D.: Semi-supervised multi-label classification - a simultaneous large-margin, subspace learning approach. In: Flach, P.A., De Bie, T., Cristianini, N. (eds.) ECML PKDD 2012, Part II. LNCS, vol. 7524, pp. 355–370. Springer, Heidelberg (2012)
9. Hazan, T., Shashua, A.: Norm-product belief propagation: Primal-dual message-passing for approximate inference. CoRR (2009)
10. Hazan, T., Urtasun, R.: Efficient learning of structured predictors in general graphical models. CoRR (2012)
11. Huang, S.J., Zhou, Z.H., Zhou, Z.H.: Multi-label learning by exploiting label correlations locally. In: AAAI (2012)
12. Joachims, T.: Transductive inference for text classification using support vector machines. In: ICML, pp. 200–209. Morgan Kaufmann (1999)
13. Joachims, T., Finley, T., Yu, C.N.J.: Cutting-plane training of structural SVMs. Machine Learning 77(1), 27–59 (2009)

14. Jojic, V., Gould, S., Koller, D.: Accelerated dual decomposition for MAP inference. In: ICML (2010)
15. Komodakis, N.: Efficient training for pairwise or higher order crfs via dual decomposition. In: CVPR (2011)
16. Komodakis, N., Paragios, N., Tziritas, G.: MRF energy minimization and beyond via dual decomposition. PAMI 33(3), 531–552 (2011)
17. Koo, T., Rush, A.M., Collins, M., Jaakkola, T., Sontag, D.: Dual decomposition for parsing with non-projective head automata. In: EMNLP (2010)
18. Kulesza, A., Pereira, F.: Structured learning with approximate inference. In: NIPS (2008)
19. Lee, C.H., Jiao, F., Wang, S., Schuurmans, D., Greiner, R.: Learning to model spatial dependency: Semi-supervised discriminative random fields. In: NIPS (2006)
20. Lindsay, B.G.: Composite likelihood methods. Contemporary Mathematics 80, 221–239 (1988)
21. Liu, D.C., Nocedal, J.: On the limited memory BFGS method for large scale optimization. Math. Program. 45(3), 503–528 (1989)
22. Liu, Y., Jin, R., Yang, L.: Semi-supervised multi-label learning by constrained non-negative matrix factorization. In: AAAI, pp. 421–426 (2006)
23. Mann, G.S., McCallum, A.: Generalized expectation criteria for semi-supervised learning with weakly labeled data. JMLR 11, 955–984 (2010)
24. Martins, A.F.T., Figueiredo, M.A.T., Aguiar, P.M.Q., Smith, N.A., Xing, E.P.: Alternating directions dual decomposition. CoRR (2012)
25. Meshi, O., Globerson, A.: An alternating direction method for dual MAP LP relaxation. In: Gunopulos, D., Hofmann, T., Malerba, D., Vazirgiannis, M. (eds.) ECML PKDD 2011, Part II. LNCS, vol. 6912, pp. 470–483. Springer, Heidelberg (2011)
26. Meshi, O., Sontag, D., Jaakkola, T., Globerson, A.: Learning efficiently with approximate inference via dual losses. In: ICML (2010)
27. Pearl, J.: Probabilistic reasoning in intelligent systems: networks of plausible inference (1988)
28. Pletscher, P., Wulff, S.: LPQP for MAP: Putting LP solvers to better use. In: ICML (2012)
29. Samdani, R., Chang, M., Roth, D.: Unified expectation maximization. In: NAACL (2012)
30. Samdani, R., Roth, D.: Efficient decomposed learning for structured prediction. In: ICML (2012)
31. Seah, C.W., Tsang, I.W., Ong, Y.S.: Transductive ordinal regression. In: TNNLS, pp. 1074–1086 (2012)
32. Sutton, C., McCallum, A.: Piecewise training for structured prediction. Machine Learning 77(2-3), 165–194 (2009)
33. Vedaldi, A.: A MATLAB wrapper of SVM^{struct} (2011), http://www.vlfeat.org/~vedaldi/code/svm-struct-matlab.html
34. Xu, L., Wilkinson, D., Schuurmans, D.: Discriminative unsupervised learning of structured predictors. In: ICML (2006)
35. Yang, Y.: An evaluation of statistical approaches to text categorization. Information Retrieval 1, 69–90 (1999)
36. Yu, C.N.: Transductive learning of structural SVMs via prior knowledge constraints. In: AISTATS (2012)
37. Yuille, A.L., Rangarajan, A.: The concave-convex procedure. Neural Computation (2003)
38. Zha, Z.J., Mei, T., Wang, J., Wang, Z., Hua, X.S.: Graph-based semi-supervised learning with multiple labels. J. Visual Communication and Image Representation 20(2), 97–103 (2009)
39. Zhang, Y., Schneider, J.: A composite likelihood view for multi-label classification. In: AISTATS (2012)
40. Zien, A., Brefeld, U., Scheffer, T.: Transductive support vector machines for structured variables. In: ICML (2007)

PSSDL: Probabilistic Semi-supervised Dictionary Learning

Behnam Babagholami-Mohamadabadi, Ali Zarghami,
Mohammadreza Zolfaghari, and Mahdieh Soleymani Baghshah

Department of Computer Engineering, Sharif University of Technology,
Tehran, Iran
{babagholami,Zarghami,mzolfaghari}@ce.sharif.edu,
soleymani@sharif.edu

Abstract. While recent supervised dictionary learning methods have attained promising results on the classification tasks, their performance depends on the availability of the large labeled datasets. However, in many real world applications, accessing to sufficient labeled data may be expensive and/or time consuming, but its relatively easy to acquire a large amount of unlabeled data. In this paper, we propose a probabilistic framework for discriminative dictionary learning which uses both the labeled and unlabeled data. Experimental results demonstrate that the performance of the proposed method is significantly better than the state of the art dictionary based classification methods.

Keywords: Dictionary learning, MAP estimation, Gibbs Random Field, Local Linear Embedding, Local Fisher Discriminant Analysis.

1   Introduction

In the recent decade, Sparse Representation (SR), and Dictionary Learning (DL) have gained much interest in the computer vision and pattern recognition areas [5]. This attention is due to the fact that many natural signals (like natural images) are sparse in their nature and can be approximated or even fully recovered by their sparse codes. A common SR formulation consists of a sparsity term and a reconstructive term as shown in the following expression

$$[\hat{A}, \hat{D}] = \underset{A,D}{\operatorname{argmin}} \sum_{i=1}^{N} \|x_i - D\alpha_i\|_2^2 + \gamma\|\alpha_i\|_1, \tag{1}$$

where x_i is the i-th input signal, D is the dictionary, $A = [\alpha_1, ..., \alpha_N]$ represents the sparse codes and γ is a regularization term. This problem is not fully convex, but by fixing A or D, and minimizing the other one, the problem can be treated as a convex problem. Methods such as K-SVD [1] can be used to find a proper dictionary and a sparse code simultaneously.

Recently, Supervised Dictionary Learning (SDL) methods [6],[23], [2], [3] have used DL for classification tasks by adding discriminative terms to the objective

H. Blockeel et al. (Eds.): ECML PKDD 2013, Part III, LNAI 8190, pp. 192–207, 2013.

function of Eq. 1. [6] added a Fisher Discriminant Analysis (FDA) term to its objective function to make the sparse codes more discriminative. The method proposed in [23] incorporated a logistic loss function into the problem definition and learned a classifier and a dictionary simultaneously. Zhang et al. [2] modified the original K-SVD method by using the classification error as a part of objective function, allowing it to apply as a sparse coding classifier. Wright et al. [3] used the training signals as dictionary atoms (basis). Using this dictionary, new signals can be represented as a sparse linear combination of the training signals. The discrimination will be performed based on the representation error caused by considering only coefficients corresponding to atoms related to each class and ignoring all other atoms.

Despite their merits, SDL methods have two main drawbacks. Firstly, the regularization parameters are usually set using cross-validation technique, which biases their cost functions toward data points that are poorly represented. Hence, they are easily affected by noisy, outlier, and mislabeled training data. Secondly, the performance of the SDL methods is highly dependent on the number of the training samples. Unfortunately, in many pattern classification problems, accessibility to a large set of the labeled data may not be possible due to the fact that labeling data is expensive and time consuming. On the other hand, unlabeled data points are easily available in abundance which have motivated machine learning researchers to develop semi-supervised learning methods which utilize a large amount of unlabeled data, along with the limited number of labeled data, to build better models for classification tasks.

One of the most well-known methods for semi-supervised classification (SSL) is Semi-Supervised Support Vector Machine (S3VM) [4], which regards the class label of unlabeled samples as extra unknowns and optimizes the classifier parameters and unknown labels simultaneously. Another popular algorithm for semi-supervised learning is Co-training [8], which assumes that features (data points) have multiple views. Based on this assumption, this algorithm utilizes the confident samples in one view to update the other view. However, in many applications such as image classification, each image has only one feature vector and hence it is difficult to apply Co-Training.

Recently, Shrivastava et al. [9] have proposed a semi-supervised dictionary learning (SSDL) algorithm for classification tasks. This algorithm uses an iterative process which goes as follows. In the first iteration, the dictionaries (one dictionary for each class) are learned using only the labeled data. Then, the class labels of the unlabeled data points are roughly inferred based on how well they are reconstructed by the dictionaries of the different classes. In the next iterations, the confident unlabeled data points are used to further refine the dictionaries.

In order to improve the discrimination power of the dictionaries, this method imposes some constraints on the DL task, in which the data samples belonging to some particular classes with high confidence, should be well represented by the corresponding dictionaries and poorly represented by other dictionaries. The

Fisher Discriminant Analysis (FDA) is also used to enhance the discrimination of the sparse codes for the labeled data.

Although the results of this method is better than the state of the art discriminative DL methods, it has several shortcomings. Firstly, due to the learning one dictionary for each class, it cannot scale to problems with large number of classes. Secondly, using FDA only for labeled data may result in overfitting due to the fact that the number of the labeled data is much smaller than that of the unlabeled data. Third, it does not consider the underlying geometrical structure of both the labeled and unlabeled data.

To overcome these shortcomings, in this paper, we propose a novel algorithm to learn discriminative dictionaries for semi-supervised classification tasks. More specifically, a single dictionary is learned jointly with a classifier in a MAP setting, by which sharing features among different classes is allowed and it leads to less computational cost and less risk of overfitting. We also introduce a new discriminative term in our probabilistic framework by combining the methods of Local Fisher Discriminant Analysis (LFDA) [7] and Locally linear Embedding (LLE) [11] which preserves the global structure of all samples in addition to enhancing the discrimination power of the dictionary. The contributions of this paper are summarized as follows:

- Our method combines the LFDA, and LLE algorithms to increase the discrimination power of the dictionary as well as preserving the geometrical structure of both labeled and unlabeled data points, by which overfitting to the labeled data is prevented.
- Our method furthermore integrates a multinomial Logistic regression classifier into the proposed probabilistic dictionary learning framework, which improves the discrimination in the sparse codes of signals, and the discrimination in the classifier construction.
- The free parameters are estimated using the MAP estimation technique which allows to avoid parameter tuning based on the cross-validation.
- The MAP parameters are efficiently estimated via the well-known feature-sign search algorithm [12].
- Our approach is validated on various well-known digit recognition, face recognition, and spoken letter classification benchmarks.

The remainder of this paper is organized as follows: The proposed probabilistic model (MAP setup) for dictionary learning is introduced in Section 2. The optimization procedure for estimating MAP parameters is discussed in Section 3. Experimental results are presented in Section 4. We conclude and discuss future work in Section 5.

2   Proposed Method

In this section, we present our probabilistic framework for dictionary learning which takes into account both the labeled and unlabeled data. Here, the intuition is to improve the discriminativeness in the dictionary and to prevent overfitting the (small-size) labeled data points by adding a classifier error term and a geometrical preserving term into the proposed MAP setting respectively.

2.1   Problem Formulation

Let $X_L = \{(x_i, y_i), i = 1, ..., N_l\}$ be the set of labeled data, and $X_U = \{x_j, j = N_l + 1, ..., N\}$ be the set of unlabeled data available for learning the dictionary, where N_l and N are the number of labeled and total samples, respectively. Here, $x_j \in R^M$ denotes the j-th sample, $y_i \in \{1, 2, ..., C\}$ is the corresponding class label of the i-th data point, C is the number of classes, and $N_u = N - N_l$ is the number of the unlabeled data points. Let $D = [d_1, ..., d_K] \in R^{M \times K}$ be the dictionary with K atoms and $A = [A_L, A_U]_{K \times N}$ be the matrix of the sparse codes, where $A_L = [\alpha_1, ..., \alpha_l]_{K \times N_l}$ and $A_U = [\alpha_{N_l+1}, ..., \alpha_N]_{K \times N_u}$ show the matrices of the sparse codes of the labeled and unlabeled data respectively.

Here, we assume that each data point $x_i (i = 1, ..., N)$ can be represented as a sparse linear combination of K dictionary atoms with additive zero-mean Gaussian noise ϵ_i ($\epsilon_i \sim \mathcal{N}(0, \sigma_i^2 I)$). Using this assumption, sparse codes can be considered as latent variables of the representation model. Consequently, the likelihood of observing a specific sample x, given the dictionary (D) and its sparse code (α) is modeled as a Gaussian:

$$P(x \mid D, \alpha, \sigma^2) \sim \mathcal{N}(D\alpha, \sigma^2 I). \tag{2}$$

To model the classification process, we use the multinomial logistic regression classifier which is defined as

$$P(y = i \mid \alpha, w_1, ..., w_C) = \frac{exp(w_i^T \alpha)}{\sum_{j=1}^{C} exp(w_j^t \alpha)}, \quad i = 1, ..., C, \tag{3}$$

where α and y are the sparse code of an ordinary sample x and its label respectively, and $W = [w_1, ..., w_C]$ shows the parameter of the classifier.

In order to further enhance the discriminative power of the dictionary, some of the previous DL methods [6], [9] have utilized the FDA algorithm, by which the trace of the within-class scatter matrix of A_L is minimized and the trace of the between-class scatter matrix of A_L is maximized.

However, in situations where the number of labeled data is small, the FDA may overfit the labeled samples. Moreover, in cases where a large set of unlabeled samples is available, FDA cannot make use of unlabeled data. Another drawback of the FDA algorithm is that its performance may be degraded if the samples in a class form several separate clusters [10]. To circumvent these shortcomings, we propose a new discrimination term based on a smooth combination of LFDA algorithm and LLE algorithm, by which the topological structure of all the data is preserved in addition to enhancing the discrimination power of the dictionary. Precisely speaking, using LFDA algorithm, within-class scatter can be computed locally, and so the within-class multimodality can be resolved. Using LLE, reliance on the global structure of all samples and information brought by labeled samples is controlled.

In LFDA method, the local between-class scatter matrix S^{lB} and the local within-class scatter matrix S^{lW} are defined as [7]

$$S^{(LB)} = \frac{1}{2} \sum_{i,j=1}^{N} W_{i,j}^{(lb)} (\alpha_i - \alpha_j)(\alpha_i - \alpha_j)^T, \tag{4}$$

$$S^{(LW)} = \frac{1}{2} \sum_{i,j=1}^{N} W_{i,j}^{(lw)} (\alpha_i - \alpha_j)(\alpha_i - \alpha_j)^T, \tag{5}$$

where $W_{i,j}^{(lb)}$ and $W_{i,j}^{(lw)}$ are the $N \times N$ matrices which are defined as

$$W_{i,j}^{(lb)} = \begin{cases} P_{i,j}(1/N_l - 1/N_{ly_i}) & \text{if } y_i = y_j \\ 1/N_l & \text{if } y_i \neq y_j \\ 0 & \text{otherwise}, \end{cases} \tag{6}$$

$$W_{i,j}^{(lw)} = \begin{cases} P_{i,j}/N_{ly_i} & \text{if } y_i = y_j \\ 0 & \text{otherwise}, \end{cases} \tag{7}$$

where N_{ly_i} denotes the number of the labeled samples in the class y_i. In the above equations, $P_{i,j}$ shows the affinity value between x_i and x_j which is defined as [7]

$$P_{i,j} = exp\left(-\frac{\|x_i - x_j\|^2}{\gamma_i \gamma_j}\right), \tag{8}$$

where the parameter γ_i represents the local scaling around x_i as

$$\gamma_i = \|x_i - x_i^k\|, \tag{9}$$

and x_i^k is the k-th nearest neighbor of x_i (a heuristic choice of k = 5 was shown to be useful through experiments).

Using LLE, we try to preserve the intrinsic topological structure of the data based on the notion of affinity preserving. In other words, by employing LLE, the geometric structure of the data is retained by maintaining locally linear relationships between sparse codes of close data points.

Given the set of the both labeled and unlabeled data points, LLE assumes that each data point in the original space can be recovered using a linearly weighted average of its neighbors. Based on this assumption, an optimal weight matrix $S^* = [s_{ij}^*]$ is reconstructed by solving the following problem:

$$S^* = \underset{S}{\operatorname{argmin}} \sum_{i=1}^{N} \|x_i - \sum_{x_j \in N_k(x_i)} s_{ij}x_j\|^2, \quad s.t. \ \forall i, \sum_{x_j \in N_k(x_i)} s_{ij} = 1, \tag{10}$$

where $N_k(x_i)$ demonstrates the set of k nearest neighbor of x_i. The above optimization problem can be solved as a constrained least-squares problem [18].

In order to utilize the information of the unlabeled data points more efficiently, we consider certain assumptions about the general geometric properties of the

Fig. 1. Left: part of a one dimensional manifold, showing the deficiency of the Euclidian distance (purple edges are shortcut edges), right: Geodesic curve between two points on a manifold (solid line shows the geodesic curve and the dashed line shows the Euclidean curve)

data. More precisely, in many applications, high dimensional data points are actually samples from a low-dimensional subspace of the actual feature space. In these cases, we can make use of the Manifold assumption which is among the most practical assumptions in semi-supervised learning tasks [19].

In the original LLE and LFDA algorithms, Euclidean distance is considered as a measure of evaluating distance between data points. However, by considering the manifold assumption, Euclidean distance measure may be misleading, since two samples having a small Euclidean distance may be located far apart on the underlying manifold of the data points (Fig. 1).

To circumvent this problem, we make use of geodesic distance as a distance measure between data points which enables us to determine the neighborhood of a data point more precisely. The geodesic distance between two sample x_i and x_j is defined as the length of shortest curve between x_i and x_j lying on the manifold of the data points (Fig. 1).

Since the underlying manifold of the samples is unknown (we only have data which are finite samples of the manifold), we cannot find the exact geodesic distance between each two data points. Hence, in this paper, we use the idea of [20] to approximate the geodesic distance between both labeled and unlabeled data points. In [20], first, a k nearest neighborhood graph of all data is constructed based on the Euclidean distance. Then, an iterative process is done to remove the shortcut edges (shortcut edges connect those points of the graph which are close to each other according to the Euclidean distance, but have large geodesic distance on the manifold [20]). After refining the constructed graph based on the idea of [20], we use an efficient shortest path algorithm to find the k nearest neighbor of each data point. After computing the optimal weight matrix S^* based on the geodesic distance, we try to minimize the following objective function in order to preserve the global structure of data in the sparse representation space.

$$\sum_{i=1}^{N} \|\alpha_i - \sum_{x_j \in N_k^G(x_i)} s_{ij}^* \alpha_j\|^2 = tr(AEA^T), \tag{11}$$

where $N_k^G(x_i)$ demonstrates the set of k nearest neighbors of x_i based on the geodesic distance, and $E = (I - S^*)^T (I - S^*)$.

Now, we define S^{RLW} and S^{RLB} as the regularized local within-class scatter matrix and the regularized local between-class scatter matrix respectively:

$$S^{RLW} = (1 - \vartheta)AL^{LW}A^T + \vartheta AEA^T \tag{12}$$

$$S^{RLB} = (1 - \vartheta)AL^{LB}A^T + \vartheta I_{K \times K} \tag{13}$$

where $\vartheta \in [0, 1]$ is a trade-off parameter, and L^{LW} and L^{LB} are the graph Laplacian matrix of the local within-scatter (S^{LW}) and the local between-class scatter (S^{LB}) matrices which are defined as

$$L^{LW} = D^{LW} - W^{(lw)}, \quad L^{LB} = D^{LB} - W^{(lb)}, \tag{14}$$

where D^{LW} and D^{LB} are diagonal $N \times N$ matrices with

$$D_{i,i}^{LW} = \sum_{j=1}^{N} W_{i,j}^{(lw)}, \quad D_{i,i}^{LB} = \sum_{j=1}^{N} W_{i,j}^{(lb)}. \tag{15}$$

Another constraint that the sparse codes should satisfy is the sparsity constraint. In order to enforce sparsity on A, we put a well-known Laplacian prior distribution on each sparse code which is shown as

$$\alpha_i \sim Lap(\alpha_i \mid b) = \frac{1}{2b}exp\left(-\frac{\|\alpha_i\|_1}{b}\right), \tag{16}$$

where b is the scale parameter of the laplacian distribution. In order to encode the sparsity constraint, the discriminative constraint by LFDA, and the global constraint by LLE into our probabilistic model, we use Gibbs Random Field (GRF). A set of random variables $\{\alpha_i\}_{i=1}^{N}$ is said to be a GRF, if and only if their joint distribution follows a Gibbs distribution. Hence, the joint distribution must take the form

$$P(\alpha_1, ..., \alpha_N) = \frac{1}{Z}exp\left(-\frac{1}{T}U(\alpha_1, ..., \alpha_N)\right), \tag{17}$$

where Z is the normalizing constant called the Partition Function, T is a constant called the temperature (in this paper its assumed to be 1), and $U(\alpha_1, ..., \alpha_N)$ is the energy function which in this paper is defined as

$$U(\alpha_1, ..., \alpha_N) = N \log 2b + \frac{1}{b} \sum_{i=1}^{N} \|\alpha_i\|_1 + tr(S^{RLW}(A)) - tr(S^{RLB}(A)). \tag{18}$$

For simplicity, we also put a Gaussian prior distribution on the dictionary and the classifier parameters. Hence we have

$$P(D \mid \sigma_d^2) \propto \prod_{i=1}^{K} \mathcal{N}(d_i; 0, \sigma_d^2 I_M), \quad P(W \mid \sigma_w^2) \propto \prod_{j=1}^{C} \mathcal{N}(w_j; 0, \sigma_w^2 I_K). \tag{19}$$

To model the prior of the parameter $\Xi = \{\sigma_i (i = 1, ..., N), b, \sigma_d, \sigma_w\}$, we choose the objective non-parametric Jeffreys prior, which has been demonstrated to perform well for regression and classification tasks [21]. So, we have

$$P(\Xi) \propto \prod_{i=1}^{N} \frac{1}{\sigma_i^2} \times \frac{1}{b} \times \frac{1}{\sigma_d^2} \times \frac{1}{\sigma_w^2}. \tag{20}$$

The prior over σ_i encourages a low variance representation which means the training data should properly fit the proposed representation model. The prior over b encourages a sparser solution for the sparse codes which is the main aim of the sparse representation based methods, and the prior over σ_d and σ_w decreases the risk of overfitting the dictionary and the classifier respectively.

After defining the prior and the likelihood distributions, the posterior distribution of the latent variables (W, D, A, Ξ) given the observations $(X_L, X_U, \{y_i\}_{i=1}^{N})$ can be computed as

$$P(W, D, A, \Xi \mid X_L, X_U, \{y_i\}_{i=1}^{N_l}) \propto \prod_{i=1}^{N} \frac{1}{\sigma_i^2} \times \frac{1}{b} \times \frac{1}{\sigma_d^2} \times \frac{1}{\sigma_w^2} \times e^{\left(-U(\alpha_1,...,\alpha_N)\right)} \times$$

$$\prod_{i=1}^{N} \mathcal{N}(x_i; D\alpha_i, \sigma_i^2) \prod_{j=1}^{N_l} \frac{exp(w_{y_j}^T \alpha_j)}{\sum_{c=1}^{C} exp(w_c^T \alpha_j)} \prod_{i=1}^{K} \mathcal{N}(d_i; 0, \sigma_d^2 I_M) \prod_{j=1}^{C} \mathcal{N}(w_j; 0, \sigma_w^2 I_K). \tag{21}$$

In order to determine the most probable point estimate for the latent variables, we compute the MAP estimation of the above posterior distribution which is easy to show that it is equal to the following minimization problem.

$$[\hat{W}, \hat{D}, \hat{A}, \hat{\Xi}] = \underset{W,D,A,\Xi}{\mathrm{argmin}} \sum_{i=1}^{N} \frac{\|x_i - D\alpha_i\|_2^2}{2\sigma_i^2} + \sum_{i=1}^{N} \log \sigma_i^{M+2} - \sum_{j=1}^{N_l} w_{y_j}^T \alpha_j$$

$$+ \sum_{j=1}^{N_l} \log \left(\sum_{c=1}^{C} exp(w_c^T \alpha_j) \right) + \frac{1}{2\sigma_w^2} \sum_{j=1}^{C} \|w_j\|_2^2 + C \log \sigma_w^{K+2}$$

$$+ \frac{1}{2\sigma_d^2} \sum_{j=1}^{K} \|d_j\|_2^2 + K \log \sigma_d^{M+2} + (N+1) \log b + \frac{1}{b} \sum_{i=1}^{N} \|\alpha_i\|_1$$

$$+ tr\left(S^{RLW}(\alpha_1, ..., \alpha_N)\right) - tr\left(S^{RLB}(\alpha_1, ..., \alpha_N)\right). \tag{22}$$

3   Optimization Procedure

In this section, we describe the optimization procedure for the proposed objective function (Eq. 22). Solving (22) is a challenging task because of two reasons. Firstly, the objective function is not convex respect to W, D, A and Ξ simultaneously. Secondly, the log-sum-exp term $(\log(\sum_{c=1}^{C} exp(w_c^T \alpha_j)))$ in the objective function prevents us using efficient methods such as feature search sign algorithm

[12] to compute the sparse codes efficiently. To address the first problem, we can easily observe that the objective function is convex with respect to each of the parameters when the others are fixed. Hence, we resort to a coordinate descent method (alternating optimization), in which unknown parameters are updated through an iterative process which updates each parameter by fixing the other parameters in each step. To circumvent the second problem, we utilize a suitable upper bound to the log-sum-exp function proposed by [22] which states that for any $\beta \in \mathbb{R}$ and $\xi_k \in [0, \infty)$, $k = 1, ..., K$

$$\log(\sum_{k=1}^{K} e^{g_k}) \leq \beta + \sum_{k=1}^{K} \left(\frac{g_k - \beta - \xi_k}{2} + \lambda(\xi_k)\left((g_k - \beta)^2 - \xi_k^2\right) + \log(1 + e^{\xi_k}) \right), \quad (23)$$

where

$$\lambda(\xi) = \frac{1}{2\xi}\left(\frac{1}{1 + e^{-\xi}} - \frac{1}{2}\right), \quad (24)$$

where β and $\{\xi_k\}_{k=1}^{K}$ are the variational parameters which can be optimized to get the tightest possible bound. So, by replacing the log-sum-exp term of the objective function with the upper bound of Eq. 23, we have

$$[\hat{W}, \hat{D}, \hat{A}, \hat{\Xi}, \hat{\Theta}] = \underset{W,D,A,\Xi,\Theta}{\mathrm{argmin}} \sum_{i=1}^{N} \frac{\|x_i - D\alpha_i\|_2^2}{2\sigma_i^2} + \sum_{i=1}^{N} \log \sigma_i^{M+2} - \sum_{j=1}^{N_l} w_{y_j}^T \alpha_j$$

$$+ \sum_{j=1}^{N_l} \beta_j + \frac{1}{2}\sum_{j=1}^{N_l}\sum_{c=1}^{C}(w_c^T \alpha_j - \beta_j - \xi_{jc}) + \sum_{j=1}^{N_l}\sum_{c=1}^{C}\lambda(\xi_{jc})\left((w_c^T \alpha_j - \beta_j)^2 - \xi_{jc}^2\right)$$

$$+ \sum_{j=1}^{N_l}\sum_{c=1}^{C}(\log(1 + e^{\xi_{jc}})) + \frac{1}{2\sigma_w^2}\sum_{j=1}^{C}\|w_j\|_2^2 + C \log \sigma_w^{K+2} + \frac{1}{2\sigma_d^2}\sum_{j=1}^{K}\|d_j\|_2^2$$

$$+ K \log \sigma_d^{M+2} + (N + 1)\log b + \frac{1}{b}\sum_{i=1}^{N}\|\alpha_i\|_1 + tr(A^T A\Gamma), \quad (25)$$

where $\Theta = \{\beta_j, \xi_{jc}\}_{j=1, c=1}^{j=N_l, c=C}$ is the set of variational parameters, and Γ is a $N \times N$ matrix which is defined as

$$\Gamma = (1 - \vartheta)L^{LW} + \vartheta E - (1 - \vartheta)L^{LB}. \quad (26)$$

Its obvious from Eq. 25 that it is convex in one parameter when the other parameters are fixed. Using the upper bound of Eq. 23, we are able to solve the above optimization problem by an efficient feature-sign search algorithm [12] which goes as follows.

Computing Sparse Codes A with Fixed W, D, Ξ and Θ: We optimize each sparse code $\alpha_i(i = 1, ..., N)$ by fixing sparse codes $\alpha_j(j \neq i)$ of other signals. Hence, for each sparse code α_i, if $x_i \in X_L$, we must solve

$$\hat{\alpha}_i = \underset{\alpha_i}{\mathrm{argmin}}\, F_L(\alpha_i), \quad (27)$$

and if $x_i \in X_U$, we must solve

$$\hat{\alpha}_i = \underset{\alpha_i}{\operatorname{argmin}} \, F_U(\alpha_i), \tag{28}$$

where

$$F_L(\alpha_i) = \frac{1}{2\sigma_i^2} \|x_i - D\alpha_i\|_2^2 - w_{y_i}^T \alpha_i + \frac{1}{2} \sum_{c=1}^{C} w_c^T \alpha_i + \sum_{c=1}^{C} \lambda(\xi_{ic})(w_c^T \alpha_i - \beta_i)^2$$

$$+ 2\alpha_i^T (A\Gamma_i) - \alpha_i^T \alpha_i \Gamma_{i,i} + \frac{1}{b} \|\alpha_i\|_1, \tag{29}$$

$$F_U(\alpha_i) = \frac{1}{2\sigma_i^2} \|x_i - D\alpha_i\|_2^2 + 2\alpha_i^T (A\Gamma_i) - \alpha_i^T \alpha_i \Gamma_{i,i} + \frac{1}{b} \|\alpha_i\|_1, \tag{30}$$

where Γ_i is the i-th column of Γ and $\Gamma_{i,i}$ is the (i,i) element of Γ.

The functions in Eqs. 29 and 30 are exactly the objective functions that the feature-sign search algorithm can minimize. This algorithm iteratively searches for the coefficient sign vector θ_i of x_i, hence (27) and (28) reduce to a standard, unconstrained quadratic optimization problem (QP). Precisely speaking, after finding the optimal coefficient sign of the sparse code α_i, $\|\alpha_i\|_1$ can be replaced by $\theta_i \alpha_i$, by which α_i can be computed analytically by setting the derivative of $F_L(\alpha_i)$ and $F_U(\alpha_i)$ respect to α_i equal to zero. The gradient of $F_L(\alpha_i)$ and $F_U(\alpha_i)$ can be calculated as

$$\frac{\partial F_L(\alpha_i)}{\partial \alpha_i} = \frac{1}{\sigma_i^2} D^T(D\alpha_i - x_i) - w_{y_i} + \frac{1}{2} \sum_{c=1}^{C} w_c - 2\beta_i \sum_{c=1}^{C} \lambda(\xi_{ic})w_c$$

$$+ 2\Big(\sum_{c=1}^{C} \lambda(\xi_{ic})w_c w_c^T\Big)\alpha_i + 2A\Gamma_i + \frac{\theta_i}{b}, \tag{31}$$

$$\frac{\partial F_U(\alpha_i)}{\partial \alpha_i} = \frac{1}{\sigma_i^2} D^T(D\alpha_i - x_i) + 2A\Gamma_i + \frac{\theta_i}{b}. \tag{32}$$

Finally, the analytic solution of α_i can be obtained when we have $\frac{\partial F_L(\alpha_i)}{\partial \alpha_i} = 0$, if $x_i \in X_L$ and $\frac{\partial F_u(\alpha_i)}{\partial \alpha_i} = 0$, if $x_i \in X_U$:

$$\hat{\alpha}_i = \Big(\frac{1}{\sigma_i^2} D^T D + 2\big(\sum_{c=1}^{C} \lambda(\xi_{ic})w_c w_c^T\big) + 2\Gamma_{i,i}I\Big)^{-1} \Big(\frac{1}{\sigma_i^2} D^T x_i + w_{y_i} - \frac{1}{2} \sum_{c=1}^{C} w_c$$

$$+ 2\beta_i \sum_{c=1}^{C} \lambda(\xi_{ic})w_c - 2\sum_{j \neq i} \Gamma_{j,i}\alpha_j - \frac{\theta_i}{b}\Big), \qquad if \; x_i \in X_L, \tag{33}$$

$$\hat{\alpha}_i = \Big(\frac{1}{\sigma_i^2} D^T D + 2\Gamma_{i,i}I\Big)^{-1} \Big(\frac{1}{\sigma_i^2} D^T x_i - 2\sum_{j \neq i} \Gamma_{j,i}\alpha_j - \frac{\theta_i}{b}\Big), \qquad if \; x_i \in X_U. \tag{34}$$

In practice, the Hessian matrices of F_L (Eq. 35), and F_U (Eq. 36) may not be positive semidifinite. So, a very small η (ηI) is added to the Hessian matrices to make them positive semidifinite, hence F_L and F_U are convex.

$$H_{F_L} = \frac{1}{\sigma_i^2} D^T D + 2 \left(\sum_{c=1}^{C} \lambda(\xi_{ic}) w_c w_c^T \right) + 2\Gamma_{i,i} I, \tag{35}$$

$$H_{F_U} = \frac{1}{\sigma_i^2} D^T D + 2\Gamma_{i,i} I. \tag{36}$$

Updating Dictionary D with Fixed A, W, Ξ and Θ: Given A, W, Ξ and Θ, the optimization problem for D can be formulated as

$$\hat{D} = \underset{D}{\operatorname{argmin}} \sum_{i=1}^{N} \frac{\|x_i - D\alpha_i\|_2^2}{2\sigma_i^2} + \frac{1}{2\sigma_d^2} \|D\|_F^2. \tag{37}$$

The above problem is an unconstrained quadratic programming, for which D can be computed analytically as

$$\hat{D} = X \Sigma A^T (A \Sigma A^T + \frac{1}{\sigma_d^2} I)^{-1}, \tag{38}$$

where Σ is a diagonal $N \times N$ matrix with

$$\Sigma_{i,i} = \frac{1}{\sigma_i^2}, \quad i = 1, 2, ..., N. \tag{39}$$

Updating the Classifier parameter W with Fixed A, D, Ξ and Θ: Without loss of generality, we assume that the first N_L^c labeled samples belong to the c-th class. So, given A, Ξ, D and Θ, the optimization problem for w_c can be formulated as

$$\hat{w}_c = \underset{w_c}{\operatorname{argmin}} \frac{1}{2} \left(\sum_{j=1}^{N_L} \alpha_j^T \right) w_c - \left(\sum_{j=1}^{N_L^c} \alpha_j^T \right) w_c + \sum_{j=1}^{N_L} \lambda(\xi_{jc}) (\alpha_j^T w_c - \beta_j)^2 + \frac{1}{2\sigma_w^2} w_c^T w_c. \tag{40}$$

By setting the derivative of the objective function of the above equation respect to w_c equal to zero, we can compute w_c analytically as

$$\hat{w}_c = \left(2 \sum_{j=1}^{N_L} \lambda(\xi_{jc}) \alpha_j \alpha_j^T + \frac{1}{\sigma_w^2} I \right)^{-1} \left(\sum_{j=1}^{N_L^c} \alpha_j + \sum_{j=1}^{N_L} (2\lambda(\xi_{jc})\beta_j - \frac{1}{2}) \alpha_j \right). \tag{41}$$

Updating the Free Parameter Ξ with Fixed A, W, D and Θ: Given A, W, D and Θ, each parameter can be computed analytically as:

$$\hat{\sigma_i}^2 = \left(\frac{1}{M+2} \right) \|x_i - D\alpha_i\|_2^2, \quad i = 1, 2, ..., N, \tag{42}$$

$$\hat{\sigma_d}^2 = \left(\frac{1}{K(M+2)}\right)\|D\|_F^2, \tag{43}$$

$$\hat{\sigma_w}^2 = \left(\frac{1}{C(K+2)}\right)\|W\|_F^2, \tag{44}$$

$$\hat{b} = \left(\frac{1}{N+1}\right)\sum_{i=1}^{N}\|\alpha_i\|_1. \tag{45}$$

Updating the Variational Parameter Θ with Fixed A, W, D and Ξ:
Given A, W, D and Ξ, we first compute the updates for $\{\beta_j\}_{j=1}^{N_L}$ by fixing other
variational parameters $(\{\xi_{jc}\}_{j=1,c=1}^{N_L,C})$ which leads to the following solution.

$$\hat{\beta}_j = \left(2\left(\sum_{c=1}^{C}\lambda(\xi_{jc})w_c^T\right)\alpha_j + \frac{C}{2} - 1\right)\bigg/2\left(\sum_{c=1}^{C}\lambda(\xi_{jc})\right), \quad j = 1, ..., N_L. \tag{46}$$

Secondly, we fix $\{\beta_j\}_{j=1}^{N_L}$, and update $\{\xi_{jc}\}_{j=1,c=1}^{N_L,C}$ by solving the following problem.

$$\hat{\xi}_{jc} = \underset{\xi_{jc}}{\operatorname{argmin}}\ \lambda(\xi_{jc})\left((w_c^T\alpha_j - \beta_j)^2 - \xi_{jc}^2\right) - \frac{1}{2}\xi_{jc} + \log(1 + e^{\xi_{jc}}). \tag{47}$$

By setting the derivative of the objective function of the above equation respect
to ξ_{jc} equal to zero, we have

$$\lambda'(\xi_{jc})\left((w_c^T\alpha_j - \beta_j)^2 - \xi_{jc}^2\right) - 2\lambda(\xi_{jc})\xi_{jc} - \frac{1}{2} + \frac{1}{1 + e^{-\xi_{jc}}} = 0. \tag{48}$$

Using the definition of $\lambda(\xi_{jc})$, the above equation can be simplified as

$$\lambda'(\xi_{jc})\left((w_c^T\alpha_j - \beta_j)^2 - \xi_{jc}^2\right) = 0. \tag{49}$$

Since $\xi_{jc} \in [0, \infty]$, $\lambda'(\xi_{jc}) \neq 0$, hence we can compute ξ_{jc} analytically as

$$\hat{\xi}_{jc} = |\ w_c^T\alpha_j - \beta_j\ |, \quad j = 1, ..., N_L, \quad c = 1, ..., C, \tag{50}$$

3.1   Class Label Prediction

After learning A, D, W and Ξ, classifying a new signal x with an unknown label
y is performed by solving the following optimization problem.

$$\hat{y} = \underset{y \in \{1,...,C\}}{\operatorname{argmax}}\ P(y \mid x, D, W, \Xi). \tag{51}$$

Using the Bayes' rule formula, the above problem can be expressed as

$$\hat{y} = \underset{y \in \{1,...,C\}}{\operatorname{argmax}}\ \iint P(y \mid \alpha, W)P(x \mid D, \alpha, \sigma^2)P(\alpha \mid b)P(\sigma)\,d\alpha\,d\sigma. \tag{52}$$

Table 1. Classification accuracy of different methods

dataset	SVM	S3VM	FDDL	SDL-G	SDL-D	S2D2	PM
MNIST	79.3 ± 1.9	83.3 ± 1.1	81.8 ± 1.8	82.1 ± 1.4	79.9 ± 2.1	86.1 ± 1.0	**87.4 ± 1.2**
USPS	80.7 ± 1.6	82.5 ± 0.9	81.1 ± 1.7	81.9 ± 1.3	80.1 ± 1.9	85.6 ± 0.9	**86.9 ± 1.0**
AR	70.4 ± 2.1	77.1 ± 1.7	74.2 ± 2.1	75.3 ± 1.5	74.1 ± 2.3	85.9 ± 1.4	**86.7 ± 1.5**
E-Yale B	72.1 ± 1.9	75.1 ± 1.7	65.9 ± 2.3	69.4 ± 1.8	67.9 ± 2.1	79.3 ± 1.8	**80.8 ± 1.8**
ISOLET	85.8 ± 1.7	87.3 ± 1.6	82.6 ± 1.9	83.4 ± 1.7	82.5 ± 1.9	89.9 ± 1.8	**91.4 ± 1.1**

where α is the sparse representation of x, and σ^2 is the representation noise variance of x (Eq. 2). By assuming that the posterior distribution over α and σ ($P(\alpha, \sigma \mid x, D, \Xi)$) can be approximated as a unit point measure at the MAP value (α_t, σ_t) of that distribution, the above problem can be replaced with the following minimization problem.

$$\hat{y} = \underset{y \in \{1,2,\dots,C\}}{argmin} \left[\underset{\alpha_t, \sigma_t}{min} \frac{\|x - D\alpha_t\|_2^2}{2\sigma_t^2} + (M+2)\log \sigma_t + \frac{1}{b}\|\alpha_t\|_1 - w_y^T \alpha_t \right.$$

$$\left. + \log\left(\sum_{c=1}^{C} exp(w_c^T \alpha_t) \right) \right]. \tag{53}$$

Again, using the upper bound of Eq. 23, we can efficiently solve the above problem (details omitted due to space limitations).

4 Experimental Results

To illustrate the efficacy of our method, we present experimental results on applications such as Face Recognition (FR), Handwritten Digit Recognition (HDR), and Letter Recognition (LR). For comparison purposes, we compare our method with some state of the art SDL methods such as FDDL [6], SDL-G [23], SDL-D [23], and two well-known classification methods SVM and S3VM [4]. We also compare our method with S2D2 [9] which is a recently introduced semi-supervised DL algorithm. In all of our experiments, the parameter ϑ is set equal to 0.5 (the results of all experiments are almost unchanged for $0.1 \le \vartheta \le 0.9$). In order to determine an appropriate number of dictionary atoms (K), and nearest neighbors of data samples (k) for computing the LLE matrix (E), Five-fold cross validation is performed to find the best pair (K, k). The tested values for K are $\{64, 128, 256, 512\}$ and for k, $\{3, 5, 7, 9, 11\}$.

Digit Recognition: We apply the proposed method on two HDR datasets MNIST [24], and USPS [25]. The MNIST dataset consists of 70,000 28 × 28 images, 60,000 for training, 10,000 for testing, each of them containing one handwritten digit. USPS is composed of 7291 training images and 2007 test images of size 16 × 16. For these datasets, 25 samples per class are randomly chosen

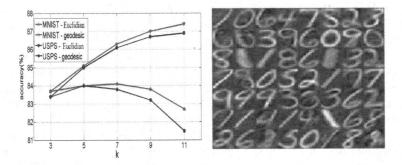

Fig. 2. Left: accuracy of the proposed method using the geodesic and the Euclidian distance for MNIST and USPS datasets, right: the learned D (K=64) for USPS dataset

from the training data as the labeled samples and the rest of the training data is used as the unlabeled data (we use the whole image as the feature vector in the digit datasets). The average recognition accuracies over 10 runs together with the standard deviation is shown in Table 1, from which we can see that the proposed method outperforms significantly the SDL methods. The improvement in performance compared to SDL methods is because of two reasons. Firstly, the number of labeled data is small, hence the SDL methods may overfit to the labeled data. Secondly, these methods cannot utilize unlabeled data for learning dictionary. Moreover, S3VM and S2D2 does not consider the topological structure of all data, hence both of them are less accurate than our method. We also provide a visualization of the learned D for USPS dataset for $K = 64$ (Fig. 2).

In order to demonstrate the superiority of the geodesic distance over the Euclidian distance, we compute the recognition performance of the proposed method on MNIST, and USPS dataset, using both the geodesic and the Euclidian distances to find the k nearest neighbors of data points. The results are presented in Fig. 2, for various number of k. The figure shows two major points. Firstly, it is obvious that using the geodesic distance leads to better performance than the Euclidian distance, because the Euclidian distance ignores the fact that the data points lie on a low dimensional manifold. Secondly, when the number of nearest neighbors grows, the recognition accuracy decreases for the Euclidian distance and increases for the geodesic distance. This is due to the fact that using Euclidian distance, by increasing k, samples from different classes are more likely to be selected as the neighbors of data points. Hence, the matrix E which captures the locality information of data points may be misleading. On the other hand, the geodesic distance considers the underlying manifold of samples, by which the neighbors of data points can be found more accurately, and hence the matrix E encodes the locality of data points more precisely.

Face Recognition: We then perform the face recognition task on the widely used E-Yale B [26], and AR [27], face databases. The E-Yale B database consists of 2,414 frontal-face images from 38 individuals (about 64 images per individual), and the AR database consists of over 4,000 frontal images from 126 individuals generated in two sessions, each of them consists of 14 images per individual (seven image for training, and seven image for testing). The E-Yale B and AR images are normalized to 54 × 48 and 60 × 40 respectively. We then perform a Principal Component Analysis (PCA) on the images to obtain 300 dimensional feature vectors. For AR dataset, we randomly choose two samples of the training session to form the labeled data and use the remaining five of that session as the unlabeled data. For E-Yale B dataset, for each class, we randomly select ten images as labeled data, twenty images as unlabeled data, and the remaining ones for testing. The average recognition accuracies over 5 runs together with the standard deviation are presented in the forth, and fifth rows of Table 1. Again, due to the small number of the labeled data, SDL methods have lower accuracy than SSDL methods. Moreover, because of considering the geometrical structure of data, our method has better performance than S3VM and S2D2 methods.

Letter Recognition: Finally, we apply our method on the ISOLET database [28], from UCI Machine Learning Repository which consists of 6238 examples and 26 classes corresponding to letters of the alphabet. We reduced the input dimensionality (originally at 617) by projecting the data onto its leading 100 principal components. For each class, We randomly select 10 samples as labeled data, 100 samples as unlabeled data, and the remaining ones for testing. The average recognition accuracies over 5 runs together with the standard deviation are presented in the last row of Table 1, from which we can see that the proposed method performs significantly better than the other algorithms.

5   Conclusion

In this paper, we proposed a probabilistic method which uses the information of unlabeled data as well as labeled data for learning discriminative dictionaries. The proposed method improves the discrimination of the dictionary and the sparse codes by incorporating a classifier error term and a discrimination term based on LFDA into the model. The topological structure of all data is also preserved based on LLE method which prevents overfitting the small labeled data. Moreover, instead of Euclidian distance, we utilized the geodesic distance which allows us to find the neighbors of data points more accurately. Experiments using various benchmark datasets demonstrate the superiority of the proposed method over the state-of-the-art SDL and SSDL methods.

References

1. Aharon, M., Elad, M., Bruckstein, A.: k-svd: An algorithm for designing dictionaries for sparse representation. IEEE Transactions on Signal Processing (2006)
2. Zhang, Q., Li, X.: Discriminative k-svd for dictionary learning in face recognition. In: CVPR (2010)

3. Wright, J., Yang, A., Ganesh, A., Sastry, S., Ma, Y.: Robust Face Recognition via Sparse Representation. IEEE PAMI (2009)
4. Sindhwani, V., Keerthi, S.S.: Large scale semi-supervised linear svms. In: ACM SIGIR (2006)
5. Wright, J., Ma, Y., Mairal, J., Sapiro, G., Yan, S.: Sparse representation for computer vision and pattern recognition. Proceedings of the IEEE (2010)
6. Yang, M., Zhang, L., Feng, X., Zhang, D.: Fisher discrimination dictionary learning for sparse representation. In: IEEE ICCV (2011)
7. Sugiyama, M.: Dimensionality reduction of multimodal labeled data by local Fisher discriminant analysis. Journal of Machine Learning Research (2007)
8. Blum, A., Mitchell, T.: Combining labeled and unlabeled data with co-training. In: ACM COLT (1998)
9. Shrivastava, A., Jaishanker, K.P., Patel, V.M., Chellappa, R.: Learning discriminative dictionaries with partially labeled data. In: IEEE ICIP (2012)
10. Sugiyama, M., Ide, T., Nakajima, S., Sese, J.: Semi-Supervised Local Fisher Discriminant Analysis for Dimensionality Reduction. J. machine Learning (2010)
11. Roweis, S.T., Saul, L.K.: Nonlinear dimensionality reduction by locally linear embedding. J. Science (2000)
12. Lee, H., Battle, A., Ng, A.Y.: Efficient sparse coding algorithms. In: NIPS (2007)
13. Huang, K., Aviyente, S.: Sparse representation for classification. In: NIPS (2007)
14. Boureau, Y., Bach, F., LeCun, Y., Ponce, J.: Learning mid-level features for recognition. In: CVPR (2010)
15. Grosse, R., Raina, R., Kwong, H., Ng, A.Y.: Shift-invariant sparse coding for audio classification. In: Conf. on Uncertainty in AI (2007)
16. Mairal, J., Leordeanu, M., Bach, F., Hebert, M., Ponce, J.: Discriminative sparse image models for class-specific edge detection and image interpretation. In: Forsyth, D., Torr, P., Zisserman, A. (eds.) ECCV 2008, Part III. LNCS, vol. 5304, pp. 43–56. Springer, Heidelberg (2008)
17. Zhang, W., Surve, A., Fern, X., Dietterich, T.: Learning non-redundant codebooks for classifying complex objects. In: ICML (2009)
18. Roweis, S.T., Saul, L.K.: Nonlinear dimensionality reduction by locally linear embedding. Science 290(5500), 2323–2326 (2000)
19. Chapelle, O., Scholkopf, B., Zien, A.: Semi-supervised learning, vol. 2. MIT Press, Cambridge (2006)
20. Ghazvininejad, M., Mahdieh, M., Rabiee, H.R., Khanipour, P., Rohban, M.H.: Isograph: Neighbourhood Graph Construction Based on Geodesic Distance for Semi-Supervised Learning. In: ICDM (2010)
21. Figueiredo, M.: Adaptive Sparseness using Jeffreys Prior. In: NIPS (2002)
22. Bouchard, G.: Efficient bounds for the softmax function. In: NIPS (2007)
23. Mairal, J., Bach, F., Ponce, J., Sapiro, G., Zisserman, A.: Supervised dictionary learning. In: NIPS (2009)
24. LeCun, Y., Bottou, L., Bengio, Y., Haffner, P.: Gradient-based learning applied to document recognition. Proc. of the IEEE (1998)
25. http://www-i6.informatik.rwth-aachen.de/~keysers/usps.html
26. Lee, K., Ho, J., Kriegman, D.: Acquiring linear subspaces for face recognition under variable lighting. IEEE TPAMI (2005)
27. Martinez, A., Benavente, R.: The AR face database. CVC Tech. Report (1998)
28. Frank, A., Asuncion, A.: UCI Machine Learning Repository (2010)

Embedding with Autoencoder Regularization

Wenchao Yu[1,2], Guangxiang Zeng[3], Ping Luo[4], Fuzhen Zhuang[1],
Qing He[1], and Zhongzhi Shi[1]

[1] The Key Laboratory of Intelligent Information Processing, Institute of Computing
Technology, Chinese Academy of Sciences, Beijing 100190, China
[2] University of Chinese Academy of Sciences, Beijing 100049, China
[3] University of Science and Technology of China, [4]HP Labs China
{yuwenchao,heqing}@ict.ac.cn, {guangxiang.zeng,ping.luo}@hp.com,
{zhuangfz,shizz}@ics.ict.ac.cn

Abstract. The problem of embedding arises in many machine learning
applications with the assumption that there may exist a small number
of variabilities which can guarantee the "semantics" of the original high-
dimensional data. Most of the existing embedding algorithms perform to
maintain the *locality-preserving* property. In this study, inspired by the
remarkable success of representation learning and deep learning, we propose a framework of embedding with autoencoder regularization (EAER
for short), which incorporates embedding and autoencoding techniques
naturally. In this framework, the original data are embedded into the
lower dimension, represented by the output of the hidden layer of the
autoencoder, thus the resulting data can not only maintain the locality-
preserving property but also easily revert to their original forms. This
is guaranteed by the joint minimization of the embedding loss and the
autoencoder reconstruction error. It is worth mentioning that instead
of operating in a batch mode as most of the previous embedding algorithms conduct, the proposed framework actually generates an *inductive* embedding model and thus supports incremental embedding efficiently. To show the effectiveness of EAER, we adapt this joint learning
framework to three canonical embedding algorithms, and apply them to
both synthetic and real-world data sets. The experimental results show
that the adaption of EAER outperforms its original counterpart. Besides, compared with the existing incremental embedding algorithms, the
results demonstrate that EAER performs incremental embedding with
more competitive efficiency and effectiveness.

Keywords: Embedding, Autoencoder, Representation Learning, Unsupervised Dimensionality Reduction.

1 Introduction

In many real-world applications, one is often confronted with overwhelmingly
complex features in the raw data and needs to obtain more useful data
representations. The manifold hypothesis, that real-world data presented in high-
dimensional spaces usually concentrate near a lower-dimensional non-linear manifold, brings a rich geometric perspective to the problem of learning meaningful

H. Blockeel et al. (Eds.): ECML PKDD 2013, Part III, LNAI 8190, pp. 208–223, 2013.
© Springer-Verlag Berlin Heidelberg 2013

representations [15]. Illustrated by this assumption, various embedding methods have been developed [19,24,1]. However, most of them only focus on the *locality-preserving* property of embedding, namely the relative distance between the points in the high dimension is preserved in the low dimension space.

Recently, deep learning and representation learning attract many research interests with its remarkable success in many applications [9,2,3,23]. In these works, autoencoder is usually adopted as a basic building block to initialize the deep neural network [3,23]. It is trained to encode the inputs into some representations so that the resulting representations can revert to their original forms. It has been shown that autoencoding is a powerful way to learn the hidden representation of the data.

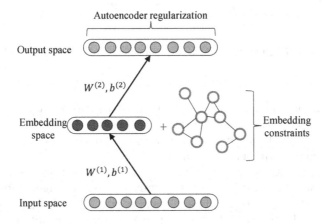

Fig. 1. Illustration of embedding with autoencoder regularization: view from autoencoders

Motivated by the remarkable success of deep learning with autoencoders, we solve the embedding problem collaboratively with autoencoding techniques. Specifically, we propose a framework of embedding with autoencoder regularization (EAER for short) with its two views from autoencoder and embedding respectively. First, from the view of autoencoder in Figure 1, EAER actually trains an autoencoder with the embedding constraints. Specifically, it contains the input layer, hidden layer and output layer from bottom to top. In this framework, besides minimizing the reconstruction error in the original autoencoder, the embedding loss at the hidden layer is also minimized simultaneously. Second, from the view of embedding in Figure 2, each data point with D dimensions is embedded into a d-dimensional space. Simultaneously, it is required that the data points in the hidden space can be recovered to their original form. In other words, the autoencoder is used here as a complementary regularizer to the embedding process. Hopefully, by this joint minimization we can derive the embedding with more semantic representations.

It should be noted that the training of the proposed framework actually generates an *inductive* embedding model, the function between the input and hidden

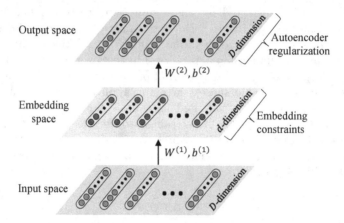

Fig. 2. Illustration of embedding with autoencoder regularization: view from embedding

layers of the autoencoder. It can directly map an instance into the low-dimensional space without accessing the original training data set. Thus, it supports incremental embedding more efficiently compared with most of the existing embedding algorithms which perform in a batch mode.

To show the effectiveness of EAER we adapt this joint learning framework to three well known embedding algorithms, namely Laplacian eigenmaps [1], multidimensional scaling [5] and margin-based embedding [7], and apply them to both the synthetic and real-world data sets. The experimental results show that the adaption of EAER outperforms its original counterpart. Also, we demonstrate that compared with the existing incremental embedding algorithms, EAER performs incremental embedding more efficiently with the competitive effectiveness.

In this paper, we describe the EAER framework, along with the details of implementation and performance on a simple synthetic example and real-world data sets. The organization of this paper is as follows: In Section 2, we review the preliminary knowledge of embedding algorithms and autoencoder. In Section 3, we describe EAER framework of learning a low dimensional mapping with autoencoder regularization. In Section 4, we illustrate the algorithm's performance by adapting this joint learning framework to three canonical embedding algorithms. In Section 5, we compare EAER framework to other unsupervised embedding algorithms and discuss several related works. Finally, in Section 6 we conclude this study and mention several directions for future work.

2   Preliminaries

In this section, the summary of general embedding algorithms will be given, followed by a brief review of autoencoders.

2.1   Embedding Algorithms

Many well known embedding algorithms can be described as a rather general form [26]: given the data set $x^{(1)}, ..., x^{(m)}$ find an embedding $f(x^{(i)})$ of each point $x^{(i)}$ by solving the following optimization problem

$$\sum_{1 \le i < j \le m} L(f(W, b; x^{(i)}), f(W, b; x^{(j)}), \varphi_{ij}) \tag{1}$$

where $f(W, b; x) \in R^d$ is the embedding result for a given input $x \in R^D$. $L(\cdot)$ is the loss function between pairs of inputs. φ_{ij} is the weight between $x^{(i)}$ and $x^{(j)}$. We define $\varphi_{ij} = 0$ if $i = j$. For certain loss function $L(\cdot)$ such as Equation (2), constraints may need to remove arbitrary factors in the embedding.

To compute the parameter set φ, one can construct an adjacency graph by putting an edge between between $x^{(i)}$ and $x^{(j)}$ if they are "similar". The similarity of any two data points can be evaluated with k-nearest neighbors (kNN) or ε-neighborhoods. Nodes $x^{(i)}$ and $x^{(j)}$ are connected by an edge if $x^{(i)}$ is among k nearest neighbors of $x^{(j)}$. If one chooses ε-neighborhoods metric, nodes are connected by an edge if $\|x^{(i)} - x^{(j)}\|^2 < \varepsilon$ where the norm is usually the Euclidean norm. The edges are weighted with Euclidean distance $\varphi_{ij} = \|x^{(i)} - x^{(j)}\|^2$ or Gaussian kernel $\varphi_{ij} = e^{-\frac{\|x^{(i)} - x^{(j)}\|^2}{\tau}}$ ($\tau \in R$) if $x^{(i)}$ and $x^{(j)}$ are connected. Another simple way for weighting the edges is to set $\varphi_{ij} = 1$ if nodes are connected, otherwise $\varphi_{ij} = 0$.

We consider the following embedding algorithms which fit into this framework.

Laplacian Eigenmaps. Laplacian eigenmaps (LE) [1] is a coherent framework for embedding by emphasizing the preservation of the locality. One constructs the adjacency graph of input samples and encodes them into d-dimensional Euclidean space. The embedding is given by the $d \times m$ matrix $\mathcal{F} = [f_1, f_2, ..., f_m]$, f_i is short for $f(W, b; x^{(i)})$. $L = D - \varphi$ is the Laplacian matrix where D is a diagonal weight matrix $D_{ii} = \sum_j \varphi_{ji}$. Then, we need to minimize

$$\sum_{i<j} L(f_i, f_j, \varphi_{ij}) = \sum_{i<j} \|f_i - f_j\|^2 \varphi_{ij} = \mathrm{Tr}(\mathcal{F} L \mathcal{F}^T) \tag{2}$$

subject to: $\mathcal{F} D \mathcal{F}^T = I$ and $\mathcal{F} D 1 = 0$.

Multidimensional Scaling. Multidimensional scaling (MDS) [5] is a canonical form of linear embedding that attempts to find an embedding from the input data into a low-dimensional space such that distances between data points are preserved. Usually, MDS is formulated as an optimization problem

$$\sum_{i<j} L(f_i, f_j, \varphi_{ij}) = \sum_{i<j} (\|f_i - f_j\| - \varphi_{ij})^2 \tag{3}$$

Kernel PCA can be interpreted as a form of metric MDS when the kernel function is isotropic [27].

Isomap [24] is one of several widely used low-dimensional embedding methods which works by defining the geodesic distance to be the sum of edge weights along the shortest path between two nodes and then performs low-dimensional embedding with classical MDS based on the pairwise distance between data points.

Margin-Based Embedding. The margin-based loss function proposed for learning a globally coherent nonlinear function that maps the data evenly to the output manifold and relies solely on neighborhood relationships [7]. The following optimization is used:

$$L(f_i, f_j, \varphi_{ij}) = \begin{cases} \|f_i - f_j\|^2 & \text{if } \varphi_{ij} = 1 \\ \max\left(0, l - \|f_i - f_j\|^2\right) & \text{if } \varphi_{ij} = 0 \end{cases} \tag{4}$$

which ensures that the data in the embedding space have a distance of at least l from each other when they are similar in the input space. In our experiments l is set to 1. The weight of edges $\varphi_{ij} = 1$ if $x^{(i)}$ and $x^{(j)}$ are connected, otherwise $\varphi_{ij} = 0$.

2.2 Autoencoders

An autoencoder neural network is an unsupervised learning algorithm that applies back-propagation [20], setting the target values to be equal to the inputs. In terms of embedding, the network learns to encode the inputs into a small number of dimensions and then decode it back into the original space. Specifically, given an unlabeled data set $x^{(i)}, i = 1, ..., m$, we want to learn representations

$$f(W^{(1)}, b^{(1)}; x^{(i)}) = \sigma(W^{(1)} x^{(i)} + b^{(1)}) \tag{5}$$

such that the output hypotheses

$$h(W, b; x^{(i)}) = \sigma(W^{(2)} f(W^{(1)}, b^{(1)}; x^{(i)}) + b^{(2)}) \tag{6}$$

is approximately $x^{(i)}$. Thus we use L_2 norm to minimize the reconstruction error $J(W, b; x)$

$$J(W, b; x) = \frac{1}{2} \sum_{i=1}^{m} \|h(W, b; x^{(i)}) - x^{(i)}\|^2 \tag{7}$$

We consider sparse autoencoder with sparsity parameter ρ and penalize it with the Kullback-Leibler (KL) divergence [10]. We then define the overall cost function to be

$$J(W, b; x) + \beta \sum_{j=1}^{d} \text{KL}\left(\rho \| \hat{\rho}_j\right) + \frac{\lambda}{2} \|W\|^2 \tag{8}$$

where $\hat{\rho}_j = \frac{1}{m} \sum_{i=1}^{m} f_j(W^{(1)}, b^{(1)}; x^{(i)})$ is the average activation of hidden unit j; m is the number of inputs and d is the number of hidden units; The last term is a weight decay term that tends to decrease the magnitude of the weights, and helps prevent over-fitting. β and λ control the weight of the corresponding penalty terms;

$$\text{KL}(\rho \| \hat{\rho}_j) = \rho \log \frac{\rho}{\hat{\rho}_j} + (1 - \rho) \log \frac{1 - \rho}{1 - \hat{\rho}_j} \tag{9}$$

is the KL divergence between Bernoulli random variables with mean ρ and $\hat{\rho}_j$ respectively.

3 Learning a Low Dimensional Mapping with Autoencoder Regularization

We consider the problem of finding a function that maps high-dimensional input data to lower-dimensional representations given the neighborhood relationships between the data in the input space.

3.1 Embedding with Autoencoder Regularization

We would like to use the ideas developed in autoencoding for embedding. The general approach we propose for EAER is to add an autoencoder regularizer to the embedding optimization function. As shown in Figure 1 and 2, we aim to simultaneously minimize the autoencoder reconstruction error at the output layer and the embedding loss in the hidden layer. The general form of this joint loss function is as follows:

$$J_{em}(W, b, \varphi; x) = \sum_{1 \leq i < j \leq m} L(f(W^{(1)}, b^{(1)}; x^{(i)}), f(W^{(1)}, b^{(1)}; x^{(j)}), \varphi_{ij})$$

$$+ \gamma J(W, b; x) + \beta \sum_{j=1}^{d} \text{KL}(\rho \| \hat{\rho}_j) + \frac{\lambda}{2} \|W\|^2 \tag{10}$$

where $L(\cdot)$ is the embedding loss function between pairs of the data (its detailed form can be any of the functions (2), (3) and (4) for different embedding algorithms). Here, $f(W^{(1)}, b^{(1)}; x^{(i)})$ actually maps $x^{(i)}$ to the lower dimension; $J(W, b; x)$ is the autoencoder reconstruction error defined by Equation (7); the last two terms are sparsity penalty term and weight decay term discussed in Section 2.2; γ, β and λ control the balance between these penalty terms. The idea that injecting an autoencoder regularization may help to guide the embedding towards better data representations, and we use a synthetic swiss roll example to illustrate this conjecture.

3.2 A Case Study on Synthetic Data

The swiss roll, considered in [1,24,19], is a flat two-dimensional sub-manifold of R^3 which is shown in Figure 3(a), and the data set of 2000 points chosen at random from the swiss roll is shown in Figure 3(b). We build the adjacency graph

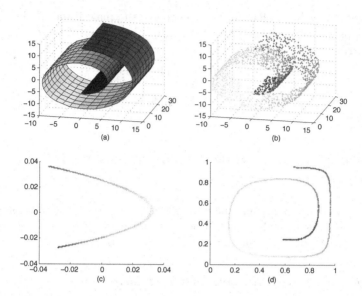

Fig. 3. A synthetic swiss roll example. (a) the synthetic swiss roll manifold, (b) 2000 points chosen at random from the swiss roll, (c) embedding result of LE, (d) embedding result of EAER-LE.

with 8 nearest neighbors, and set the weight $\varphi_{ij} = 1$ if node i is among the 8 nearest neighbors of node j, otherwise $\varphi_{ij} = 0$. For EAER-LE (the EAER framework applied to the LE algorithm), we have the following parameter settings in Equation (10): $\gamma = 0.65$, $\lambda = 0.003$, and $\beta = 0$. We compute the two-dimensional representations by LE and EAER-LE, and their results are respectively shown in Figure 3(c) and Figure 3(d).

The curve in Figure 3(c) is only a half ellipse, while the curve in Figure 3(d) maintains the roll in the two-dimension space. Thus, it is obvious that the result of EAER-LE is more "semantic" since it preserves the curve of the original manifold together with the locality properties. As shown in Section 4, we can also obtain such meaningful results when EAER is applied to the other embedding algorithms.

3.3   Model Learning

Our goal is to minimize $J_{em}(W, b, \varphi; x)$ as a function of W, b and φ. One can construct a weighted graph by the method of k-nearest neighbors (kNN) or ε-neighborhoods to compute φ and then train this regularized neural network parameterized by W and b. The key step of model learning is computing the partial derivatives $\frac{\partial}{\partial W^{(l)}} J_{em}(W, b, \varphi; x)$ and $\frac{\partial}{\partial b^{(l)}} J_{em}(W, b, \varphi; x)$ with respect to the input x. We will use an efficient way to compute the partial derivatives in light of the intuition behind the back-propagation algorithm [20]. In order to measure how much the nodes of the same layer is "responsible" for the errors of the output hypotheses, we introduce an "error term" vector δ, and define $\delta^{(1)}$ and $\delta^{(2)}$ for the hidden layer and output layer respectively. To incorporate the KL-divergence term into the derivative calculation, $\delta^{(1)}$ and $\delta^{(2)}$ are computed as follows [16]:

$$\delta^{(1)} = \left(\left(W^{(2)} \right)^T \delta^{(2)} + \beta \frac{\hat{\rho} - \rho_0}{\hat{\rho}(1 - \rho_0)} \right) \cdot \sigma'(z^{(1)}) \tag{11}$$

$$\delta^{(2)} = \frac{\partial}{\partial z^{(2)}} \frac{1}{2} \| h(W, b; x) - x \|^2 = - (h(W, b; x) - x) \cdot \sigma'(z^{(2)}) \tag{12}$$

where $\hat{\rho} = \frac{1}{m} \sum_{i=1}^{m} f(W^{(1)}, b^{(1)}; x^{(i)})$ is the average activation of embedding layer; $\rho_0 \in R^d$ is a vector with all entries ρ; "." denotes the element-wise product operator; $z^{(1)} = W^{(1)} x + b^{(1)}$ and $z^{(2)} = W^{(2)} f(W^{(1)}, b^{(1)}; x) + b^{(2)}$. In detail, the procedure can be described in Algorithm 1.

Algorithm 1. Partial Derivatives Computation

Input: The input sample x

Output: Partial derivatives of the function defined by Equation (10): $\frac{\partial}{\partial W^{(l)}} J_{em}(W, b, \varphi; x)$ and $\frac{\partial}{\partial b^{(l)}} J_{em}(W, b, \varphi; x)$.

1. Randomly initialized $W^{(l)}$ and $b^{(l)}$, ($l = 1, 2$).
2. Perform a feedforward pass, computing the activations for the embedding layer and output layer.
3. For the output layer, compute $\delta^{(2)}$ by Equation (12).
4. Compute $\delta^{(1)}$ by Equation (11).
5. Compute the partial derivatives:

$$\frac{\partial}{\partial W^{(2)}} J_{em}(W, b, \varphi; x) = \gamma \delta^{(2)} \left(f(W^{(1)}, b^{(1)}; x) \right)^T + \lambda W^{(2)};$$

$$\frac{\partial}{\partial b^{(2)}} J_{em}(W, b, \varphi; x) = \gamma \delta^{(2)};$$

$$\frac{\partial}{\partial W^{(1)}} J_{em}(W, b, \varphi; x) = \frac{\partial}{\partial W^{(1)}} \sum_{ij} L(\cdot) + \gamma \delta^{(1)} x^T + \lambda W^{(1)};$$

$$\frac{\partial}{\partial b^{(1)}} J_{em}(W, b, \varphi; x) = \frac{\partial}{\partial b^{(1)}} \sum_{ij} L(\cdot) + \gamma \delta^{(1)} \text{ where}$$

$$L(\cdot) = L \left(f(W^{(1)}, b^{(1)}; x^{(i)}), f(W^{(1)}, b^{(1)}; x^{(j)}), \varphi_{ij} \right).$$

In Step 5 of Algorithm 1, we compute $\frac{\partial}{\partial W^{(1)}} \sum_{ij} L(\cdot)$ according to its concrete form of different embedding algorithms, such as LE, MDS and margin-based embedding. For certain embedding loss function, the gradient can be incorporated into the "error term" $\delta^{(1)}$ in order to speed up the algorithm. Pseudocode of the full approach is given in Algorithm 2, where α is the learning rate. To train this model, we can now repeatedly take steps of gradient descent to reduce our cost function $J_{em}(W, b, \varphi; x)$. Note that EAER is a general framework, which can be adapted for different embedding algorithms. In the experiments we adapt the framework to LE, MDS and the margin-based embedding algorithm for comparison.

Algorithm 2. Algorithm for EAER Framework

Input: The input data set $\{x^{(i)}\}_{i=1}^m$.
Output: Results of embedding layer $f(W^{(1)}, b^{(1)}; x)$.

1. Construct the adjacency graph of $\{x^{(i)}\}_{i=1}^m$ and compute φ_{ij}.
2. **while** not stopping criterion **do**
3.   Set $\Delta W^{(l)} = 0$, $\Delta b^{(l)} = 0$ for all $l = 1, 2$.
4.   Use Algorithm 1 to compute $\frac{\partial}{\partial W^{(l)}} J_{em}(W, b, \varphi; x^{(i)})$ and $\frac{\partial}{\partial b^{(l)}} J_{em}(W, b, \varphi; x^{(i)})$
      for all $x^{(i)}$.
5.   Compute $\Delta W^{(l)} = \sum_{i=1}^m \frac{\partial}{\partial W^{(l)}} J_{em}(W, b, \varphi; x^{(i)})$;
6.   Compute $\Delta b^{(l)} = \sum_{i=1}^m \frac{\partial}{\partial b^{(l)}} J_{em}(W, b, \varphi; x^{(i)})$.
7.   Update: $W^{(l)} = W^{(l)} - \alpha \left(\frac{1}{m} \Delta W^{(l)} \right)$, $b^{(l)} = b^{(l)} - \alpha \left(\frac{1}{m} \Delta b^{(l)} \right)$.
8. **end while**
9. Compute the embedding results $f(W^{(1)}, b^{(1)}; x)$.

3.4   Incremental Embedding with EAER

Most of the embedding algorithms operate in a "batch" mode. That is, all data need to be available during training. If the new data come, the naive method is to re-run the training on the union of the original and new data, which prohibitively involves with expensive computing. To address this point, some incremental version of embedding algorithms, such as incremental Isomap [12] and incremental locally linear embedding (LLE) [22], were proposed. Their basic idea is to identify the k-nearest neighbors in the original training data for the new point and use these neighbors to represent the new one in the embedding space. It is clear that the original training data must be accessed in these methods of incremental embedding.

As to the EAER framework it is naturally an inductive embedding model. After all the model parameters are learned, any instance x can be embedded to $f(W^{(1)}, b^{(1)}; x)$ in the lower dimension directly without accessing the original training data. It should be noted that, when we use the Sigmoid function as the activation function, the input data need to be normalized to the range $[0, 1]$. Therefore, it is convenient to apply incremental embedding to those data sets with known data intervals, say MNIST [13], with each element ranging from 0 to 255.

Assume that we have m points for training and n for testing. EAER only takes linear time $O(n)$ to compute the low-dimensional embedding for the new n points. However, the incremental versions of the existing embedding algorithms need to compute its k-nearest neighbors among the training set. Thus, the overall time complexity of these algorithms is $O(mn)$ (assuming the linear scanning method for k-nearest neighbors is adopted here). Therefore, EAER dramatically reduces the running time for incremental embedding. The experiments in Section 4 also show the effectiveness of EAER for incremental embedding.

4   Experimental Evaluation

To verify the embedding performance of EAER framework, we conduct the experiments on various kinds of benchmark data sets. We begin with a simple synthetic example to intuitively show the embedding performance of EAER in Section 3. In this section, we build softmax classifiers [6] based on the embedded features of the real-world data sets and check the performance on classification accuracy. Then, we compare EAER with the incremental version of the baseline methods.

4.1   Benchmark Data Sets and Baseline Methods

The benchmark data sets are summarized in Table 1. One is MNIST [13] and the other six are taken from the UCI repository [4]. All these data sets are provided with the class labels.

Table 1. Data sets description

#	Data Sets	#Instances	#Attributes	#Classes
1	Iris	150	4	3
2	Wine	178	13	3
3	Glass	214	9	6
4	Diabetes	768	8	2
5	Segment	2310	19	7
6	Satimage	6435	36	6
7	MNIST	70000	784	10

In our experiments, we adapt the proposed EAER framework to the following embedding algorithms.

- **LE**: Laplacian Eigenmaps with the loss function defined by Equation (2) [1];
- **MDS**: Classical multidimensional scaling [5] with the loss function in Equation (3), where the norm adopted is the Euclidean distance;
- **Margin-based**: Embedding with the margin-based loss function defined by Equation (4) [7].

The embedding algorithms adapted to EAER are denoted as EAER-LE, EAER-MDS and EAER-Margin respectively.

4.2   Experimental Results on Real-World Data Sets

The experiments are conducted on 7 real-world data sets listed in Table 1. For each data set we first apply any embedding algorithm to it and then build the classification model on the resultant features. The classification accuracy on the training data is used as the evaluation measure for embedding.

The model parameters are set as follows. In the joint loss function of Equation (10), we set $\rho = 0.1$, $\beta = 0.2$, $\lambda = 0.003$, and W, b are randomly initialized. We set the weight $\varphi_{ij} = 1$ if node i is among the nearest neighbors of node j, otherwise $\varphi_{ij} = 0$. Then, for the rest three parameters, namely the number d of hidden units, the number k of the nearest neighbors in the adjacency graph and the weight γ of the autoencoder regularizer, their ranges are given as follows. γ is sampled from 0.1 to 1 with the interval of 0.1, k varies from 2 to 10 with the interval of 1, and d is set as $2 \leq d \leq \frac{D+1}{2}$ for the UCI data sets and $2 \leq d \leq 20$ for the MNIST data set, where D is the dimension of the original data sets. The data values are all normalized on each feature.

Table 2. Results on the data sets described in Table 1. We report the best accuracy for each method over the parameter ranges. For MNIST, we randomly select 1000 and 5000 samples and evaluate the performances on them respectively. The last column "Original" is the classification results on the original data sets.

Methods	LE	EAER-LE	MDS	EAER-MDS	Margin-based	EAER-Margin	Original
Iris	0.8933	0.9467	0.9400	0.9733	0.9667	**0.9800**	0.9467
Wine	0.9663	0.9719	0.9831	**0.9944**	0.9775	0.9888	**0.9944**
Glass	0.5654	0.5374	0.5935	0.6916	0.6355	**0.6963**	0.6075
Diabetes	0.6680	0.6576	0.7083	0.7552	0.7578	0.7669	**0.7813**
Segment	0.7758	0.8823	0.6390	0.7643	0.9028	**0.9443**	0.9130
Satimage	0.8362	0.8410	0.7984	0.8413	0.8522	**0.8738**	0.8578
MNIST^{1k}	0.8140	0.9520	0.8320	0.9020	0.9510	0.9840	**1.0000**
MNIST^{5k}	0.8370	0.9340	0.8520	0.9260	0.9370	0.9900	**1.0000**
Average	0.7945	**0.8404**	0.7933	**0.8560**	0.8726	**0.9030**	0.8876

For each embedding algorithm we train the softmax classifier on the resultant features and the original data sets, and record its classification accuracy on these training data. Among the parameter ranges, the results with best training accuracy are reported in Table 2. It shows that the EAER adaption is usually better than its original counterpart except that on the two data sets of *Glass* and *Diabetes*, EAER-LE is slightly worse than LE. On the whole, EAER increases the average accuracy over all the data sets by 4.59%, 6.27%, 3.04% compared with the corresponding three baseline methods respectively. It is worth mentioning that the average accuracy of EAER-Margin is better than the classification performance with original data sets, which again verifies EAER can learn better semantic representations.

Figure 4 shows the average accuracy of EAER-LE, EAER-MDS and EAER-Margin (denoted as "EAER" in the figure) and baseline methods (denoted as "Embedding" in the figure) when the embedding layer dimensionality d varies. The parameter setting is identical to previous experiment. Figure 4 demonstrates that when the embedding layer dimensionality d increases, the embedding accuracy increases. Yet we also observed that embedding methods adapted to

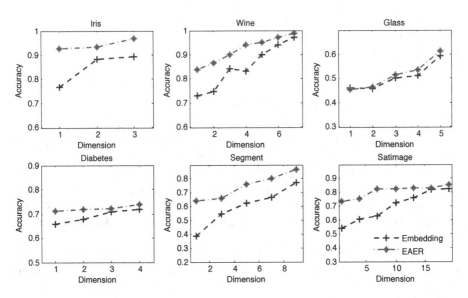

Fig. 4. The average accuracy of EAER-LE, EAER-MDS and EAER-Margin (denoted as "EAER" in the figure) and baseline methods (denoted as "Embedding" in the figure) when the embedding layer dimensionality d varies. The experiment was conducted on six UCI data sets.

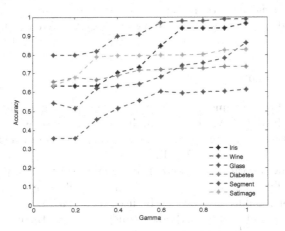

Fig. 5. The average accuracy of embedding methods adapted to EAER (EAER-LE, EAER-MDS and EAER-Margin) on each data set when γ changes. γ is sampled from 0.1 to 1 with the interval of 0.1, the experiment was conducted on six UCI data sets.

EAER framework led to comparatively higher accuracy even when d is small. So to speak, embedding methods adapted to EAER framework is able to preserve more data information. Figure 5 shows the average accuracy of embedding methods adapted to EAER when γ changes. As can be seen from the figure, when γ is small, the accuracy increases as γ goes up. When γ reaches a certain value (usually 0.6 to 0.8, the value varies according to different data sets), the result becomes comparatively stable. However, larger γ does not indicate better result. When its value exceeds a certain value (say 3 or 5), the accuracy begins to fall off (not shown in this figure).

4.3   Incremental Embedding Results

The EAER framework is an inductive embedding model. After all the model parameters are learned, any instance x can be embedded to a lower-dimensional space via $f(W^{(1)}, b^{(1)}; x)$. Thus, EAER can naturally handle incremental embedding. Since LE, MDS and Margin-based methods all perform in a batch mode, we adapt them to handle new data as follows, similar to the method in [22]. For a new instance we first identify its k-nearest neighbors in the original data and then construct the output by combining the embedded features of these neighbors.

To evaluate the performance of incremental embedding, each data set is randomly divided into two parts for training and testing respectively. On the training data we first apply the embedding algorithm and then train a classifier based on the resultant features. Next, for each instance in the testing data we apply the incremental embedding on it and then use the classifier to test on its embedded features. Thus, the classification accuracy on the testing data can be used as the evaluation measure for incremental embedding. We randomly sample the training and testing data for 10 times and average the testing accuracy values for each method. All the results are shown in Figure 6. We can find that EAER achieves highest accuracy among 16 of the whole 18 cases, and 1 tie-break.

5   Related Work

Two canonical forms of linear embedding are eigenvector methods of principle component analysis (PCA) [11] and multidimensional scaling (MDS) [5]. Recently there have been lots of nonlinear approaches to compute a low-dimensional embedding, including Isomap [24], LLE [19], Laplacian eigenmaps [1], margin-based embedding algorithms [7] and their variants [17,8]. The proposed EAER framework is different from these methods in the following aspects. First, with the autoencoder regularization it can be used as a general approach to complement any existing embedding method. Second, it embeds the new data using the resultant inductive model without accessing the original training data.

Autoencoders are usually used as basic building blocks to train the deep neural network [3] and currently variants of autoencoder have been investigated, such as contractive autoencoders [18] and denoising autoencoders [25]. These are often called regularized autoencoders, where some regularization terms are proposed to improve the data reconstruction performance. However, in the proposed framework, an autoencoder as a whole is used as a regularizer to improve the embedding algorithms.

There are also some other works related to EAER. First, in [26], the embedding-based regularizer is plugged into the layers of deep architectures as

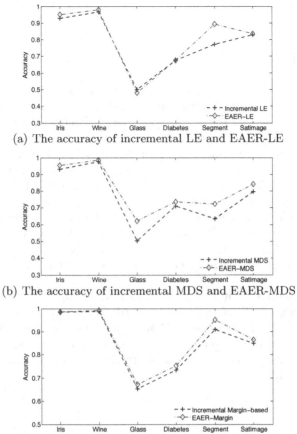

(a) The accuracy of incremental LE and EAER-LE

(b) The accuracy of incremental MDS and EAER-MDS

(c) The accuracy of incremental Margin-based embedding and EAER-Margin

Fig. 6. The accuracy of LE, MDS, Margin-based embedding and the corresponding methods adapted to EAER

an auxiliary task for semi-supervised embedding. The focus of this work is on semi-supervised learning, while ours is for unsupervised embedding with the autoencoder regularizer. Second, in [14] the stacked restricted Boltzmann machines (RBMs) are pre-trained and then fine-tuned with the supervised embedding constraints. It is clear that this approach is a disjoint way of autoencoding and embedding while our method trains them jointly. In [21], a multilayer neural network is pre-trained and fine-tuned to learning a nonlinear embedding by preserving class neighborhood structure. However, EAER is a general framework which compatible with different types of embedding algorithms.

Often we need to generalize the embedding results for the new data. LLE [19] was extended to its incremental version [22] by identifying the k-nearest neighbors of the new input and construct its output with its neighbors. Also, the incremental version of Isomap was proposed [12]. EAER naturally generates an inductive embedding model, whereas the methods mentioned embed new data in a transductive way.

6   Conclusion

In this paper we proposed an *embedding with autoencoder regularization* (EAER) framework for unsupervised nonlinear dimensionality reduction. By minimizing the embedding loss and the autoencoder reconstruction error simultaneously, EAER can learn more semantic representations of the inputs. We adapt the framework to the embedding algorithms of Laplacian eigenmaps, multidimensional scaling and margin-based method, and the results demonstrate that the embedding methods adapted to EAER outperform the original counterparts when applying the embedding codes to the classification tasks. The EAER framework proposed in the paper is naturally an inductive model, thus can embed the new data more efficiently. We plan to further investigate the performance of EAER by extending it to deep architectures or combining the advanced autoencoders, such as contractive autoencoders and denoising autoencoders.

Acknowledgments. This work is supported by the National Natural Science Foundation of China (No. 61175052, 61203297, 60933004, 61035003), National High-tech R&D Program of China (863 Program) (No. 2013AA01A606, 2012AA011003), and National Program on Key Basic Research Project (973 Program) (No. 2013CB329502). Thanks to the internship program of Hewlett Packard Labs China.

References

1. Belkin, M., Niyogi, P.: Laplacian eigenmaps for dimensionality reduction and data representation. In: Neural Computation, pp. 1373–1396 (2003)
2. Bengio, Y., Courville, A., Vincent, P.: Unsupervised feature learning and deep learning: A review and new perspectives. In: CoRR (2012)
3. Bengio, Y., Lamblin, P., Popovici, D., Larochelle, H.: Greedy layer-wise training of deep networks. In: Advances in Neural Information Processing Systems, pp. 153–160 (2007)
4. Blake, C., Merz, C.: Uci repository of machine learning databases (1998)
5. Cox, T., Cox, M.: Multidimensional scaling. Chapman & Hall, London (1994)
6. Greene, W., Zhang, C.: Econometric analysis. Prentice Hall, Upper Saddle River (1997)
7. Hadsell, R., Chopra, S., LeCun, Y.: Dimensionality reduction by learning an invariant mapping. In: IEEE Conference on Computer Vision and Pattern Recognition, pp. 1735–1742 (2006)
8. He, X., Cai, D., Yan, S., Zhang, H.: Neighborhood preserving embedding. In: Tenth IEEE International Conference on Computer Vision, pp. 1208–1213 (2005)
9. Hinton, G., Salakhutdinov, R.: Reducing the dimensionality of data with neural networks. Science, 504–507 (2006)
10. Hinton, G.: A practical guide to training restricted boltzmann machines. Momentum (2010)

11. Jolliffe, I.: Principal component analysis. Springer, New York (1986)
12. Law, M., Jain, A.: Incremental nonlinear dimensionality reduction by manifold learning. IEEE Trans. Pattern Analysis and Machine Intelligence, 377–391 (2006)
13. LeCun, Y., Bottou, L., Bengio, Y., Haffner, P.: Gradient-based learning applied to document recognition. Proceedings of the IEEE, 2278–2324 (1998)
14. Min, R., van der Maaten, L., Yuan, Z., Bonner, A., Zhang, Z.: Deep supervised t-distributed embedding. In: Proceedings of the 27th International Conference on Machine Learning (2010)
15. Narayanan, H., Mitter, S.: Sample complexity of testing the manifold hypothesis. In: Advances in Neural Information Processing Systems, pp. 1786–1794 (2010)
16. Ng, A.: Cs294a lecture notes: Sparse autoencoder. Stanford University (2010)
17. Niyogi, X.: Locality preserving projections. In: Advances in Neural Information Processing Systems, pp. 153–160 (2004)
18. Rifai, S., Vincent, P., Muller, X., Glorot, X., Bengio, Y.: Contractive auto-encoders: Explicit invariance during feature extraction. In: Proceedings of the 28th International Conference on Machine Learning (2011)
19. Roweis, S., Saul, L.: Nonlinear dimensionality reduction by locally linear embedding. Science, 2323–2326 (2000)
20. Rumelhart, D., Hintont, G., Williams, R.: Learning representations by back-propagating errors. Nature, 533–536 (1986)
21. Salakhutdinov, R., Hinton, G.: Learning a nonlinear embedding by preserving class neighbourhood structure. In: AI and Statistics (2007)
22. Saul, L., Roweis, S.: Think globally, fit locally: unsupervised learning of low dimensional manifolds. The Journal of Machine Learning Research, 119–155 (2003)
23. Socher, R., Pennington, J., Huang, E.H., Ng, A.Y., Manning, C.D.: Semi-supervised recursive autoencoders for predicting sentiment distributions. In: Proceedings of the Conference on Empirical Methods in Natural Language Processing, pp. 151–161 (2011)
24. Tenenbaum, J., De Silva, V., Langford, J.: A global geometric framework for non-linear dimensionality reduction. Science 2319–2323 (2000)
25. Vincent, P., Larochelle, H., Bengio, Y., Manzagol, P.: Extracting and composing robust features with denoising autoencoders. In: Proceedings of the 25th International Conference on Machine Learning (2008)
26. Weston, J., Ratle, F., Collobert, R.: Deep learning via semi-supervised embedding. In: Proceedings of the 25th International Conference on Machine Learning (2008)
27. Williams, C.: On a connection between kernel pca and metric multidimensional scaling. In: Machine Learning, pp. 11–19. Springer (2002)

Reduced-Rank Local Distance Metric Learning

Yinjie Huang[1], Cong Li[1], Michael Georgiopoulos[1],
and Georgios C. Anagnostopoulos[2]

[1] University of Central Florida, Department of Electrical Engineering & Computer
Science, 4000 Central Florida Blvd, Orlando, Florida, 32816, USA
yhuang@eecs.ucf.edu, licong1112@gmail.com, michaelg@ucf.edu
[2] Florida Institute of Technology, Department of Electrical and Computer
Engineering, 150 W University Blvd, Melbourne, FL 32901, USA
georgio@fit.edu

Abstract. We propose a new method for local metric learning based
on a conical combination of Mahalanobis metrics and pair-wise similarities between the data. Its formulation allows for controlling the rank
of the metrics' weight matrices. We also offer a convergent algorithm
for training the associated model. Experimental results on a collection
of classification problems imply that the new method may offer notable
performance advantages over alternative metric learning approaches that
have recently appeared in the literature.

Keywords: Metric Learning, Local Metric, Proximal Subgradient Descent, Majorization Minimization.

1 Introduction

Many Machine Learning problems and algorithms entail the computation of
distances with prime examples being the k-nearest neighbor (KNN) decision rule
for classification and the k-Means algorithm for clustering problems. Also, when
computing distances, the use of the Euclidean distance metric, or a weighted
variation of it, the Mahalanobis metric, are most often encountered because of
their simplicity and geometric interpretation. However, employing these metrics
for computing distances may not necessarily perform well for all problems. Early
on, attention was directed to data-driven approaches in order to infer the best
metric for a given problem (*e.g.* [1] and [2]). This is accomplished by taking
advantage of the data's distributional characteristics or other *side information*,
such as similarities between samples. In general, such paradigms are referred to as
metric learning techniques. A typical instance of such approaches is the learning
of the weight matrix that determines the Mahalanobis metric. This particular
task can equivalently be viewed as learning a decorrelating linear transformation
of the data in their native space and computing Euclidean distances in the range
space of the learned linear transform (feature space). When the problem at hand
is a classification problem, a KNN algorithm based on the learned metric is
eventually employed to label samples.

H. Blockeel et al. (Eds.): ECML PKDD 2013, Part III, LNAI 8190, pp. 224–239, 2013.

This paper focuses on metric learning methods for classification tasks, where the Mahalanobis metric is learned with the assistance of pair-wise sample similarity information. In our context, two samples will be deemed similar, if they feature the same class label. The goal of such approaches is to map similar samples close together and to map dissimilar samples far apart as measured by the learned metric. This is done so that an eventual application of a KNN decision rule exhibits improved performance over an application of KNN using a Euclidean metric.

Many such algorithms show significant improvements over the case of KNN that uses Euclidean metrics. For example, [1] poses similarity-based metric learning as a convex optimization problem, while [3] builds a trainable system to map similar faces to low dimensional spaces using a convolutional network to address geometric distortions. Moreover, [2] provides an online algorithm for learning a Mahalanobis metric based on kernel operators. Another approach, Neighborhood Components Analysis (NCA) [4], maximizes the leave-one-out performance on the training data based on stochastic nearest neighbors. Furthermore, in Large Margin Nearest Neighbor (LMNN) [5], the metric is learned so that the k-nearest neighbors of each sample belong to the same class, while others are separated by a large margin. Finally, [6] formulates the problem using information entropy and proposes the Information Theoretic Metric Learning (ITML) technique. In specific, ITML minimizes the differential relative entropy between two multivariate Gaussian distributions with distance metric constraints.

A common thread of the aforementioned methods is the use of a single, global metric, *i.e.*, a metric that is used for all distance computations. However, learning a global metric may not be well-suited in some settings that entail multi-modality or non-linearities in the data. To illustrate this point, Figure 1 displays a toy dataset consisting of 4 samples drawn from two classes. Sub-figure (a) shows the samples in their native space and sub-figure (b) depicts their images in the feature space resulting from learning a global metric. Finally, sub-figure (c) depicts the transformed data, when a local metric is learned, that takes into account the location and similarity characteristics of the data involved. We'll refer to such metrics as *local metrics*. Unlike the results obtained via the use of a global metric, one can (somewhat, due to the 3-dimensional nature of the depiction) observe in sub-figure (c) that images of similar samples (in this case, of the same class label) have been mapped closer to each other, when a local metric is learned. This may potentially result into improving 1-NN classification performance, when compared to the sample distributions in the other two cases.

Much work has been already performed on local metric learning. For example, [7] defines "local" as nearby pairs. In particular, they develop a model that aims to co-locate similar pairs and to separate dissimilar pairs. Additionally, their probabilistic framework is solved using an Expectation-Maximization-like algorithm. [8] learns local metrics through reducing neighborhood distances in directions that are orthogonal to the local decision boundaries, while expanding those parallel to the boundaries. In [9], the authors of LMNN also developed the LMNN-Multiple Metric (LMNN-MM) technique. When LMNN-MM is applied

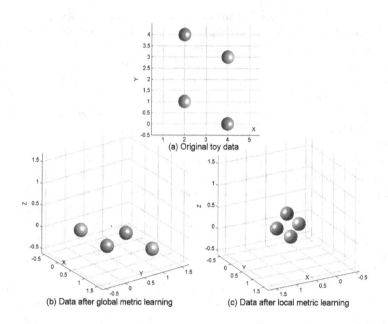

Fig. 1. Toy dataset that illustrates the potential advantages of learning a local metric instead of a global one. (a) Original data distribution. (b) Data distribution in the feature space obtained by learning a global metric. (c) Data distribution in the feature space obtained by learning a local metric.

in a classification context, the number of metrics utilized equals the number of classes. [10] introduced a similar approach, in which a metric is defined for each cluster. Moreover, in [11], the authors proposed Generative Local Metric Learning (GLML), which learns local metrics through NN classification error minimization. Their model employs a rather strong assumption, namely, they assume that the data has been drawn from a Gaussian mixture. Furthermore, in [12], the authors propose Parametric Local Metric Learning (PLML), in which each local metric is defined in relation to an anchor point of the instance space. Next, they use a linear combination of the resulting metric-defining weight matrices and employ a projected gradient method to optimize their model.

In this paper, we propose a new local metric learning approach, which we will be referring to as Reduced-Rank Local Metric Learning (R^2LML). As detailed in Section 2, for our method, the local metric is modeled as a conical combination of Mahalanobis metrics. Both the Mahalonobis metric weight matrices and the coefficients of the combination are learned from the data with the aid of pair-wise similarities in order to map similar samples close to each other and dissimilar samples far from each other in the feature space. Furthermore, the proposed problem formulation is able to control the rank of the involved linear mappings through a sparsity-inducing matrix norm. Additionally, in Section 3 we supply an algorithm for training our model. We then show that the set of fixed points of our algorithm includes the Karush-Kuhn-Tucker (KKT) points of our minimization problem. Finally, in Section 4 we demonstrate the capabilities of R^2LML with respect to classification tasks. When compared to other recent global or local metric learning methods, R^2LML exhibits the best classification accuracy in 7 out of the 9 datasets we considered.

2   Problem Formulation

Let $\mathbb{N}_M \triangleq \{1, 2, \ldots, M\}$ for any positive integer M. Suppose we have a training set $\{\boldsymbol{x}_n \in \mathbb{R}^D\}_{n \in \mathbb{N}_N}$ and corresponding pair-wise sample similarities arranged in a matrix $\boldsymbol{S} \in \{0, 1\}^{N \times N}$ as side information with the convention that, if $\boldsymbol{x}_m$ and $\boldsymbol{x}_n$ are similar, then $s_{mn} = 1$; if otherwise, then $s_{mn} = 0$. In a classification scenario, two samples can be naturally deemed similar (or dissimilar), if they feature the same (or different) class labels.

Now, the Mahalanobis distance between two samples $\boldsymbol{x}_n$ and $\boldsymbol{x}_m$ is defined as $d_{\boldsymbol{A}}(\boldsymbol{x}_m, \boldsymbol{x}_n) \triangleq \sqrt{(\boldsymbol{x}_m - \boldsymbol{x}_n)^T \boldsymbol{A}(\boldsymbol{x}_m - \boldsymbol{x}_n)}$, where $\boldsymbol{A} \in \mathbb{R}^{D \times D}$ is a positive semi-definite matrix (denoted as $\boldsymbol{A} \succeq 0$), which we will refer to as the *weight matrix of the metric*. Obviously, when $\boldsymbol{A} = \boldsymbol{I}$, the previous metric becomes the usual Euclidean distance. Being positive semi-definite, the weight matrix can be expressed as $\boldsymbol{A} = \boldsymbol{L}^T \boldsymbol{L}$, where $\boldsymbol{L} \in \mathbb{R}^{P \times D}$ with $P \leq D$. Hence, the previously defined distance can be expressed as $d_{\boldsymbol{A}}(\boldsymbol{x}_m, \boldsymbol{x}_n) = \|\boldsymbol{L}(\boldsymbol{x}_m - \boldsymbol{x}_n)\|_2$. Evidently, this last observation implies that the Mahalanobis distance based on $\boldsymbol{A}$ between two points in the native space can be viewed as the Euclidean distance between the images of these points in a feature space obtained through the linear transformation $\boldsymbol{L}$.

In metric learning, we are trying to learn $\boldsymbol{A}$ so to minimize the distances between pairs of similar points, while maintaining above a certain threshold (if not maximizing) the distances between dissimilar points in the feature space. Such a problem could be formulated as follows:

$$\min_{\boldsymbol{A} \succeq 0} \sum_{m,n} s_{mn} d_{\boldsymbol{A}}(\boldsymbol{x}_m, \boldsymbol{x}_n) \tag{1}$$

$$s.t. \sum_{m,n} (1 - s_{mn}) d_{\boldsymbol{A}}(\boldsymbol{x}_m, \boldsymbol{x}_n) \geq 1$$

Problem (1) is a semi-definite programming problem involving a global metric based on $\boldsymbol{A}$. There are several methods for learning a single global metric like the ones used for LMNN, ITML and NCA. However, as we have shown in Figure 1, use of a global metric may not be advantageous under all circumstances.

In this paper, we propose R^2LML, a new local metric approach, which we delineate next. Our formulation assumes that the metric involved is expressed as a conical combination of $K \geq 1$ Mahalanobis metrics. We also define a vector $\boldsymbol{g}^k \in \mathbb{R}^N$ for each local metric k. The nth element g_n^k of this vector may be regarded as a measure of how important metric k is, when computing distances involving the nth training sample. We constrain the vectors $\boldsymbol{g}^k$ to belong to $\Omega_g \triangleq$

$\left\{ \{g_k\}_{k \in \mathbb{N}_K} \in [0, 1]^N : g^k \succeq 0, \ \sum_k g^k = 1 \right\}$, where '$\succeq$' denotes component-wise ordering. The fact that the g^k's need to sum up to the all-ones vector $\mathbf{1}$ forces at least one metric to be relevant, when computing distances from each training sample. Note that, if $K = 1$, $g^1 = \mathbf{1}$, which corresponds to learning a single global metric.

Based on what we just described, the weight matrix for each pair (m, n) of training samples is given as $\sum_k A^k g_m^k g_n^k$. Observe that the distance between every pair of points features a different weight matrix. Motivated by Problem (1), one could consider the following formulation:

$$\min_{L^k, g^k \in \Omega_g, \xi_{m,n}^k \geq 0} \sum_k \sum_{m,n} s_{mn} \left\| L^k \Delta x_{mn} \right\|_2^2 g_n^k g_m^k + \tag{2}$$
$$+ C \sum_k \sum_{m,n} (1 - s_{mn}) \xi_{mn}^k + \lambda \sum_k \text{rank}(L^k)$$
$$s.t. \ \left\| L^k \Delta x_{mn} \right\|_2^2 \geq 1 - \xi_{mn}^k, \quad m, n \in \mathbb{N}_N, \ k \in \mathbb{N}_K$$

where $\Delta x_{mn} \triangleq x_m - x_n$ and $\text{rank}(L^k)$ denotes the rank of matrix L^k. The first term of the objective function attempts to minimize the measured distance between similar samples, while the second term along with the first set of soft constraints (due to the presence of slack variables ξ_{mn}^k) encourage distances between pairs of dissimilar samples to be larger than 1. Evidently, $C > 0$ controls the penalty of violating the previous desiteratum and can be chosen via a validation procedure. Finally, the last term penalizes large ranks of the linear transformations L^k. Therefore, the regularization parameter $\lambda \geq 0$, in essence, controls the dimensionality of the feature space.

Problem (2) can be somewhat reformulated by first eliminating the slack variables. Let $[\cdot]_+ : \mathbb{R} \to \mathbb{R}_+$ be the hinge function defined as $[u]_+ \triangleq \max\{u, 0\}$ for all $u \in \mathbb{R}$. It is straightforward to show that $\xi_{mn}^k = \left[1 - \left\| L^k \Delta x_{mn} \right\|_2^2 \right]_+$, which can be substituted back into the objective function. Next, we note that $\text{rank}(L^k)$ is a non-convex function w.r.t. L^k and is, therefore, hard to optimize. Following the approaches of [13] and [14], we replace $\text{rank}(L^k)$ with its convex envelope, i.e., the nuclear norm L^k, which is defined as the sum of L^k's singular values. These considerations lead to the following problem:

$$\min_{L^k, g^k \in \Omega_g} \sum_k \sum_{m,n} s_{mn} \left\| L^k \Delta x_{mn} \right\|_2^2 g_n^k g_m^k + \tag{3}$$
$$+ C(1 - s_{mn}) \left[1 - \left\| L^k \Delta x_{mn} \right\|_2^2 \right]_+ + \lambda \sum_k \left\| L^k \right\|_*$$

where $\|\cdot\|_*$ denotes the nuclear norm; in specific, $\left\| L^k \right\|_* \triangleq \sum_{s=1}^P \sigma_s(L^k)$, where σ_s is a singular value of L^k.

3   Algorithm

Problem (3) reflects a minimization over two sets of variables. When the g^k's are considered fixed, the problem is non-convex w.r.t. L^k, since the second term in Eq. (3) is the combination of a convex function (hinge function) and a non-monotonic function, $1 - \left\| L^k \Delta x_{mn} \right\|_2^2$, w.r.t. L^k. On the other hand, if the L^k's are considered fixed, the problem is also non-convex w.r.t g^k, since the similarity matrix S is almost always indefinite as it will be argued in the sequel. This implies that the objective function may have multiple minima. Therefore, an iterative procedure seeking to minimize it may have to be started multiple times with different initial estimates of the unknown parameters in order to find its global minimum. In what follows, we discuss a two-step, block-coordinate descent algorithm that is able to perform the minimization in question.

3.1   Two-Step Algorithm

For the first step, we fix g^k and try to solve for each L^k. In this case, Problem (3) becomes an unconstrained minimization problem. We observe that the objective function is of the form $f(w) + r(w)$, where w is the parameter we are trying to minimize over, $f(w)$ is the hinge loss function, which is non-differentiable, and $r(w)$ is a non-smooth, convex regularization term. If $f(w)$ were smooth, one could employ a proximal gradient method to find a minimum. As this is clearly not the case with the objective function at hand, in our work we resort to using a Proximal Subgradient Descent (PSD) method in a similar fashion to what has been done in [15] and [16]. Moreover, our approach is a special case of [17], based on which we show that our PSD steps converge (see Section 3.2).

Correspondingly, for the second step we assume the L^k's to be fixed and minimize w.r.t. each g^k vector. Consider a matrix $\bar{S}^k$ associated to the kth metric, whose (m, n) element is defined as:

$$\bar{s}_{mn}^k \triangleq s_{mn} \left\| L^k \Delta x_{mn} \right\|_2^2, \quad m, n \in \mathbb{N}_N \tag{4}$$

Then Problem (3) becomes:

$$\min_{g^k \in \Omega_g} \sum_k (g^k)^T \bar{S}^k g^k \tag{5}$$

Let $g \in \mathbb{R}^{KN}$ be the vector that results from concatenating all individual g^k vectors into a single vector and define the matrix

$$\tilde{S} \triangleq \begin{bmatrix} \bar{S}^1 & 0 & \dots & 0 \\ 0 & \bar{S}^2 & \dots & 0 \\ \vdots & \vdots & \ddots & \vdots \\ 0 & \dots & 0 & \bar{S}^K \end{bmatrix} \in \mathbb{R}^{KN \times KN} \tag{6}$$

Based on the previous definitions, the cost function becomes $\boldsymbol{g}^T \tilde{\boldsymbol{S}} \boldsymbol{g}$ and $\boldsymbol{g}$'s constraint set becomes $\Omega_g = \left\{\boldsymbol{g} \in [0,1]^{KN} : \boldsymbol{g} \succeq \boldsymbol{0}, \ \boldsymbol{Bg} = \boldsymbol{1}\right\}$, where $\boldsymbol{B} \triangleq \boldsymbol{1}^T \otimes \boldsymbol{I}_N$, $\otimes$ denotes the Kronecker product and $\boldsymbol{I}_N$ is the $N \times N$ identity matrix. Hence, the minimization problem for the second step can be re-expressed as:

$$\min_{\boldsymbol{g} \in \Omega_g} \ \boldsymbol{g}^T \tilde{\boldsymbol{S}} \boldsymbol{g} \tag{7}$$

Problem (7) is non-convex, since $\tilde{\boldsymbol{S}}$ is almost always indefinite. This stems from the fact that $\tilde{\boldsymbol{S}}$ is a block diagonal matrix, whose blocks are Euclidean Distance Matrices (EDMs). It is known that EDMs feature exactly one positive eigenvalue (unless all of them equal 0). Since each EDM is a hollow matrix, its trace equals 0. This, in turn, implies that its remaining eigenvalues must be negative [18]. Hence, $\tilde{\boldsymbol{S}}$ will feature negative eigenvalues.

In order to obtain a minimizer of Problem (7), we employ a Majorization Minimization (MM) approach [19], which first requires identifying a function of $\boldsymbol{g}$ that majorizes the objective function at hand. Let $\mu \triangleq -\lambda_{max}(\tilde{\boldsymbol{S}})$, where $\lambda_{max}(\tilde{\boldsymbol{S}})$ is the largest eigenvalue of $\tilde{\boldsymbol{S}}$. As the latter matrix is indefinite, $\lambda_{max}(\tilde{\boldsymbol{S}}) > 0$. Then, $\boldsymbol{H} \triangleq \tilde{\boldsymbol{S}} + \mu \boldsymbol{I}$ is negative semi-definite. Let $q(\boldsymbol{g}) \triangleq \boldsymbol{g}^T \tilde{\boldsymbol{S}} \boldsymbol{g}$ be the cost function in Eq. (7). Since $(\boldsymbol{g} - \boldsymbol{g}')^T \boldsymbol{H}(\boldsymbol{g} - \boldsymbol{g}') \leq 0$ for any $\boldsymbol{g}$ and $\boldsymbol{g}'$, we have that $q(\boldsymbol{g}) < -\boldsymbol{g}'^T \boldsymbol{H} \boldsymbol{g}' + 2\boldsymbol{g}'^T \boldsymbol{H} \boldsymbol{g} - \mu \|\boldsymbol{g}\|_2^2$ for all $\boldsymbol{g} \neq \boldsymbol{g}'$ and equality, only if $\boldsymbol{g} = \boldsymbol{g}'$. The right hand side of the aforementioned inequality constitutes q's majorizing function, which we will denote as $q(\boldsymbol{g}|\boldsymbol{g}')$. The majorizing function is used to iteratively optimize $\boldsymbol{g}$ based on the current estimate $\boldsymbol{g}'$. So we have the following minimization problem, which is convex w.r.t $\boldsymbol{g}$:

$$\min_{\boldsymbol{g} \in \Omega_g} \ 2\boldsymbol{g}'^T \boldsymbol{H} \boldsymbol{g} - \mu \|\boldsymbol{g}\|_2^2 \tag{8}$$

This problem is readily solvable, as the next theorem implies.

Theorem 1. *Let $\boldsymbol{g}, \boldsymbol{d} \in \mathbb{R}^{KN}$, $\boldsymbol{B} \triangleq \boldsymbol{1}^T \otimes \boldsymbol{I}_N \in \mathbb{R}^{N \times KN}$ and $c > 0$. The unique minimizer $\boldsymbol{g}^*$ of*

$$\min_{\boldsymbol{g}} \ \frac{c}{2} \|\boldsymbol{g}\|_2^2 + \boldsymbol{d}^T \boldsymbol{g} \tag{9}$$

$$s.t. \ \boldsymbol{Bg} = \boldsymbol{1}, \ \boldsymbol{g} \succeq \boldsymbol{0}$$

has the form

$$g_i^* = \frac{1}{c} \left[(\boldsymbol{B}^T \boldsymbol{\alpha})_i - d_i \right]_+, \quad i \in \mathbb{N}_{KN} \tag{10}$$

where g_i is the ith element of $\boldsymbol{g}$ and $\boldsymbol{\alpha} \in \mathbb{R}^N$ is the Lagrange multiplier vector associated to the equality constraint.

Proof. The Lagrangian of Problem (9) is expressed as:

$$L(\boldsymbol{g}, \boldsymbol{\alpha}, \boldsymbol{\beta}) = \frac{c}{2} \boldsymbol{g}^T \boldsymbol{g} + \boldsymbol{d}^T \boldsymbol{g} + \boldsymbol{\alpha}^T (\boldsymbol{1} - \boldsymbol{Bg}) - \boldsymbol{\beta}^T \boldsymbol{g} \tag{11}$$

Algorithm 1. Minimization of Problem (3)

Input: Data $\boldsymbol{X} \in R^{D \times N}$, number of metrics K
Output: $\boldsymbol{L}^k, \boldsymbol{g}^k \quad k \in \mathbb{N}_K$
01. Initialize $\boldsymbol{L}^k, \boldsymbol{g}^k$ for all $k \in \mathbb{N}_K$
02. **While** not converged **Do**
03.   Step 1: Use a PSD method to solve Problem (3) for each $\boldsymbol{L}^k$
04.   Step 2:
05. $\tilde{\boldsymbol{S}} \leftarrow Eq.$ (6)
06. $\mu \leftarrow -\lambda_{max}(\tilde{\boldsymbol{S}})$
07. $\boldsymbol{H} \leftarrow \tilde{\boldsymbol{S}} + \mu \boldsymbol{I}$
08.   **While** not converged **Do**
09.   Apply binary search to obtain each $\boldsymbol{g}^k$ using Eq. (10)
10.   **End While**
11. **End While**

---

where $\boldsymbol{\alpha} \in \mathbb{R}^N$ and $\boldsymbol{\beta} \in \mathbb{R}^{KN}$ with $\boldsymbol{\beta} \succeq \boldsymbol{0}$ are Lagrange multiplier vectors. If we set the partial derivative of $L(\boldsymbol{g}, \boldsymbol{\alpha}, \boldsymbol{\beta})$ with respect to $\boldsymbol{g}$ to $\boldsymbol{0}$, we readily obtain that

$$g_i = \frac{1}{c} \left((\boldsymbol{B}^T \boldsymbol{\alpha})_i + \beta_i - d_i \right), \quad i \in \mathbb{N}_{KN} \tag{12}$$

Let $\gamma_i \triangleq (\boldsymbol{B}^T \boldsymbol{\alpha})_i - d_i$. Combining Eq. (12) with the complementary slackness condition $\beta_i g_i = 0$, one obtains that, if $\gamma_i \leq 0$, then $\beta_i = -\gamma_i$ and $g_i = 0$, while, when $\gamma_i > 0$, then $\beta_i = 0$ and, evidently, $g_i = \frac{1}{c}\gamma_i$. These two observations can be summarized into $g_i = \frac{1}{c} [\gamma_i]_+$, which completes the proof. □

In order to exploit the result of Theorem 1 for obtaining a concrete solution to Problem (8), we ought to point out that the (unknown) optimal values of the Lagrange multipliers α_i can be found via binary search, so they satisfy the equality constraint $\boldsymbol{Bg} = \boldsymbol{1}$.

In conclusion, the entire algorithm for solving Problem (3) can be recapitulated as follows: For step 1, the $\boldsymbol{g}^k$ vectors are assumed fixed and a PSD is being employed to minimize the cost function of Eq. (3) w.r.t. each weight matrix $\boldsymbol{L}^k$. For step 2, all $\boldsymbol{L}^k$'s are held fixed to the values obtained after completion of the previous step and the solution offered by Theorem 1 along with binary searches for the α_i's are used to compute the optimal $\boldsymbol{g}_k$'s by iteratively solving Problem (8) via a MM scheme. Note that these two main steps are repeated until convergence is established; the whole process is depicted in Algorithm 1.

3.2 Analysis

In this subsection, we investigate the convergence of our proposed algorithm. Suppose that a PSD method is employed to minimize the function $f(\boldsymbol{w}) + r(\boldsymbol{w})$, where both f and r are non-differentiable. Denote ∂f as the subgradient of f and define $\|\partial f(\boldsymbol{w})\| \triangleq \sup_{g \in \partial f(\boldsymbol{w})} \|g\|$; the corresponding quantities for r are similarly defined. Like in [20] and [21], we assume that the subgradients are bounded, *i.e.*:

$$\|\partial f(\boldsymbol{w})\|^2 \le Af(\boldsymbol{w}) + G^2, \quad \|\partial r(\boldsymbol{w})\|^2 \le Ar(\boldsymbol{w}) + G^2 \qquad (13)$$

where A and G are scalars. Let $\boldsymbol{w}^*$ be a minimizer of $f(\boldsymbol{w}) + r(\boldsymbol{w})$. Then we have the following lemma for the problem under consideration.

Lemma 2. *Suppose that a PSD method is employed to solve $\min_{\boldsymbol{w}} f(\boldsymbol{w}) + r(\boldsymbol{w})$. Assume that 1) f and r are lower-bounded; 2) the norms of any subgradients ∂f and ∂r are bounded as in Eq. (13); 3) $\|\boldsymbol{w}^*\| \le D$ for some $D > 0$; 4) $r(\boldsymbol{0}) = 0$. Let $\eta_t \triangleq \frac{D}{\sqrt{8TG}}$, where T is the number of iterations of the PSD algorithm. Then, for a constant $c \le 4$, such that $(1 - cA\frac{D}{\sqrt{8T}D}) > 0$, and initial estimate of the solution $\boldsymbol{w}_1 = \boldsymbol{0}$, we have:*

$$\min_{t \in \{1...T\}} f(\boldsymbol{w}_t) + r(\boldsymbol{w}_t) \le \frac{1}{T} \sum_{t=1}^{T} f(\boldsymbol{w}_t) + r(\boldsymbol{w}_t) \le$$

$$\le \frac{4\sqrt{2}DG}{\sqrt{T}(1 - \frac{cAD}{G\sqrt{8T}})} + \frac{f(\boldsymbol{w}^*) + r(\boldsymbol{w}^*)}{1 - \frac{cAD}{G\sqrt{8T}}} \qquad (14)$$

The proof of Lemma 2 is straightforward as it is based on [17] and, therefore, is omitted here. Lemma 2 implies that, as T grows, the PSD iterates approach $\boldsymbol{w}^*$.

Theorem 3. *Algorithm 1 yields a convergent, non-increasing sequence of cost function values relevant to Problem (3). Furthermore, the set of fixed points of the iterative map embodied by Algorithm 1 includes the KKT points of Problem (3).*

Proof. We first prove that each of the two steps in our algorithm decreases the objective function value. This is true for the first step, according to Lemma 2. For the second step, since a MM algorithm is used, we have the following relationships

$$q(\boldsymbol{g}^*) = q(\boldsymbol{g}^*|\boldsymbol{g}^*) \le q(\boldsymbol{g}^*|\boldsymbol{g}') \le q(\boldsymbol{g}'|\boldsymbol{g}') = q(\boldsymbol{g}') \qquad (15)$$

This implies that the second step always decreases the objective function value. Since the objective function is lower bounded, our algorithm converges.

Next, we prove that the set of fixed points of the proposed algorithm includes the KKT points of Problem (3). Towards this purpose, suppose the algorithm has converged to a KKT point $\left\{\boldsymbol{L}^{k*}, \boldsymbol{g}^{k*}\right\}_{k \in \mathbb{N}_K}$; then, it suffices to show that this point is also a fixed point of the algorithm's iterative map. For notational brevity, let $f_0(\boldsymbol{L}^k, \boldsymbol{g}^k)$, $f_1(\boldsymbol{g}^k)$ and $h_1(\boldsymbol{g}^k)$ be the cost function, inequality constraint and equality constraint of Problem (3) respectively. By definition, the KKT point will satisfy

$$\boldsymbol{0} \in \partial_{\boldsymbol{L}^k} f_0(\boldsymbol{L}^{k*}, \boldsymbol{g}^{k*}) + \nabla_{\boldsymbol{g}^k} f_0(\boldsymbol{L}^{k*}, \boldsymbol{g}^{k*})$$

$$- (\boldsymbol{\beta}^k)^T \nabla_{\boldsymbol{g}^k} f_1(\boldsymbol{g}^{k*}) + \boldsymbol{\alpha}^T \nabla_{\boldsymbol{g}^k} h_1(\boldsymbol{g}^{k*}) \quad k \in \mathbb{N}_K \qquad (16)$$

In relation to Problem (7), which step 2 tries to solve, the KKT point will satisfy the following equality (gradient of the problem's Lagrangian set to $\mathbf{0}$):

$$2\tilde{\boldsymbol{S}}\boldsymbol{g}^* - \boldsymbol{\beta} - \boldsymbol{B}^T\boldsymbol{\alpha} = \boldsymbol{0} \tag{17}$$

Problem (8) can be solved based on Eq. (12) of Theorem 1; in specific, we obtain that

$$\boldsymbol{g} = -\frac{1}{2\mu}(\boldsymbol{B}^T\boldsymbol{\alpha} + \boldsymbol{\beta} - 2\boldsymbol{H}\boldsymbol{g}^*) \tag{18}$$

Substituting Eq. (17) and $\boldsymbol{H} = \tilde{\boldsymbol{S}} + \mu\boldsymbol{I}$ into Eq. (18), one immediately obtains that

$$\boldsymbol{g} = -\frac{1}{2\mu}(\boldsymbol{B}^T\boldsymbol{\alpha} + \boldsymbol{\beta} - 2\boldsymbol{H}\boldsymbol{g}^*) = -\frac{1}{2\mu}(2\tilde{\boldsymbol{S}}\boldsymbol{g}^* - 2\tilde{\boldsymbol{S}}\boldsymbol{g}^* - 2\mu\boldsymbol{g}^*) = \boldsymbol{g}^* \tag{19}$$

In other words, step 2 will not update the solution. Now, if we substitute Eq. (17) back into Eq. (16), we obtain $\boldsymbol{0} \in \partial_{\boldsymbol{L}^k} f_0(\boldsymbol{L}^{k*}, \boldsymbol{g}^{k*})$ for all k, which is the optimality condition for the subgradient method; the PSD step (step 1 of our algorithm) will also not update the solution. Thus, a KKT point of Problem (3) is a fixed point of our algorithm. □

Table 1. Details of benchmark data sets. For the Letter and Pendigits datasets, only 4 and 5 classes were considered respectively.

	#D	#CLASSES	#TRAIN	#VALIDATION	#TEST
ROBOT	4	4	240	240	4976
LETTER A-D	16	4	200	400	2496
PENDIGITS 1-5	16	5	200	1800	3541
WINEQUALITY	12	2	150	150	6197
TELESCOPE	10	2	300	300	11400
IMGSEG	18	7	210	210	1890
TWONORM	20	2	250	250	6900
RINGNORM	20	2	250	250	6900
IONOSPHERE	34	2	80	50	221

4   Experiments

In this section, we performed experiments on 9 datasets, namely, Robot Navigation, Letter Recognition, Pendigits, Wine Quality, Gamma Telescope, Ionosphere datasets from the *UCI machine learning repository*[1], and Image Segmentation, Two Norm, Ring Norm datasets from the *Delve Dataset Collection*[2]. Some characteristics of these datasets are summarized in Table 1. We first explored how

---

[1] http://archive.ics.uci.edu/ml/datasets.html
[2] http://www.cs.toronto.edu/~delve/data/datasets.html

the performance of R^{2}LML[1] varies with respect to the number of local metrics. Then, we compared R^{2}LML with other global or local Metric Learning algorithms, including ITML, LMNN, LMNN-MM, GLML and PLML.

The computation of the distances between some test sample x and the training samples x_n according to our formulation requires the value of g corresponding to x. One option to assign a value to g would be to utilize transductive learning. However, as such an approach could prove computationally expensive, we opted instead to assign g the value of the corresponding vector associated to x's nearest (in terms of Euclidean distance) training sample as was done in [12].

4.1   Number of Local Metrics

In this subsection, we show how the performance of R^{2}LML varies with respect to the number of local metrics K. In [9], the authors set K equal to the number of classes for each dataset, which might not necessarily be the optimal choice. In our experiments, we let K vary from 1 to 7. This range covers the maximum number of classes in the datasets that are considered in our experiments. As we will show, the optimal K is not always the same as the number of classes.

Besides K, we held the remaining parameters (refer to Eq. (2)) fixed: the penalty parameter C was set to 1 and the nuclear norm regularization parameter λ to 0.1. Moreover, we terminated our algorithm, if it reached 10 epochs or when the difference of cost function values between two consecutive iterations was less than 10^{-4}. In each epoch, the PSD inner loop ran for 500 iterations. The PSD step length was fixed to 10^{-5} for the Robot and Ionosphere datasets, to 10^{-6} for the Letter A-D, Two norm and Ring Norm datasets, to 10^{-8} for the Pendigits 1-5, Wine Quality and Image Segmentation datasets and to 10^{-9} for the Gamma Telescope dataset. The MM loop was terminated, if the number of iterations reached 3000 or when difference of cost function values between two consecutive iterations was less than 10^{-3}. The relation between number of local metrics and the classification accuracy for each dataset is reported in Figure 2.

Several observations can be made based on Figure 2. First of all, our method used as a local metric learning method (when $K \geq 2$) performs much better than when used with a single global metric (when $K = 1$) for all datasets except the Ring Norm dataset. For the latter dataset, the classification performance deteriorates with increasing K. Secondly, one cannot discern a deterministic relationship between the classification accuracy and the number of local metrics utilized that is suitable for all datasets. For example, for the Robot dataset, the classification accuracy is almost monotonically increasing with respect to K. For the remaining datasets, the optimal K varies in a non-apparent fashion with respect to their number of classes. For example, in the case of the Ionosphere dataset (2-class problem), $K = 3, 6, 7$ yield the best generalization results. All these observations suggest that validation over K is needed to select the best performing model.

---

[1] https://github.com/yinjiehuang/R2LML/archive/master.zip

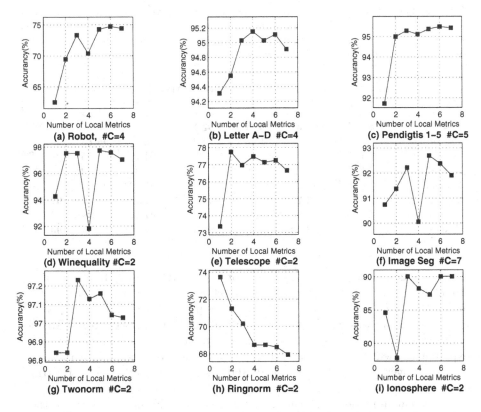

Fig. 2. R^2LML classification accuracy results on the 9 benchmark datasets for varying number K of local metrics. $\#C$ indicates the number of classes of each dataset.

4.2 Comparisons

We compared R^2LML with several other metric learning algorithms, including Euclidean metric KNN, ITML [6], LMNN [5], LMNN-MM [9], GLML [11] and PLML [12]. Both ITML and LMNN learn a global metric, while LMNN-MM, GLML and PLML are local metric learning algorithms. After the metrics are learned, the KNN classifier is utilized for classification with k (number of nearest neighbors) set to 5.

For our experiments we used LMNN, LMNN-MM[1], ITML[2] and PLML[3] implementations that we found available online. For ITML, a good value of γ is found via cross-validation. Also, for LMNN and LMNN-MM, the number of attracting neighbors during training is set to 1. Additionally, for LMNN, at most 500 iterations were performed and 30% of training data were used as a validation

---

[1] http://www.cse.wustl.edu/~kilian/code/code.html
[2] http://www.cs.utexas.edu/~pjain/itml/
[3] http://cui.unige.ch/~wangjun/papers/PLML.zip

Table 2. Percent accuracy results of 7 algorithms on 9 benchmark datasets. For each dataset, the statistically best and comparable results for a family-wise significance level of 0.05 are highlighted in boldface. All algorithms are ranked from best to worst; algorithms share the same rank, if their performance is statistically comparable.

	EUCLIDEAN	ITML	LMNN	LMNN-MM	GLML	PLML	R^2LML
ROBOT	65.31^{2nd}	65.86^{2nd}	66.10^{2nd}	66.10^{2nd}	62.28^{3rd}	61.03^{3rd}	$\mathbf{74.16^{1st}}$
LETTER A-D	88.82^{2nd}	$\mathbf{93.39^{1st}}$	$\mathbf{93.79^{1st}}$	$\mathbf{93.83^{1st}}$	89.30^{2nd}	$\mathbf{94.43^{1st}}$	$\mathbf{95.07^{1st}}$
PENDIGITS 1-5	88.31^{4th}	93.17^{2nd}	91.19^{3rd}	91.27^{3rd}	88.37^{4th}	$\mathbf{95.88^{1st}}$	$\mathbf{95.43^{1st}}$
WINEQUALITY	86.12^{7th}	96.11^{3rd}	94.43^{4th}	93.38^{5th}	91.79^{6th}	$\mathbf{98.55^{1st}}$	97.53^{2nd}
TELESCOPE	70.31^{3rd}	71.42^{2nd}	72.16^{2nd}	71.45^{2nd}	70.31^{3rd}	$\mathbf{77.52^{1st}}$	$\mathbf{77.97^{1st}}$
IMGSEG	80.05^{4th}	90.21^{2nd}	90.74^{2nd}	89.42^{2nd}	87.30^{3rd}	90.48^{2nd}	$\mathbf{92.59^{1st}}$
TWONORM	96.54^{2nd}	$\mathbf{96.78^{1st}}$	96.32^{2nd}	96.30^{2nd}	96.52^{2nd}	$\mathbf{97.32^{1st}}$	$\mathbf{97.23^{1st}}$
RINGNORM	55.84^{7th}	77.35^{2nd}	59.36^{6th}	59.75^{5th}	$\mathbf{97.09^{1st}}$	75.68^{3rd}	73.73^{4th}
IONOSPHERE	75.57^{3rd}	$\mathbf{86.43^{1st}}$	82.35^{2nd}	82.35^{2nd}	71.95^{3rd}	78.73^{3rd}	$\mathbf{90.50^{1st}}$

set. The maximum number of iterations for LMNN-MM was set to 50 and a step size of 10^{-7} was employed. For GLML, we chose γ by maximizing performance over a validation set. Finally, the PLML hyperparameter values were chosen as in [12], while α_1 was chosen via cross-validation. With respect to R^2LML, for each dataset we used K's optimal value as established in the previous series of experiments, while the regularization parameter λ was chosen via a validation procedure over the set $\{0.01, 0.1, 1, 10, 100\}$. The remaining parameter settings of our method were the same as the ones used in the previous experiments.

For pair-wise model comparisons, we employed McNemar's test. Since there are 7 algorithms to be compared, we used Holm's step-down procedure as a multiple hypothesis testing method to control the Family-Wise Error Rate (FWER) [22] of the resulting pair-wise McNemar's tests. The experimental results for a family-wise significance level of 0.05 are reported in Table 2.

It is observed that R^2LML achieves the best performance on 7 out of the 9 datasets, while GLML, ITML and PLML outperform our model on the Ring Norm dataset. GLML's surprisingly good result for this particular dataset is probably because GLML assumes a Gaussian mixture underlying the data generation process and the Ring Norm dataset is a 2-class recognition problem drawn from a mixture of two multivariate normal distributions. Even though not being the best model for this dataset, R^2LML is still highly competitive compared to LMNN, LMNN-MM and Euclidean KNN. Next, PLML performs best in 5 out of 9 datasets, even outperforming R^2LML on the Wine Quality dataset. However, PLML gives poor results on some datasets like Robot or Ionosphere. Also, PLML does not show much improvements over KNN and may even perform worse like for the Robot dataset. Note, that R^2LML is still better for the Image Segmentation, Robot and Ionosphere datasets. Additionally, ITML is ranked first for 3 datasets and even outperforms R^2LML on the Ring Norm dataset. Often, ITML ranks at least 2nd and seems to be suitable for low dimensional datasets. However, R^2LML still performs better than ITML for 5 out of the 9 datasets. Finally,

GLML rarely performs well; according to Table 2, GLML only achieves 3rd or 4th ranks for 6 out of the 9 datasets.

Another general observation that can be made is the following: employing metric learning is almost always a good choice, since the classification accuracy of utilizing a Euclidean metric is almost always the lowest among all the 7 methods we considered. Interestingly, LMNN-MM, even though being a local metric learning algorithm, does not show any performance advantages over LMNN (a global metric method); for some datasets, it even obtained lower classification accuracy than LMNN. It is possible that fixing the number of local metrics to the number of classes present in the dataset curtails LMNN-MM's performance. According to the obtained results, R^2LML yields much better performance for all datasets compared to LMNN-MM. This consistent performance advantage may not only be attributed to the fact that K was selected via a validation procedure, since, for cases where the optimal K equaled the number of classes (*e.g.* Letter A-D dataset), R^2LML still outperformed LMNN-MM.

5 Conclusions

In this paper, we proposed a new local metric learning model, namely Reduced-Rank Local Metric Learning (R^2LML). It learns K Mahalanobis-based local metrics that are conically combined, such that similar points are closer to each other, while the separation between dissimilar ones is encouraged to increase. Additionally, a nuclear norm regularizer is adopted to obtain low-rank weight matrices for calculating metrics. In order to solve our proposed formulation, a two-step algorithm is showcased, which iteratively solves two sub-problems in an alternating fashion; the first sub-problem is minimized via a Proximal Subgradient Descent (PSD) approach, while the second one via a Majorization Minimization (MM) procedure. Moreover, we have demonstrated that our algorithm converges and that its fixed points include the Karush-Kuhn-Tucker (KKT) points of our proposed formulation.

In order to show the merits of R^2LML, we performed a series of experiments involving 9 benchmark classification problems. First, we varied the number of local metrics K and discussed the influence of K on classification accuracy. We concluded that there is no obvious relation between K and the classification accuracy. Furthermore, the obtained optimal K does not necessarily equal the number of classes of the dataset under consideration. Finally, in a second set of experiments, we compared R^2LML to several other metric learning algorithms and demonstrated that our proposed method is highly competitive.

Acknowledgments. Y. Huang acknowledges partial support from a UCF Graduate College Presidential Fellowship and National Science Foundation (NSF) grant No. 1200566. C. Li acknowledges partial support from NSF grants No. 0806931 and No. 0963146. Furthermore, M. Georgiopoulos acknowledges partial support from NSF grants No. 1161228 and No. 0525429, while G. C. Anagnostopoulos acknowledges partial support from NSF grant No. 1263011. Note that any opinions, findings, and conclusions or recommendations expressed in this material are those of the authors and do not necessarily reflect the views of the NSF. Finally, the authors would like to thank the 3 anonymous reviewers of this manuscript for their helpful comments.

References

1. Xing, E.P., Ng, A.Y., Jordan, M.I., Russell, S.: Distance metric learning with application to clustering with side-information. In: Neural Information Processing Systems Foundation (NIPS), pp. 505–512. MIT Press (2002)
2. Shalev-Shwartz, S., Singer, Y., Ng, A.Y.: Online and batch learning of pseudometrics. In: International Conference on Machine Learning (ICML). ACM (2004)
3. Chopra, S., Hadsell, R., Lecun, Y.: Learning a similarity metric discriminatively, with application to face verification. In: Computer Vision and Pattern Recognition (CVPR), pp. 539–546. IEEE Press (2005)
4. Goldberger, J., Roweis, S., Hinton, G., Salakhutdinov, R.: Neighbourhood components analysis. In: Neural Information Processing Systems Foundation (NIPS), pp. 513–520. MIT Press (2004)
5. Weinberger, K.Q., Blitzer, J., Saul, L.K.: Distance metric learning for large margin nearest neighbor classification. In: Neural Information Processing Systems Foundation (NIPS). MIT Press (2006)
6. Davis, J.V., Kulis, B., Jain, P., Sra, S., Dhillon, I.S.: Information-theoretic metric learning. In: International Conference on Machine Learning (ICML), pp. 209–216. ACM (2007)
7. Yang, L., Jin, R., Sukthankar, R., Liu, Y.: An efficient algorithm for local distance metric learning. In: AAAI Conference on Artificial Intelligence (AAAI). AAAI Press (2006)
8. Hastie, T., Tibshirani, R.: Discriminant adaptive nearest neighbor classification. IEEE Transactions on Pattern Analysis and Machine Intelligence 18(6), 607–616 (1996)
9. Weinberger, K., Saul, L.: Fast solvers and efficient implementations for distance metric learning. In: International Conference on Machine Learning (ICML), pp. 1160–1167. ACM (2008)
10. Bilenko, M., Basu, S., Mooney, R.J.: Integrating constraints and metric learning in semi-supervised clustering. In: International Conference on Machine Learning (ICML), pp. 81–88. ACM (2004)
11. Noh, Y.K., Zhang, B.T., Lee, D.D.: Generative local metric learning for nearest neighbor classification. In: Neural Information Processing Systems Foundation (NIPS). MIT Press (2010)
12. Wang, J., Kalousis, A., Woznica, A.: Parametric local metric learning for nearest neighbor classification. In: Neural Information Processing Systems Foundation (NIPS), pp. 1610–1618. MIT Press (2012)
13. Candès, E.J., Tao, T.: The power of convex relaxation: Near-optimal matrix completion. CoRR abs/0903.1476 (2009)
14. Candès, E.J., Recht, B.: Exact matrix completion via convex optimization. CoRR abs/0805.4471 (2008)
15. Rakotomamonjy, A., Flamary, R., Gasso, G., Canu, S.: lp-lq penalty for sparse linear and sparse multiple kernel multi-task learning. IEEE Transactions on Neural Networks 22, 1307–1320 (2011)

16. Chen, X., Pan, W., Kwok, J.T., Carbonell, J.G.: Accelerated gradient method for multi-task sparse learning problem. In: International Conference on Data Mining (ICDM), pp. 746–751. IEEE Computer Society (2009)
17. Duchi, J., Singer, Y.: Efficient online and batch learning using forward backward splitting. Journal of Machine Learning Research (JMLR) 10, 2899–2934 (2009)
18. Balaji, R., Bapat, R.: On euclidean distance matrices. Linear Algebra and its Applications 424(1), 108–117 (2007)
19. Hunter, D.R., Lange, K.: A tutorial on mm algorithms. The American Statistician 58(1) (2004)
20. Langford, J., Li, L., Zhang, T.: Sparse online learning via truncated gradient. Journal of Machine Learning Research (JMLR) 10, 777–801 (2009)
21. Shalev-Shwartz, S., Tewari, A.: Stochastic methods for l1-regularized loss minimization. Journal of Machine Learning Research (JMLR) 12, 1865–1892 (2011)
22. Hochberg, Y., Tamhane, A.C.: Multiple comparison procedures. John Wiley & Sons, Inc., New York (1987)

Learning Exemplar-Represented Manifolds in Latent Space for Classification

Shu Kong and Donghui Wang *

College of Computer Science and Technology
Zhejiang University
Hangzhou, Zhejiang 310027, China
{aimerykong,dhwang}@zju.edu.cn

Abstract. Intrinsic manifold structure of a data collection is valuable information for classification task. By considering the manifold structure in the data set for classification and with the sparse coding framework, we propose an algorithm to: (1) find exemplars from each class to represent the class-specific manifold structure, in which way the object-space dimensionality is reduced; (2) simultaneously learn a latent feature space to make the mapped data more discriminative according to the class-specific manifold measurement. We call the proposed algorithm Exemplar-represented Manifold in Latent Space for Classification (EMLSC). We also present the nonlinear extension of EMLSC based on kernel tricks to deal with highly nonlinear situations. Experiments on synthetic and real-world datasets demonstrate the merit of the proposed method.

Keywords: Sparse Coding, Dimensionality Reduction, Manifold, Exemplar Selection, Classification.

1 Introduction

Among various areas of machine learning, information retrieval and signal processing, one needs to deal with high-dimensional data collections ahead of specific tasks, such as classification focused on in this paper. This has motivated a lot of work in dimensionality reduction, whose goal is to find compact representations of the data that can save memory and computational time, and also enhance the performance of algorithms that deal with the data.

Since datasets often consist of high-dimensional data, most dimensionality reduction methods aim at reducing the feature-space dimension for all the data, *e.g.* PCA [1], LLE [2] and Isomap [3], etc. Among these methods, geometrically motivated approaches are shown to be effective in discovering the geometrical structure in the data. Meanwhile, manifold-based methods, such as LLE and Isomap and their variants, have attracted considerable attention in data analysis [4,5,6], and have achieved very encouraging performances in clustering, classification and data visualization. However, these methods separate the manifold-motivated dimensionality reduction and classifier learning apart, in which way, further improved classification performance is prevented.

---

* Corresponding author. This work is supported by Natural Science Foundations (No.61071218) of China and 973 Program (No.2010CB327904).

H. Blockeel et al. (Eds.): ECML PKDD 2013, Part III, LNAI 8190, pp. 240–255, 2013.
© Springer-Verlag Berlin Heidelberg 2013

On the other hand, since datasets usually contain a large number of data, dimensionality reduction in the object space is a desirable solution [7]. This can be achieved either by learning an adaptive dictionary [8,9] or finding exemplars [7]. Learning a compact dictionary to represent data (see [10] and therein) and the problem of learning a supervised dictionary for classification have been well studied in literature [9,11]. But such learned dictionaries intrinsically ignore the data manifold structures. Because, the dictionary atoms almost never coincide with the original data [12,11]. Specifically, for example, the negative sign of some atoms are hard to interpret, and the unit Euclidean length of the atoms means they just act as bases for reconstruction of data points but not for representing them. This intrinsic problem in dictionary learning has also been recognized in [7]. Therefore, the learned dictionary atoms cannot be considered as good representatives for the collection of data points when meeting various tasks such as classification. In contrast, one can find a small subset of the data to appropriately represent the whole data collection owing to the self-expressiveness property, which has been studied for subspace clustering using sparse representation [13,7] and low-rank representation [14]. The selected exemplars can naturally represent the manifold structure of the dataset, and thus reduce the dimensionality in the object space. Actually, finding exemplars is of particular significance in large-scale dataset summarization and visualization, and improves memory requirement and computational time of classification on such large-scale datasets. Nevertheless, merely selecting exemplars in the original space is insufficient to cover all the data points for classification task, since these data points distribute along complex manifolds and the exemplars may be neighbors to the data points from different classes. For this reason, it is desirable to learn a latent space in which the selecting exemplars can better serve the classification purpose.

By considering the two ways of dimensionality reduction and their limitations in classification presented above, we propose an algorithm to implement dimensionality reduction along the two directions by considering the manifold structure of each class. The proposed algorithm simultaneously does the following:

- find exemplars from each class to represent the class-specific manifold structure, in which way the object-space dimensionality is reduced;
- learn a latent space in the feature space to make the mapped data more discriminative according to the class-specific manifold measurement.
- carry out classification under a simple sparse coding framework.

We call this algorithm Exemplar-represented Manifolds in Latent Space for Classification (EMLSC). In the classification stage, EMLSC adopts a simple sparse coding framework for classification in a way like Sparse Representation-based Classification (SRC) [15]. But different from employing all training data in SRC, EMLSC only uses the selected exemplars as the bases. As the sparse coding is done in the lower-dimensional latent space and the number bases is far smaller than the whole training set, it is anticipated that the classification is performed faster than the original SRC method, in which the whole training set is used for classification. Furthermore, we present a nonlinear extension of EMLSC via kernel tricks (K-EMLSC) to deal with highly nonlinear situations. Through experimental validation, we can see that (K-)EMLSC not only reduces the dimension of the data and the scale of the dataset, but also improves classification performance.

2 Notations and Related Work

Let $\mathbf{X} \in \mathbb{R}^{p \times N}$ denote a training data set which consists of N data points from C classes, and $\mathbf{X}_c \in \mathbb{R}^{p \times N_c}$ denote the subset of data from the cth class, where $N = \sum_{c=1}^{C} N_c$. $\mathbf{I}$ is an identity matrix with appropriate size.

Under SRC framework [15], Ngugen *et al.* propose a unified method called Latent Sparse Embedding Residual Classifier (LASERC) to learn dictionary and a latent space [16]. In detail, for each class, LASERC jointly learn an adaptive dictionary $\mathbf{D}_c$ and a latent space defined by a projection $\mathbf{W}_c$ through:

$$\min_{\mathbf{W}_c, \mathbf{D}_c, \mathbf{A}_c} \|\mathbf{W}_c^T \mathbf{X}_c - \mathbf{D}_c \mathbf{A}_c\|_F^2 + \lambda \|\mathbf{X}_c - \mathbf{W}_c \mathbf{W}_c^T \mathbf{X}_c\|_F^2,$$

$$\text{s.t. } \mathbf{W}_c^T \mathbf{W}_c = \mathbf{I}, \|\mathbf{A}_c\|_1 \leq T, \tag{1}$$

where $\mathbf{A}_c$ is the coefficient matrix and $\|\mathbf{A}_c\|_1$ is the sum of ℓ_1 norms of all columns in $\mathbf{A}_c$. LASERC uses the projection to reduce the dimensionality of the data, and adopts a reconstruction error based classifier for the final classification. However, even though the method can be extended to nonlinear version via kernel tricks, it fails to consider the discrimination power among the separately learned class-specific dictionaries $\mathbf{D}_c$'s, such that it is not guaranteed to produce improved classification performance.

Elhamifar *et al.* propose to find exemplars in the dataset to reduce the dimensionality in the object space [17], so that computational cost and memory requirements are significantly reduced. They use nearest neighbor to do classification, and achieve comparable results with exemplars to that with all the training data. In [7], the authors also propose Sparse Modeling Representative Selection (SMRS) to find exemplars for classification with different classifiers. SMRS first selects exemplars by solving the following objective function over row-sparse coefficient matrix $\mathbf{A}$:

$$\min_{\mathbf{A}} \|\mathbf{X} - \mathbf{X}\mathbf{A}\|_F^2, \quad \text{s.t. } \|\mathbf{A}\|_{1,q} \leq \tau, \mathbf{1}^T \mathbf{A} = \mathbf{1}^T, \tag{2}$$

where $\|\mathbf{A}\|_{1,q} = \sum_{i=1}^{N} \|\mathbf{a}^i\|_q$ denotes the sum of ℓ_q norms[1] of the rows of coefficient matrix $\mathbf{A} = [\mathbf{a}^1; \ldots; \mathbf{a}^N] \in \mathbb{R}^{N \times N}$; $\tau > 1$ is an appropriately chosen parameter to make the optimization program in Eq. 2 convex; and the affine constraint $\mathbf{1}^T \mathbf{A} = \mathbf{1}^T$ means invariance of the selected exemplars w.r.t global translation of the data. As the $\ell_{1,q}$-norm vanishes rows of $\mathbf{A}$, the exemplars can be found according to nonzero rows in $\mathbf{A}$. SMRS learns different classifiers over the selected exemplars, and experimental results demonstrate that well-chosen exemplars can not only reduce the scale of the training set, but also produce very good classification performances with far fewer data points. Despite the effectiveness of SMRS, it separately finds the exemplars in the original space and learns the classifier. Therefore, the learned classifiers are not optimal for classification based on the selected exemplars. Moreover, SMRS simply selects exemplars from all the classes, hence using the exemplars as a subset of the training data in the original space can significantly change the inner and intra class distances of the training data, such that good classification performance is not guaranteed as discussed in [7].

---

[1] In this paper, we merely set $q = 2$, *i.e.* using an $\ell_{1,2}$-norm regularizer in the objective functions presented latter.

3   Exemplar-Represented Manifold in Latent Space for Classification (EMLSC)

As reviewed previously, finding exemplars in the original space directly from all classes is not optimal for classification, and separately learning class-specific dictionaries also limits the discrimination power of the dictionaries. Since we understand the importance of selecting exemplars opposed to learning adaptive dictionaries in representing the class-specific manifold structure, it is worth simultaneously finding exemplars in each class and learning a latent space with consideration of the discrimination power.

3.1   Derivation of EMLSC Objective Function

In SMRS, solving Eq. 2 means finding exemplars from all classes in the original space, and it cannot serve classification purpose well. Therefore, it is desirable to finding exemplars in a simultaneously learned latent space, in which the exemplars can effectively represent the data points according to their class-specific manifold structure. Suppose a linear projection $\mathbf{W} \in \mathbb{R}^{p \times m}$ defines this m-dimensional latent space, then we have the new data in the latent space as $\mathbf{W}^T \mathbf{X}$. By replacing the original data set $\mathbf{X}$ in Eq. 2 with $\mathbf{W}^T \mathbf{X} \in \mathbb{R}^{m \times N}$, and constraining $\mathbf{W}^T \mathbf{W} = \mathbf{I}$, we have:

$$\min_{\mathbf{A}, \mathbf{W}} \|\mathbf{W}^T \mathbf{X} - \mathbf{W}^T \mathbf{X} \mathbf{A}\|_F^2,$$

$$\text{s.t. } \mathbf{W}^T \mathbf{W} = \mathbf{I}, \|\mathbf{A}\|_{1,q} \leq \tau, \mathbf{1}^T \mathbf{A} = \mathbf{1}^T. \tag{3}$$

The constraint of $\mathbf{W}^T \mathbf{W} = \mathbf{I}$ not only leads to a computationally efficient scheme for optimization as we see in the next subsection, but also allows the extension of our proposed method to the nonlinear version as demonstrated in Section 4.

Moreover, it is essential to guarantee that the exemplars from a specific class can well represent all the data of this class. Specifically, for the cth class $\mathbf{X}_c$, we should also minimize the following constraint:

$$\|\mathbf{W}^T \mathbf{X}_c - \mathbf{W}^T \mathbf{X}_c \mathbf{A}_c^{(c)}\|_F^2, \tag{4}$$

where $\mathbf{A}_c^{(c)}$ is the cth part of coefficient matrix $\mathbf{A}_c = [\mathbf{A}_c^{(1)}; \ldots; \mathbf{A}_c^{(i)}; \ldots; \mathbf{A}_c^{(C)}]$ corresponding to $\mathbf{X}_c$, i.e. $\mathbf{X}_c \approx \mathbf{X} \mathbf{A}_c = \sum_{i=1}^C \mathbf{X}_i \mathbf{A}_c^{(i)}$. For brevity, we introduce a selection operator $\mathbf{Q}_c = [\mathbf{q}_1^c, \ldots, \mathbf{q}_j^c, \ldots, \mathbf{q}_{K_c}^c] \in \mathbb{R}^{N \times N_c}$, in which the jth column of $\mathbf{Q}_c$ is of the following form:

$$\mathbf{q}_j^c = [\underbrace{0, \ldots, 0}_{\sum_{i=1}^{c-1} N_i}, \overbrace{\underbrace{0, \ldots, 0, 1, 0, \ldots, 0}_{N_c}}^{j-1}, \underbrace{0, \ldots, 0}_{\sum_{i=c+1}^{C} N_i}]^T. \tag{5}$$

Therefore, we have $\mathbf{Q}_c^T \mathbf{Q}_c = \mathbf{I}$, $\mathbf{X}_c = \mathbf{X} \mathbf{Q}_c$, and $\mathbf{A}_j^{(c)} = \mathbf{Q}_c^T \mathbf{A}_j \in \mathbb{R}^{N_j}$ means the cth part of coefficient matrix $\mathbf{A}_j$ corresponding to $\mathbf{X}_c$. Now, we can rewrite Eq. 4 as:

$$\|\mathbf{W}^T \mathbf{X}_c - \mathbf{W}^T \mathbf{X} \mathbf{Q}_c \mathbf{Q}_c^T \mathbf{A}_c\|_F^2. \tag{6}$$

Let $\tilde{\mathbf{Q}}_c = [\mathbf{Q}_1, \ldots, \mathbf{Q}_{c-1}, \mathbf{Q}_{c+1}, \ldots, \mathbf{Q}_C]$, then we have $\mathbf{X}\tilde{\mathbf{Q}}_c = [\mathbf{X}_1, \ldots, \mathbf{X}_{c-1}, \mathbf{X}_{c+1}, \ldots, \mathbf{X}_C]$, and $\tilde{\mathbf{Q}}_c^T \mathbf{A}_c = [\mathbf{A}_c^{(1)}; \ldots; \mathbf{A}_c^{(c-1)}; \mathbf{A}_c^{(c+1)}; \ldots; \mathbf{A}_c^{(C)}]$. To guarantee that exemplars from other classes do not contribute to representing the data from class c, we should also minimize the following:

$$\|\mathbf{W}^T \mathbf{X}\tilde{\mathbf{Q}}_c \tilde{\mathbf{Q}}_c^T \mathbf{A}_c\|_F^2. \tag{7}$$

This term measures how much the unrelated exemplars (from undesirable classes) contribute to the representation of the data points from a specific class. Thus, minimizing this term means drawing apart the exemplars belonging to different classes [11], in which way the data points from different classes are better separated.

Considering the above three terms, *i.e.* Eq. 3, Eq. 6 and Eq. 7, we have our objective function as below:

$$\min_{\mathbf{A}_c, \mathbf{W}} \sum_{c=1}^{C} \left\{ \|\mathbf{W}^T \mathbf{X}_c - \mathbf{W}^T \mathbf{X}\mathbf{A}_c\|_F^2 + \alpha \|\mathbf{W}^T \mathbf{X}_c - \mathbf{W}^T \mathbf{X}\mathbf{Q}_c \mathbf{Q}_c^T \mathbf{A}_c\|_F^2 \right.$$
$$\left. + \beta \|\mathbf{W}^T \mathbf{X}\tilde{\mathbf{Q}}_c \tilde{\mathbf{Q}}_c^T \mathbf{A}_c\|_F^2 \right\} \tag{8}$$
$$\text{s.t. } \mathbf{W}^T \mathbf{W} = \mathbf{I}, \|\mathbf{A}_c\|_{1,q} \le s, \mathbf{1}^T \mathbf{A}_c = \mathbf{1}^T, \text{ for } \forall c.$$

In the objective function, α and β are two parameters to balance relative importance of the three terms, and s denotes the sparse level of the coefficient $\mathbf{A}_c$. There are other possible ways to add discrimination power to the latent space and exemplars, such as the methods based on Linear Discriminative Analysis [18] and Maximum Margin Criterion [19]. But it is worth noting the way of improving discrimination in Eq. 8 has an intrinsic relation to the classifier adopted in this paper. As described in Section 5, since our classifier is based on sparse coding technique, this discrimination-enhancing method in Eq. 8 can benefit the classifier a lot.

3.2 Numerical Solution

Even though the optimization problem in Eq. 8 is a non-convex problem with two matrix variables $\mathbf{W}$ and $\mathbf{A} = [\mathbf{A}_1, \ldots, \mathbf{A}_C]$, we still can derive effective solutions through iterative minimization, as demonstrated by experiments in Section 6. In this subsection, we present the detailed optimization of each variable matrix.

Update Projection W. By omitting the terms which are independent to $\mathbf{W}$, we have the following:

$$\mathbf{W}^* = \operatorname*{argmin}_{\mathbf{W}} \|\mathbf{W}^T (\mathbf{X} - \mathbf{X}\mathbf{A})\|_F^2 + \alpha \|\mathbf{W}^T (\mathbf{X} - \mathbf{X}\hat{\mathbf{A}})\|_F^2 + \beta \|\mathbf{W}^T (\mathbf{X}\tilde{\mathbf{A}})\|_F^2,$$
$$\text{s.t. } \mathbf{W}^T \mathbf{W} = \mathbf{I}, \tag{9}$$

where $\hat{\mathbf{A}} = [\mathbf{Q}_1 \mathbf{Q}_1^T \mathbf{A}_1, \ldots, \mathbf{Q}_c \mathbf{Q}_c^T \mathbf{A}_c, \ldots, \mathbf{Q}_C \mathbf{Q}_C^T \mathbf{A}_C]$, $\tilde{\mathbf{A}} = [\tilde{\mathbf{Q}}_1 \tilde{\mathbf{Q}}_1^T \mathbf{A}_1, \ldots, \tilde{\mathbf{Q}}_c \tilde{\mathbf{Q}}_c^T \mathbf{A}_c, \ldots, \tilde{\mathbf{Q}}_C \tilde{\mathbf{Q}}_C^T \mathbf{A}_C]$, and $\mathbf{A} = [\mathbf{A}_1, \ldots, \mathbf{A}_c, \ldots, \mathbf{A}_C]$. Through simple derivation, we have the following concise function:

$$\mathbf{W}^* = \operatorname*{argmin}_{\mathbf{W}} \operatorname{tr}(\mathbf{W}^T \boldsymbol{\Omega} \mathbf{W}), \quad \text{s.t. } \mathbf{W}^T \mathbf{W} = \mathbf{I}, \tag{10}$$

where $\Omega = (\mathbf{X} - \mathbf{X}\mathbf{A})(\mathbf{X} - \mathbf{X}\mathbf{A})^T + \alpha(\mathbf{X} - \mathbf{X}\hat{\mathbf{A}})(\mathbf{X} - \mathbf{X}\hat{\mathbf{A}})^T + \beta(\mathbf{X}\tilde{\mathbf{A}})(\mathbf{X}\tilde{\mathbf{A}})^T$.
Therefore, to derive the optimal $\mathbf{W}^*$ with fixed $\mathbf{A}$, we can simply solve this eigenvalue decomposition problem, and choose the eigenvectors w.r.t the m smallest eigenvalues as the columns of $\mathbf{W}^*$.

Update the Coefficient Matrix $\mathbf{A}_c$. Specifically, by fixing the projection $\mathbf{W}$, we focus on updating $\mathbf{A}_c$ for demonstration as below:

$$
\begin{aligned}
\mathbf{A}_c^* &= \underset{\mathbf{A}_c}{\arg\min} \|\mathbf{W}^T\mathbf{X}_c - \mathbf{W}^T\mathbf{X}\mathbf{A}_c\|_F^2 + \alpha\|\mathbf{W}^T\mathbf{X}_c - \mathbf{W}^T\mathbf{X}\mathbf{Q}_c\mathbf{Q}_c^T\mathbf{A}_c\|_F^2 \\
&\quad + \beta\|\mathbf{W}^T\mathbf{X}\tilde{\mathbf{Q}}_c\tilde{\mathbf{Q}}_c^T\mathbf{A}_c\|_F^2 \\
&= \underset{\mathbf{A}_c}{\arg\min} \|\bar{\mathbf{X}}_c - \bar{\mathbf{X}}\mathbf{A}_c\|_F^2
\end{aligned}
\tag{11}
$$

$$
\text{s.t. } \|\mathbf{A}_c\|_{1,q} \leq s, \mathbf{1}^T\mathbf{A}_c = \mathbf{1}^T,
$$

where $\bar{\mathbf{X}}_c = \begin{pmatrix} \mathbf{W}^T\mathbf{X}_c \\ \sqrt{\alpha}\mathbf{W}^T\mathbf{X}_c \\ 0 \end{pmatrix}$ and $\bar{\mathbf{X}} = \begin{pmatrix} \mathbf{W}^T\mathbf{X} \\ \sqrt{\alpha}\mathbf{W}^T\mathbf{X}\mathbf{Q}_c\mathbf{Q}_c^T \\ \sqrt{\beta}\mathbf{W}^T\mathbf{X}\tilde{\mathbf{Q}}_c\tilde{\mathbf{Q}}_c^T \end{pmatrix}$. In this paper, by

using Lagrange multipliers on $\|\mathbf{A}_c\|_{1,q}$, we turn to an Alternating Direction Method of Multipliers optimization framework [20] to solve the above problem.

In sum, the overall optimization procedure iterates the two steps, updating $\mathbf{W}$ in Eq. 10 and updating $\mathbf{A}$ in Eq. 11. It stops when meeting a predefined condition, *i.e.* reaching a maximum number of iterations or the difference between two consecutive the projection $\mathbf{W}$ is small enough. Finally, we choose the data as selected exemplars according to nonzero rows of the coefficient matrix $\mathbf{A}$.

4 Kernel EMLSC

Even if the proposed EMLSC exploit the data manifolds in the learned latent space based on the selected exemplars, it may fail to discover the intrinsic geometry when the data manifold is highly nonlinear. In this section, we discuss how to perform EMLSC in Reproducing Kernel Hilbert Space (RKHS), which gives rise to kernel version of EMLSC, denoted as K-EMLSC.

4.1 An Equivalent Objective Function

Before deriving the K-EMLSC, we provide an equivalent objective function to the original one in Eq. 8. We first present the following proposition.

Proposition 1. *With fixed $\mathbf{A}$, there exists an optional solution $\mathbf{W}^*$ to Eq. 8 that has the following form:*

$$
\mathbf{W}^* = \mathbf{X}\mathbf{P}
\tag{12}
$$

for some $\mathbf{P} \in \mathbb{R}^{N \times m}$.

The proof of this proposition is given in Appendix A. As a corollary of Proposition 1, it is sufficient to seek an optimal solution for the optimization in Eq. 8 through $\mathbf{P}$ and coefficient matrix $\mathbf{A}$. By substituting Eq. 12 into Eq. 8, we have:

$$\min_{\mathbf{A},\mathbf{P}} \|\mathbf{P}^T\mathbf{K}(\mathbf{I}-\mathbf{A})\|_F^2 + \alpha\|\mathbf{P}^T\mathbf{K}(\mathbf{I}-\hat{\mathbf{A}})\|_F^2 + \beta\|\mathbf{P}^T\mathbf{K}\tilde{\mathbf{A}}\|_F^2,$$

$$\text{s.t. } \mathbf{P}^T\mathbf{K}\mathbf{P} = \mathbf{I}, \|\mathbf{A}_c\|_{1,q} \le s, \mathbf{1}^T\mathbf{A}_c = \mathbf{1}^T, \text{ for } \forall c, \tag{13}$$

where $\mathbf{K} = \mathbf{X}^T\mathbf{X}$, $\mathbf{A} = [\mathbf{A}_1, \ldots, \mathbf{A}_c, \ldots, \mathbf{A}_C]$, $\hat{\mathbf{A}} = [\mathbf{Q}_1\mathbf{Q}_1^T\mathbf{A}_1, \ldots, \mathbf{Q}_c\mathbf{Q}_c^T\mathbf{A}_c, \ldots,$ $\mathbf{Q}_C\mathbf{Q}_C^T\mathbf{A}_C]$, and $\tilde{\mathbf{A}} = [\tilde{\mathbf{Q}}_1\tilde{\mathbf{Q}}_1^T\mathbf{A}_1, \ldots, \tilde{\mathbf{Q}}_c\tilde{\mathbf{Q}}_c^T\mathbf{A}_c, \ldots, \tilde{\mathbf{Q}}_C\tilde{\mathbf{Q}}_C^T\mathbf{A}_C]$.

To derive the optimal $\mathbf{P}$, we have the following proposition.

Proposition 2. *The optimal solution of Eq. 13 when $\mathbf{A}$ is fixed is:*

$$\mathbf{P}^* = \mathbf{U}\mathbf{S}^{-\frac{1}{2}}\mathbf{G}^*, \tag{14}$$

where $\mathbf{U}$ and $\mathbf{S}$ come from the SVD of $\mathbf{K} = \mathbf{U}\mathbf{S}\mathbf{U}^T$, and $\mathbf{G} \in \mathbb{R}^{N \times m}$ is the optimal solution of the following minimum eigenvalue problem:

$$\mathbf{G}^* = \underset{\mathbf{G}}{\arg\min} \operatorname{tr}\mathbf{G}^T\tilde{\mathbf{H}}\mathbf{G}, \quad s.t. \ \mathbf{G}^T\mathbf{G} = \mathbf{I}, \tag{15}$$

where $\tilde{\mathbf{H}} = \mathbf{S}^{\frac{1}{2}}\mathbf{U}^T\mathbf{H}\mathbf{U}\mathbf{S}^{\frac{1}{2}}$ in which $\mathbf{H} = (\mathbf{I}-\mathbf{A})(\mathbf{I}-\mathbf{A})^T + \alpha(\mathbf{I}-\hat{\mathbf{A}})(\mathbf{I}-\hat{\mathbf{A}})^T + \beta\tilde{\mathbf{A}}\tilde{\mathbf{A}}^T$.

The proof of this proposition is provided in Appendix B. From the equivalence illustrated by Proposition 2, we can derive the optimal $\mathbf{W} = \mathbf{X}\mathbf{P}$ after having the optimal $\mathbf{P}$. It is worth noting the following remark.

Remark 1. With fixed coefficient matrix $\mathbf{A}$, we can derive the optimal projection $\mathbf{W}$ either through solving Eq. 10 or Eq. 12 (Eq. 14 and Eq. 15 are used). The difference between these two ways can be beneficial in different situations. Particularly, when the number of training data $N \gg p$, which stands for the dimensionality of the data, we can choose the first way to derive the optimal $\mathbf{W}$, as the complexity of the eigenvalue problem is $\mathcal{O}(p^3)$. When $p \gg N$, we can use the second strategy to calculate the optimal $\mathbf{W}$ with the $\mathcal{O}(N^3)$ complexity of the eigenvalue problem .

4.2 Derivation of Kernel EMLSC

Since we have $\mathbf{x}_i \in \mathbb{R}^p$ where $\mathbf{x}_i$ is the ith training sample, we consider the problem in a feature space $\mathcal{H}$ induced by some nonlinear mappings:

$$\phi : \mathbb{R}^p \to \mathcal{H}. \tag{16}$$

For a proper chosen ϕ, an inner produce $\langle \cdot, \cdot \rangle$ can be defined on $\mathcal{H}$ which makes for a reproducing kernel Hilbert space (RKHS). More specifically,

$$\langle \phi(\mathbf{x}), \phi(\mathbf{y}) \rangle = \mathcal{K}(\mathbf{x}, \mathbf{y}) \tag{17}$$

holds where $\mathcal{K}(\cdot, \cdot)$ is a positive semi-definite kernel function. Several popular kernel functions are Gaussian kernel $\mathcal{K}(\mathbf{x}, \mathbf{y}) = \exp(-\|\mathbf{x} - \mathbf{y}\|_2^2/\sigma^2)$, polynomial kernel $\mathcal{K}(\mathbf{x}, \mathbf{y}) = (1 + \mathbf{x}^T\mathbf{y})^\alpha$, and Sigmoid kernel $\mathcal{K}(\mathbf{x}, \mathbf{y}) = \tanh(\mathbf{x}^T\mathbf{y} + \alpha)$.

Let $\boldsymbol{\Phi}$ denote the data matrix in RKHS:

$$\boldsymbol{\Phi} = [\phi(\mathbf{x}_i), \ldots, \phi(\mathbf{x}_N)]. \tag{18}$$

Now, the problem Eq. 13 in RKHS can be written as below:

$$\min_{\mathbf{A},\mathbf{P}} \|\mathbf{P}^T\boldsymbol{\Phi}^T\boldsymbol{\Phi}(\mathbf{I} - \mathbf{A})\|_F^2 + \alpha\|\mathbf{P}^T\boldsymbol{\Phi}^T\boldsymbol{\Phi}(\mathbf{I} - \hat{\mathbf{A}})\|_F^2 + \beta\|\mathbf{P}^T\boldsymbol{\Phi}^T\boldsymbol{\Phi}\tilde{\mathbf{A}}\|_F^2,$$
$$\text{s.t. } \mathbf{P}^T\boldsymbol{\Phi}^T\boldsymbol{\Phi}\mathbf{P} = \mathbf{I}, \|\mathbf{A}_c\|_{1,q} \leq s, \mathbf{1}^T\mathbf{A}_c = \mathbf{1}^T, \text{ for } \forall c. \tag{19}$$

Denote the kernel matrix by $\mathcal{K} = \boldsymbol{\Phi}^T\boldsymbol{\Phi}$, in which $\mathcal{K}_{ij} = \mathcal{K}(\mathbf{x}_i, \mathbf{x}_j) = \langle \phi(\mathbf{x}_i), \phi(\mathbf{x}_j) \rangle$. Then, we have:

$$\min_{\mathbf{A},\mathbf{P}} \|\mathbf{P}^T\mathcal{K}(\mathbf{I} - \mathbf{A})\|_F^2 + \alpha\|\mathbf{P}^T\mathcal{K}(\mathbf{I} - \hat{\mathbf{A}})\|_F^2 + \beta\|\mathbf{P}^T\mathcal{K}\tilde{\mathbf{A}}\|_F^2,$$
$$\text{s.t. } \mathbf{P}^T\mathcal{K}\mathbf{P} = \mathbf{I}, \|\mathbf{A}_c\|_{1,q} \leq s, \mathbf{1}^T\mathbf{A}_c = \mathbf{1}^T, \text{ for } \forall c. \tag{20}$$

The resulting kernelized objective function in Eq. 20 can be solved in the same way as in the linear case. Note that in the nonlinear case, the dimension m of the output space can be higher than the dimension p of the input space, and is only upper bounded by the number of training samples. For a sample datum $\mathbf{x}$ either from the training set or a testing one, we have the corresponding mapped point $\mathbf{z}$ in RKHS as $\mathbf{z} = \mathbf{P}^T\mathcal{K}(\mathbf{X}, \mathbf{x})$.

5   Classification Scheme

After learning the projection and selecting the exemplars, we have the linear or nonlinear mapped exemplars as $\mathbf{Z} = [\mathbf{Z}_1, \ldots, \mathbf{Z}_C] \in \mathbb{R}^{m \times M}$, in which $\mathbf{Z}_c \in \mathbb{R}^{m \times M_c}$ is the mapped exemplars selected from the cth class ($\sum_{c=1}^{C} M_c = M$). Specifically, in the linear version, we have $\mathbf{Z}_c = \mathbf{W}^T\mathbf{X}_c$, while in nonlinear situation, we have $\mathbf{Z}_c = \mathbf{P}^T\mathcal{K}(\mathbf{X}, \mathbf{X}_c)$. When comes a query datum $\mathbf{x}$, we have the mapped point $\mathbf{z} \in \mathbb{R}^m$. To classify the query, we follow the classification framework of SRC [15]. In detail, we first solve the following sparse coding problem:

$$\boldsymbol{\alpha}^* = \operatorname*{argmin}_{\boldsymbol{\alpha}} \|\mathbf{z} - \mathbf{Z}\boldsymbol{\alpha}\|_F^2 + \gamma\|\boldsymbol{\alpha}\|_1. \tag{21}$$

Then, we calculate the reconstruction error of each class by: $e_c = \|\mathbf{z} - \mathbf{Z}_c\boldsymbol{\alpha}_c^*\|_F^2$, where $\boldsymbol{\alpha}_c$ is the part in coefficient $\boldsymbol{\alpha}^*$ corresponding to $\mathbf{Z}_c$. Finally, we classify the query to class c^* such that $c^* = \operatorname{argmin}_c e_c$.

6   Experimental Results

In this section, we evaluate the performance of EMLSC with its kernel version denoted by K-EMLSC on both synthetic and two real-world datasets. We compare our

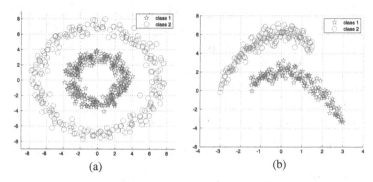

Fig. 1. Two synthetic datasets and each one has two classes: (a) circle-like distributed data and (b) parabola-like distributed data

proposed (K-)EMLSC with several state-of-the-art methods that are based on Sparse Representation-based Classification (SRC) framework for classification on the learned dictionaries or the selected exemplars. The standard SRC method (SRC) acts as a baseline method, which uses all the training data without learning dictionaries or finding exemplars. As our EMLSC can also learn a project along the feature-space dimension, for comparing the effect of the dimensionality on the feature-space dimension, we first apply random projection, PCA and LPP [6] to reduce the feature-space dimensionality, and then use SRC for classification. We call these schemes rSRC, pSRC and lSRC respectively. Moreover, we compare two closely related methods for comparison, they are Sparse Modeling Representative Selection (SMRS) [7] and Latent Sparse Embedding Residual Classifier (LASERC) [16]. SMRS merely selects exemplars in the original space, and LASERC simultaneously learns the projection and adaptive dictionaries for each class. Both SMRS and LASERC use SRC framework for classification. For all experiments, we simply set $\alpha = \beta = 1$ for EMLSC and K-EMLSC. As for K-EMLSC, we choose Gaussian kernel with $\sigma = 1.7$.

6.1   Experiments with Synthetic Data

We first evaluate K-EMLSC for its ability of discriminating class-specific manifolds in the learned latent space. We synthesize two datasets of two-dimensional data points, and each set includes two classes of data. These two datasets consist of circle-like and parabola-like distributed data as illustrated by Fig. 1. We can easily see there are highly nonlinear manifold structures in the data. Our K-EMLSC jointly learns a latent space and selects exemplars, therefore, we can draw the mapped data points of the two classes in the learned latent space. Among the compared methods, only LASERC can jointly learn a latent space, but learns dictionary for each class individually. Therefore, we focus on the comparison of K-EMLSC and LASERC, and choose Gaussian kernel with $\sigma = 1.7$ for both of the methods.

Fig. 2 displays the mapped data of the synthetic datasets in the latent space by LASERC and K-EMLSC. (a) to (d) are the mapped data from the circle-like and parabola-like datasets by LASERC and K-EMLSC, respectively; as well, we also plot

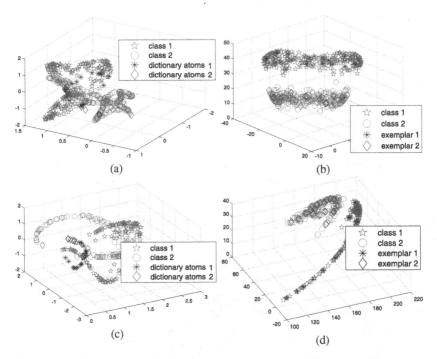

Fig. 2. Synthetic example: circle-like distributed data in the original 2D space and the latent space with the learned dictionary atoms or selected exemplars are presented in (a) and (b), respectively; circle-like distributed data in the original 2D space and the latent space with the learned atoms or selected exemplars are illustrated by (c) and (d), respectively

the learned dictionary atoms of LASERC and the selected exemplars of our K-EMLSC. Since LASERC learns the class-specific dictionaries and projection matrices of each class individually, we plot the data points of the two classes in one figure, as illustrated in (a) and (c). Then, we can see the data are mixed up, as well as the dictionary atoms. This means the dictionary atoms cannot represent the data points well. This observation coincides with what we analyze previously — dictionary atoms act bases to reconstruct data points, and thus the atoms cannot be directly used to represent the data. On the contrary, our K-EMLSC separates the two classes and preserves the class-specific manifolds clearly as expected as demonstrated in (b) and (d), because K-EMLSC considers the discrimination power and representation power of the selected exemplars in the mapped space. Moreover, the selected exemplars can fully represent the data of each class, specifically, the exemplars in the learned latent space can reflect the class-specific data manifolds clearly.

6.2   Experiments with Real Data

Now, we examine the classification performances of the proposed method on two real-world datasets, USPS digit database [21] and Extended-YaleB face database [22]. We

Table 1. USPS digit recognition accuracy (%) with different reduced feature-space dimension

	10	20	40	60	80
rSRC	89.3 ± 0.6	90.1 ± 0.7	92.5 ± 0.4	93.8 ± 0.4	95.6 ± 0.5
pSRC	93.2 ± 0.4	95.9 ± 0.5	97.3 ± 0.6	98.1 ± 0.3	98.6 ± 0.4
LASERC	85.6 ± 1.6	86.3 ± 1.3	86.9 ± 1.2	87.2 ± 1.1	87.9 ± 0.9
lSRC	93.6 ± 0.6				
SRC	98.9 ± 0.7				
SMRS	91.7 ± 0.6				
EMLSC	95.8 ± 0.7	96.2 ± 0.6	97.1 ± 0.7	97.8 ± 0.5	98.2 ± 0.4
K-EMLSC	96.1 ± 0.3	96.5 ± 0.4	97.3 ± 0.8	97.9 ± 0.4	98.4 ± 0.5

Table 2. Extended-YaleB face recognition accuracy (%) with different reduced feature-space dimension

	30	60	100	150	200
rSRC	82.7 ± 1.3	91.6 ± 1.4	94.6 ± 1.1	95.8 ± 1.2	96.4 ± 0.9
pSRC	86.4 ± 0.7	91.8 ± 0.7	93.4 ± 0.9	93.8 ± 0.8	94.6 ± 0.6
LASERC	83.3 ± 1.8	87.5 ± 1.5	89.8 ± 1.4	91.4 ± 1.6	92.1 ± 1.3
lSRC	87.4 ± 0.6				
SRC	98.2 ± 0.8				
SMRS	93.1 ± 0.7				
EMLSC	93.6 ± 0.8	95.5 ± 0.6	96.3 ± 0.5	97.9 ± 0.5	98.5 ± 0.3
K-EMLSC	94.2 ± 0.3	95.9 ± 0.4	96.7 ± 0.4	98.2 ± 0.6	98.7 ± 0.5

show that class-specific manifolds commonly exist in real-world data sets, and our (K-)EMLSC can achieve very promising classification results by simultaneously learning the latent space and selecting the exemplars with consideration of the class-specific manifolds.

In USPS/Extended-YaleB dataset, we randomly select 1000 (USPS) / 51 (YaleB) samples of each class for training, and restrict our (K-)EMLSC to select 20 (USPS) / 7 (YaleB) exemplars in each class. As well, SMRS also selects the same number of exemplars as (K-)EMLSC, and LASERC learns the same number of dictionary atoms for each class. The recognition accuracy of each method is averaged over 10 runs. As for dimensionality reduction along the feature-space dimension, rSRC, pSRC, LASERC and our (K-)EMLSC reduce the data to the same dimension; while lSRC reduces the data to $C - 1$ dimension, where C is the number of classes; SRC and SMRS directly perform on the original data without dimensionality reduction process. In this evaluation, only K-EMLSC uses a Gaussian kernel.

Table 1 and Table 2 report the averaged classification accuracies with standard deviations for USPS and Extended-YaleB respectively, and the results are obtained from each method after running 10 times. Fig.3 and Fig. 4 illustrate the two-dimensional mapped

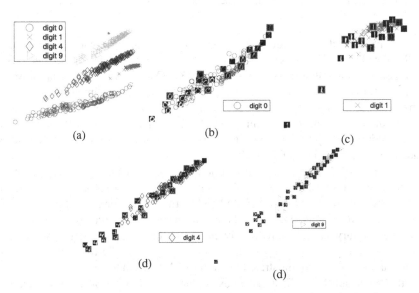

Fig. 3. Four digits from USPS are projected into the 2D latent space shown in (a). (b), (c) and (d) are three digits zoomed in with the selected exemplars highlighted in red box (best seen in color).

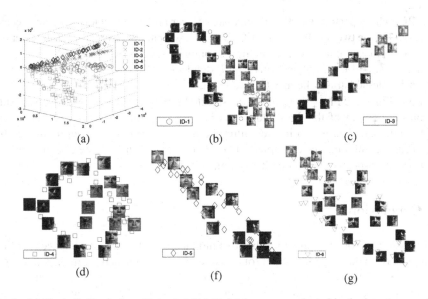

Fig. 4. (a) Six individuals from Extended YaleB database are projected in the latent space. (b), (c) and (d) present facial images of three persons, and the images in red box are the selected exemplars (best seen in color).

images and exemplars of the two databases, respectively. From these results, we make the following remarks:

1. rSRC, pSRC and lSRC reduce the dimension in the feature space and then apply SRC for classification in separate stages. As seen in the tables, these methods do not generate favorable results compared to (K-)EMLSC. In contrast, (K-)EMLSC jointly learns the latent space for reducing the feature-space dimension and selects the exemplars to represent class-specific manifolds. Therefore, with lower feature-space dimension and less training samples (exemplars) for classification, it can still produce very promising results. When kernel tricks are adopted, K-EMLSC achieves better results than EMLSC.

2. Even if LASERC jointly learns projection and dictionaries, the learned dictionary atoms cannot represent the data points but act as bases for reconstructing the data as argued previously. Moreover, LASERC learns the dictionary and projection for each class individually, therefore, classification performance cannot be guaranteed. On the contrary, (K-)EMLSC jointly considers the discrimination power and the representation power of the exemplars in the class-specific manifold viewpoint. Fig.3 and Fig. 4 clearly illustrate the merit of EMLSC on this respect.

3. Both SMRS and (K-)EMLSC select exemplars, but (K-)EMLSC obtains higher accuracy, even with much lower feature-space dimension. This is because (K-)EMLSC considers the discrimination power of the mapped exemplars in the latent space, and these exemplars can better represent the class-specific manifolds in a discriminative way.

4. SRC, which performs in the original space over all training data, produces decent results in the two databases. But (K-)EMLSC, with much fewer and much lower dimensional exemplars as bases, achieves comparable performances to SRC on USPS database, and even outperforms it in Extended-YaleB database. This observation further demonstrates the merit of (K-)EMLSC.

5. From Fig.3 and Fig. 4, we can see the selected exemplars by EMLSC mainly reside on the contour of the manifolds. This is a good result, because it is reasonable to anticipate data which locate within the manifold can be well approximated by linearly combining the exemplars. This phenomenon validates the effectiveness of using exemplars to represent the data manifold structure.

7 Conclusion

In this paper, we propose an algorithm to simultaneously learn a latent space and find exemplars from the mapped dataset for classification. The selected exemplars can naturally represent the data manifolds, and our method analyze the manifold structure in a discriminative way. Therefore, the exemplar-based class-specific manifolds of the classes are driven to be discriminative in the mapped latent space. We further extend this method to nonlinear version with kernel tricks, therefore, the kernelized method can deal with highly nonlinear cases. Through experiments, we demonstrate the merit of our proposed method. In face of big-data era, as a future work, it is worth extending our method to online learning framework to deal with large-scale situations.

References

1. Jolliffe, I.: Principal Component Analysis. Springer (2002)
2. Roweis, S., Saul, L.: Nonlinear dimensionality reduction by locally linear embedding. Science 290(5500), 2323–2326 (2000)
3. Tenenbaum, J.B.: A global geometric framework for nonlinear dimensionality reduction. Science 290(5500), 2319–2323 (2000)
4. Belkin, M., Niyogi, P.: Laplacian eigenmaps and spectral techniques for embedding and clustering. In: NIPS (2001)
5. Cai, D., He, X., Zhou, K., Han, J., Bao, H.: Locality sensitive discriminant analysis. In: IJCAI (2007)
6. He, X., Niyogi, P.: Locality preserving projections. In: NIPS (2003)
7. Elhamifar, E., Sapiro, G., Vidal, R.: See all by looking at a few: Sparse modeling for finding representative objects. In: CVPR (2012)
8. Aharon, M., Elad, M., Bruckstein, A.M.: The k-svd: An algorithm for designing of overcomplete dictionaries for sparse representation. IEEE Trans. Signal Process. 54(11), 4311–4322 (2006)
9. Mairal, J., Bach, F., Ponce, J., Sapiro, G., Zisserman, A.: Supervised dictionary learning. In: NIPS (2008)
10. Elad, M.: Sparse and Redundant Representations - From Theory to Applications in Signal and Image Processing. Springer (2010)
11. Kong, S., Wang, D.: A dictionary learning approach for classification: Separating the particularity and the commonality. In: Fitzgibbon, A., Lazebnik, S., Perona, P., Sato, Y., Schmid, C. (eds.) ECCV 2012, Part I. LNCS, vol. 7572, pp. 186–199. Springer, Heidelberg (2012)
12. Mairal, J., Bach, F., Ponce, J., Sapiro, G., Zisserman, A.: Discriminative learned dictionaries for local image analysis. In: CVPR (2008)
13. Elhamifar, E., Vidal, R.: Sparse subspace clustering. In: CVPR (2009)
14. Liu, G., Lin, Z., Yu, Y.: Robust subspace segmentation by low-rank representation. In: ICML (2010)
15. Wright, J., Yang, A., Ganesh, A., Sastry, S., Ma, Y.: Robust face recognition via sparse representation. IEEE Trans. Pattern Anal. Mach. Intell. 31(2), 210–227 (2009)
16. Nguyen, H.V., Patel, V.M., Nasrabadi, N.M., Chellappa, R.: Sparse embedding: A framework for sparsity promoting dimensionality reduction. In: Fitzgibbon, A., Lazebnik, S., Perona, P., Sato, Y., Schmid, C. (eds.) ECCV 2012, Part VI. LNCS, vol. 7577, pp. 414–427. Springer, Heidelberg (2012)
17. Elhamifar, E., Sapiro, G., Vidal, R.: Finding exemplars from pairwise dissimilarities via simultaneous sparse recovery. In: NIPS (2012)
18. Kong, S., Wang, X., Wang, D., Wu, F.: Multiple feature fusion for face recognition. In: FG (2013)
19. Kong, S., Wang, D.: Learning individual-specific dictionaries with fused multiple features for face recognition. In: FG (2013)
20. Gabay, D., Mercier, B.: A dual algorithm for the solution of nonlinear variational problems via finite element approximation. Comp. Math. Appl. 2(1), 17–40 (1976)
21. Hull, J.J.: A database for handwritten text recognition research. IEEE Trans. Pattern Anal. Mach. Intell. 16(5), 550–554 (1994)
22. Lee, K.C., Ho, J., Kriegman, D.J.: Acquiring linear subspaces for face recognition under variable lighting. IEEE Trans. Pattern Anal. Mach. Intell. 27(5), 684–698 (2005)

Appendix A: Proof of Proposition 1

Denote the optimal solution of $\mathbf{W}$ by $\mathbf{W}^*$. we show that $\mathbf{W}^*$ must have the form $\mathbf{W}^* = \mathbf{XP}$, for some $\mathbf{P} \in \mathbb{R}^{N \times m}$.

In advance, we rewrite the objective function into a compact form:

$$\min_{\mathbf{A},\mathbf{W}} \|\mathbf{W}^T\mathbf{X} - \mathbf{W}^T\mathbf{XA}\|_F^2 + \alpha\|\mathbf{W}^T\mathbf{X} - \mathbf{W}^T\mathbf{X}\hat{\mathbf{A}}\|_F^2$$
$$+ \beta\|\mathbf{W}^T\mathbf{X}\tilde{\mathbf{A}}\|_F^2 + \lambda\|\mathbf{A}\|_{1,2} \tag{22}$$
$$\text{s.t. } \mathbf{W}^T\mathbf{W} = \mathbf{I}.$$

where $\mathbf{A} = [\mathbf{A}_1, \ldots, \mathbf{A}_c, \ldots, \mathbf{A}_C]$, $\hat{\mathbf{A}} = [\mathbf{Q}_1\mathbf{Q}_1^T\mathbf{A}_1, \ldots, \mathbf{Q}_c\mathbf{Q}_c^T\mathbf{A}_c, \ldots, \mathbf{Q}_C\mathbf{Q}_C^T\mathbf{A}_C]$, and $\tilde{\mathbf{A}} = [\tilde{\mathbf{Q}}_1\tilde{\mathbf{Q}}_1^T\mathbf{A}_1, \ldots, \tilde{\mathbf{Q}}_c\tilde{\mathbf{Q}}_c^T\mathbf{A}_c, \ldots, \tilde{\mathbf{Q}}_C\tilde{\mathbf{Q}}_C^T\mathbf{A}_C]$.

Using the orthogonal decomposition of $\mathbf{W}^*$, we have:

$$\mathbf{W}^* = \mathbf{W}_\| + \mathbf{W}_\perp, \quad \text{where } \mathbf{W}_\| = \mathbf{XP}, \text{ and } \mathbf{W}_\perp\mathbf{X} = \mathbf{0} \tag{23}$$

for some appropriate $\mathbf{P} \in \mathbb{R}^{N \times m}$. Columns of $\mathbf{W}_\|$ and $\mathbf{W}_\perp$ are in and orthogonal to the column subspace of $\mathbf{X}$, respectively. Substituting Eq. 23 back into objective function Eq. 8, we can rewrite the first term in the following:

$$\|\mathbf{W}^T\mathbf{X} - \mathbf{W}^T\mathbf{XA}\|_F^2 = \|(\mathbf{W}_\| + \mathbf{W}_\perp)^T(\mathbf{X} - \mathbf{XA})\|_F^2 = \|\mathbf{W}_\|^T(\mathbf{X} - \mathbf{XA})\|_F^2$$
$$= \text{tr}\{\mathbf{W}_\|^T\mathbf{X}(\mathbf{I} - \mathbf{A})(\mathbf{I} - \mathbf{A})^T\mathbf{X}^T\mathbf{W}_\|\}, \tag{24}$$

the second term as:

$$\|\mathbf{W}^T\mathbf{X} - \mathbf{W}^T\mathbf{X}\hat{\mathbf{A}}\|_F^2 = \|(\mathbf{W}_\| + \mathbf{W}_\perp)^T(\mathbf{X} - \mathbf{X}\hat{\mathbf{A}})\|_F^2 = \|\mathbf{W}_\|^T(\mathbf{X} - \mathbf{X}\hat{\mathbf{A}})\|_F^2$$
$$= \text{tr}\{\mathbf{W}_\|^T\mathbf{X}(\mathbf{I} - \hat{\mathbf{A}})(\mathbf{I} - \hat{\mathbf{A}})^T\mathbf{X}^T\mathbf{W}_\|\}, \tag{25}$$

and the third term as below:

$$\|\mathbf{W}^T\mathbf{X}\tilde{\mathbf{A}}\|_F^2 = \|(\mathbf{W}_\| + \mathbf{W}_\perp)^T\mathbf{X}\tilde{\mathbf{A}}\|_F^2 = \|\mathbf{W}_\|^T\mathbf{X}\tilde{\mathbf{A}}\|_F^2$$
$$= \text{tr}\{\mathbf{W}_\|^T\mathbf{X}\tilde{\mathbf{A}}\tilde{\mathbf{A}}^T\mathbf{X}^T\mathbf{W}_\|\}. \tag{26}$$

Let $\mathbf{X}^T\mathbf{X} = \mathbf{K}$, by putting Eq. 23, Eq. 24, Eq. 25 and Eq. 26 together, and omitting the unrelated term to $\mathbf{W}$, we have:

$$\min_{\mathbf{W}} \text{tr}\{\mathbf{W}_\|^T\mathbf{X}(\mathbf{I} - \mathbf{A})(\mathbf{I} - \mathbf{A})^T\mathbf{X}^T\mathbf{W}_\|\}$$
$$+ \alpha\text{tr}\{\mathbf{W}_\|^T\mathbf{X}(\mathbf{I} - \hat{\mathbf{A}})(\mathbf{I} - \hat{\mathbf{A}})^T\mathbf{X}^T\mathbf{W}_\|\} + \beta\text{tr}\{\mathbf{W}_\|^T\mathbf{X}\tilde{\mathbf{A}}\tilde{\mathbf{A}}^T\mathbf{X}^T\mathbf{W}_\|\}, \tag{27}$$
$$\text{s.t. } \mathbf{W}^T\mathbf{W} = \mathbf{I}.$$

Let the singular value decomposition (SVD) of $\mathbf{K} = \mathbf{USU}^T = \mathbf{US}^{\frac{1}{2}}\mathbf{S}^{\frac{1}{2}}\mathbf{U}^T$, and $\mathbf{H} = (\mathbf{I} - \mathbf{A})(\mathbf{I} - \mathbf{A})^T + \alpha(\mathbf{I} - \hat{\mathbf{A}})(\mathbf{I} - \hat{\mathbf{A}})^T + \beta\tilde{\mathbf{A}}\tilde{\mathbf{A}}^T$, we have:

$$\text{tr}\{\mathbf{W}_\|^T\mathbf{XHX}^T\mathbf{W}_\|\} = \text{tr}\{\mathbf{P}^T\mathbf{US}^{\frac{1}{2}}\mathbf{S}^{\frac{1}{2}}\mathbf{U}^T\mathbf{HUS}^{\frac{1}{2}}\mathbf{S}^{\frac{1}{2}}\mathbf{U}^T\mathbf{P}\},$$
$$= \text{tr}\{\mathbf{G}^T\tilde{\mathbf{H}}\mathbf{G}\}, \tag{28}$$

where $\tilde{\mathbf{H}} = \mathbf{S}^{\frac{1}{2}} \mathbf{U}^T \mathbf{H} \mathbf{U} \mathbf{S}^{\frac{1}{2}}$ and $\mathbf{G} = \mathbf{S}^{\frac{1}{2}} \mathbf{U}^T \mathbf{P}$. Therefore, we have:

$$\mathrm{tr}\{\mathbf{G}^T \tilde{\mathbf{H}} \mathbf{G}\} \geq \sum_{i=1}^{m} \sigma_i, \tag{29}$$

where σ_i is the ith smallest eigenvalue of $\tilde{\mathbf{H}}$. In order for the objective function to achieve its minimum, columns of $\mathbf{G}$ have to be the same with eigenvectors corresponding to the smallest eigenvalues of $\tilde{\mathbf{H}}$. Therefore we have $\mathbf{G}^T \mathbf{G} = \mathbf{I}$. Equivalently, we have the constraint:

$$\begin{aligned}
\mathbf{W}^T \mathbf{W} = \mathbf{I} &= (\mathbf{W}_{\parallel} + \mathbf{W}_{\perp})^T (\mathbf{W}_{\parallel} + \mathbf{W}_{\perp}) \\
&= \mathbf{W}_{\parallel}^T \mathbf{W}_{\parallel} + \mathbf{W}_{\perp}^T \mathbf{W}_{\perp} \\
&= \mathbf{P}^T \mathbf{K} \mathbf{P} + \mathbf{W}_{\perp}^T \mathbf{W}_{\perp} \\
&= \mathbf{G}^T \mathbf{G} + \mathbf{W}_{\perp}^T \mathbf{W}_{\perp},
\end{aligned} \tag{30}$$

which means $\mathbf{W}_{\perp} = \mathbf{0}$. In short, the optimal solution of $\mathbf{W}$ has the form:

$$\mathbf{W}^* = \mathbf{W}_{\parallel} = \mathbf{X} \mathbf{P}. \tag{31}$$

This completes the proof.

Appendix B: Proof of Proposition 2

When fixing $\mathbf{A}$, by omitting unrelated terms, we derive from objective function Eq. 13 as below:

$$\begin{aligned}
&\|\mathbf{P}^T \mathbf{K} (\mathbf{I} - \mathbf{A})\|_F^2 + \alpha \|\mathbf{P}^T \mathbf{K} (\mathbf{I} - \hat{\mathbf{A}})\|_F^2 + \beta \|\mathbf{P}^T \mathbf{K} \tilde{\mathbf{A}}\|_F^2 \\
&= \mathrm{tr}\Big\{ \mathbf{P}^T \mathbf{K} \big((\mathbf{I} - \mathbf{A})(\mathbf{I} - \mathbf{A})^T + \alpha (\mathbf{I} - \hat{\mathbf{A}})(\mathbf{I} - \hat{\mathbf{A}})^T + \beta \tilde{\mathbf{A}} \tilde{\mathbf{A}}^T \big) \mathbf{K} \mathbf{P} \Big\} \\
&= \mathrm{tr}\Big\{ \mathbf{P}^T \mathbf{K} \mathbf{H} \mathbf{K} \mathbf{P} \Big\}, \\
&\text{s.t. } \mathbf{P}^T \mathbf{K} \mathbf{P} = \mathbf{I},
\end{aligned} \tag{32}$$

where $\mathbf{H} = (\mathbf{I} - \mathbf{A})(\mathbf{I} - \mathbf{A})^T + \alpha (\mathbf{I} - \hat{\mathbf{A}})(\mathbf{I} - \hat{\mathbf{A}})^T + \beta \tilde{\mathbf{A}} \tilde{\mathbf{A}}^T$. Let SVD of $\mathbf{K} = \mathbf{U} \mathbf{S} \mathbf{U}^T = \mathbf{U} \mathbf{S}^{\frac{1}{2}} \mathbf{S}^{\frac{1}{2}} \mathbf{U}^T$, then we have:

$$\begin{aligned}
\mathrm{tr}\Big\{ \mathbf{P}^T \mathbf{K} \mathbf{H} \mathbf{K} \mathbf{P} \Big\} &= \mathrm{tr}\Big\{ \mathbf{P}^T \mathbf{U} \mathbf{S}^{\frac{1}{2}} \mathbf{S}^{\frac{1}{2}} \mathbf{U}^T \mathbf{H} \mathbf{U} \mathbf{S}^{\frac{1}{2}} \mathbf{S}^{\frac{1}{2}} \mathbf{U}^T \mathbf{P} \Big\} \\
&= \mathrm{tr}\Big\{ (\mathbf{S}^{\frac{1}{2}} \mathbf{U}^T \mathbf{P})^T \mathbf{S}^{\frac{1}{2}} \mathbf{U}^T \mathbf{H} \mathbf{U} \mathbf{S}^{\frac{1}{2}} (\mathbf{S}^{\frac{1}{2}} \mathbf{U}^T \mathbf{P}) \Big\} \\
&= \mathrm{tr}\Big\{ \mathbf{G}^T \tilde{\mathbf{H}} \mathbf{G} \Big\} \\
&\text{s.t. } (\mathbf{S}^{\frac{1}{2}} \mathbf{U}^T \mathbf{P})^T (\mathbf{S}^{\frac{1}{2}} \mathbf{U}^T \mathbf{P}) = \mathbf{I},
\end{aligned} \tag{33}$$

where $\mathbf{G} = \mathbf{S}^{\frac{1}{2}} \mathbf{U}^T \mathbf{P}$ and $\tilde{\mathbf{H}} = \mathbf{S}^{\frac{1}{2}} \mathbf{U}^T \mathbf{H} \mathbf{U} \mathbf{S}^{\frac{1}{2}}$. And the constraint can also be simplified as $\mathbf{G}^T \mathbf{G} = \mathbf{I}$.

Now, we can see the equivalence of optimization in Eq. 13 and Eq. 33. And the optimal solution $\mathbf{P}^*$ can be recovered as in Eq. 14, *i.e.* $\mathbf{P}^* = \mathbf{U} \mathbf{S}^{-\frac{1}{2}} \mathbf{G}^*$.

Note that since $\mathbf{K}$ is a positive semidefinite matrix, the diagonal matrix $\mathbf{S}$ has non-negative entries. $\mathbf{S}^{-\frac{1}{2}}$ is obtained by setting non-zero entries along the diagonal of $\mathbf{S}$ to the inverse of their square root and keeping zero elements the same.

This completes the proof.

Locally Linear Landmarks
for Large-Scale Manifold Learning

Max Vladymyrov and Miguel Á. Carreira-Perpiñán

Electrical Engineering and Computer Science, University of California, Merced, USA
{mvladymyrov,mcarreira-perpinan}@ucmerced.edu

Abstract. Spectral methods for manifold learning and clustering typically construct a graph weighted with affinities from a dataset and compute eigenvectors of a graph Laplacian. With large datasets, the eigendecomposition is too expensive, and is usually approximated by solving for a smaller graph defined on a subset of the points (landmarks) and then applying the Nyström formula to estimate the eigenvectors over all points. This has the problem that the affinities between landmarks do not benefit from the remaining points and may poorly represent the data if using few landmarks. We introduce a modified spectral problem that uses all data points by constraining the latent projection of each point to be a local linear function of the landmarks' latent projections. This constructs a new affinity matrix between landmarks that preserves manifold structure even with few landmarks, allows one to reduce the eigenproblem size, and defines a fast, nonlinear out-of-sample mapping.

Keywords: manifold learning, spectral methods, optimization.

1 Introduction

Manifold learning algorithms have long been used for exploratory analysis of a high-dimensional dataset, to reveal structure such as clustering, or as a preprocessing step to extract some low-dimensional features that are useful for classification or other tasks. Here we focus on the well-known class of spectral manifold learning algorithms [1]. The input to these algorithms is a symmetric positive (semi)definite matrix $\mathbf{A}_{N \times N}$ (affinity matrix, graph Laplacian, etc.) that contains information about the similarity between pairs of data points $\mathbf{Y} \in \mathbb{R}^{D \times N}$, and a symmetric positive definite matrix $\mathbf{B}_{N \times N}$ that usually sets the scale of the solution. Given these two matrices, we seek a solution $\mathbf{X} \in \mathbb{R}^{d \times N}$ to the following *generalized spectral problem*:

$$\min_{\mathbf{X}} \operatorname{tr} \left(\mathbf{X} \mathbf{A} \mathbf{X}^T \right) \text{ s.t. } \mathbf{X} \mathbf{B} \mathbf{X}^T = \mathbf{I}. \tag{1}$$

Within this framework it is possible to represent manifold learning methods such as Laplacian Eigenmaps (LE) [2], Kernel PCA [3], MDS [4], ISOMAP [5] and LLE [6], as well as spectral clustering [7].

H. Blockeel et al. (Eds.): ECML PKDD 2013, Part III, LNAI 8190, pp. 256–271, 2013.

The solution of the spectral problem (1) is given by $\mathbf{X} = \mathbf{U}_d^T \mathbf{B}^{-\frac{1}{2}}$, where $\mathbf{U}_d = (\mathbf{u}_1, \ldots, \mathbf{u}_d)$ are the d trailing eigenvectors of the matrix $\mathbf{C} = \mathbf{B}^{-\frac{1}{2}} \mathbf{A} \mathbf{B}^{-\frac{1}{2}}$. In large problems (large N), the computational cost means the matrices $\mathbf{A}$, $\mathbf{B}$ and $\mathbf{C}$ have to be sparse, and these eigenvectors are found with numerical linear algebra techniques such as restarted Arnoldi iterations [8]. The resulting cost is still large when N and d are large. The primary goal of this paper is to find fast, approximate solutions to the spectral problem (1) (and thus to LE, spectral clustering, etc.). We propose a method we call *Locally Linear Landmarks (LLL)*, based on the idea of selecting a subset of $L \ll N$ landmarks $\widetilde{\mathbf{Y}}_{L \times N}$ from the data, approximating the data manifold by a globally nonlinear but locally linear manifold around these landmarks, and then constraining the solution $\mathbf{X}$ to follow this locally linear structure. The locally linear mapping is given by a projection matrix $\mathbf{Z} \in \mathbb{R}^{L \times N}$ that satisfies

$$\mathbf{Y} \approx \widetilde{\mathbf{Y}} \mathbf{Z} \tag{2}$$

in the high-dimensional space, and by enforcing it in the low-dimensional space, we can re-express the problem (1) as a new spectral problem on a smaller number of variables L. This reduces the cost of the eigendecomposition dramatically and, as we will show, constructs affinity matrices that preserve much manifold information because the problem still involves the entire dataset. Note that LLL is not a new manifold learning method, but a fast, approximate way to solve an existing method of the form (1).

The LLL algorithm can be used for purposes beyond fast solutions of spectral problems. First, it is particularly useful for model selection. The similarity matrices $\mathbf{A}$ and $\mathbf{B}$ are usually constructed using some meta-parameters, such as a bandwidth σ of Gaussian affinities and a sparsity level K_W (number of neighbors). In practice, a user has to tune these parameters to the dataset by hand by solving the spectral problem for each parameter value. This is extremely costly with large datasets. As we will show, we can run LLL with very few landmarks so that the *shape* of the model selection curve (especially its minimum) is preserved well. This way we can identify the optimal meta-parameters much faster and then solve the spectral problem (possibly using more landmarks). Second, LLL solves the out-of-sample problem for the spectral problem (1) (which projects only the training points) by providing a natural, explicit mapping to project new points, which does not exist in the original spectral problem. Finally, we observe that the gain of LLL is much bigger when the number of eigenvectors d is large, which makes it very attractive as a preprocessing step for classification to other machine learning tasks.

Related Work. The most widespread method to find an approximate, fast solution of the spectral problem is the Nyström method [9,10,11,12]. It approximates the eigendecomposition of a large positive semidefinite matrix using the eigendecomposition of a much smaller matrix of landmarks. It can be seen as an out-of-sample extension where we first solve for the landmarks separately from the non-landmark points, and then use it to project the non-landmark points. Since, during the projection of the landmarks, the Nyström method does not use the data from the non-landmark points, which is available from the beginning, it can result in large approximation errors if the number of landmarks is low.

It is possible to redefine the affinities between landmarks so that they use information from all points, for example by using a commute distance (the expected time it takes a random walk to travel from the first to the second node and back). Besides the fact that this solves a different spectral problem, computing these distances is costly, it provides no out-of-sample mapping, and commute distances have been shown to be problematic with large datasets in high dimensions [13]. As we will show, in LLL the affinities between landmarks use naturally the information in non-landmarks without us having to define new affinities.

Other landmark-based methods can be seen as forms of a Nyström approach. De Silva and Tenenbaum [14] suggested to run the metric MDS algorithm on a subset of the data, while the rest of the points can be located through a distance-based triangulation process. The same idea can be applied to a graph of geodesic distances (instead of Euclidean ones) which leads to the Landmark Isomap algorithm [15]. This algorithm is able to give better results due to its ability to deal with nonlinear manifolds. These approaches have been shown [10,16] to be a Nyström approximation combined with classical MDS or Isomap.

The idea of representing points by linear coding as in eq. (2) has been used in many different domains of machine learning, such as image classification [17,18], manifold learning [19,20,21], supervised [22] and semi-superwised [23] learning. In addition to linearity, many of above algorithms try to find local, sparse representations of the data, so that points are reconstructed using only nearby landmarks. An early work is the LLE method for manifold learning [19], which computes the matrix $\mathbf{Z}$ that best reconstructs each data point from a set of nearby points. Variations exist, such as using multiple local weight vectors in constructing $\mathbf{Z}$ in the MLLE algorithm [24]. However, these works use local linear mappings to define a new spectral problem, while LLL uses them to approximate an existing spectral problem. The AnchorGraph algorithm [23] uses local coding in the graph Laplacian regularization term of a semi-supervised learning problem. The problem it solves is different from (1), and does not generalize beyond the Laplacian regularizer, compared to the more general approach that we propose here. Chen and Cai [25] propose a landmarks-based approximation for spectral clustering. However, the affinities they construct entirely ignore the original affinity matrix and thus cannot be seen as approximating the target problem. Landmark SDE [20] proposes to reconstruct kernel matrix using much smaller matrix of inner products between the landmarks only. This problem is also different to ours.

Two approaches exist to construct out-of-sample mappings for spectral problems such as Laplacian eigenmaps: Bengio el al. [26] apply the Nyström formula using the affinity kernel that defined the problem. Carreira-Perpiñán and Lu [27] augment the spectral problem with the test point and solve it subject to not changing the points already embedded, which results in a kernel regression mapping. In LLL, the out-of-sample mapping is a natural subproduct of assuming each low-dimensional point to be a local linear mapping of the landmark projections associated with it.

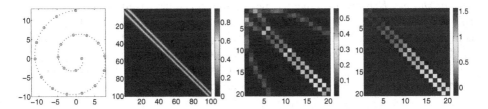

Fig. 1. Affinity matrices for landmarks in a spiral dataset. *From left to right:* 100 points along the spiral (in red) with 20 landmarks selected uniformly (in blue); the affinity matrix **W** used by LE constructed using all the points; the affinity matrix **W** built using just landmarks; the learned affinity matrix **C** of LLL using the whole dataset.

2   Solving Spectral Problems with Locally Linear Landmarks

The fundamental assumption in LLL is that the local dependence of points on landmarks that occurs in high-dimensional space, eq. (2), is preserved in the low-dimensional space:

$$\mathbf{X} \approx \widetilde{\mathbf{X}}\mathbf{Z}. \tag{3}$$

Substituting this into the spectral problem (1) gives the following *reduced spectral problem* (on dL parameters):

$$\min_{\widetilde{\mathbf{X}}} \operatorname{tr}\left(\widetilde{\mathbf{X}}\widetilde{\mathbf{A}}\widetilde{\mathbf{X}}^T\right) \text{ s.t. } \widetilde{\mathbf{X}}\widetilde{\mathbf{B}}\widetilde{\mathbf{X}}^T = \mathbf{I}, \tag{4}$$

where the matrices

$$\widetilde{\mathbf{A}} = \mathbf{Z}\mathbf{A}\mathbf{Z}^T, \qquad \widetilde{\mathbf{B}} = \mathbf{Z}\mathbf{B}\mathbf{Z}^T. \tag{5}$$

are of $L \times L$. The solution for the reduced problem is given by $\widetilde{\mathbf{X}} = \widetilde{\mathbf{U}}_d^T \widetilde{\mathbf{B}}^{-\frac{1}{2}}$, where $\widetilde{\mathbf{U}}_d$ are d trailing eigenvectors of the matrix $\widetilde{\mathbf{C}} = \widetilde{\mathbf{B}}^{-\frac{1}{2}}\widetilde{\mathbf{A}}\widetilde{\mathbf{B}}^{-\frac{1}{2}}$. After the solution for the landmarks is found, the values of **X** can be recovered by applying the formula (3) once again.

We can see the reduced problem (4) as a spectral problem for just the landmark points using a similarity matrix $\widetilde{\mathbf{A}}$ that incorporates information from the whole dataset. For example, in Laplacian Eigenmaps (see Section 4) the matrix **A** is given by the graph Laplacian of a matrix **W** of affinities (typically Gaussian). Using LLL we can dramatically improve the quality of **W** over that of constructing **W** using only the landmarks, by including additional information from the whole dataset. Fig. 1 shows the affinity matrix constructed in the usual way for a spiral dataset in the full case (using all 100 points) and using 20 landmark points versus the affinity matrix learned using LLL. The latter one (right plot) is almost perfectly banded with uniform entries. This means the connectivity pattern proceeds along the spiral, respecting its geometry, rather than across it. However, when affinities are constructed directly on landmarks that are quite distant from each other, undesirable interactions across the spiral occur.

Out-of-sample Extension. Given a new point $\mathbf{y}_0 \in \mathbb{R}^D$ that is not a part of the original dataset, we find its projection on the low-dimensional space by computing a new projection vector $\mathbf{z}_0$ for that point using K_Z landmarks around $\mathbf{y}_0$. The embedding of $\mathbf{y}_0$ is found from a linear combination of the landmark projections $\mathbf{x}_0 = \widehat{\mathbf{X}}\mathbf{z}_0$. The cost of the out-of-sample is $\mathcal{O}(DK_Z^2 + Ld)$, which is linear for all the parameters except for K_Z, which is usually low.

Construction of the Projection Matrix $\mathbf{Z}$. Let us define the landmarks as a set $\widehat{\mathbf{Y}} = (\widetilde{\mathbf{y}}_1, \ldots, \widetilde{\mathbf{y}}_L) \in \mathbb{R}^{D \times L}$ of L points in the same space as the high-dimensional input $\mathbf{Y}$. Now each datapoint $\mathbf{y}_n$ can be expressed as a linear combination of nearby landmark points: $\mathbf{y}_n = \sum_{k=1}^{L} \widetilde{\mathbf{y}}_k z_{nk}$ where $\mathbf{z}_n$ is a local projection vector for the point $\mathbf{y}_n$. There are multiple ways to make this projection local. One can consider choosing K_Z landmarks closest to $\mathbf{y}_n$ or ϵ-balls centered around $\mathbf{y}_n$. Moreover, the choice of landmarks can be different for every n. In our experiments, we keep only the K_Z landmarks that are closest to $\mathbf{y}_n$ and use the same value of K_Z for all the points. Therefore, the projection matrix $\mathbf{Z} = (\mathbf{z}_1, \ldots, \mathbf{z}_N) \in \mathbb{R}^{L \times N}$ has only K_Z nonzero elements for every column. This matrix intuitively corresponds to the proximity of the points in the dataset to the nearby landmarks and it should be invariant to rotation, rescaling and translation. The invariance to rotation and rescaling is given by the linearity of the reconstructing matrix $\widetilde{\mathbf{Y}}\mathbf{Z}$ with respect to $\widetilde{\mathbf{Y}}$, whereas translation invariance must be enforced by constraining columns of $\mathbf{Z}$ to sum to one. This leads to the following optimization problem:

$$\min_{\mathbf{Z}} \|\mathbf{Y} - \widetilde{\mathbf{Y}}\mathbf{Z}\|^2 \text{ s.t. } \mathbf{1}^T\mathbf{Z} = \mathbf{1}^T. \qquad (6)$$

Following the approach proposed in the LLE algorithm [19] we introduce a pointwise Gram matrix $\mathbf{G} \in \mathbb{R}^{L \times L}$ with elements

$$g_{ij} = (\mathbf{y}_n - \widetilde{\mathbf{y}}_i)^T (\mathbf{y}_n - \widetilde{\mathbf{y}}_j) \qquad (7)$$

for every $n = 1, \ldots, N$. Now, the solution to problem (6) is found by solving a linear system $\sum_{k=1}^{L} g_{jk} z_{nk} = 1$ and rescaling the weights so they sum to one.

Computational Complexity. This algorithm reduces the number of computations for the eigendecomposition in the solution to the problem (1). However, we also need to perform extra computations to evaluate $\mathbf{Z}$, compute auxiliary matrices (5) and perform the final multiplication (3) to recover the full embedding.

The computation of $\mathbf{Z}$ consists of computing the pointwise Gram matrix $\mathbf{G}$ and solving the linear system. $\mathbf{G}$ is sparse and has only K_Z nonzero elements in each row, so it takes $\mathcal{O}(NDK_Z^2)$ to compute it. The linear system also should be solved just for K_Z unknowns, so it takes $\mathcal{O}(NK_Z^3)$. Among the two, the computation of $\mathbf{G}$ matrix dominates because $K_Z < D$, as we will show below. Note this step is independent of the number of landmarks L. The cost of computing $\widetilde{\mathbf{A}}$ and $\widetilde{\mathbf{B}}$ is $\mathcal{O}(K_Z N^2)$ with dense matrices and $\mathcal{O}(K_Z Nc)$ with sparse matrices, where c is some constant that depends on the sparsity of the matrices $\mathbf{A}$ and $\mathbf{B}$ and on the particular location of the nonzero elements in $\mathbf{Z}$. Computing $\widetilde{\mathbf{C}}$ and performing the eigendecomposition both take $\mathcal{O}(L^3)$, and recovering the final embedding takes $\mathcal{O}(NLd)$. Overall, the complexity of LLL is $\mathcal{O}\big(K_Z N^2 + N(Ld + DK_Z^2) + L^3)\big)$ with dense inputs and $\mathcal{O}\big(N(K_Z c + Ld + DK_Z^2) + L^3)\big)$ with sparse inputs, which is asymptotically much faster than the cost of the eigendecomposition if $L \ll N$.

The computational cost of the out-of-sample mapping is $\mathcal{O}(DK_Z^2)$ to find a projection vector $\mathbf{z}_0$ and $\mathcal{O}(Ld)$ for a reprojection in the low-dimensional space, hence $\mathcal{O}(DK_Z^2 + Ld)$ overall.

3   Choice of Parameters

Number of Landmarks L. One should use as many landmarks as one can afford computationally, because the more landmarks the better the approximation. As L increases, the results look more and more similar to the solution of the original spectral problem, which is recovered when $L = N$.

Number of Landmarks K_Z around Each Point. Each point should be a local linear reconstruction of nearby landmarks. Thus it is important that there are enough landmarks around each point so that its nearest landmarks are chosen along the manifold. These landmarks will have nonzero weights in the reconstruction, thus achieving locally and linearity. Non-local weights may not work unless the manifold is globally linear.

Using weights that are nonzero only for the nearest K_Z landmarks implies that the low-dimensional space is partitioned into regions where $\mathbf{X}$ is piecewise linear as a function of the corresponding subset of landmarks. If the K_Z landmarks are in general position, they span a linear manifold of dimension $K_Z - 1$. Therefore, we need no more than $D + 1$ landmarks, since $K_Z = D + 1$ of them reconstruct any point in D dimensions perfectly. On the other hand, using $K_Z > D + 1$ makes the weights non-unique and we need to add a regularization term to (7) to penalize the weight norm by adding a small positive amount to the diagonal of the linear system. However, the manifold learning assumption implies that the intrinsic dimensionality of the manifold is lower than D. For example, if the manifold is linear with dimension $\hat{d}$ then the number of landmarks needed to reconstruct any point is $K_Z = \hat{d} + 1$ by the same argument as above. However, if the manifold is nonlinear with local dimension $\hat{d}$, then $K_Z = \hat{d} + 1$ landmarks reconstruct the point approximately (near its projection on the tangent plane). Thus, overall the number of landmarks around each point should be between $\hat{d}+1$ (which may have a certain reconstruction error, particularly if the landmarks are not in general position) and $D+1$ (which achieves perfect reconstruction). If the reconstruction is imperfect, we introduce an additional error on the embedding, by implicitly replacing each original data point with its projection on landmarks. Thus, K_Z is a user parameter with values in $[\hat{d} + 1, D + 1]$: the larger K_Z the smaller the error and the larger the computational cost. In practice, K_Z can be estimated so a desired reconstruction error $\|\mathbf{Y} - \widetilde{\mathbf{Y}}\mathbf{Z}\|$ is achieved, but it should not be much bigger than $\hat{d} + 1$. Note $\hat{d}$ in this context is an intrinsic local dimensionality of the manifold and not the dimensionality of the low-dimensional output d, which may or many not match $\hat{d}$.

The Location of Landmarks. Kumar et al. [28] provide a formal analysis of different types of sampling and show that, at least for the Nyström approximation, uniform sampling works best. Our experiments confirm this as well. However, we should not spend much computation on selecting landmarks, so as to introduce as little computational overhead as possible. Based on this, we investigated three general methods on how to compute the location of the landmarks.

First, we can always choose the landmarks at random from a set of existing points. This method requires almost no computational resources. However, the result can vary dramatically, especially when only a small number of landmarks is available. We can apply an additional heuristic to make the landmark location as close to uniform as possible: we select $K + M$ landmarks at random, find M pairs of closest landmarks and then discard one landmark from each pair. This heuristic is also useful because the distances are already given to us from the adjacency matrix. Even when the adjacency matrix is sparse, it is usually the largest distances that are missing. Thus, we can always identify closest landmarks to each other.

Second, we can select the landmarks by running a clustering algorithm with L clusters and choose each landmark in the middle of the clusters. For instance, one can run k-means and set the landmarks to the points that are closest to the centroids of the clusters. One problem with this approach is that the clustering is usually quite expensive. Another is that, for data with a nonconvex manifold structure, the landmarks can end up in between branches of the manifold (although this could be avoided with a k-modes algorithm that places landmarks in high-density regions of the data [29]). In our experiments we avoid dealing with landmarks that are not part of the dataset.

Finally, the landmarks can be also selected using other heuristics so they span the manifold as uniformly as possible. It has been proposed [14] to use a MinMax algorithm which chooses landmarks one by one by maximizing the mutual distance between the new landmark and the existing set of landmarks. However, this requires having the mutual distances between all the points, which in case of a large number of points N is unavailable.

4   Locally Linear Landmarks for Laplacian Eigenmaps

A particular case of the spectral method for which we can apply LLL is the Laplacian Eigenmaps (LE) algorithm [2]. The general embedding formulation is recovered using $\mathbf{A}$ as a graph Laplacian matrix $\mathbf{L} = \mathbf{D} - \mathbf{W}$ defined on a symmetric affinity matrix $\mathbf{W}$ with degree matrix $\mathbf{D} = \text{diag}\left(\sum_{m=1}^{N} w_{nm}\right)$ and $\mathbf{B} = \mathbf{D}$. The objective function is thus

$$\min_{\mathbf{X}} \text{tr}\left(\mathbf{X}\mathbf{L}\mathbf{X}^T\right) \text{ s.t. } \mathbf{X}\mathbf{D}\mathbf{X}^T = \mathbf{I}, \mathbf{X}\mathbf{D}\mathbf{1} = \mathbf{0}. \tag{8}$$

Note that adding the second constraint does not alter the general formulation of the spectral solution, but just removes the first eigenvector, which is constant and equal to $\mathbf{D}^{-\frac{1}{2}}\mathbf{1}$ with eigenvalue 1.

The matrices in the reduced spectral problem (4) are then:

$$\widetilde{\mathbf{A}} = \mathbf{Z}\mathbf{L}\mathbf{Z}^T, \qquad \widetilde{\mathbf{B}} = \mathbf{Z}\mathbf{D}\mathbf{Z}^T. \qquad (9)$$

Similarly to the case of the original LE, the second constraint is satisfied by discarding the first eigenvector. We can see this by noticing that $\widetilde{\mathbf{A}}\mathbf{1} = \mathbf{0}$ and looking at the eigendecomposition of $\widetilde{\mathbf{C}}$:

$$\widetilde{\mathbf{B}}^{-\frac{1}{2}}\widetilde{\mathbf{A}}\widetilde{\mathbf{B}}^{-\frac{1}{2}}\widetilde{\mathbf{u}}_1 = \widetilde{\mathbf{B}}^{-\frac{1}{2}}\widetilde{\mathbf{A}}\widetilde{\mathbf{x}}^T = \lambda_1\widetilde{\mathbf{u}}_1.$$

Therefore, the solution corresponding to the eigenvalue $\lambda_1 = 0$ is trivial.

The affinity matrix $\mathbf{W}$ for LE is usually computed using a Gaussian kernel with a bandwidth parameter σ (or a separate bandwidth per point [30])). The affinities are also sparsified by retaining only the K_W biggest values for every row. The performance of LE depends crucially on the choice of those parameters and they have to be tuned quite carefully in order to achieve good results. Unfortunately, in most cases there is no procedure to check the quality of the affinity matrix without running LE itself. However, instead of solving multiple, expensive LE problems, we can tune those parameters by running LLL. This gives a much cheaper runtime, especially considering that the matrix $\mathbf{Z}$ is independent of the choice of σ and K_W and, thus, is computed only once.

5   Experimental Evaluation

We compared LLL for LE to three natural baselines. (1) "Exact LE" runs LE on the full dataset and gives the optimal embedding by definition, but the runtime is large. (2) "LE ($\mathbf{Z}$)" runs LE only on a set of landmark points and then projects non-landmark points using the projection matrix $\mathbf{Z}$, which gives a locally linear (but globally nonlinear) out-of-sample mapping. (3) "LE (Nys.)" runs LE only on a set of landmark points and uses the Nyström out-of-sample formula. The latter two Landmark LE baselines give faster performance, but the embedding quality can be worse because non-landmark points are completely ignored in solving the spectral problem. For all our experiments we used Matlab's `eigs` function to compute the partial eigendecomposition of a sparse matrix.

Role of the Number of Landmarks. We used 60 000 MNIST digits with sparsity $K_W = 200$ and bandwidth $\sigma = 200$ to build the affinity matrix and reduced the dimensionality to $d = 50$. For LLL, we set $K_Z = 50$ and increased the number of landmarks logarithmically from 50 to 60 000. We chose landmarks at random and repeated the experiment 5 times for different random initialization to see the sensibility of the results to the random choice of the landmarks. To quantify the error with respect to Exact LE we first used Procrustes alignment [4, ch. 5] to align the embeddings of the methods and then computed the relative error between the aligned embeddings.

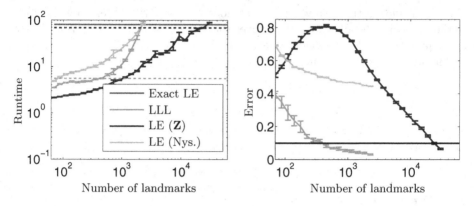

Fig. 2. Performance of LLL (green), Landmark LE with **Z** as an out-of-sample (blue) and Landmark LE with Nyström as an out-of-sample (cyan). *Left*: runtime as the number of landmarks changes. The green and blue dashed lines correspond to the runtime that gives 10% error with respect to Exact LE for LLL and Landmark LE using **Z**, respectively. *Right*: error with respect to Exact LE. The black line corresponds to 10% error. Note the log scale in most of the axes.

Fig. 2 shows the error as well as the overall runtime for different algorithms as the number of landmarks increases. Our first indicator of performance is to see which algorithm can attain an error of 10% faster. LLL needed 451 landmarks and 5.5 seconds (shown by a dashed green line in the left plot). This is 14 times faster compared to Exact LE, which takes 80 seconds. Landmark LE with **Z** as out-of-sample mapping attains the same error with 23 636 landmarks and the runtime of 69 seconds (1.15 speedup, blue dashed line in the right plot). Landmark LE with Nyström is not able to attain an error smaller than 50% with any number of landmarks. Note the deviation from the mean for 5 runs of randomly chosen landmarks is rather small, suggesting the algorithm is robust to different locations of landmarks. Fig. 3 shows the embedding of Exact LE and the embedding of LLL with 451 randomly selected landmarks. The embedding of LLL is very similar to the one of Exact LE, but the runtime is 15 times faster (5.5 seconds compared to 80 seconds). Using more landmarks only decreases the error further and for 3 000 landmarks, where the runtime of LLL matches the runtime of Exact LE, the mean error among 5 runs drops to 3%. Landmark LE with **Z** as an out-of-sample attained the same error only by using 23 636 landmarks and a runtime of 69 seconds (1.15 speedup, blue dashed line in the right plot). Landmark LE with Nyström is not able to attain an error smaller than 50% for any number of landmarks. Note the deviation from the mean for 5 runs of randomly chosen landmarks is rather small, suggesting the algorithm is relatively robust to different locations of landmarks.

Model Selection. We evaluated the use of LLL to select the parameters of the affinity matrix. We used 4 000 points from the Swissroll dataset and ran the

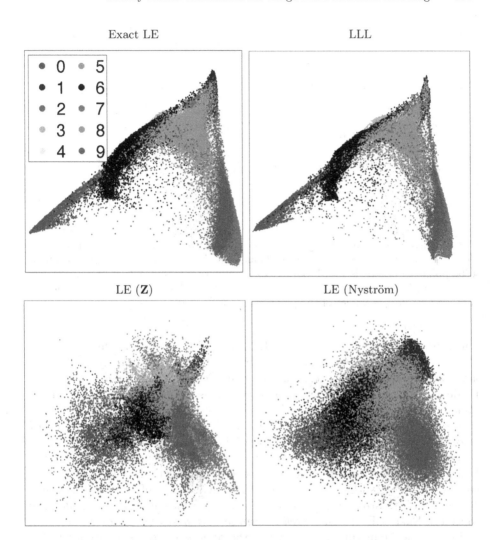

Fig. 3. Embedding 60 000 MNIST digits using the first two dimensions. *Left to right*: Exact LE ($t = 80$ s), LLL ($t = 5.5$ s, 451 landmarks), Landmark LE with $\mathbf{Z}$ as out-of-sample mapping ($t = 5.5$ s, 1 144 landmarks), and LE with Nyström as out-of-sample mapping ($t = 5.5$ s, 88 landmarks).

methods varying different parameters of the algorithm. We ran LLL and Landmark LE 5 times using different random initializations to show the general behavior of the algorithm. Experimentally we discovered that the best results are obtained with a bandwidth $\sigma = 1.6$, a number of landmarks L no less than 300 and a sparsity level K_W around 150. We then fixed two out of these three parameters and changed the third one to see how the error curve changes. Fig. 4 shows the results. First, for different σ values the error curve of Exact LE is

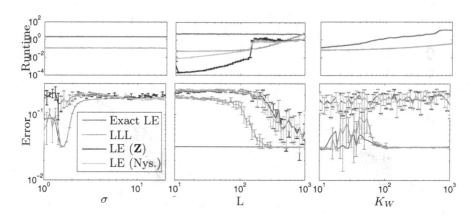

Fig. 4. Quality of the embedding with respect to the ground truth for different values of the bandwidth σ, number of landmarks L, and sparsity level K_W. The dataset contains 4 000 points from a Swissroll. From left to right: vary σ for fixed $L = 300$, $K_W = 150$; vary L for fixed $\sigma = 1.6$, $K_W = 150$; vary K_W for fixed $L = 300$, $\sigma = 1.6$. *Top row:* runtime for different values of the parameters. *Bottom row:* error.

much more similar to the one from LLL, but LLL is able to achieve it about 18× faster (top plot). Compared to that, Landmark LE definitely needs more landmarks in order to show a similar behavior. Second, the number of landmarks needed to achieve the same error as Exact LE is much lower for LLL than for Landmark LE. Using 300 landmarks the error of LLL is about 3% and it is also 18 times faster than Exact LE. Landmark LE is never able to achieve a 10% error for any set of landmarks up to 1 000. Third, changing the sparsity level parameter K_W, the error curve is again very similar between Exact LE and LLL, but very different between Exact LE and Landmark LE. The speedup of LLL compared to Exact LE varies between 2 for small values of K_W to 40 for large K_W. Note that, although LLL is not able to reproduce an error curve identical to that of Exact LE, *it does match the minima of these curves* (for σ and K_W), and of course the minima correspond to the parameter values we are interested in. That is, LLL can be used as a fast way to find good parameter values for Exact LE. This suggests a practical procedure to set the parameters of Exact LE: we run LLL to obtain the values of σ and K_W that give the (approximately) minimum error and then run Exact LE using those values.

Classification. Here our goal was to find a good set of parameters to achieve a low 1-nearest neighbor classification error for the full 70 000 MNIST digits dataset. We first split the dataset into three independent sets: 50 000 digits as a training set, 10 000 digits as a test set and 10 000 digits for out-of-sample mapping. We then projected training and test sets (overall 60 000 points) to 500 dimensions using LLL with 1 000 landmarks selected using k-means with $K_Z = 50$. We did this a number of times for different values of K_W from 1 to 200 and σ from 4.6 to 1000. Note the $\mathbf{Z}$ matrix is independent from the affinities and depends only on

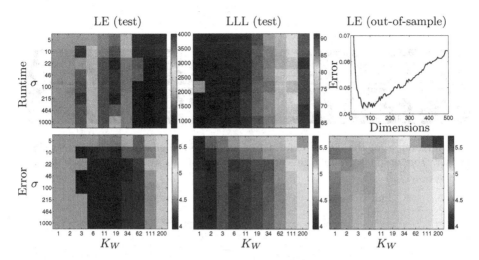

Fig. 5. 1-nearest neighbor classification error of MNIST digits after applying LLL, for different values of σ and K_W. See the main text for details. Gray areas corresponds to Matlab's `eigs` routine not converging. *Top two plots*: runtime for Exact LE and LLL. The color corresponds to the runtime in seconds. *Bottom three plots*: error of Exact LE, LLL and out-of-sample set, respectively. The color corresponds to percent classification error. The out-of-sample runtime is constant and equal to 30 seconds for all values of K_W and σ. *Top right plot*: 1-nearest neighbor classification error for different dimensions for the test subset with $\sigma = 10$ and $K_W = 1$.

the choice of landmark points, so we can save 30 seconds' runtime for each point by precomputing that matrix and using it for all variations of the parameters. Given the location of the embedding points $\widetilde{\mathbf{X}}$, we also computed the out-of-sample projection matrix $\mathbf{Z}_{OOS}$ to find the embedding of the out-of-sample set as well. We computed the 1-nearest neighbor classification for different number of dimensions separately and reported the smallest error, for both the test and the out-of-sample sets. Fig. 5 shows the results. The smallest error is achieved for very small values of K_W. There is also little discrepancy between the error for test and out-of-sample sets, which indicate our out-of-sample mapping is accurate. The top right corner shows the variation of the error as we change the dimensionality. The results are shown for $\sigma = 10$ and $K_W = 1$, but the curve is very similar for other sets of parameters as well. Note the runtime of LLL is less than two minutes for the embedding of as many as 60 000 MNIST points.

We also tried to repeat the same experiment for Exact LE to compare the results with LLL, but we found many complications. First of all, it turns out that for small values of K_W the graph Laplacian is not connected and Matlab's `eigs` routine does not converge (at least not all 500 requested eigenvalues). For larger values of K_W `eigs` converges, but takes many iterations, which increase the runtime dramatically to almost 4 000 seconds. Note it is exactly for those values that both Exact LE and LLL give the smallest error (in fact the smallest

Fig. 6. Example of the elastic deformation of MNIST digits in the infinite MNIST dataset. *Top*: original. *Bottom*: one of the 16 deformations we applied to each digit.

Full embedding   Only landmarks

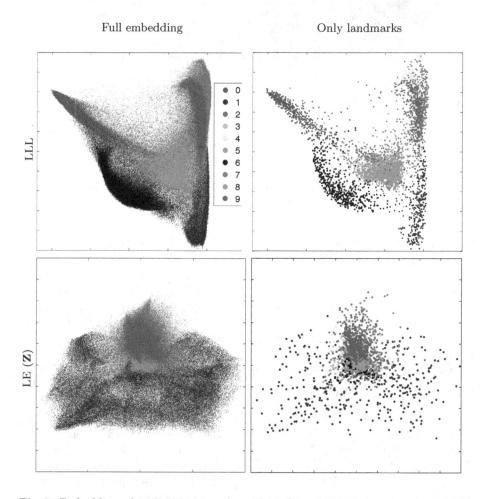

Fig. 7. Embedding of 1 020 000 points from the infinite MNIST dataset using 10 000 landmarks. *Top*: LLL, *bottom*: LE ($\mathbf{Z}$). *Left*: full dataset $\mathbf{X}$, *right*: landmarks $\tilde{\mathbf{X}}$ only.

error for LLL is achieved for $K_W = 1$, for which Exact LE did not even converge). Increasing K_W improves the connectivity of the graph Laplacian, but the runtime of eigs did not decrease much below 1 000 seconds, which means LLL is 15–40× faster depending on the particular set of parameters. Finally, the general pattern of variation and values of the error is almost the same for Exact LE, LLL and the out-of-sample set. The error gradually increases from the lower left corner to the upper right in all three cases.

Large-Scale Experiment. We used the infinite MNIST dataset [31], where we generated 1 020 000 handwritten digits using elastic deformations of the original MNIST dataset (see examples of the deformations in Fig. 6). We reduce the dimensionality to two with 10 000 randomly selected landmarks and $K_Z = 5$ nearest landmarks. LLL took 4.2 minutes to compute the projection matrix $\mathbf{Z}$ and 14 minutes to compute the embedding. We also run LE ($\mathbf{Z}$) on the same 10 000 landmarks. Fig. 7 shows the resulting embeddings. In the embedding of LLL, zeros, sixes and ones are separated from the rest of the digits, and nines, fours and sevens form their own group (all those digits contain in them a straight vertical line). The embedding for For LE ($\mathbf{Z}$) shows far ess structure. Only ones and a group containing sevens and nines can be separated. The rest of the points are trapped in the center of the figure.

6 Conclusion

Spectral methods for manifold learning and clustering often give good solutions to problems involving nonlinear manifolds or complex clusters, and are in widespread use. However, scaling them up to large datasets (large N) and non-trivial numbers of eigenvectors (d) requires approximations. The Locally Linear Landmarks (LLL) method proposes a reduced formulation of the original spectral problem that optimizes only over a small set of landmarks, while retaining structure of the whole data. The algorithm is well defined theoretically and has better performance than the Nyström method, allowing users to scale up applications to larger dataset sizes. LLL also defines a natural out-of-sample extension that is cheaper and better than the Nyström method. This paper has focused on the case of Laplacian eigenmaps, where LLL was able to achieve 10×–20× speedups with small approximation error.

The basic framework of LLL, where we replace the low-dimensional projections by a fixed linear function of only a few of the projections, is applicable to any spectral method. However, the best choice of the linear function is an interesting topic of future research. In particular for spectral clustering, the input data need not have manifold structure, but the cluster label of a point may be well approximated by a function of some of its neighboring landmarks.

Acknowledgment. Work funded in part by NSF CAREER award IIS-0754089.

References

1. Saul, L.K., Weinberger, K.Q., Ham, J.H., Sha, F., Lee, D.D.: Spectral methods for dimensionality reduction. In: Chapelle, O., Schölkopf, B., Zien, A. (eds.) Semi-Supervised Learning. Adaptive Computation and Machine Learning Series, pp. 293–308. MIT Press (2006)
2. Belkin, M., Niyogi, P.: Laplacian eigenmaps for dimensionality reduction and data representation. Neural Computation 15(6), 1373–1396 (2003)
3. Schölkopf, B., Smola, A., Müller, K.R.: Nonlinear component analysis as a kernel eigenvalue problem. Neural Computation 10(5), 1299–1319 (1998)
4. Cox, T.F., Cox, M.A.A.: Multidimensional Scaling. Chapman & Hall, London (1994)
5. Tenenbaum, J.B., de Silva, V., Langford, J.C.: A global geometric framework for nonlinear dimensionality reduction. Science 290(5500), 2319–2323 (2000)
6. Saul, L.K., Roweis, S.T.: Think globally, fit locally: Unsupervised learning of low dimensional manifolds. J. Machine Learning Research 4, 119–155 (2003)
7. Shi, J., Malik, J.: Normalized cuts and image segmentation. IEEE Trans. Pattern Analysis and Machine Intelligence 22(8), 888–905 (2000)
8. Lehoucq, R.B., Sorensen, D.C.: Deflation techniques for an implicitly restarted Arnoldi iteration. SIAM J. Matrix Anal. and Apps. 17(4), 789–821 (1996)
9. Williams, C.K.I., Seeger, M.: Using the Nyström method to speed up kernel machines. In: Leen, T.K., Dietterich, T.G., Tresp, V. (eds.) Advances in Neural Information Processing Systems (NIPS), vol. 13, pp. 682–688. MIT Press, Cambridge (2001)
10. Bengio, Y., Paiement, J.F., Vincent, P., Delalleau, O., Le Roux, N., Ouimet, M.: Out-of-sample extensions for LLE, Isomap, MDS, Eigenmaps, and spectral clustering. In: Thrun, S., Saul, L.K., Schölkopf, B. (eds.) Advances in Neural Information Processing Systems (NIPS), vol. 16. MIT Press, Cambridge (2004)
11. Drineas, P., Mahoney, M.W.: On the Nyström method for approximating a Gram matrix for improved kernel-based learning. J. Machine Learning Research 6, 2153–2175 (2005)
12. Talwalkar, A., Kumar, S., Rowley, H.: Large-scale manifold learning. In: Proc. of the 2008 IEEE Computer Society Conf. Computer Vision and Pattern Recognition (CVPR 2008), Anchorage, AK, June 23-28 (2008)
13. von Luxburg, U., Radl, A., Hein, M.: Getting lost in space: Large sample analysis of the resistance distance. In: Lafferty, J., Williams, C.K.I., Shawe-Taylor, J., Zemel, R., Culotta, A. (eds.) Advances in Neural Information Processing Systems (NIPS), vol. 23, pp. 2622–2630. MIT Press, Cambridge (2010)
14. de Silva, V., Tenenbaum, J.B.: Sparse multidimensional scaling using landmark points (June 30, 2004)
15. de Silva, V., Tenenbaum, J.B.: Global versus local approaches to nonlinear dimensionality reduction. In: Becker, S., Thrun, S., Obermayer, K. (eds.) Advances in Neural Information Processing Systems (NIPS), vol. 15, pp. 721–728. MIT Press, Cambridge (2003)
16. Platt, J.: FastMap, MetricMap, and landmark MDS are all Nyström algorithms. In: Cowell, R.G., Ghahramani, Z. (eds.) Proc. of the 10th Int. Workshop on Artificial Intelligence and Statistics (AISTATS 2005), Barbados, January 6-8, pp. 261–268 (2005)

17. Gao, S., Tsang, I.W.H., Chia, L.T., Zhao, P.: Local features are not lonely — Laplacian sparse coding for image classification. In: Proc. of the 2010 IEEE Computer Society Conf. Computer Vision and Pattern Recognition (CVPR 2010), San Francisco, CA, June 13-18, pp. 3555–3561 (2010)

18. Wang, F.Y., Chi, C.Y., Chan, T.H., Wang, Y.: Nonnegative least-correlated component analysis for separation of dependent sources by volume maximization. IEEE Trans. Pattern Analysis and Machine Intelligence 32(5), 875–888 (2010)

19. Roweis, S.T., Saul, L.K.: Nonlinear dimensionality reduction by locally linear embedding. Science 290(5500), 2323–2326 (2000)

20. Weinberger, K., Packer, B., Saul, L.: Nonlinear dimensionality reduction by semidefinite programming and kernel matrix factorization. In: Cowell, R.G., Ghahramani, Z. (eds.) Proc. of the 10th Int. Workshop on Artificial Intelligence and Statistics (AISTATS 2005), Barbados, January 6-8, pp. 381–388 (2005)

21. Yu, K., Zhang, T., Gong, Y.: Nonlinear learning using local coordinate coding. In: Bengio, Y., Schuurmans, D., Lafferty, J., Williams, C.K.I., Culotta, A. (eds.) Advances in Neural Information Processing Systems (NIPS), vol. 22. MIT Press, Cambridge (2009)

22. Ladický, Ľ., Torr, P.H.S.: Locally linear support vector machines. In: Getoor, L., Scheffer, T. (eds.) Proc. of the 28th Int. Conf. Machine Learning (ICML 2011), Bellevue, WA, June 28-July 2, pp. 985–992 (2011)

23. Liu, W., He, J., Chang, S.F.: Large graph construction for scalable semi-supervised learning. In: Fürnkranz, J., Joachims, T. (eds.) Proc. of the 27th Int. Conf. Machine Learning (ICML 2010), Haifa, Israel, June 21-25 (2010)

24. Zhang, Z., Wang, J.: MLLE: Modified locally linear embedding using multiple weights. In: Schölkopf, B., Platt, J., Hofmann, T. (eds.) Advances in Neural Information Processing Systems (NIPS), vol. 19, pp. 1593–1600. MIT Press, Cambridge (2007)

25. Chen, X., Cai, D.: Large scale spectral clustering with landmark-based representation. In: Proc. of the 25th National Conference on Artificial Intelligence (AAAI 2011), San Francisco, CA, August 7-11, pp. 313–318 (2011)

26. Bengio, Y., Delalleau, O., Le Roux, N., Paiement, J.F., Vincent, P., Ouimet, M.: Learning eigenfunctions links spectral embedding and kernel PCA. Neural Computation 16(10), 2197–2219 (2004)

27. Carreira-Perpiñán, M.Á., Lu, Z.: The Laplacian Eigenmaps Latent Variable Model. In: Meilă, M., Shen, X. (eds.) Proc. of the 11th Int. Workshop on Artificial Intelligence and Statistics (AISTATS 2007), San Juan, Puerto Rico, March 21-24, pp. 59–66 (2007)

28. Kumar, S., Mohri, M., Talwalkar, A.: Sampling methods for the Nyström method. J. Machine Learning Research (2012)

29. Carreira-Perpiñán, M.Á., Wang, W.: The K-Modes algorithm for clustering. arXiv:1304.6478 (April 23, 2013) (unpublished manuscript)

30. Vladymyrov, M., Carreira-Perpiñán, M.Á.: Entropic affinities: Properties and efficient numerical computation. In: Proc. of the 30th Int. Conf. Machine Learning (ICML 2013), Atlanta, GA, June 16-21, pp. 477–485 (2013)

31. Loosli, G., Canu, S., Bottou, L.: Training invariant support vector machines using selective sampling. In: Bottou, L., Chapelle, O., DeCoste, D., Weston, J. (eds.) Large Scale Kernel Machines. Neural Information Processing Series, pp. 301–320. MIT Press (2007)

Discovering Skylines of Subgroup Sets

Matthijs van Leeuwen[1] and Antti Ukkonen[2]

[1] Department of Computer Science, KU Leuven, Belgium
[2] Helsinki Institute for Information Technology HIIT, Aalto University, Finland
matthijs.vanleeuwen@cs.kuleuven.be, antti.ukkonen@aalto.fi

Abstract. Many tasks in exploratory data mining aim to discover the top-k results with respect to a certain interestingness measure. Unfortunately, in practice top-k solution sets are hardly satisfactory, if only because redundancy in such results is a severe problem. To address this, a recent trend is to find *diverse sets of high-quality patterns*. However, a 'perfect' diverse top-k cannot possibly exist, since there is an inherent trade-off between quality and diversity.

We argue that the best way to deal with the quality-diversity trade-off is to *explicitly consider the Pareto front, or skyline, of non-dominated solutions*, i.e. those solutions for which neither quality nor diversity can be improved without degrading the other quantity. In particular, we focus on k-pattern set mining in the context of Subgroup Discovery [6]. For this setting, we present two algorithms for the discovery of skylines; an exact algorithm and a levelwise heuristic.

We evaluate the performance of the two proposed skyline algorithms, and the accuracy of the levelwise method. Furthermore, we show that the skylines can be used for the objective evaluation of subgroup set heuristics. Finally, we show characteristics of the obtained skylines, which reveal that different quality-diversity trade-offs result in clearly different subgroup sets. Hence, the discovery of skylines is an important step towards a better understanding of 'diverse top-k's'.

1 Introduction

"Find me the k highest scoring solutions to this problem." This phrase describes the goal of many common tasks in the fields of both exploratory data mining and information retrieval. Consider for example document retrieval, where the task is to return the top-k documents that are most relevant for a given query. Also in pattern mining, it is quite common to ask for the top-k patterns with respect to a given interestingness or quality measure. As a result, many efficient algorithms for finding top-k's have been proposed in the literature.

Unfortunately, in practice top-k solution sets are hardly satisfactory to the user, which is due to two reasons. First, it is hard to formalise interestingness and therefore it is unlikely that a used quality measure completely matches perceived interestingness. Second, the top-k results are often very redundant, which can have different causes. Focusing on pattern mining, the main cause lies in the use of expressive pattern languages in which many different patterns

H. Blockeel et al. (Eds.): ECML PKDD 2013, Part III, LNAI 8190, pp. 272–287, 2013.

describe the same structure in the data. Consequently, there are many patterns with almost the same interestingness or quality. When ranking the complete set of patterns according to interestingness and taking the top-k, this results in a clearly redundant result set: the top-k contains many variations of the same theme, while many potentially interesting patterns fall outside the top-k and are thus completely ignored.

Acknowledging this problem, a trend in recent years has been to move away from mining *individual patterns*, and towards mining *pattern sets* [1]. The main idea of pattern set mining is that one should mine *a diverse set of high-quality patterns*, where quality and diversity depend on the specific task. In other words, pattern set mining aims at finding a diverse top-k rather than the top-k. Note that result diversification is also very common in e.g. document retrieval.

An important observation is that a single, 'perfect' diverse top-k cannot possibly exist: whenever we replace elements from the top-k with other elements to improve diversity, it is no longer the top-k with regard to quality. This is inherent to the problem, and can be compared to the risk-return trade-off that forms the core of modern portfolio theory [11]. Consequently, many instances of pattern set mining have a trade-off between quality and diversity, even if this is often obfuscated by parameters that need tuning. However, this also implies that there exists a *Pareto front*, or *skyline*, of *non-dominated pattern sets*. In other words, there must be a set of pattern sets such that no other pattern sets exist that have both higher quality and diversity.

Subgroup Discovery. Although most contributions can be generalised to other instances of pattern set mining, in the remainder of this paper we focus on Subgroup Discovery (SD) [6]. SD is an instance of pattern mining that is concerned with finding regions in the data that stand out with respect to a particular target variable. As an example, consider a dataset from the medical domain with patient information, in which we treat hemoglobin concentration as the target. By performing subgroup discovery, we could identify patterns such as $sex = male \rightarrow high$, implying that men tend to have a higher hemoglobin

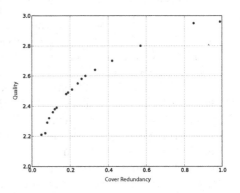

Fig. 1. The effect of varying the diversity parameter of cover-based subgroup selection [10] on quality and cover redundancy ('inverse diversity') (*Car*)

concentration than the overall population. Compared to other pattern mining tasks, a notable advantage is that it is pattern type agnostic and can thus deal with almost any type of data.

We recently proposed several heuristic methods for selecting *diverse top-k subgroup sets* from large sets of candidate subgroups [10]. Diversity and quality can be balanced by the user by setting several parameters. To demonstrate this effect and the resulting trade-off between quality and diversity, we conducted a series of selection experiments in which we varied the 'trade-off' parameter. The qualities and diversities of the resulting subgroup sets are plotted in Figure 1. A higher quality implies that individual subgroups have a higher quality on average, and a lower cover redundancy implies that the subgroups cover more diverse parts of the data. The figure clearly shows the trade-off: higher quality comes at the cost of less diversity and vice versa.

Approach and Contributions. We argue that the best way to deal with the quality-diversity trade-off is by *explicitly considering the Pareto front of optimal solutions*, i.e. those solutions for which neither quality nor diversity can be improved without degrading the other quantity. For this we have the following three arguments. First, maximising only quality or diversity clearly does not give satisfactory results, but neither is it possible to determine in advance what trade-off is desirable for a user. Second, existing heuristic methods yield a single subgroup set, but one cannot possibly know if and where it resides on the Pareto front. Neither theoretical nor empirical arguments have addressed whether the discovered subgroup sets are on or even near the Pareto front. Third and last, the trade-off is usually implicit and hidden by parameters that are hard to tune. This tuning is dataset-dependent and small parameter changes can have a big impact on the resulting trade-off. Hence, only by explicitly considering subgroup set skylines, we can learn more about its characteristics and properties, and find out what subgroup sets are preferred by users.

The approach and contributions of this paper can be summarised as follows:

1. We introduce the concept of the quality-diversity skyline for k-pattern sets, and argue that it is important to investigate its shape. Such analysis helps in making more principled choices with regard to the quality-diversity trade-off, e.g. interactively or by generalising frequent behaviour, but may also result in insights that lead to novel heuristics for pattern set mining.
2. For developing our theory and methodology, we focus on instances of the k-pattern set mining problem for which quality of the pattern set is the *sum of the individual qualities*, and diversity is quantified using *joint entropy*.
3. For this setting, we present two algorithms that compute the skyline given a set of candidate patterns.
   (a) The first is a branch-and-bound algorithm that computes the exact Pareto front. Properties of joint entropy are used to prune the search space, which results in considerable speedups and makes skyline computation feasible for modestly sized candidate sets.
   (b) The second is a heuristic that takes a levelwise approach; the Pareto front of size i is approximated by augmenting the Pareto front of subgroup sets of size $i - 1$. Although this method is a heuristic, it approximates the results of the exact method very well and is much faster.
4. We perform experiments to investigate how the methods perform in the context of subgroup sets. We study runtimes of the two skyline algorithms, and investigate how accurate the approximations of the heuristic method are. In addition, we show that the skylines can be used for the objective evaluation of

subgroup set heuristics. Finally, we show characteristics of the obtained skylines, which reveals that subgroup sets with very different quality-diversity trade-offs exist. Hence, explicitly computing and considering skylines is an important step towards a better understanding of 'diverse top-k's'.

After discussing related work in Section 2, we introduce notation and preliminaries in Section 3. Section 4 introduces Pareto optimal subgroup sets, after which Sections 5 and 6 introduce the skyline discovery algorithms. Section 7 presents the experiments and we round up with conclusions in Section 8.

2   Related Work

Subgroup Discovery has been around since the late 90s [6], and is closely related to Contrast Set Mining and Emerging Pattern Mining. Subgroup Discovery is probably the most generic of these, as the general task makes very few assumptions about the data.

Diverse subgroup set discovery, introduced in [10], resembles supervised pattern set selection methods [12,2,1], but quantifying quality separately from diversity makes it substantially different. This decoupling has clear advantages, as the two can be independently varied and it becomes more apparent how each of the two perform. Supervised pattern set mining methods often aim to achieve good classification, which does not match the *exploratory* goal of Subgroup Discovery.

Entropy was used to quantify diversity of k-sized pattern sets by Knobbe and Ho [8], but our overall approach and aims are rather different. Knobbe and Ho's *maximally informative k-itemsets* maximize entropy and are therefore maximally diverse, but no other interestingness or quality measure is taken into account. Skylines consisting of individual patterns were previously studied in [13].

Our problem can also be regarded as an instance of a bicriteria or multiobjective combinatorial optimisation (MOCO) problem. Quite some approaches to MOCO problems have been proposed, both exact [4] and heuristic [5]. The main property that distinguishes our problem is that given a point in decision space, it is very hard to determine whether a feasible solution in criteria space exists, let alone to compute this solution. This is due to the large size and complex structure of the criteria space, i.e. the set of all possible pattern sets. Consequently, searching through criteria space is required to find a skyline, and this cannot be accomplished with standard MOCO approaches.

3   Preliminaries

We assume that the tuples to be analysed are described by k (≥ 1) *description attributes* A and a target attribute Y. All attributes A_i (and Y) have a domain of possible values $\text{Dom}(A_i)$ (resp. $\text{Dom}(Y)$). A dataset $\mathcal{D}$ is a bag of tuples t over the complete set of attributes $\{A_1, \ldots, A_k, Y\}$. The central concept is the *subgroup*, which consists of a *description* and a corresponding *cover*. A *subgroup cover* is a bag of tuples $G \subseteq \mathcal{D}$ and $|G|$ denotes its size, also called *subgroup size* or *coverage*. A *subgroup description* is a formula s, consisting of a conjunction of conditions on the description attributes, and its corresponding subgroup cover is the set of tuples that satisfy the formula, i.e. $G_s = \{t \in \mathcal{D} \mid t \vDash s\}$.

A *subgroup quality measure* is a function $\varphi : 2^Y \mapsto \mathbb{R}$ that assigns a numeric value to any subgroup based on its target values $\pi_Y(G)$. The traditional Subgroup Discovery (SD) task is to find the top-k ranking subgroups according to φ. Depending on the data and task, either exhaustive search or beam search can be used. Several parameters influence the search, e.g. a minimum coverage threshold requires subgroup covers to consist of at least *mincov* tuples, and the maximum depth parameter *maxdepth* imposes a maximum on the pattern length.

k**-subgroup sets** A k-*subgroup set* S is a set consisting of exactly k subgroups. Subgroup set quality, denoted by $q(S)$, is simply the sum of the individual qualities of the k subgroups, i.e. $q(S) = \sum_{G \in S} \varphi(G)$. Any suitable quality measure can be used for $\varphi(G)$; in this paper we use Weighted Relative Accuracy [6], probably the best-known quality measure for subgroups.

To quantify diversity among the subgroup covers of a subgroup set, we use *joint entropy*. Joint entropy, denoted by H, is obtained by computing the entropy over the binary features defined by the subgroups in the set.

Definition 1 (Joint Entropy). *Given a k-subgroup set $S = \{G_1, \ldots, G_k\}$, and let $B = (b_1, \ldots, b_k) \in \{0, 1\}^k$ be a tuple of binary values. Let $p(s_{G_1} = b_1, \ldots, s_{G_k} = b_k)$ denote the fraction of tuples $t \in \mathcal{D}$ such that $s_{G_1}(t) = b_1 \wedge \ldots \wedge s_{G_k}(t) = b_k$. The* joint entropy *of S is defined as:*

$$H(S) = -\sum_{B \in \{0,1\}^k} p(s_{G_1} = b_1, \ldots, s_{G_k} = b_k) \log_2 p(s_{G_1} = b_1, \ldots, s_{G_k} = b_k).$$

H is measured in bits, and each subgroup contributes at most 1 bit of information, so that $H(S) \leq |S|$. A higher entropy indicates higher diversity. Note that we decided not to use Cover redundancy [10], because joint entropy is more widely known and used, and can be used to prune the search space.

4   Skylines of Pareto Optimal k-Subgroup Sets

In this section we formally state the problem that we address in this paper, i.e. that of finding skylines of Pareto optimal k-subgroup sets. Before discussing the problem in more detail, we define the notion of *dominance*.

Definition 2. *Given points x and y from some set Ω, and a set of functions $\mathcal{F} = \{f_1, \ldots, f_n\}$, $f_i : \Omega \to \mathbb{R}$ for all $f_i \in \mathcal{F}$, we say that point x dominates point y in terms of $\mathcal{F}$, denoted $x \succ_{\mathcal{F}} y$, if and only if $f_i(x) \geq f_i(y)$ for all $f_i \in \mathcal{F}$ and there exists at least one function $f_i \in \mathcal{F}$ for which $f_i(x) > f_i(y)$.*

The Pareto front, or skyline, of a set of points are those points that are not being dominated by any other point. Finding the Pareto front is in general not hard. For a given set of n points it can be computed in time $O(n^2)$ by a brute-force method, and in time $O(n \log n)$ by using an improved algorithm [9]. However, in our case n is very large, because the set of points we want to compute the Pareto front of is in fact the *power set* of the set C of candidate subgroups, denoted 2^C. That is, we are addressing the following problem:

Problem 1 (k-subset skyline). Given a discrete set C, an integer k, a set of functions $\mathcal{F} = \{f_1, \ldots, f_n\}$, with $f_i : 2^C \to \mathbb{R}$ for all $f_i \in \mathcal{F}$, find the set

$$P_k = \{S \in 2^C : |S| = k \text{ and } \not\exists S' \in 2^C \text{ st. } S' \succ_{\mathcal{F}} S\}.$$

Note that we only consider subsets of C of a fixed size k. The brute-force approach to Problem 1 simply materialises all k-sized subsets of C, and then runs a standard Pareto front algorithm on this. Clearly this is not going to work unless C is very small.

In our k-subgroup set application the set $\mathcal{F}$ contains two functions, one for subgroup set quality, and another for diversity. For the rest of the paper we let $\mathcal{F} = \{q, d\}$, where q is a quality measure of the set S, defined as $q(S) = \sum_{G \in S} \varphi(G)$, while d is joint entropy as given in Definition 1. Although in principle d could be any diversity measure, the exact method presented in the next section exploits properties of joint entropy for pruning the search space.

Problem 2 (k-subgroup set skyline). Given a set of subgroups C, an integer k, the set of functions $\mathcal{F} = \{q, d\}$, with $q(S) = \sum_{G \in S} \varphi(G)$ and $d(S) = H(S)$, find the k-subset skyline (as defined by Problem 1).

5   Exact Algorithm

In this section we present an exact algorithm for solving Problem 2 that combines an efficient subset enumeration scheme with results from [7] to prune a substantial part of the search space.

As a first step in designing the exact algorithm, suppose that we have a list L with all k-sized subsets of C sorted in non-increasing order of $q(S)$. Clearly the first element of L must belong to the skyline as it has the highest quality of all subsets. Assign its diversity to $d_{\max}$. A simple algorithm to construct the skyline is to scan over L until we find a subset with diversity larger than $d_{\max}$. This subset is added to the skyline, we set $d_{\max}$ equal to the diversity of this subset, and continue scanning L. When the algorithm reaches the end of L we have found the exact skyline.[1]

Our exact algorithm works in the same way, but it does this *without materialising the complete list* L. Instead, we enumerate the subsets in decreasing order of quality in an online fashion with *polynomial delay*, meaning that the computation required to obtain the next subset in the sequence is polynomial in

---

[1] To be precise, this is only true in the absence of ties in $q(S)$. If some subsets all have the same quality, and ties were broken at random when sorting the subsets, this algorithm may include some subsets in the skyline that are dominated. However, removing these is a simple post-processing step.

Algorithm 1. Outline of an algorithm that enumerates all $S \subset C$ with $|S| = k$ in non-increasing order of $q(S)$

1. $Q \leftarrow$ empty priority queue, insert $X = \{1, 2, \ldots k\}$ into Q with priority $q(C(X))$.
2. While Q is not empty:
   (a) Pop the highest priority index set X from Q, and output $C(X)$.
   (b) Insert every $X' \in N(X)$ into Q with $q(C(S'))$ as the priority *unless* X' has already been inserted into Q.

$|C|$ and k. In Subsections 5.1 and 5.2 we describe how to make the enumeration efficient and how to prune the search space, we now continue with the main idea.

Without loss of generality, assume that the candidate subgroups $G \in C$ are sorted in non-increasing order of $\varphi(G)$, and let $C(i)$ denote the subgroup at position i. That is, we have $\varphi(C(i)) \geq \varphi(C(j))$ for every $i < j$. Let $X \subset \{1, \ldots, |C|\}$, $|X| = k$, denote a set of indices to C that induce the k-subgroup set $C(X)$. To simplify notation, we sometimes write $f(X)$ in place of $f(C(X))$ for $f \in \{q, d\}$.

Definition 3. *The i-neighbour of X, denoted $n_i(X)$, is a copy of X with the index at position i incremented by one. More formally:*

$$n_i(X) = \begin{cases} \{X_1, \ldots, X_i + 1, \ldots X_k\} & \text{if } i < k \text{ and } X_i + 1 < X_{i+1}, \\ \{X_1, \ldots, X_k + 1\} & \text{if } i = k \text{ and } X_k + 1 \leq |C|, \\ \emptyset & \text{otherwise.} \end{cases}$$

The neighbourhood of X, denoted $N(X)$, is the set $\{n_i(X) \mid i = 1, \ldots, k\}$. Finally, the set X is a parent of the set X' whenever $X' \in N(X)$.

For example, if $X = \{1, 3, 5\}$, its neighbourhood contains the sets $\{2, 3, 5\}$, $\{1, 4, 5\}$, and $\{1, 3, 6\}$, while for $X = \{1, 2, 3\}$ we have $N(X) = \{\{1, 2, 4\}\}$[2]. Observe that the neighbourhood $N(X)$ of a set X *only contains sets having at most the same quality as X,* i.e. we have $q(X) \geq q(X')$ for every $X' \in N(X)$.

Using this, we can generate the list L on the fly by following the procedure shown in Algorithm 1. The algorithm starts from $\{1, \ldots, k\}$ and maintains a priority queue of subsets with $q(S)$ as the priority value.

Proposition 1. *Algorithm 1 is both complete and correct in the sense that it enumerates all k-sized subsets in non-increasing order of $q(S)$.*

Proof. Correctness: We show that a subset output by the algorithm can not have a *larger* quality than any subset that was output before. By the mechanics of the algorithm, every index set X' in priority queue Q must belong to the neighbourhood $N(X)$ of some $C(X)$ that was already output. And by definition, all $X' \in N(X)$ have quality at most $q(X)$. Hence Q can only contain subsets with qualities that are upper bounded by qualities of the already generated subsets. This implies that the subsets are output in non-increasing order of $q(S)$.

---

[2] The empty sets in $N(X)$ are not considered.

Completeness: The algorithm outputs every k-sized subset of C. Clearly every subset that enters Q is eventually output. Algorithm 1 fails to output a subset S if and only if none of its parents is output, because whenever a parent of S is output, S is put into Q. Consider one possible chain of parents from any index set X until we reach the set $\{1, \ldots, k\}$. Because Q is initialised with $\{1, \ldots, k\}$, this chain must exist, meaning that every subset in the chain has at least one parent in Q. This implies that X has at least one parent in Q, and thus $S = C(X)$ is found.   □

Notice that the amount of computation needed between two subsets is $O(k|C|)$. Size of the priority queue Q is trivially upper bounded by $2^{|C|}$, which means that insertions and extractions to Q are linear in the size of C using e.g. a binomial heap to implement Q. The size of $N(X)$ is upper bounded by k, meaning we need at most k insertions and one extraction.

5.1 Efficient Subset Enumeration

A problem with the enumeration scheme given in Algorithm 1 is that it generates the entire neighbourhood $N(X)$ for every X, and these neighbourhoods are partly overlapping. Consider for example the set $\{1, 3, 5\}$, which is both the 2-neighbour of $\{1, 2, 5\}$, and the 3-neighbour of the set $\{1, 3, 4\}$. Algorithm 1 would thus generate the set $\{1, 3, 5\}$ twice. To avoid this, it must keep track of every set that was inserted into Q at some point. This is clearly undesirable, as the amount of space needed is $O(2^{|C|})$. We thus need an algorithm that generates every subset of size k once and only once without additional bookkeeping.

Enumerating subsets once and only once is of course a known problem. However, our situation has the additional challenge that we want to generate the subsets in decreasing order of $q(S)$ with polynomial delay (preferably $O(k|C|)$). An approach that combines the priority queue with a simple subset enumeration scheme, e.g. depth-first traversal, does not satisfy this property. Instead, we use a modification to Algorithm 1 that maintains its computational properties, but avoids duplicates. The idea is to insert only *a subset of $N(X)$* to Q on every iteration. This subset can be chosen so that Proposition 1 still holds. To this end, we arrange the search space of all k-sized subsets in a directed graph T.

Definition 4. *Given the set C and the integer k, the directed graph T has the node (X, j) for every $X \subset \{1, \ldots, |C|\}$, $|X| = k$. Here $j \in \{1, \ldots, k\}$ is the position associated with index set X in the given node. Node (X, j) has neighbours*

$$\{(n_i(X), i) \mid i = 1, \ldots, j\}. \tag{1}$$

That is, the neighbours of (X, j) in T are those i-neighbours of X where the modifications take place in the first j positions of X.

Proposition 2. *The graph T is a tree rooted at $(\{1, \ldots, k\}, k)$.*

Proof. First, observe that $(\{1, \ldots, k\}, k)$ can have no incoming edges, because by Definitions 3 and 4 such an edge should come from a node with a value smaller

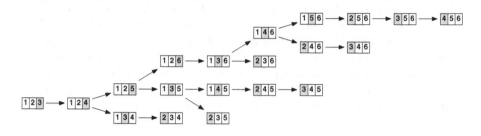

Fig. 2. Example of the subset graph T for 3-sized subsets of the set $\{1,2,\ldots,6\}$. The *position* of a node (see Definition 4) is indicated by a light grey background.

than k at position k. Clearly such a node cannot exist, because k is the smallest value that a node may have at position k. Second, every node (X,i) except the root has one, and only one incoming edge. Suppose this edge comes from the node (X',j), with $j \geq i$. By Definition 3, the i-neighbour of a set X' (if it exists) is identical to X' except at position i which is incremented by one. This means that the set X of the node (X,i) must have $X_i = X'_i + 1$ and $X_h = X'_h$ for every $h \neq i$. Clearly there is only one $X' \subset \{1,\ldots,|C|\}$ that satisfies this, and since by definition of T no two nodes of T may contain the same subset, there can be only one node with an outgoing edge to (X,i).   □

An example of such a tree for all 3-sized subsets of $\{1,\ldots,6\}$ is shown in Figure 2. Any algorithm that traverses T will generate every k-sized subset once and only once. We can thus enumerate the subsets in a way that avoids duplicates by making Algorithm 1 traverse the tree T. This is accomplished simply by keeping nodes of T in the priority queue Q, and only inserting those nodes to Q that are neighbours of (X,j) in T (which can be determined without materialising T).

5.2   Entropy-Based Pruning of the Search Space

Algorithm 1 scans over a sorted list L of all possible subsets, but in practice L is too long already with small C and k. Next we discuss how to use a property of joint entropy and the tree T to *prune* parts of list L. Recall that at every stage of the algorithm we know that $d_{\max}$ is the largest diversity observed so far. If we can show that no subset below a given node of T can have diversity larger than $d_{\max}$, we can skip the entire subtree. From Figure 2 we can make the following observations:

Observation 1: For any (X,j) of T, the suffix starting at position $(j+1)$, denoted R^X_{j+1}, is the same in every node of the subtree rooted at (X,j). For example, all nodes below node $(\{1,3,6\},2)$ contain the element 6 at position 3.

Observation 2: For any (X,j) of T with $j < k$, in every node that appears in the subtree rooted at (X,j), the elements at positions $1,\ldots j$ all have a value less than X_{j+1}. For example, all nodes below $(\{1,3,5\},2)$ only contain elements that are less than 5 at positions 1 and 2.

Algorithm 2. EXACT(C, k)

$X \leftarrow \{1, 2, \ldots, k\}$, $Q \leftarrow$ empty priority queue, $d_{\max} \leftarrow d(X)$, $P_k \leftarrow \emptyset$
push (X, k) into Q with priority $q(X)$
while Q is not empty **do**
    $(X, j) \leftarrow$ highest priority item from Q
    if $d(X) \geq d_{\max}$ **then**
        $P_k \leftarrow P_k \cup \{C(X)\}$
        $d_{\max} \leftarrow d$
    for every $(X', i) \in \{(n_i(X), i) \mid i = 1, \ldots, j\}$ **do**
        if $\tilde{d}(X') \geq d_{\max}$ **then**
            insert (X', i) into Q with priority $q(X')$
return P_k

These observations are general properties of the tree T, and can be shown to follow from Definitions 3 and 4. Together with the following proposition we can use these to prune entire subtrees of T.

Proposition 3. *(Prop. 4 of [7]): Let $S \subset C$ denote a subgroup set, and let H as in Def. 1. Suppose that $\{B_1, \ldots, B_m\}$ is a partition of S. Then $H(S) \leq \sum_{i=1}^{m} H(B_i)$.*

For any (X, j) in T, let $\hat{X}$ denote any index set that appears in the subtree rooted at (X, j). Given (X, j), we must find an upper bound for $d(\hat{X}) = H(\hat{X})$. From Obs. 1 we know that the suffix R_{j+1}^X is the same in X and $\hat{X}$. We have thus $H(R_{j+1}^X) = H(R_{j+1}^{\hat{X}})$ for every $\hat{X}$. Obs. 2 tells us that only certain elements may occur in the first j positions of $\hat{X}$. We can compute the singleton entropies $d(G) = H(G)$ for every subgroup $G \in C$. For $j < k$, let $Z(X, j)$ denote the set of j highest entropy subgroups in the first $X_{j+1} - 1$ positions of C. Now, by construction and Prop. 3, the sum $\sum_{G \in Z(X, j)} H(G)$ must be an upper bound for the entropy of the first j positions of any $\hat{X}$. This means that for all $\hat{X}$ we have

$$d(\hat{X}) \leq d(R_{j+1}^X) + \sum_{G \in Z(X, j)} d(G) = \tilde{d}(X),$$

which gives the desired upper bound. If $\tilde{d}(X) < d_{\max}$, the entire subtree rooted at (X, j) can be pruned.

The algorithm EXACT, shown in Algorithm 2, implements all details discussed in this section. It combines the improved subset enumeration scheme with the pruning results. While this is a substantial improvement over the basic scheme of Alg. 1, the size of the priority queue can still be exponential in $|C|$, as it is proportional to the size of a *cut* of the subset tree T.

6   A Greedy Levelwise Algorithm

The exact algorithm we described above has two drawbacks: 1) it is optimised for the joint entropy diversity function of Definition 1, and 2) it requires exponential

---

Algorithm 3. LEVELWISE(C, k)

$P_2 \leftarrow$ SKYLINE$(C \times C)$, $i \leftarrow 3$
while $i \leq k$ **do**
   $P_i \leftarrow \emptyset$
   for $S \in P_{i-1}$ **do**
      $P_i \leftarrow$ SKYLINE$(P_i \cup \{S \cup c \mid c \in \{C \setminus S\}\})$
   $i \leftarrow i + 1$
return P_k

---

space. Of these point 1) makes the method ill-suited for some applications, while point 2) rules out large candidate sets C and subgroup set sizes k. Because of this we also introduce a greedy heuristic for finding the skyline.

In general the main computational bottleneck is caused by the large number of points that must be considered when computing the skyline. However, the resulting skyline itself is very likely going to be orders of magnitude smaller than $\binom{|C|}{k}$. Our algorithm will exploit this property. Moreover, consider the Pareto fronts for sets of size $i - 1$ and i, denoted P_{i-1} and P_i, respectively. It seems unlikely that a subset of size $i - 1$ that is very far from P_{i-1} in terms of the functions $\mathcal{F}$ would have a superset that belongs to P_i. On the other hand, subsets in P_{i-1} might be more likely to have a superset that belongs to P_i.

We propose an algorithm that constructs an approximate skyline P_k one level at a time starting from P_2. Given the set P_{i-1}, we define P_i as the skyline of the points that are obtained by combining every point in P_{i-1} with every unused candidate in C. More formally, we can define P_i recursively as follows:

$$
P_i = \begin{cases} \text{SKYLINE}\left(\{(S, c) \mid S \in P_{i-1}, c \in \{C \setminus S\}\}\right) & \text{if } i > 2, \\ \text{SKYLINE}\left(\{(c_1, c_2) \in \{C \times C\} \mid c_1 \neq c_2\}\right) & \text{if } i = 2, \end{cases} \tag{2}
$$

Here SKYLINE is any algorithm that computes skylines in two dimensions.

We put these ideas together in the LEVELWISE algorithm shown in Algorithm 3. It first computes P_2 exactly by considering all subgroup pairs in $C \times C$. In subsequent steps the algorithm applies Equation 2 until it reaches P_k. In practice we obtain better performance by not materialising the entire Cartesian product of P_{i-1} and C in one step, but by incrementally "growing" the set P_i.

7 Experiments

In this section we empirically evaluate the proposed approach and methods.

Datasets: Table 1 presents the datasets that we use, which were all taken from the UCI Machine Learning repository[3]. For each dataset we give the number of tuples, the number of discrete resp. numeric attributes, and the domain size y of the target attribute. Target attributes with more than two classes are treated as binary by considering the majority class as target.

---

[3] http://archive.ics.uci.edu/ml/

Candidate Sets: Candidate sets C of subgroups are generated using exhaustive search. As conditions for the discrete attributes, $A_i = c$ and $A_i \neq c$ for all constants $c \in Dom(A_i)$ are considered. For numeric attributes, conditions $A_i > c$ and $A_i < c$ are considered, where the 'split' values are determined by local binning of occurring values into 6 equal-sized bins.

Search parameters are chosen to result in short (and thus simple) subgroup descriptions, substantial subgroup sizes, and reasonably sized candidate sets. We set $maxdepth = 2$ and $mincov = 5\% \times |\mathcal{D}|$ (except for Adult and Mushroom: $10\% \times |\mathcal{D}|$). For those experiments for which a p-value is given, a permutation test [3] that aims to eliminate false discoveries is used to prune the candidate set. For the remaining experiments, all subgroups found are used as candidates.

Table 1. Dataset properties

dataset	$|\mathcal{D}|$	$|A^{\mathrm{disc}}|$	$|A^{\mathrm{num}}|$	y
Adult	48842	8	6	2
Car	1728	6	0	4
Cmc	1473	7	2	3
Credit-A	690	9	6	2
Credit-G	1000	13	7	2
Mushroom	8124	22	0	2
Pima	768	0	8	2
Tictactoe	958	9	0	2

Evaluation: We need measures to compare two skylines, P and P'. Intuitively, P is better than P' if there are more points in P that dominate points in P' than vice versa. We denote the fraction of points in P' that are dominated by at least one point in P by $\#\{P \succ P'\}$. For these points, we also measure by how much a skyline dominates another skyline. This is expressed by the quantity $\Delta_f(P \succ P')$, defined as the median of the set $\{(f(S) - f(S'))/f(S') \mid S \in P, S' \in P', S \succ S'\}$, i.e. the median of the relative differences between dominated sets and the sets that dominate it.

7.1 Exact and Levelwise Skyline Discovery

Table 2 presents the results obtained on all datasets with the EXACT and LEVEL-WISE algorithms. The candidate sets for the first six datasets were pruned using the aforementioned permutation test. Due to long runtimes the exact method was only used with $k = 5$, the levelwise method was also used with $k = 10$. The last two datasets are too large to be used with the exact method, but for the levelwise approach no pruning of the candidate set was needed.

Runtimes greatly vary depending on the dataset and desired subgroup set size. This can be explained by the large variation in the number of 'points' in the search space that need to be explored: for Car with LEVELWISE and $k = 5$ only 4380 subgroup sets are considered, but for Credit-A with EXACT and $k = 5$ a staggering amount of 2.7×10^8 points is considered. Although a large number, this is still only a small fraction of the total search space: 10^{-2}. The exact method explored the same fraction for all datasets, implying that its pruning is effective: 99% of the search space is pruned. Despite this, runtimes are still quite long.

The greedy, levelwise approach explores much smaller parts of the search space: fractions between 10^{-23} and 10^{-4} are reported. The natural question is whether the resulting skylines approximate the exact solutions well. Looking at

Table 2. Results with EXACT and LEVELWISE algorithms. For each experiment, we give dataset, used p-value for the permutation test (if any), candidate set size $|C|$, algorithm, and subgroup set size k. Then follow runtime, the number of points in the search space considered, the fraction of the complete search space considered, the size of the resulting skyline, and corresponding quality and diversity ranges [min, max].

dataset	p	$|C|$	method	k	time	#points	fraction	$|S|$	$q(S)$	$d(S)$
Car	10^{-3}	71	LEVEL	5	< 1s	4380	10^{-4}	55	[0.23, 0.41]	[3.49, 4.90]
			EXACT	5	92s	179219	10^{-2}	128	[0.21, 0.41]	[3.49, 5.00]
			LEVEL	10	2s	25211	10^{-8}	99	[0.56, 0.74]	[5.56, 8.43]
Cmc	10^{-3}	98	LEVEL	5	< 1s	10908	10^{-4}	58	[0.18, 0.26]	[1.57, 4.61]
			EXACT	5	680s	1545129	10^{-2}	71	[0.17, 0.26]	[1.57, 4.61]
			LEVEL	10	5s	57105	10^{-9}	147	[0.35, 0.50]	[3.39, 7.05]
Credit-A	10^{-7}	232	LEVEL	5	10s	126516	10^{-5}	305	[0.42, 0.89]	[1.41, 4.56]
			EXACT	5	20.5h	274696613	10^{-2}	460	[0.41, 0.89]	[1.41, 4.56]
			LEVEL	10	139s	616899	10^{-12}	613	[0.89, 1.75]	[2.13, 6.87]
Credit-G	10^{-7}	114	LEVEL	5	1s	20493	10^{-4}	133	[0.22, 0.37]	[1.61, 4.74]
			EXACT	5	3347s	10149837	10^{-2}	137	[0.22, 0.37]	[1.61, 4.74]
			LEVEL	10	34s	145703	10^{-9}	435	[0.46, 0.73]	[1.88, 7.77]
Pima	10^{-7}	166	LEVEL	5	1s	28509	10^{-5}	66	[0.31, 0.49]	[2.62, 4.77]
			EXACT	5	1h	12221188	10^{-2}	107	[0.29, 0.49]	[2.62, 4.84]
			LEVEL	10	10s	127252	10^{-11}	248	[0.60, 0.96]	[2.85, 7.75]
Tictactoe	10^{-2}	90	LEVEL	5	2s	22303	10^{-4}	78	[0.20, 0.37]	[3.36, 4.94]
			EXACT	5	390s	1107493	10^{-2}	281	[0.19, 0.37]	[3.36, 4.94]
			LEVEL	10	4s	53181	10^{-9}	75	[0.41, 0.67]	[5.68, 7.89]
Adult		5116	LEVEL	5	55h	15870624	10^{-10}	252	[0.10, 0.48]	[1.48, 4.99]
			LEVEL	10	85h	37774186	10^{-23}	2040	[0.28, 0.96]	[1.83, 9.45]
Mushroom		1617	LEVEL	5	785s	2439272	10^{-8}	601	[0.27, 1.09]	[1.85, 4.97]
			LEVEL	10	2940s	9230365	10^{-19}	1310	[0.50, 2.12]	[2.36, 9.06]

the skyline sizes, we observe that skylines generally consist of modest numbers of subgroup sets (in particular given the total number of subgroup sets). LEVELWISE tends to find slightly smaller skylines, which is perhaps unsurprising as it explores a relatively small part of the search space. However, are the subgroup sets that it does find on the 'true' skyline?

The minimum-maximum values of subgroup set quality and diversity indicate that the levelwise skylines span almost the same ranges as the exact skylines. Table 3 shows a more elaborate comparison. The second column shows the fraction of points on the levelwise skyline that are domi-

Table 3. Comparison of the exact (E) and levelwise (L) skylines with $k = 5$

dataset	$\#\{E \succ L\}$	$\Delta_q(E \succ L)$	$\Delta_d(E \succ L)$
Car	22.73%	6.48%	1.49%
Cmc	51.72%	0.92%	0.69%
Credit-A	64.59%	1.04%	0.46%
Credit-G	5.31%	0.10%	0.19%
Pima	39.39%	0.78%	0.33%
Tictactoe	0%	–	–

nated by any point on the exact skyline. This reveals that substantial parts of the approximation are on the Pareto front. For those points that are dominated,

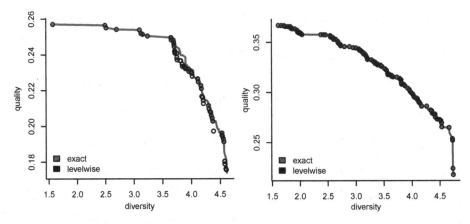

Fig. 3. Comparing exact and levelwise skylines (k=5); Cmc (left) and Credit-G (right)

it is of interest to investigate by how much. This is shown in the rightmost two columns, for both quality and diversity. These numbers are very small and indicate that the levelwise skylines approximate the exact skylines very well. To further illustrate this, consider the skylines plotted in Figure 3. The Credit-G approximate skyline is virtually identical to the exact one, and the Cmc approximation, one of the worst according to Table 3, is still a close approximation for all practical purposes.

The results obtained on Adult and Mushroom, shown at the bottom of Table 2, demonstrate that LEVELWISE can handle moderately sized datasets and candidate sets, although runtimes may increase substantially. However, note that the levelwise approach finds all skylines up to k – *not just for k*. For Adult, for example, 85 hours of runtime gives all skylines for $k = 2$ up to and including $k = 10$. Provisional results can be inspected at any time, and search can be terminated when the user thinks k is large enough. Such an approach is impossible with the exact method, as it only enumerates subgroup sets of exactly size k.

Finally, it is important to note that the quality and diversity ranges spanned by the skylines are often quite large (see Table 2 and Figure 3). For example, for both Cmc and Credit-G diversity ranges from 1.5 up to almost 5 bits – taking into account that 5 bits is the maximum for $k = 5$, this implies large differences between the corresponding subgroup sets. This demonstrates that very different trade-offs between quality and diversity are possible, and hence investigating these skylines is useful. For Cmc, for example, Figure 3 shows that diversity can be increased from 1.5 to 3 bits without affecting quality much. Such knowledge is likely to influence the preference of the end user.

7.2 Evaluating a Heuristic Subgroup Set Selection Method

In the introduction, we argued that disadvantages of existing heuristic subgroup selection methods are that 1) it is unknown whether the resulting subgroup sets are Pareto optimal, and 2) it is hard to tune the quality-diversity trade-off. To illustrate this, we evaluate a heuristic using LEVELWISE skylines.

Table 4. Comparing LEVELWISE (L) with the entropy-based selection heuristic (H)

dataset	k	$\#\{H \succ L\}$	$\Delta_q(H \succ L)$	$\Delta_d(H \succ L)$	$\#\{L \succ H\}$	$\Delta_q(L \succ H)$	$\Delta_d(L \succ H)$
Car	5	3.45%	0.00%	0.01%	75.86%	16.36%	7.29%
	10	1.79%	0.01%	0.00%	86.21%	12.65%	6.37%
Cmc	5	15.38%	1.67%	1.52%	65.52%	12.06%	2.07%
	10	0.43%	0.00%	0.01%	82.76%	6.03%	1.12%
Credit-A	5	0.00%	-	-	86.21%	1.99%	1.06%
	10	0.09%	0.00%	0.01%	79.31%	2.36%	1.94%
Credit-G	5	0.00%	-	-	68.97%	4.32%	0.27%
	10	0.00%	-	-	82.76%	6.16%	0.71%
Mushroom	5	0.67%	0.00%	0.01%	79.31%	10.92%	5.18%
	10	0.00%	-	-	93.10%	6.35%	4.68%
Pima	5	0.91%	0.00%	0.01%	51.72%	2.65%	0.36%
	10	0.59%	0.00%	0.01%	89.66%	2.36%	0.62%
Tictactoe	5	0.00%	-	-	62.07%	9.03%	1.08%
	10	0.00%	-	-	93.10%	4.73%	1.92%

The recently proposed cover-based subgroup set selection heuristic [10] selects a diverse k-subgroup set from a candidate set, but preliminary experiments revealed that it performed badly in terms of our diversity measure, i.e. joint entropy. We therefore slightly modified it to use entropy as selection criterion:

Given a candidate set C, first order it descending by subgroup quality (φ). Initialise subgroup set S to contain only the highest-quality subgroup and remove it from C. Then, iteratively add subgroups until $|S| = k$. In each iteration, pick that subgroup $G \in C$ that maximises $\alpha\varphi(G)+(1-\alpha)H(S\cup\{G\})$, where $\alpha \in [0,1]$ is a parameter. The selected subgroup is added to S and removed from C.

With this entropy-based heuristic, a single subgroup set can be found. By carefully varying the α parameter, we obtain a 'skyline' consisting of 29 subgroup sets for each dataset (except for Adult, for which running this heuristic many times took too long). Table 4 presents a comparison of this skyline to the one obtained with LEVELWISE. The heuristic rarely finds better solutions than the levelwise method, but most of the solutions found by the heuristic are dominated by the levelwise skyline. Also when considering the relative differences in quality and diversity for those points that are dominated, it is clear that the levelwise method often finds much better skylines than the heuristic.

We conclude that even when carefully tuning the parameter of a pattern set selection method, this does not guarantee that we discover a set of Pareto optimal solutions. Furthermore, this comparison also demonstrates that skyline discovery can be a useful tool in the evaluation of (existing) heuristics.

8   Conclusions

We have argued that whenever there is a quality-diversity trade-off in a k-subset selection task, it is important to explicitly consider the skyline of Pareto optimal solutions. In this paper we focused on the task of pattern set selection in the context of Subgroup Discovery, but many similar 'diverse top-k' tasks exist, not only in exploratory data mining but also for example in information retrieval. We proposed two algorithms for discovering skylines of k-subgroup sets. If we use joint entropy as diversity measure and use its properties for pruning, the EXACT method can be used with modest candidate sets and k. LEVELWISE performs a greedy, levelwise search and is therefore considerably faster. Furthermore, experiments showed that the obtained skylines closely approximate the exact solutions. Finally, we demonstrated that the skyline discovery algorithms can be used for the objective evaluation of heuristic selection methods.

One might argue that having multiple subgroup sets instead of one only complicates the situation, but observe that this skyline *always exists*; the problem is that users may not be aware of this. Therefore, the explicit discovery of skylines is an important step towards a better understanding of 'diverse top-k's'. Skylines of subgroup sets can be interactively explored, allowing the user to make informed choices based on both the subgroup sets and the shape of the skyline.

Acknowledgements. Matthijs van Leeuwen is supported by a Rubicon grant of the Netherlands Organisation for Scientific Research (NWO).

References

1. Bringmann, B., Nijssen, S., Tatti, N., Vreeken, J., Zimmermann, A.: Mining sets of patterns: Next generation pattern mining. In: Tutorial at ICDM 2011 (2011)
2. Bringmann, B., Zimmermann, A.: The chosen few: On identifying valuable patterns. In: Proceedings of the ICDM 2007, pp. 63–72 (2007)
3. Duivesteijn, W., Knobbe, A.: Exploiting false discoveries – statistical validation of patterns and quality measures in subgroup discovery. In: Proceedings of the ICDM 2011, pp. 151–160 (2011)
4. Ehrgott, M., Gandibleux, X.: A survey and annoted bibliography of multiobjective combinatorial optimization. OR Spektrum (2000)
5. Ehrgott, M., Gandibleux, X.: Approximative solution methods for multiobjective combinatorial optimization. TOP: An Official Journal of the Spanish Society of Statistics and Operations Research 12(1), 1–63 (2004)
6. Klösgen, W.: Explora: A Multipattern and Multistrategy Discovery Assistant. In: Advances in Knowledge Discovery and Data Mining, pp. 249–271 (1996)
7. Knobbe, A., Ho, E.K.Y.: Maximally informative k-itemsets and their efficient discovery. In: Proceedings of the KDD 2006, pp. 237–244 (2006)
8. Knobbe, A.J., Ho, E.K.Y.: Pattern teams. In: Fürnkranz, J., Scheffer, T., Spiliopoulou, M. (eds.) PKDD 2006. LNCS (LNAI), vol. 4213, pp. 577–584. Springer, Heidelberg (2006)
9. Kung, H.T., Luccio, F., Preparata, F.P.: On finding the maxima of a set of vectors. J. ACM 22(4), 469–476 (1975)
10. van Leeuwen, M., Knobbe, A.: Diverse subgroup set discovery. Data Mining and Knowledge Discovery 25, 208–242 (2012)
11. Markowitz, H.: Portfolio selection. The Journal of Finance 7(1), 77–91 (1952)
12. Peng, H., Long, F., Ding, C.: Feature selection based on mutual information: Criteria of max-dependency, max-relevance, and min-redundancy. IEEE Transactions on Pattern Analysis and Machine Intelligence 27(8), 1226–1238 (2005)
13. Soulet, A., Raïssi, C., Plantevit, M., Crémilleux, B.: Mining dominant patterns in the sky. In: Proceedings of the ICDM 2011, pp. 655–664 (2011)

Difference-Based Estimates
for Generalization-Aware Subgroup Discovery

Florian Lemmerich, Martin Becker, and Frank Puppe

Artificial Intelligence and Applied Computer Science Group,
University of Würzburg, D-97074 Würzburg, Germany
{lemmerich, becker, puppe}@informatik.uni-wuerzburg.de

Abstract. For the task of subgroup discovery, generalization-aware interesting measures that are based not only on the statistics of the patterns itself, but also on the statistics of their generalizations have recently been shown to be essential. A key technique to increase runtime performance of subgroup discovery algorithms is the application of optimistic estimates to limit the search space size. These are upper bounds for the interestingness that any specialization of the currently evaluated pattern may have. Until now these estimates are based on the anti-monotonicity of instances, which are covered by the current pattern. This neglects important properties of generalizations. Therefore, we present in this paper a new scheme of deriving optimistic estimates for generalization aware subgroup discovery, which is based on the instances by which patterns differ in comparison to their generalizations. We show, how this technique can be applied for the most popular interestingness measures for binary as well as for numeric target concepts. The novel bounds are incorporated in an efficient algorithm, which outperforms previous methods by up to an order of magnitude.

1 Introduction

Subgroup discovery [17] is a key technique for data mining and machine learning. It aims at identifying descriptions for subsets of instances in a dataset, which have an interesting deviation with respect to the distribution of a predefined concept of interest. This task has been studied under different terminology such as contrast set mining [7], emerging pattern mining [11], correlated itemset mining [21], discriminative pattern mining [10] or association rule mining with a fixed consequent [20]. While the specific goal of these tasks may vary, the algorithmic challenges and approaches are very closely related, see [18,26].

The selection of patterns in the search space is commonly based on an interestingness measure. These measures use statistics derived from the instances covered by a pattern to determine a score for the pattern. The best patterns according to this score are then returned to the user. As an example, consider a dataset of patients and their medical data. Let the target concept be surgery successful, which is true for 30% of the patients. Then a pattern like gender=male $\wedge$ smoker=false with a higher rate of successful surgeries, e.g. 50%, receives a higher score and is more likely to be included in the result.

H. Blockeel et al. (Eds.): ECML PKDD 2013, Part III, LNAI 8190, pp. 288–303, 2013.

Practical applications have shown that results for traditional interestingness measures often contain variants of the same pattern multiple times. To avoid this problem, several authors postulated that a pattern should not only be evaluated with respect to its own statistics, but also with respect to the statistics of its generalizations, see for instance [8,4,5,19]. Considering the example above the pattern gender=male ∧ smoker=false would be rated as less interesting if it can be explained by one of its generalizations alone, e.g., if the pattern smoker=false already describes a set of patients with a 50% surgery success rate. While the practical use of such generalization-aware interestingness measures has been widely acknowledged, the efficient mining in this setting has received little attention. A key technique to improve runtime performance of subgroup discovery in general is the application of optimistic estimates, that is, upper bounds for the interestingness of any specialization of the currently evaluated pattern. Although research has shown that improving the tightness of the utilized bounds improves the runtime performance substantially [15], there has been no extensive research so far concerning upper bounds for generalization aware interestingness measures beyond the trivial transfer of bounds for traditional measures.

In this paper we propose a novel method to exploit specific properties of generalization-aware measures to derive additional optimistic estimate bounds, which allow to speed-up the search. Unlike previous approaches, the bounds are not exclusively based on the instances that are contained in the currently examined subgroup, but on the instances that were excluded in comparison to generalizations of the current pattern. We show, how this general concept can be applied to exemplary interestingness measures in different setting, i.e., for subgroup discovery with binary target concepts and with numeric target concepts using a mean-based interestingness measure. The bounds are incorporated in a novel *apriori*-based algorithm that allows efficient propagation of the required statistics. Experiments show that exploiting the presented bounds results in substantial runtime improvements. The optimistic estimates are especially effective in tasks that incorporate selectors, which cover a majority of the dataset.

The rest of the paper is structured as follows: Section 2 provides background on subgroup discovery and the used terminology. Then, related work is discussed in Section 3. Next, the new scheme to derive optimistic estimate bounds and its application to different interestingness measures is presented in Section 4. Afterwards, we explain, how the new optimistic estimate bounds can efficiently be exploited in an algorithm in Section 5. Section 6 presents experimental results, before we conclude in Section 7.

2   Background

Let A be an *attribute space* $A = A_1 \times \ldots \times A_m$, where each set A_i represents an *attribute*. A *dataset* is a tuple $D = (I, A)$ with $I \subseteq A$. Each $i \in I$ is called a *data instance*. *Selectors sel* (also called basic patterns) are boolean functions $sel : I \to \{false, true\}$ defined by selection expressions on the set of attributes. In the case of nominal attributes typical selection expressions are given by attribute-value pairs, in the case of numeric attributes by intervals. For example, the selector $age =]12; \infty[$ is true, iff the attribute age has a value greater than 12. A (complex) *pattern* (also called subgroup description) combines selectors into a boolean formula. For a typical conjunctive description language, on which we focus in this paper, a pattern $P = \{sel_1, \ldots, sel_k\}$ is defined by a set of selectors sel_j, which are interpreted as a conjunction, i.e., $P = sel_1 \wedge \ldots \wedge sel_k$. Thus, an instance $i \in I$ is *covered* by pattern P, iff $(\forall sel \in P : sel(i) = true)$ or short $P(i) = true$. A subgroup $sg(P)$ is given by the set of individuals covered by the pattern P: $sg(P) = \{i \in I | P(i) = true\}$. For short notation, $i_P = |sg(P)|$ is the number of individuals covered by a pattern P. Furthermore we denote as $\Delta(A, B) = sg(A) \setminus sg(B)$ the instances, which are covered by A, but not by B. We call a pattern G a *generalization* of its *specialization* S, iff $G \subset S$.

A *pattern mining task* is specified by a 5-tuple (D, T, q, Σ, k). D is a dataset. The target concept T assigns a target value $tc(i)$ to each instance. It can either be defined by a pattern (binary case) or by a single numeric attribute (numeric case). In the binary case, we write p_P (n_P) for all individuals with a true (false) target concept. $q : 2^\Sigma \to \mathbb{R}$ is a quality function that measures the interest-ingness of a pattern with respect to the chosen target concept T. Σ defines the search space by providing a set of selectors to build conjunctive patterns from. k specifies the number of patterns contained in the result set. The overall task is then to identify the best k patterns in the search space 2^Σ according to the quality function q $(q \gg 0)$.

A huge amount of quality functions has been proposed in literature, cf. [17,13]. While the general approach of this paper could also be applied to other qual-ity functions, we especially focus on the following popular measures: The most popular interestingness measures trade-off the covered instances i_P of a pattern versus the deviation of the target share $\tau_P - \tau_\emptyset$, where $\tau_P = \frac{p_P}{p_P + n_P}$ is the ratio of positive instances versus all instances in pattern P and $\tau_\emptyset$ is the same ratio for the overall population. This is formalized as:

$$q_{bin}^a(P) = i_P^a \cdot (\tau_P - \tau_0), a \in [0; 1]$$

This includes for example the weighted relative accuracy for the size parameter $a = 1$, a simplified binomial function for $a = 0.5$, or the added value for $a = 0$. For numeric target concepts this can easily adapted by replacing the target share for the pattern and the overall population with the respective mean values μ_P and $\mu_\emptyset$ of the target attribute:

$$q_{num}^a(P) = i_P^a \cdot (\mu_P - \mu_\emptyset), a \in [0; 1].$$

This definition includes the mean test quality function [17] for $a = 0.5$ and the impact quality function [23] for $a = 1$.

Consider a pattern P with an interestingly high target share τ_P. If another selector sel with $sel \notin P$ is added to the pattern P, which does either cover a majority of the instances of P $(sg(P) \approx sg(P \wedge sel))$ or is statistically independent from P and the target concept, then the pattern $P \wedge sel$ will have roughly the same target share as pattern P. Thus, this pattern may also receive a high score according to the previously presented quality measures due to its high target share. However, it should not be presented to users in the result set, since the additional selector *sel* does not contribute to the increased target share. To avoid such redundant output the *minimum improvement constraint* has been introduced, see [8]. By using this additional filter, all patterns with a target share that is lower or equal to the target share of any of its generalizations are removed from the result set. Nonetheless, patterns that improve the target share only by a small margin, e.g., due to noise, will still be contained in the result. Therefore, more recent approaches incorporate the comparison of pattern statistics with the statistics of its generalizations directly into the interestingness measure [5,14,19] resulting in *generalization-aware interesting measures*. The target share (or the mean value in case of numeric targets) within the pattern is not compared to the target share (mean value) of the overall population but to the maximum target share (mean value) of all its generalizations:

$$r^a_{bin}(P) = i^a_P \cdot (\tau_P - \max_{H \subset P} \tau_H), a \in [0; 1]$$

$$r^a_{num}(P) = i^a_P \cdot (\mu_P - \max_{H \subset P} \mu_H), a \in [0; 1]$$

Thus, a pattern is only regarded as interesting if its target share (mean value) is considerably higher than it is in all of its generalizations. Although other interestingness measures can be adapted accordingly, we focus on these two families of generalization-aware measures in this paper, since they are the only ones, which have been described in previous literature and applied in practical applications. We will also not argue about advantages of these functions in comparison to traditional measures or other methods that avoid redundant output, such as closed pattern [12], but focus on efficient mining for these generalization-aware measures by introducing novel, difference-based optimistic estimates.

The concept of optimistic estimates has been introduced in order to speed up the subgroup discovery task, see [22,25]. The basic idea of optimistic estimates is as following: if one can guarantee that no specialization of the currently evaluated pattern will have a quality, which is good enough to include the respective pattern into the result set, then we can safely omit these patterns from the search. In doing so we can substantially reduce the number of patterns, which have to be evaluated, while maintaining the optimality of the results. In this regard, we aim at as strict as possible bounds to reduce the remaining search space and thus to speed up the search process. Formally, given a pattern P and an interestingness measure q an optimistic estimate function $oe_q(P)$ is a function such that for each specialization $S \supset P$ of P the quality is lower than the value of the optimistic estimate function for pattern P: $\forall S \supset P : q(S) \leq oe_q(P)$.

3 Related Work

Subgroup discovery is a long studied field [17]. An essential technique for efficient discovery showed to be pruning based on optimistic estimates [22,25]. As Grosskreutz et al. showed, the efficiency of the pruning is strongly influenced by the *tightness* of the bounds [15]. A more general method to derive optimistic estimates for a whole class of interestingness measures, that is, *convex* measures, was introduced in [20] and later extended in [26]. In this paper, we provide a different technique to determine optimistic estimates to another family of interestingness measures, i.e., generalization-aware measures.

The necessity to consider also generalizations of patterns in selection criteria has been recognized in [8,4]. These early approaches used a *minimum improvement constraint*, which is applied only as a post-processing operation after the mining algorithm. Webb and Zhang presented an efficiency improvement in mining with this constraint in the context of association rules [24] by introducing a pruning condition based on the difference in covering. While the method of Webb and Zhang requires *full* coverage on all instances, the method presented in this work can also be applied with only partial coverage. In addition our method is used to derive upper bounds for interestingness measures instead of exploiting constraints and is also applied in settings with numeric target concepts.

Recent approaches incorporate differences with respect to generalizations directly in the interestingness measure. This showed positive results in descriptive [14,19] as well as predictive settings [6] for both binary and numeric target concepts. However, these papers focus more on which patterns are to be selected and not on efficient mining through pruning. As an exception, Batal and Hausknecht utilized a pruning scheme in an Apriori-based algorithm that is based exclusively on the positives covered by a subgroup [5]. This algorithm is used for comparison in the evaluation section. Utilizing pruning in settings with numeric concepts of interest is more challenging than in the binary case [2]. While for the impact measure q_{num}^1 an optimistic estimate has been employed [2,23] in the standard subgroup setting, to the authors knowledge no other pruning bounds for numeric generalization-aware measures have been proposed so far.

4   Estimates for Generalization-Aware Subgroup Mining

In this section, we introduce a novel scheme to derive optimistic estimates for generalization-aware interestingness measures. These optimistic estimates help to improve the runtime performance of algorithms by pruning the search space. We start by generalizing estimates that have been previously presented for this task to outline the conventional approach to derive estimates. Then, we present the core idea of our new scheme to derive upper bounds: difference-based optimistic estimates. Next, we show how this concept can be exploited by deriving estimates for quality functions in the binary and the numeric case using the quality functions r_{bin}^a and r_{num}^a.

4.1   Optimistic Estimates Based on Covered Positive Instances

Traditionally, optimistic estimates for subgroup discovery are based only on the anti-monotonicity of instance coverage. That is, when adding an additional selector to a pattern P, then the resulting pattern only covers a subset of the instances covered by P. To give an example for this traditional approach, the following theorem generalizes the optimistic estimate bounds for r_{bin}^a used in [5], which covers only the special case using the parameter $a = 0.5$.

Theorem 1. *Let p_P be the number of all positive instances covered by the currently evaluated pattern P and $\max_{H \subset P}(\tau_H)$ the maximum of the target shares for P and any of its generalizations. Then, optimistic estimate bounds for the family of quality functions r^a_{bin} are given by: $oe_{r^a_{bin}} = (p_P)^a \cdot (1 - \max_{H \subseteq P} \tau_H)$.*

Proof. We first show that the quality of any specialization S does not decrease, if all negatives are removed. Let n_s be the number negatives in S. Then,

$$r^a_{bin}(S) = (p_S + n_S)^a \cdot \left(\frac{p_S}{p_S + n_S} - \max_{H \subset S} \tau_H \right) = \frac{p_S}{(p_S + n_S)^{1-a}} - (p_S + n_S)^a \cdot \max_{H \subset S} \tau_H$$

We examine this term as a function of n_S, $n_S \geq 0$: The first summand decreases with increasing n_S, since $1 - a \geq 0$. The second, negative summand increases with increasing n_S, as $max_t \geq 0$. Thus, the maximum is reached for $n_S = 0$. We can conclude that:

$$r^a_{bin}(S) = (p_S + n_S)^a \cdot \left(\frac{p_S}{p_S + n_S} - \max_{H \subset S} \tau_H \right)$$

$$\leq (p_S)^a \cdot \left(\frac{p_S}{p_S} - \max_{H \subset S} \tau_H \right) \leq (p_P)^a \cdot (1 - \max_{H' \subseteq P} \tau_{H'}),$$

as the number of positives in the specialization S is smaller than the number of positives in the more general pattern P, and the generalizations of S include all generalizations of P. □

As has been exemplified in [5] this bound can already achieve significant runtime improvements. Note, that these bounds use only the anti-monotonicity of the covered positive instances. In contrast, we will show in the next sections, how we can exploit additional information on the difference of negative instances between patterns and their generalizations to derive additional bounds.

4.2 Difference-Based Pruning

Next, we provide the core idea for our novel scheme to derive optimistic estimates. It utilizes that the instances by which a pattern and its specialization differ are – in a certain way – anti-monotonic. More specifically, we will exploit the following lemma to derive optimistic estimates:

Lemma 1. *Let $P = A \wedge B$ be any pattern with A, B potentially being a conjunction of patterns themselves and $B \neq \emptyset$. Then for any specialization $S \supset P$ there exists a generalization $\gamma(S) \subset S$, such that $\Delta(\gamma(S), S) \subseteq \Delta(A, B)$.*

Proof. Consider for any specialization $S = A \wedge B \wedge X$ (X being potentially a conjunction itself) the pattern $\gamma(S) = A \wedge X$, which is a real generalization of S, since $B \neq \emptyset$. Then, $\Delta(\gamma(S), S) = sg(A \wedge X) \setminus sg(A \wedge B \wedge X) = (sg(A) \cap sg(X)) \setminus (sg(A) \cap sg(B) \cap sg(X)) = sg(X) \cap (sg(A) \setminus (sg(A) \cap sg(B))) = sg(X) \cap (sg(A) \setminus sg(B)) = sg(X) \cap \Delta(A, B)$, which is a subset of $\Delta(A, B)$. □

The subset property implies directly that for each specialization S the generalization $\gamma(S)$ contains at most $i_{sg(S)} + i_{\Delta(A,B)}$ instances. Additionally, in the case of a binary target, we can estimate the number of negative instances in this generalization: $n_{\gamma(S)} \leq n_S + n_{\Delta(A,B)}$. Furthermore, in the case of a numeric target, the minimum target value of $\Delta(\gamma(S), S)$ is higher than the minimum target value in $\Delta(A, B)$. In mining algorithms, statistics for $\Delta(A, B)$ can be computed with almost no additional effort. For instance, n_A and $n_{A \wedge B}$ are both required anyway in order to evaluate the pattern $A \wedge B$ with r_{bin}^a. Then, $n_{\Delta(A,B)}$ is given by $n_{\Delta(A,B)} = n_A - n_{A \wedge B}$.

As an example, assume that the pattern A covers 20 positive and 10 negative instances and the evaluation of the pattern $A \wedge B$ shows that this pattern also covers 10 negative instances. That is, B covers all negative instances, which are covered by A, $n_{\Delta(A,B)} = 0$. Now consider any specialization S of this pattern. According to the lemma, S has another generalization $\gamma(S)$ that contains the same number of negative instances as S since $n_{\gamma(S)} \leq n_S + n_{\Delta(A,B)}$. As S (as a specialization of $\gamma(S)$) additionally has no more positive instances than S, the target share in S is equal or smaller than for its generalization $\gamma(S)$. Thus, the quality of S according to any generalization-aware measure r_{bin}^a is ≤ 0. Since this is the case for any specialization of $A \wedge B$, specializations of $A \wedge B$ can be pruned from the search space without influencing the results.

This is an extreme example: *all* negative instances of A are also covered by $A \wedge B$. Now assume that $A \wedge B$ had covered only 8 negative instance, thus $n_{\Delta(A,B)} = 10 - 8 = 2$. In this case the lemma guarantees that S has a generalization $\gamma(S)$ with *at most 2* negative instances more than S. If S itself covers a decent amount of instances, the target share in S cannot be much higher than in $\gamma(S)$. Thus, either S is small or there is only a small increase (or a decrease) in the target share comparing S and its generalization $\gamma(S)$. In both cases, the interestingness of S according to r_{bin}^a is low.

Overall we conclude that, if the difference of covered instances between A and $A \wedge B$ is small, then the interestingness score for all specializations is limited. In the next sections we formalize these considerations by deriving formal optimistic estimate bounds that can be used to prune the search space.

4.3   Difference-Based Optimistic Estimates for Binary Targets

Following, we provide for generalization-aware measures $r_{bin}^a = i_P^a \cdot (\tau_P - \max_{H \subset P} \tau_H)$ with binary targets new optimistic estimates, which are based on the difference of pattern coverage in comparison to the coverage of generalizations.

Theorem 2. *Consider the pattern P with p_P positive instances. $P' \subseteq P$ is either P itself or one of its generalizations and $P'' \subset P'$ a generalization of P'. Let $n_\Delta = n_{P''} - n_{P'}$ be the difference in coverage of negative instances between these patterns. Then, an optimistic estimate of P for r_{bin}^a is given by:*

$$oe_{r_{bin}^a}(P) = \begin{cases} \frac{p_P \cdot n_\Delta}{p_P + n_\Delta}, & \text{if } a = 1 \\ \frac{n_\Delta}{1 + n_\Delta}, & \text{if } a = 0 \\ \frac{\hat{p}^a \cdot n_\Delta}{\hat{p} + n_\Delta}, \text{ with } \hat{p} = \min(\frac{a \cdot n_\Delta}{1-a}, p_P), & \text{else} \end{cases}$$

Proof. Let S be any specialization of P and $G = \gamma(S)$ the generalization with $\Delta(G,S) \subseteq \Delta(P',P'')$, which exists according to the previous lemma, since S is also a specialization of P'. The number of negatives in G is equal to the number of negatives covered by S plus the number of negatives, which are covered by G, but not by S: $n_G = n_S + n_{\Delta(G,S)}$. By construction it holds that $n_{\Delta(G,S)} \leq n_\Delta$. Additionally, we can assume $p_S > 0$, that is, S contains at least one positive instance, since $r_{bin}^a(S) \leq 0$ otherwise.

In the proof, we will first derive an upper bound that depends on the number of positives in the specialization S, which is unknown at the time P is evaluated. In a second step we therefore determine the maximum value of this function. The quality of S is given by:

$$r_{bin}^a(S) = (p_S + n_S)^a \cdot (\tau_S - \max_{H \subset S} \tau_H) \tag{1}$$

$$\leq (p_S + n_S)^a \cdot (\tau_S - \tau_G) \tag{2}$$

$$= (p_S + n_S)^a \cdot \left(\frac{p_S}{p_S + n_S} - \frac{p_G}{p_G + n_S + n_{\Delta(G,S)}} \right) \tag{3}$$

$$\leq (p_S + n_S)^a \cdot \left(\frac{p_S}{p_S + n_S} - \frac{p_S}{p_S + n_S + n_{\Delta(G,S)}} \right) \tag{4}$$

$$= (p_S + n_S)^a \cdot \left(\frac{p_S \cdot (p_S + n_S + n_{\Delta(G,S)}) - (p_S \cdot (p_S + n_S))}{(p_S + n_S)(p_S + n_S + n_{\Delta(G,S)})} \right) \tag{5}$$

$$= \frac{p_S \cdot n_{\Delta(G,S)}}{(p_S + n_S)^{1-a}(p_S + n_S + n_{\Delta(G,S)})} \tag{6}$$

$$\leq \frac{p_S \cdot n_{\Delta(G,S)}}{(p_S)^{1-a}(p_S + n_{\Delta(G,S)})} \tag{7}$$

$$= \frac{p_S^a \cdot n_{\Delta(G,S)}}{(p_S + n_{\Delta(G,S)})} \tag{8}$$

$$\leq \frac{p_S^a \cdot n_\Delta}{(p_S + n_\Delta)} := f^a(p_S) \tag{9}$$

The transformation to line 2 is possible, since $G \subset S$. In line 4 it is used that $p_S \leq p_G$, as the positives of S are a subset of the positive of its generalization G. In line 7 it is exploited that the denominator is strictly increasing with increasing n_S, because $1 - a \in [0;1]$. Therefore, the smallest denominator and thus the largest value for the overall term is achieved by setting $n_S = 0$. The term in line 8 is strictly increasing as a function of $n_{\Delta(G,S)}$. Since $n_{\Delta(G,S)} \leq n_\Delta$, line 9 follows.

In the final line 9, the function $f^a(p_S)$ is defined, which provides an upper bound on the interestingness of P that depends on the number of positives within the specialization. This number is not known, when the pattern P is evaluated. Intuitively, for large number of positives in the specialization removing n_Δ negative instances will not change the target share in the subgroup much, therefore the interestingness of the generalization is limited. On the other hand, for small numbers of positive instances S is overall small and possibly not interesting for that reason. p_S is at least 1, since S otherwise is not interesting anyway and at most p_P, as the number of positives for S is smaller than for its generalization P. Next, we analyze for which value of p_S the function $f^a(p_S)$ of line 9 reaches its maximum in the interval $[1; p_P]$. This depends on the parameter a of the interestingness measure:

1. For $a = 1$ it holds that $f^1(p_S) = \frac{p_S \cdot n_\Delta}{p_S + n_\Delta}$. This function is strictly increasing in p_S. That is, the more positive instances are contained in S, the higher is the derived upper bound. The maximum is reached at highest value in the domain of definition: $\max(f^1(p_S)) = f^1(p_P) = \frac{p_P \cdot n_\Delta}{p_P + n_\Delta}$.

2. In contrast for $a = 0$, $f^0(p_S) = \frac{n_\Delta}{p_S + n_\Delta}$ is strictly decreasing. Thus, the maximum value of f^0 is reached for $p_S = 1$, the minimum possible value of p_S: $\max(f^0(p_S)) = f^0(1) = \frac{n_\Delta}{1 + n_\Delta}$.

3. For $0 < a < 1$, f^a reaches a maximum for a certain value p^* within the domain of definition. To determine that, we compute the first derivative of f^a using the quotient rule.

$$\frac{d}{dp_S} f^a(p_S) = n_\Delta \cdot \frac{d}{dp_S} \frac{p_S^a}{p_S + n_\Delta}$$

$$= n_\Delta \frac{(n_\Delta + p_S) \cdot a \cdot p_S^{a-1} - p_S^a}{(n_\Delta + p_S)^2}$$

$$= n_\Delta \cdot p_S^{a-1} \frac{a n_\Delta + a \cdot p_S - p_S}{(n_\Delta + p_S)^2} := (f^a)'$$

The only root of this derivative is at $p^* := \frac{a \cdot n_\Delta}{1-a}$. As can be easily shown, $(f^a)'(p_S)$ is greater than zero for p_S smaller than p^* and lower than zero for p_S greater than p^*. Therefore, p^* is the only maximum of $f^a(p_S)$. Thus, if $p_P > p^*$, then p^* is the maximum value of f^a, otherwise the maximum is reached at the highest value of the domain of definition: $\max(f^a(p_S)) = f^a(\hat{p}) = \frac{\hat{p}^a \cdot n_\Delta}{\hat{p} + n_\Delta}$, with $\hat{p} = \min(\frac{a \cdot n_\Delta}{1-a}, p_P)$.

Overall, for any specialization S it holds that $r^a_{bin}(S) \leq f^a(p_S) \leq \max f^a(p_S) = oe_{r^a_{bin}}(P)$, with the function maxima as described above, therefore $oe_{r^a_{bin}}(P)$ as defined in the theorem is a correct optimistic estimate.   □

For any pair of generalizations of P (P' and P'') as well as for any pair of P ($P' = P$) and one of its generalization (P''), this theorem provides an optimistic estimate of P. The optimistic estimate bound is dependent on the number of positives in the subgroup and the difference of negative instances between P' and P''. It is low, if either there are only few positives in P or the difference of negative instances between the pair of generalizations is small (or a combination of both). Since the number of positives in P is independent of the chosen pair P', P'', the pair with the minimum difference of negative instances implies the tightest upper bound, which should be used to maximize the effects of pruning.

As a special case the theorem includes that the interestingness of any pattern is ≤ 0, if n_Δ is 0. To the authors knowledge, it is the first measure that includes these differences in optimistic estimate bounds for subgroup discovery.

4.4 Difference-Based Optimistic Estimates for Numeric Targets

Next, we will show that a related approach can be used to obtain optimistic estimates for generalization-aware interestingness $r^a_{num} = i^a_P \cdot (\mu_P - \max_{H \subset P} \mu_H)$ in settings with numeric target concepts.

Theorem 3. *In a task with a numeric target concept, consider the pattern P with i_P instances and a maximum target value of $\max_P$. $P' \subseteq P$ is either P itself or one of its generalizations and $P'' \subset P'$ is a generalization of P'. Let $i_\Delta = |\Delta(P'', P')|$ be the number of instances contained in P'', but not in P' and $\min_\Delta$ the minimum target value contained in $\Delta(P'', P')$. Then, an optimistic estimate of P for the generalization aware quality function r^a_{num} is given by:*

$$oe_{r^a_{num}}(P) = \max(0, oe'_{r^a_{num}}(P)),$$

$$oe_{r^a_{num}}(P)' = \begin{cases} \frac{i_\Delta \cdot i_P}{i_P + i_\Delta} \cdot (\max_P - \min_\Delta), & \text{if } a = 1 \\ \frac{i_\Delta}{1 + i_\Delta} \cdot (\max_P - \min_\Delta), & \text{if } a = 0 \\ \frac{\hat{i}^a \cdot i_\Delta}{\hat{i} + i_\Delta} \cdot (\max_P - \min_\Delta), \text{ with } \hat{i} = \min(\frac{a \cdot i_\Delta}{1-a}, i_P), & \text{else} \end{cases}$$

Proof. We consider any specialization $S \supset P$ and its generalization $G = \gamma(S)$ according to Lemma 1. Then we can estimate the interestingness of S:

$$r^a_{num}(S) = i_S{}^a \cdot (\mu_S - \max_{H \subset S} \mu_H) \tag{1}$$

$$\leq i_S{}^a \cdot (\mu_S - \mu_G) \tag{2}$$

$$= i_S{}^a \cdot \left(\frac{\sum\limits_{i \in sg(S)} tc(i)}{i_S} - \frac{\sum\limits_{i \in sg(S)} tc(i) + \sum\limits_{j \in \Delta(G,S)} tc(j)}{i_S + i_{\Delta(G,S)}} \right) \tag{3}$$

$$= i_S{}^{a-1} \cdot \left(\sum\limits_{i \in sg(S)} tc(i) - \frac{i_S \cdot \left(\sum\limits_{i \in sg(S)} tc(i) + \sum\limits_{j \in \Delta(G,S)} tc(j) \right)}{i_S + i_{\Delta(G,S)}} \right) \tag{4}$$

$$= i_S{}^{a-1} \cdot \left(\frac{i_{\Delta(G,S)} \sum\limits_{i \in sg(S)} tc(i) - i_S \sum\limits_{j \in \Delta(G,S)} tc(j)}{i_S + i_{\Delta(G,S)}} \right) \tag{5}$$

$$\leq i_S{}^{a-1} \cdot \left(\frac{i_{\Delta(G,S)} \cdot i_S \cdot \max\limits_{i \in S} tc(i) - i_S \cdot i_{\Delta(G,S)} \cdot \min\limits_{j \in \Delta(G,S)} tc(j)}{i_S + i_{\Delta(G,S)}} \right) \tag{6}$$

$$= \frac{i_{\Delta(G,S)} \cdot i_S{}^a}{i_S + i_{\Delta(G,S)}} \cdot \left(\max\limits_{i \in S} tc(i) - \min\limits_{j \in \Delta(G,S)} tc(j) \right) \tag{7}$$

$$\leq \frac{i_\Delta \cdot i_S{}^a}{i_S + i_\Delta} \cdot \left(\max\limits_P - \min\limits_\Delta \right) = f(i_S) \cdot \left(\max\limits_P - \min\limits_\Delta \right) \tag{8}$$

In line 2 it is used that G is a generalization of S, then it is exploited that $sg(G) = sg(S) \cup \Delta(G, S), S \cap \Delta(G, S) = \emptyset$. In line 6 we utilize that the sum of any set of values is bigger than the minimum appearing value times the size of the set, but smaller than the maximum appearing value times the size of the set. Line 8 uses that $i_{\Delta(G,S)} \leq i_\Delta$.

f^a is a function over the unknown number of all instances in the specialization, which can be any number in $[1; i_P]$. f^a is always positive. Therefore, if $(\max_P - \min_\Delta) \leq 0$, the optimistic estimate is given by 0. Else, the maxima of f^a, which have already been derived in the proof of Theorem 2, determine the bound: $f^a(i_S)$ is strictly increasing for $a = 1$, strictly decreasing for $a = 0$ and reaches a maximum at $\frac{a \cdot i_\Delta}{1-a}$ or at i_P otherwise. Thus: $r^a_{num}(S) \leq (f^a(i_S)) \cdot (\max_P - \min_\Delta) \leq \max(f^a(i_S) \cdot (\max_P - \min_\Delta)$. The bounds follow directly from the inserting the resp. maxima values. Since this holds for any specialization S of P, $oe_{r^a_{num}}(P)$ is a correct optimistic estimate for P.   □

Similar to the optimistic estimate in the binary case, the derived optimistic estimate is low, if either the number of instances covered by P is low, or if the difference in the number of instances covered between the generalizations P'' and P' is low (or a combination of both). However additionally, the bound also considers the range of the target variable in these patterns, that is, the maximum occurring target value in P and the minimum target value in the difference set of instances. As a result, the bound gets zero, if the minimum target value removed by adding a selector to a generalization of P was higher than the maximum remaining target value in P.

5   Algorithm

The presented optimistic estimates can in general be applied in combination with any search strategy. In this paper we focus on adapting an exhaustive algorithm, i.e., apriori [1,16]. This approach is especially suited for the task of generalization-aware subgroup discovery, since its levelwise search strategy guarantees that specializations are always evaluated after their generalizations and the highest target share found in generalizations can efficiently be propagated from generalizations to specializations, see [5]. Therefore, and for better comparability with previous approaches, we chose apriori as a basis for our novel algorithm. Using the following adaptations the algorithm is not only capable of determining the proposed optimistic estimates. The algorithm also propagates the required information very efficiently. Due to limited space, we will not describe the base algorithm, which has been extensively described in literature [1,20,16,5], but instead focus only on the differences. We start by describing the binary case.

Apriori performs a levelwise search, where new candidate patterns are generated from the last level of more general patterns. In our adaptation of the algorithm additional information is stored for each candidate. This includes the maximum target share in generalizations of this pattern, the minimum number of negatives covered by any generalization and the minimum number of negatives that were removed in generalizations of this pattern. After the evaluation of a pattern the number of positives, the number of negatives and the resulting target share are additionally saved in each candidate. The minimum number of negative instance in a generalization is required to compute the minimum number of instances, which are contained in the pattern, but not in a generalization. The other statistics are directly required to compute either the quality or the optimistic estimates of the pattern. Whenever a new candidate pattern P is generated in apriori, it is checked for all its direct generalizations G, if it is contained in the last levels candidate set. During this check, the statistics for the maximum target share in generalizations, the minimum number of negatives in a generalization and the minimum number of negatives that were removed in any generalization of this pattern can be computed by using the information stored in the generalizations and simple minimum/maximum functions. In doing so, the statistics required to compute the quality of the pattern and the optimistic estimates are propagated very efficiently from one level of patterns to the next level of more specific patterns.

In the evaluation phase (the counting phase in classical apriori) each candidate is evaluated. This requires to determine the coverage of the pattern. Combined with previously computed statistics about generalizations this is used to compute the interestingness according to the chosen generalization-aware measure. Subgroups with sufficient high score are placed in the result set, potentially replacing others in a top-k approach. Afterwards the target share in generalizations and the minimum number of removed negative instances are updated by using the statistics of the current patterns coverage. After the evaluation of a pattern all optimistic estimates, that is, traditional estimates (see theorem 1) and difference-based estimates are computed from the information stored for a candidate. If any optimistic estimate is lower than the threshold given by the result set for a top-k pattern, then the pattern is removed from the list of current candidates. Thus, no specializations of this pattern are explored in the next level of search.

The approach for numeric target concepts is very similar, except that minimum/maximum and mean target values as well as overall instance counts of the candidate patterns are stored instead of counts of positives and negatives. When determining the pruning bounds, a pattern is compared with all its direct generalizations. For each generalization an optimistic estimate bound is computed based on the difference of instances between the generalization and the specialization and the stored minimum/maximum target values. The tightest bound can be applied for pruning.

For the experiments, the algorithm was implemented in the open-source environment VIKAMINE [3]. The implementation utilizes an efficient bitset-based data structure to determine the coverage of patterns efficiently.

6   Evaluation

In this section, we show the effectiveness of the presented approach in experiments using well-known datasets from the UCI [9] repository. As a baseline algorithm we use a variant of the MPR-algorithm presented in [5], as this is the most recently proposed algorithm for this task. The algorithm was slightly modified to support top-k mining and to incorporate the bounds of Theorem 1 for any a. Since this algorithm follows the same search strategy as our novel algorithm, that is, apriori, it allows to determine the improvements that originate directly from the advanced pruning bounds presented in this paper. Results below are shown for $k = 20$, a realistic number for practical applications, which was also used for example as beam size in [26]. Different choices of k lead to similar results. For the numeric attributes an equal-frequency discretization was used, using all half-open intervals from the cutpoints as selectors. The experiments were performed on an office PC with 2.8 Ghz and 6 GB RAM.

In the first part of the evaluation we investigated the setting of a binary target concept using different generalization-aware quality functions r_{bin}^a. We compared the runtimes of the presented algorithm with traditional pruning only and with the novel generalization-aware bounds. The results show, that utilizing difference-based pruning leads to significant runtime improvements in almost all tasks, see Table 1. The improvements range from a factor of about 2 to over 20 in the datasets hypothyroid, audiology and spammer. For a more detailed analysis we investigated these tasks more closely. It turned out that the search space for these datasets contained multiple selectors that covered a vast majority of the instances. Conjunctive combinations of subsets of these selectors still cover a large part of the dataset and especially of the positive instances. As traditional optimistic estimates are based on this number of covered positive instance, pruning cannot be applied on these combinations efficiently. In contrast, since the number of negative instances, by which those patterns differ from generalizations, is often very low in these cases, such combination can be pruned often using the difference-based optimistic estimates presented in this paper. This leads to the massive improvements. We can conclude that our new pruning scheme is especially efficient, if many selectors cover a majority of the dataset. In some cases the algorithms did not finish due to out of memory errors despite the large amount of available memory. This does occur less often using the novel bounds, see for example the results for the vehicle dataset, since less candidates are generated in apriori, if more advanced bounds are applied.

In the second part of the evaluation the interestingness measure $r_{bin}^{0.5}$, a generalization-aware variant of the binomial-test, was further analyzed by comparing the runtimes for different search depth (maximum number of selectors in a pattern), see Table 2. As before, almost all tasks finished earlier using the novel difference-based pruning. While the improvement is only moderate for low search depth, massive speedups can be observed for $d = 5$ and $d = 6$. For $d = 6$ many algorithms with only traditional pruning did not finish because of limited memory. When additionally using the novel bounds, this happened only in two datasets, as less candidates were generated.

In the last part of the evaluation the improvements in a setting with numeric target concepts and quality functions q_{num}^a were examined. For subgroup discovery with numeric targets and generalization-aware quality functions no optimistic estimates have been proposed so far. To allow for a comparison nonetheless, we use the optimistic estimate bound $\bar{o}e_{num}^1 = \sum_{x:tc(x)>\mu_\emptyset}(tc(x) - \mu_\emptyset)$, which has been shown to be a correct optimistic estimate for q_{num}^1. Since $r_{num}^a(P) \leq r_{num}^1(P) \leq q_{num}^1(P)$ this can also be used as a (non-tight) optimistic estimate for any generalization-aware quality function r_{num}^a. Results are shown in Table 3. Since

Table 1. Runtime comparison (in s) of the base algorithm with traditional pruning based on the positives (std) and the novel algorithm with additional difference-based pruning (dbp) using different size parameters a for quality functions r_{bin}^a. The maximum number describing selectors was limited to $d = 5$. "-" indicates that the algorithm did not finish due to lack of memory.

a	0.0		0.1		0.5		1.0	
pruning	dpb	std	dpb	std	dpb	std	dpb	std
adults	1.1	1.0	17.8	48.7	1.6	8.1	1.0	1.7
audiology	0.2	62.3	24.9	51.6	0.6	51.7	0.1	57.4
census-kdd	16.4	16.2	-	-	107.9	2954.3	18.5	94.0
colic	<0.1	<0.1	1.7	4.8	0.4	5.1	0.1	1.2
credit-a	<0.1	<0.1	2.6	4.1	1.2	3.6	<0.1	0.4
credit-g	0.2	0.2	24.4	42.5	4.0	35.2	0.4	4.6
diabetes	1.0	3.8	5.9	12.6	1.2	9.3	<0.1	0.7
hepatitis	1.5	11.8	2.3	4.9	0.8	3.3	<0.1	0.5
hypothyroid	0.1	1.2	2.0	37.1	1.7	39.0	<0.1	21.2
spammer	4.3	5.5	133.0	-	29.3	172.2	0.5	27.6
vehicle	2.3	2.7	-	-	15.6	-	0.9	-

Table 2. Runtime comparison (in s) of the base algorithm with traditional pruning based on the positives (std) and the novel algorithm with additional difference-based pruning (dbp) using different maximum numbers d of describing selectors in a pattern. As quality functions the generalization-aware mean test $r_{bin}^{0.5}$ was used. "-" indicates that the algorithm did not finish due to lack of memory.

d	3		4		5		6	
pruning	dpb	std	dpb	std	dpb	std	dpb	std
adults	1.0	1.1	0.9	1.8	1.6	8.1	1.7	30.2
audiology	0.1	0.1	0.1	2.8	0.6	51.7	-	-
census-kdd	17.9	20.6	37.2	99.8	107.9	2954.3	267.5	-
colic	0.1	0.2	0.3	1.1	0.4	5.1	0.4	16.4
credit-a	0.1	0.1	0.3	0.7	1.2	3.6	1.2	12.9
credit-g	0.2	0.2	1.5	4.0	4.0	35.2	7.0	-
diabetes	0.1	0.1	0.5	1.3	1.2	9.3	2.0	67.1
hepatitis	<0.1	0.1	0.2	0.6	0.8	3.3	0.3	11.9
hypothyroid	0.1	0.2	0.5	2.7	1.7	39.0	-	-
spammer	1.3	1.6	5.7	15.5	29.3	172.2	88.3	-
vehicle	1.0	1.3	4.8	57.8	15.6	-	-	-

the applied traditional bound is tight for $a = 1$, the runtimes in this case are relatively low already for the studied datasets, leaving only little room for improvement. For lower values of a, significant runtime improvements can be observed, which reach a full order of magnitude (e.g., for the datasets concrete_data and housing). The relative runtime improvement is on average highest for $a = 0.5$. This can be explained by the fact that for lower values of a even small subgroups can be considered as interesting. This makes it more difficult to exclude subgroups by pruning also when using the difference-based bounds.

Table 3. Runtime comparison (in s) of the base algorithm with traditional pruning based on the positives (std) and the novel algorithm with additional difference-based pruning (dbp) using different size parameters a for quality functions r_{num}^a for numeric target concepts. The maximum number describing selectors was limited to $d = 5$.

a	0.0		0.1		0.5		1.0	
pruning	dpb	std	dpb	std	dpb	std	dpb	std
adults	19.6	92.5	22.5	89.6	14.8	64.7	3.9	14.9
concrete_data	4.7	20.8	6.2	20.1	1.3	11.2	0.1	0.3
credit-a	3.7	14.1	5.1	13.8	3.1	9.7	0.4	0.8
credit-g	6.6	53.0	8.5	54.5	7.9	40.3	0.5	1.0
diabetes	5.6	20.4	8.5	18.6	5.2	15.0	0.3	0.7
forestfires	2.3	10.4	3.4	11.1	2.7	9.6	2.4	6.5
heart-c	3.5	17.5	5.6	17.5	2.9	13.2	0.2	0.5
housing	2.0	28.2	3.4	26.4	1.7	23.8	0.1	3.0
yeast	3.1	14.3	3.5	13.9	1.5	8.4	0.1	0.8

7   Conclusions

In this paper we proposed a new scheme of deriving optimistic estimates bounds for subgroup discovery with interesting measures that take statistics of generalizations into account. In contrast to previous approaches the bounds are not only based on the anti-monotonicity of instances, which are contained within the subgroup, but also on the number of instance that are covered by a pattern, but not by its generalization. The optimistic estimates have been incorporated in an efficient algorithm that outperforms previous approaches by up to an order of magnitude. The speed-up is especially high, if the dataset contains selection expressions that cover a large part of the dataset.

In the future we plan to extend this approach to explore novel interestingness measures that take generalizations into account. Furthermore, an analysis of different search strategies, e.g., reverse-depth-first search, for this task is an interesting direction.

References

1. Agrawal, R., Imielienski, T., Swami, A.: Mining association rules between sets of items in large databases. ACM SIGMOD Record, 1–10 (May 1993)
2. Atzmueller, M., Lemmerich, F.: Fast subgroup discovery for continuous target concepts. In: Rauch, J., Raś, Z.W., Berka, P., Elomaa, T. (eds.) ISMIS 2009. LNCS, vol. 5722, pp. 35–44. Springer, Heidelberg (2009)
3. Atzmueller, M., Lemmerich, F.: VIKAMINE–Open-Source Subgroup Discovery, Pattern Mining, and Analytics. In: Flach, P.A., De Bie, T., Cristianini, N. (eds.) ECML PKDD 2012, Part II. LNCS, vol. 7524, pp. 842–845. Springer, Heidelberg (2012)
4. Aumann, Y., Lindell, Y.: A statistical theory for quantitative association rules. In: Knowledge Discovery and Data Mining, pp. 261–270 (1999)
5. Batal, I., Hauskrecht, M.: A concise representation of association rules using minimal predictive rules. In: Balcázar, J.L., Bonchi, F., Gionis, A., Sebag, M. (eds.) ECML PKDD 2010, Part I. LNCS, vol. 6321, pp. 87–102. Springer, Heidelberg (2010)

6. Batal, I., Hauskrecht, M.: Constructing classification features using minimal predictive patterns. In: Proceedings of the 19th ACM International Symposium on High Performance Distributed Computing, pp. 869–877 (2010)
7. Bay, S., Pazzani, M.: Detecting change in categorical data: Mining contrast sets. In: Proceedings of the Fifth ACM SIGKDD Int. Conf. on KDD (1999)
8. Bayardo, R.: Efficiently mining long patterns from databases. ACM SIGMOD Record, 85–93 (1998)
9. Blake, C., Merz, C.J.: {UCI} Repository of machine learning databases (1998)
10. Cheng, H., Yan, X., Han, J., Yu, P.: Direct discriminative pattern mining for effective classification. In: ICDE 2008, Proceedings of the 2008 IEEE 24th International Conference on Data Engineering, pp. 169–178 (April 2008)
11. Dong, G., Li, J.: Efficient mining of emerging patterns: Discovering trends and differences. In: Proceedings of the Fifth ACM SIGKDD International Conference on Knowledge Discovery and Data Mining, pp. 1–11 (1999)
12. Garriga, G., Kralj, P., Lavrac, N.: Closed sets for labeled data. The Journal of Machine Learning Research 9, 559–580 (2008)
13. Geng, L., Hamilton, H.J.: Interestingness measures for data mining. ACM Computing Surveys 38(3), 9–es (2006)
14. Grosskreutz, H., Boley, M., Krause-Traudes, M.: Subgroup discovery for election analysis: a case study in descriptive data mining. Disc. Science, 57–71 (2010)
15. Grosskreutz, H., Rüping, S., Wrobel, S.: Tight optimistic estimates for fast subgroup discovery. In: Daelemans, W., Goethals, B., Morik, K. (eds.) ECML PKDD 2008, Part I. LNCS (LNAI), vol. 5211, pp. 440–456. Springer, Heidelberg (2008)
16. Kavšek, B., Lavrač, N.: Apriori-Sd: Adapting Association Rule Learning To Subgroup Discovery 20 (September 2006)
17. Klösgen, W.: Explora: A multipattern and multistrategy discovery assistant. In: Advances in Knowledge Discovery and Data Mining, pp. 249–271. American Association for Artificial Intelligence (1996)
18. Kralj Novak, P., Lavrač, N., Webb, G.I.: Supervised Descriptive Rule Discovery: A Unifying Survey of Contrast Set. Emerging Pattern and Subgroup Mining 10, 377–403 (2009)
19. Lemmerich, F., Puppe, F.: Local Models for Expectation-Driven Subgroup Discovery. In: 2011 IEEE 11th International Conference on Data Mining, pp. 360–369 (2011)
20. Morishita, S., Sese, J.: Traversing Itemset Lattices with Statistical Metric Pruning. In: Proc. of ACM SIGMOD, pp. 226–236 (2000)
21. Nijssen, S., Guns, T., Raedt, L.D.: Correlated itemset mining in roc space: a constraint programming approach. In: Proceedings of the 15th ACM SIGKDD International Conference on Knowledge Discovery and Data Mining (2009)
22. Webb, G.I.: OPUS: An efficient admissible algorithm for unordered search. arXiv preprint cs/9512101 3, 431–465 (1995)
23. Webb, G.I.: Discovering associations with numeric variables. In: Proceedings of the seventh ACM SIGKDD Int. Conf. on Knowledge Discovery and Data Mining (2001)
24. Webb, G.I., Zhang, S.: Removing trivial associations in association rule discovery. In: Proceedings of the First International NAISO Congress on Autonomous Intelligent Systems. NAISO Academic Press, Geelong (2002)
25. Wrobel, S.: An algorithm for multi-relational discovery of subgroups. In: Komorowski, J., Żytkow, J.M. (eds.) PKDD 1997. LNCS, vol. 1263, pp. 78–87. Springer, Heidelberg (1997)
26. Zimmermann, A., Raedt, L.D.: From Subgroup Discovery to Clustering. Machine Learning 77(1), 125–159 (2009)

Local Outlier Detection with Interpretation

Xuan Hong Dang[1], Barbora Micenková[1], Ira Assent[1], and Raymond T. Ng[2]

[1] Aarhus University, Denmark
{dang,barbora,ira}@cs.au.dk
[2] University of British Columbia, Canada
rng@cs.ubc.ca

Abstract. Outlier detection aims at searching for a small set of objects that are inconsistent or considerably deviating from other objects in a dataset. Existing research focuses on outlier identification while omitting the equally important problem of outlier interpretation. This paper presents a novel method named LODI to address both problems at the same time. In LODI, we develop an approach that explores the quadratic entropy to adaptively select a set of neighboring instances, and a learning method to seek an optimal subspace in which an outlier is maximally separated from its neighbors. We show that this learning task can be solved via the matrix eigen-decomposition and its solution contains essential information to reveal features that are most important to interpret the exceptional properties of outliers. We demonstrate the appealing performance of LODI via a number of synthetic and real world datasets and compare its outlier detection rates against state-of-the-art algorithms.

1 Introduction

Data mining aims at searching for novel and actionable knowledge from data. Mining techniques can generally be divided into four main categories: clustering, classification, frequent pattern mining and anomalies detection. Unlike the first three main tasks whose objective is to find patterns that characterize for majority data, the fourth one aims at finding patterns that only represent the minority data. Such kind of patterns usually do not fit well to the mechanisms that have generated the data and are often referred to as outliers, anomalies or surprising patterns. Mining that sort of rare patterns therefore poses novel issues and challenges. Yet, they are of interest and particularly important in a number of real world applications ranging from bioinformatics [28], direct marketing [18], to various types of fraud detection [4].

Outlying patterns may be divided into two types: global and local outliers. A global outlier is an object which has a significantly large distance to its k-th nearest neighbor (usually greater than a global threshold) whereas a local outlier has a distance to its k-th neighbor that is large *relatively to* the average distance of its neighbors to their own k-th nearest neighbors [6]. Although it is also possible to create a ranking of global outliers (and select the top outliers), it is noted in [6,3] that the notion of local outliers remains more general than that of

H. Blockeel et al. (Eds.): ECML PKDD 2013, Part III, LNAI 8190, pp. 304–320, 2013.
© Springer-Verlag Berlin Heidelberg 2013

global outliers and, usually, a global outlier is also a local one but not vice versa, making the methods to discover local outliers typically more computationally expensive. In this study, our objective is to focus on mining and interpreting *local* outliers.

Although there is a large number of techniques for discovering global and local anomalous patterns [29,26], most attempts focus solely on the aspect of outlier *identification*, ignoring the equally important problem of outlier *interpretation*. For many application domains, especially those with data described by a large number of features, the description/intepretation of outliers is essential. As such, an outlier should be explained clearly and compactly, like a subset of features, that shows its exceptionality. This knowledge obviously assists the user to evaluate the validity of the uncovered outliers. More importantly, it offers him/her a facility to gain insights into why an outlier is exceptionally different from other regular objects. To our best knowledge, the study developed in [13] is the only attempt that directly addresses this issue, yet for global outliers but not for the more challenging patterns of local outliers (shortly reviewed in Section 2).

In this work, we introduce a novel approach that achieves both objectives of local outlier detection and interpretation at the same time. We propose a technique relying on the information theoretic measure of entropy to select an appropriate set of neighboring objects of an outlier candidate. Unlike most existing methods which often select the k closest objects as neighbors, our proposed technique goes further by requiring strong interconnections (or high entropy) amongst all neighboring members. This helps to remove irrelevant objects that can be nearby outliers or the objects coming from other distributions, and thus ensures all remaining objects to be truly normal inliers generated by the same distribution (illustrated via examples later). This characteristic is crucial since the statistical properties of the neighborhood play an essential role in our explanation of the outlierness. We then develop a method, whose solution firmly relies on the matrix eigen-decomposition, to learn an optimal one-dimensional subspace in which an outlier is most distinguishable from its neighboring set. The basic idea behind this approach is to consider the local outlier detection problem as a binary classification and thus ensure that a single dimension is sufficient to discriminate an outlier from its vicinity. The induced dimension is in essence a linear combination of the original features and thus contains all intrinsic information to reveal which original features are the most important to explain outliers. A visualization associated with the outlier interpretation is provided for intuitive understanding. Our explanation form not only shows the relevant features but also ranks objects according to their outlierness.

2   Related Work

Studies in outlier detection can generally be divided into two categories stemming from: (i) statistics and (ii) data mining. In the statistical approach, most methods assume that the observed data are governed by some statistical process to which a standard probability distribution (e.g., Binomial, Gaussian, Poisson etc.) with appropriate parameters can be fitted to. An object is identified as an outlier based on how unlikely it could have been generated by that distribution [2]. Data mining techniques, on the other hand, attempt to avoid model assumptions; relying on the concepts of distance and density, as stated earlier. For most distance-based methods [12,27], two parameters called distance d and data fraction p are required. Following that, an outlier has at least fraction p of all instances farther than d from it [12]. As both d and p are parameters defined over the entire data, methods based on distance can only find *global* outliers. Techniques relying on density, in contrast, attempt to seek *local* outliers, whose outlying degrees ("local outlier factor"—LOF) are defined w.r.t. their neighborhoods rather than the entire dataset [3,6]. There are several recent studies that attempt to find outliers in spaces with reduced dimensionality. Some of them consider every single dimension [10] or every combination of two dimensions [7] as the reduced dimensional subspaces, others [19,11] go further in refining the number of relevant subspaces. While the work in [19] makes assumptions that outliers can only exist in subspaces with non-uniform distributions, the method developed in [11] assumes that outliers only appear in subspaces showing high dependencies amongst their related dimensions. These studies, exploring either subspace projections [19,11] or subspace samplings [18,10,7], appear to be appropriate for the purpose of outlier interpretation. Nonetheless, as the outlier score of an object is aggregated from multiple spaces, it remains unclear which subspace should be selected to interpret its outlierness property. In addition, the number of explored subspaces for every object should be large in order to obtain good outlier ranking results. These techniques are hence closer to outlier ensembles [25] rather than outlier interpretation. The recent SOD method [14] pursues a slightly different approach in which it seeks an axis-parallel hyperplane (w.r.t. an object) as one spanned by the attributes with the highest data variances. The anomaly degree of the object is thus computed in the space orthogonal to this hyperplane. This technique also adopts an approach based on the shared neighbors between two objects to measure their similarity, which alleviates the almost equi-distance effect among all instances in a high dimensional space and thus can achieve better selection for neighboring sets. SOD was demonstrated to be effective in uncovering outliers that deviate from the most variance attributes yet it seems somewhat limited in searching outliers having extreme values in such directions. A similar approach is adopted in [16] where the subspace can be arbitrarily oriented (not only axis-parallel) and a form of outlier characterization based on vector directions have been proposed. ABOD [15] pursues a different approach where variance of angles among objects is taken into account to compute outlierness, making the method suitable for high dimensional data. In terms of outlier detection, we provide experimental comparisons with state-of-the-art algorithms in Section 4.

3   Our Approach

In this work, we consider $\mathcal{X} = \{\mathbf{x}_1, \mathbf{x}_2, \ldots, \mathbf{x}_N\}$ a dataset of N instances and each $\mathbf{x}_i \in \mathcal{X}$ is represented as a vector in a D-dimensional space. Each dimension represents a feature f_1 to f_D. We aim for an algorithm that can rank the objects in $\mathcal{X}$ w.r.t. their outlier degrees with the most outlying objects on the top. Having been queried for M outliers in $\mathcal{X}$, the algorithm returns the top M outliers and for a threshold $\lambda \in (0,1)$ (to be clear in Section 3.3), each outlier $\mathbf{x}_i$ is associated with a small set of features $\{f_1^{(\mathbf{x}_i)}, \ldots, f_d^{(\mathbf{x}_i)}\}, d \ll D$ explaining why the object is exceptional. The value of d may vary across different outliers. In addition, $f_1^{(\mathbf{x}_i)}, \ldots, f_d^{(\mathbf{x}_i)}$ are also weighted according to the degree to which they contribute to discriminate $\mathbf{x}_i$ as an outlier.

3.1   Neighboring Set Selection

Compared to global anomalous patterns, mining local outliers is generally harder and more challenging since it has to further deal with the problem of locally different densities in the data distribution. An outlier is considered anomalous if its density value is significantly different from the average density computed from the neighboring objects. The anomalous property of an outlier is thus decided by the local density distribution rather than the global knowledge derived from the entire distribution. For most existing studies [3,14], the set of k nearest neighboring objects (kNNs) is used. Nonetheless, this approach has not been thoroughly investigated and may be misleading for outlier explanation. The difficulty comes from the fact that identifying a proper value of k is not only a non-trivial task [22,3] but such a set of k closest neighbors might also contain nearby outliers or inliers from several distributions, which both strongly affect the statistical properties of the neighboring set. To give an illustration, we borrow a very popular data set from subspace clustering [23,17] which includes four clusters in a 3-dimensional space with 20 outliers randomly added as shown in Figure 1(a). Each cluster is only visible in 2-dimensional subspace [17] and each outlier is considered anomalous w.r.t. its closest cluster. Now taking the outlier $\mathbf{o}_1$ as an example, regardless of how small k is selected, other nearby outliers such as $\mathbf{o}_2, \mathbf{o}_3$ or $\mathbf{o}_4$ are included in its neighbors since they are amongst the closest objects (see Figure 1(a)). On the other hand, increasing k to include more inliers from the upper distribution can alleviate the effect of these outliers on the $\mathbf{o}_1$'s anomalous property. Unfortunately, such a large setting also comprises instances from the lower right distribution as shown in Figure 1(b). To cope with these issues, our objective is to ensure that all $\mathbf{o}_1$'s neighbors are truly inliers coming from a single closest distribution and thus $\mathbf{o}_1$ can be considered as its local outlier. Our proposed approach to handle this issue stems from the well-studied concept of entropy in information theory. The technique is adaptive by not fixing the number of neighboring inliers k. Instead, we only use k as a lower bound to ensure that the number of final nearby inliers is no less than k.

In information theory, entropy is used to measure the uncertainty (or disorder) of a stochastic event. Following the definition by Shannon, the entropy of that event is defined by $H(X) = -\int p(\mathbf{x}) \log p(\mathbf{x}) d\mathbf{x}$, of which X is the stochastic event or more specifically, a continuous random variable, and $p(\mathbf{x})$ is its corresponding probability distribution. If the entropy of X is large, its purity is low, or equivalently, X's uncertainty is high. Therefore, it is natural to exploit entropy for our task of selecting neighboring inliers. Intuitively, for the entropy computed with respect to this set, we would expect its value to be small in order to infer that objects within the set are all similar (i.e., high purity) and thus there is a high possibility that they are being generated from the same statistical mechanism or distribution. Nonetheless, computing entropy in Shannon's definition is not an easy task since it requires $p(\mathbf{x})$ to be known. We thus utilize a more general form, the Renyi entropy [24], which enables a straightforward computation. Mathematically, given α as an order, Renyi entropy is defined as:

$$H_{R_\alpha}(X) = \frac{1}{1-\alpha} \log \int p(\mathbf{x})^\alpha d\mathbf{x}, \text{ for } \alpha > 0, \ \alpha \neq 1. \tag{1}$$

in which Shannon entropy is a special case when α is approaching 1 (i.e., $\lim_{\alpha \to 1} H_{R_\alpha}(X) = H(X)$ [24]). However, in order to ensure the practical computation and impose no assumption regarding the probability distribution $p(\mathbf{x})$, we select $\alpha = 2$, yielding the quadratic form of entropy, and use the non-parametric Parzen window technique to estimate $p(\mathbf{x})$. More specifically, let us denote $R(\mathbf{o}) = \{\mathbf{x}_1, \mathbf{x}_2, \ldots, \mathbf{x}_s\}$ as the initial set of nearest neighboring instances closest to an outlier candidate $\mathbf{o}$. Following the Parzen window technique, we approximate $p(\mathbf{x})$ w.r.t. this set via the sum of kernels placed at each $\{\mathbf{x}_i\}_{i=1}^s$ and it follows that:

$$p(\mathbf{x}) = s^{-1} \sum_i G(\mathbf{x} - \mathbf{x}_i, \sigma^2) \tag{2}$$

where $G(\mathbf{x} - \mathbf{x}_i, \sigma^2) = (2\pi\sigma)^{-D/2} \exp\left\{-\frac{||\mathbf{x}-\mathbf{x}_i||^2}{2\sigma^2}\right\}$ is the Gaussian in the D-dimensional space used as the kernel function. In combination with setting $\alpha = 2$, this leads to a direct computation of the local quadratic Renyi entropy as follows:

$$QE(R(\mathbf{o})) = -\ln \int \left(\frac{1}{s} \sum_{i=1}^s G(\mathbf{x} - \mathbf{x}_i, \sigma^2)\right) \left(\frac{1}{s} \sum_{j=1}^s G(\mathbf{x} - \mathbf{x}_j, \sigma^2)\right)$$

$$= -\ln \frac{1}{s^2} \sum_i^s \sum_j^s G(\mathbf{x}_i - \mathbf{x}_j, 2\sigma^2) \tag{3}$$

Notice that, unlike Shannon entropy, the above computation removes burden of the computation of the numerical integration due to the advantages of the quadratic form and the convolution property of two Gaussian functions. Essentially, the sum within the logarithm operation can be interpreted as the local information potential. Each term in the summation satisfies the positivity and increases as the distance between $\mathbf{x}_i$ and $\mathbf{x}_j$ decreases, very much analogous to the potential energy between two physical particles. As such, our objective of minimizing the entropy is equivalent to maximizing the information potential within the neighboring set. The higher the information potential of the set is, the more similar the elements within the set are.

Having the way to capture the local quadratic entropy, an appropriate set of nearest neighbors can be selected adaptively as follows. We begin by setting

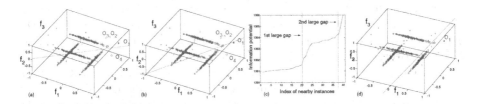

Fig. 1. Neighbors selection: object under consideration is o_1 and circle points are its nearest neighbors (figures are best visualized in colors).

the number of initial nearest neighbors to s (in our experiments, a setting of $s = 2k$ often gives good results), and aim to find an optimal subset of no less than k instances with maximum local information potential. Obviously, a naive way to find such an optimal set may require computing all $\sum_{i=k}^{s} \binom{s}{i}$ possible combinations, which is prohibitively expensive. We thus make use of an heuristic approach to select such a subset. Specifically, removing an object from the neighboring set will lead to a decrement in the total information potential. Those instances resulting in the most decrement are important ones whereas those causing least decrement tend to be irrelevant for the neighboring set. With the latter objects, their potential energy is minor as they loosely interact with the rest of neighboring objects and thus excluding them makes the neighboring set less uncertain or more pure. These objects in fact can be either other outliers or part of nearby distributions. Our method thus ranks the total information potential left in the increasing order and removes objects behind the first significant gap as long as the number of remaining instances is no less than k. A significant gap is defined to have a value larger than the average gap.

For illustration, we plot in Figure 1(c) the total information potential left (ordered increasingly) after excluding each of nearest neighboring objects represented in Figure 1(b). One may observe that there are two remarkably large gaps in the plot (noted by the red vertical lines in Figure 1(c)), which indeed reflect the nature of local distribution surrounding outlier o_1. In particular, the first large gap signifies the information decrement in removing instances from the lower right distribution whereas the second one corresponds to the removal of nearby outliers. By excluding these irrelevant objects from the set of o_1's neighboring instances, the remaining ones are true inliers coming from the same and closest distribution shown as blue points in Figure 1(d).

3.2 Anomaly Degree Computation

Given a way to compute the neighboring (or "reference") set above, we develop a method to calculate the anomaly degree for each object in the dataset $\mathcal{X}$. Essentially, directly computing that measure in the original multidimensional data space is often less reliable since many features may not be relevant for the task of identifying local outliers. We thus exploit an approach of a *local dimensionality reduction*. For the remaining discussion, let us denote o as an outlier candidate under consideration, $R(\mathbf{o})$ as its neighboring inliers found by the entropy-based technique presented in the previous section and $\mathbf{R}$ as the matrix form of $R(\mathbf{o})$. Each neighboring inlier $\mathbf{x}_i \in R(\mathbf{o})$ corresponds to a column in $\mathbf{R}$ and together with $\mathbf{o}$, they are all vectors in the $\mathbb{R}^D$ space.

Essentially, we view the local outlier detection as a binary classification problem in the sense that the outlier candidate $\mathbf{o}$ should be distinguished from its neighbors $R(\mathbf{o})$. By dimensionality reduction, this objective is equivalent to the objective of learning an optimal subspace such that $\mathbf{o}$ is maximally separated from every object in $R(\mathbf{o})$. More specifically, $\mathbf{o}$ needs to be strongly deviating from $R(\mathbf{o})$ while at the same time $R(\mathbf{o})$ shows high density or low variance in that induced subspace. Following this approach, we denote the optimal 1-dimensional subspace as $\mathbf{w}$ and in order to achieve our goal, data variance is obviously an important statistical measure to explore. Toward this goal, we define the first variance of all neighboring objects projected onto $\mathbf{w}$ as follows:

$$Var(R(\mathbf{o})) = \mathbf{w}^T \left(\mathbf{R} - \mathbf{Ree}^T/N_\mathbf{o}\right) \left(\mathbf{R} - \mathbf{Ree}^T/N_\mathbf{o}\right)^T \mathbf{w} = \mathbf{w}^T \mathbf{A}\mathbf{A}^T \mathbf{w} \quad (4)$$

where $\mathbf{A} = \left(\mathbf{R} - \mathbf{Ree}^T/N_\mathbf{o}\right)$, $N_\mathbf{o}$ is the number of neighboring instances in $R(\mathbf{o})$ and $\mathbf{e}$ is the vector with all entries equal to 1.

Another important statistic in our approach is the distance between $\mathbf{o}$ and every object in $R(\mathbf{o})$. This resembles an *average proximity* in a hierarchical clustering technique[9] where all pairwise data distances are taken into account. Compared to the two extremes of using minimum or maximum distance, this measure often shows better stability. We hence formulate their variance in the projected dimension $\mathbf{w}$ as the following quantity:

$$D_{(\mathbf{o}, R(\mathbf{o}))} = \mathbf{w}^T \left(\sum (\mathbf{o} - \mathbf{x}_i)(\mathbf{o} - \mathbf{x}_i)^T\right) \mathbf{w} = \mathbf{w}^T \mathbf{B}\mathbf{B}^T \mathbf{w}, \quad (5)$$

where $\mathbf{x}_i \in R(\mathbf{o})$ and $\mathbf{B}$ is defined as the matrix whose each column corresponds to a vector $(\mathbf{o} - \mathbf{x}_i)$. Intuitively, in order to achieve the goal of optimally distinguishing $\mathbf{o}$ from its neighboring reference inliers, we want to learn a direction for $\mathbf{w}$ such that the variance of $R(\mathbf{o})$ projected onto it is minimized whereas the variance between $\mathbf{o}$ and $R(\mathbf{o})$ also projected on that direction is maximized. One possible way to do that is to form an objective function resembling Rayleigh's quotient which maximizes the ratio between $D_{(\mathbf{o}, R(\mathbf{o}))}$ and $R(\mathbf{o})$ as follows:

$$\arg\max_{\mathbf{w}} J(\mathbf{w}) = \frac{D_{(\mathbf{o}, R(\mathbf{o}))}}{Var(R(\mathbf{o}))} = \frac{\mathbf{w}^T \mathbf{B}\mathbf{B}^T \mathbf{w}}{\mathbf{w}^T \mathbf{A}\mathbf{A}^T \mathbf{w}}. \quad (6)$$

It is obvious that setting the derivative of $J(\mathbf{w})$ w.r.t. $\mathbf{w}$ equal to 0 results in $(\mathbf{w}^T \mathbf{B}\mathbf{B}^T \mathbf{w})\mathbf{A}\mathbf{A}^T \mathbf{w} = (\mathbf{w}^T \mathbf{A}\mathbf{A}^T \mathbf{w})\mathbf{B}\mathbf{B}^T \mathbf{w}$, which is in essence equivalent to solving the following generalized eigensystem:

$$J(\mathbf{w})\mathbf{A}\mathbf{A}^T \mathbf{w} = \mathbf{B}\mathbf{B}^T \mathbf{w}. \quad (7)$$

In dealing with this objective function, note that $\mathbf{A}\mathbf{A}^T$, though symmetric, may not be full rank as the number of neighbors can be smaller than the number of features. This matrix is thus not directly invertible. Moreover, the size of $\mathbf{AA}^T$ can be large and quadratically proportional to the feature number which makes its eigendecomposition computationally expensive. To alleviate this problem, we propose to approximate $\mathbf{A}$ via its singular value decomposition and consequently $\mathbf{w}$ can be computed using the pseudo inversion of $\mathbf{AA}^T$.

Specifically, since $\mathbf{A}$ in general is a rectangular matrix, it can be decomposed into three matrices $\mathbf{A} = \mathbf{U\Sigma V}^T$ of which $\mathbf{U}$ and $\mathbf{V}$ are matrices whose columns are $\mathbf{A}$'s left and right singular eigenvectors and $\mathbf{\Sigma}$ is the diagonal matrix of its singular values. In essence, as our objective is to compute matrix inversion, we remove singular values which are very close to 0 and approximate $\mathbf{A}$ by its set of leading singular values and vectors. More concretely, we estimate $\mathbf{A} = \sum_\ell \mathbf{u}_\ell \sigma_\ell \mathbf{v}_\ell^T$ such that the sum over keeping singular values σ_ℓ's explains for 95% (as demonstrated in our experimental studies) of the total values in the diagonal matrix $\mathbf{\Sigma}$. Additionally, we compute $\mathbf{U}$ via the eigendecomposition of $\mathbf{A}^T\mathbf{A}$ which has a lower dimensionality. Particularly, we can see that:

$$\mathbf{A}^T\mathbf{A} = \mathbf{V\Sigma}^2\mathbf{V}^T. \tag{8}$$

Then, taking the square of both sides and pre-multiplying with $\mathbf{\Sigma}^{-1}\mathbf{V}^T$ and post-multiplying with $\mathbf{V\Sigma}^{-1}$, we obtain:

$$\mathbf{\Sigma}^{-1}\mathbf{V}^T\mathbf{A}^T(\mathbf{AA}^T)\mathbf{AV\Sigma}^{-1} = \mathbf{\Sigma}^2$$
$$\mathbf{U}^T(\mathbf{AA}^T)\mathbf{U} = \mathbf{\Sigma}^2. \tag{9}$$

This implies that columns in $\mathbf{U}$ are the eigenvectors of $\mathbf{AA}^T$ and they can be computed via the eigenvectors of the smaller matrix $\mathbf{A}^T\mathbf{A}$, i.e., $\mathbf{U} = \mathbf{AV\Sigma}^{-1}$. Thus, the final pseudo inversion $(\mathbf{AA}^T)^\dagger$ can be simply approximated by $\mathbf{U\Sigma}^{-2}\mathbf{U}^T$. Plugging this value into our objective function in Eq.(7), it is straight-forward to see that the optimal direction for $\mathbf{w}$ is the first eigenvector of the matrix $\mathbf{U\Sigma}^{-2}\mathbf{U}^T\mathbf{BB}^T$ of which $J(\mathbf{w})$ achieves the maximum value as the largest eigenvalue of this matrix.

Given the optimal direction $\mathbf{w}$ uncovered by the technique developed above, the statistical distance between $\mathbf{o}$ and $R(\mathbf{o})$ can be calculated in terms of the standard deviation as follows:

$$AD(\mathbf{o}) = \max\left\{ \sqrt{\frac{(\mathbf{w}^T\mathbf{o} - \sum_i \frac{\mathbf{w}^T\mathbf{x}_i}{N_o})^2}{Var(\mathbf{w}^T R(\mathbf{o}))}}, \sqrt{Var(\mathbf{w}^T R(\mathbf{o}))} \right\} \tag{10}$$

where the second term in the max operation is added to ensure that the projection of $\mathbf{o}$ is not too close to the center of the projected neighboring instances (calculated in the first term). Notice that unlike most techniques that find multiple subspaces and have to deal with the problem of dimensionality bias [20], our approach naturally avoids this issue since it learns a 1-dimensional subspace and thus directly enables a comparison across objects. Therefore, with the objective of generating an outlier ranking over all objects, the relative difference between the statistical distance of an object $\mathbf{o}$ defined above and that of its neighboring objects is used to define its local anomalous degree:

$$LAD(\mathbf{o}) = AD(\mathbf{o}) \times \left(\sum AD(\mathbf{x}_i)/N_\mathbf{o} \right)^{-1}. \tag{11}$$

For this relative outlier measure, it is easy to see that if $\mathbf{o}$ is a regular object embedded in a cluster, its local anomaly degree is close to 1 whereas if it is a true outlier, the value will be greater than 1.

3.3   Outlier Interpretation

In interpreting the anomaly degree of an outlier, it is possible to rely on the correlation between the projected data in $\mathbf{w}$ and those in each of the original dimensions (i.e., $\mathbf{R}$'s rows). Features with highest absolute values can be used to interpret the anomaly degree of $\mathbf{o}$ since values of $\mathbf{o}$ and its referenced objects on these features are correlated to those projected onto $\mathbf{w}$. Nonetheless, this approach requires computing correlations with respect to all original features. A better and more direct approach is to exploit the optimal direction $\mathbf{w}$ directly. Recall that the projection of $R(\mathbf{o})$ over $\mathbf{w}$ is equivalent to the local linear combination of the original features. Consequently, coefficients within the eigenvector $\mathbf{w}$ are truly the weights of the original features. The feature corresponding to the largest absolute coefficient is the most important in determining $\mathbf{o}$ as an outlier. Analogously, the second important feature is the one corresponding to the second $\mathbf{w}$'s largest absolute component and so on. In this way, we are not only able to figure out which original features are crucial in distinguishing $\mathbf{o}$ but also show how important they are via the weights of the corresponding components in $\mathbf{w}$.

Generally, we can provide the user with a parameter λ, whose values are between $(0,1)$, to control the number of features used to interpret the anomaly degree. We select $\{f_i\}_{i=1}^{d}$ as the set of features that correspond to the top d largest absolute coefficients in $\mathbf{w}$ and s.t. $\sum_{i=1}^{d} |w_i| \geq \lambda \times \sum_{j=1}^{D} |w_j|$. The degree of importance of each respective f_i can be further computed as the ratio $|w_i|/\sum_{j=1}^{D} |w_j|$. An object $\mathbf{o}$ therefore can be interpreted as an outlier in the d-subspace $\{f_1, ..., f_d\}$ with the corresponding feature importance degrees. An illustration is given in Figure 1 where $\mathbf{w}$ is plotted as the green line whose coefficients in the rightmost subgraph are $(0.11, 4.63, 5.12)$ (or in terms of importance degrees $(0.03, \mathbf{0.46}, \mathbf{0.51})$) which obviously reveals $\{f_2, f_3\}$ being two important features to explain $\mathbf{o}_1$ as an outlier. Note that the corresponding values of $\mathbf{w}$ (green lines) in Figures 1(a) and (b) are respectively $(6.12, 5.17, 0.59)$ and $(2.91, 4.72, 2.01)$, which tend to select $\{f_1, f_2\}$ and $\{f_1, f_2, f_3\}$ as the subspaces for $\mathbf{o}_1$ due to the influence of nearby *irrelevant* instances. The advantage of our entropy-based neighbor selection is thus demonstrated here where only the direction of $\mathbf{w}$ in Figures 1(d) is in parallel to the relevant subspace $\{f_2, f_3\}$ (compared to the slant lines of $\mathbf{w}$ shown in Figures 1(a) and (b)).

3.4   Algorithm Complexity

We name our algorithm LODI which stands for Local Outlier Detection with Interpretation and its computation complexity is analyzed as follows. LODI requires the calculation of the neighboring set as well as the local quadratic Renyi entropy. Both these steps take $O(DN \log N)$ with the implementation of

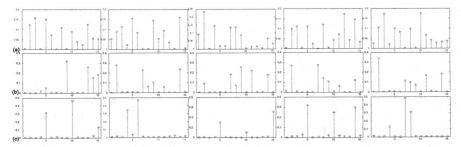

Fig. 2. Feature visualization over 5 top ranking outliers found in Syn1, Syn2 and Syn3 datasets (x- and y-axis are respectively the features' index and importance degree).

the $k-d$ tree data structure. The size of the matrix $\mathbf{A}^T\mathbf{A}$ is $s \times s$ and thus its eigen-decomposition is $O(Ds \log s)$ using the Lanczos method [8]. Similarly, computing the eigen-decomposition of $\mathbf{U\Sigma}^{-2}\mathbf{U}^T\mathbf{BB}^T$ amounts to $O(D^2 \log D)$. We compute these steps for all instances to render the outlier ranking list so these computations take $O(DN(s \log s + D \log D))$. The overall complexity is thus at most $O(DN(\log N + s \log s + D \log D))$.

4   Experimental Results

In this section, we provide experimental results on both synthetic and real-world datasets. We compare LODI against the following algorithms: LOF (density-based technique) [3], ABOD (angle-based) [15] and SOD (axis-parallel subspaces) [14]. The last two algorithms are adapted from the ELKI package[1] with some small changes in their output formats. Unless specified differently, we use $k = 20$ as the lower bound for the number of kNNs used in LODI. We also vary the number of neighbors, like $minPts$ in LOF or reference points in SOD, between 10 and 40 and report the best results. With SOD, we further set $\alpha = 0.8$ as recommended by the authors [14].

4.1   Synthetic Data

Data Description. We generate three synthetic datasets Syn1, Syn2 and Syn3, each consists of 50K data instances generated from 10 normal distributions. For each dimension ith of a normal distribution, the center μ_i is randomly selected from $\{10, 20, 30, 40, 50\}$ while variance σ_i is taken from either of two (considerably different) values 10 and 100. Such a setting aims to ensure that if the dimension ith of a distribution takes the large variance, its corresponding generated data will spread out in almost entire data space and thus an outlier close to this distribution can be hard to uncover in the ith dimension due to the strongly overlapping values projected onto this dimension. We set the percentage of the large variance to 40%, 60% and 80%, respectively, to generate Syn1, Syn2 and Syn3. For each dataset, we vary 1%, 2%, 5% and 10% of the whole data as the

---

[1] http://elki.dbs.ifi.lmu.de/

number of randomly generated outliers within the range of the data space and also vary the dimensionality of each dataset from 15 to 50.

Outlier Explanation. In Figure 2, we provide a feature visualization of the 5 top-ranked outliers returned by our LODI algorithm on the three datasets. For each graph in the figure, the x-axis shows the index of features while the y-axis shows their degree of importance. For the purpose of visualization, we plot the results where three datasets are generated with 5% outlier percentage and in 15 dimensions. The results for higher dimensionalities and other outlier percentages are very similar to those plotted here and thus were omitted to save space (yet, they are summarized in Table 1 and will be soon discussed). As observed from these graphs, the number of relevant features used to explain the anomalous property of each outlier is varied considerably across the three datasets. In Syn1 (Figure 2(a)), each identified outlier can be interpreted in a large number of dimensions since the percentage of the large variance used to generate this dataset is small, only 40%. When increasing the percentage to 60% in Syn2 and to 80% in Syn3 (Figures 2(b-c)), the number of relevant features reduces accordingly. In Syn3 dataset, generally only 3 features are needed to interpret its outliers. These results have been anticipated and quite intuitive since once the number of dimensions with large variance increases, the dimensionality of the subspaces in which an outlier can be found and explained will be narrowed down. This is due to the wide overlapping of outliers and regular objects projected onto these (large variance) dimensions.

For comparison against other techniques, we select the SOD algorithm. Recall that SOD is not directly designed for outlier interpretation, yet its uncovered axis-parallel subspaces might be used to select outliers' relevant features. For these experiments, we select Syn3 dataset and vary the outlier percentage from 1% to 10%, and the data dimensionality from 15 to 50 features. Table 1 reports the average subspace's dimensionality of LODI and SOD computed from their top ranking outliers. The first column shows the outlier percentages while D15, D30 and D50 denote the data dimensionality. We set $\lambda = 0.8$ (see Section 3.3) for LODI and $\alpha = 0.8$ for SOD to ensure their good performance. As one can observe, LODI tends to select subspaces with dimensionality close to the true one whereas the dimensionality of the axis-parallel subspaces in SOD is often higher. For example, at D15, LODI uses around 3 original features to explain each outlier, which is quite consistent with the percentage of 80% of the large variance while it is approximately 8 features for SOD. It can further be observed that the number of relevant features uncovered by SOD also greatly varies, which is indicated by the high standard deviation. Additionally, it tends to increase as the percentage of outliers increases. In contrast, our method performs better and the relevant subspace dimensionality is less sensitive to the variation of the outlier percentages as well as to the number of original features.

Outlier Detection. For comparison of outlier detection rates, we further include the angle-based ABOD and the density-based LOF techniques. The receiver operating characteristic (ROC) is used to evaluate the performance of all algorithms. It was observed that all methods performed quite competitively in

Table 1. Average dimensionality of the subspaces selected for outlier explanation in LODI and SOD in Syn3 dataset (values after ± are standard deviations).

Outlier %	D15		D30		D50	
	LODI	SOD	LODI	SOD	LODI	SOD
1%	3.12±0.84	8.35±1.61	6.34±1.27	16.05±2.46	10.92±2.15	26.50±3.95
2%	3.20±0.72	8.40±1.68	6.41±1.14	16.13±2.56	11.03±2.07	27.57±4.15
5%	3.15±0.81	8.16±1.69	6.70±1.18	16.20±2.69	10.87±2.21	26.62±4.31
10%	3.14±0.96	7.84±1.85	6.42±1.23	15.87±3.05	11.08±2.31	25.87±4.81

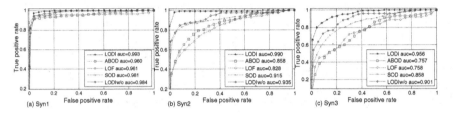

Fig. 3. Outlier detection rate of all algorithms on three synthetic datasets (D=50)

the low dimensionality yet their performances were more divergent on higher dimensional data. We hence report in Figure 3 the outlier detection rates of all methods in D50, setting for all 3 datasets. As observed from these graphs, the outlier detection performance of all algorithms is likely to be decreased as the the large variance percentage used to generate the data increases. However, while the detection rates decrease vastly for other methods, our technique LODI remains stable from Syn1 to Syn2 dataset, and only slightly reduces in Syn3. Nonetheless, its area under the ROC (AUC) is still around 96% for this dataset. Amongst other techniques, the AUCs of LOF are the lowest. This could be explained through its density-based approach which often makes LOF's performance deteriorated in high dimensional data. The performances of both ABOD and SOD are quite competitive yet their ROC curves are still lower than that of LODI for all three examined datasets. In Figure 3, we also report the performance of LODI not using the entropy-based approach in kNNs selection (denoted as LODIw/o). Instead, k is varied from 10 to 40 and the best result is reported. As seen in Figure 3, the AUC of LODIw/o in all cases are smaller than that of LODI, which highlights the significance of the entropy-based approach for kNNs selection. However, compared to other techniques, LODIw/o's outlier detection rate is still better, demonstrating the appealing approach of computing outlier degrees in subspaces learnt from the objective function developed in Eq.(6).

Parameters Sensitivity. To provide more insights into the performance of our LODI technique, we further test its detection rates with various parameter settings. In Figure 4(a), we plot its AUC performance on the Syn3 dataset when the data dimensionality increases from D15 to D50 and the outlier percentage varies from 1% to 10%. The lower threshold k for the neighboring set remains

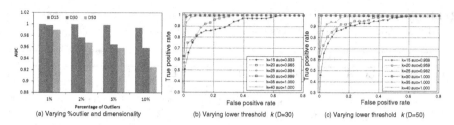

Fig. 4. Performance of LODI on Syn3 dataset with varying % outliers and threshold k

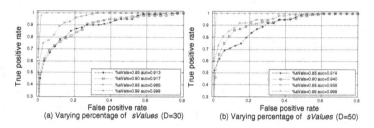

Fig. 5. Performance of LODI on Syn3 dataset with varying % singular values

at 20. One may see that LODI's performance slightly deteriorates as the number of outliers generated in the dataset increases. This happens since once the number of outliers increases, there are higher chances for them to be included in other instances' neighboring sets. Recall that LODI has alleviated this issue by excluding those with low information potential via the use of quadratic entropy. And in order to gain insights into this matter, we further test the case when the lower threshold for the neighboring set is varied. Figures 4(b-c) show the algorithm's ROC curves when k is changed from 15 to 40 for two cases of $D30$ and $D50$. As expected, once k increases, LODI has more capability in excluding irrelevant instances from the neighboring sets and its overall performance increases. As visualized from Figures 4(a-b), a general setting of k around 20 or 25 often leads to competitive results. We finally provide the impact of the total number of singular values used in our matrix approximation. In Figure 5, our algorithm's ROC curves are plotted as the percentage of keeping singular values is varied from 85%, 90%, 95% to 99%. We use Syn3 dataset for these experiments with the data dimensionality at 30 and 50. It is clearly seen that LODI performs better for higher percentages of singular values and in order to keep it at high performance, this parameter should be set around 90% or 95%.

4.2   Real World Data

In this section, we provide the experimental results of all algorithms on three real-world datasets selected from the UCI repository [1]. The first dataset is the image segmentation data which includes 2 310 instances of outdoor images {brickface, sky, foliage, cement, window, path, grass} classified into 7 classes. Each instance is a 3×3 region described by 19 attributes. However, we remove

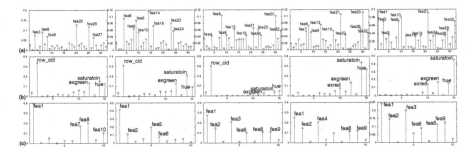

Fig. 6. Feature visualization over 5 top outliers found in: Ionosphere data (a), Image segmentation data (b) and Vowel data (c) (relevant features are shown with labels)

three features 5,7 and 9 from this data as they are known to be repetitive with the attributes 4,6 and 8 [1]. The second dataset is the vowel data consisting of 990 instances and is described by 11 variables (low pass filtered signals), of which the last one is the class label corresponding to 11 different English vowels {hid, hId, hEd, hAd, hYd, had, hOd, hod, hUd, hud, hed}. The third dataset is the ionosphere data containing 351 instances and being described by 32 features (electromagnetic signals). Instead of randomly generating artificial outliers and adding them to these datasets, it is more natural to directly downsample several classes and treat them as hidden outliers (as suggested in [19,11]). Specifically, we keep instances from two randomly selected classes of segmentation data as regular objects and downsample five remaining classes, each to 2 instances to represent hidden outliers. Likewise with the vowel dataset, we keep one class of regular objects and randomly sample 10 instances from the remaining classes to represent outliers. With the 2-class ionosphere data, we select instances from the second class as outliers since its number of objects is much lower than that of the first class.

Unlike the synthetic data where we can manage the data distributions and report the average subspace sizes for all outliers, it is harder to perform such analysis for the real-world datasets since different outliers may have relevant subspaces of different sizes. However, in an attempt to interpret the results of LODI, we plot in Figures 6(a-c) the original features' important degrees of 5 top-ranking outliers respectively selected from the ionosphere, image segmentation and vowel datasets. Figure 6(a) reveals that, for each outlier, there are only few features having high importance degrees and they can be selected as the subspace to interpret the abnormal property of the outlier. However, as this dataset has a large number of outliers, the subspaces do not have many features in common. It is thus more interesting to observe the feature visualization for the two other datasets. Looking at the the 5 top outliers of the image segmentation dataset in Figure 6(b), one can see that out of 15 original features, only a few are suitable to interpret the outliers. For example, the space spanned by {$row\_icd, exgreen, saturatioin, hue$} attributes is suitable to interpret the exceptional property of the first 3 outliers while the space spanned by {$exred, exgreen, saturatioin, hue$} is appropriate to explain the last 2 outliers.

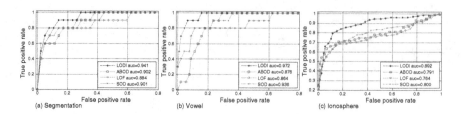

Fig. 7. Performance of all algorithms on three UCI real datasets.

Taking a closer look, we find out that these two types of outliers are indeed exceptional with respect to the 2 main distributions which correspond to the outdoor imaging instances of 2 classes (number 3rd and 6th) in the segmentation data. In the last dataset, vowel, shown in 6(c), few prominent features stand out for outlier interpretation, yet the features vary across different outliers (using $\lambda = 0.8$). Nevertheless, a common and interesting point is that the first attribute always has the highest value across all outliers, indicating it is the most important feature. Recall that for this dataset, we keep instances from only a single vowel (by random selection it is "hYd") as normal objects while randomly downsample one from each of the remaining vowels as hidden outliers. This might also justify the diversity of the other prominent features across the 5 outliers shown here.

We now compare the performance of LODI and the other algorithms through their outlier detection rates. In Figure 7, we report the ROC curves of all algorithms over the three datasets. As observed, LODI shows the best detection performance compared to all three techniques. In the segmentation data, LOF is less successful with its AUC value around 88% though we have tried to optimize its parameter $minPts$. The detection rates of ABOD and SOD are quite competitive and achieve 90% AUC which yet is still lower than LODI's 94%. Moreover, LODI is also likely to uncover all true outliers earlier than the other techniques. As observed in Figure 7, its false positive rate is only at 24% when all outliers are found compared to that of 43% for SOD or 60% for LOF. With the vowel dataset, we observe a similar behavior. Nevertheless, in the ionosphere where the number of outliers is considerably larger, none of the algorithms can discover all outliers before their false positive rate reaches 100%. However, it is seen that while both SOD and ABOD can uncover at most 70% of true outliers when the false positive rate is at 20%, LODI retrieves 86% at the same level. Its overall area under the curve is 89% which is clearly better than the other algorithms.

5   Conclusion

In this work, we developed the LODI algorithm to address outlier identification and explanation at the same time. In achieving this twin-objective, our method makes use of an approach firmly rooted from information theory to select appropriate sets of neighboring objects. We developed an objective function to learn subspaces in which outliers are most separable from their nearby inliers. We showed that the optimization problem can be optimally solved from the matrix eigen-decomposition of which relevant features are obtained to understand exceptional properties of outliers. Our thorough evaluation on both synthetic and real-world datasets demonstrated the appealing performance of LODI and its interpretation form over outliers is intuitive and meaningful. Nonetheless, LODI has some limitations. First, its computation is rather expensive (quadratic in the dimensionality), making LODI less suitable for very large and high dimensional datasets. In dealing with this issue, approaches based on features' sampling [21] seem to be potential; yet they also lead to some information loss. The challenge is thus to compromise the trade-off between these two criteria. Second, LODI made an assumption that an outlier can be *linearly* separated from inliers. This assumption may not be practical if distributions of inliers exhibit non-convex shapes. Though several learning techniques based on nonlinear dimensionality reduction can be applied to uncover such outliers [5], this, however, still leaves open to the difficult question of what can be an appropriate form to interpret these "*nonlinear*" outliers. We consider these challenges as the immediate issues for our future work.

Acknowledgements. Part of this work has been supported by the Danish Council for Independent Research - Technology and Production Sciences (FTP), grant 10-081972.

References

1. Asuncion, A., Newman, D.: UCI machine learning repository (2007)
2. Barnett, V., Lewis, T.: Outliers in statistical data, 3rd edn. John Wiley & Sons Ltd. (1994)
3. Breunig, M.M., Kriegel, H.P., Ng, R.T., Sander, J.: LOF: Identifying density-based local outliers. In: SIGMOD (2000)
4. Chandola, V., Banerjee, A., Kumar, V.: Anomaly detection: A survey. ACM Computing Surveys 41(3) (2009)
5. Dang, X.H., Micenková, B., Assent, I., Ng, R.T.: Outlier detection with space transformation and spectral analysis. In: SIAM-SDM (2013)
6. de Vries, T., Chawla, S., Houle, M.E.: Finding local anomalies in very high dimensional space. In: ICDM, pp. 128–137 (2010)
7. Foss, A., Zaïane, O.R., Zilles, S.: Unsupervised class separation of multivariate data through cumulative variance-based ranking. In: ICDM (2009)
8. Golub, G., Loan, C.: Matrix Computations, 3rd edn. The Johns Hopkins University Press (1996)
9. Han, J., Kamber, M., Pei, J.: Data Mining: Concepts and Techniques, 3rd edn. Morgan Kaufmann Publishers Inc. (2012)
10. He, Z., Deng, S., Xu, X.: A unified subspace outlier ensemble framework for outlier detection. In: Fan, W., Wu, Z., Yang, J. (eds.) WAIM 2005. LNCS, vol. 3739, pp. 632–637. Springer, Heidelberg (2005)
11. Keller, F., Müller, E., Böhm, K.: Hics: High contrast subspaces for density-based outlier ranking. In: ICDE (2012)

12. Knorr, E.M., Ng, R.T.: Algorithms for mining distance-based outliers in large datasets. In: VLDB (1998)
13. Knorr, E.M., Ng, R.T.: Finding intensional knowledge of distance-based outliers. The VLDB Journal 8, 2111–2222 (1999)
14. Kriegel, H.-P., Kröger, P., Schubert, E., Zimek, A.: Outlier detection in axis-parallel subspaces of high dimensional data. In: Theeramunkong, T., Kijsirikul, B., Cercone, N., Ho, T.-B. (eds.) PAKDD 2009. LNCS, vol. 5476, pp. 831–838. Springer, Heidelberg (2009)
15. Kriegel, H., Schubert, M., Zimek, A.: Angle-based outlier detection in high-dimensional data. In: SIGKDD (2008)
16. Kriegel, H.-P., Kröger, P., Schubert, E., Zimek, A.: Outlier detection in arbitrarily oriented subspaces. In: ICDM, pp. 379–388 (2012)
17. Kriegel, H.-P., Kröger, P., Zimek, A.: Clustering high-dimensional data: A survey on subspace clustering, pattern-based clustering, and correlation clustering. TKDD 3(1) (2009)
18. Lazarevic, A., Kumar, V.: Feature bagging for outlier detection. In: SIGKDD, pp. 157–166 (2005)
19. Müller, E., Schiffer, M., Seidl, T.: Statistical selection of relevant subspace projections for outlier ranking. In: ICDE, pp. 434–445 (2011)
20. Nguyen, H.V., Gopalkrishnan, V., Assent, I.: An unbiased distance-based outlier detection approach for high-dimensional data. In: Yu, J.X., Kim, M.H., Unland, R. (eds.) DASFAA 2011, Part I. LNCS, vol. 6587, pp. 138–152. Springer, Heidelberg (2011)
21. Olken, F., Rotem, D.: Random sampling from databases - a survey. Statistics and Computing 5, 25–42 (1994)
22. Papadimitriou, S., Kitagawa, H., Gibbons, P.B., Faloutsos, C.: Loci: Fast outlier detection using the local correlation integral. In: ICDE, pp. 315–326 (2003)
23. Parsons, L., Haque, E., Liu, H.: Subspace clustering for high dimensional data: a review. SIGKDD Explorations 6(1), 90–105 (2004)
24. Renyi, A.: On measures of entropy and information. In: Proc. Fourth Berkeley Symp. Math., Statistics, and Probability, pp. 547–561 (1960)
25. Schubert, E., Wojdanowski, R., Zimek, A., Kriegel, H.-P.: On evaluation of outlier rankings and outlier scores. In: SDM (2012)
26. Schubert, E., Zimek, A., Kriegel, H.-P.: Local outlier detection reconsidered: a generalized view on locality with applications to spatial, video, and network outlier detection. In: Data Mining and Knowledge Discovery, pp. 1–48 (2012)
27. Tao, Y., Xiao, X., Zhou, S.: Mining distance-based outliers from large databases in any metric space. In: SIGKDD (2006)
28. Tibshirani, R., Hastie, T.: Outlier sums for differential gene expression analysis. Biostatistics 8(1), 2–8 (2007)
29. Zimek, A., Schubert, E., Kriegel, H.-P.: A survey on unsupervised outlier detection in high-dimensional numerical data. Statistical Analysis and Data Mining 5(5), 363–387 (2012)

Anomaly Detection in Vertically Partitioned Data by Distributed Core Vector Machines

Marco Stolpe[1], Kanishka Bhaduri[2], Kamalika Das[3], and Katharina Morik[1]

[1] TU Dortmund, Computer Science, LS 8, 44221 Dortmund, Germany
[2] Netflix Inc., Los Gatos, CA 94032, USA
[3] UARC, NASA Ames, CA 94035, USA

Abstract. Observations of physical processes suffer from instrument malfunction and noise and demand data cleansing. However, rare events are not to be excluded from modeling, since they can be the most interesting findings. Often, sensors collect features at different sites, so that only a subset is present (vertically distributed data). Transferring all data or a sample to a single location is impossible in many real-world applications due to restricted bandwidth of communication. Finding interesting abnormalities thus requires efficient methods of distributed anomaly detection.

We propose a new algorithm for anomaly detection on vertically distributed data. It aggregates the data directly at the local storage nodes using RBF kernels. Only a fraction of the data is communicated to a central node. Through extensive empirical evaluation on controlled datasets, we demonstrate that our method is an order of magnitude more communication efficient than state of the art methods, achieving a comparable accuracy.

Keywords: 1-class learning, core vector machine, distributed features, communication efficiency.

1 Introduction

Outlier or anomaly detection [8] refers to the task of identifying abnormal or inconsistent patterns in a dataset. It is a well studied problem in the literature of data mining, machine learning, and statistics. While outliers are, in general, deemed as undesirable entities, their identification and further analysis can be crucial to many tasks such as fraud and intrusion detection [7], climate pattern discovery in Earth sciences [29], quality control in manufacturing processes [15], and adverse event detection in aviation safety applications [11].

Large amounts of data are accumulated and stored in an entirely decentralized fashion. In cases where storage, bandwidth, or power limitations prohibit the transfer of all data to a central node for analysis, distributed algorithms are needed that are communication efficient, but nevertheless accurate. For example, the high amount of data in large scale applications such as Earth sciences makes it infeasible to store all the data at a central repository. Communication is one of

H. Blockeel et al. (Eds.): ECML PKDD 2013, Part III, LNAI 8190, pp. 321–336, 2013.
© Springer-Verlag Berlin Heidelberg 2013

the most expensive operations in wireless networks of battery-powered sensors [3] and mobile devices [6].

Research has focused on *horizontally partitioned data* [24], where each node stores a subset of all observations. In contrast, many applications such as the detection of outliers in spatio-temporal data have the observations *vertically partitioned, i.e.* each node stores different feature sets of the (same set of) observations. For instance, at the NASA's Distributed Active Archive Centers (DAAC), all precipitation data for all locations on the earth's surface can be at one data center while humidity observations for the same locations can be at another data center. Another example are oceanic water levels and weather conditions recorded by spatially distributed sensors over one day before a Tsunami. Only few communication efficient algorithms have been proposed for the vertically partitioned scenario. This task is particularly challenging, if the analysis depends on a combination of features from more than a single data location.

In this paper, we introduce a distributed method for 1-class learning which works in the vertically partitioned data scenario and has low communication costs. In particular, the contributions of this paper are:

- A new method for distributed 1-class learning on vertically partitioned data is proposed, the $\underline{V}$ertically $\underline{D}$istributed $\underline{C}$ore $\underline{V}$ector $\underline{M}$achine (VDCVM).
- It is theoretically proven and empirically demonstrated that the VDCVM can have an order of magnitude lower communication cost compared to a current state of the art method for distributed 1-class learning [10].
- The anomaly detection accuracy of VDCVM is systematically assessed on synthetic and real world datasets of varying difficulty. It is demonstrated that VDCVM can have similar accuracy as the state of the art [10] while reducing communication cost.

The rest of this paper is organized as follows. In the next section (Section 2) we discuss some work related to the task of outlier detection. Relevant background material concerning traditional and distributed 1-class learning is discussed in Section 3. Details about the core algorithmic contribution of VDCVM are presented in Section 4. We demonstrate the performance of VDCVM in Section 5. Finally, we conclude the paper in Section 6.

2   Related Work

Several researchers have developed methods for parallelizing or distributing the optimization problem in SVMs. This is particularly useful when the datasets are large and the computation cannot be executed on a single machine. The cascade SVM by Graf *et al.* [14] uses a cascade of binary SVMs arranged in an inverted binary tree topology to train a global model. This method is guaranteed to reach a global optimum. In a different method, Chang *et al.* [9] present a parallel SVM formulation which reduces memory use through performing a row-based, approximate matrix factorization, and which loads only essential data to each machine to perform parallel computation. The solution to the optimization problem is achieved using a parallel interior point method (IPM) which computes the update rules in a distributed fashion. Hazan *et al.* [16] presents a method for parallel SVM learning based on the parallel Jacobi block update scheme derived from the convex conjugate Fenchel duality. Unfortunately, this method cannot guarantee optimality. In a related method, Flouri *et al.* [12] and Lu *et al.* [22] have proposed techniques in which the computation is done by the local nodes and then the central node performs aggregation of the results. In their method, SVMs are learned at each node independently and then the SVs are passed onto the other nodes for updating the models of the other nodes. This process has to be repeated for a few iterations to ensure convergence. Another interesting technique is the ADMM-based consensus SVM method proposed by Forero *et al.* [13] where the authors build a global SVM model in a sensor network without any central authority. The proposed algorithm is asynchronous in which messages are exchanged only among the neighboring nodes.

The aforementioned methods distribute the SVM problem, but do not focus on outlier detection. Those can be detected using *unsupervised, supervised,* or *semi-supervised* techniques [17,8]. In the field of distributed anomaly detection, researchers have mainly focused on the horizontally distributed scenario. In the PBay algorithm by Lozano and Acuna [21], a master node first splits the data into separate chunks for each processor. Then the master node loads each block of test data and broadcasts it to each of the worker nodes. Each worker node then executes a distance based outlier detection technique using its local database and the test block. Hung and Cheung [18] present a parallel version of the basic nested loop algorithm which is not suitable for distributed computation since it requires all the dataset to be exchanged among all the nodes. Otey *et al.* [24] present a distributed algorithm for mixed attribute datasets. Angiulli *et al.* [1] present a distributed distance-based outlier detection algorithm based on the concept of solving set which can be viewed as a compact representation of the original dataset. The solving set is such that by comparing any data point to only the elements of the solving set, it can be concluded if the point is an outlier or not. More recently, Bhaduri *et al.* [2] have developed a distributed method by using an efficient parallel pruning rule. For the vertically partitioned scenario, Brefeld *et al.* [4] use co-regularisation and block coordinate descent for least squares regression. Lee *et al.* [20] have proposed a separable primal formulation of the SVM training local SVM models and combining their predictions. While their algorithm can be extended to anomaly detection, their main focus is on supervised learning. A technique more focused on anomaly detection in the vertically partitioned scenario is proposed by Das *et al.* [10] for Earth science datasets. It trains local 1-class models and reduces communication by a pruning rule. This technique is used for comparisons and discussed in details in Sec. 3.3.

3   The Problem of 1-Class Learning

The next sections introduce important notations by giving some background information on 1-class learning and the problem of vertically partitioned data.

3.1   Support Vector Data Description

The task of data description, or 1-class classification [23], is to find a model that well describes a training set of observations. The model can then be used to check whether new observations are similar or dissimilar to the previously seen data points and mark dissimilar points as anomalies or outliers. It has been shown by Tax and Duin [27] that instead of estimating the data distribution based on a training set, it is more efficient to compute a spherical boundary around the data. This method, called the Support Vector Data Description (SVDD), allows for choosing the diameter of an enclosing ball in order to control the volume of the training data that falls within the ball. Observations inside the ball are then classified as normal whereas those outside the ball are treated as outliers.

Given a vector space X and a set $\mathcal{S} = \{\mathbf{x}_1, \ldots, \mathbf{x}_n\} \subseteq X$ of training instances, the primal problem is to find a minimum enclosing ball (MEB) with radius R and center $\mathbf{c}$ around all data points $\mathbf{x}_i \in \mathcal{S}$:

$$\min_{R, \mathbf{c}} R^2 : ||\mathbf{c} - \mathbf{x}_i||^2 \leq R^2, \ i = 1, \ldots, n$$

When the input space is not a vector space or the decision boundary is non-spherical, observations may be mapped by $\varphi : X \to \mathcal{F}$ to a feature space $\mathcal{F}$ for which an inner product is defined. The explicit computation of this mapping to an (possibly) infinite dimensional space can be avoided by use of a kernel function $k : X \times X \to \mathbb{R}$, which computes the inner product in $\mathcal{F}$ between the input observations. The dual problem after the kernel transformation then becomes

$$\max_{\alpha} \sum_{i=1}^{n} \alpha_i k(\mathbf{x}_i, \mathbf{x}_i) - \sum_{i,j=1}^{n} \alpha_i \alpha_j k(\mathbf{x}_i, \mathbf{x}_j) \qquad (1)$$

$$\text{s.t.} \quad \alpha_i \geq 0, \quad i = 1, \ldots, n, \quad \sum_{i=1}^{n} \alpha_i = 1$$

Here $\alpha = (\alpha_1, \ldots, \alpha_n)^T$ is a vector of Lagrange multipliers and the primal variables can be recovered using

$$\mathbf{c} = \sum_{i=1}^{n} \alpha_i \varphi(\mathbf{x}_i), \quad R = \sqrt{\alpha^T \text{diag}(\mathbf{K}) - \alpha^T \mathbf{K} \alpha}$$

where $\mathbf{K}_{n \times n} = (k(\mathbf{x}_i, \mathbf{x}_j))$ is the kernel matrix. After optimal αs are found, the model consists of all data points for which $\alpha_i > 0$, called the support vectors (set SV), and the corresponding αs. An observation $\mathbf{x}$ is said to belong to the training set distribution if its distance from the center $\mathbf{c}$ is smaller than the radius R, where the distance is expressed in terms of the support vectors and the kernel function as

$$||\mathbf{c} - \varphi(\mathbf{x})||^2 = k(\mathbf{x}, \mathbf{x}) - 2 \sum_{i=1}^{|SV|} \alpha_i k(\mathbf{x}, \mathbf{x}_i) + \sum_{i,j=1}^{|SV|} \alpha_i \alpha_j k(\mathbf{x}_i, \mathbf{x}_j) \leq R^2$$

It can be shown [28] that for kernels $k(\mathbf{x}, \mathbf{x}) = \kappa$ (κ constant), that map all input patterns to a sphere in feature space, (1) can be simplified to the optimization problem

$$\max_{\alpha} -\alpha^T \mathbf{K}\alpha : \alpha \geq \mathbf{0}, \alpha^T \mathbf{1} = 1 \tag{2}$$

where $\mathbf{0} = (0, \ldots, 0)^T$ and $\mathbf{1} = (1, \ldots, 1)^T$.

Whenever the kernel satisfies $k(\mathbf{x}, \mathbf{x}) = \kappa$, any problem of the form (2) is an MEB problem. For example, Schölkopf [25] proposed the 1-class ν-SVM that, instead of minimizing an enclosing ball, separates the normal data by a hyperplane with maximum margin from the origin in feature space. If $k(\mathbf{x}, \mathbf{x}) = \kappa$, the optimization problems of the SVDD and the 1-class ν-SVM with $C = 1/(\nu n)$ are equivalent and yield identical solutions.

3.2   Core Vector Machine (CVM)

Bǎdoiu and Clarkson [5] have shown that a $(1+\varepsilon)$-approximation of the MEB can be computed with constant time and space requirements. Their algorithm only depends on ε, but not on the dimension m or the number of training examples n. Tsang et al. [28] have adopted this algorithm for kernel methods like the SVDD.

Let $\mathcal{S}$ be the training data as described in Section 3.1. For an $\varepsilon > 0$, the ball $B(\mathbf{c}, (1 + \varepsilon)R)$ with center $\mathbf{c}$ and radius R is an $(1 + \varepsilon)$-approximation of the MEB($\mathcal{S}$), the minimum enclosing ball that contains all data points of $\mathcal{S}$. A subset $Q \subseteq \mathcal{S}$ is called the core set of $\mathcal{S}$ if the expansion of MEB(Q) by the factor $(1 + \varepsilon)$ contains $\mathcal{S}$.

The Core Vector Machine (CVM) algorithm shown in Figure 1 starts with an empty core set and extends it consecutively by the furthest point from the current center in feature space until all data is contained in an approximate MEB. The algorithm uses a modified kernel function $\tilde{k}$ for the reason that optimization problem (1) yields a hard margin solution, but can be transformed into a soft margin problem [19] by introducing a 2-norm error on the slack variables, i.e. by replacing $C \sum_{i=1}^{n} \xi_i$ with $C \sum_{i=1}^{n} \xi_i^2$, and replacing the original kernel function k with a new kernel function $\tilde{k} : \tilde{\varphi} \to \tilde{\mathcal{F}}$, where

$$\tilde{k}(\mathbf{x}_i, \mathbf{x}_i) = k(\mathbf{x}_i, \mathbf{x}_j) + \frac{\delta_{ij}}{C}, \quad \delta_{ij} = \begin{cases} 1 : i = j \\ 0 : i \neq j \end{cases} \tag{3}$$

The new kernel again satisfies $\tilde{k}(\mathbf{z}, \mathbf{z}) = \tilde{\kappa}$ with $\tilde{\kappa}$ being constant.

The furthest point calculation in step 2 takes $O(|\mathcal{S}_t|^2 + n|\mathcal{S}_t|)$ time for the tth iteration. However, as is mentioned by Schölkopf [26], the furthest point obtained from a randomly sampled subset $\mathcal{S}' \subset \mathcal{S}$ of size 59 already has a probability of 95% to be among the furthest 5% points in the whole dataset $\mathcal{S}$. By using this probabilistic speed-up strategy, i.e. determining the furthest point on a small sampled subset of points in each iteration, the running time for the furthest point calculation can be reduced to $O(|\mathcal{S}_t|^2)$. As shown by Tsang et al. [28], with probabilistic speed-up and a standard QP solver, the CVM reaches a $(1+\varepsilon)^2$-approximation of the MEB with high probability. The total number of

---

Core Vector Machine (CVM)

$\mathcal{S}$: training set, consisting of n examples
$\mathcal{S}_t \subseteq \mathcal{S}$: core set of $\mathcal{S}$ at iteration t
$\mathbf{c}_t$: center of the MEB around $\mathcal{S}_t$ in feature space
R_t: radius of the MEB

1. **Initialization:** Uniformly at random choose a point $\mathbf{z} \in \mathcal{S}$. Determine a point $\mathbf{z}_a \in \mathcal{S}$ that is furthest away from $\mathbf{z}$ in feature space, then a point $\mathbf{z}_b \in \mathcal{S}$ that is furthest away from $\mathbf{z}_a$. Set $\mathcal{S}_0 := \{\mathbf{z}_a, \mathbf{z}_b\}$ and the initial radius

$$R_0 := \frac{1}{2}\sqrt{2\tilde{\kappa} - 2\tilde{\kappa}(\mathbf{z}_a, \mathbf{z}_b)}$$

2. **Furthest point calculation:** Find $\mathbf{z} \in \mathcal{S}$ such that $\tilde{\phi}(\mathbf{z})$ is furthest away from $\mathbf{c}_t$. The new core set becomes $\mathcal{S}_{t+1} = \mathcal{S}_t \cup \{\mathbf{z}\}$. The squared distance of any point from the center in $\tilde{\mathcal{F}}$ can be calculated using the kernel function

$$\|\mathbf{c}_t - \tilde{\phi}(\mathbf{z}_\ell)\|^2 = \sum_{z_i, z_j \in \mathcal{S}_t} \alpha_i \alpha_j \tilde{k}(\mathbf{z}_i, \mathbf{z}_j) - 2 \sum_{z_i \in \mathcal{S}_t} \alpha_i \tilde{k}(\mathbf{z}_i, \mathbf{z}_\ell) + \tilde{k}(\mathbf{z}_\ell, \mathbf{z}_\ell)$$

3. **Termination check:** Terminate if all training points are inside the $(1 + \varepsilon)$-ball $B(\mathbf{c}_t, (1 + \varepsilon)R_t)$ in feature space, *i.e.* $\|\mathbf{c}_t - \tilde{\phi}(\mathbf{z})\| \le R_t(1 + \varepsilon)$.
4. **MEB calculation:** Find a new MEB$(\mathcal{S}_{t+1})$ by solving the QP problem

$$\max_\alpha -\alpha^T \tilde{\mathbf{K}} \alpha \ : \ \alpha \ge \mathbf{0}, \alpha^T \mathbf{1} = 1, \ \tilde{\mathbf{K}} = [\tilde{k}(\mathbf{z}_i, \mathbf{z}_j)]$$

on all points of the core set. Set $R_{t+1} := \sqrt{\tilde{\kappa} - \alpha^T \tilde{\mathbf{K}} \alpha}$.
5. $t := t + 1$, then go to step 2.

Fig. 1. Core Vector Machine (CVM) algorithm by Tsang *et al.* [28]

iterations is bounded by $O(1/\varepsilon^2)$, the running time by $O(1/\varepsilon^8)$, and the space complexity by $O(1/\varepsilon^4)$. The running time and resulting core set size are thus constant and *independent* of the size of the whole dataset.

The CVM already seems to be better suited for a network setting than the 1-class ν-SVM, as it works incrementally and could sample only as much data as needed from the storage nodes. However, it is no distributed algorithm. Section 4 discusses how the CVM can be turned into a distributed algorithm that is even more communication efficient.

3.3   Vertically Distributed 1-Class Learning

In the vertically partitioned data scenario, each data site has all observations, but only a subset of the features. Let $P_0, ..., P_k$ be a set of nodes where P_0 is designated as the central node and the others are denoted as the data nodes. For the rest of the paper, it is assumed that all nodes can also be used for computations. Let the dataset at node P_i ($\forall i > 0$) be denoted by $\mathcal{S}_i = [\mathbf{x}_1^{(i)} \ ... \ \mathbf{x}_n^{(i)}]^T$ consisting

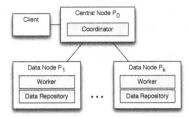

Fig. 2. Components of the VDCVM

of n rows where $\mathbf{x}_j^{(i)} \in \mathbb{R}^{m_i}$ and m_i is the number of variables in the ith site. Here, each row corresponds to an observation and each column corresponds to a variable (feature). There should be a one-to-one mapping between the rows across the different nodes. There exist crossmatching techniques that can be used to ensure that. The global set of features A is the vertical concatenation of all the $m = \sum_{i=1}^{k} m_i$ features over all nodes and is defined as $A = [A_1 \, A_2 \, \ldots \, A_k]$ (using Matlab notation). Hence, the global data $\mathcal{S}$ is the $n \times m$ matrix defined as the union of all data over all nodes, $i.e.$ $\mathcal{S} = [\mathbf{x}_1 \, \ldots \, \mathbf{x}_n]^T$ with $\mathbf{x}_j \in \mathbb{R}^m$. The challenge is to learn an accurate 1-class model without transferring all data to a central node.

Das $et\ al.$ [10] have proposed a synchronized distributed anomaly detection algorithm (called VDSVM in this paper) for vertically partitioned data based on the 1-class ν-SVM. At each data node P_i, a local 1-class model is trained. Points identified as local outliers are sent to the central node P_0, together with a small sample of all observations. At the central node, a global model is trained on the sample and used to decide if the outlier candidates sent from the data nodes are global outliers or not. The VDSVM cannot detect outliers which are global due to a combination of attributes. However, the algorithm shows good performance if global outliers are also local outliers. Moreover, in the application phase, the algorithm is highly communication efficient, since the number of outlier candidates is often only a small fraction of the data. A major drawback is that the fixed-size sampling approach gives no guarantees or bounds on the correctness of the global model. For a user, it is therefore difficult to set the sampling size correctly, in advance. Moreover, during training, no other strategies than sampling are used for reducing communication costs. We address these issues in this paper by developing a distributed version of the Core Vector Machine (VDCVM) which is more communication-efficient in the training phase and samples only as many points as needed, with known bounds for the correctness of the global model.

4   Vertically Distributed CVM (VDCVM)

In this section, we introduce the Vertically Distributed Core Vector Machine (VDCVM). It consists of the components shown in Figure 2. The *Coordinator* communicates with *Worker* components that reside locally on the data nodes and can access the values of the local feature subsets directly, without any network communication. While the termination check and the QP optimization (steps 3 and 4 of the CVM algorithm in Figure 1) are still done centrally by the *Coordinator*, the sampling and furthest point calculations are combined in a single step and done in parallel by the *Worker* components, as described in the next sections.

4.1   Distributed Furthest Point Calculation

In any iteration t, the original CVM algorithm (Figure 1) with probabilistic speedup draws a fixed-sized sample of data points from the whole dataset. Let V_t denote the sample drawn at iteration t and $|V_t|$ denote its size. From the sample, the CVM determines the point $\mathbf{z}_t$ furthest away from the current center $\mathbf{c}_t$ in feature space by calculating the squared distance $||\mathbf{c}_t - \tilde{\phi}(\mathbf{z}_\ell)||^2$ for each sample point $\mathbf{z}_\ell \in V_t$. Since $\tilde{k}(\mathbf{z}_\ell, \mathbf{z}_\ell) = \tilde{\kappa}$ is constant and the sum $\sum_{z_i, z_j \in S_t} \alpha_i \alpha_j \tilde{k}(\mathbf{z}_i, \mathbf{z}_j)$ does not depend on the sampled points, the furthest point calculation at iteration t can be simplified to

$$\mathbf{z}_t = \text{argmax}_{\mathbf{z}_\ell \in V_t} \left[-\sum_{z_j \in S_t} \alpha_j \tilde{k}(\mathbf{z}_j, \mathbf{z}_\ell) \right] \tag{4}$$

Let $\mathbf{z}[i]$ denote the ith component of vector $\mathbf{z}$. With the linear dot product kernel $k(\mathbf{z}_i, \mathbf{z}_j) = \langle \mathbf{z}_i, \mathbf{z}_j \rangle$, the sum in (4) could be written as

$$\mathbf{z}_t = \text{argmin}_{\mathbf{z}_\ell \in V_t} \sum_{z_j \in S_t} \alpha_j \langle \mathbf{z}_j, \mathbf{z}_\ell \rangle = \text{argmin}_{\mathbf{z}_\ell \in V_t} \sum_{i=1}^{m} \sum_{z_j \in S_t} \alpha_j \mathbf{z}_j[i] \mathbf{z}_\ell[i] \tag{5}$$

Since the dot product kernel multiplies each component i of the $\mathbf{z}_\ell$ and $\mathbf{z}_j$ vectors independently, we only need the index i of the vector component and all α values for calculating the partial inner sums separately on each node

$$v_\ell^{(p)} = \sum_{i=1}^{m_p} \sum_{z_j^{(p)} \in S_t} \alpha_j \mathbf{z}_j^{(p)}[i] \, \mathbf{z}_\ell^{(p)}[i] \tag{6}$$

over its subset of attributes A_p for all random indices $\ell \in I_t$ and send these partial sums back to the coordinator. The coordinator then aggregates the sums and determines the index $\ell_{\max} \in I_t$ of the furthest point:

$$\ell_{\max} = \text{argmin}_{\ell \in I_t} \sum_{p=1}^{k} v_\ell^{(p)} \tag{7}$$

Each data node thus only transmits a *single* numerical value for each point of the random sample, instead of sending *all* attribute values of the sampled points to the central node.

VDCVM Coordinator

---

on *workerInitialized()*:
>   **if** received message from all workers **then**
>>   Determine random index set I_0 for data points.
>>   Send *getPartialSums(I_0)* to all workers.

on *getPartialSumsAnswer($v_\ell^{(p)}$ $\forall \ell \in I_t$)*:
>   Store partial sums received from worker p.
>   **if** received message from all workers **then**
>> $\ell_{\max} = \mathrm{argmin}_{\ell \in I_t} \sum_{p=1}^{k} v_\ell^{(p)}$
>>   Send *getData($\{\ell\}$)* to all repositories.

on *getDataAnswer($\{\mathbf{z}_t[1 \dots m_p]\}$)*:
>   Store attribute values received from *Data Repository p*.
>   **if** received message from all repositories **then**
>>   Construct furthest point $\mathbf{z}_t$ from attribute values.
>>   **if** $\|\mathbf{c}_t - \phi(\mathbf{z}_t)\| \leq (1 + \varepsilon) \cdot R_t$ **then**
>>>   Broadcast *stop* and return model.
>>   **else**
>>> $\mathcal{S} := \mathcal{S} \cup \{\mathbf{z}_t\}$.
>>>   Calculate new $\mathrm{MEB}(\mathcal{S}_{t+1})$. (Solve QP problem.)
>>> $R_{t+1} := \sqrt{\tilde{\kappa} - \alpha^T \tilde{\mathbf{K}} \alpha}$.
>>>   Determine random index set I_t for data points.
>>>   Send *getPartialSums(I_t, αs)* to workers.
>>> $t := t + 1$.

Fig. 3. Operations of the VDCVM *Coordinator*

Splitting (4) into partial sums can be done for the linear kernel, but it is impossible or at least non-trivial for non-linear mappings. For example, the SVDD is usually used with the RBF kernel $k(\mathbf{z}_i, \mathbf{z}_j) = e^{-\gamma \|\mathbf{z}_i - \mathbf{z}_j\|^2}$ and the CVM requires $k(\mathbf{x}, \mathbf{x})$ to be constant, which holds for the RBF kernel. One possible choice for a kernel is the summation of kernels defined on the local attributes only, like a combination of RBF kernels (see also Lee *et al.* [20]):

$$k(\mathbf{z}_i, \mathbf{z}_j) = \sum_{p=1}^{k} e^{-\gamma_p \|\mathbf{z}_i^{(p)} - \mathbf{z}_j^{(p)}\|^2} \tag{8}$$

In Section 5 it is empirically shown that such a combination yields a similar accuracy as VDSVM on most of the tested datasets.

4.2   The VDCVM Algorithm

The *Coordinator* retrieves meta information attached to the datasets from all *Data Repository* components. As initial point $\mathbf{z}$, the algorithm takes the mean vector of the minimum and maximum attribute values. Thereby, $\mathbf{z}$ does not need to be sampled from the network. The constant $\tilde{\kappa}$ is calculated and all data

---

VDCVM Worker

---

on *initializeWorker(parameters, $\mathbf{z}$, $\tilde{\kappa}$):*
    Store C, γ, $\mathbf{z}$ and $\tilde{\kappa}$.
    Set $\mathcal{S} = \emptyset$, $t := 0$.
    Send *workerInitialized* to *Coordinator*.

on *getPartialSums(I_t, αs):*
    if $t \geq 2$ **then** store new αs.
    $\mathcal{S} := \mathcal{S} \cup \{\mathbf{z}_{\ell_{\max}}[1 \ldots m_p]\}$.
    Calculate $v_\ell^{(p)}$ $\forall \ell \in I_t$ — see (6) and (7)

on *stop():*
    Free all resources.

Fig. 4. Operations of the VDCVM *Worker*

structures (the core set) are initialized. These are transmitted together with the parameters C and $\gamma_1, \ldots, \gamma_p$ to all *Worker* components.

The main part of the *Coordinator* is shown in Figure 3. The indices of $|V_t|$ random data points are sampled and sent to the workers in a request for the partial sums v_ℓ. When the *Coordinator* has received all partial sums, it can calculate the index $\ell_{\max}$ of the furthest point $\mathbf{z}_t$ and ask the repositories for their feature values. If the termination criterion is not fulfilled, the coordinator goes on with solving the QP problem and calculates the new radius R_{t+1}. It then determines a new random index set I_t and requests the next partial sums from the workers. It furthermore transmits all updated α values.

Based on the updated αs, each *Worker* gets the local components of points $\mathbf{z}_t$ by its furthest index $\ell_{\max}$. It then calculates $v_\ell^{(p)}$ for all random indices $\ell \in I_t$ received from the *Coordinator*, according to Equation (7). The partial sums are then sent back to the *Coordinator* which continues with the main algorithm.

4.3   Analysis of Running Time and Communication Costs

The VDCVM performs exactly the same calculations as the original CVM algorithm. It therefore inherits all properties of the CVM, including the constant bound on the total number of iterations (see Section 3.2) and the $(1 + \varepsilon)^2$-approximation guarantee for the calculated MEB.

Regarding communication costs, we assume that messages can be broadcast to all workers, that training point indices are represented by 4 bytes and real numbers by 8 bytes. The total number of bytes transferred (excluding initialization and message headers) when sending all m attributes of n points in a sample to a central server for training (as does VDSVM) is

$$B_{\text{central}}(n) = n \cdot 4 + n \cdot m \cdot 8$$

Table 1. Numbers of iterations up to which VDCVM is more communication efficient than VDSVM, for different numbers of nodes k and attributes m, $s = 59$

m	$k=1$	$k=2$	$k=5$	$k=10$	$k=25$
10	1,042	924	570	0	-
25	2,782	2,664	2,310	1,720	0
50	5,682	5,564	5,210	4,620	2,850
100	11,482	11,364	11,010	10,420	8,650

In contrast, the bytes transferred by VDCVM up to iteration T are

$$B_{\text{VDCVM}}(T) = [T \cdot s \cdot 4 + T \cdot s \cdot k \cdot 8] + [T \cdot 4 + T \cdot m \cdot 8] + \left[\frac{T(T+1)}{2} \cdot 8 \right]$$

The coordinator at the central node first broadcasts s index values to all data nodes and receives partial kernel sums for each, from k workers (first term in brackets). Then, the index value of the furthest point is broadcast to all data nodes and the coordinator receives its m feature values (second term). The total number of αs transmitted is quadratic in the number of iterations (last term).

The break even point T_{worse}, *i.e.* the iteration from when on the VDCVM has worse communication costs than central sampling, can be calculated by setting $B_{\text{central}}(n)$ with $n = Ts$ equal to $B_{\text{VDCVM}}(T)$ and solving for T:

$$T_{\text{worse}} = 2 \cdot m \cdot (s - 1) - 2 \cdot k \cdot s \tag{9}$$

According to (9), the communication efficiency of VDCVM depends on the number of attributes per node. Table 1 contains values of T_{worse} for different numbers of data nodes k and attributes m. The number of iterations occuring in practice is often much lower than those in Table 1 (cf. Section 5).

5   Experimental Evaluation

In this section we demonstrate the performance of VDCVM on a variety of datasets and compare it to VDSVM and a single central model. In 1-class learning, the ground truth about the outliers is often not available. For a systematic performance evaluation of the algorithms, synthetic data containing known outliers was therefore generated. In addition, the methods also have been evaluated on three real world datasets with known binary class labels.

Synthetic Data. Figure 5 visualizes the generated datasets for two dimensions. The points were generated randomly in a unit hypercube of m dimensions (for $m = 2, 4, 8, 16, 32, 64$). The different types of data pose varying challenges to the algorithms when vertically partitioned among network nodes. The easiest scenario is the one in which each attribute reveals all information about the label, represented by SepBox. For Gaussian, the means $\mu_{+,-}$ and standard deviations

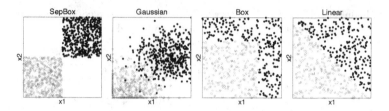

Fig. 5. Generated normal data (grey) and outliers (black) in two dimensions

Table 2. Number of data points (total, training, test and validation set)

Dataset	Total	Training	Test		Validation	
			normal	outliers	normal	outliers
Random datasets	60,000	20,000	10,000	10,000	10,000	10,000
letter	1,000	400	150	150	150	150
kddcup-99	60,000	20,000	10,000	10,000	10,000	10,000
face	20,000	10,000	2,500	2,500	2,500	2,500

$\sigma_{+,-}$ of two Gaussians were chosen randomly and independently for each attribute (with $\mu_{+,-} \in [0.1, 0.9]$ and $\sigma_{+,-} \in [0, 0.25]$). If the Gaussians overlap in each single dimension, they may nevertheless become separable by a combination of attributes. In the Box dataset, an outlier is a point for which $\exists \mathbf{x}[i] > \rho$ with $\rho = 0.5^{(1/m)}$ (*i.e.* the normal data lies in half the volume of the m-dimensional unit hypercube). Separation is only given by all dimensions in conjunction. The same is true for the Linear dataset, where the normal data is separated from the outliers by the hyperplane $h = \{\mathbf{x} \mid \mathbf{x}/\|m\| - 0.5\|m\| = 0\}$.

Real World Data. All real world data was taken from the CVM authors' web site[1]. The letter dataset consists of 20,000 data points for the 26 letters of the latin alphabet, each represented by 16 attributes. For the experiments, 773 examples of the letter G were taken as normal data and 796 of letter T extracted as outliers. The KDDCUP-99 data consists of 5,209,460 examples of network traffic described by 127 features. The task is to differentiate between normal and bad traffic patterns. The extended MIT face dataset contains 513,455 images consisting of 19x19 (361) grey scale values. The task is to decide if the image contains a human face or not.

5.1   Experimental Setup

VDCVM was implemented in Java using the Spread Toolkit[2]. VDSVM was implemented in Python using LibSVM.

Table 2 shows that 60,000 points were generated for each of the random datasets. From each of the real-world datasets, only a random sample was taken

---

[1] http://c2inet.sce.ntu.edu.sg/ivor/cvm.html
[2] http://www.spread.org

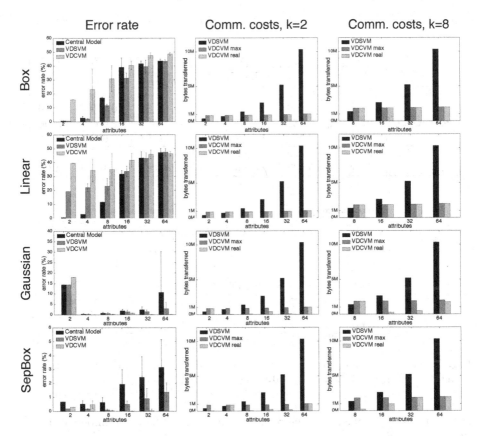

Fig. 6. Performance of VDCVM, VDSVM and a central model with standard RBF kernel on the generated datasets

(column *Total*). These sets were randomly splitted further into independent sets for training, testing (*i.e.* parameter optimization) and validation, with sizes as shown. The central and local VDSVM models were trained on the whole training set, while VDCVM was allowed to sample up to the same amount of data. The methods require different parameters γ and ν (or C), since VDSVM uses a standard RBF kernel and the 1-norm on its slack variables, while VDCVM uses the 2-norm and a combination of local kernels. For VDSVM, 75 random parameter combinations were tested, and for VDCVM 100 combinations, alternatingly conducting a local and global random search. All error rates shown in the next section result from a single run on the validation set, with tuned parameters.

5.2   Results

The plots in Figure 6 compare the performance of VDCVM to a single central 1-class model with standard RBF kernel and to VDSVM, *i.e.* local 1-class models which communicate only outlier candidates to the central model for testing. The

Table 3. Results on real world datasets (*m*: attributes, *k*: nodes, *err*: error rate in %, *bytes*: amount of bytes transferred)

Dataset	m	k	Central model		VDSVM		VDCVM	
			err	Kbytes	err	Kbytes	err	Kbytes
letter	16	2	10.000	54	9.333	54	6.500	9
		4	11.000	54	9.333	54	10.500	16
kddcup-99	127	2	0.285	20,401	0.220	20,401	0.000	1,206
		4	0.450	20,401	0.290	20,401	0.002	1,526
face	361	2	6.220	29,006	7.900	29,006	4.940	808
		4	5.580	29,006	6.880	29,006	5.100	969

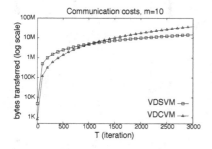 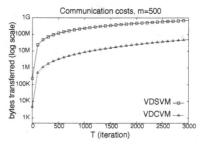

Fig. 7. Bytes transferred (log scale) by VDSVM and VDCVM with a growing number of iterations (T), for 10 (left) and 500 (right) attributes, $s = 59$, $k = 1$

error rates are averaged over the results obtained for different numbers of nodes (2, 4, 8, 16, 32).

All methods, including the central 1-class model, have difficulties to separate the `Linear` and `Box` datasets in higher dimensions. In low dimensions, the combined RBF kernel has worse error rates than a standard RBF kernel and VDSVM's ensemble of local classifiers. VDCVM shows similar or even slightly *better performance on the* `Gaussian` *and* `SepBox` *datasets*, whose attribute values provide more information about the label locally. Even for the maximum number of 20,000 points allowed to sample (*VDCVM max*), it already has much *lower communication costs* than VDSVM in most cases. When the data is easy to separate (`Gaussian` and `SepBox`), the real number of sampled points is often lower, resulting in *even less communication* (see *VDCVM real*).

All methods achieve similar error rates on the real world datasets (see Table 3), with VDCVM being more *communication efficient*. The central 1-class model performing worse in some cases can be explained by VDSVM using an ensemble of (local and global) classifiers, increasing chances by at least one of them making the correct prediction. The better performance of VDCVM might be explained by better parameters found with the random tuning strategy.

The plots in Figure 7 show how the number of transmitted bytes grows with the number of iterations, for a fixed number of features. As shown in the left figure for 10 features, the crossover occurs at 1,000 iterations. The right figure plots the transmitted bytes for 500 features. Here, the VDCVM is at least an order of magnitude more communication efficient for all plotted iterations of the algorithm. In general, the more attributes are stored at each data node, the more can be saved in comparison to transmitting all data to the central node.

6 Conclusion

In this paper we have developed VDCVM – a distributed algorithm for anomaly detection from vertically distributed datasets. VDCVM is based on the recently proposed Core Vector Machine algorithm which uses a minimum enclosing ball formulation to describe the data. The algorithm solves local problems which are iteratively coordinated with the central server to compute the final solution. The proposed algorithm can be as accurate as state of the art methods, with an order of magnitude less communication overhead in the training phase. Extensive experiments on datasets containing ground truth demonstrate the validity of the claims. In future work, we want to explore other combinations of local kernels and learning tasks such as distributed support vector clustering using VDCVM. Moreover, an implementation on embedded devices remains to be done, together with measuring real energy consumption.

Acknowledgment. This work has been supported by the DFG, Collaborative Research Center SFB 876 (http://sfb876.tu-dortmund.de/), project B3.

References

1. Angiulli, F., Basta, S., Lodi, S., Sartori, C.: A Distributed Approach to Detect Outliers in Very Large Data Sets. In: D'Ambra, P., Guarracino, M., Talia, D. (eds.) Euro-Par 2010, Part I. LNCS, vol. 6271, pp. 329–340. Springer, Heidelberg (2010)

2. Bhaduri, K., Matthews, B.L., Giannella, C.: Algorithms for speeding up distance-based outlier detection. Proc. of KDD 2011, 859–867 (2011)

3. Bhaduri, K., Stolpe, M.: Distributed data mining in sensor networks. In: Aggarwal, C.C. (ed.) Managing and Mining Sensor Data. Springer, Heidelberg (2013)

4. Brefeld, U., Gärtner, T., Scheffer, T., Wrobel, S.: Efficient co-regularised least squares regression. In: Proc. of the 23rd Int. Conf. on Machine Learning, ICML 2006, pp. 137–144. ACM, New York (2006)

5. Bădoiu, M., Clarkson, K.: Optimal core sets for balls. In: DIMACS Workshop on Computational Geometry (2002)

6. Carroll, A., Heiser, G.: An analysis of power consumption in a smartphone. In: Proc. of the 2010 USENIX Conf. on USENIX Ann. Technical Conf., USENIXATC 2010. USENIX Association, Berkeley (2010)

7. Chan, P., Fan, W., Prodromidis, A., Stolfo, S.: Distributed Data Mining in Credit Card Fraud Detection. IEEE Intelligent Systems 14, 67–74 (1999)

8. Chandola, V., Banerjee, A., Kumar, V.: Anomaly detection: A survey. ACM Comp. Surveys 41(3), 1–58 (2009)

9. Chang, E.Y., Zhu, K., Wang, H., Bai, H., Li, J., Qiu, Z., Cui, H.: Psvm: Parallelizing support vector machines on distributed computers. In: NIPS (2007)

10. Das, K., Bhaduri, K., Votava, P.: Distributed anomaly detection using 1-class SVM for vertically partitioned data. Stat. Anal. Data Min. 4(4), 393–406 (2011)

11. Das, S., Matthews, B., Srivastava, A., Oza, N.: Multiple kernel learning for heterogeneous anomaly detection: algorithm and aviation safety case study. In: Proc. of KDD 2010, pp. 47–56 (2010)

12. Flouri, K., Beferull-Lozano, B., Tsakalides, P.: Optimal gossip algorithm for distributed consensus svm training in wireless sensor networks. In: Proceedings of DSP 2009, pp. 886–891 (2009)

13. Forero, P.A., Cano, A., Giannakis, G.B.: Consensus-based distributed support vector machines. J. Mach. Learn. Res. 99, 1663–1707 (2010)

14. Graf, H., Cosatto, E., Bottou, L., Durdanovic, I., Vapnik, V.: Parallel support vector machines: The cascade svm. In: NIPS (2004)

15. Harding, J., Shahbaz, M., Srinivas, K.A.: Data mining in manufacturing: A review. Manufacturing Science and Engineering 128(4), 969–976 (2006)

16. Hazan, T., Man, A., Shashua, A.: A parallel decomposition solver for svm: Distributed dual ascend using fenchel duality. In: CVPR 2008, pp. 1–8 (2008)

17. Hodge, V., Austin, J.: A survey of outlier detection methodologies. A. I. Review 22(2), 85–126 (2004)

18. Hung, E., Cheung, D.: Parallel Mining of Outliers in Large Database. Distrib. Parallel Databases 12, 5–26 (2002)

19. Keerthi, S., Shevade, S., Bhattacharyya, C., Murthy, K.: A fast iterative nearest point algorithm for support vector machine classifier design. IEEE Transactions on Neural Networks 11(1), 124–136 (2000)

20. Lee, S., Stolpe, M., Morik, K.: Separable approximate optimization of support vector machines for distributed sensing. In: Flach, P.A., De Bie, T., Cristianini, N. (eds.) ECML PKDD 2012, Part II. LNCS, vol. 7524, pp. 387–402. Springer, Heidelberg (2012)

21. Lozano, E., Acuna, E.: Parallel algorithms for distance-based and density-based outliers. In: ICDM 2005, pp. 729–732 (2005)

22. Lu, Y., Roychowdhury, V.P., Vandenberghe, L.: Distributed Parallel Support Vector Machines in Strongly Connected Networks. IEEE Transactions on Neural Networks 19(7), 1167–1178 (2008)

23. Moya, M., Koch, M., Hostetler, L.: One-class classifier networks for target recognition applications. In: Proc. World Congress on Neural Networks, pp. 797–801. International Neural Network Society (1993)

24. Otey, M., Ghoting, A., Parthasarathy, S.: Fast Distributed Outlier Detection in Mixed-Attribute Data Sets. Data Min. Knowl. Discov. 12, 203–228 (2006)

25. Schölkopf, B., Platt, J.C., Shawe-Taylor, J.C., Smola, A.J., Williamson, R.C.: Estimating the support of a high-dimensional distribution. Neural Comp. 13(7), 1443–1471 (2001)

26. Schölkopf, B., Smola, A.J.: Learning with Kernels. MIT Press (2002)

27. Tax, D.M.J., Duin, R.P.W.: Support vector data description. Mach. Learn. 54, 45–66 (2004)

28. Tsang, I., Kwok, J., Cheung, P.: Core Vector Machines: Fast SVM Training on Very Large Data Sets. J. Mach. Learn. Res. 6, 363–392 (2005)

29. Zhang, J., Roy, D., Devadiga, S., Zheng, M.: Anomaly detection in MODIS land products via time series analysis. Geo-Spat. Inf. Science 10, 44–50 (2007)

Mining Outlier Participants: Insights Using Directional Distributions in Latent Models

Didi Surian[1] and Sanjay Chawla[2]

[1,2]University of Sydney and [1,2]NICTA
dsur5833@uni.sydney.edu.au, sanjay.chawla@sydney.edu.au

Abstract. In this paper we will propose a new probabilistic topic model to score the expertise of participants on the projects that they contribute to based on their previous experience. Based on each participant's score, we rank participants and define those who have the lowest scores as *outlier participants*. Since the focus of our study is on outliers, we name the model as Mining **O**utlier **P**articipants from **P**rojects (MOPP) model. MOPP is a topic model that is based on directional distributions which are particularly suitable for outlier detection in high-dimensional spaces. Extensive experiments on both synthetic and real data sets have shown that MOPP gives better results on both topic modeling and outlier detection tasks.

1 Introduction

We present a new topic model to capture the interaction between participants and the projects that they participate in. We are particularly interested in outlier projects, i.e., those which include participants who are unlikely to join in based on their past track record.

Example: Consider the following example. Three authors $A1, A2$ and $A3$ come together to write a research paper. The authors and the paper profiles are captured by a "category" vector as shown in Table 1. A category can be a "word" or "topic" and is dependent upon the model we use.

Table 1. Example: "category" vectors for authors and paper profiles. The dot products determine the "outlierness" of the authors to the paper

Entity	Category 1	Category 2	Category 3	Dot.P
Paper	0.1	0.1	0.8	
A1	0.1	0.2	0.7	0.59
A2	0.2	0.2	0.6	0.52
A3	0.8	0.1	0.1	0.17

Then we can compute the dot products $< Paper, A1 >$, $< Paper, A2 >$ and $< Paper, A3 >$ as shown in the last column. Based on the dot product, $A3$ is the most outlier participant in the paper.

H. Blockeel et al. (Eds.): ECML PKDD 2013, Part III, LNAI 8190, pp. 337–352, 2013.

The challenge we address in this paper is to develop a new topic model to accurately form categories which can be used to discover outlier behavior as illustrated in Table 1.

A natural approach is to use Latent Dirichlet Allocation (LDA) to model the track record of the participants and also the project descriptions. The advantage of using LDA is that we carry out the analysis in the "topic" space, which is known to be more robust compared to a word-level analysis. However, LDA is not particularly suitable for outlier detection as we illustrate in the following example.

Assume that we want to cluster five documents into two clusters ($C1$, $C2$) and there are three unique words in the vocabulary (i.e. the dimension is three). Let $d = [n_1, n_2, n_3]$ represents number of occurrences of word w_1, w_2, and w_3 in document d. Assume we have: $d_1 = [3,0,0]$, $d_2 = [0,8,3]$, $d_3 = [0,9,2]$, $d_4 = [0,2,10]$, and $d_5 = [0,2,7]$ (illustrated in Fig. 1(a)). It is likely that d_1 does not share any similarity with the other documents because w_1 in d_1 does not appear in the other documents. On the other hand, d_2, d_3, d_4 and d_5 have some common words, i.e., w_2 and w_3. Intuitively, d_1 should be clustered separately ($C1$), while d_2, d_3, d_4 and d_5 should be clustered together (in $C2$). However, because LDA is mainly affected by the word counts, d_2 will be clustered together with d_3 (in $C1$), and d_4 with d_5 (in $C2$); while d_1 will be clustered either to $C1$ *or* $C2$. Figure 1(b) shows the results of running ten consecutive trials with LDA[1]. As we can observe that **none** of the ten consecutive trials follows our first intuition (i.e. d_1 in a separate cluster). On the other hand, our proposed model gives a better solution which is shown in Fig. 1(c). Notice from Fig. 1(d) that if we represent the documents as unit vectors on a sphere, document d_1 is well separated from d_2 and d_3 or d_3 and d_4.

As the above example illustrates, the weakness of LDA is that it is not fully sensitive to the directionality of data and is essentially governed by word counts. Our proposed approach extends LDA by integrating directional distribution and treating the observations in a vector space. Specifically, we represent very high dimensional (and often sparse) observations as unit vectors, where direction plays a pivotal role in distinguishing one entity from another.

We highlight the importance of our proposed model from two perspectives. First, due to the integration of directional distribution, the resulting clusters are more robust against outliers and it could potentially give a better clustering solution. Secondly, because of the robustness, the outliers are well separated from the rest of the data, which could be used as a base for outlier detection. We present more details about the directional distribution that we use, the von Mises-Fisher (vMF) distribution, in Sect. 2. We summarize our contributions as follows:

1. We introduce a novel problem for discovering outlier participants in projects based on their previous working history.

---

[1] As LDA treats a document as a finite mixture over a set of topics, we assume a topic with the largest proportion as the topic of a document.

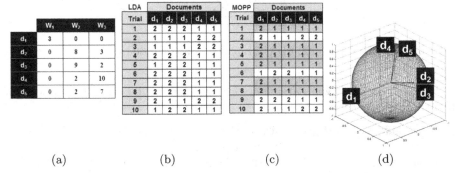

(a)   (b)   (c)   (d)

Fig. 1. Example: (a) the word counts for five documents; Topic assignments for the five documents: (b) LDA (c) MOPP; (d) Distribution of the documents on the unit sphere. Notice from (c) that MOPP separates d_1 from the other documents 6 out of 10 times.

2. We model the proposed problem using a topic-based hierarchical generative model based on the von Mises-Fisher (vMF) directional distribution. The model is named MOPP: **M**ining **O**utlier **P**articipants from **P**rojects.

3. We have implemented MOPP and compared it with a variation of Latent Dirichlet Allocation (LDA) on several synthetic and real data sets. We show that MOPP improves both the ability to detect outliers and form high quality clusters compared to LDA.

2   Related Work

The outlier or anomaly detection problem has been extensively researched in the data mining, machine learning and statistical communities. The survey by Chandola et. al. [1], provides an overview of contemporary data mining methods used for outlier detection. Our proposed model is mainly inspired by the concept of hierarchical structure in topic model used in Latent Dirichlet Allocation (LDA) [2]. In this section we present LDA and some work in directional distributions.

2.1   Latent Dirichlet Allocation

Latent Dirichlet Allocation (LDA) is a generative probabilistic model proposed by Blei et al. [2]. LDA describes the generative process and captures the latent structure of topics in a text corpus. LDA is now widely used for the clustering and topic modeling tasks. The graphical representation of LDA is shown in Fig. 2.

The plate M represents documents and the plate Nm represents words in document m. $W_{m,n}$ represents the observed word n in document m. $Z_{m,n}$ represents topic assignment of word n in document m. θ_m represents topic mixture in document m. β represents the underlying latent topics of those documents. Both α

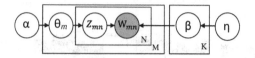

Fig. 2. LDA: Graphical Representation

and η represent the hyperparameters for the model. The generative process of LDA is summarized as follows:

Topic mixture: $\theta_m|\alpha \sim \text{Dirichlet}(\alpha)$, $m \in M$
Topic: $\beta_k|\eta \sim \text{Dirichlet}(\eta)$, $k \in K$
Topic assignment: $Z_{m,n}|\theta_m \sim \text{Multinomial}(\theta_m)$, $n \in Nm$
Word: $W_{m,n}|\beta_{Z_{m,n}} \sim \text{Multinomial}(\beta_{Z_{m,n}})$

2.2 Directional Distribution

Mardia [3,4] and Fisher [5] discussed von Mises-Fisher (vMF) distribution as a natural distribution and the simplest parametric distribution for directional statistics. The vMF distribution has support on $\mathbb{S}^{d-1}$ or unit $(d\text{-}1)$-sphere embedded in $\mathbb{R}^d$, and has properties analogous to the multi-variate Gaussian distribution. More details about vMF distribution can be found in [3]. The probability density function of vMF distribution is described as follows:

$$f(\mathbf{x}|\mu,\kappa) = \frac{\kappa^{d/2-1}}{(2\pi)^{d/2}I_{d/2-1}(\kappa)}e^{\kappa\mu^T\mathbf{x}} \tag{1}$$

where μ is called *mean direction*, $\|\mu\| = 1$; κ is called *concentration parameter* and characterizes how strongly the unit random vectors are concentrated[2] about the mean direction (μ), and $\kappa \geq 0$. $I_r(\cdot)$ represents the modified Bessel function of the first kind of order r.

A body of work has shown the effectiveness of directional distribution for modeling text data. Zhong et al. [6] shown that vMF distribution gives superior results in high dimensions comparing to Euclidean distance-based measures. Banerjee et al. [7,8] proposed the use of EM algorithm for a mixture of von Mises-Fisher distributions (movMF). Banerjee et al. [9] also observed the connection between vMF distributions in a generative model and the spkmeans algorithm [10] which is superior for clustering high-dimensional text data. Reisinger et al. [11] proposed a model named SAM that decomposes spherically distributed data into weighted combinations of component vMF distributions. However, both movMF and SAM lack a hierarchical structure and cannot be scaled-up for domains involving multiple levels of structure.

The effectiveness of vMF distribution has also been studied in many outlier detection studies. Ide et al. [12] proposed an eigenspace-based outliers detection

---

[2] Specifically, if $\kappa = 0$, the distribution is uniform and, if $\kappa \to \infty$, the distribution tends to concentrate on one density.

in computer systems, especially in the application layer of Web-based systems. Fujimaki et al. [13] proposed the use of vMF distribution for spacecraft outliers detection. Both of these two papers use a single vMF distribution and compute the angular difference of two vectors to determine outliers. Kriegel et al. [14] proposed an approach to detect outliers based on angular deviation. Our proposed model uses a mixture of K-topics as latent factors that underlie the generative process of observations to detect outlier participants from projects.

3   Mining Outlier Participants

3.1   The Proposed Model

Figure 3(a) shows the graphical model of our proposed model. We use the following assumption: rectangular with solid line represents replication of plates, rectangular with dashed-line represents a single plate (named as a dummy plate), a shaded circle represents an observable variable, an unshaded circle represents latent/unobservable variable, and a directed arrow among circles represents a dependency among them.

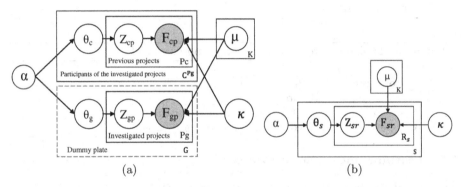

(a)   (b)

Fig. 3. Graphical representation: (a) MOPP, (b) The simplified MOPP

Recall that each project has a profile vector and each participant has a set of profile vectors. We refer the project as the *investigated project*. Each investigated project is represented by an L_2-normalized TF-IDF unit vector and each participant of the investigated project at time t has a number of previous projects before t. The subscripts on the participants and the investigated projects plates in Fig. 3(a) are used to differentiate between the two plates. We keep a record about the respective subscripts and merge them to simplify MOPP. We call the new model as simplified MOPP and use it in the learning and inference process. The record will be used later after the learning and inference process to recover information about which plates the learnt latent values originally belong to. We show the simplified MOPP in Fig. 3(b). Notice that in the simplified MOPP, we "stack" the dummy plate G (with P_g) and the participants of the investigated projects plates C^{P_g} (with P_c). We then refer the stacked G and C^{P_g} as S. P_g and P_c on the respective G and C^{P_g} plates are referred as R_s. We also rename the subscripts c and g as s, and the subscripts p as r. Table 2 summarizes the main symbols we use in this work. We present the generative process of the proposed model in Algorithm 1.

Table 2. Main symbols and their definitions used in this work

Symbol	Definition	Symbol	Definition
α	Hyperparameter	θ_g	Topic proportions for dummy plate
K	Numb. of topics	P_g	Investigated projects
μ	Mean direction of vMF	$|P_g|$	Numb. of P_g
κ	Concentration parameter of vMF	Z_{gp}	Topic assignment of project $p \in P_g$
V	Dimension of unit vector F	F_{gp}	L_2-normalized TF-IDF unit vector
C^{P_g}	Participants (of the investigated		of project $p \in P_g$
	projects at snapshot t) plate	S	C + a dummy plate, $C = \forall C^{P_g}$
$|C^{P_g}|$	Numb. of C^{P_g}	$|S|$	$|C| + 1$
θ_c	Topic proportions for participant c	R_s	Projects on plate s, $R_s \in \{\forall P_g, \forall P_c\}$
P_c	Projects of participant c before t	$|R_s|$	Numb. of R_s
$|P_c|$	Numb. of P_c	θ_s	Topic proportions for s, $s \in S$
Z_{cp}	Topic assignment of project $p \in P_c$	Z_{sr}	Topic assignment of project $r \in R_s$
F_{cp}	L_2-normalized TF-IDF unit vector of	F_{sr}	L_2-normalized TF-IDF unit vector
	project $p \in P_c$		of project $r \in R_s$
G	Dummy plate		

Algorithm 1. Generative process of the proposed model

---

for $s=1$ to $|S|$, $s \in S$ **do**

    Choose topic proportions $\theta_s \sim Dir(\alpha)$

    for·$r=1$ to $|R_s|$, $r \in R_s$ **do**

        Choose a topic μ_k, $\{1..K\} \ni Z_{sr} \sim Multinomial(\theta_s)$

        Compute L_2-normalized TF-IDF unit vector $||F_{sr}||_2 \sim \text{vMF}(\mu_k, \kappa)$

3.2 Learning and Inference Process

The joint distribution for a plate s is given as follows:

$$p(\theta, Z, F | \alpha, \kappa, \mu) = p(\theta|\alpha) \prod_{r=1}^{|R_s|} p(Z_r|\theta) p(F_r|Z_r, \mu, \kappa) \quad (2)$$

We introduce variational parameters γ and ϕ in the following variational distribution q for the inference step in (3).

$$q(\theta, Z | \gamma, \phi) = q(\theta|\gamma) \prod_{r=1}^{|R_s|} q(Z_r|\phi_r) \quad (3)$$

Application of Jensen's inequality for a plate s [2] results in:

$$\log p(F|\alpha, \mu, \kappa) = L(\gamma, \phi; \alpha, \mu, \kappa) + KL(q(\theta, Z|\gamma, \phi)||p(\theta, Z|F, \alpha, \mu, \kappa)) \quad (4)$$

where KL represents the Kullback-Leibler divergence notation. By using the factorization of p and q, we then expand the lower bound $L(\gamma, \phi; \alpha, \mu, \kappa)$ in (5):

$$L(\gamma, \phi; \alpha, \mu, \kappa) = E_q[\log p(\theta|\alpha)] + E_q[\log p(Z|\theta)] + \boxed{E_q[\log p(F|Z, \mu, \kappa)]}$$
$$- E_q[\log q(\theta)] - E_q[\log q(Z)] \tag{5}$$

The derivation from (4) to (5) is similar with the derivation from (13) to (14) in ([2] p.1019) except for the third terms on the right hand side of (5) (highlighted), which now includes the vMF distribution. Due to the space constraint, interested readers should refer to [2] for the expansion details of the first, second, fourth, and fifth terms on the right hand side of (5).

The variational parameter γ_k is calculated by maximizing (5) w.r.t. variational parameter γ_k. Using the same approach, the variational parameter ϕ is computed by maximizing (5) w.r.t. variational parameter ϕ_{rk} and introducing Lagrange multiplier, $\sum_{k=1}^{K} \phi_{rk} = 1$. Both of these two steps result in (6).

$$\gamma_k^* = \alpha_k + \sum_{r=1}^{|R_s|} \phi_{rk}$$

$$\phi_{rk}^* \propto exp((\Psi(\gamma_k) - \Psi(\sum_{k=1}^{K} \gamma_k)) + \log p(F_r|\mu_k, \kappa)) \tag{6}$$

where Ψ is the digamma function (the first derivative of the log Gamma function). To compute μ, we need to calculate:

$$\mu^* = \arg\max_{\mu_k} \sum_{s=1}^{|S|} \sum_{r=1}^{|R_s|} \sum_{k=1}^{K} \phi_{srk} \log p(F_{sr}|\mu_k, \kappa) \tag{7}$$

Equation (7) is the same as fitting von Mises-Fisher distributions in a mixture of von Mises-Fisher distributions [7], where ϕ_{srk} is the mixture proportions. The complete process for variational EM algorithm in the learning and inference process includes the following iterative steps:

E-step: Compute the optimized values of γ and ϕ for each plate s using (6).
M-step: Maximize the lower bound w.r.t. to the model parameters α and μ described in the standard LDA model [2] and (7) respectively.

3.3 Scoring the Expertise to Project's Topic

The learning and inference process (Sect. 3.2) assigns a topic to each project on R_s. Recall from Section 3.1 that we keep a record about the subscripts in MOPP but use the simplified MOPP for the learning and inference process. After the learning and inference process, we need to *reverse* map the learnt latent values back to the plate that they originally belong to. This process is crucial because our goal is to retrieve the topic proportions of each participant (θ_c) and the topic assignment of the investigated project (Z_{gp}). The topic proportions in θ_c based on the investigated project's topic will become the score of each participant in that project. An end-to-end pseudo-code that summarizes the steps to score each participant's expertise to projects and mine the outlier participants is shown in Algorithm 2.

Algorithm 2. Mining outlier participants algorithm

Input: 1) Investigated projects' L_2-normalized unit vectors, 2) Each participant from the investigated projects with their L_2-normalized unit vectors, 3) O: number of outlier participants

Output: $topO$: List of Top-O [<outlier participants, his project, his score>] outlier participants

Steps:

1: Let $lPart \leftarrow \emptyset$, $topO \leftarrow \emptyset$

2: Translate MOPP to simplified MOPP (Sect. 3.1)
3: Do the learning and inference process (Sect. 3.2)
4: Reverse map the learnt latent values to the respective projects. (Sect. 3.3)

5: **for** every investigated project $p \in P_g$
6:     Get its topic assignment Z_{gp}
7:     **for** every participant $c \in C^{P_g}$
8:         Get the topic proportions θ_c
9:         Get his score, $partScore \leftarrow \theta_c[Z_{gp}]$
10:         $lPart \leftarrow <c, p, partScore>$
11: Sort $lPart$ in an ascending order based on $partScore$ values
12: $topO \leftarrow$ take the participants in the first Top-O lowest scores in $lPart$
13: Output $topO$

We translate MOPP to simplified MOPP (Line 2) following our description in Sect. 3.1. Line 3 refers to the learning and inference process, which is summarized in the E-step and M-step (Sect. 3.2). Line 4: after we learn the model, we translate back the simplified MOPP to MOPP. Lines 5–10: for every project p in P_g, we infer its topic assignment (Z_{gp}) and score each participant based on his topic proportion in Z_{gp}. Lines 11–13: we label $topO$ lowest scores participants as outlier participants.

4   Experiments

In this section we present our experiments to evaluate the performance of our proposed model. The proposed model is implemented in Matlab and we conduct the experiments on a machine with Intel® Core(TM) Duo CPU T6400 @2.00 GHz, 1.75 GB of RAM.

4.1   Baseline Methods

For our baseline methods, we use a cosine similarity test and a latent topic model Latent Dirichlet Allocation (LDA). We specifically measure the cosine similarity between the TF-IDF vectors of each participant and his/her current project. We form the TF-IDF vector of each participant from the words of all his/her previous projects, while the TF-IDF vector for a project is extracted from the words in the title. The cosine similarity is defined as follows:

$$cos(vec_1, vec_2) = \frac{vec_1 \cdot vec_2}{|vec_1||vec_2|} \qquad (8)$$

The original LDA model (Sect. 2.1) is not suitable to be used directly for our purpose. We introduce the modified LDA for our purpose in Fig. 4. Following the translation for the simplified MOPP in Sect. 3.1, we keep a record of the subscripts. We perform the reverse mapping after the learning and inference process. The remaining steps are the same as step 5-13 in Algorithm 2. We set $\eta = 0.1$ and $\alpha = 50/K$, where K is the number of topics [15]. Because LDA returns topic mixture for each project and each participant, we use (9) to score the participants. We sort the scores in an ascending order to list the outlier participants, where t_p is the topic mixture of project p and $s_{i,p}$ is the topic mixture of a participant i in project p.

$$< t_p \cdot s_{i,p} > \qquad (9)$$

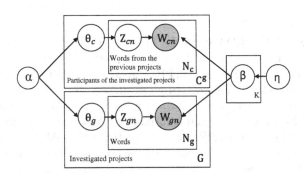

Fig. 4. The modified LDA for mining outlier participants

4.2 Semi Synthetic and Synthetic Data Sets

Our experiments are divided into two parts: using semi-synthetic and synthetic data sets. For the semi-synthetic data set, we use data from the Arxiv HEP-TH (high energy physics theory) network[3]. This data set was originally released as a part of 2003 KDD Cup. We analyze the publications in year 2003 and extract the authors. We then take a number of authors from DBLP[4] and their publications. These DBLP authors will act as the outlier participants. For all publications extracted from HEP-TH and DBLP, we use words from the title. We then inject the authors from DBLP to the HEP-TH randomly. We have 1,212 HEP-TH authors in 554 projects. We vary the number of outlier participants and the results are the average values over ten trials. To evaluate the performance of our proposed model and the baseline methods, we use precision, recall and the F1 score. We show the results in Fig. 5.

---

[3] http://snap.stanford.edu/data/cit-HepTh.html
[4] www.informatik.uni-trier.de/ ley/db/

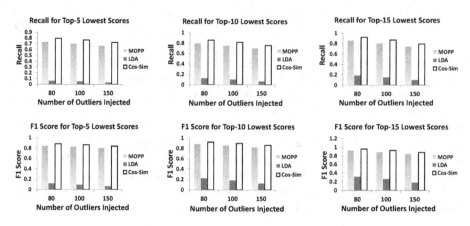

Fig. 5. Recall and F1 score for the first Top-5, 10 and 15 lowest scores in MOPP, LDA, and cosine similarity (Cos-Sim)

Figure 5 shows that LDA gives the lowest performance in detecting outlier participants (both the recall and F1 score are very low), while cosine similarity seems to be slightly better than MOPP. This is intuitive because the nature of the data set itself (HEP-TH and DBLP) is almost well-separated (the words in computer science are less likely to appear in the physics publications)[5].

We use the Normalized Mutual Information (NMI) measure to compare the performance of LDA and MOPP in reconstructing the underlying label distribution in the data set. NMI [17] is used to evaluate the clustering result [18,19]. NMI is defined as follows:

$$NMI(X,Y) = \frac{I(X,Y)}{\sqrt{H(X)H(Y)}} \qquad (10)$$

where X represents cluster assignments, Y represents true labels on the data set, I and H represent mutual information and marginal entropy. Table 3 presents the result and shows that MOPP gives a better cluster quality than LDA does.

Table 3. NMI score: MOPP vs. LDA. The best values are highlighted, where NMI score close to 0 represents bad clustering quality and NMI = 1 for perfect clustering quality

NMI	Number of Outliers Injected		
	80	100	150
LDA	0.308	0.404	0.626
MOPP	**0.803**	**0.811**	**0.815**

---

[5] In our initial experiments, we also included a classical outlier detection method Local Outlier Factor (LOF) [16]. Unfortunately LOF fails to detect outlier participants in any settings so we do not include the result here.

In the second part of the experiment, we form a small synthetic data set that represents a scenario illustrated in Fig. 6(a). This scenario is often observed in the real word that the extracted words from the participants may not appear in the investigated projects' extracted words. For example in Fig. 6(a) the word W_1 and W_2 appear in the investigated project 2, but do not appear in the previous projects of participant P_4. Because cosine similarity compares directly between TF-IDF vector of a participant and an investigated project, obviously P_1 and P_4 in Fig. 6(a) will be marked as outlier participants in project 1 and 2 respectively. However if we analyze in the topic space, the *true* outlier participant should be P_1, because P_4's words are likely to share same topic with the investigated project 2 (through words in P_5 and P_6). We generate the synthetic data set by first generating words with random occurrence for the investigated projects. We then generate words for the participants with also random occurrence[6]. We generate three types of participant: "normal", "spurious-outlier", and "true-outlier" participants. P_2, P_3, P_5 and P_6 are examples of the normal participants, P_4 is an example of the spurious-outlier participant, and P_1 is an example of the true-outlier participant. We randomly assign all the participants to the investigated projects and inject the true-outlier participants.

The table in Fig. 6(b) shows that MOPP outperforms both cosine similarity and LDA[7] in this scenario. As we can observe MOPP returns all the true-outlier participants and all true-outlier participants have the lowest score. The cosine similarity returns all the true-outlier participants together with the spurious-outlier participants (low precision and high recall score). On the other hand, LDA correctly returns the true-outlier participants (high precision score), but misses many true outliers (low recall score).

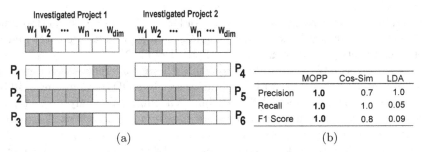

(a)   (b)

Fig. 6. (a) Scenario used in the synthetic experiment: *dim* is the dimensionality of vocabulary, W represents word, P represents participant, shaded box of W_n represents a certain number of appearances of word W_n and unshaded box of W_n means word W_n does not appear (b) Precision, recall and F1 score of MOPP, cosine similarity (Cos-Sim) and LDA

---

[6] To fit the scenario, we generate words with occurrence > 0.

[7] For MOPP and cosine similarity, we consider results from the lowest returned score, while for LDA we take the results from the Top-5 lowest returned scores.

Running time: MOPP vs. LDA. In this section we compare the running time of LDA and MOPP w.r.t. the dimension of data. We form synthetic data sets with various dimensions: 100, 500, 1,000, 1,500, 2,000, 2,500, 3,000, 3,500, 4,000, 4,500, and 5,000. We then run LDA and MOPP with various number (5, 10, 15 and 20) of topics/clusters. Figure 7 presents the results of the running time for 1,000 iterations of the respective model. It is clear that MOPP scales linearly as the dimensionality of the data set increases. As the number of topics and dimension keep increasing, MOPP can run under 500 seconds for 1,000 iterations or less than half second per iteration.

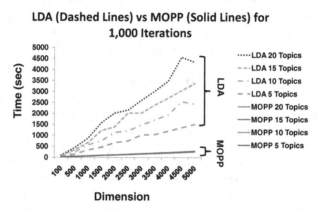

Fig. 7. Running time: MOPP vs. LDA for synthetic data set with various dimensions

4.3   Real Data set

Experimental Settings. We use a subset of DBLP to evaluate the performance of MOPP. In a bibliographic setting, a publication represents a project and an author represents a participant in our proposed model[8].

We take projects from year 2005 to 2011 of four conferences that represent four main research fields, i.e. VLDB (databases), SIGKDD (data mining), SIGIR (information retrieval), and NIPS (machine learning). We analyze the title of each project. We remove words which are *too common* or *too rare*[9] from our analysis. We have 141,999 projects in total.

We only consider projects which have at least two participants who have more than nine previous[10] projects. We assume that a participant who has at least ten previous projects in his/her profile has already "matured" his/her research direction. We refer these filtered projects as the *investigated projects*.

---

[8] Henceforth we use the term a project for a publication and a participant for an author.

[9] We determine words that appear less than 100 times are too rare and more than 100,000 times are too often.

[10] This includes projects before year 2005 as well.

At the end of process, we have 792 investigated projects, 1,424 participants and the dimensionality of data (size of vocabulary) is 2,108. In average, each participant has 28.71 previous projects before he/she joins in the investigated project. The number of topics K that we use is five, i.e. four main research topics and one for the other), κ for MOPP = 2,500, and α for MOPP is initialized to 1 to represent a non-informative prior [20,21].

Convergence Rate. We now show our empirical verification that the variational EM of MOPP is able to converge. The convergence rate of α and variational EM for DBLP data set are shown in Fig. 8(a) and Fig. 8(b) respectively.

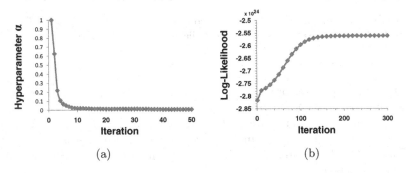

(a)   (b)

Fig. 8. DBLP: Rate of convergence of (a) α value (b) variational EM

Experimental Results. From MOPP and the baseline methods, we aim to mine those participants who have the lowest scores. We examine these three cases for our analysis and present the results in the following paragraphs (Fig. 9, 10 and 11):

Case 1: Participant who has the lowest score from the cosine similarity only
Case 2: Participant who has the lowest score from LDA only
Case 3: Participant who has the lowest θ_c from MOPP only

Case 1. The cosine similarity measures the correlation between two vectors. We focus on the participants who have cosine similarity with the investigated project equal to 0. Figure 9 shows the extracted words from an investigated project and previous projects of a participant (*ID 16544*)[11]. The cosine similarity of this participant's TF-IDF vector and the investigated project's TF-IDF vector is zero. This participant is marked as an outlier participant by cosine similarity. However, we can observe that the words from his previous projects (*multiprocessor, microprocessor, chip, pipeline, architecture*) have a strong relation with the the word from the investigated project (i.e. *multicore*). The score from LDA is 0.23 and from MOPP is 0.047. Both of these scores are not the lowest scores in the respective models.

---

[11] We use an anonymous ID for a participant.

Words extracted from the investigated project:	map-reduce, machine, learning, multicore
Venue of this investigated project:	NIPS – 2006
Participant ID:	16544

Words extracted from participant's previous projects and the frequency:
multiprocessor (6), caching (4), microprocessor (4), application (3), design (3), java (3), programs (3), system (3), transaction (3), chip (2), clustering (2), coherent (2), consistent (2), data (2), dynamic (2), implemented (2), optimal (2), parallel (2), pipeline (2), spaces (2), speculative (2), address (1), alternate (1), analysis (1), approach (1), architecture (1), associated (1), bandwidth (1), benefits (1), circuits (1), clock (1), concurrent (1), considerations (1), correct (1), efficient (1), embedded (1), environment (1), evaluation (1), exploitation (1), explorer (1), extraction (1), filtering (1), framework (1), hardware-software (1), high-performance (1), impact (1), increasing (1), investigation (1), language (1), memory (1), on-chip (1), performance (1), polymorphic (1), porting (1), primary (1), profiles (1), prototype (1), real-time (1), shared-memory (1), sharing (1), specification (1), support (1), synchronizing (1), testing (1), threading (1), timed (1), verifiably (1), verification (1)

Fig. 9. Case 1: participant with the lowest score from cosine similarity only (Cos-Sim score: 0). Highlight the weakness of cosine similarity at the word level.

Case 2. In case 2, we present a participant (*ID 10261*) which LDA marks as an outlier participant (Figure 10). However, as we can see here too that the words from his previous projects (*internet, network, malicious, online*) are related to the extracted words in the investigated project (*spammer, online, social, networks*). The scores from LDA, cosine-similarity and MOPP are 0.095, 0.097, and 0.167 respectively.

Words extracted from the investigated project:	detecting, spammers, content, promoters, online, video, social, networks
Venue of this investigated project:	SIGIR – 2009
Participant ID:	10261

Words extracted from participant's previous projects and the frequency:
analyzing (3), characterization (3), content (3), interactions (3), internet (3), management (3), media (3), network (3), stream (3), adaptive (1), analysis (1), architecture (1), auctions (1), behavior (1), clients (1), comparative (1), distributed (1), dynamic (1), education (1), energy (1), graphs (1), incentives (1), live (1), malicious (1), methodology (1), mobile (1), online (1), p2p (1), peer (1), perspectives (1), placement (1), protocols (1), quality (1), resource (1), search (1), security (1), self-adaptive (1), server (1), service-oriented (1), sharing (1), summary (1), system (1), theoretical (1), tradeoffs (1), traffic (1), understanding (1), user (1), video (1), wide-area (1), workload (1)

Fig. 10. Case 2: participant with the lowest score from LDA only (LDA score: 0.095)

Case 3. The last case that we consider is when MOPP gives the lowest score. Figure 11 presents the results for participant *ID 116471*. From the words extracted from this participant's previous projects, it seems that this participant focus more on web graph analysis. However, the words from the investigated project suggest that this project presents method in text analysis algorithms. This participant has a score 0.197 from LDA and 0.0828 from cosine similarity.

Words in title of the investigated publication:	variable, latent, semantic, indexing
Venue of this investigated publication:	KDD – 2005
Participant ID:	116471
Words extracted from participant's previous projects and the frequency:	

web (6), analysis (3), graphs (3), models (3), algorithm (2), caching (2), information (2), prefetchers (2), random (2), semantic (2), system (2), abstraction (1), algebraic (1), annotating (1), application (1), automated (1), based (1), bootstrap (1), bounded (1), clock (1), comparison (1), computing (1), connection (1), discovery (1), extending (1), extraction (1), fast (1), hypertext (1), interval (1), knowledge (1), large-scale (1), linear (1), markov (1), measuring (1), method (1), metrics (1), minimality (1), mining (1), online (1), parallel (1), probabilistic (1), recommendation (1), retrieval (1), scheduler (1), segmentation (1), skewed (1), sub-graph (1), targeted (1), tasks (1), teaching (1), transition (1), tree (1), understanding (1), walk (1), zero (1)

Fig. 11. Case 3: participant with the lowest score from MOPP only (MOPP score: 0)

5 Conclusion

In this paper, we introduce the **M**ining **O**utlier **P**articipants from **P**rojects (MOPP) model, to address the problem of scoring and ranking participant's expertise to their projects. Each participant is scored based on their project working history to the project's topic. Participants who have the lowest scores are marked as outlier participants, which means that these participants have different topic interests compare to the projects that they are working on. MOPP incorporates the structure and nature of hierarchical generative model and directional distribution, the von Mises-Fisher distribution. Experiments on semi-synthetic and synthetic data sets show that MOPP outperforms baseline methods. We also present the result from real data set extracted from DBLP. The proposed model consistently gives more meaningful and semantically correct results from the bibliographic network DBLP. For future work, we would like to extend the model to non-parametric model and compare its performance to other non-parametric topic models. We also plan to implement MOPP in different domains.

References

1. Chandola, V., Banerjee, A., Kumar, V.: Anomaly detection: A survey. ACM Computing Surveys 41 (2009)
2. Blei, D., Ng, A., Jordan, M.: Latent dirichlet allocation. Journal of Machine Learning Research (JMLR) 3, 993–1022 (2003)
3. Mardia, K.: Statistical of directional data (with discussion). Journal of the Royal Statistical Society 37(3), 390 (1975)
4. Mardia, K., Jupp, P.: Directional Statistics. John Wiley and Sons, Ltd. (2000)
5. Fisher, N., Lewis, T., Embleton, B.: Statistical Analysis of Spherical Data. Cambridge University Press (1987)
6. Zhong, S., Ghosh, J.: Generative model-based document clustering: A comparative study. Knowledge and Information Systems 8(3), 374–384 (2005)
7. Banerjee, A., Dhillon, I., Ghosh, J., Sra, S.: Clustering on the unit hypersphere using von mises-fisher distributions. Journal of Machine Learning Research (JMLR) 6, 1345–1382 (2005)

8. Banerjee, A., Dhillon, I., Ghosh, J., Sra, S.: Generative model-based clustering of directional data. In: ACM Conference on Knowledge Discovery and Data Mining, SIGKDD (2003)
9. Banerjee, A., Ghosh, J.: Frequency sensitive competitive learning for clustering on high-dimensional hyperspheres. In: Proceedings International Joint Conference on Neural Networks, vol. 15, pp. 1590–1595 (2002)
10. Dhillon, I.S., Modha, D.S.: Concept decompositions for large sparse text data using clustering. Machine Learning 42, 143–175 (2001)
11. Reisinger, J., Waters, A., Silverthorn, B., Mooney, R.J.: Spherical topic models. In: International Conference on Machine Learning, ICML (2010)
12. Ide, T., Kashima, H.: Eigenspace-based anomaly detection in computer systems. In: ACM Conference on Knowledge Discovery and Data Mining, SIGKDD (2004)
13. Fujimaki, R., Yairi, T., Machida, K.: An approach to spacecraft anomaly detection problem using kernel feature space. In: ACM Conference on Knowledge Discovery and Data Mining, SIGKDD (2005)
14. Kriegel, H.P., Schubert, M., Zimek, A.: Angle-based outlier detection in high-dimensional data. In: ACM Conference on Knowledge Discovery and Data Mining, SIGKDD (2008)
15. Griffith, T., Steyvers, M.: Finding scientific topics. PNAS 101, 5228–5235 (2004)
16. Breunig, M., Kriegel, H., Ng, R., Sander, J.: Lof: Identifying density-based local outliers. In: IEEE International Conference on Data Mining, ICDM (2000)
17. Strehl, A., Ghosh, J.: Cluster ensembles - a knowledge reuse framework for combining multiple partitions. Journal of Machine Learning Research (JMLR) 3, 583–617 (2002)
18. Zhong, S., Ghosh, J.: Generative model-based document clustering: a comparative study. Knowledge and Information Systems 8, 374–384 (2005)
19. Hu, X., Zhang, X., Lu, C., Park, E., Zhou, X.: Exploiting wikipedia as external knowledge for document clustering. In: ACM Conference on Knowledge Discovery and Data Mining, SIGKDD (2009)
20. DeGroot, M.H.: Probability and Statistics, 2nd edn. Addison-Wesley (1986)
21. Xu, Z., Ke, Y., Wang, Y., Cheng, H., Cheng, J.: A model-based approach to attributed graph clustering. In: SIGMOD (2012)

Anonymizing Data with Relational and Transaction Attributes

Giorgos Poulis[1], Grigorios Loukides[2], Aris Gkoulalas-Divanis[3],
and Spiros Skiadopoulos[1]

[1] University of Peloponnese
{poulis,spiros}@uop.gr
[2] Cardiff University
g.loukides@cs.cf.ac.uk
[3] IBM Research - Ireland
arisdiva@ie.ibm.com

Abstract. Publishing datasets about individuals that contain both relational and transaction (i.e., set-valued) attributes is essential to support many applications, ranging from healthcare to marketing. However,
preserving the privacy and utility of these datasets is challenging, as it
requires *(i)* guarding against attackers, whose knowledge spans both attribute types, and *(ii)* minimizing the overall information loss. Existing
anonymization techniques are not applicable to such datasets, and the
problem cannot be tackled based on popular, multi-objective optimization strategies. This work proposes the first approach to address this
problem. Based on this approach, we develop two frameworks to offer
privacy, with bounded information loss in one attribute type and minimal information loss in the other. To realize each framework, we propose
privacy algorithms that effectively preserve data utility, as verified by
extensive experiments.

1 Introduction

Privacy-preserving data mining has emerged to address privacy concerns related
to the collection, analysis, and sharing of data and aims at preventing the disclosure of individuals' private and sensitive information from the published data.
Publishing datasets containing both relational and transaction attributes, *RT-
datasets* for short, is essential in many real-world applications. Several marketing
studies, for example, need to find product combinations that appeal to specific
types of customers. Consider the *RT*-dataset in Fig. 1a, where each record corresponds to a customer. Age, Origin and Gender are relational attributes, whereas
Purchased-products is a transaction attribute that contains a *set* of *items*, representing commercial transactions. Such studies may require finding all customers
below 30 years old who purchased products E and F. Another application is in
healthcare, where several medical studies require analyzing patient demographics
and diagnosis information together. In such *RT*-datasets, patients features (e.g.,
demographics) are modeled as relational attributes and diagnosis as a transaction attribute. In all these applications, the privacy protection of data needs

H. Blockeel et al. (Eds.): ECML PKDD 2013, Part III, LNAI 8190, pp. 353–369, 2013.

Id	Name	Relational attributes			Transaction attribute
		Age	Origin	Gender	Purchased-products
0	John	19	France	Male	E F B <u>G</u>
1	Steve	22	Greece	Male	E F D <u>H</u>
2	Mary	28	Germany	Female	B C E <u>G</u>
3	Zoe	39	Spain	Female	F D <u>H</u>
4	Ann	70	Algeria	Female	E <u>G</u>
5	Jim	55	Nigeria	Male	A F <u>H</u>

Id	Relational attributes			Transaction attribute
	Age	Origin	Gender	Purchased-products
0	[19:22]	Europe	Male	E F (A,B,C,D) <u>G</u>
1	[19:22]	Europe	Male	E F (A,B,C,D) <u>H</u>
2	[28:39]	Europe	Female	E (A B,C D) <u>G</u>
3	[28:39]	Europe	Female	F (A,B,C,D) <u>H</u>
4	[55:70]	Africa	All	E <u>G</u>
5	[55:70]	Africa	All	F (A,B,C,D) <u>H</u>

(a)   (b)

Fig. 1. (a) An RT-dataset with patient demographics and IDs of purchased products, and (b) a 2-anonymous dataset with respect to relational attributes and 2^2-anonymous with respect to the transaction attribute. Identifiers Id and Name are not published.

to performed without adding *fake* or removing *truthful* information [5,16]. This precludes the application of ϵ-*differential privacy* [3], which only allows releasing noisy answers to user queries or noisy summary statistics, as well as *suppression* [19], which deletes values prior to data release.

A plethora of methods can be used to preserve the privacy of datasets containing only relational or only transaction attributes [9,12,15,18]. However, there are currently no methods for anonymizing RT-datasets, and simply anonymizing each attribute type separately, using existing methods (e.g., [9,12,15,18]), is not enough. This is because information concerning *both* relational and transaction attributes may lead to *identity disclosure* (i.e., the association of an individual to their record) [15]. Consider, for example, the dataset in Fig. 1a which is anonymized by applying the methods of [18] and [8] to the relational and transaction attributes, as shown in Fig. 1b. An attacker, who knows that Jim is a 55-year-old Male from Nigeria who purchased F, can associate Jim with record 5 in Fig. 1b. Thwarting identity disclosure is essential to comply with legislation, e.g., HIPAA, and to help future data collection. At the same time, many applications require preventing *attribute disclosure* (i.e., the association of an individual with sensitive information). In medical data publishing, for example, this ensures that patients are not associated with sensitive diagnoses [17].

Furthermore, anonymized RT-datasets need to have minimal information loss in relational and in transaction attributes. However, these two requirements are conflicting, and the problem is difficult to address using multi-objective optimization strategies [4]. In fact, these strategies are either inapplicable or incur excessive information loss, as we show in Section 3.

CONTRIBUTIONS. Our work makes the following specific contributions:

- We introduce the problem of anonymizing RT-datasets and propose the first approach to tackle it. Our privacy model prevents an attacker, who knows the set of an individual's values in the relational attributes and up to m items in the transaction attribute, from linking the individual to their record.
- We develop an approach for producing (k, k^m)-anonymous RT-datasets with bounded information loss in one attribute type and minimal information loss in the other. Following this approach, we propose two frameworks which employ *generalization* [15] and are based on a three-phase process: *(i)* creating k-anonymous clusters with respect to the relational attributes, *(ii)* merging these clusters in a way that helps anonymizing RT-datasets with low information loss, and *(iii)* enforcing (k, k^m)-anonymity to each merged cluster.

Relational attributes			Transaction attribute
Id	Age	Origin	Gender	Purchased-products
0	[19:22]	Europe	Male	D E (B,D) G
1	[19:22]	Europe	Male	E E (B,D) H
2	[28:39]	Europe	Female	(B,C,F) (D,E) G
3	[28:39]	Europe	Female	(B,C,F) (D,E) H
4	[55:70]	Africa	All	(A,E,F) G
5	[55:70]	Africa	All	(A,E,F) H

(a)

Relational attributes			Transaction attribute
Id	Age	Origin	Gender	Purchased-products
0	[19:70]	All	All	E F (A,B,C,D) G
1	[19:70]	All	All	E F (A,B,C,D) H
2	[19:70]	All	All	E (A,B,C,D) G
3	[19:70]	All	All	F (A,B,C,D) H
4	[19:70]	All	All	E G
5	[19:70]	All	All	F (A,B,C,D) H

(b)

Relational attributes			Transaction attribute
Id	Age	Origin	Gender	Purchased-products
0	[19:39]	Europe	All	E F (B,C,D) G
1	[19:39]	Europe	All	E F (B,C,D) H
2	[19:39]	Europe	All	E (B,C,D) G
3	[19:39]	Europe	All	F (B,C,D) H
4	[55:70]	Africa	All	(A,E,F) G
5	[55:70]	Africa	All	(A,E,F) H

(c)

Relational attributes			Transaction attribute
Id	Age	Origin	Gender	Purchased-products
0	[19:70]	All	All	E F (A,B,D) G
1	[19:70]	All	All	E F (A,B,D) H
4	[19:70]	All	All	E G
5	[19:70]	All	All	F (A,B,D) H
2	[28:39]	Europe	Female	(B,C,F) (D,E) G
3	[28:39]	Europe	Female	(B,C,F) (D,E) H

(d)

Fig. 2. The $(2, 2^2)$-anonymous datasets from applying (a) RFIRST, and (b) TFIRST to the dataset of Fig. 1a, and (c) RMERGE$_R$, and (d) RMERGE$_T$, to the clusters of Fig. 2a

- We propose a family of algorithms to implement the second phase in each framework. These algorithms operate by building clusters, which can be made (k, k^m)-anonymous with minimal information loss, and preserve different aspects of data utility.
- We investigate the effectiveness of our approach by conducting experiments on two real-world RT-datasets. Our results verify that the proposed approach is effective at preserving data utility.

PAPER ORGANIZATION. Section 2 defines concepts used in this work. Section 3 clarifies why popular multi-objective optimization strategies are unsuited for enforcing (k, k^m)-anonymity and formulates the target problems. Sections 4 and 5 present our approach and an instance of it. Sections 6 and 7 present experiments and discuss related work, and Section 8 concludes the paper.

2 RT-Datasets and Their Anonymity

RT-DATASETS. An RT-dataset D consists of records containing relational attributes $R_1, \ldots, R_v$, which are single-valued, and a transaction attribute T, which is set-valued. For convenience, we consider that: (i) identifiers have been removed from D, and (ii) a single transaction attribute T is contained in D^1.

(k, k^m)-ANONYMITY. We propose (k, k^m)-anonymity to guard against *identity disclosure*. To prevent both identity and attribute disclosure, (k, k^m)-anonymity can be extended, as we explain in Section 5.

Before defining (k, k^m)-anonymity, we associate each record r in an RT-dataset D with a *group* of records $G(r)$, as shown below.

---

[1] Multiple transaction attributes $T_1, \ldots, T_u$ can be transformed to a single transaction attribute T, whose domain contains every item in the domain of $T_1, \ldots, T_u$, preceded by the domain name, i.e., $dom(T) = \{d.t \mid d = T_i \text{ and } t \in dom(T_i), i \in [1, u]\}$.

Definition 1. *For a record $r \in D$, its group $G(r)$ is a set of records that contains r and each record $q \in D$, such that $q[R_1, \ldots, R_v] = r[R_1, \ldots, R_v]$ and $q[T] \cap I = r[T] \cap I$, where I is any set of m or fewer items of $r[T]$*.[2]

Group $G(r)$ contains r and all records that are indistinguishable from r to an attacker, who knows the values of r in relational attributes *and* up to m items in the transaction attribute. The size of $G(r)$, denoted with $|G(r)|$, represents the risk of associating an individual with a record r. Thus, to provide privacy, we may lower-bound $|G(r)|$. This idea is captured by (k, k^m)-anonymity.

Definition 2. *A group of records $G(r)$ is (k, k^m)-anonymous, if and only if $|G(r)| \geq k$, for each record r in $G(r)$. An RT-dataset D is (k, k^m)-anonymous, if and only if the group $G(r)$ of each record $r \in D$ is (k, k^m)-anonymous.*

For example, in Fig. 2a groups $\{0,1\}$ ($=G(0)=G(1)$), $\{2,3\}$ ($=G(2)=G(3)$) and $\{4,5\}$ ($=G(4)=G(5)$) are $(2, 2^2)$-anonymous, rendering the whole dataset $(2, 2^2)$-anonymous. Note that in each group, all records have the same values in the relational attributes, as required by Definition 1, but do not necessarily have the same items in the transaction attribute Purchased-products (see Fig. 2b).

The notion of (k, k^m)-anonymity for RT-datasets extends and combines relational k-anonymity [15] and transactional k^m-anonymity [17].

Proposition 1. *Let $D[R_1, \ldots, R_v]$ and $D[T]$ be the relational and transaction part of an RT-dataset D, respectively. If D is (k, k^m)-anonymous, then $D[R_1, \ldots, R_v]$ is k-anonymous and $D[T]$ is k^m-anonymous.*

Proposition 1 shows that (k, k^m)-anonymity provides the same protection as k-anonymity [15], for relational attributes, and as k^m-anonymity [17], for transaction attributes. *Unfortunately, the inverse does not hold.* That is, an RT-dataset may be k and k^m but not (k, k^m)-anonymous. For instance, let D be the dataset of Fig. 1b. Note that $D[\text{Age, Origin, Gender}]$ is 2-anonymous and $D[\text{Purchased-products}]$ is 2^2-anonymous, but D is not $(2, 2^2)$-anonymous.

GENERALIZATION. We employ the generalization functions defined below.

Definition 3. *A relational generalization function $\mathcal{R}$ maps a value v in a relational attribute R to a generalized value $\tilde{v}$, which is a range of values, if R is numerical, or a collection of values, if R is categorical.*

Definition 4. *A transaction generalization function $\mathcal{T}$ maps an item u in the transaction attribute T to a generalized item $\tilde{u}$. The generalized item $\tilde{u}$ is a non-empty subset of items in T that contains u.*

The way relational values and transactional items are generalized is fundamentally different, as they have different semantics [19]. Specifically, a generalized value bears *atomic* semantics and is interpreted as a *single value* in a range or a collection of values, whereas a generalized item bears *set* semantics and is interpreted as *any non-empty subset* of the items mapped to it [12]. For instance,

---

[2] Expression $r[A]$ is a shortcut for the projection $\pi_A(r)$.

the generalized value [19:22] in Age, in the record 0 in Fig. 2a, means that the actual Age is in $[19, 22]$. Contrary, the generalized item (B, D) in Purchased-products means that B, or D, or both products were bought. Given a record r, the function $\mathcal{R}$ is applied to a single value $v \in R$, and all records in the k-anonymous group $G(r)$ must have the same generalized value in R. On the other hand, the function $\mathcal{T}$ is applied to one of the potentially many items in T, and the records in the k^m-anonymous $G(r)$ may not have the same generalized items.

DATA UTILITY MEASURES. In this work, we consider two general data utility measures; **Rum**, for relational attributes, and **Tum**, for the transaction attribute. These measures satisfy Properties 1, 2 and 3.

Property 1. Lower values in **Rum** and **Tum** imply better data utility.

Property 2. **Rum** is *monotonic* to subset relationships. More formally, given two groups G and G' having at least k records, and a relational generalization function $\mathcal{R}$, it holds that $\mathbf{Rum}(\mathcal{R}(G) \cup \mathcal{R}(G')) \leq \mathbf{Rum}(\mathcal{R}(G \cup G'))$.

Property 2 suggests that data utility is preserved better, when we generalize the relational values of small groups, and is consistent with prior work on relational data anonymization [2,6]. Intuitively, this is because the group $G \cup G'$ contains more distinct values in a relational attribute R than G or G', and thus more generalization is needed to make its values indistinguishable.

A broad class of measures, such as *NCP*, the measures expressed as Minkowski norms [6], *Discernability* [1], and the *Normalized average equivalence class size metric* [9], satisfy Property 2 [6], and can be used as **Rum**.

Property 3. **Tum** is *anti-monotonic* to subset relationships. More formally, given two groups G and G' having at least k records, and a transaction generalization function $\mathcal{T}$ that satisfies Definition 4 and *(i)* maps each item in the group it is applied to a generalized item that is not necessarily unique, and *(ii)* constructs the mapping with the minimum **Tum**, it holds that $\mathbf{Tum}(\mathcal{T}(G) \cup \mathcal{T}(G')) \geq \mathbf{Tum}(\mathcal{T}(G \cup G'))$.

Property 3 suggests that generalizing large groups can preserve transaction data utility better, and is consistent with earlier works [12,17]. Intuitively, this is because, all mappings between items and generalized items constructed by $\mathcal{T}$ when applied to G and G' separately (Case I) can also be constructed when $\mathcal{T}$ is applied to $G \cup G'$ (Case II), but there can be mappings that can only be considered in Case II. Thus, the mapping with the minimum **Tum** in Case I cannot have lower **Tum** than the corresponding mapping in Case II.

3   Challenges of Enforcing (k, k^m)-Anonymity

LACK OF OPTIMAL SOLUTION. Constructing a (k, k^m)-anonymous RT-dataset D with minimum information loss is far from trivial. Lemma 1 follows from Theorem 1 and shows that there is no (k, k^m)-anonymous version of D with minimum (i.e., optimal) **Rum** and **Tum**, for any D of realistic size.

Theorem 1. *Let $\mathcal{D}_R$ and $\mathcal{D}_T$ be the optimal (k, k^m)-anonymous version of an RT-dataset D with respect to* **Rum** *and* **Tum**, *respectively. Then, no group in $\mathcal{D}_R$ contains more than $2k - 1$ records, and $\mathcal{D}_T$ is comprised of a single group.*

Proof. (Sketch) The proof that no group in $\mathcal{D}_R$ contains more than $2k - 1$ records is based on Property 2, and has been given in [6]. The proof that $\mathcal{D}_T$ is comprised of a single group is similar and, it is based on Property 3.

Lemma 1. *There is no optimal (k, k^m)-anonymous version $\mathcal{D}$ of an RT-dataset D with respect to both* **Rum** *and* **Tum**, *unless $|D| \in [k, 2k - 1]$.*

INADEQUACY OF POPULAR OPTIMIZATION STRATEGIES. Constructing useful (k, k^m)-anonymous RT-datasets requires minimizing information loss with respect to both **Rum** and **Tum**. Such multi-objective optimization problems are typically solved using the *lexicographical*, the *conventional weighted-formula*, or the *Pareto optimal* approach [4]. We will highlight why these approaches are not adequate for our problem.

Lexicographical. In this approach, the optimization objectives are ranked and optimized in order of priority. In our case, we can prioritize the lowering of information loss in *(i)* the relational attributes (i.e., minimal **Rum**), or *(ii)* the transaction attribute (i.e., minimal **Tum**).

Given an RT-dataset D and anonymization parameters k and m, an algorithm that implements strategy *(i)* is RFIRST. This algorithm partitions D into a set of k-anonymous groups $\mathcal{C}$, with respect to the relational attributes (e.g., using [18]), and applies $\mathcal{T}$ to generalize items in each group of records in $\mathcal{C}$, separately (e.g., using [17]). Symmetrically, to implement strategy *(ii)*, we may use an algorithm TFIRST, which first partitions D into a set of k^m-anonymous groups (e.g., using the LRA algorithm [17]), and then applies a relational generalization function (see Definition 3) to each relational attribute, in each group.

Both RFIRST and TFIRST enforce (k, k^m)-anonymity, but produce vastly different results. For instance, Figs. 2a and 2b show $(2, 2^2)$-anonymous versions of the dataset in Fig. 1a, produced by RFIRST and TFIRST, repetively. Observe that RFIRST did not generalize the relational attributes as heavily as TFIRST but applied more generalization to the transaction attribute. This is because, RFIRST constructs small groups, and does not control the grouping of items. Contrary, the groups created by TFIRST contain records, whose items are not heavily generalized, unlike their values in the relational attributes. In either case, the purpose of producing anonymized RT-datasets that allow meaningful analysis of relational and transaction attributes together, is defeated.

Conventional weighted-formula. In this approach, all objectives are combined into a single one, using a weighted formula. The combined objective is then optimized by a single-objective optimization algorithm. For example, a clustering-based algorithm [13] would aim to minimize the weighted sum of **Rum** and **Tum**. However, this approach works only for *commensurable* objectives [4]. This is not the case for **Rum** and **Tum**, which are fundamentally different and have different properties (see Section 2). Therefore, this approach is not suitable.

Algorithm: Rum-BOUND

```
// Initial cluster formation
1 {C₁,...,Cₙ} := CLUSTERFORMATION(D, k)
2 D := {C₁,...,Cₙ}
3 if Rum(D) > δ then return false

// Cluster merging
4 D := RMERGE(D, T, δ)

// (k, kᵐ)-anonymization
5 for each cluster C ∈ D do
6 ⌊ D := (D \ C) ∪ T(C)

7 return D
```

Algorithm: Tum-BOUND

```
// Initial cluster formation
1 {C₁,...,Cₙ} := CLUSTERFORMATION(D, k)
2 D := {C₁,···,Cₙ}
3 if Tum(T(D)) ≤ δ then return D

// Cluster merging
4 D := TMERGE(D, T, δ)

// (k, kᵐ)-anonymization
5 for each cluster C ∈ D do
6 ⌊ D := (D \ C) ∪ T(C)

7 if Tum(D) > δ then return false
8 return D
```

Pareto optimal. This approach finds a set of solutions that are *non-dominated* [4], from which the most appropriate solution is selected by the data publisher, according to their preferences. However, the very large number of non-dominated solutions that can be constructed by flexible generalization functions, such as those in Definitions 3 and 4, render this approach impractical.

PROBLEM FORMULATION. To construct a (k, k^m)-anonymous version of an RT-dataset, we *either* upper-bound the information loss in relational attributes and seek to minimize the information loss in the transaction attribute (Problem 1), *or* upper-bound the information loss in the transaction attribute and seek to minimize the information loss in relational attributes (Problem 2).

Problem 1. Given an RT-dataset D, data utility measures **Rum** and **Tum**, parameters k and m, and a threshold δ, construct a (k, k^m)-anonymous version $\mathcal{D}$ of D, such that $\mathbf{Rum}(\mathcal{D}) \leq \delta$ and $\mathbf{Tum}(\mathcal{D})$ is minimized.

Problem 2. Given an RT-dataset D, data utility measures **Rum** and **Tum**, parameters k and m, and a threshold δ, construct a (k, k^m)-anonymous version $\mathcal{D}$ of D, such that $\mathbf{Tum}(\mathcal{D}) \leq \delta$ and $\mathbf{Rum}(\mathcal{D})$ is minimized.

Threshold δ must be specified by data publishers, as in [6]. Thus, constructing $\mathcal{D}$ might be infeasible for an arbitrary δ. Solving Problem 1 or Problem 2 ensures that $\mathcal{D}$ preserves privacy and utility, but it is NP-hard (proof follows from [12]).

4   Anonymization Approach

We propose an approach that overcomes the deficiencies of the aforementioned optimization approaches and works in three phases:

Initial cluster formation: k-anonymous clusters with respect to relational attributes, which incur low information loss, are formed.

Cluster merging: Clusters are merged until the conditions set by Problems 1 or 2 are met.

(k, k^m)-*anonymization*: Each cluster becomes (k, k^m)-anonymous, by generalizing the its items with low **Tum**.

Based on our approach, we developed two anonymization frameworks, **Rum**-BOUND and **Tum**-BOUND, which address Problems 1 and 2, respectively. **Rum**-BOUND seeks to produce a dataset with minimal **Tum** and acceptable **Rum**, and implements the phases of our approach, as follows.

Initial cluster formation (Steps 1–3): Algorithm **Rum**-BOUND clusters D, using a function CLUSTERFORMATION, which can be implemented by any generalization-based k-anonymity algorithm [9,18,2]. This function produces a set of k-anonymous clusters $C_1, \ldots, C_n$, from which a dataset $\mathcal{D}$ containing $C_1, \ldots, C_n$, is created (Step 2). The dataset $\mathcal{D}$ must have a lower **Rum** than δ, since subsequent steps of the algorithm cannot decrease **Rum** (see Property 2). If the dataset $\mathcal{D}$ does not satisfy this condition, it cannot be a solution to Problem 1, and false is returned (Step 3).

Cluster merging (Step 4): This phase is the crux of our framework. It is performed by a function **R**MERGE, which merges the clusters of $\mathcal{D}$ to produce a version that can be (k, k^m)-anonymized with minimal **Tum** and without violating δ. To implement **R**MERGE we propose three algorithms, namely **R**MERGE$_R$, **R**MERGE$_T$ and **R**MERGE$_{RT}$, which aim at minimizing **Tum** using different heuristics.

(k, k^m)-*anonymization (Steps 5–7):* In this phase, $\mathcal{D}$ is made (k, k^m)-anonymous, by applying a transaction generalization function $\mathcal{T}$ to each of its clusters.

Tum-BOUND, on the other hand, focuses on Problem 2 and aims at creating a dataset with minimal **Rum** and acceptable **Tum**. This framework has the following major differences from **Rum**-BOUND.

- At Step 3, after the formation of $\mathcal{D}$, **Tum**-BOUND checks if $\mathcal{D}$ has lower **Tum** than the threshold δ. In such case, $\mathcal{D}$ is a solution to Problem 2.
- At Step 4, function **T**MERGE merges clusters until the **Tum** threshold is reached, or no more merging is possible. To implement **T**MERGE we propose three algorithms: **T**MERGE$_R$, **T**MERGE$_R$ and **T**MERGE$_{RT}$, which aim at minimizing **Rum** using different heuristics.
- At Step 7, **Tum**-BOUND checks if $\mathbf{Tum}(\mathcal{D}) > \delta$; in this case, we cannot satisfy Problem 2 conditions and, thus, return false.

CLUSTER-MERGING ALGORITHMS. We now present three algorithms that implement function **R**MERGE, which is responsible for the merging phase of **Rum**-BOUND (Step 4). Our algorithms are based on different merging heuristics. Specifically, **R**MERGE$_R$ merges clusters with similar relational values, **R**MERGE$_T$ with similar transaction items and **R**MERGE$_{RT}$ takes a middle line between these two algorithms. In all cases, relational generalization is performed by a set of functions $\mathcal{G} = \{\mathcal{L}_1, \ldots, \mathcal{L}_v\}$, one for each relational attribute (Definition 3) and transaction generalization is performed by function $\mathcal{T}$ (Definition 4).

RMERGE$_R$ selects the cluster C with the minimum $\mathbf{Rum}(C)$ as a seed (Step 2). Cluster C contains relational values that are not highly generalized and is expected to be merged with a low relational utility loss. The algorithm locates the cluster C' with the most similar relational values to C (Step 3) and constructs a temporary dataset $\mathcal{D}_{tmp}$ that reflects the merging of C and C' (Step 4). If $\mathcal{D}_{tmp}$ does not violate the **Rum** threshold, it is assigned to $\mathcal{D}$ (Step 5).

Algorithm: $\mathbf{R}\textsc{merge}_R$

1. **while** $\mathcal{D}$ *changes* **do**
2.   Select, as a seed, the cluster $C \in \mathcal{D}$ with minimum $\mathbf{Rum}(C)$
3.   Find the cluster $C' \in \mathcal{D}$ that minimizes $\mathbf{Rum}(\mathcal{G}(C \cup C'))$.
4. $\mathcal{D}_{tmp} := ((\mathcal{D} \setminus C) \setminus C') \cup \mathcal{G}(C \cup C')$
5.   **if** $\mathbf{Rum}(\mathcal{D}_{tmp}) \leq \delta$ **then**
           $\mathcal{D} := \mathcal{D}_{tmp}$
6. **return** $\mathcal{D}$

Algorithm: $\mathbf{R}\textsc{merge}_T$

1. **while** $\mathcal{D}$ *changes* **do**
2.   Select, as a seed, the cluster $C \in \mathcal{D}$ with minimum $\mathbf{Rum}(C)$
       `// Find the appropriate cluster C' to be merged with C`
3.   Let $\{C_1, \ldots, C_t\}$ be the set of clusters in $\mathcal{D} \setminus C$ ordered by increasing $\mathbf{Tum}(\mathcal{T}(C \cup C_i))$, $i \in [1, t)$
4.   **for** $i := 1$ **to** t **do**   `// Test if C' = C_i`
5. $\mathcal{D}_{tmp} := ((\mathcal{D} \setminus C) \setminus C_i) \cup \mathcal{G}(C \cup C_i)$
6.   **if** $\mathbf{Rum}(\mathcal{D}_{tmp}) \leq \delta$ **then**   `// C' is C_i`
7. $\mathcal{D} := \mathcal{D}_{tmp}$
8.   **exit** the **for** loop
9. **return** $\mathcal{D}$

Algorithm: $\mathbf{R}\textsc{merge}_{RT}$

1. **while** $\mathcal{D}$ *changes* **do**
2.   Select, as a seed, the cluster $C \in \mathcal{D}$ with minimum $\mathbf{Rum}(C)$
3.   Let $\{C_1, \ldots, C_t\}$ (resp. $\{\hat{C}_1, \ldots, \hat{C}_t\}$) be the set of clusters in $\mathcal{D} \setminus C$ ordered by increasing $\mathbf{Rum}(\mathcal{G}(C \cup C_i))$ (resp. $\mathbf{Tum}(\mathcal{T}(C \cup \hat{C}_i)))$, $i \in [1, t)$
       `// Find the appropriate cluster C' to be merged with C`
4.   **for** $i := 1$ **to** t **do**
5.   Find cluster C', that has the i-th minimum sum of indices $u + v$ s.t. $C_u \in \{C_1, \ldots, C_t\}$ and $C_v \in \{\hat{C}_1, \ldots, \hat{C}_t\}$
6. $\mathcal{D}_{tmp} := ((\mathcal{D} \setminus C) \setminus C_i) \cup \mathcal{G}(C \cup C_i)$
7.   **if** $\mathbf{Rum}(\mathcal{D}_{tmp}) \leq \delta$ **then**
8. $\mathcal{D} := \mathcal{D}_{tmp}$
9.   **exit** the **for** loop
10. **return** $\mathcal{D}$

$\mathbf{R}\textsc{merge}_T$ starts by selecting the same seed C as $\mathbf{R}\textsc{merge}_R$ (Step 2) and seeks a cluster C' that contains similar transaction items to C and, when merged with C, results in a dataset with $\mathbf{Rum}$ no higher than δ. To this end, $\mathbf{R}\textsc{merge}_T$ merges C with every other cluster C_i in $\mathcal{D} \setminus C$ and orders the clusters by increasing $\mathbf{Tum}(\mathcal{T}(C \cup C_i))$ (Step 3). This allows efficiently finding the best merging for minimizing $\mathbf{Tum}$ that does not violate $\mathbf{Rum}(\mathcal{D}) \leq \delta$. The algorithm considers the clusters with increasing $\mathbf{Tum}(\mathcal{T}(C \cup C_i))$ scores. The first cluster that gives a dataset with acceptable $\mathbf{Rum}$ is used for merging (Steps 4–5).

$\mathbf{R}\textsc{merge}_{RT}$ combines the benefits of $\mathbf{R}\textsc{merge}_R$ and $\mathbf{R}\textsc{merge}_T$. It selects the same seed cluster C as $\mathbf{R}\textsc{merge}_T$, and constructs two orderings, which sort the generalized merged clusters in ascending order of $\mathbf{Rum}$ and $\mathbf{Tum}$, respectively (Step 3). Then, a cluster C' that is as close as possible to C, based on both orderings (i.e., it has the i-th minimum sum $(u + v)$, where u and v are the indices of C' in the $\{C_1, \ldots, C_t\}$ and orderings $\{\hat{C}_1, \ldots, \hat{C}_t\}$ repsectively), is found (Step 5). The next steps of $\mathbf{R}\textsc{merge}_{RT}$ are the same as in $\mathbf{R}\textsc{merge}_T$.

We now discuss $\mathbf{T}\textsc{merge}_R$, $\mathbf{T}\textsc{merge}_R$, and $\mathbf{T}\textsc{merge}_{RT}$, used in $\mathbf{Tum}\text{-}\textsc{bound}$. These algorithms perform cluster merging, until $\mathcal{D}$ satisfies the $\mathbf{Tum}$ threshold, or all possible mergings have been considered. The pseudocode of $\mathbf{R}\textsc{merge}_R$ is the same as that of $\mathbf{T}\textsc{merge}_R$, except that Step 5 in $\mathbf{R}\textsc{merge}_R$ is replaced by the following steps. Note that D is returned if it satisfies the $\mathbf{Tum}$ threshold, because $\mathbf{Rum}$ cannot be improved by further cluster merging (Property 2).

The pseudocode of $\textbf{TMERGE}_R$ and $\textbf{TMERGE}_{RT}$ can be derived by replacing the same steps with Steps 5 and 7 in $\textbf{TMERGE}_R$ and $\textbf{TMERGE}_{RT}$, respectively.

```
5  if Tum(𝒟_tmp) ≤ δ then
■      𝒟 := 𝒟_tmp
■      return 𝒟
```

The runtime cost of anonymization is $O(\mathcal{F} + |\mathcal{C}|^2 \cdot (\mathcal{K}_\mathcal{R} + \mathcal{K}_\mathcal{T}))$, where $\mathcal{F}$ is the cost for initial cluster formation, $|\mathcal{C}|$ the number of clusters in $\mathcal{D}$, and $\mathcal{K}_\mathcal{R}$ and $\mathcal{K}_\mathcal{T}$ the cost of generalizing the relational and transaction part of a cluster.

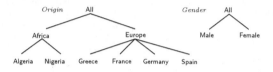

Fig. 3. Hierarchies for the dataset of Fig. 1a

5   Instantiating and Extending the Frameworks

Our frameworks can be parameterized by generalization functions, data utility measures, and initial cluster formation algorithms. This section presents such instantiations and strategies to improve their efficiency, as well as extensions of our frameworks to prevent both identity and attribute disclosure.

GENERALIZATION FUNCTIONS. We employ the *local recoding* [18] and *set-based generalization* [8,12]. As an example, the dataset in Fig. 1b has been created by applying these functions to the dataset in Fig. 1a, using the hierarchies in Fig. 3.

DATA UTILITY MEASURES. To measure data utility in relational and transaction attributes, we used *Normalized Certainty Penalty (NCP)* [18] and *Utility Loss (UL)* [12], respectively. The *NCP* for a generalized value $\tilde{v}$, a record r, and an *RT*-dataset D, is defined as: $NCP_R(\tilde{v}) = \begin{cases} 0, |\tilde{v}| = 1 \\ |\tilde{v}|/|R|, \text{otherwise} \end{cases}$, $NCP(r) = \sum_{i \in [1,v]} w_i \cdot NCP_{R_i}(r[R_i])$ and $NCP(D) = \frac{\sum_{r \in D} NCP(r)}{|D|}$ resp., where $|R|$ denotes the number of leaves in the hierarchy for a categorical attribute R (or domain size for a numerical attribute R), $|\tilde{v}|$ denotes the number of leaves of the subtree rooted at $\tilde{v}$ in the hierarchy for a categorical R (or the length of the range for a numerical R), and $w_i \in [0,1]$ is a weight that measures the importance of an attribute. The *UL* for a generalized item $\tilde{u}$, a record r, and an *RT*-dataset D, is defined as: $UL(\tilde{u}) = (2^{|\tilde{u}|} - 1) \cdot w(\tilde{u})$, $UL(r) = \frac{\sum_{\forall \tilde{u} \in r} UL(\tilde{u})}{2^{\sigma(r)} - 1}$ and $UL(D) = \frac{\sum_{\forall r \in D} UL(r)}{|D|}$ resp., where $|\tilde{u}|$ is the number of items mapped to $\tilde{u}$, $w(\tilde{u}) \in [0,1]$ a weight reflecting the importance of $\tilde{u}$ [12], and $\sigma(r)$ the sum of sizes of all generalized items in r.

INITIAL CLUSTER FORMATION WITH CLUSTER. The initial cluster formation phase should be implemented using algorithms that create many small clusters, with low **Rum**, because this increases the chance of constructing a (k, k^m)-anonymous dataset with good data utility. Thus, we employ CLUSTER, an algorithm that is instantiated with NCP and local recoding, and it is inspired by the algorithm in [2]. The time complexity of CLUSTER is $O(\frac{|D|^2}{k} \cdot \log(|D|))$.

Algorithm: CLUSTER

```
1  C := ∅
   // Create clusters of size k
2  while |D| ≥ k do
3      Select, as a seed, a random record s from D
4      Add s and each record r ∈ D having one of the lowest k−1 values in NCP(G({s,r})) to
        cluster C
5      Add cluster C to C and remove its records from D

   // Accommodate the remaining |D| mod k records
6  for each record r ∈ D do
7      Add r to the cluster C ∈ C that minimizes NCP(G(C ∪ r))

8  Apply G to the relational values of each cluster in C
   // Extend clusters
9  for each cluster C ∈ C do
10     Let S be the set of clusters in C with the same values in relational attributes as C.
11     Extend C with the records of S and remove each cluster in S from C.

12 return C
```

EFFICIENCY OPTIMIZATION STRATEGIES. To improve the efficiency of cluster-merging algorithms, we compute $\mathbf{Rum}(\mathcal{D}_{tmp})$ incrementally, thereby avoiding to access all records in $\mathcal{D}_{tmp}$, after a cluster merging. This can be performed for all measures in Section 2, but we illustrate it for NCP. We use a list λ of tuples $<|C|, NCP(r_c))>$, for each cluster C in $\mathcal{D}_{tmp}$ and any record r_c in C, which is initialized based on $\mathcal{D}$. Observe that $NCP(\mathcal{D}_{tmp}) = \frac{\sum_{\forall C \in \mathcal{D}_{tmp}}(|C| \cdot NCP(r_c))}{|D|}$, and it can be updated, after C and C' are merged, by adding: $\frac{(|C|+|C'|) \cdot NCP(r_{c \cup c'})}{|D|} - \frac{|C| \cdot NCP(r_c) - |C'| \cdot NCP(r_{c'})}{|D|}$. This requires accessing only the records in $C \cup C'$.

The efficiency of RMERGE$_T$, RMERGE$_{RT}$, TMERGE$_R$, and TMERGE$_{RT}$ can be further improved by avoiding computing $\mathbf{Tum}(\mathcal{T}(C \cup C_1)), \ldots, \mathbf{Tum}(\mathcal{T}(C \cup C_t))$. For this purpose, we merge clusters using *Bit-vector Transaction Distance* (*BTD*). The *BTD* for records r_1, r_2 is defined as $BTD(r_1, r_2) = \frac{ones(b_1 \veebar b_2)+1}{ones(b_1 \wedge b_2)+1}$. $ones(b_1 \vee b_2)$, where b_1 and b_2 are the bit-vector based representations of $r_1[T]$ and $r_2[T]$, $\veebar$, $\wedge$ and $\vee$ are the Boolean operators, for XOR, AND, and OR, and the function *ones* counts the number of 1 bits in a bit-vector. The *BTD* of a cluster C is defined as $BTD(C) = \max\{BTD(r_1, r_2)|$ for all $r_1, r_2 \in C\}$. *BTD* helps enforcing (k, k^m)-anonymity with minimal **Tum**, as it favors the grouping of records with a small number of items, many of which are common.

PREVENTING BOTH IDENTITY AND ATTRIBUTE DISCLOSURE. To prevent both types of disclosure, we propose the concept of (k, ℓ^m)-diversity, defined below.

Let $G(r)$ be a group of records and $G(r')$ be a group with the same records as $G(r)$ projected over $\{R_1, \ldots, R_v, T'\}$, where T' contains only the nonsensitive items in T. $G(r)$ is (k, ℓ^m)-*diverse*, if and only if $G(r')$ is (k, k^m)-anonymous, and an attacker, who knows up to m nonsensitive items about an individual, cannot associate any record in $G(r)$ to any combination of sensitive items, with

Table 1. Description of the datasets

Dataset	$|D|$	Rel. att.	$\|dom(T)\|$	Max, Avg # items/record
INFORMS	36553	5	619	17, 4.27
YOUTUBE	131780	6	936	37, 6.51

a probability greater than $\frac{1}{\ell}$. An RT-dataset D is (k, ℓ^m)-diverse, if and only if the group $G(r)$ of each record $r \in D$ is (k, ℓ^m)-diverse.

(k, ℓ^m)-diversity forestalls identity disclosure, and, additionally, the inference of any combination of sensitive items, based on ℓ^m-diversity [17]. Extending our anonymization frameworks to enforce (k, ℓ^m)-diversity requires: *(i)* applying **Tum** to nonsensitive items, and *(ii)* replacing the transaction generalization function $\mathcal{T}$, which enforces k^m-anonymity to each cluster, with one that applies ℓ^m-diversity. The ℓ^m-diversity version of AA [17] was used as such a function.

6   Experimental Evaluation

In this section, we evaluate our algorithms in terms of data utility and efficiency, and demonstrate the benefit of choices made in their design.

EXPERIMENTAL SETUP. We implemented all algorithms in C++ and applied them to INFORMS (https://sites.google.com/site/informsdataminingcontest) and YOUTUBE (http://netsg.cs.sfu.ca/youtubedata) datasets, whose characteristics are shown in Table 1[1]. The default parameters were k=25, m=2, and δ=0.65, and hierarchies were created as in [17]. Our algorithms are referred to in abbreviated form (e.g., **RM**$_R$ for **RMERGE**$_R$) and were not compared against prior works, since they cannot (k, k^m)-anonymize RT-datasets. The algorithms that enforce (k, ℓ^m)-diversity are named after those based on (k, k^m)-anonymity. All experiments ran on an Intel i5 at 2.7 GHz with 8 GB of RAM.

DATA UTILITY. We evaluated data utility on INFORMS and YOUTUBE using k=25 and k=100, respectively, and varied δ in $[X, 1)$, where X is the NCP of the dataset produced by CLUSTER, for **Rum**-BOUND, or the UL, for **Tum**-BOUND. Data utility is captured using ARE [9,12,16], which is invariant of the way our algorithms work and reflects the average number of records that are retrieved incorrectly, as part of query answers. We used workloads of 100 queries, involving relational, transaction, or both attribute types, which retrieve random values and/or sets of 2 items by default [9,12]. Low ARE scores imply that anonymized data can be used to accurately estimate the number of co-occurrences of relational values and items. This statistic is an important building block of several data mining models.

Figs. 4a to 4g demonstrate the conflicting objectives of minimizing information loss in relational and transaction attributes, and that **Rum**-BOUND can produce useful data. By comparing Fig. 4a with 4c, and Fig. 4d with 4g, it can be seen

---

[1] INFORMS contains the relational attributes {*month of birth, year of birth, race, years of education, income*}, and the transaction attribute *diagnosis_codes*. YOUTUBE contains the relational attributes {*age, category, length, rate, #ratings, #comments*}, and the transaction attribute *related_videos*.

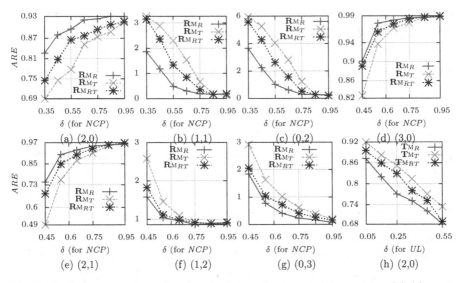

Fig. 4. ARE for queries involving (x, y) relational values and items. Figs. (a)-(c) are for INFORMS; (d)-(g) for YOUTUBE (**Rum**-BOUND). Fig. (h) is for INFORMS (**Tum**-BOUND).

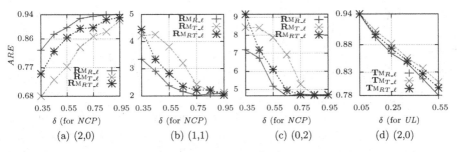

Fig. 5. ARE for queries involving (x, y) relational values and items. Figs. (a)-(c) are for INFORMS (**Rum**-BOUND); Fig. (d) is for INFORMS (**Tum**-BOUND).

that a small δ forces all algorithms to incur low information loss in the relational attributes, whereas a large δ favors the transaction attribute. Also, *NCP* is at most δ, in all tested cases, and data remain useful for queries involving both attribute types (see Figs. 4b, 4e, and 4f). We performed the same experiments for the **Tum**-BOUND and present a subset of them in Fig. 4h. Note that, increasing δ (i.e., the bound for *UL*), favors relational data, and that the information loss in the transaction attribute is low. Similar observations can be made for the (k, l^m)-diversity algorithms (see Fig. 5).

Next, we compared $\mathbf{R}M_R$, $\mathbf{R}M_T$, and $\mathbf{R}M_{RT}$. As shown in Fig. 4, $\mathbf{R}M_R$ incurred the lowest information loss in the transaction attribute, and the highest in the relational attributes, and $\mathbf{R}M_T$ had opposite trends. $\mathbf{R}M_{RT}$ allows more accurate query answering than $\mathbf{R}M_R$, in relational attributes, and than $\mathbf{R}M_T$, in the transaction attribute, as it merges clusters, based on both attribute types.

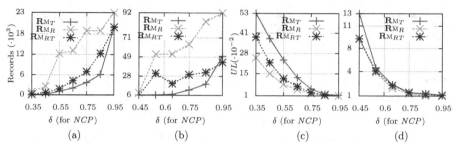

Fig. 6. Max. cluster size for (a) INFORMS, and (b) YOUTUBE, and UL for (c) INFORMS and (d) YOUTUBE (**Rum**-BOUND)

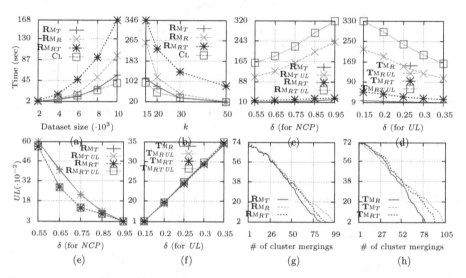

Fig. 7. Runtime (a) vs. $|D|$ and (b) vs. k. Impact of using BTD on runtime for (c) **Rum**-BOUND and (d) (**Tum**-BOUND), and on UL for (e) **Rum**-BOUND and (f) **Tum**-BOUND. UL vs. number of cluster mergings for (g) **Rum**-BOUND, and (h) **Tum**-BOUND

Similar results were obtained for YOUTUBE (see Figs. 4d-4g), from comparing $\mathbf{T}_{M_T}$, $\mathbf{T}_{M_R}$, and $\mathbf{T}_{M_{RT}}$ (see e.g., Fig. 4h), and from comparing the (k, l^m)-diversity algorithms (see Figs. 5). Figs. 6a and 6b show the size of the largest cluster created by $\mathbf{R}_{M_R}$, $\mathbf{R}_{M_T}$, and $\mathbf{R}_{M_{RT}}$, for varying δ. $\mathbf{R}_{M_R}$ created the largest clusters, as it merges many clusters with similar relational values. These clusters have low UL, as shown in Figs. 6c and 6d. Furthermore, Figs. 6a and 6c, show that $\mathbf{R}_{M_{RT}}$ created slightly larger clusters than $\mathbf{R}_{M_T}$, which have lower UL scores. The results for $\mathbf{T}_{M_T}$, $\mathbf{T}_{M_R}$, and $\mathbf{T}_{M_{RT}}$ and the (k, l^m)-diversity algorithms were similar.

EFFICIENCY. We studied the impact of dataset size using random subsets of INFORMS, whose records were contained in all larger sets. As can be seen in Fig. 7a, $\mathbf{R}_{M_T}$ outperformed $\mathbf{R}_{M_R}$ and $\mathbf{R}_{M_{RT}}$, and it was more scalable, due to the use of the BTD measure. $\mathbf{RM}_{RT}$ was the slowest, because it computes two cluster orderings. $\mathbf{TM}_T$, $\mathbf{TM}_R$, and $\mathbf{TM}_R$ perform similarly to $\mathbf{RM}_R$, $\mathbf{RM}_T$, and $\mathbf{RM}_{RT}$ (their results were omitted). Fig. 7a shows the cost of CL. We also studied the impact of k using the largest dataset of the previous experiment. Fig. 7b shows that the runtime of $\mathbf{RM}_R$, $\mathbf{RM}_T$, and $\mathbf{RM}_{RT}$ improves with k, as fewer clusters are merged. $\mathbf{RM}_T$ was up to 2.2 times more efficient than $\mathbf{RM}_R$ and $\mathbf{RM}_{RT}$ was the least efficient. Fig. 7b shows that the runtime of CL improves with k. The cost of the (k, l^m)-diverse algorithms was similar (omitted).

BENEFITS OF ALGORITHMIC CHOICES. To show that BTD helps efficiency without degrading data utility, we developed the baseline algorithms $\mathbf{RM}_{T\,UL}$, $\mathbf{RM}_{T\,UL}$, $\mathbf{TM}_{T\,UL}$, and $\mathbf{TM}_{RT\,UL}$, which do not perform the optimization of Section 5. Due to their high runtime, a subset of INFORMS with $4K$ records was used. Observe in Figs. 7c and 7e that $\mathbf{RM}_T$ and $\mathbf{RM}_{RT}$ have the same UL scores with their corresponding baseline algorithms, but are at least 10 times more efficient and scalable with respect to δ. Similar observations can be made from Figs. 7d and 7f, for $\mathbf{TM}_R$ and $\mathbf{TM}_{RT}$. Last, we show that UL decreases monotonically, as our algorithms merge clusters. Figs. 7g-7h show the results with $\delta = 1$ for the dataset used in the previous experiment. The fact that UL never increases shows that avoiding to compute $UL(\mathcal{T}(\mathcal{D}_{tmp}))$ after a cluster merge does not impact data utility but helps efficiency. The (k, l^m)-diversity algorithms performed similarly.

7   Related Work

Preventing identity disclosure is crucial in many real-world applications [5,11] and can be achieved through k-anonymity [15]. This privacy principle can be enforced through various generalization-based algorithms (see [5] for a survey). Thwarting *attribute disclosure* may additionally be needed [14,19,17], and this can be achieved by applying other privacy models, such as l-diversity [14], together with k-anonymity.

Privacy-preserving transaction data publishing requires new privacy models and algorithms, due to the high dimensionality and sparsity of transaction data [19,7,17]. k^m-anonymity is a model for protecting transaction data against attackers, who know up to m items about an individual [17]. Under this condition, which is often satisfied in applications [17,16,11], an individual cannot be associated with fewer than k records in the dataset. k^m-anonymity can be enforced using several algorithms [17,12,8], which can be incorporated into our frameworks. However, k^m-anonymity does not guarantee protection against stronger attackers, who know that an individual is associated with exactly certain items [17,16]. This is because, by excluding records that have exactly these items from consideration, the attackers may be able to increase the probability of associating an individual with their record to greater than $\frac{1}{k}$ (although not necessarily 1). A recent method [20] can guard against such attackers while preserving data utility based on a *nonreciprocal recoding* anonymization scheme. To thwart both identity and attribute disclosure in transaction data publishing, [17] proposes ℓ^m-diversity, which we also employ in our frameworks.

Our frameworks employ generalization, which incurs lower information loss than suppression [17] and helps preventing identity disclosure, contrary to bucketization [7]. Also, we seek to publish record-level and truthful data. Thus, we do not employ ϵ-differential privacy [3], nor disassociation [16]. However, the relationship between (k, k^m)-anonymization and relaxed differential privacy definitions is worth investigating to strengthen protection. For instance, Li et al. [10] proved that *safe* k-anonymization algorithms, which perform data grouping and recoding in a differentially private way, can satisfy a relaxed version of differential privacy when preceded by a random sampling step.

8   Conclusions

In this paper, we introduced the problem of anonymizing RT-datasets and proposed the first approach to protect such datasets, along with two frameworks for enforcing it. Three cluster-merging algorithms were developed, for each framework, which preserve different aspects of data utility. Last, we showed how our approach can be extended to prevent both identity and attribute disclosure.

Acknowledgements. G. Poulis is supported by the Research Funding Program: Heraclitus II. G. Loukides is partly supported by a Research Fellowship from the Royal Academy of Engineering. S. Skiadopoulos is partially supported by EU/Greece the Research Funding Program: Thales.

References

1. Bayardo, R.J., Agrawal, R.: Data privacy through optimal k-anonymization. In: ICDE, pp. 217–228 (2005)
2. Byun, J.-W., Kamra, A., Bertino, E., Li, N.: Efficient k-anonymization using clustering techniques. In: Kotagiri, R., Radha Krishna, P., Mohania, M., Nantajeewarawat, E. (eds.) DASFAA 2007. LNCS, vol. 4443, pp. 188–200. Springer, Heidelberg (2007)
3. Dwork, C.: Differential privacy. In: Bugliesi, M., Preneel, B., Sassone, V., Wegener, I. (eds.) ICALP 2006. LNCS, vol. 4052, pp. 1–12. Springer, Heidelberg (2006)
4. Freitas, A.A.: A critical review of multi-objective optimization in data mining: a position paper. SIGKDD Explorations 6(2), 77–86 (2004)
5. Fung, B.C.M., Wang, K., Chen, R., Yu, P.S.: Privacy-preserving data publishing: A survey on recent developments. ACM Comput. Surv. 42 (2010)
6. Ghinita, G., Karras, P., Kalnis, P., Mamoulis, N.: A framework for efficient data anonymization under privacy and accuracy constraints. TODS 34(2) (2009)
7. Ghinita, G., Tao, Y., Kalnis, P.: On the anonymization of sparse high-dimensional data. In: ICDE, pp. 715–724 (2008)
8. Gkoulalas-Divanis, A., Loukides, G.: Utility-guided clustering-based transaction data anonymization. Trans. on Data Privacy 5(1), 223–251 (2012)
9. LeFevre, K., DeWitt, D.J., Ramakrishnan, R.: Mondrian multidimensional k-anonymity. In: ICDE, p. 25 (2006)

10. Lii, N., Qardaji, W., Su, D.: On sampling, anonymization, and differential privacy or, k-anonymization meets differential privacy. In: ASIACCS, pp. 32–33 (2012)
11. Loukides, G., Gkoulalas-Divanis, A., Malin, B.: Anonymization of electronic medical records for validating genome-wide association studies. Proceedings of the National Academy of Sciences 17, 7898–7903 (2010)
12. Loukides, G., Gkoulalas-Divanis, A., Malin, B.: COAT: Constraint-based anonymization of transactions. Knowledge and Information Systems 28(2), 251–282 (2011)
13. Loukides, G., Shao, J.: Clustering-based K-anonymisation algorithms. In: Wagner, R., Revell, N., Pernul, G. (eds.) DEXA 2007. LNCS, vol. 4653, pp. 761–771. Springer, Heidelberg (2007)
14. Machanavajjhala, A., Gehrke, J., Kifer, D., Venkitasubramaniam, M.: l-diversity: Privacy beyond k-anonymity. In: ICDE, p. 24 (2006)
15. Samarati, P., Sweeney, L.: Generalizing data to provide anonymity when disclosing information (abstract). In: PODS, p. 188 (1998)
16. Terrovitis, M., Liagouris, J., Mamoulis, N., Skiadopoulos, S.: Privacy preservation by disassociation. PVLDB 5(10), 944–955 (2012)
17. Terrovitis, M., Mamoulis, N., Kalnis, P.: Local and global recoding methods for anonymizing set-valued data. VLDB J. 20(1), 83–106 (2011)
18. Xu, J., Wang, W., Pei, J., Wang, X., Shi, B., Fu, A.W.-C.: Utility-based anonymization using local recoding. In: KDD, pp. 785–790 (2006)
19. Xu, Y., Wang, K., Fu, A.W.-C., Yu, P.S.: Anonymizing transaction databases for publication. In: KDD, pp. 767–775 (2008)
20. Xue, M., Karras, P., Raïssi, C., Vaidya, J., Tan, K.: Anonymizing set-valued data by nonreciprocal recoding. In: KDD, pp. 1050–1058 (2012)

Privacy-Preserving Mobility Monitoring
Using Sketches of Stationary Sensor Readings

Michael Kamp, Christine Kopp, Michael Mock, Mario Boley, and Michael May

Fraunhofer IAIS, Schloss Birlinghoven,
St. Augustin, Germany
{michael.kamp,christine.kopp,michael.mock,mario.boley,
michael.may}@iais.fraunhofer.de

Abstract. Two fundamental tasks of mobility modeling are (1) to track the number of distinct persons that are present at a location of interest and (2) to reconstruct flows of persons between two or more different locations. Stationary sensors, such as Bluetooth scanners, have been applied to both tasks with remarkable success. However, this approach has privacy problems. For instance, Bluetooth scanners store the MAC address of a device that can in principle be linked to a single person. Unique hashing of the address only partially solves the problem because such a pseudonym is still vulnerable to various linking attacks. In this paper we propose a solution to both tasks using an extension of linear counting sketches. The idea is to map several individuals to the same position in a sketch, while at the same time the inaccuracies introduced by this overloading are compensated by using several independent sketches. This idea provides, for the first time, a general set of primitives for privacy preserving mobility modeling from Bluetooth and similar address-based devices.

1   Introduction

Advanced sensor technology and spread of mobile devices allows for increasingly accurate mobility modeling and monitoring. Two specific tasks are crowd monitoring, i.e., counting the number of mobile entities in an area, and flow monitoring between locations, i.e., counting the number of entities moving from one place to another within a given time interval.[1] Both have several applications in event surveillance and marketing [10, 16]. Moreover, matrices containing the flow between every pair of locations (origin-destination, or OD-matrices) are an important tool in many GIS applications, notably traffic planning and management [3].

Today's sensor technologies such as GPS, RFID, GSM, and Bluetooth have revolutionized data collection in this area, although significant problems remain to be solved. One of those problems are privacy concerns. They mandate that, while the count of groups of people can be inferred, inference on an individual person remains infeasible. Directly tracing IDs through the sensors violates this

---

[1] In this paper, we use the term 'flow' always as a short-hand for 'flow between two or more locations'.

H. Blockeel et al. (Eds.): ECML PKDD 2013, Part III, LNAI 8190, pp. 370–386, 2013.

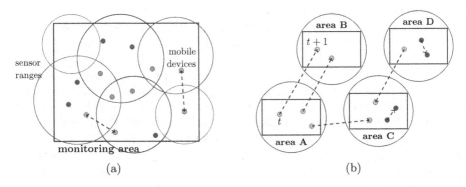

Fig. 1. The two mobility modeling tasks addressed in this paper: (a) *crowd monitoring* and (b) *flow monitoring*

privacy constraint, because the amount of information stored allows for linking attacks [12]. In such an attack, sensor information is linked to additional knowledge in order to identify a person and infer upon her movement behavior. Hence, application designers have to design and use new, privacy preserving methods.

The contribution of this paper is to provide a general set of primitives for privacy-preserving mobility monitoring and modeling using stationary sensor devices. Following the *privacy-by-design* paradigm [17, 21], we present a method that stores just enough information to perform the desired inference task and discards the rest. Thereby, privacy constraints are much easier to enforce.

Technically, the method we propose is based on Linear Counting sketches [26], a data structure that allows to probabilistically count the distinct amount of unique items in the presence of duplicates. Linear Counting not only obfuscates the individual entities by hashing, but furthermore provides a probabilistic form of k-anonymity. This form of anonymity guarantees that, by having access to all stored information, an attacker cannot be certain on a single individual but can at most infer upon k individuals as a group. Furthermore, Linear Counting is an efficient and easy to implement method that outperforms other approaches in terms of accuracy and privacy on the cost of higher memory usage [19].

The rest of the paper is structured as follows. After discussing related work in section 2, we describe the application scenarios in section 3. In section 4 we present our extension to the linear counting method and give a theoretical analysis of the error. Subsequently, we analyze the privacy of our method in section 5. In section 6 we conduct extensive experiments on the accuracy of Linear Counting and flow estimation under different privacy requirements to test our approach. These experiments have been carried out on a a real-world simulation. Section 7 concludes with a discussion of the results and pointers to future directions.

2   Related Work

The basic tool we are using in this paper are sketches (see Cormode et al. [9] for a good general introduction). Sketches are summaries of possibly huge data collections that, on the one hand, discard some information for the sake of space efficiency or privacy, but that, on the other hand, still contain enough information to restore the current value of certain variables of interest. Sketches are a very universal tool and have been successfully applied for inferring heavy hitters, moments of a distribution, distinct elements, and more. The general idea of using sketches for privacy is described in Aggarwal and Yu [1] and Mir et al. [20]. The first relates the privacy of a sketch to the variance of the estimator. The latter discusses privacy paradigms that go beyond differential privacy [11] and address security as well. They use various techniques for achieving privacy, notably adding noise. In our approach, we do not employ noise as a source of privacy. However, due to the probabilistic nature of our method, noise has no excessive impact on the accuracy. Hence, adding noise to improve privacy can be combined with our method.

The crucial task in this paper is to count the number of distinct objects at a location (see Gibbons [14] for a general overview on approaches for this problem). The method discussed in our paper, Linear Counting, is first described in [26]. This method can be seen as a special case of Bloom filters for counting [6]. So far, it has received relatively little attention, because it is not as space efficient as log-space approaches such as FM sketches, and traditionally sketches have mainly been used to provide short summaries of huge data collections. However, in the context of privacy preservation in the mobility modeling scenarios of this paper, space is not so much an issue as is *accuracy*. To this end, recently very positive results are reported for comparisons of Linear Counting with FM sketches and other methods [19, 15]. Especially for smaller set sizes that appear in mobility mining, Linear Counting has often an advantage in terms of accuracy. Hence, for our scenario, it is a promising choice.

The idea of using Linear Counting for stationary sensors, specifically Bluetooth measurements, recently has also been described by Gonçalves et al. [15] and similar work for mobile devices is reported in Bergamini et al. [4]. However, here, for the first time, we describe extensions of the basic Linear Count sketches that can be used to monitor flows between locations as well as to compensate for the precision loss incurred for raising privacy.

Our approach for tracking flows is based on the ability to compute the intersection of sketches. A general method for computing general set expressions from sketches is described in Ganguly et al. [13]. Though highly general, the approach is ruled out for our application because the scheme requires to store information at the coordinator which could be used for identifying persons and thus is not privacy aware—it should be noted that it was not build for that purpose.

3   Application Scenarios

In the following, the two application scenarios in mobility monitoring covered by this paper are described. The general setup in both applications is using stationary sensor devices centralizing their sensor readings at a coordinator.

In general, a well-studied approach to monitoring people in an area is to use sensors that count the number of mobile devices in their sensor radius [14,

15], such as Bluetooth, or GSM scanners. From the amount of mobile devices, the actual amount of people can be accurately estimated by assuming a stable average fraction of people carrying such a device [18].

A stationary sensor S scans periodically for devices in its sensor range. Each device is identifiable by its address a from a **global address space** $\mathcal{A}$. The stream of **sensor readings** of a sensor S is defined by $\mathcal{R}_S \subseteq \mathcal{A} \times \mathbb{R}_+$ where $(a, t) \in \mathcal{R}_S$ means that S has read a device address a at time stamp t. For a measurement interval $T = [b, e] \subseteq \mathbb{R}_+$, the readings of sensor S in this time interval is denoted by $\mathcal{R}_S^T = \{a \in \mathcal{A} | \exists t \in T : (a, t) \in \mathcal{R}_S\}$. In both application scenarios, the sensor readings are evaluated to solve the application-specific problem.

3.1 Crowd Monitoring

In major public events, such as concerts, sport events or demonstrations, continuously monitoring the amount of people in certain areas is a key tool for maintaining security. It is vital for the prevention of overcrowding as well as for allocating security and service personnel to the places they are needed most.

To monitor an area using stationary sensor devices, a set of sensors $\mathcal{S} = \{S_1, ..., S_k\}$ is distributed over the area such that the union of their sensor ranges covers the complete area (see Fig. 1(a)). For a single sensor S, the **count distinct** of unique entities that have been present in its range during a time interval T is $|\mathcal{R}_S^T|$. The task is then to continuously monitor at a central site the count distinct of the **union** of all sensor ranges. That is, we aim to monitor $|\mathcal{R}_{S_1}^{T_i} \cup ... \cup \mathcal{R}_{S_k}^{T_i}|$ for consecutive measurement intervals $T_0, T_1, T_2...$ of a fixed time resolution.

Note that the problem of duplicates in the sensor readings cannot be avoided in practice because of several reasons: Covering an area with circular sensor ranges requires overlap and the radius of each sensor range cannot be accurately estimated beforehand. Furthermore, entities can move between sensor ranges during a measurement interval. Thus, independent of the specific application design, summing the distinct counts of individual sensor readings usually overestimate the true number of devices. Without privacy restrictions, this problem can be solved by centralizing the read device addresses or a unique hash of them to eliminate duplicates. However, when addresses or unique identifiers are' centralized, devices can be tracked and linked to real persons, thereby violating common privacy constraints.

To solve this problem, we use **Linear Counting** that has been introduced as an accurate and privacy preserving method for estimating the number of distinct items in the presence of duplicates. Linear counting is a sketching technique, i.e., the vector of sensor readings is compressed to a lower dimensional vector, the so called sketch, in such a way that a desired quantity can still be extracted sufficiently accurate from the sketch. The linear count sketch maintains privacy by deliberately compressing the sensor readings with a certain amount of collisions, such that a deterministic inference from a sketch to an address is impossible. A detailed analysis of the privacy aspect is provided in section 5. The distinct amount of people in an area covered by several overlapping sensors is estimated by combining the individual sketches to a sketch of the union of sensor ranges. The method is described in section 4.1. For that, only the sketches have to be stored and send to the coordinator so that privacy is preserved at a very basic level. This provides two general primitives for monitoring the amount of entities in an area: linear count sketches for privacy-preserving estimation of the **count distinct** of mobile entities in a sensor radius and estimation of the distinct count of entities in an area defined as the **union** of sensor ranges.

3.2   Flow Monitoring

The flow of mobile entities, such as pedestrians, cars or trains, is a key quantity in traffic planning. Streets, rails and public transportation networks are optimized for handling these flows. For a given set of locations, all flows between the locations can be combined in the form of an origin-destination matrix. These matrices are an important tool in traffic management [2].

Flows of mobile entities between a set of locations can be estimated using stationary sensors [3]. For that, given a set of locations of interest, a sensor is placed at each location (e.g., see Fig. 1(b)). For a given time period, the flow between two locations denotes the amount of mobile entities that have been present at one location at the beginning of that time period and present at the other location at the end of the period. An entity is present at a location, if it is staying within the range of the sensor placed at that location. Thus, given a time interval at the beginning of the period, T_b, and one at the end, T_e, the **flow** between two sensors S and S' is defined as $v(S, S') = |\{a \in \mathcal{A} \mid a \in \mathcal{R}_S^{T_b} \wedge a \in \mathcal{R}_{S'}^{T_e}\}|$. For convenience, we assume that sensor ranges do not overlap. In the case of overlap, this notion of flow has to be extended so that the number of mobile devices that stayed in the intersection is handled separately.

The existing approaches to flow monitoring with stationary sensors rely on tracing unique identifiers through the sensor network. Hence, identifying and tracking a specific device, i.e., a specific person, is possible in monitoring systems as soon as the identifier can be linked to a person. Again, this violates common privacy restrictions. In order to monitor flows in a privacy-preserving manner using linear count sketches, the definition of flow is modified to be able to express the flow as the intersection of sensor readings of different time intervals. Therefore, let T_b and T_e be disjoint time intervals as in the aforementioned definition of flow, then the flow can be expressed as $v(S, S') = |\mathcal{R}_S^{T_b} \cap \mathcal{R}_{S'}^{T_e}|$. In section 4.2, we show how local linear count sketches can be combined in a privacy-preserving manner to estimate the size of an intersection.

The extension to paths of arbitrary length is straightforward. Given three consecutive, disjoint time intervals T_1, T_2, T_3, the flow on a path $S_1 \to S_2 \to S_3$ can be represented as $|\mathcal{R}_{S_1}^{T_1} \cap \mathcal{R}_{S_2}^{T_2} \cap \mathcal{R}_{S_3}^{T_3}|$. However, the accuracy of the flow estimation is highly dependent on the size of the intersection compared to the original sets. Moreover, the accuracy drops drastically in the number of intersections, thereby limiting the length of paths that can be monitored. To boost the accuracy and thereby ensure applicability, we present an improved estimator that uses a set of intermediate estimators and their mean value in section 4.3.

An OD-matrix L for a set of locations $\{l_1, ..., l_k\}$ is defined as the flow between each pair of locations, i.e., $L \in \mathbb{R}^{k \times k}$ with $L_{ij} = v(l_i, l_j)$. By placing a sensor at each location, that is, sensor S_i is placed at location l_i, an OD-matrix L can be estimated by $L_{ij} = v(S_i, S_j)$. This provides another mobility mining primitive: the estimation of flows, paths and OD-matrices based on the **intersection** of sensor readings.

4   Extending Linear Counting

In this section, we present the technical solution to the application scenarios introduced in section 3. We start by recapitulating Linear Counting sketches, which serve as fundamental tool. In particular, for the flow-monitoring it is necessary to extend this sketching technique to monitoring the size of an intersection of two or more sensor readings. For both application scenarios, privacy preservation demands that we use basic sketches at the sensors with relatively high variance in their estimates. This variance even increases when estimating intersections. In order to increase the accuracy again on the output layer, we present an improved estimator that reduces the variance by combining several independent sketches.

4.1   From Sensors to Sketches

Given the sensor readings $\mathcal{R}_S^T = \{a \in \mathcal{A} \mid \exists t \in T : (a,t) \in \mathcal{R}_S\}$ of a sensor S during a time interval T, the goal is to represent the number of distinct devices observed without explicitly storing their addresses. A problem of this kind is referred to as **count distinct problem**, which can be tackled by Linear Counting sketches [26]. They have originally been introduced to estimate the number of unique elements within a table of a relational database.

In our scenario, this means that, instead of storing all readings within a measurement interval T, a sensor just maintains a binary **sketch** $\mathrm{sk}(\mathcal{R}_S^T) \in \{0,1\}^m$ of some fixed length m. The sketch is determined by a random **hash map** $h : \mathcal{A} \to \{0, \ldots, m-1\}$ such that the following **uniformity** property holds: for all $a \in \mathcal{A}$ and all $k \in \{0, \ldots, m-1\}$ it holds that $\mathbb{P}(h(a) = k) = 1/m$. For practical purposes this can be approximately achieved by choosing, e.g., $h(a) = ((va + w) \bmod p) \bmod m$, with uniform random numbers $v, w \in \mathbb{N}$ and a fixed large prime number p. Other choices of hash functions are possible (see e.g., Preneel [22]).

A sensor maintains its sketch as follows. At the beginning of the measurement interval it starts with an empty sketch $(0, \ldots, 0)$, and on every address $a \in \mathcal{A}$ read, until the end of the interval, the sketch is updated by setting the $h(a)$-th position to 1. For the whole measurement interval this results in a sketch

$$\mathrm{sk}(\mathcal{R}_S^T)[k] = \begin{cases} 1 & , \text{ if } \exists a \in \mathcal{R}_S^T, \, h(a) = i \\ 0 & , \text{ otherwise} \end{cases}.$$

In our application scenarios, a **global population** of mobile entities $\mathcal{P} \subseteq \mathcal{A}$ of size $|\mathcal{P}| = n$ is monitored with a set of sensors. The size of the global population

n is an upper bound to the number of mobile entities in a sensor reading. In each sensor reading, a subset of the global population is captured, i.e., $\mathcal{R}_S^T \subseteq \mathcal{P}$, thus $|\mathcal{R}_S^T| \leq |\mathcal{P}|$, or $n_S \leq n$ (from now on we denote $|\mathcal{R}_S^T|$ as n_S).

The number of distinct addresses within a sensor reading can then be estimated based on the sketch as follows. Assume a sensor reading $\mathcal{R}_S^T$ with $|\mathcal{R}_S^T| = n_S$ and the respective sketch $\mathrm{sk}(\mathcal{R}_S^T)$. Let $\mathbf{u}_S$ denote the number of zeros in the sketch and $\mathbf{v}_S = \mathbf{u}_S/m$ the relative zero count. Now, the maximum likelihood **count estimator** for the number of distinct items n_S is $\widehat{n}_S = -m \ln \mathbf{v}_S$. Whang et al. [26] shows that the expected value, and the variance of this estimator are asymptotically well-behaved. Here, asymptotically refers to the **limit** for increasing n while the **loadfactor** $t = n/m$ and the **relative size** $c_S = n_S/n$ of S are kept constant. With this notion of limit, that we simply denote by lim for the remainder of this paper, the expected value and the variance can be expressed as

$$\lim \mathbb{E}[\widehat{n}_S] = n_S + (e^{n_S/m} - n_S/m - 1)/2 = n_S \tag{1}$$

$$\lim \mathbb{V}(\widehat{n}_S) = m \left(e^{n_S/m} - n_S/m - 1 \right) , \tag{2}$$

respectively. Hence, asymptotically, the estimator is unbiased and has a bounded variance. Standard concentration inequalities can be used to convert this result into probabilistic error guarantees.

In the crowd monitoring scenario, the size of the global population n can be estimated as the size of the union of all individual sensor readings. By construction of the sketches, it is possible to build a sketch of the union of sensor readings $\mathcal{R}_S^T$ and $\mathcal{R}_{S'}^T$ by combining the individual sketches with the point-wise binary OR operation (i.e., $\mathrm{sk}(\mathcal{R}_S^T)[k] \vee \mathrm{sk}(\mathcal{R}_{S'}^T)[k]$ is equal to 1 if and only if $\mathrm{sk}(\mathcal{R}_S^T)[k] = 1$ or $\mathrm{sk}(\mathcal{R}_{S'}^T)[k] = 1$). The following statement holds:

Proposition 1 (Whang et al. [26]). *Let $\mathcal{R}_{S_1}, ..., \mathcal{R}_{S_k}$ be readings of a set of sensors $S = S_1, ..., S_k$. The sketch constructed from the union of these readings can be obtained by calculating the binary or of the individual sketches. That is,*

$$\mathrm{sk}\left(\bigcup_{i=1}^{k} \mathcal{R}_{S_i} \right) = \bigvee_{i=1}^{k} \mathrm{sk}(\mathcal{R}_{S_i}) .$$

This is already sufficient to continuously track the total number of distinct addresses in the crowd monitoring scenario: for a pre-determined time resolution, the sensor nodes construct sketches of their readings, send them to a monitoring coordinator, and start over with new sketches. As required by the application, the coordinator can then compute the estimate of distinct counts of mobile entities based on the OR-combination of all local sketches. However, for the flow-monitoring scenario we have to be able to compute the number of distinct addresses in the intersection of sensor readings. Therefore, we have to extend the Linear Counting approach.

4.2   Intersection Estimation

In the following, a method is presented for estimating the intersection of two sets using linear count sketches. The sketch of the intersection cannot be constructed from the individual sketches (note that using the binary 'and' operation on the two sketches does in general not result in the correct sketch of the intersection). Therefore, this method is based on the inclusion-exclusion formula for sets. That is, we can express the size of the intersection of two sets A, B as $|A \cap B| = |A| + |B| - |A \cup B|$. The estimator for the intersection of two sensor ranges is defined in a similar way, using the estimators for each sensor and their union. Let $\hat{n}_S, \hat{n}_{S'}$ denote the estimator for $|\mathcal{R}_S^T|$, respectively $|\mathcal{R}_{S'}^T|$. Let furthermore $\hat{n}_{S \cup S'}$ denote the estimator based on the sketch of the union of the sensor readings $\mathcal{R}_S^T$ and $\mathcal{R}_{S'}^T$ as defined in proposition 1. Then the **intersection estimator** is defined as

$$\tilde{n}_{S,S'} = \hat{n}_S + \hat{n}_{S'} - \hat{n}_{S \cup S'} \tag{3}$$

It turns out that also this estimator asymptotically is unbiased and has a bounded variance. The first follows directly from the linearity of the expected value. Thus, we can note:

Proposition 2. *For a constant loadfactor t and constant fractions c_S, c_S', the estimator $\tilde{n}_{S,S'}$ is asymptotically unbiased, i.e., $\lim \mathbb{E}[\tilde{n}_{S,S'}]/|S \cap S'| = 1$.*

Furthermore, we can bound the variance of our estimator by the variance of the union estimator. This implies that resulting probabilistic error guarantees become tighter the closer the ratio $|S \cap S'|/|S \cup S'|$ is to one.

Proposition 3. *Asymptotically, the variance of the intersection estimator $\tilde{n}_{S,S'}$ is bounded by the variance of the count estimator for the union, i.e., $\lim \mathbb{V}(\tilde{n}_{S,S'}) \leq \lim \mathbb{V}(\hat{n}_{S \cup S'})$.*

Proof (sketch). For some subset $A \subseteq \mathcal{P}$ and a fixed sketch position $k \in \{0, \ldots, m - 1\}$ let us denote by $p_A = \mathbb{P}(sk(A)[k] = 0)$ the probability that the sketch of A has entry 0 at position k. Due to the uniformity of the hash function h it holds that $p_A = (1 - 1/m)^{n_A}$. The limit of this probability $p_A^* = \lim p_A$ is equal to $\lim (1 - t/n)^{n c_A} = e^{-t \cdot c_A}$. The variance of the intersection estimator can be re-expressed in terms of the covariances σ of the individual count estimators:

$$\begin{aligned}
\mathbb{V}(\tilde{n}_{S,S'}) &= \mathbb{V}(\hat{n}_S + \hat{n}_{S'} - \hat{n}_{S \cup S'}) \\
&= \mathbb{V}(\hat{n}_S) + \mathbb{V}(\hat{n}_{S'}) + \mathbb{V}(\hat{n}_{S \cup S'}) + 2\sigma(\hat{n}_S, \hat{n}_{S'}) \\
&\quad - 2\sigma(\hat{n}_S, \hat{n}_{S \cup S'}) - 2\sigma(\hat{n}_{S'}, \hat{n}_{S \cup S'}) \ . \tag{4}
\end{aligned}$$

In order to determine the limit of the covariances for the count estimators $\hat{n}_A$ and $\hat{n}_B$ for some arbitrary subsets $A, B \subseteq \mathcal{P}$ we can use Whang et al. [26, Eq. (8)] and the bi-linearity of the covariance

$$\begin{aligned}
\lim \sigma(\hat{n}_A, \hat{n}_B) &= \sigma(m \, (tc_A - \mathbf{v}_A/p_A^* - 1), m \, (tc_B - \mathbf{v}_B/p_B^* - 1)) \\
&= m^2 \sigma(\mathbf{v}_A, \mathbf{v}_B)/(p_A^* p_B^*) = m^2 \sigma(\mathbf{u}_A/m, \mathbf{u}_B/m)/(p_A^* p_B^*) \\
&= \sigma(\mathbf{u}_A, \mathbf{u}_B)/(p_A^* p_B^*) \ . \tag{5}
\end{aligned}$$

Let $\overline{\mathrm{sk}}(A)\,[k]$ denote the binary negation of sketch position k. The co-variances of the absolute number of zero entries $\mathbf{u}_A$ and $\mathbf{u}_B$ is

$$\sigma(\mathbf{u}_A, \mathbf{u}_B) = \sum_{k=1}^{m}\sum_{l=1}^{m} \sigma\big(\overline{\mathrm{sk}}(A)\,[k], \overline{\mathrm{sk}}(B)\,[l]\big) \tag{6}$$

$$= \sum_{k=1}^{m} \sigma\big(\overline{\mathrm{sk}}(A)\,[k], \overline{\mathrm{sk}}(B)\,[k]\big) + \sum_{\substack{k,l=1 \\ k \neq l}}^{m} \sigma\big(\overline{\mathrm{sk}}(A)\,[k], \overline{\mathrm{sk}}(B)\,[l]\big) \ .$$

Let $U = A \cup B$ and $I = A \cap B$. We state without proof that the co-variances of the individual sketch positions are given by

$$\sigma\big(\overline{\mathrm{sk}}(A)\,[k], \overline{\mathrm{sk}}(B)\,[k]\big) = p_U - p_A p_B$$

$$\sigma\big(\overline{\mathrm{sk}}(A)\,[k], \overline{\mathrm{sk}}(B)\,[l]\big) = p_{A\backslash B}\, p_{A\backslash B}(1 - 2/m)^{|I|} - p_A p_B$$

for the cases $k = l$ and $k \neq l$, respectively. From this and Eq. (6) it follows that

$$\lim \sigma(\mathbf{u}_A, \mathbf{u}_B) = m\left(p_U^* - p_A^* p_B^*\right) + m(m-1)\left(p_{A\backslash B}^* p_{B\backslash A}^*(1 - 2/m)^{|I|} - p_A^* p_B^*\right)$$

$$= m\left(e^{-t(c_A + c_B - c_I)} - e^{-t(c_A + c_B)} - c_I t e^{-t(c_A + c_B)}\right) \ ,$$

where the second equality follows from several steps of elementary calculus that we omit here. By Eq. (5) we can then conclude

$$\lim \sigma(\hat{n}_A, \hat{n}_B) = m\left(e^{t c_I} - t c_I - 1\right) = \lim \mathbb{V}(\hat{n}_I)$$

Inserting this result in Eq. (4), and noting that for fixed $c_A \leq c_B$ it holds that $\lim \mathbb{V}(\hat{n}_A) \leq \lim \mathbb{V}(\hat{n}_B)$ (see eq. (2)), in particular $\mathrm{Var}[\hat{n}_{S \cap S'}] \leq \mathrm{Var}[\hat{n}_S]$ as well as $\mathrm{Var}[\hat{n}_{S \cap S'}] \leq \mathrm{Var}[\hat{n}_{S'}]$, yields

$$\lim \mathbb{V}(\tilde{n}_{S,S'}) = \lim(\mathbb{V}(\hat{n}_{S \cup S'}) + 2\mathbb{V}(\hat{n}_{S \cap S'}) - \mathbb{V}(\hat{n}_S) - \mathbb{V}(\hat{n}_{S'}))$$

$$\leq \lim(\mathbb{V}(\hat{n}_{S \cup S'}) + 2\mathbb{V}(\hat{n}_{S \cap S'}) - \mathbb{V}(\hat{n}_{S \cap S'}) - \mathbb{V}(\hat{n}_{S \cap S'}))$$

$$= \lim \mathbb{V}(\hat{n}_{S \cup S'})$$

$\square$

When estimating the flow, the intersection of the readings of sensor S in a time interval T_b are intersected with the readings of sensor S' in consecutive time interval T_e. Let ΔT denote the time period between those two intervals. Then we denote the estimator for the flow between S and S' for this time period as $\tilde{n}_{S,S'}^{\Delta T}$. This method can straight-forwardly be extended to paths. The flow on a path $S_1 \to S_2 \to S_3$ can be represented as $|\mathcal{R}_{S_1}^{T_1} \cap \mathcal{R}_{S_2}^{T_3} \cap \mathcal{R}_{S_3}^{T_3}|$. This quantity can again be estimated using the inclusion-exclusion formula.

$$\tilde{n}_{S_1 S_2, S_3}^{\Delta T} = \hat{n}_{S_1} + \hat{n}_{S_2} + \hat{n}_{S_3} - \hat{n}_{S_1 \cup S_2} - \hat{n}_{S_1 \cup S_3} - \hat{n}_{S_2 \cup S_3} + \hat{n}_{S_1 \cup S_2 \cup S_3}$$

The drawback of estimating the flow on paths is that the accuracy decreases drastically in the number of nodes on the path. In conclusion, we now have two major sources of high variance. A high loadfactor that is necessary to comply with high privacy requirements and a large number of intersections, required to monitor long paths. In the following a method for reducing the variance of the estimators is presented that improves the estimation of count distinct at a single sensor, as well as union and intersection estimation. Through this improvement, a higher loadfactor can be chosen to increase privacy while maintaining the same estimation accuracy. Furthermore, this improvement allows for monitoring the flow on longer paths with sufficient accuracy.

4.3   Improved Estimator

The improved estimator is based on the idea that the average of independent estimations of the same quantity is again an equally biased estimator with lower variance [7]. Hence, at each sensor, not one sketch is constructed, but r different sketches using r different and independent hash functions. This yields r different intermediate estimates, $\widehat{n}^1, ..., \widehat{n}^r$. The improved estimator is then defined as the mean of these intermediate estimates, i.e., $\widehat{\eta} = \frac{1}{r}\sum_{i=1}^{r} \widehat{n}^i$. The $\widehat{n}^1, ..., \widehat{n}^r$ are maximum likelihood estimators for count distinct and as such they are normally distributed and independent with common mean and variance [24], i.e., for all $i \in \{1, ..., r\}$ it holds that $\widehat{n}^i \sim \mathcal{N}(\mathbb{E}[\widehat{n}], \mathbb{V}(\widehat{n}))$. Thus, the improved estimator is normally distributed with $\widehat{\eta} \sim \mathcal{N}\left(\mathbb{E}[\widehat{n}], \frac{1}{r}\mathbb{V}(\widehat{n})\right)$. The improved estimator has the same expected value as the intermediate estimates, that is, it is asymptotically unbiased, whereas the variance of the improved estimator is reduced by a factor of $1/r$. Furthermore, because the intermediate estimators are normally distributed and asymptotically unbiased, the improved estimator based on their mean is not only again a maximum likelihood estimator for the count distinct, it is also the uniformly minimum variance unbiased estimator and the minimum risk invariant estimator [23].

However, in the pathological event that a sketch becomes full, i.e., $\mathbf{u}_n = 0$, the estimate for the count distinct based on this sketch is infinity. If only one of the r sketches runs full, the estimator fails. This drawback can be circumvented by using the median of the intermediate results instead of their mean. The median is very robust to outliers but has also weaker error guarantees, i.e., to guarantee an error not larger than ϵ with probability $1 - \delta$, the mean estimator requires $r \geq z_{1-\delta}\sqrt{\mathbb{V}(\widehat{n})}/\epsilon$, the median method requires $r \geq \log(1/\delta)/\epsilon^2$ [8] intermediate estimators. Consequently, for $\epsilon < 1$, the mean estimator requires less intermediate estimates to be as accurate as the median method.

5   Privacy Analysis

The main threat to privacy in the presented application scenarios is the so called linking attack, i.e., an attacker infiltrates or takes over the monitoring system and links this knowledge to background information in order to draw novel conclusions. For example, in a standard monitoring system that distributes the sensor readings, i.e., the device addresses, an attacker that knows the device address of a certain person as background knowledge, and furthermore infiltrates the monitoring system, is able to track this person throughout the monitored area.

Sketching prevents these linking attacks in two ways, obfuscation and k-anonymity. Obfuscation is accomplished by hashing the device address to sketch positions. Hence, before an attacker is able to re-identify a device, she has to infer the employed hash function. However, this very basic obfuscation technique can be vanquished using statistical analysis on sensor readings. The second anonymization technique is accomplished by the natural property of sketches to compress the address space, implicating collisions of addresses when mapped to sketch positions. Whereas these collisions entail a loss in accuracy, they create a form of anonymity, because an attacker can only infer upon a set of devices whose addresses are all mapped to that very same sketch position.

Formally, a monitoring system guarantees k-anonymity (see Sweeney [25]), if every monitored entity is indistinguishable from at least k other entities. Using linear count sketches with a loadfactor t results in t collisions per bucket on expectation, as implied by the uniformity property of the hash function, i.e.,

$$\forall i \in \{0, ..., m-1\} : \mathbb{E}\left[|\{a \in \mathcal{R}_S^T : h(a) = i\}|\right] = t .$$

Hence, the expected level of anonymity is t. We denote this form of anonymity **expected k-anonymity**, because the number of collisions is not deterministically guaranteed as required by regular k-anonymity. For a mathematical derivation of a similar probabilistic guarantee in the context of Bloom filters the reader is referred to Bianchi et al. [5].

The union of sensor readings is estimated by the binary or of the individual sketches. The binary or of a set of sketches has a loadfactor at least as high as the individual sketches themselves. Therefore, the level of expected k-anonymity is at least as high.

The intersection of sensor readings can contain far less device addresses than the individual readings. A sketch that is created on the readings of the intersection has thus a lower loadfactor. Even so, the intersection estimator presented in this paper is based on the estimators of the individual sketches and their union; the sketch of the intersection is not constructed at all. Therefore, the level of k-anonymity of this estimator for the intersection is again at least as high as the anonymity of the individual sketches.

6   Experiments

In this section the empirical analysis of our method is presented. The general set up of experiments is as follows. A set of n addresses is randomly sampled out of a pre-defined address range ($\mathcal{A} = \{1, \cdots, 5 \times 10^7\}$). Out of this set we repeatedly sample with duplicates. The set is partitioned into k subsets $S_1, \cdots, S_k$ where S_i represents the sensor readings of sensor i. For each sensor S_i, a sketch sk_i of size m is generated using a global hash function h for all sensors. The estimate of sketches, their unions and intersections are then calculated as explained above.

6.1   Properties of the Estimator

For the first experiment, we simulate 3 sensors and vary the number of persons inside the sensor range from 500 to 250,000. Results for the average ratio of estimator and true value and the standard deviation of the ratio are shown in Fig. 2(a) and Fig. 2(b), respectively. The estimate is highly accurate—the error is always below 1%. Compared to the error introduced by the inference of the number of persons present in the area from the number of active Bluetooth devices [18], the error is negligible.

For simple Linear Counting, these results confirm existing expectations from theory and experimental studies. We need not go into a detailed comparison with other sketching methods in this paper, since two recent studies [19, 15] have done that already. One of the basic findings in these studies is that Linear Counting gives, using a suitable loadfactor, much more accurate estimates of the number of distinct objects than other sketching methods, e.g. FM-sketches or sampling based methods. This holds especially for small set sizes—where a number of 10,000 might already be considered small. For our application scenario this is important, since the size—especially of the intersections—can decrease to a few hundred persons. The experiment goes beyond the existing studies by showing that for the intersection of two sets the error can also be very low with a suitable loadfactor and that we can always set up a very accurate estimator using Linear Counting. A significant error of the estimator comes in only because we deliberately trade privacy against accuracy. As shown in section 5 the basic mechanism responsible for privacy is increasing the loadfactor. We analyze this trade-off, i.e., we investigate the impact of the loadfactor on the accuracy of estimates of one sensor as well as of intersections of up to five sensors. We simulate 5 sensors, and vary the loadfactor and the number of intersections. We average the results over 2,000 runs. The results are depicted in Fig. 3(a).

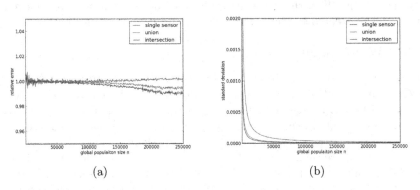

(a)   (b)

Fig. 2. Average estimate $\hat{n}$ (a) and average standard deviation (b) relative to the true value n for a loadfactor of 1

The results confirm that the standard error increases with the loadfactor (Fig. 3(a), upper part), and even more rapidly with the number of intersections (Fig. 3(a), lower part). From this experiment we conclude that simple Linear Counting is indeed suitable for loadfactors smaller than 2 and intersection of at most two sensors. But for higher loadfactors or more intersections the trade-off can become unacceptable. This finding motivates the improved estimator investigated in the following.

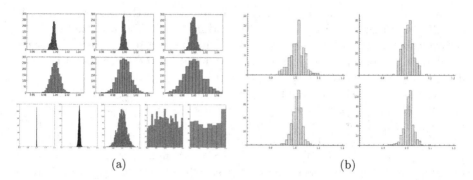

(a)   (b)

Fig. 3. (a) Distribution of estimates for loadfactors $0.5, 1, 2, 3, 4, 5$ (upper part) and numbers of sensors intersected, $1, 2, 3, 4, 5$ with a constant loadfactor of 5 (lower part). (b) Distribution of estimates using improved estimator with number of intermediate estimates r set to 5 (upper left), 15, 30 and 50 (lower right).

For that, we concentrate on a loadfactor of 5, because in regular k-anonymity 5 is a common value to ensure privacy. For each sensor, we take $r \in \{5, 15, 30, 50\}$ different random initializations of the hash function, resulting in r different sketches per sensor. Results, averaged over 500 runs, are shown in Fig. 3(b). The mean of estimates reduces the variance and is close to the true value. A good trade-off between the increase in accuracy and the higher memory requirements for storing multiple sketches is in the range between 15-30 sketches, since for higher numbers of sketches the variance reduction per additional sketch becomes insignificant. With these results we have demonstrated how to achieve a good trade-off between accuracy, privacy level, and memory consumption.

6.2   Real-World Simulation

To investigate the flow and crowd monitoring in a more realistic setting, we implemented a simulation environment as follows. A random graph of k nodes with Bernoulli Graph Distribution ($p = 0.4$) is generated; the position of nodes in 2D-space is calculated using an automatic graph layout method. A number s of node locations is attached with sensors with a predefined range. In general, a sensor may cover more than one node and several edges. A number of n objects, i.e. the global population, is created. For each object a random sequence of tour stops (nodes of the graph) is generated. For every pair of tour stops the shortest path is determined using Dijkstra's method and inserted into the sequence between the stops. Finally, to each object a velocity, starting time and a step size is assigned (the latter because objects are not only at the node positions, but travel along the edges). During the simulation, for each time step the objects follow the tour with the assigned velocity and starting time, and their position along the edges is calculated. Each sensor monitors at each time step the objects in its sensor range. For each sensor and time period a new sketch is calculated and stored. As a ground truth, also the object address are stored. The simulation stops when the last object has completed its tour.

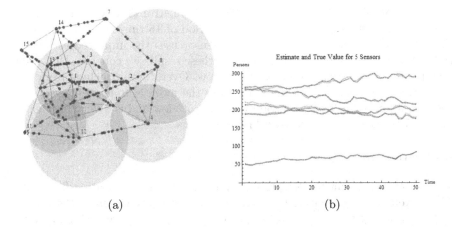

(a)   (b)

Fig. 4. (a)Crowd simulation with Random Bernoulli Graph, 1000 persons (red dots), 5 Bluetooth sensors with overlapping range. (b) Estimated value (green dashed) and true value (red dotted) for each sensor for the first 50 time steps.

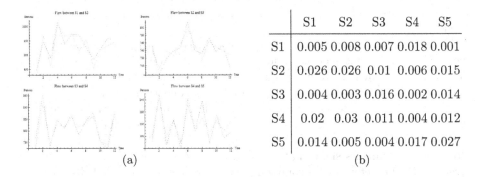

	S1	S2	S3	S4	S5
S1	0.005	0.008	0.007	0.018	0.001
S2	0.026	0.026	0.01	0.006	0.015
S3	0.004	0.003	0.016	0.002	0.014
S4	0.02	0.03	0.011	0.004	0.012
S5	0.014	0.005	0.004	0.017	0.027

(a)   (b)

Fig. 5. Flow estimation (green) and true values (red) between four sensors (a) as well as relative error of an OD-matrix between five sensors (b). Results are from flow simulation with 20000 persons and 5 sensors.

For the crowd monitoring scenario, overlapping sensors are simulated. Fig. 4(a) shows a snapshot from a simulation run. For the flow monitoring we use non-overlapping areas. The main difference compared to the experiments discussed in the last section is that the distribution of objects at nodes is not independent from each other because of flow constraints along the graph. The distribution is generated by a process very similar to real traffic flow, so that we have realistic flow properties over time. Fig. 4(b) shows an example for crowd monitoring with 15 nodes, 5 sensors, 1000 objects and a loadfactor of 5. Evidently, the estimates closely track the true values, as expected from the theoretical analysis and the experiments reported in the last section. For the flow monitoring scenario, Fig. 5(a) shows the estimate (green, dashed) and true (red, dotted) value of the flows between 4 sensors over time. Next we simulated the OD-matrix construction problem. Table 5(b) shows the relative error of an OD-matrix construction for 20,000 persons moving over a period of 15 time steps in a system with 12 nodes, tracked by 5 sensors. Each sensor uses the improved estimator with 30 sketches and a loadfactor of 5. From these results we conclude that even for moderately sized sets, the error is very low. Overall, we conclude from the experiments that Linear Counting behaves in the more complex setting of the simulation as expected from a theoretical point of view and is a very promising approach for deployment in real-world applications.

7   Discussion

In this paper we present a new, privacy aware method for estimating flows and tracks as well as for estimating OD-matrices. This way, we extended the Linear Counting approach to a general set of primitives for privacy-preserving mobility modeling. We show theoretically and empirically that two challenging application scenarios can be solved using and combining this set of primitives. To compensate the accuracy deficit of Linear Counting for strict privacy requirements we present a method for increasing the accuracy, while maintaining the privacy. This method is also applicable to boost accuracy in flow estimation, allowing to monitor even tracks.

In contrast to many privacy-preserving approaches, this one is easy to implement, has excellent accuracy and can be implemented efficiently. Our experiments suggest that it is immensely useful in a practical settings and can have a real impact on how stationary sensor based data collection is done.

While being accurate on count distinct and flow estimation, even for strict privacy, the accuracy of our method drops drastically with the length of monitored tracks. The improved estimator can compensate this drop to a certain level. However, experiments show that estimating tracks of length greater 5 leads to large errors. Therefore, we recommend using our method on count distinct and flows. When monitoring tracks, depending on their length, a user might have to reduce privacy requirements in order to maintain a certain accuracy.

The main drawback of Linear Counting when compared to other sketching techniques is the memory usage. Most sketching techniques, e.g., FM Sketches, use memory logarithmic in the number of items it estimates. The linear count sketches, however, have linear memory usage, leading to potentially large sketches. Fortunately, stationary sensors usually can be equipped with large memory (e.g., 32GB flash memory). Hence, this is unproblematic for our application scenarios. Still, the memory footprint can become problematic, because communication is in general costly. If large sketches have to be send very frequently, communication costs can become significant, or sketch sizes might even exceed network capacities.

In follow up research, we want to tackle the general problem of communication costs when using stationary sensors. However, when monitoring non-linear functions, like the union or intersection of sets, this task is not trivial. The LIFT-approach provides a framework for communication reduction in distributed systems, allowing communication efficient monitoring of non-linear functions. We want to apply the LIFT-approach to our monitoring system and test the benefits of employing the LIFT-approach in a real-world experiment.

Acknowledgements. This research has been supported by the EU FP7/2007-2013 under grant 255951 (LIFT) and by the German Science Foundation under 'GA 1615/2-1'.

References

[1] Aggarwal, C.C., Yu, P.S.: A general survey of privacy-preserving data mining models and algorithms. In: Privacy-Preserving Data Mining, pp. 11–52 (2008)

[2] Ashok, K., Ben-Akiva, M.E.: Dynamic origin-destination matrix estimation and prediction for real-time traffic management systems. In: International Symposium on the Theory of Traffic Flow and Transportation (1993)

[3] Barceló, J., Montero, L., Marquès, L., Carmona, C.: Travel time forecasting and dynamic origin-destination estimation for freeways based on bluetooth traffic monitoring. Transportation Research Record: Journal of the Transportation Research Board 2175(1), 19–27 (2010)

[4] Bergamini, L., Becchetti, L., Vitaletti, A.: Privacy-preserving environment monitoring in networks of mobile devices. In: Casares-Giner, V., Manzoni, P., Pont, A. (eds.) NETWORKING 2011 Workshops. LNCS, vol. 6827, pp. 179–191. Springer, Heidelberg (2011)

[5] Bianchi, G., Bracciale, L., Loreti, P.: "Better than nothing" privacy with bloom filters: To what extent? In: Domingo-Ferrer, J., Tinnirello, I. (eds.) PSD 2012. LNCS, vol. 7556, pp. 348–363. Springer, Heidelberg (2012)

[6] Broder, A., Mitzenmacher, M.: Network applications of bloom filters: A survey. Internet Mathematics 1(4), 485–509 (2004)

[7] Cormode, G., Garofalakis, M.: Sketching probabilistic data streams. In: Proceedings of the 2007 ACM SIGMOD International Conference on Management of Data, pp. 281–292. ACM (2007)

[8] Cormode, G., Datar, M., Indyk, P., Muthukrishnan, S.: Comparing data streams using hamming norms (how to zero in). IEEE Transactions on Knowledge and Data Engineering 15(3), 529–540 (2003)

[9] Cormode, G., Muthukrishnan, S., Yi, K.: Algorithms for distributed functional monitoring. ACM Transactions on Algorithms (TALG) 7(2), 21 (2011)

[10] Davies, A.C., Yin, J.H., Velastin, S.A.: Crowd monitoring using image processing. Electronics & Communication Engineering Journal 7(1), 37–47 (1995)

[11] Dwork, C.: Differential privacy. In: Bugliesi, M., Preneel, B., Sassone, V., Wegener, I. (eds.) ICALP 2006. LNCS, vol. 4052, pp. 1–12. Springer, Heidelberg (2006)

[12] Friedman, A., Wolff, R., Schuster, A.: Providing k-anonymity in data mining. The VLDB Journal 17(4), 789–804 (2008)

[13] Ganguly, S., Garofalakis, M., Rastogi, R.: Tracking set-expression cardinalities over continuous update streams. The VLDB Journal 13(4), 354–369 (2004)

[14] Gibbons, P.B.: Distinct-values estimation over data streams. In: Data Stream Management: Processing High-Speed Data. Springer (2009)

[15] Gonçalves, N., José, R., Baquero, C.: Privacy preserving gate counting with collaborative bluetooth scanners. In: Meersman, R., Dillon, T., Herrero, P. (eds.) OTM 2011 Workshops. LNCS, vol. 7046, pp. 534–543. Springer, Heidelberg (2011)

[16] Heikkila, J., Silvén, O.: A real-time system for monitoring of cyclists and pedestrians. In: Second IEEE Workshop on Visual Surveillance (VS 1999), pp. 74–81. IEEE (1999)

[17] Langheinrich, M.: Privacy by design - principles of privacy-aware ubiquitous systems. In: Abowd, G.D., Brumitt, B., Shafer, S. (eds.) UbiComp 2001. LNCS, vol. 2201, pp. 273–291. Springer, Heidelberg (2001)

[18] Liebig, T., Xu, Z., May, M., Wrobel, S.: Pedestrian quantity estimation with trajectory patterns. In: Flach, P.A., De Bie, T., Cristianini, N. (eds.) ECML PKDD 2012, Part II. LNCS, vol. 7524, pp. 629–643. Springer, Heidelberg (2012)

[19] Metwally, A., Agrawal, D., El Abbadi, A.: Why go logarithmic if we can go linear?: Towards effective distinct counting of search traffic. In: Proceedings of the 11th International Conference on Extending Database Technology: Advances in Database Technology, EDBT, pp. 618–629. ACM (2008)

[20] Mir, D., Muthukrishnan, S., Nikolov, A., Wright, R.N.: Pan-private algorithms via statistics on sketches. In: Proceedings of the 30th Symposium on Principles of Database Systems of Data, pp. 37–48. ACM (2011)

[21] Monreale, A.: Privacy by Design in Data Mining. PhD thesis, University Pisa (2011)

[22] Preneel, B.: Analysis and design of cryptographic hash functions. PhD thesis, Katholieke Universiteit te Leuven (1993)

[23] Rusu, F., Dobra, A.: Statistical analysis of sketch estimators. In: Proceedings of the 2007 ACM SIGMOD International Conference on Management of Data, pp. 187–198. ACM (2007)

[24] Self, S.G., Liang, K.Y.: Asymptotic properties of maximum likelihood estimators and likelihood ratio tests under nonstandard conditions. Journal of the American Statistical Association 82(398), 605–610 (1987)

[25] Sweeney, L.: k-anonymity: A model for protecting privacy. International Journal of Uncertainty, Fuzziness and Knowledge-Based Systems 10(5), 557–570 (2002)

[26] Whang, K.Y., Vander-Zanden, B.T., Taylor, H.M.: A linear-time probabilistic counting algorithm for database applications. ACM Transactions on Database Systems (TODS) 15(2), 208–229 (1990)

Evasion Attacks against Machine Learning at Test Time

Battista Biggio[1], Igino Corona[1], Davide Maiorca[1], Blaine Nelson[2],
Nedim Šrndić[3], Pavel Laskov[3], Giorgio Giacinto[1], and Fabio Roli[1]

[1] Dept. of Electrical and Electronic Engineering, University of Cagliari,
Piazza d'Armi, 09123 Cagliari, Italy
{battista.biggio,igino.corona,davide.maiorca,giacinto,
roli}@diee.unica.it
http://prag.diee.unica.it/
[2] Institut für Informatik, Universität Potsdam,
August-Bebel-Straße 89, 14482 Potsdam, Germany
bnelson@cs.uni-potsdam.de
[3] Wilhelm Schickard Institute for Computer Science, University of Tübingen,
Sand 1, 72076 Tübingen, Germany
{nedim.srndic,pavel.laskov}@uni-tuebingen.de

Abstract. In security-sensitive applications, the success of machine learning depends on a thorough vetting of their resistance to adversarial data. In one pertinent, well-motivated attack scenario, an adversary may attempt to evade a deployed system at test time by carefully manipulating attack samples. In this work, we present a simple but effective gradient-based approach that can be exploited to systematically assess the security of several, widely-used classification algorithms against evasion attacks. Following a recently proposed framework for security evaluation, we simulate attack scenarios that exhibit different risk levels for the classifier by increasing the attacker's knowledge of the system and her ability to manipulate attack samples. This gives the classifier designer a better picture of the classifier performance under evasion attacks, and allows him to perform a more informed model selection (or parameter setting). We evaluate our approach on the relevant security task of malware detection in PDF files, and show that such systems can be easily evaded. We also sketch some countermeasures suggested by our analysis.

Keywords: adversarial machine learning, evasion attacks, support vector machines, neural networks.

1 Introduction

Machine learning is being increasingly used in security-sensitive applications such as spam filtering, malware detection, and network intrusion detection [3, 5, 9, 11, 14–16, 19, 21]. Due to their intrinsic adversarial nature, these applications differ from the classical machine learning setting in which the underlying data distribution is assumed to be *stationary*. To the contrary, in security-sensitive

H. Blockeel et al. (Eds.): ECML PKDD 2013, Part III, LNAI 8190, pp. 387–402, 2013.
© Springer-Verlag Berlin Heidelberg 2013

applications, samples (and, thus, their distribution) can be actively manipulated by an intelligent, adaptive adversary to confound learning; *e.g.*, to avoid detection, spam emails are often modified by obfuscating common spam words or inserting words associated with legitimate emails [3, 9, 16, 19]. This has led to an arms race between the designers of learning systems and their adversaries, which is evidenced by the increasing complexity of modern attacks and countermeasures. For these reasons, classical performance evaluation techniques are not suitable to reliably assess the security of learning algorithms, *i.e.*, the performance degradation caused by carefully crafted attacks [5].

To better understand the security properties of machine learning systems in adversarial settings, paradigms from security engineering and cryptography have been adapted to the machine learning field [2, 5, 14]. Following common security protocols, the learning system designer should use *proactive* protection mechanisms that anticipate and prevent the adversarial impact. This requires (*i*) finding potential vulnerabilities of learning *before* they are exploited by the adversary; (*ii*) investigating the impact of the corresponding attacks (*i.e.*, evaluating classifier security); and (*iii*) devising appropriate countermeasures if an attack is found to significantly degrade the classifier's performance.

Two approaches have previously addressed security issues in learning. The min-max approach assumes the learner and attacker's loss functions are antagonistic, which yields relatively simple optimization problems [10, 12]. A more general game-theoretic approach applies for non-antagonistic losses; *e.g.*, a spam filter wants to accurately identify legitimate email while a spammer seeks to boost his spam's appeal. Under certain conditions, such problems can be solved using a Nash equilibrium approach [7, 8]. Both approaches provide a *secure* counterpart to their respective learning problems; *i.e.*, an optimal anticipatory classifier.

Realistic constraints, however, are too complex and multi-faceted to be incorporated into existing game-theoretic approaches. Instead, we investigate the vulnerabilities of classification algorithms by deriving *evasion attacks* in which the adversary aims to avoid detection by manipulating malicious test samples.[1] We systematically assess classifier security in attack scenarios that exhibit increasing risk levels, simulated by increasing the attacker's knowledge of the system and her ability to manipulate attack samples. Our analysis allows a classifier designer to understand how the classification performance of each considered model degrades under attack, and thus, to make more informed design choices.

The problem of evasion at test time was addressed in prior work, but limited to linear and convex-inducing classifiers [9, 19, 22]. In contrast, the methods presented in Sections 2 and 3 can generally evade linear or non-linear classifiers using a gradient-descent approach inspired by Golland's discriminative directions technique [13]. Although we focus our analysis on widely-used classifiers such as Support Vector Machines (SVMs) and neural networks, our approach is applicable to any classifier with a differentiable discriminant function.

---

[1] Note that other kinds of attacks are possible, *e.g.*, if the adversary can manipulate the training data. A comprehensive taxonomy of attacks can be found in [2, 14].

This paper is organized as follows. We present the evasion problem in Section 2 and our gradient-descent approach in Section 3. In Section 4 we first visually demonstrate our attack on the task of handwritten digit recognition, and then show its effectiveness on a realistic application related to the detection of PDF malware. Finally in Section 5, we summarize our contributions, discuss possibilities for improving security, and suggest future extensions of this work.

2   Optimal Evasion at Test Time

We consider a classification algorithm $f : \mathcal{X} \mapsto \mathcal{Y}$ that assigns samples represented in some feature space $\mathbf{x} \in \mathcal{X}$ to a label in the set of predefined classes $y \in \mathcal{Y} = \{-1, +1\}$, where -1 $(+1)$ represents the legitimate (malicious) class. The classifier f is trained on a dataset $\mathcal{D} = \{\mathbf{x}_i, y_i\}_{i=1}^{n}$ sampled from an underlying distribution $p(\mathbf{X}, Y)$. The label $y^c = f(\mathbf{x})$ given by a classifier is typically obtained by thresholding a continuous discriminant function $g : \mathcal{X} \mapsto \mathbb{R}$. In the sequel, we use y^c to refer to the label assigned by the classifier as opposed to the true label y. We further assume that $f(\mathbf{x}) = -1$ if $g(\mathbf{x}) < 0$, and $+1$ otherwise.

2.1   Adversary Model

To motivate the optimal attack strategy for evasion, it is necessary to disclose one's assumptions of the adversary's knowledge and ability to manipulate the data. To this end, we exploit a general model of the adversary that elucidates specific assumptions about adversary's goal, knowledge of the system, and capability to modify the underlying data distribution. The considered model is part of a more general framework investigated in our recent work [5], which subsumes evasion and other attack scenarios. This model can incorporate application-specific constraints in the definition of the adversary's capability, and can thus be exploited to derive practical guidelines for developing the optimal attack strategy.

Adversary's goal. As suggested by Laskov and Kloft [17], the adversary's goal should be defined in terms of a utility (loss) function that the adversary seeks to maximize (minimize). In the evasion setting, the attacker's goal is to manipulate a single (without loss of generality, positive) sample that should be misclassified. Strictly speaking, it would suffice to find a sample $\mathbf{x}$ such that $g(\mathbf{x}) < -\epsilon$ for any $\epsilon > 0$; *i.e.*, the attack sample only just crosses the decision boundary.[2] Such attacks, however, are easily thwarted by slightly adjusting the decision threshold. A better strategy for an attacker would thus be to create a sample that is misclassified with high confidence; *i.e.*, a sample minimizing the value of the classifier's discriminant function, $g(\mathbf{x})$, subject to some feasibility constraints.

Adversary's knowledge. The adversary's knowledge about her targeted learning system may vary significantly. Such knowledge may include:

---

[2] This is also the setting adopted in previous work [9, 19, 22].

- the training set or part of it;
- the feature representation of each sample; *i.e.*, how *real* objects such as emails, network packets are mapped into the classifier's feature space;
- the type of a learning algorithm and the form of its decision function;
- the (trained) classifier model; *e.g.*, weights of a linear classifier;
- or feedback from the classifier; *e.g.*, classifier labels for samples chosen by the adversary.

Adversary's capability. In the evasion scenario, the adversary's capability is limited to modifications of test data; *i.e.*altering the training data is not allowed. However, under this restriction, variations in attacker's power may include:

- modifications to the input data (limited or unlimited);
- modifications to the feature vectors (limited or unlimited);
- or independent modifications to specific features (the semantics of the input data may dictate that certain features are interdependent).

Most of the previous work on evasion attacks assumes that the attacker can arbitrarily change every feature [8, 10, 12], but they constrain the degree of manipulation, *e.g.*, limiting the number of modifications, or their total cost. However, many real domains impose stricter restrictions. For example, in the task of PDF malware detection [20, 24, 25], removal of content is not feasible, and content addition may cause correlated changes in the feature vectors.

2.2   Attack Scenarios

In the sequel, we consider two attack scenarios characterized by different levels of adversary's knowledge of the attacked system discussed below.

Perfect knowledge (PK). In this setting, we assume that the adversary's goal is to minimize $g(\mathbf{x})$, and that she has perfect knowledge of the targeted classifier; *i.e.*, the adversary knows the feature space, the type of the classifier, and the trained model. The adversary can transform attack points in the test data but must remain within a maximum distance of $d_{\max}$ from the original attack sample. We use $d_{\max}$ as parameter in our evaluation to simulate increasingly pessimistic attack scenarios by giving the adversary greater freedom to alter the data.

The choice of a suitable distance measure $d : \mathcal{X} \times \mathcal{X} \mapsto \mathbb{R}^+$ is application specific [9, 19, 22]. Such a distance measure should reflect the adversary's effort required to manipulate samples or the cost of these manipulations. For example, in spam filtering, the attacker may be bounded by a certain number of words she can manipulate, so as not to lose the semantics of the spam message.

Limited knowledge (LK). Here, we again assume that the adversary aims to minimize the discriminant function $g(\mathbf{x})$ under the same constraint that each transformed attack point must remain within a maximum distance of $d_{\max}$ from the corresponding original attack sample. We further assume that the attacker knows the feature representation and the type of the classifier, but does not know either the learned classifier f or its training data $\mathcal{D}$, and hence can not directly compute $g(\mathbf{x})$. However, we assume that she can collect a surrogate dataset $\mathcal{D}' = \{\hat{\mathbf{x}}_i, \hat{y}_i\}_{i=1}^{n_q}$ of n_q samples drawn from the same underlying distribution $p(\mathbf{X}, Y)$ from which $\mathcal{D}$ was drawn. This data may be collected by an adversary in several ways; *e.g.*, by sniffing some network traffic during the classifier operation, or by collecting legitimate and spam emails from an alternate source.

Under this scenario, the adversary proceeds by approximating the discriminant function $g(\mathbf{x})$ as $\hat{g}(\mathbf{x})$, where $\hat{g}(\mathbf{x})$ is the discriminant function of a surrogate classifier $\hat{f}$ learnt on $\mathcal{D}'$. The amount of the surrogate data, n_q, is an attack parameter in our experiments. Since the adversary wants her surrogate $\hat{f}$ to closely approximate the targeted classifier f, it stands to reason that she should learn $\hat{f}$ using the labels assigned by the targeted classifier f, when such feedback is available. In this case, instead of using the true class labels $\hat{y}_i$ to train $\hat{f}$, the adversary can query f with the samples of $\mathcal{D}'$ and subsequently learn using the labels $\hat{y}_i^c = f(\hat{\mathbf{x}}_i)$ for each $\mathbf{x}_i$.

2.3   Attack Strategy

Under the above assumptions, for any target malicious sample $\mathbf{x}^0$ (the adversary's desired instance), an optimal attack strategy finds a sample $\mathbf{x}^*$ to minimize $g(\cdot)$ or its estimate $\hat{g}(\cdot)$, subject to a bound on its distance[3] from $\mathbf{x}^0$:

$$\mathbf{x}^* = \arg\min_{\mathbf{x}} \quad \hat{g}(\mathbf{x}) \tag{1}$$

$$\text{s.t.} \quad d(\mathbf{x}, \mathbf{x}^0) \le d_{\max}.$$

Generally, this is a non-linear optimization problem. One may approach it with many well-known techniques, like gradient descent, or quadratic techniques such as Newton's method, BFGS, or L-BFGS. We choose a gradient-descent procedure. However, $\hat{g}(\mathbf{x})$ may be non-convex and descent approaches may not achieve a global optima. Instead, the descent path may lead to a flat region (local minimum) outside of the samples' support (*i.e.*, where $p(\mathbf{x}) \approx 0$) where the attack sample may or may not evade depending on the behavior of g in this unsupported region (see left and middle plots in Figure 1).

Locally optimizing $\hat{g}(\mathbf{x})$ with gradient descent is particularly susceptible to failure due to the nature of a discriminant function. Besides its shape, for many classifiers, $g(\mathbf{x})$ is equivalent to a posterior estimate $p(y^c = -1 | \mathbf{x})$; *e.g.*, for neural networks, and SVMs [23]. The discriminant function does not incorporate the evidence we have about the data distribution, $p(\mathbf{x})$, and thus, using gradient descent to optimize Eq. 1 may lead into unsupported regions ($p(\mathbf{x}) \approx 0$). Because of the insignificance of these regions, the value of g is relatively unconstrained by criteria such as risk minimization. This problem is compounded by our finite

---

[3] One can also incorporate additional application-specific constraints on the attack samples. For instance, the box constraint $0 \le x_f \le 1$ can be imposed if the fth feature is normalized in $[0, 1]$, or $x_f^0 \le x_f$ can be used if the fth feature of the target $\mathbf{x}^0$ can be only incremented.

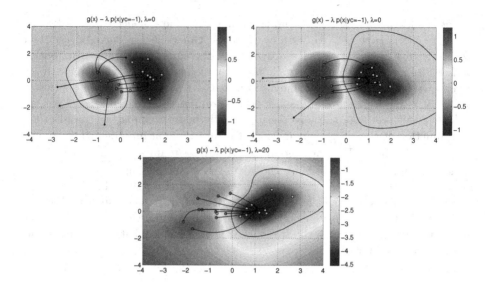

Fig. 1. Different scenarios for gradient-descent-based evasion procedures. In each, the function $g(\mathbf{x})$ of the learned classifier is plotted with a color map with high values (red-orange-yellow) for the malicious class, and low values (green-cyan-blue) for the legitimate class. The decision boundary is shown in black. For every malicious sample, we plot the gradient descent path against a classifier with a closed boundary around the malicious class (**top-left**) and against a classifier with a closed boundary around the benign class (**top-right**). Finally, we plot the modified objective function of Eq. (2) and the resulting descent paths against a classifier with a closed boundary around the benign class (**bottom**).

(and possibly small) training set, since it provides little evidence in these regions to constrain the shape of g. Thus, when our gradient descent procedure produces an evasion example in these regions, the attacker cannot be confident that this sample will actually evade the corresponding classifier. Therefore, to increase the probability of successful evasion, the attacker should favor attack points from densely populated regions of legitimate points, where the estimate $\hat{g}(\mathbf{x})$ is more reliable (closer to the real $g(\mathbf{x})$), and tends to become negative in value.

To overcome this shortcoming, we introduce an additional component into our attack objective, which estimates $p(\mathbf{x}|y^c = -1)$ using a density estimator. This term acts as a penalizer for $\mathbf{x}$ in low density regions and is weighted by a parameter $\lambda \geq 0$ yielding the following modified optimization problem:

$$\arg\min_x F(\mathbf{x}) = \hat{g}(\mathbf{x}) - \frac{\lambda}{n} \sum_{i|y_i^c = -1} k\left(\frac{\mathbf{x} - \mathbf{x}_i}{h}\right) \tag{2}$$

$$\text{s.t. } d(\mathbf{x}, \mathbf{x}^0) \leq d_{\max} , \tag{3}$$

where h is a bandwidth parameter for a kernel density estimator (KDE), and n is the number of benign samples ($y^c = -1$) available to the adversary. This

Algorithm 1. Gradient-descent evasion attack

Input: $\mathbf{x}^0$, the initial attack point; t, the step size; λ, the trade-off parameter; $\epsilon > 0$ a small constant.

Output: $\mathbf{x}^*$, the final attack point.

1: $m \leftarrow 0$.
2: **repeat**
3:     $m \leftarrow m + 1$
4:     Set $\nabla F(\mathbf{x}^{m-1})$ to a unit vector aligned with $\nabla g(\mathbf{x}^{m-1}) - \lambda \nabla p(\mathbf{x}^{m-1}|y^c = -1)$.
5:     $\mathbf{x}^m \leftarrow \mathbf{x}^{m-1} - t\nabla F(\mathbf{x}^{m-1})$
6:     **if** $d(\mathbf{x}^m, \mathbf{x}^0) > d_{\max}$ **then**
7:         Project $\mathbf{x}^m$ onto the boundary of the feasible region.
8:     **end if**
9: **until** $F(\mathbf{x}^m) - F(\mathbf{x}^{m-1}) < \epsilon$
10: **return:** $\mathbf{x}^* = \mathbf{x}^m$

alternate objective trades off between minimizing $\hat{g}(\mathbf{x})$ (or $p(y^c = -1|\mathbf{x})$) and maximizing the estimated density $p(\mathbf{x}|y^c = -1)$. The extra component favors attack points that imitate features of known legitimate samples. In doing so, it reshapes the objective function and thereby biases the resulting gradient descent towards regions where the negative class is concentrated (see the bottom plot in Fig. 1). This produces a similar effect to that shown by *mimicry* attacks in network intrusion detection [11].[4] For this reason, although our setting is rather different, in the sequel we refer to this extra term as the *mimicry* component.

Finally, we point out that, when mimicry is used ($\lambda > 0$), our gradient descent clearly follows a suboptimal path compared to the case when only $g(\mathbf{x})$ is minimized ($\lambda = 0$). Therefore, more modifications may be required to reach the same value of $g(\mathbf{x})$ attained when $\lambda = 0$. However, as previously discussed, when $\lambda = 0$, our descent approach may terminate at a local minimum where $g(\mathbf{x}) > 0$, without successfully evading detection. This behavior can thus be qualitatively regarded as a trade-off between the probability of evading the targeted classifier and the number of times that the adversary must modify her samples.

3   Gradient Descent Attacks

Algorithm 1 solves the optimization problem in Eq. 2 via gradient descent. We assume $g(\mathbf{x})$ to be differentiable almost everywhere (subgradients may be used at discontinuities). However, note that if g is non-differentiable or insufficiently smooth, one may still use the mimicry / KDE term of Eq. (2) as a search heuristic. This investigation is left to future work.

---

[4] Mimicry attacks [11] consist of camouflaging malicious network packets to evade anomaly-based intrusion detection systems by mimicking the characteristics of the legitimate traffic distribution.

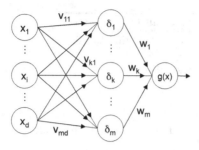

Fig. 2. The architecture of a multi-layer perceptron with a single hidden layer

3.1   Gradients of Discriminant Functions

Linear classifiers. Linear discriminant functions are $g(\mathbf{x}) = \langle \mathbf{w}, \mathbf{x} \rangle + b$ where $\mathbf{w} \in \mathbb{R}^d$ is the feature weights and $b \in \mathbb{R}$ is the bias. Its gradient is $\nabla g(\mathbf{x}) = \mathbf{w}$.

Support vector machines. For SVMs, $g(\mathbf{x}) = \sum_i \alpha_i y_i k(\mathbf{x}, \mathbf{x}_i) + b$. The gradient is thus $\nabla g(\mathbf{x}) = \sum_i \alpha_i y_i \nabla k(\mathbf{x}, \mathbf{x}_i)$. In this case, the feasibility of our approach depends on whether the kernel gradient $\nabla k(\mathbf{x}, \mathbf{x}_i)$ is computable as it is for many numeric kernels. For instance, the gradient of the RBF kernel, $k(\mathbf{x}, \mathbf{x}_i) = \exp\{-\gamma \|\mathbf{x} - \mathbf{x}_i\|^2\}$, is $\nabla k(\mathbf{x}, \mathbf{x}_i) = -2\gamma \exp\{-\gamma\|\mathbf{x} - \mathbf{x}_i\|^2\}(\mathbf{x} - \mathbf{x}_i)$, and for the polynomial kernel, $k(\mathbf{x}, \mathbf{x}_i) = (\langle \mathbf{x}, \mathbf{x}_i \rangle + c)^p$, it is $\nabla k(\mathbf{x}, \mathbf{x}_i) = p(\langle \mathbf{x}, \mathbf{x}_i \rangle + c)^{p-1}\mathbf{x}_i$.

Neural networks. For a multi-layer perceptron with a single hidden layer of m neurons and a sigmoidal activation function, we decompose its discriminant function g as follows (see Fig. 2): $g(\mathbf{x}) = (1+e^{-h(\mathbf{x})})^{-1}$, $h(\mathbf{x}) = \sum_{k=1}^{m} w_k \delta_k(\mathbf{x})+b$, $\delta_k(\mathbf{x}) = (1 + e^{-h_k(\mathbf{x})})^{-1}$, $h_k(\mathbf{x}) = \sum_{j=1}^{d} v_{kj}x_j + b_k$. From the chain rule, the ith component of $\nabla g(\mathbf{x})$ is thus given by:

$$\frac{\partial g}{\partial x_i} = \frac{\partial g}{\partial h}\sum_{k=1}^{m}\frac{\partial h}{\partial \delta_k}\frac{\partial \delta_k}{\partial h_k}\frac{\partial h_k}{\partial x_i} = g(\mathbf{x})(1 - g(\mathbf{x}))\sum_{k=1}^{m} w_k \delta_k(\mathbf{x})(1 - \delta_k(\mathbf{x}))v_{ki} \ .$$

3.2   Gradients of Kernel Density Estimators

Similarly to SVMs, the gradient of kernel density estimators depends on the kernel gradient. We consider generalized RBF kernels of the form $k\left(\frac{\mathbf{x}-\mathbf{x}_i}{h}\right) = \exp\left(-\frac{d(\mathbf{x},\mathbf{x}_i)}{h}\right)$, where $d(\cdot, \cdot)$ is any suitable distance function. Here we use the same distance $d(\cdot, \cdot)$ defined in Eq. (3), but, in general, they can be different. For ℓ_2- and ℓ_1-norms (*i.e.*, RBF and Laplacian kernels), the KDE (sub)gradients are respectively given by:

$$-\frac{2}{nh}\sum_{i|y_i^c=-1}\exp\left(-\frac{\|\mathbf{x}-\mathbf{x}_i\|_2^2}{h}\right)(\mathbf{x} - \mathbf{x}_i) \ ,$$

$$-\frac{1}{nh}\sum_{i|y_i^c=-1}\exp\left(-\frac{\|\mathbf{x}-\mathbf{x}_i\|_1}{h}\right)(\mathbf{x} - \mathbf{x}_i) \ .$$

Note that the scaling factor here is proportional to $O(\frac{1}{nh})$. Therefore, to influence gradient descent with a significant mimicking effect, the value of λ in the objective function should be chosen such that the value of $\frac{\lambda}{nh}$ is comparable with (or higher than) the range of values of the discriminant function $\hat{g}(\mathbf{x})$.

3.3   Descent in Discrete Spaces

In discrete spaces, gradient approaches travel through infeasible portions of the feature space. In such cases, we need to find a feasible neighbor $\mathbf{x}$ that maximally decrease $F(\mathbf{x})$. A simple approach to this problem is to probe F at every point in a small neighborhood of $\mathbf{x}$, which would however require a large number of queries. For classifiers with a differentiable decision function, we can instead select the neighbor whose change best aligns with $\nabla F(\mathbf{x})$ and decreases the objective function; *i.e.*, to prevent overshooting a minimum.

4   Experiments

In this section, we first report a toy example from the MNIST handwritten digit classification task [18] to visually demonstrate how the proposed algorithm modifies digits to mislead classification. We then show the effectiveness of the proposed attack on a more realistic and practical scenario: the detection of malware in PDF files.

4.1   A Toy Example on Handwritten Digits

Similar to Globerson and Roweis [12], we consider discriminating between two distinct digits from the MNIST dataset [18]. Each digit example is represented as a gray-scale image of 28×28 pixels arranged in raster-scan-order to give feature vectors of $d = 28 \times 28 = 784$ values. We normalized each feature (pixel) $\mathbf{x} \in [0, 1]^d$ by dividing its value by 255, and we constrained the attack samples to this range. Accordingly, we optimized Eq. (2) subject to $0 \le x_f \le 1$ for all f.

We only consider the perfect knowledge (PK) attack scenario. We used the *Manhattan* distance (ℓ_1-norm), d, both for the kernel density estimator (*i.e.*, a Laplacian kernel) and for the constraint $d(\mathbf{x}, \mathbf{x}^0) \le d_{\max}$ in Eq. (3), which bounds the total difference between the gray level values of the original image $\mathbf{x}^0$ and the attack image $\mathbf{x}$. We used $d_{\max} = \frac{5000}{255}$ to limit the total gray-level change to 5000. At each iteration, we increased the ℓ_1-norm value of $\mathbf{x} - \mathbf{x}^0$ by $\frac{10}{255}$, or equivalently, we changed the total gray level by 10. This is effectively the gradient step size. The targeted classifier was an SVM with the linear kernel and $C = 1$. We randomly chose 100 training samples and applied the attacks to a correctly-classified positive sample.

In Fig. 3 we illustrate gradient attacks in which a "3" is to be misclassified as a "7". The left image shows the initial attack point, the middle image shows the first attack image misclassified as legitimate, and the right image shows the attack point after 500 iterations. When $\lambda = 0$, the attack images exhibit only

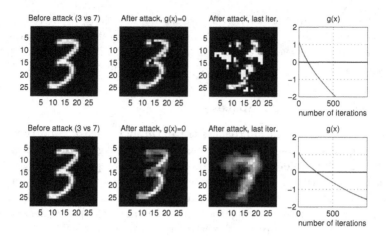

Fig. 3. Illustration of the gradient attack on the digit data, for $\lambda = 0$ (**top row**) and $\lambda = 10$ (**bottom row**). Without a mimicry component ($\lambda = 0$) gradient descent quickly decreases g but the resulting attack image does not resemble a "7". In contrast, the attack minimizes g slower when mimicry is applied ($\lambda = 0$) but the final attack image closely resembles a mixture between "3" and "7", as the term "mimicry" suggests.

a weak resemblance to the target class "7" but are, nevertheless, reliably misclassified. This is the same effect demonstrated in the top-left plot of Fig. 1: the classifier is evaded by making the attack sample sufficiently dissimilar from the malicious class. Conversely, when $\lambda = 10$, the attack images strongly resemble the target class because the mimicry term favors samples that are more similar to the target class. This is the same effect seen in the bottom plot of Fig. 1.

Finally note that, as expected, $g(\mathbf{x})$ tends to decrease more gracefully when mimicry is used, as we follow a suboptimal descent path. Since the targeted classifier can be easily evaded when $\lambda = 0$, exploiting the mimicry component would not be the optimal choice in this case. However, in the case of limited knowledge, as discussed at the end of Section 2.3, mimicry may allow us to trade for a higher probability of evading the targeted classifier, at the expense of a higher number of modifications.

4.2   Malware Detection in PDF Files

We now focus on the task of discriminating between legitimate and malicious PDF files, a popular medium for disseminating malware [26]. PDF files are excellent vectors for malicious-code, due to their flexible *logical structure*, which can described by a hierarchy of interconnected objects. As a result, an attack can be easily hidden in a PDF to circumvent file-type filtering. The PDF format further allows a wide variety of resources to be embedded in the document including JavaScript, Flash, and even binary programs. The type of the embedded object is specified by *keywords*, and its content is in a *data stream*. Several recent works proposed machine-learning techniques for detecting malicious PDFs using the file's logical structure to accurately identify the malware [20, 24, 25]. In this case study, we use the feature representation of Maiorca *et al.* [20] in which each feature corresponds to the tally of occurrences of a given keyword.

The PDF structure imposes natural constraints on attacks. Although it is difficult to *remove* an embedded object (and its keywords) from a PDF without corrupting the PDF's file structure, it is rather easy to *insert* new objects (and, thus, keywords) through the addition of a new *version* to the PDF file [1]. In our feature representation, this is equivalent to allowing only feature increments, *i.e.*, requiring $\mathbf{x}^0 \leq \mathbf{x}$ as an additional constraint in the optimization problem given by Eq. (2). Further, the total difference in keyword counts between two samples is their *Manhattan* distance, which we again use for the kernel density estimator and the constraint in Eq. (3). Accordingly, $d_{\max}$ is the maximum number of additional keywords that an attacker can add to the original $\mathbf{x}^0$.

Experimental setup. For experiments, we used a PDF corpus with 500 malicious samples from the *Contagio* dataset[5] and 500 benign samples collected from the web. We randomly split the data into five pairs of training and testing sets with 500 samples each to average the final results. The features (keywords) were extracted from each training set as described in [20]. On average, 100 keywords were found in each run. Further, we also bounded the maximum value of each feature to 100, as this value was found to be close to the 95th percentile for each feature. This limited the influence of outlying samples.

We simulated the *perfect* knowledge (PK) and the *limited* knowledge (LK) scenarios described in Section 2.1. In the LK case, we set the number of samples used to learn the surrogate classifier to $n_g = 100$. The reason is to demonstrate that even with a dataset as small as the 20% of the original training set size, the adversary may be able to evade the targeted classifier with high reliability. Further, we assumed that the adversary uses feedback from the *targeted* classifier f; *i.e.*, the labels $\hat{y}_i^c = f(\hat{\mathbf{x}}_i)$ for each surrogate sample $\hat{\mathbf{x}}_i \in \mathcal{D}'$.[6]

As discussed in Section 3.2, the value of λ is chosen according to the scale of the discriminant function $g(\mathbf{x})$, the bandwidth parameter h of the kernel density estimator, and the number of legitimate samples n in the surrogate training set. For computational reasons, to estimate the value of the KDE at $\mathbf{x}$, we only consider the 50 nearest (legitimate) training samples to $\mathbf{x}$; therefore, $n \leq 50$ in our case. The bandwidth parameter was set to $h = 10$, as this value provided a proper rescaling of the Manhattan distances observed in our dataset for the KDE. We thus set $\lambda = 500$ to be comparable with $O(nh)$.

For each targeted classifier and training/testing pair, we learned five surrogate classifiers by randomly selecting n_g samples from the test set, and we averaged their results. For SVMs, we sought a surrogate classifier that would correctly match the labels from the targeted classifier; thus, we used parameters $C = 100$, and $\gamma = 0.1$ (for the RBF kernel) to heavily penalize training errors.

---

[5] http://contagiodump.blogspot.it

[6] Similar results were also obtained using the true labels (without relabeling), since the targeted classifiers correctly classified almost all samples in the test set.

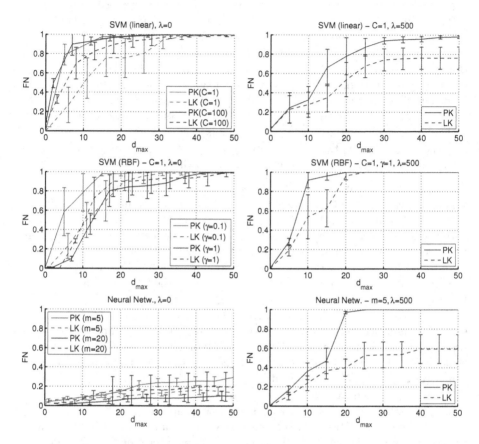

Fig. 4. Experimental results for SVMs with linear and RBF kernel (first and second row), and for neural networks (third row). We report the FN values (attained at FP=0.5%) for increasing d_{max}. For the sake of readability, we report the average FN value $\pm$ half standard deviation (shown with error bars). Results for perfect (PK) and limited (LK) knowledge attacks with $\lambda = 0$ (without mimicry) are shown in the first column, while results with $\lambda = 500$ (with mimicry) are shown in the second column. In each plot we considered different values of the classifier parameters, *i.e.*, the regularization parameter C for the linear SVM, the kernel parameter γ for the SVM with RBF kernel, and the number of neurons m in the hidden layer for the neural network, as reported in the plot title and legend.

Experimental results. We report our results in Figure 4, in terms of the false negative (FN) rate attained by the targeted classifiers as a function of the maximum allowable number of modifications, $d_{max} \in [0, 50]$. We compute the FN rate corresponding to a fixed false positive (FP) rate of FP= 0.5%. For $d_{max} = 0$, the FN rate corresponds to a standard performance evaluation using unmodified PDFs. As expected, the FN rate increases with d_{max} as the PDF is increasingly modified. Accordingly, a more secure classifier will exhibit a more graceful increase of the FN rate.

Results for $\lambda = 0$. We first investigate the effect of the proposed attack in the PK case, without considering the mimicry component (Figure 4, first column), for varying parameters of the considered classifiers. The linear SVM (Figure 4, top-left plot) is almost always evaded with as few as 5 to 10 modifications, independent of the regularization parameter C. It is worth noting that attacking a linear classifier amounts to always incrementing the value of the same highest-weighted feature (corresponding to the /Linearized keyword in the majority of the cases) until it reaches its upper bound. This continues with the next highest weighted non-bounded feature until termination. This occurs simply because the gradient of $g(\mathbf{x})$ does not depend on $\mathbf{x}$ for a linear classifier (see Section 3.1). With the RBF kernel (Figure 4, middle-left plot), SVMs exhibit a similar behavior with $C = 1$ and various values of its γ parameter,[7] and the RBF SVM provides a higher degree of security compared to linear SVMs (*cf.* top-left plot and middle-left plot in Figure 4). Interestingly, compared to SVMs, neural networks (Figure 4, bottom-left plot) seem to be much more robust against the proposed evasion attack. This behavior can be explained by observing that the decision function of neural networks may be characterized by flat regions (*i.e.*, regions where the gradient of $g(\mathbf{x})$ is close to zero). Hence, the gradient descent algorithm based solely on $g(\mathbf{x})$ essentially stops after few attack iterations for most of the malicious samples, without being able to find a suitable attack.

In the LK case, without mimicry, classifiers are evaded with a probability only *slightly* lower than that found in the PK case, even when only $n_g = 100$ surrogate samples are used to learn the surrogate classifier. This aspect highlights the threat posed by a skilled adversary with incomplete knowledge: only a small set of samples may be required to successfully attack the target classifier using the proposed algorithm.

Results for $\lambda = 500$. When mimicry is used (Figure 4, second column), the success of the evasion of linear SVMs (with $C = 1$) decreases both in the PK (*e.g.*, compare the blue curve in the top-left plot with the solid blue curve in the top-right plot) and LK case (*e.g.*, compare the dashed red curve in the top-left plot with the dashed blue curve in the top-right plot). The reason is that the computed direction tends to lead to a slower descent; *i.e.*, a less direct path that often requires more modifications to evade the classifier. In the non-linear case (Figure 4, middle-right and bottom-right plot), instead, mimicking exhibits some beneficial aspects for the attacker, although the constraint on feature addition

---

[7] We also conducted experiments using $C = 0.1$ and $C = 100$, but did not find significant differences compared to the presented results using $C = 1$.

may make it difficult to properly mimic legitimate samples. In particular, note
how the targeted SVMs with RBF kernel (with $C = 1$ and $\gamma = 1$) in the PK case
(*e.g.*, compare the solid blue curve in the middle-left plot with the solid blue curve
in the middle-right plot) is evaded with a significantly higher probability than
in the case of $\lambda = 0$. The reason is that, as explained at the end of Section 2.3, a
pure descent strategy on $g(\mathbf{x})$ may find local minima (*i.e.*, attack samples) that
do not evade detection, while the mimicry component biases the descent towards
regions of the feature space more densely populated by legitimate samples, where
$g(\mathbf{x})$ eventually attains lower values. For neural networks, this aspect is even more
evident, in both the PK and LK settings (compare the dashed/solid curves in
the bottom-left plot with those in the bottom-right plot), since $g(\mathbf{x})$ is essentially
flat far from the decision boundary, and thus pure gradient descent on g can not
even commence for many malicious samples, as previously mentioned. In this
case, the mimicry term is thus critical for finding a reasonable descent path to
evasion.

Discussion. Our attacks raise questions about the feasibility of detecting malicious PDFs solely based on logical structure. We found that /Linearized,
/OpenAction, /Comment, /Root and /PageLayout were among the most commonly manipulated keywords. They indeed are found mainly in legitimate PDFs,
but can be easily added to malicious PDFs by the versioning mechanism. The
attacker can simply insert comments inside the malicious PDF file to augment
its /Comment count. Similarly, she can embed *legitimate* OpenAction code to add
/OpenAction keywords or add new pages to insert /PageLayout keywords.

5   Conclusions, Limitations and Future Work

In this work we proposed a simple algorithm for evasion of classifiers with differentiable discriminant functions. We investigated the attack effectiveness in
the case of perfect and limited knowledge of the attacked system, and empirically showed that very popular classification algorithms (in particular, SVMs
and neural networks) can still be evaded with high probability even if the adversary can only learn a copy of the classifier from a small surrogate dataset. Thus,
our investigation raises important questions on whether such algorithms can be
reliably employed in security-sensitive applications.

We believe that the proposed attack formulation can be extended to classifiers
with non-differentiable discriminant functions as well, such as decision trees and
k-nearest neighbors; *e.g.*, by defining suitable search heuristics similar to our
mimicry term to minimize $g(\mathbf{x})$.

Interestingly our analysis also suggests improvements for classifier security.
From Fig. 1, it is clear that a tighter *enclosure* of the legitimate samples increasingly forces the adversary to mimic the legitimate class, which may not always
be possible; *e.g.*, malicious network packets or PDF files must contain a valid
exploit for the attack to be successful. Accordingly, more secure classifiers can be
designed by employing regularization terms that promote enclosure of the legitimate class; *e.g.*, by penalizing "blind spots" - regions with low $p(\mathbf{x})$ - classified

as legitimate. Alternatively, one may explicitly model the attack distribution, as in [4]; or add the generated attack samples to the training set. Nevertheless, improving security probably must be balanced with a higher FP rate.

In our example applications, the feature representations could be *inverted* to obtain a corresponding real-world objects (*e.g.*, spam emails, or PDF files); *i.e.*, it is straightforward to manipulate the given real-world object to obtain the desired feature vector $\mathbf{x}^*$ of the *optimal* attack. However, in practice some complex feature mappings can not be easily inverted; *e.g.*, n-gram features [11]. Another idea would be to modify the real-world object at each step of the gradient descent to obtain a sample in the feature space which is as close as possible to the sample that would be obtained at the next attack iteration. A similar technique has been already exploited by [6] to overcome the pre-image problem.

Other interesting extensions of our work may be to (*i*) consider more effective strategies such as those proposed by [19, 22] to build a small but representative set of surrogate data; and (*ii*) improve the classifier estimate $\hat{g}(\mathbf{x})$. To this end, one may exploit ensemble techniques such as bagging or the random subspace method to train several classifiers and then average their output.

Acknowledgments. This work has been partly supported by the project CRP-18293, L.R. 7/2007, Bando 2009, and by the project "Advanced and secure sharing of multimedia data over social networks in the future Internet" (CUP F71J11000690002), both funded by Regione Autonoma della Sardegna. Davide Maiorca gratefully acknowledges Regione Autonoma della Sardegna for the financial support of his PhD scholarship. Blaine Nelson thanks the Alexander von Humboldt Foundation for providing additional financial support. The opinions expressed in this paper are solely those of the authors and do not necessarily reflect the opinions of any sponsor.

References

1. Adobe: PDF Reference, sixth edn. version 1.7
2. Barreno, M., Nelson, B., Sears, R., Joseph, A.D., Tygar, J.D.: Can machine learning be secure? In: ASIACCS 2006: Proc. of the 2006 ACM Symp. on Information, Computer and Comm. Security, pp. 16–25. ACM, New York (2006)
3. Biggio, B., Fumera, G., Roli, F.: Multiple classifier systems for robust classifier design in adversarial environments. Int'l J. of Machine Learning and Cybernetics 1(1), 27–41 (2010)
4. Biggio, B., Fumera, G., Roli, F.: Design of robust classifiers for adversarial environments. In: IEEE Int'l Conf. on Systems, Man, and Cybernetics (SMC), pp. 977–982 (2011)
5. Biggio, B., Fumera, G., Roli, F.: Security evaluation of pattern classifiers under attack. IEEE Trans. on Knowl. and Data Eng. 99(PrePrints), 1 (2013)
6. Biggio, B., Nelson, B., Laskov, P.: Poisoning attacks against support vector machines. In: Langford, J., Pineau, J. (eds.) 29th Int'l Conf. on Mach. Learn. (2012)
7. Brückner, M., Scheffer, T.: Stackelberg games for adversarial prediction problems. In: Knowl. Disc. and D. Mining (KDD), pp. 547–555 (2011)

8. Brückner, M., Kanzow, C., Scheffer, T.: Static prediction games for adversarial learning problems. J. Mach. Learn. Res. 13, 2617–2654 (2012)
9. Dalvi, N., Domingos, P., Mausam, S.S., Verma, D.: Adversarial classification. In: 10th ACM SIGKDD Int'l Conf. on Knowl. Discovery and Data Mining (KDD), pp. 99–108 (2004)
10. Dekel, O., Shamir, O., Xiao, L.: Learning to classify with missing and corrupted features. Mach. Learn. 81, 149–178 (2010)
11. Fogla, P., Sharif, M., Perdisci, R., Kolesnikov, O., Lee, W.: Polymorphic blending attacks. In: Proc. 15th Conf. on USENIX Sec. Symp. USENIX Association, CA (2006)
12. Globerson, A., Roweis, S.T.: Nightmare at test time: robust learning by feature deletion. In: Cohen, W.W., Moore, A. (eds.) Proc. of the 23rd Int'l Conf. on Mach. Learn., vol. 148, pp. 353–360. ACM (2006)
13. Golland, P.: Discriminative direction for kernel classifiers. In: Neu. Inf. Proc. Syst (NIPS), pp. 745–752 (2002)
14. Huang, L., Joseph, A.D., Nelson, B., Rubinstein, B., Tygar, J.D.: Adversarial machine learning. In: 4th ACM Workshop on Art. Int. and Sec (AISec 2011), Chicago, IL, USA, pp. 43–57 (2011)
15. Kloft, M., Laskov, P.: Online anomaly detection under adversarial impact. In: Proc. of the 13th Int'l Conf. on Art. Int. and Stats (AISTATS), pp. 405–412 (2010)
16. Kolcz, A., Teo, C.H.: Feature weighting for improved classifier robustness. In: Sixth Conf. on Email and Anti-Spam (CEAS), Mountain View, CA, USA (2009)
17. Laskov, P., Kloft, M.: A framework for quantitative security analysis of machine learning. In: AISec 2009: Proc. of the 2nd ACM Works. on Sec. and Art. Int., pp. 1–4. ACM, New York (2009)
18. LeCun, Y., Jackel, L., Bottou, L., Brunot, A., Cortes, C., Denker, J., Drucker, H., Guyon, I., Müller, U., Säckinger, E., Simard, P., Vapnik, V.: Comparison of learning algorithms for handwritten digit recognition. In: Int'l Conf. on Art. Neu. Net., pp. 53–60 (1995)
19. Lowd, D., Meek, C.: Adversarial learning. In: Press, A. (ed.) Proc. of the Eleventh ACM SIGKDD Int'l Conf. on Knowl. Disc. and D. Mining (KDD), Chicago, IL, pp. 641–647 (2005)
20. Maiorca, D., Giacinto, G., Corona, I.: A pattern recognition system for malicious pdf files detection. In: MLDM, pp. 510–524 (2012)
21. Nelson, B., Barreno, M., Chi, F.J., Joseph, A.D., Rubinstein, B.I.P., Saini, U., Sutton, C., Tygar, J.D., Xia, K.: Exploiting machine learning to subvert your spam filter. In: LEET 2008: Proc. of the 1st USENIX Work. on L.-S. Exp. and Emerg. Threats, pp. 1–9. USENIX Association, Berkeley (2008)
22. Nelson, B., Rubinstein, B.I., Huang, L., Joseph, A.D., Lee, S.J., Rao, S., Tygar, J.D.: Query strategies for evading convex-inducing classifiers. J. Mach. Learn. Res. 13, 1293–1332 (2012)
23. Platt, J.: Probabilistic outputs for support vector machines and comparison to regularized likelihood methods. In: Smola, A., Bartlett, P., Schölkopf, B., Schuurmans, D. (eds.) Adv. in L. M. Class, pp. 61–74 (2000)
24. Smutz, C., Stavrou, A.: Malicious pdf detection using metadata and structural features. In: Proc. of the 28th Annual Comp. Sec. App. Conf., pp. 239–248 (2012)
25. Šrndić, N., Laskov, P.: Detection of malicious pdf files based on hierarchical document structure. In: Proc. 20th Annual Net. & Dist. Sys. Sec. Symp. (2013)
26. Young, R.: 2010 IBM X-force mid-year trend & risk report. Tech. rep., IBM (2010)

The Top-k Frequent Closed Itemset Mining Using Top-k SAT Problem

Said Jabbour, Lakhdar Sais, and Yakoub Salhi

CRIL - CNRS, Université d'Artois, France
F-62307 Lens Cedex, France
{jabbour,sais,salhi}@cril.fr

Abstract. In this paper, we introduce a new problem, called Top-k SAT, that consists in enumerating the Top-k models of a propositional formula. A Top-k model is defined as a model with less than k models preferred to it with respect to a preference relation. We show that Top-k SAT generalizes two well-known problems: the partial Max-SAT problem and the problem of computing minimal models. Moreover, we propose a general algorithm for Top-k SAT. Then, we give the first application of our declarative framework in data mining, namely, the problem of enumerating the Top-k frequent closed itemsets of length at least min ($\mathcal{FCIM}^k_{min}$). Finally, to show the nice declarative aspects of our framework, we encode several other variants of $\mathcal{FCIM}^k_{min}$ into the Top-k SAT problem.

Keywords: Data Mining, Itemset Mining, Satisfiability.

1 Introduction

The problem of mining frequent itemsets is well-known and essential in data mining, knowledge discovery and data analysis. It has applications in various fields and becomes fundamental for data analysis as datasets and datastores are becoming very large. Since the first article of Agrawal [1] on association rules and itemset mining, the huge number of works, challenges, datasets and projects show the actual interest in this problem (see [2] for a recent survey of works addressing this problem). Important progress has been achieved for data mining and knowledge discovery in terms of implementations, platforms, libraries, etc. As pointed out in [2], several works deal with designing highly scalable data mining algorithms for large scale datasets. An important problem of itemset mining and data mining problems, in general, concerns the huge size of the output, from which it is difficult for the user to retrieve relevant informations. Consequently, for practical data mining, it is important to reduce the size of the output, by exploiting the structure of the itemsets data. Computing for example, closed, maximal, condensed, discriminative itemset patterns are some of the well-known and useful techniques. Most of the works on itemset mining require the specification of a minimum support threshold λ. This constraint allows the user to control at least to some extent the size of the output by mining only itemsets covering at least λ transactions. However, in practice, it is difficult for users to provide an appropriate threshold. As pointed out in [3], a too small threshold may lead to the generation of a huge number of itemsets, whereas a too

H. Blockeel et al. (Eds.): ECML PKDD 2013, Part III, LNAI 8190, pp. 403–418, 2013.

high value of the threshold may result in no answer. In [3], based on a total ranking between patterns, the authors propose to mine the n most interesting itemsets of arbitrary length. In [4], the proposed task consists in mining Top-k frequent closed itemsets of length greater than a given lower bound min, where k is the desired number of frequent closed itemsets to be mined, and min is the minimal length of each itemset. The authors demonstrate that setting the minimal length of the itemsets to be mined is much easier than setting the usual frequency threshold. Since the introduction of Top-k mining, several research works investigated its use in graph mining (e.g. [5,6]) and other datamining tasks (e.g. [7,8]). This new framework can be seen as a nice way to mine the k preferred patterns according to some specific constraints or measures. Starting from this observation, our goal in this paper is to define a general logic based framework for enumerating the Top-k preferred patterns according to a predefined preference relation.

The notion of preference has a central role in several disciplines such as economy, operations research and decision theory in general. Preferences are relevant for the design of intelligent systems that support decisions. Modeling and reasoning with preferences play an increasing role in Artificial Intelligence (AI) and its related fields such as nonmonotonic reasoning, planning, diagnosis, configuration, constraint programming and other areas in knowledge representation and reasoning. For example, in nonmonotonic reasoning the introduction of preferential semantics by Shoham [9] gives an unifying framework where nonmonotonic logic is reduced to a standard logic with a preference relation (order) on the models of that standard logic. Several models for representing and reasoning about preferences have been proposed. For example, soft constraints [10] are one of the most general way to deal with quantitative preferences, while CP-net (Conditional Preferences networks) [11] is most convenient for qualitative preferences. There is a huge literature on preferences (see [12,13,14] for a survey at least from the AI perspective). In this paper we focus on qualitative preferences defined by a preference relation on the models of a propositional formula. Preferences in propositional satisfiability (SAT) has not received a lot of attention. In [15], a new approach for solving satisfiability problems in the presence of qualitative preferences on literals (defined as partial ordered set) is proposed. The authors particularly show how DPLL procedure can be easily adapted for computing optimal models induced by the partial order. The issue of computing optimal models using DPLL has also been investigated in SAT [16]. paper we propose a new framework, where the user is able to control through a parameter k the output by searching only for the top-k preferred models.

The contribution of this paper is twofold. Firstly, we propose a generic framework for dealing with qualitative preferences in propositional satisfiability. Our qualitative preferences are defined using a reflexive and transitive relation (preorder) over the models of a propositional formula. Such preference relation on models is first used to introduce a new problem, called Top-k SAT, defined as the problem of enumerating the Top-k models of a propositional formula. Here a Top-k model is defined as a model with no more than k-1 models preferred to it with respect to the considered preference relation. Then, we show that Top-k SAT generalizes the two well-known problems, the partial Max-SAT problem and the problem of generating minimal models. We also define a particular preference relation that allows us to introduce a general algorithm for computing the Top-k models.

Secondly, we introduce the first application of our declarative framework to data mining. More precisely, we consider the problem of mining Top-k frequent closed itemsets of minimum length min [17]. In this problem, the minimum support threshold usually used in frequent itemset mining is not known, while the minimum length can be set to 0 if one is interested in itemsets of arbitrary length. In itemset mining, the notion of Top-k frequent itemsets is introduced as an alternative to finding the appropriate value for the minimum support threshold. It is also an elegant way to control the size of the output. Consequently, itemset mining is clearly a nice application of our new defined Top-k SAT problem. In this paper, we provide a SAT encoding and we show that computing the Top-k closed itemsets of length at least min corresponds to computing the Top-k models of the obtained propositional formula. Finally, to show the nice declarative aspects of our framework, we encode several other variants of this data mining problem as Top-k SAT problems. Finally, preliminary experiments on some datasets show the feasibility of our proposed approach.

2   Preliminary Definitions and Notations

In this section, we describe the Boolean satisfiability problem (SAT) and some necessary notations. We consider the conjunctive normal form (CNF) representation for the propositional formulas. A *CNF formula* Φ is a conjunction of clauses, where a *clause* is a disjunction of literals. A *literal* is a positive (p) or negated ($\neg p$) propositional variable. The two literals p and $\neg p$ are called *complementary*. A CNF formula can also be seen as a set of clauses, and a clause as a set of literals. Let us recall that any propositional formula can be translated to CNF using linear Tseitin's encoding [18]. We denote by $Var(\Phi)$ the set of propositional variables occuring in Φ.

An *interpretation* $\mathcal{M}$ of a propositional formula Φ is a function which associates a value $\mathcal{M}(p) \in \{0, 1\}$ (0 corresponds to *false* and 1 to *true*) to the variables p in a set V such that $Var(\Phi) \subseteq V$. A *model* of a formula Φ is an interpretation $\mathcal{M}$ that satisfies the formula. The *SAT problem* consists in deciding if a given CNF formula admits a model or not.

We denote by $\bar{l}$ the complementary literal of l. More precisely, if $l = p$ then $\bar{l}$ is $\neg p$ and if $l = \neg p$ then $\bar{l}$ is p. For a set of literals L, $\bar{L}$ is defined as $\{\bar{l} \mid l \in L\}$. Moreover, we denote by $\overline{\mathcal{M}}$ ($\mathcal{M}$ is an interpretation over $Var(\Phi)$) the clause $\bigvee_{p \in Var(\Phi)} s(p)$, where $s(p) = p$ if $\mathcal{M}(p) = 0$, $\neg p$ otherwise. Let Φ be a CNF formula and $\mathcal{M}$ an interpretation over $Var(\Phi)$. We denote by $\mathcal{M}(\Phi)$ the set of clauses satisfied by $\mathcal{M}$. Let us now consider a set X of propositional variables such that $X \subseteq Var(\Phi)$. We denote by $\mathcal{M} \cap X$ the set of variables $\{p \in X | \mathcal{M}(p) = 1\}$. Moreover, we denote by $\mathcal{M}_{|X}$ the restriction of the model $\mathcal{M}$ to X.

3   Preferences and Top-k Models

Let Φ be a propositional formula and Λ_Φ the set of all its models. A preference relation $\succeq$ over Λ_Φ is a reflexive and transitive binary relation (a preorder). The statement $\mathcal{M} \succeq$

$\mathcal{M}'$ means that $\mathcal{M}$ is at least as preferred as $\mathcal{M}'$. We denote by $P(\varPhi, \mathcal{M}, \succeq)$ the subset of $\Lambda_\varPhi$ defined as follows:

$$P(\varPhi, \mathcal{M}, \succeq) = \{\mathcal{M}' \in \Lambda_\varPhi \mid \mathcal{M}' \succ \mathcal{M}\}$$

where $\mathcal{M}' \succ \mathcal{M}$ means that $\mathcal{M}' \succeq \mathcal{M}$ holds but $\mathcal{M} \succeq \mathcal{M}'$ does not. It corresponds to all models that are strictly preferred to $\mathcal{M}$.

We now introduce an equivalence relation $\approx_X$ over $P(\varPhi, \mathcal{M}, \succeq)$, where X is a set of propositional variables. It is defined as follows:

$$\mathcal{M}' \approx_X \mathcal{M}'' \text{ iff } \mathcal{M}' \cap X = \mathcal{M}'' \cap X$$

Thus, the set $P(\varPhi, \mathcal{M}, \succeq)$ can be partitioned into a set of equivalence classes by $\approx_X$, denoted by $[P(\varPhi, \mathcal{M}, \succeq)]^X$. In our context, this equivalence relation is used to take into consideration only a subset of propositional variables. For instance, we introduce new variables in Tseitin's translation [18] of propositional formula to CNF, and such variables are not important in the case of some preference relations.

Definition 1 (Top-k Model). *Let $\varPhi$ be a propositional formula, $\mathcal{M}$ a model of $\varPhi$, $\succeq$ a preference relation over the models of $\varPhi$ and X a set of propositional variables. $\mathcal{M}$ is a Top-k model w.r.t. $\succeq$ and X iff $|[P(\varPhi, \mathcal{M}, \succeq)]^X| \leq k - 1$.*

Let us note that the number of the Top-k models of a formula is not necessarily equal to k. Indeed, it can be strictly greater or smaller than k. For instance, if a formula is unsatisfiable, then it does not have a Top-k model for any $k \geq 1$. Furthermore, if the considered preference relation is a total order, then the number of Top-k models is always smaller than or equal to k.

It is easy to see that we have the following *monotonicity property*: if $\mathcal{M}$ is a Top-k model and $\mathcal{M}' \succeq \mathcal{M}$, then $\mathcal{M}'$ is also a Top-k model.

Top-k SAT problem. Let $\varPhi$ be propositional formula, $\succeq$ a preference relation over the models of $\varPhi$, X a set of propositional variables and k a strictly positive integer. The Top-k SAT problem consists in computing a set $\mathcal{L}$ of Top-k models of $\varPhi$ with respect to $\succeq$ and X satisfying the two following properties:

1. for all $\mathcal{M}$ Top-k model, there exists $\mathcal{M}' \in \mathcal{L}$ such that $\mathcal{M} \approx_X \mathcal{M}'$; and
2. for all $\mathcal{M}$ and $\mathcal{M}'$ in $\mathcal{L}$, if $\mathcal{M} \neq \mathcal{M}'$ then $\mathcal{M} \not\approx_X \mathcal{M}'$.

The two previous properties come from the fact that we are only interested in the truth values of the variables in X. Indeed, the first property means that, for all Top-k model, there is a model in $\mathcal{L}$ equivalent to it with respect to $\approx_X$. Moreover, the second property means that $\mathcal{L}$ does not contain two equivalent Top-k models.

In the following definition, we introduce a particular type of preference relations, called δ-preference relation, that allows us to introduce a general algorithm for computing Top-k models.

Definition 2. *Let Φ be a formula and $\succeq$ a preference relation on the models of Φ. Then $\succeq$ is a δ-preference relation, if there exists a polytime function $f_{\succeq}$ from Boolean interpretations to the set of CNF formulae such that, for all $\mathcal{M}$ model of Φ and for all $\mathcal{M}'$ Boolean interpretation, $\mathcal{M}'$ is a model of $\Phi \wedge f_{\succeq}(\mathcal{M})$ iff $\mathcal{M}'$ is a model of Φ and $\mathcal{M} \not\succeq \mathcal{M}'$.*

Note that, given a model $\mathcal{M}$ of a CNF formula Φ, $f_{\succeq}(\mathcal{M})$ is a formula such that when added to Φ together with $\overline{\mathcal{M}}$, the models of the resulting formula are different from $\mathcal{M}$ and they are at least as preferred as $\mathcal{M}$. Intuitively, this can be seen as a way to introduce a lower bound during the enumeration process. From now, we only consider δ-preference relations.

3.1 Top-k SAT and Partial MAX-SAT

In this section, we show that the Top-k SAT problem generalizes the Partial MAX-SAT problem (e.g. [19]). In Partial MAX-SAT each clause is either relaxable (soft) or non-relaxable (hard). The objective is to find an interpretation that satisfies all the hard clauses together with the maximum number of soft clauses. The MAX-SAT problem is a particular case of Partial MAX-SAT where all the clauses are relaxable.

Let $\Phi = \Phi_h \wedge \Phi_s$ be a partial MAX-SAT instance such that Φ_h is the hard part and Φ_s the soft part. The relation denoted by $\succeq_{\Phi_s}$ corresponds to preference relation defined as follows: for all $\mathcal{M}$ and $\mathcal{M}'$ models of Φ_h defined over $Var(\Phi_h \wedge \Phi_s)$, $\mathcal{M} \succeq_{\Phi_s} \mathcal{M}'$ if and only if $|\mathcal{M}(\Phi_s)| \geq |\mathcal{M}'(\Phi_s)|$.

Note that $\succeq_{\Phi_s}$ is a δ-preference relation. Indeed, we can define $f_{\succeq_{\Phi_s}}$ as follows:

$$f_{\succeq_{\Phi_s}}(\mathcal{M}) = (\bigwedge_{C \in \Phi_s} p_C \leftrightarrow C) \wedge \sum_{C \in \Phi_s} p_C \geq |\mathcal{M}(\Phi_s)|$$

where p_C for $C \in \Phi_s$ are fresh propositional variables.

The Top-1 models of Φ_h with respect to $\succeq_{\Phi_s}$ and $Var(\Phi)$ correspond to the set of all solutions of Φ in Partial Max-SAT. Naturally, they are the most preferred models with respect to $\succeq_{\Phi_s}$, and that means they satisfy Φ_h and satisfy the maximum number of clauses in Φ_s. Thus, the Top-k SAT problem can be seen as a generalization of Partial MAX-SAT.

The formula $f_{\succeq_{\Phi_s}}(\mathcal{M})$ involves the well-known cardinality constraint (0/1 linear inequality). Several polynomial encodings of this kind of constraints into a CNF formula have been proposed in the literature. The first linear encoding of general linear inequalities to CNF has been proposed by Warners [20]. Recently, efficient encodings of the cardinality constraint to CNF have been proposed, most of them try to improve the efficiency of constraint propagation (e.g. [21,22]).

3.2 Top-k SAT and X-minimal Model Generation Problem

Let $\mathcal{M}$ and $\mathcal{M}'$ be two Boolean interpretations and X a set of propositional variables. Then, $\mathcal{M}$ is said to be smaller than $\mathcal{M}'$ with respect to X, written $\mathcal{M} \preceq_X \mathcal{M}'$, if

---

Algorithm 1: Top-k

Input: a CNF formula Φ, a preorder relation $\succeq$, an integer $k \geq 1$, and a set X of Boolean variables
Output: A set of Top-k models $\mathcal{L}$

```
1  Φ' ← Φ;
2  L ← ∅;                                              /* Set of all Top-k models */
3  while (solve(Φ')) do                                /* M is a model of Φ' */
4      if (∃M' ∈ L.M ≈_X M'  &  M ≻ M') then
5          │  replace(M, M', L);
6      else if (∀M' ∈ L.M ≉_X M'  &  |preferred(M, L)| < k) then
7          │  S ← min_top(k, L);
8          │  add(M, L);
9          │  remove(k, L);
10         │  S ← min_top(k, L) \ S;
11         │  Φ' ← Φ' ∧ ⋀_{M'∈S} f_{⪰}(M');
12     else
13         │  Φ' ← Φ' ∧ f_{⪰}(M)
14     Φ' ← Φ' ∧ M̄;
15 return L;
```

$\mathcal{M} \cap X \subseteq \mathcal{M}' \cap X$. We now consider $\preceq_X$ as a preference relation, i.e., $\mathcal{M} \preceq_X \mathcal{M}'$ means that $\mathcal{M}$ is at least as preferred as $\mathcal{M}'$.

We now show that $\preceq_X$ is a δ-preference relation. We can define $f_{\preceq_X}$ as follows:

$$f_{\preceq_X}(\mathcal{M}) = \left(\bigvee_{p \in \mathcal{M} \cap X} \overline{p} \right) \vee \bigwedge_{p' \in X \setminus \mathcal{M}} \overline{p'}$$

Absolutely, $\mathcal{M}'$ is a model of a formula $\Phi \wedge \overline{\mathcal{M}} \wedge f_{\preceq_X}(\mathcal{M})$ if and only if $\mathcal{M}'$ is a model of Φ, $\mathcal{M}' \neq \mathcal{M}$, and either $\mathcal{M}' \cap X = \mathcal{M} \cap X$ or $(\mathcal{M} \cap X) \setminus (\mathcal{M}' \cap X) \neq \emptyset$. The two previous statements mean that $\mathcal{M} \not\prec_X \mathcal{M}'$. In fact, if $\mathcal{M}'$ satisfies $\bigwedge_{p' \in X \setminus \mathcal{M}} \overline{p'}$, then $\mathcal{M}' \cap X \subseteq \mathcal{M} \cap X$ holds. Otherwise, $\mathcal{M}'$ satisfies $\bigvee_{p \in \mathcal{M} \cap X} \overline{p}$ and that means that $(\mathcal{M} \cap X) \setminus (\mathcal{M}' \cap X) \neq \emptyset$. This latter statement expresses that either $\mathcal{M}' \cap X \subset \mathcal{M} \cap X$ or $\mathcal{M}$ and $\mathcal{M}'$ are incomparable with respect to $\preceq_X$.

Let Φ be a propositional formula, X a set of propositional variables and $\mathcal{M}$ a model of Φ. Then $\mathcal{M}$ is said to be an X-*minimal model* of Φ if there is no model strictly smaller than $\mathcal{M}$ with respect to $\preceq_X$. In [23], it was shown that finding an X-minimal model is $P^{NP[O(log(n))]}$-hard, where n is the number of propositional variables.

The set of all X-minimal models corresponds to the set of all top-1 models with respect to $\preceq_X$ and $Var(\Phi)$. Indeed, if $\mathcal{M}$ is a top-1 model, then there is no model $\mathcal{M}'$ such that $\mathcal{M}' \prec_X \mathcal{M}$, and that means that $\mathcal{M}$ is an X-minimal model. In this context, let us note that computing the set of Top-k models for $k \geq 1$ can be seen as a generalization of X-minimal model generation problem.

3.3 An Algorithm for Top-k SAT

In this section, we describe our algorithm for computing Top-k models in the case of the δ-preference relations (Algorithm 1). The basic idea is simply to use the formula $f_{\succeq}(\mathcal{M})$ associated to a model $\mathcal{M}$ to obtain models that are at least as preferred as $\mathcal{M}$. This algorithm takes as input a CNF formula Φ, a preference relation $\succeq$, a strictly positive integer k, and a set X of propositional variables allowing to define the equivalence relation $\approx_X$. It has as output a set $\mathcal{L}$ of Top-k models of Φ satisfying the two properties given in the definition of the Top-k SAT problem.

Algorithm Description. In the while-loop, we use lower bounds for finding optimal models. These lower bounds are obtained by using the fact that the preorder relation considered is a δ-preference relation. In each step, the lower bound is integrated by using the formula:

$$\bigwedge_{\mathcal{M}' \in S} f_{\succeq}(\mathcal{M}')$$

- **Lines 4 – 5.** Let us first mention that the procedure replace$(\mathcal{M}, \mathcal{M}', \mathcal{L})$ replaces $\mathcal{M}'$ with $\mathcal{M}$ in $\mathcal{L}$. We apply this replacement because there exists a model $\mathcal{M}'$ in $\mathcal{L}$ which is equivalent to $\mathcal{M}'$ and $\mathcal{M}$ allows to have a better bound.
- **Lines 6 – 11.** In the case where $\mathcal{M}$ is not equivalent to any model in $\mathcal{L}$ and the number of models in $\mathcal{L}$ preferred to it is strictly less than k ($|\text{preferred}(\mathcal{M}, \mathcal{L})|$ $< k$), we add $\mathcal{M}$ to $\mathcal{L}$ (add$(\mathcal{M}, \mathcal{L})$). Note that S contains first the models of $\mathcal{L}$ before adding $\mathcal{M}$ that have exactly $k - 1$ models preferred to them in this set. After adding $\mathcal{M}$ to $\mathcal{L}$, we remove from $\mathcal{L}$ the models that are not Top-k, i.e., they have more than $k - 1$ models in $\mathcal{L}$ that are strictly preferred to them (remove$(k, \mathcal{L})$). Next, we modify the content of S. Note that the elements of S before adding $\mathcal{M}$ are used as bounds in the previous step. Hence, in order to avoid adding the same bound several times, the new content of S corresponds to the models in $\mathcal{L}$ that have exactly $k - 1$ models preferred to them in $\mathcal{L}$ (min_top$(k, \mathcal{L})$) deprived of the elements of the previous content of S. In line 11, we integrate lower bounds in Φ' by using the elements of S. Indeed, for all model $\mathcal{M}$ of a formula $\Phi' \wedge \bigwedge_{\mathcal{M}' \in S} f_{\succeq}(\mathcal{M}')$, $\mathcal{M}' \not\succ \mathcal{M}$ holds, for any $\mathcal{M}' \in S$.
- **Lines 12 – 13.** In the case where $\mathcal{M}$ is not a Top-k model, we integrate its associated lower bound.
- **Line 14.** This instruction enables us to avoid finding the same model in two different steps of the while-loop.

Proposition 1. *Algorithm 1 (Top-k) is correct.*

Proof. The proof of the partial correctness is based on the definition of the δ-preference relation. Indeed, the function $f_{\succeq}$ allows us to exploit bounds to systematically improve the preference level of the models. As the number of models is bounded, adding the negation of the found model at each iteration leads to an unsatisfiable formula. Consequently the algorithm terminates.

As explained in the algorithm description, we use lower bounds for finding optimal models. These bounds are obtained by using the function $f_{\succeq}$.

4   Total Preference Relation

We here provide a second algorithm for computing Top-k models in the case of the total δ-preference relations (Algorithm 2). Let us recall that a δ-preference relation $\succeq$ is total if, for all models $\mathcal{M}$ and $\mathcal{M}'$, we have $\mathcal{M} \succeq \mathcal{M}'$ or $\mathcal{M}' \succeq \mathcal{M}$.

Our algorithm in this case is given in Algorithm 2:

- **Lines 3 – 8.** In this part, we compute a set $\mathcal{L}$ of k different models of Φ such that, for all $\mathcal{M}, \mathcal{M}' \in \mathcal{L}$ with $\mathcal{M} \neq \mathcal{M}'$, we have $\mathcal{M} \not\approx_X \mathcal{M}'$. Indeed, if $\mathcal{M}$ is a model of Φ and $\mathcal{M}'$ is a model of $\Phi \wedge \overline{\mathcal{M}^1_{|X}} \wedge \cdots \wedge \overline{\mathcal{M}^n_{|X}} \wedge \overline{\mathcal{M}_{|X}}$, then it is trivial that $\mathcal{M} \not\approx_X \mathcal{M}'$.

- **Line 9.** Note that the set $min(\mathcal{L})$ corresponds to the greatest subset of $\mathcal{L}$ satisfying the following property: for all $\mathcal{M} \in min(\mathcal{L})$, there is no model in $\mathcal{L}$ which is strictly less preferred than $\mathcal{M}$. The assignment in this line allows us to have only models that are at least as preferred as an element of $min(\mathcal{L})$. Indeed, we do not need to consider the models that are less preferred than the elements of $min(\mathcal{L})$ because it is clear that they are not Top-k models. Note that all the elements of $min(\mathcal{L})$ are equivalent with respect to the equivalence relation $\approx$ induced by $\succeq$, since this preorder relation is total.

- **Line 10 – 21.** This while-loop is similar to that in Algorithm 1 (Top-k). We only remove the condition $|preferred(\mathcal{M}, \mathcal{L})| < k$ and replace $min\_top(k, \mathcal{L})$ with $min(\mathcal{L})$. In fact, since the preference relation $\succeq$ is a total preorder, it is obvious that we have $|preferred(\mathcal{M}, \mathcal{L})| < k$ because of the lower bounds added previously. Moreover, as $\succeq$ is total, the set of removed models by $remove(k, \mathcal{L})$ (Line 16) is either the empty set or $min(\mathcal{L})$.

Proposition 2. *Algorithm 2 (Top-k^T) is correct.*

Correctness of this algorithm is obtained from that of the algorithm Top-k and the fact that the considered δ-preference relation is total.

5   An Application of Top-k SAT in Data Mining

The problem of mining frequent itemsets is well-known and essential in data mining [1], knowledge discovery and data analysis. Note that several data mining tasks are closely related to the itemset mining problem such as the ones of association rule mining, frequent pattern mining in sequence data, data clustering, etc. Recently, De Raedt et al. in [24,25] proposed the first constraint programming (CP) based data mining framework for itemset mining. This new framework offers a declarative and flexible representation model. It allows data mining problems to benefit from several generic and efficient CP solving techniques. This first study leads to the first CP approach for itemset mining displaying nice declarative opportunities.

In itemset mining problem, the notion of Top-k frequent itemsets is introduced as an alternative to finding the appropriate value for the minimum support threshold. In this section, we propose a SAT-based encoding for enumerating all closed itemsets. Then we use this encoding in the Top-k SAT problem for computing all Top-k frequent closed itemsets.

Algorithm 2: Top-k^T

Input: a CNF formula Φ, a total preorder relation $\succeq$, an integer $k \geq 1$, and a set X of Boolean variables
Output: the set of all Top-k models $\mathcal{L}$

1 $\Phi' \leftarrow \Phi$;
2 $\mathcal{L} \leftarrow \emptyset$;   /* Set of all Top-$k$ models */
3 **for** $(i \leftarrow 0$ to $k - 1)$ **do**
4 | **if** (solve(Φ')) **then**
5 | | add($\mathcal{M}, \mathcal{L}$);   /* $\mathcal{M}$ is a model of Φ' */
6 | | $\Phi' \leftarrow \Phi' \wedge \overline{\mathcal{M}_{|X}}$;
7 | **else**
8 | | **return** $\mathcal{L}$;
9 $\Phi' \leftarrow \Phi \wedge \bigwedge_{\mathcal{M} \in \mathcal{L}} \overline{\mathcal{M}} \wedge \bigwedge_{\mathcal{M}' \in min(\mathcal{L})} f_{\succeq}(\mathcal{M}')$;
10 **while** (solve(Φ')) **do**   /* $\mathcal{M}$ is a model of Φ' */
11 | **if** ($\exists \mathcal{M}' \in \mathcal{L}.\mathcal{M} \approx_X \mathcal{M}'$ & $\mathcal{M} \succ \mathcal{M}'$) **then**
12 | | replace($\mathcal{M}, \mathcal{M}', \mathcal{L}$);
13 | **else if** ($\forall \mathcal{M}' \in \mathcal{L}.\mathcal{M} \not\approx_X \mathcal{M}'$) **then**
14 | | $S \leftarrow min(\mathcal{L})$;
15 | | add($\mathcal{M}, \mathcal{L}$);
16 | | remove($k, \mathcal{L}$);
17 | | $S \leftarrow min(\mathcal{L}) \setminus S$;
18 | | $\Phi' \leftarrow \Phi' \wedge \bigwedge_{\mathcal{M}' \in S} f_{\succeq}(\mathcal{M}')$;
19 | **else**
20 | | $\Phi' \leftarrow \Phi' \wedge f_{\succeq}(\mathcal{M})$
21 | $\Phi' \leftarrow \Phi' \wedge \overline{\mathcal{M}}$;
22 **return** $\mathcal{L}$;

5.1 Problem Statement

Let $\mathcal{I}$ be a set of *items*. A *transaction* is a couple (tid, I) where tid is the *transaction identifier* and I is an *itemset*, i.e., $I \subseteq \mathcal{I}$. A *transaction database* is a finite set of transactions over $\mathcal{I}$ where, for all two different transactions, they do not have the same transaction identifier. We say that a transaction (tid, I) *supports* an itemset J if $J \subseteq I$.

The *cover* of an itemset I in a transaction database $\mathcal{D}$ is the set of transaction identifiers in $\mathcal{D}$ supporting I: $\mathcal{C}(I, \mathcal{D}) = \{tid \mid (tid, J) \in \mathcal{D}, I \subseteq J\}$. The *support* of an itemset I in $\mathcal{D}$ is defined by: $\mathcal{S}(I, \mathcal{D}) = \mid \mathcal{C}(I, \mathcal{D}) \mid$. Moreover, the frequency of I in $\mathcal{D}$ is defined by: $\mathcal{F}(I, \mathcal{D}) = \frac{\mathcal{S}(I, \mathcal{D})}{|\mathcal{D}|}$.

For instance, consider the following transaction database $\mathcal{D}$:

tid	itemset
1	a, b, c, d
2	a, b, e, f
3	a, b, c, m
4	a, c, d, f, j
5	j, l
6	d
7	d, j

In this database, we have $\mathcal{S}(\{a, b, c\}, \mathcal{D}) = |\{1, 3\}| = 2$ and $\mathcal{F}(\{a, b\}, \mathcal{D}) = \frac{3}{7}$.

Let $\mathcal{D}$ be a transaction database over $\mathcal{I}$ and λ a minimum support threshold. The *frequent itemset mining problem* consists in computing the following set:

$$\mathcal{FIM}(\mathcal{D}, \lambda) = \{I \subseteq \mathcal{I} \mid \mathcal{S}(I, \mathcal{D}) \geq \lambda\}$$

Definition 3 (Closed Itemset). *Let $\mathcal{D}$ be a transaction database (over $\mathcal{I}$) and I an itemset ($I \subseteq \mathcal{I}$) such that $\mathcal{S}(I, \mathcal{D}) \geq 1$. I is closed if for all itemset J such that $I \subset J$, $\mathcal{S}(J, \mathcal{D}) < \mathcal{S}(I, \mathcal{D})$.*

One can easily see that all frequent itemsets can be obtained from the closed frequent itemsets by computing their subsets. Since the number of closed frequent itemsets is smaller than or equal to the number of frequent itemsets, enumerating all closed itemsets allows us to reduce the size of the output.

In this work, we mainly consider the problem of mining Top-k frequent closed itemsets of minimum length min. In this problem, we consider that the minimum support threshold λ is not known.

Definition 4 ($\mathcal{FCIM}_{min}^{k}$). *Let k and min be strictly positive integers. The problem of mining Top-k frequent closed itemsets consists in computing all closed itemsets of length at least min such that, for each one, there exist no more than $k - 1$ closed itemsets of length at least min with supports greater than its support.*

5.2   SAT-Based Encoding for $\mathcal{FCIM}_{min}^{k}$

We now propose a Boolean encoding of $\mathcal{FCIM}_{min}^{k}$. Let $\mathcal{I}$ be a set of items, $\mathcal{D} = \{(0, t_i), \ldots, (n - 1, t_{n-1})\}$ a transaction database over $\mathcal{I}$, and k and min are strictly positive integers. We associate to each item a appearing in $\mathcal{D}$ a Boolean variable p_a. Such Boolean variables encode the candidate itemset $I \subseteq \mathcal{I}$, i.e., $p_a = true$ iff $a \in I$. Moreover, for all $i \in \{0, \ldots, n - 1\}$, we associate to the i-th transaction a Boolean variable b_i.

We first propose a constraint allowing to consider only the itemsets of length at least min. It corresponds to a cardinality constraint:

$$\sum_{a \in \mathcal{I}} p_a \geq min \tag{1}$$

We now introduce a constraint allowing to capture all the transactions where the candidate itemset does not appear:

$$\bigwedge_{i=0}^{n-1} (b_i \leftrightarrow \bigvee_{a \in \mathcal{I} \backslash t_i} p_a) \tag{2}$$

This constraint means that b_i is true if and only if the candidate itemset is not in t_i.

By the following constraint, we force the candidate itemset to be closed:

$$\bigwedge_{a \in \mathcal{I}} (\bigwedge_{i=0}^{n-1} \overline{b_i} \rightarrow a \in t_i) \rightarrow p_a \tag{3}$$

Intuitively, this formula means that if $\mathcal{S}(I) = \mathcal{S}(I \cup \{a\})$ then $a \in I$ holds. Thus, it allows us to obtain models that correspond to closed itemsets.

In this context, computing the Top-k closed itemsets of length at least min corresponds to computing the Top-k models of (1), (2) and (3) with respect to $\succeq_B$ and $X = \{p_a | a \in \mathcal{I}\}$, where $B = \{b_0, \dots, b_{n-1}\}$ and $\succeq_B$ is defined as follows: $\mathcal{M} \succeq_B \mathcal{M}'$ if and only if $|\mathcal{M}(B)| \leq |\mathcal{M}'(B)|$. This preorder relation is a δ-preference relation. Indeed, one can define $f_{\succeq_B}$ as follows:

$$f_{\succeq_B}(\mathcal{M}) = (\sum_{i=0}^{n-1} b_i \leq |\mathcal{M}(B)|)$$

Naturally, this formula allows us to have models corresponding to closed itemsets with supports greater or equal to the support of the closed itemset obtained from $\mathcal{M}$.

5.3 Some Variants of $\mathcal{FCIM}_{min}^k$

In this section, our goal is to illustrate the nice declarative aspects of our proposed framework. To this end, we simply consider slight variations of the problem, and show that their encodings can be obtained by simple modifications.

Variant 1 ($\mathcal{FCIM}_{max}^k$). In this variant, we consider the problem of mining Top-k closed itemsets of length at most max. Our encoding in this case is obtained by adding to (2) and (3) the following constraint:

$$\sum_{a \in \mathcal{I}} p_a \leq max \tag{4}$$

In this case, we use the δ-preference relation $\succeq_B$ defined previously.

Variant 2 ($\mathcal{FCIM}_{\lambda}^k$). Let us now propose an encoding of the problem of mining Top-k closed itemsets of supports at least λ (minimal support threshold). In this context, a Top-k closed itemset is a closed itemset such that, for each one, there exist no more than k - 1 closed itemsets of length greater than its length. Our encoding in this case is obtained by adding to (2) and (3) the following constraint:

$$\sum_{i=0}^{n} \overline{b_i} \geq \lambda \tag{5}$$

The preference relation used in this case is $\succeq_I$ defined as follows: $\mathcal{M} \succeq_I \mathcal{M}'$ if and only if $|\mathcal{M}(I)| \geq |\mathcal{M}'(I)|$. It is a δ-preference relation because $f_{\succeq_I}$ can be defined as follows:

$$f_{\succeq_I}(\mathcal{M}) = \sum_{a \in I} p_a \geq |\mathcal{M}(I)|$$

Variant 3 ($\mathcal{FMIM}_\lambda^k$). We consider here a variant of the problem of mining maximal frequent itemsets. It consists in enumerating Top-k maximal itemsets of supports at least λ and for each one, there exist no more than k - 1 maximal itemsets of length greater than its length. Our encoding of this problem consists of (2) and (5). We use in this case the δ-preference relation $\succeq_I$.

6   Experiments

This section evaluates the performance of our Algorithm for Top-k SAT empirically. The primary goal is to assess the declarativity and the effectiveness of our proposed framework. For this purpose, we consider the problem $\mathcal{FCIM}_{min}^k$ of computing the Top-k frequent closed itemsets of minimum length min described above.

For our experiments, we implemented the Algorithm 1 (Top-k) on the top of the state-of-the-art SAT solver MiniSAT 2.2 [1]. In our SAT encoding of $\mathcal{FCIM}_{min}^k$, we used the sorting networks, one of the state-of-the-art encoding of the cardinality constraint (0/1 linear inequality) to CNF proposed in [26].

We considered a variety of datasets taken from the FIMI repository [2] and CP4IM [3]. All the experiments were done on Intel Xeon quad-core machines with 32GB of RAM running at 2.66 Ghz. For each instance, we used a timeout of 4 hours of CPU time.

The table 1 details the characteristics of the different transaction databases ($\mathcal{D}$). The first column mentions the name of the considered instance. In the second and third column, we give the size of $\mathcal{D}$ in terms of number of transactions (#trans) and number of items (#items) respectively. The fourth column shows the density (dens) of the transaction database, defined as the percentage of 1's in $\mathcal{D}$. The panel of datasets ranges from sparse (e.g. mushroom) to dense ones (e.g. Hepatitis). Finally, in the two last columns, we give the size of the CNF encoding (#vars, #clauses) of $\mathcal{FCIM}_{min}^k$. As we can see, our proposed encoding leads to CNF formula of reasonable size. The maximum size is obtained for the instance *connect* (67 815 variables and 5 877 720 clauses).

In order to analyze the behavior of our Top-k algorithm on $\mathcal{FCIM}_{min}^k$, we conducted two kind of experiments. In the first one, we set the minimum length min of the itemsets to 1, while the value of k is varied from 1 to 10000. In the second experiment, we fix the parameter k to 10, and we vary the minimal length min from 1 to the maximum size of the transactions.

Results for a representative set of datasets are shown in Figure 1 (log scale). The other instances present similar behavior. As expected, the CPU time needed for computing the Top-k models increase with k. For the *connect* dataset, our algorithm fails to compute the Top-k models for higher value of $k > 1000$ in the time limit of 4 hours. This figure clearly shows that finding the Top-k models (the most interesting ones) can be computed efficiently for small values of k. For example, on all datasets the top-10 models are computed in less than 100 seconds of CPU time. When a given instance contains a huge number of frequent closed itemsests, the Top-k problem offers an alternative to the user to control the size of the output and to get the most preferred models.

---

[1] MiniSAT: http://minisat.se/

[2] FIMI: http://fimi.ua.ac.be/data/

[3] CP4IM: http://dtai.cs.kuleuven.be/CP4IM/datasets/

Table 1. Characteristics of the datasets

instance	#trans	#items	$dens(\%)$	#vars	#clauses
zoo-1	101	36	44	173	2196
Hepatitis	137	68	50	273	4934
Lymph	148	68	40	284	6355
audiology	216	148	45	508	17575
Heart-cleveland	296	95	47	486	15289
Primary-tumor	336	31	48	398	5777
Vote	435	48	33	531	14454
Soybean	650	50	32	730	22153
Australian-credit	653	125	41	901	48573
Anneal	812	93	45	990	39157
Tic-tac-toe	958	27	33	1012	18259
german-credit	1000	112	34	1220	73223
Kr-vs-kp	3196	73	49	3342	121597
Hypothyroid	3247	88	49	3419	143043
chess	3196	75	49	3346	124797
splice-1	3190	287	21	3764	727897
mushroom	8124	119	18	8348	747635
connect	67558	129	33	67815	5877720

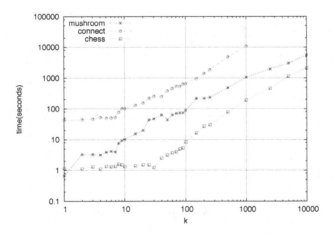

Fig. 1. $\mathcal{FCIM}_1^k$ results for different values of k

In Figure 2, we show the results obtained on the hardest instance from Table 1. On splice-1, the algorithm fails to solve the problem under the time limit for $k > 20$.

In our second experiment, our goal is to show the behavior of our algorithm when varying the minimum length. In Figure 3, we give the results obtained on the three representative datasets (mushroom, connect and chess) when k is fixed to 10 and min is varied from 1 to the maximum size of the transactions. The problem is easy at both the under-constrained (small values of min - many Top-k models) and the over-constrained (high values of min - small number of Top-k models) regions. For the connect dataset, the algorithm fails to solve the problem for $min > 15$ under the time limit. For all the other datasets, the different curves present a pick of difficulty for medium values of the minimal length of the itemsets.

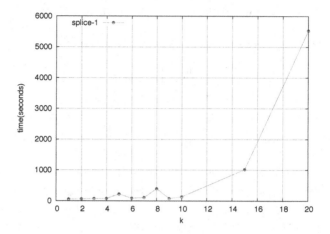

Fig. 2. Hardest $\mathcal{FCIM}_1^k$ instance

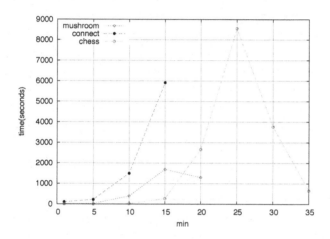

Fig. 3. $\mathcal{FCIM}_{min}^{10}$ results for different values of min

7   Conclusion and Perspectives

In this paper, we introduce a new problem, called Top-k SAT, defined as the problem of enumerating the Top-k models of a propositional formula. A Top-k model is a model having no more than k-1 models preferred to it with respect to the considered preference relation. We also show that Top-k SAT generalizes the two well-known problems: the partial Max-SAT problem and the problem of computing minimal models. A general algorithm for this problem is proposed and evaluated on the problem of enumerating top-k frequent closed itemsets of length at least min.

While our new problem of computing the Top-k preferred models in Boolean satisfiability is flexible and declarative, there are a number of questions that deserve further research efforts. One direction is the study of (preferred/Top-k) model enumeration algorithm so as to achieve a further speedup of the runtime. This fundamental problem has not received a lot of attention in the SAT community, except some interesting works on enumerating minimal/preferred models.

Acknowledgments. This work was supported by the French Research Agency (ANR) under the project DAG "Declarative Approaches for Enumerating Interesting Patterns" - Défis program 2009.

References

1. Agrawal, R., Imielinski, T., Swami, A.N.: Mining association rules between sets of items in large databases. In: ACM SIGMOD International Conference on Management of Data, pp. 207–216. ACM Press, Baltimore (1993)
2. Tiwari, A., Gupta, R., Agrawal, D.: A survey on frequent pattern mining: Current status and challenging issues. Inform. Technol. J 9, 1278–1293 (2010)
3. Fu, A.W.-C., Kwong, R.W.-W., Tang, J.: Mining N-most Interesting Itemsets. In: Ohsuga, S., Raś, Z.W. (eds.) ISMIS 2000. LNCS (LNAI), vol. 1932, pp. 59–67. Springer, Heidelberg (2000)
4. Han, J., Wang, J., Lu, Y., Tzvetkov, P.: Mining top-k frequent closed patterns without minimum support. In: Proceedings of the 2002 IEEE International Conference on Data Mining (ICDM 2002), pp. 211–218. IEEE Computer Society (2002)
5. Ke, Y., Cheng, J., Yu, J.X.: Top-k correlative graph mining. In: Proceedings of the SIAM International Conference on Data Mining (SDM 2009), pp. 1038–1049 (2009)
6. Valari, E., Kontaki, M., Papadopoulos, A.N.: Discovery of top-k dense subgraphs in dynamic graph collections. In: Ailamaki, A., Bowers, S. (eds.) SSDBM 2012. LNCS, vol. 7338, pp. 213–230. Springer, Heidelberg (2012)
7. Lam, H.T., Calders, T.: Mining top-k frequent items in a data stream with flexible sliding windows. In: Proceedings of the 16th ACM SIGKDD International Conference on Knowledge Discovery and Data Mining (KDD 2010), pp. 283–292 (2010)
8. Lam, H.T., Calders, T., Pham, N.: Online discovery of top-k similar motifs in time series data. In: Proceedings of the Eleventh SIAM International Conference on Data Mining, SDM 2011, pp. 1004–1015 (2011)
9. Shoham, Y.: Reasoning about change: time and causation from the standpoint of artificial intelligence. MIT Press, Cambridge (1988)
10. Meseguer, P., Rossi, F., Schiex, T.: 9. In: Soft Constraints. Elsevier (2006)
11. Boutilier, C., Brafman, R.I., Domshlak, C., Poole, D.L., Hoos, H.H.: CP-nets: A Tool for Representing and Reasoning with Conditional Ceteris Paribus Preference Statements. Journal of Artificial Intelligence Research (JAIR) 21, 135–191 (2004)
12. Walsh, T.: Representing and reasoning with preferences. AI Magazine 28(4), 59–70 (2007)
13. Brafman, R.I., Domshlak, C.: Preference Handling - An Introductory Tutorial. AI Magazine 30(1), 58–86 (2009)
14. Domshlak, C., Hüllermeier, E., Kaci, S., Prade, H.: Preferences in AI: An overview. Artificial Intelligence 175(7-8), 1037–1052 (2011)
15. Rosa, E.D., Giunchiglia, E., Maratea, M.: Solving satisfiability problems with preferences. Constraints 15(4), 485–515 (2010)
16. Castell, T., Cayrol, C., Cayrol, M., Berre, D.L.: Using the davis and putnam procedure for an efficient computation of preferred models. In: ECAI, pp. 350–354 (1996)

17. Wang, J., Han, J., Lu, Y., Tzvetkov, P.: TFP: An efficient algorithm for mining top-k frequent closed itemsets. IEEE Transactions on Knowledge Data Engineering 17(5), 652–664 (2005)
18. Tseitin, G.: On the complexity of derivations in the propositional calculus. In: Structures in Constructive Mathematics and Mathematical Logic, Part II, pp. 115–125 (1968)
19. Fu, Z., Malik, S.: On Solving the Partial MAX-SAT Problem. In: Biere, A., Gomes, C.P. (eds.) SAT 2006. LNCS, vol. 4121, pp. 252–265. Springer, Heidelberg (2006)
20. Warners, J.P.: A linear-time transformation of linear inequalities into conjunctive normal form. Information Processing Letters (1996)
21. Bailleux, O., Boufkhad, Y.: Efficient CNF Encoding of Boolean Cardinality Constraints. In: Rossi, F. (ed.) CP 2003. LNCS, vol. 2833, pp. 108–122. Springer, Heidelberg (2003)
22. Sinz, C.: Towards an optimal CNF encoding of boolean cardinality constraints. In: van Beek, P. (ed.) CP 2005. LNCS, vol. 3709, pp. 827–831. Springer, Heidelberg (2005)
23. Cadoli, M.: On the complexity of model finding for nonmonotonic propositional logics. In: 4th Italian Conference on Theoretical Computer Science, pp. 125–139 (1992)
24. Raedt, L.D., Guns, T., Nijssen, S.: Constraint programming for itemset mining. In: ACM SIGKDD, pp. 204–212 (2008)
25. Guns, T., Nijssen, S., De Raedt, L.: Itemset mining: A constraint programming perspective. Artificial Intelligence 175(12-13), 1951–1983 (2011)
26. Eén, N., Sörensson, N.: Translating pseudo-boolean constraints into SAT. JSAT 2(1-4), 1–26 (2006)

A Declarative Framework
for Constrained Clustering

Thi-Bich-Hanh Dao, Khanh-Chuong Duong, and Christel Vrain

Univ. Orléans, ENSI de Bourges, LIFO, EA 4022, F-45067, Orléans, France
{thi-bich-hanh.dao,khanh-chuong.duong,christel.vrain}@univ-orleans.fr

Abstract. In recent years, clustering has been extended to constrained clustering, so as to integrate knowledge on objects or on clusters, but adding such constraints generally requires to develop new algorithms. We propose a declarative and generic framework, based on Constraint Programming, which enables to design clustering tasks by specifying an optimization criterion and some constraints either on the clusters or on pairs of objects. In our framework, several classical optimization criteria are considered and they can be coupled with different kinds of constraints. Relying on Constraint Programming has two main advantages: the declarativity, which enables to easily add new constraints and the ability to find an optimal solution satisfying all the constraints (when there exists one). On the other hand, computation time depends on the constraints and on their ability to reduce the domain of variables, thus avoiding an exhaustive search.

1 Introduction

Clustering is an important task in Data Mining and many algorithms have been designed for it. It has been extended to semi-supervised clustering, so as to integrate previous knowledge on objects that must be or cannot be in the same cluster, and most algorithms have been adapted to handle such information. Other kinds of constraints could be specified by the user, as for instance the sizes of the clusters or their diameters, but classical frameworks are not designed to integrate different types of knowledge. Yet, in the context of an exploratory process, it would be important to be able to express constraints on the task at hand, tuning the model for getting finer solutions. Constrained clustering aims at integrating constraints in the clustering process, but the algorithms are usually developed for handling one kind of constraints. Developing general solvers with the ability of handling different kinds of constraints is therefore of high importance for Data Mining. We propose a declarative and generic framework, based on Constraint Programming, which enables to design a clustering task by specifying an optimization criterion and some constraints either on the clusters or on pairs of objects.

Relying on Constraint Programming (CP) has two main advantages: the declarativity, which enables to easily add new constraints and the ability to find an optimal solution satisfying all the constraints (when there exists one).

H. Blockeel et al. (Eds.): ECML PKDD 2013, Part III, LNAI 8190, pp. 419–434, 2013.

Recent progress in CP have made this paradigm more powerful and several work [1,2,3] have already shown its interest for Data Mining.

In a recent work[4], we have proposed a CP model for constrained clustering, aiming at finding a partition of data that minimizes the maximal diameter of classes. In this paper, we generalize the model with more optimization criteria, namely maximizing the margin between clusters and minimizing the Within-Cluster Sums of Dissimilarities (WCSD). Clustering with WCSD criterion is NP-Hard, since one instance of this problem is the weighted max-cut problem, which is NP-Complete. Recent work [5] has addressed the problem of finding an exact optimum but the size of the database must be quite small, all the more when k is high. We have developed propagation algorithms for the WCSD problem, and experiments show that we are able of finding the optimal solution for small to medium databases. Moreover, adding constraints allows to reduce the computation time.

The main contribution of our paper is a general framework for Constrained Clustering, which integrates different kinds of optimization criteria and finds a global optimum. Moreover, we show that coupling optimization with some types of constraints allows to handle larger databases and can be interesting for the users.

The paper is organized as follows. In Section 2, we give background notions on clustering, constrained clustering and Constraint Programming. Section 3 presents related work. Section 4 is devoted to the model and Section 5 to experiments. A discussion on future work is given in Section 6.

2   Preliminaries

2.1   Clustering

Clustering is the process of grouping data into classes or clusters, so that objects within a cluster have high similarity but are very dissimilar to objects in other clusters. More formally, we consider a database of n objects $\mathcal{O} = \{o_1, \ldots, o_n\}$ and a dissimilarity measure $d(o_i, o_j)$ between two objects o_i and o_j of $\mathcal{O}$. Clustering is often seen as an optimization problem, i.e., finding a partition of the objects which optimizes a given criterion. Optimized criteria may be, among others: (the first four criterion must be minimized whereas the last one must be maximized)

• Within-Cluster Sum of Dissimilarities (WCSD) criterion:
$$E = \sum_{c=1}^{k} \sum_{o_i, o_j \in C_c} d(o_i, o_j)^2$$

• Within-Cluster Sum of Squares (WCSS) criterion, also called the least square criterion (m_c denotes the center of cluster C_c):
$$E = \sum_{c=1}^{k} \sum_{o_i \in C_c} d(m_c, o_i)^2$$

• Absolute-error criterion (r_c denotes a representative object of the cluster C_c):
$$E = \sum_{c=1}^{k} \sum_{o_i \in C_c} d(o_i, r_c)$$

• Diameter-based criterion: $E = max_{c \in [1,k], o_i, o_j \in C_c}(d(o_i, o_j))$. E represents the maximum diameter of the clusters, where the diameter of a cluster is the maximum distance between any two of its objects.

• Margin-based criterion: $E = min_{c < c' \in [1,k], o_i \in C_c, o_j \in C_{c'}}(d(o_i, o_j))$. E is the minimal margin between clusters, where the margin between two clusters $C_c, C_{c'}$ is the minimum value of the distances $d(o_i, o_j)$, with $o_i \in C_c$ and $o_j \in C_{c'}$.

We do not detail here well-known classical clustering algorithms, such as k-means that finds a local optimum of the WCSS criterion, or k-medoids for the absolute-error criterion. The FPF (Furthest Point First) method introduced in [6] is a very efficient method (complexity $O(kn)$) for finding a local optimum of the maximum diameter criterion. Moreover theoretical bounds are given and we show in Section 4 how such bounds can be used to reduce the complexity, when modeling the problem in CP.

Some algorithms do not rely on an optimization algorithm, as for instance DBSCAN [7], based on the notion of density. Parameters are needed to adjust the notion of density. Although our model does not currently allow to simulate the behavior of DBSCAN, the notion of density can be integrated as a constraint on the clustering task.

2.2 Constraint-Based Clustering

Most clustering methods rely on an optimization criterion, and because of the inherent complexity search for a local optimum. Several optima may exist, some may be closer to the one expected by the user. In order to better model the task, but also in the hope of reducing the complexity, user-specified constraints are added, leading to Constraint-based Clustering that aims at finding clusters that satisfy user-specified constraints. User constraints can be classified into cluster-level constraints, specifying requirements on the clusters, or instance-level constraints, specifying requirements on pairs of objects.

Most of the attention has been put on instance-level constraints, first introduced in [8]. Commonly, two kinds of constraints are used. A must-link constraint specifies that two objects o_i and o_j have to appear in the same cluster: $\forall c \in [1,k], o_i \in C_c \Leftrightarrow o_j \in C_c$. A cannot-link constraint specifies that two objects must not be in the same cluster: $\forall c \in [1,k], \neg(o_i \in C_c \wedge o_j \in C_c)$.

Cluster-level constraints impose requirements on the clusters. We give some examples of such constraints that have been integrated to our model.

The minimum capacity constraint requires that each cluster has a number of objects greater than a given threshold α: $\forall c \in [1,k], |C_c| \geq \alpha$, whereas the maximum capacity constraint requires each cluster to have a number of objects inferior to a predefined threshold β: $\forall c \in [1,k], |C_c| \leq \beta$.

The maximum diameter constraint specifies an upper bound on the diameter of the clusters: $\forall c \in [1,k], \forall o_i, o_j \in C_c, d(o_i, o_j) \leq \gamma$ (γ is a given parameter). The minimum margin constraint, also called the δ-constraint in [9], requires the distance between any two points of different clusters to be superior to a given threshold δ: $\forall c \in [1,k], \forall c' \neq c, \forall o_i \in C_c, o_j \in C_{c'}, d(o_i, o_j) \geq \delta$.

The ϵ-constraint introduced in [9] requires for each point o_i to have in its neighborhood of radius ϵ at least another point of the same cluster:

$$\forall c \in [1,k], \forall o_i \in C_c, \exists o_j \in C_c, o_j \neq o_i \text{ and } d(o_i, o_j) \leq \epsilon.$$

This constraint tries to capture the notion of density, introduced in DBSCAN. We propose a new density-based constraint, stronger than the ϵ-constraint: it requires that for each point o_i, its neighborhood of radius ϵ contains at least *MinPts* points belonging to the same cluster as o_i.

In the last ten years, many works have been done to extend classical algorithms for handling must-link and cannot-link constraints, as for instance an extension of COBWEB [8], of k-means [10,11], hierarchical non supervised clustering [12] or spectral clustering [13,14], etc. This is achieved either by modifying the dissimilarity measure, or the objective function or the search strategy. However, to the best of our knowledge there is no general solution to extend traditional algorithms to different types of constraints. Our framework relying on Constraint Programming allows to add directly user-specified constraints.

2.3   Constraint Programming

Constraint Programming is a powerful paradigm to solve combinatorial problems, based on Artificial Intelligence or Operational Research methods. A *Constraint Satisfaction Problem (CSP)* is a triple $\langle X, D, C \rangle$ where $X = \{x_1, x_2, \ldots, x_n\}$ is a set of variables, $D = \{D_1, D_2, \ldots, D_n\}$ is a set of domains ($x_i \in D_i$), $C = \{C_1, C_2, \ldots, C_t\}$ is a set of constraints where each constraint C_i expresses a condition on a subset of X. A solution of a CSP is a complete assignment of values from D_i to each variable x_i that satisfies all the constraints of C. A *Constraint Optimization Problem (COP)* is a CSP with an objective function to be optimized. An optimal solution of a COP is a solution of the CSP that optimizes the objective function. In general, solving a CSP is NP-hard. Nevertheless, the methods used by the solvers enable to efficiently solve a large number of real applications. They rely on constraint propagation and search strategies.

Constraint propagation operates on a constraint c and removes all the values that cannot be part of a solution from the domains of the variables of c. A set of propagators is associated to each constraint, they depend on the kind of consistency required for this constraint (e.g. arc consistency removes all the inconsistent values, while bound consistency modifies only the bounds of the domain). Consistency is chosen by the programmer when the constraint is established. Let us notice that a formula or a mathematic relation can be a constraint in CP only if a set of propagators can be defined on it.

In a CP solver, two steps, constraint propagation and branching, are repeated until a solution is found. Constraints are propagated until a stable state, in which the domains of the variables are reduced as much as possible. If the domains of all the variables are reduced to singletons then a solution is found. If the domain of a variable becomes empty, then there exists no solution with the current partial assignment and the solver backtracks. In the other cases, the solver chooses a variable whose domain is not reduced to a singleton and splits its domain into different parts, thus leading to new branches in the search tree. The solver then explores each branch, activating constraint propagation since the domain of a variable has been modified.

The search strategy can be determined by the programmer. When using a depth-first strategy, the solver orders branches, following the order given by the programmer and explores in depth each branch. For an optimization problem, a branch-and-bound strategy can be integrated to depth-first search: each time a solution, i.e. a complete assignment of variables satisfying the constraints, is found, the value of the objective function for this solution is computed and a new constraint is added, expressing that a new solution must be better than this one. This implies that only the first best solution found is returned by the solver. The solver performs a complete search, pruning only branches that cannot lead to solutions and therefore finds an optimal solution. The choice of variables and of values at each branching is very important, since it may drastically reduce the search space and therefore computation time. For more details, see [15].

Example 1. Let us illustrate by the following COP: find an assignment of letters to digits such that $SEND + MOST = MONEY$, which maximizes $MONEY$. This problem can be modeled by a COP with eight variables S, E, N, D, M, O, T, Y, of the domain the set of digits $\{0, \ldots, 9\}$. Constraints for this problem are:

- the digits for S and M are different from 0: $S \neq 0$, $M \neq 0$
- the values of the variables are pairwise different: alldifferent(S, E, N, D, M, O, T, Y). Let us notice that instead of using a constraint $\neq$ for each pair of variables, the constraint *alldifferent* on a set of variables is used. This is a global constraint in CP, as the following linear constraint.
- $(1000 \times S + 100 \times E + 10 \times N + D) + (1000 \times M + 100 \times O + 10 \times S + T) = 10000 \times M + 1000 \times O + 100 \times N + 10 \times E + Y$
- maximize($10000 \times M + 1000 \times O + 100 \times N + 10 \times E + Y$).

The initial constraint propagation leads to a stable state, with the domains: $D_S = \{9\}$, $D_E = \{2, 3, 4, 5, 6, 7\}$, $D_M = \{1\}$, $D_O = \{0\}$, $D_N = \{3, 4, 5, 6, 7, 8\}$ and $D_D = D_T = D_Y = \{2, 3, 4, 5, 6, 7, 8\}$. Since some domains are not reduced to singletons, branching is then performed. At the end of the search, we get the optimal solution with the assignment $S = 9, E = 7, N = 8, D = 2, M = 1, O = 0, T = 4, Y = 6$, leading to $MONEY = 10876$.

Strategies specifying the way branching is performed are very important. When variables are chosen in the order S, E, N, D, M, O, T, Y and when values are chosen following an increasing order, the search tree is composed of 29 nodes and 7 intermediary solutions (solutions satisfying all the constraints, better than the previous ones found but not optimal). When variables are chosen in the order S, T, Y, N, D, E, M, O, the search tree has only 13 nodes and 2 intermediary solutions.

3   Related Work

Recent work [16,17] has proposed to use Constraint Programming for conceptual clustering. The problem is then formalized as the search of frequent, pairwise non overlapping k-patterns that cover the whole dataset. Several optimization criteria are considered as maximizing the minimal size of the clusters or minimizing the difference between the sizes of classes. These approaches can only be applied to qualitative databases, whereas our approach can handle all kinds of data, as soon as a dissimilarity measure is defined on data. Another approach is based on Integer Linear Programming [18,19], where a set of candidate clusters must be known beforehand and the model searches for the best clustering among the subset of clusters. It has been experimented in the context of conceptual clustering, based on frequent patterns. This framework is less convenient for clustering in general since finding a good set of candidate clusters is difficult as the number of candidate clusters is exponential in the number of objects. A SAT framework [20] is proposed for constrained clustering, but only for a 2-class problem ($k = 2$). Several kinds of constraints are considered: must-link and cannot-link constraints on instances, constraints on cluster diameters and margins. Based on SAT, the algorithm allows to obtain a global optimum. Our approach is more general, since the number of classes is not limited to 2, and several optimization criteria as well as a larger class of constraints are considered.

Clustering with the presented criteria is NP-Hard, most algorithms are heuristics. For instance, k-means finds a local optimum for the WCSS criterion. There are few exact algorithms for the WCSD and WCSS criteria: they rely on lower bounds, which must be computed in a reasonable time and finding such bounds is a difficult subtask. The best known exact method for both WCSD and the maximum diameter criterion is a repetitive branch-and-bound algorithm [5]. This algorithm is efficient when the number k of groups is small; it solves the problem first with $k + 1$ objects, then with $k + 2$ objects and so on, until all n objects are considered. When solving large problems, smaller problems are solved for quickly calculating good lower bounds. The authors give the size n of the databases that can be handled: $n = 250$ for the minimum diameter criterion, $n = 220$ for the WCSS criterion, and only $n = 50$ with k up to 5 or 6 for the WCSD criterion. For the WCSS criterion, the best known exact method is a recent column generation algorithm [21]. The method solves problems with $n = 2300$, however, the number of objects per group (n/k) must be small, roughly equal to 10, in order to have a reasonable computation time. To the best of our knowledge, there exists no exact algorithm for WCSD or WCSS criterion that integrates user-constraints.

4   A CP Framework for Constrained Clustering

We present a CP model for constrained clustering. As input, we have a dataset of n points and a dissimilarity measure between pairs of points, denoted by $d(i, j)$. Without loss of generality, we suppose that points are indexed and named by their index. The number of clusters is fixed by the user and we aim at finding a partition of data into k clusters, satisfying a set of constraints specified by the user and optimizing a given criterion.

4.1   A Constraint-Based Model

Variables. For each cluster $c \in [1, k]$, the point with the smallest index is considered as the representative point of the cluster[1]. An integer variable $I[c]$ is introduced, its value is the index of the representative point of c; the domain of $I[c]$ is therefore the interval $[1, n]$. Assigning a point to a cluster becomes assigning the point to the representative of the cluster. Therefore, for each point $i \in [1, n]$, an integer variable $G[i] \in [1, n]$ is introduced: $G[i]$ is the representative point of the cluster which contains the point i.

Let us for instance suppose that we have 7 points $o_1, \ldots, o_7$ and that we have 2 clusters, the first one composed of o_1, o_2, o_4 and the second one composed of the remaining points. The points are denoted by their integer (o_1 is denoted by 1, o_2 by 2 and so on). Then $I[1] = 1$ and $I[2] = 3$ (since 1 is the smallest index among $\{1, 2, 4\}$ and 3 is the smallest index among $\{3, 5, 6, 7\}$), $G[1] = G[2] = G[4] = 1$ (since 1 is the representative of the first cluster) and $G[3] = G[5] = G[6] = G[7] = 3$ (since 3 is the representative of the second cluster).

A variable is introduced for representing the optimization criterion. It is denoted by D for the maximal diameter, S for the minimal margin and V for the Within-Cluster Sum of Dissimilarities. It is a real-valued variable, since distance are real numbers. The domains of D and S are the interval whose lower (upper) bound is the minimal (maximal, resp.) distance between any two points. The domain of V is upper-bounded by the sum of the distances between all pairs of points. The clustering task is represented by the following constraints.

Constraints on the representation.

- Each representative belongs to its cluster: $\forall c \in [1, k], G[I[c]] = I[c]$.
- Each point is assigned to a representative: $\forall i \in [1, n], \bigvee_{c \in [1,k]} (G[i] = I[c])$.
  This relation can be expressed by a cardinality constraint in CP:
  $$\forall i \in [1, n], \quad \#\{c \mid I[c] = G[i]\} = 1.$$
- The representative of a cluster is the point in this cluster with the minimal index; in other words, the index i of a point is greater or equal to the index of its representative given by $G[i]$: $\forall i \in [1, n], \quad G[i] \leq i$.

A set of clusters could be differently represented, depending on the order of clusters. For instance, in the previous example, we could have chosen $I[1] = 3$ and $I[2] = 1$, thus leading to another representation of the same set of clusters. To avoid this symmetry, the following constraints are added:

- Representatives are sorted in increasing order: $\forall c < c' \in [1, k], I[c] < I[c']$.
- The representative of the first cluster is the first point: $I[1] = 1$.

Modeling different objective criteria. When minimizing the maximal diameter:

- Two points at a distance greater than the maximal diameter must be in different clusters: $\forall i < j \in [1, n], d(i, j) > D \rightarrow (G[i] \neq G[j])$. $(*)$

---

[1] It allows to have a single representation of a cluster. It must not be confused with the notion of representative in the medoid approach.

- The maximal diameter is minimized: minimize D.

When maximizing the minimal margin between clusters:

- Two points at a distance less than the minimal margin must be in the same cluster: $\forall i < j \in [1, n], d(i, j) < S \rightarrow G[i] = G[j]$.
- The minimal margin is maximized: maximize S.

When minimizing the Within-Cluster Sum of Dissimilarities (WCSD):

- $V = \sum_{i,j \in [1,n]} (G[i] == G[j]) d(i, j)^2$. $(**)$
- The sum value is minimized: minimize V.

Modeling user-defined constraints. All popular user-defined constraints may be straightforwardly integrated:

- Minimal size α of clusters: $\forall c \in [1, k], \#\{i \mid G[i] = I[c]\} \geq \alpha$.
- Maximal size β of clusters: $\forall c \in [1, k], \#\{i \mid G[i] = I[c]\} \leq \beta$.
- A δ-constraint expresses that the margin between two clusters must be at least δ. Therefore, for each $i < j \in [1, n]$ satisfying $d(i, j) < \delta$, we put the constraint: $G[i] = G[j]$.
- A diameter constraint expresses that the diameter of each cluster must be at most γ, therefore for each $i < j \in [1, n]$ such that $d(i, j) > \gamma$, we put the constraint: $G[i] \neq G[j]$.
- A density constraint that we have introduced expresses that each point must have in its neighborhood of radius ϵ, at least $MinPts$ points belonging to the same cluster as itself. So, for each $i \in [1, n]$, the set of points in its ϵ-neighborhood is computed and a constraint is put on its cardinality:
  $$\#\{j \mid d(i, j) \leq \epsilon, G[j] = G[i]\} \geq MinPts.$$
- A must-link constraint on two points i and j is expressed by: $G[i] = G[j]$.
- A cannot-link constraint on i and j is expressed by: $G[i] \neq G[j]$.

Adding such constraints involves other constraints on D or S, as for instance $G[i] = G[j]$ implies $D \geq d(i, j)$ and $G[i] \neq G[j]$ implies $S \leq d(i, j)$.

Search strategy. Let us recall that a solver iterates two steps: constraint propagation and branching when needed. In our model, variables $I[c]$ ($c \in [1, k]$) are instantiated before variables $G[i]$ ($i \in [1, n]$). This means that cluster representatives are first instantiated, allowing constraint propagation to assign some points to clusters; when all the $I[c]$ are instantiated, the variables $G[i]$ whose domains are not singletons are instantiated.

Variables $I[c]$ are chosen from $I[1]$ to $I[k]$. Since the representative is the one with the minimal index in the cluster, values for instantiating each $I[c]$ are chosen in an increasing order. Variables $G[i]$ are chosen so that the ones with the smallest remaining domain are chosen first. For instantiating $G[i]$, the index of the closest representative is chosen first.

4.2 Model Improvement

Using constraint propagation to reduce the search space by deleting values in the domain of variables that cannot lead to a solution, CP solvers perform an exhaustive search, allowing to find an optimal solution. In order to improve the efficiency of the system, different aspects are considered.

Improvement by ordering the points. To instantiate the variables $I[c]$, the values (which are point indices) are chosen in an increasing order. The way points are indexed is therefore really important. Points are then ordered and indexed, so that points that are probably representatives have small index. In order to achieve this, we rely on FPF algorithm, introduced in Section 2. The algorithm is applied with $k = n$ (as many classes as points): a first point is chosen, the second point is the furthest from this point, the third one is the furthest from the two first and so on until all points have been chosen.

Improvement when minimizing the maximal diameter. Let us consider first the case where no user-defined constraints are put in the model. In [6], it is proven that if d_{FPF} represents the maximal diameter of the partition computed by FPF, then it satisfies $d_{opt} \leq d_{FPF} \leq 2d_{opt}$, with d_{opt} the maximal diameter of the optimal solution. This knowledge gives bounds on D: $D \in [d_{FPF}/2, d_{FPF}]$. Moreover, for each pair of points i, j :

- if $d(i, j) < d_{FPF}/2$, the reified constraint (*) on i, j is no longer put,
- if $d(i, j) > d_{FPF}$, the constraint (*) is replaced by: $G[i] \neq G[j]$.

Such a result allows to remove several reified constraints, without modifying the semantics of the model, and thus allows to improve the efficiency of the model, since handling reified constraints requires to introduce new variables.

In the case where user constraints are added, this result is no longer true, since the optimal diameter is in general greater than the optimal diameter d_{opt} obtained without user-constraints. The upper bound is no longer satisfied but the lower bound, namely $d_{FPF}/2$, still holds. Therefore for each pair of points i, j, if $d(i, j) < d_{FPF}/2$, the constraint (*) on i, j is not put.

Improvement when minimizing WCSD. Computing WCSD (**) requires to use a linear constraint on boolean variables ($G[i] == G[j]$). However, a partial assignment of points to clusters does not allow to filter the domain of the remaining values, thus leading to an inefficient constraint propagation. We have proposed a new method for propagating this constraint and filtering the domain of remaining variables. It is out of the scope of this paper, for more details, see [22]. Experiments in Section 5 show that it enables to handle databases that are out of reach of the most recent exact algorithms.

5 Experiments

5.1 Datasets and Methodology

Eleven datasets are used in our experiments. They vary significantly in their size, number of attributes and number of clusters. Nine datasets are from the

Table 1. Properties of data sets used in the experimentation

Dataset	# Objects	# Attributes	# Clusters
Iris	150	4	3
Wine	178	13	3
Glass	214	9	7
Ionosphere	351	34	2
GR431	431	2	not available
GR666	666	2	not available
WDBC	569	30	2
Letter Recognition	600	16	3
Synthetic Control	600	60	6
Vehicle	846	18	4
Yeast	1484	8	10

UCI repository [23]: Iris, Wine, Glass, Ionosphere, WDBC, Letter Recognition, Synthetic Control, Vehicle, Yeast. For the dataset Letter Recognition, only 600 objects are considered from the 20.000 objects of the initial data set, they are composed of the first 200 objects of each class. The datasets GR431 and GR666 are obtained from the library TSPLIB [24]; they contain coordinates of 431 and 666 European cities [25]. These two datasets do not contain information about the number of clusters and we choose $k = 3$ for the tests. Table 1 summarizes information about these datasets. There are few systems aiming at reaching a global optimum. In Subsection 5.2, our model without user-constraints is compared to the Repetitive Branch-and-Bound Algorithm (RBBA) [5][2]. To the best of our knowledge, it is the best exact algorithm for the maximal diameter and WCSD criteria but it does not integrate user-constraints. The distance between objects is the Euclidean distance and the dissimilarity is measured as the squared Euclidean distance. As far as we know, there is no work optimizing the criteria presented in the paper and integrating user-constraints (with $k > 2$). In Subsections 5.3, 5.4 and 5.5, we show the ability of our model to handle different kinds of user-constraints.

Our model is implemented with the Gecode 4.0.0 library[3]. In this version released in 2013, float variables are supported. This property is important to obtain exact optimal value. All the experiments (our model and RBBA) are performed on a PC Intel core i5 with 3.8 GHz and 8 GB of RAM. The time limit for each test is 2 hours.

5.2   Minimizing Maximum Diameter without User-Constraint

Table 2 shows the results for the maximal diameter criterion. The first column gives the datasets, the second column reports the optimal values of the diameter. They are the same for both our model and the RBBA approach, since both

---

[2] The program can be found in http://mailer.fsu.edu/~mbrusco/

[3] http://www.gecode.org

Table 2. Comparison of performance with the maximal diameter criterion

Dataset	Optimal Diameter	CP Framework	RBBA
Iris	2.58	0.03s	1.4s
Wine	458.13	0.3s	2.0s
Glass	4.97	0.9s	42.0s
Ionosphere	8.60	8.6s	> 2 hours
GR431	141.15	0.6s	> 2 hours
GR666	180.00	31.7s	> 2 hours
WDBC	2377.96	0.7s	> 2 hours
Letter Recognition	18.84	111.6s	> 2 hours
Synthetic Control	109.36	56.1s	> 2 hours
Vehicle	264.83	14.9s	> 2 hours
Yeast	0.67	2389.9s	> 2 hours

approaches find the global optimal. The third and fourth columns give the total CPU times (in seconds) for each approach.

The results show that RBBA finds the optimal diameter only for the first three datasets. In [5], the authors mention that their algorithm is effective for databases with less than 250 objects. Table 2 shows that our model is able to find the optimal diameter for a data set with up to $n = 1484$ and $k = 10$. The performance does not only depend on the number of objects n and on the number of groups k, but also on the margin between objects and on the database features. The behavior of our model differs from classical models: for instance, when k increases, the search space is larger and in many approaches, solving such a problem takes more time. Indeed, since there are more clusters, the maximum diameter is smaller, and propagation of the diameter constraint is more effective, thus explaining that in some cases, it takes less computation time. As already mentioned, they may exist several partitions with the same optimal diameter; because of the search strategy of Constraint Optimization Problem in CP, our model finds only one partition with the optimal diameter.

5.3   Minimizing Maximum Diameter with User-Constraints

Let us consider now the behavior of our system with user-constraints considering the Letter Recognition dataset. Figure 1 presents the results we obtain when must-link constraints (generated from the true classes of objects) and a margin constraint δ (the margin between two clusters must be at least δ) is used. The number of must-link constraints varies from 0.01% to 1% the total number of pairs of objects where δ ranges from 3% to 12% the maximum distance between two objects. Regarding to the results, both must-link and margin constraints boost the performance for this data set.

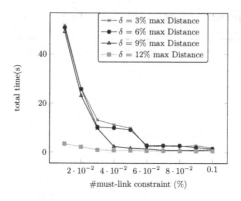

Fig. 1. Must-link and margin constraints

5.4   Minimizing Within-Cluster Sum of Dissimilarities

Minimizing the Within-Cluster Sum of Dissimilarities (WCSD) is a difficult task since the propagation of the sum constraint is less efficient than the propagation of the diameter constraint. Without users-constraints, both our model and the RBBA approach can find the optimal solutions only with the Iris dataset. Our model needs 4174s to complete the search whereas the RBBA takes 3249s. However, with appropriate user-constraints, the performance of our model can be significantly improved.

WCSD and the margin constraint. Let us add a margin constraint δ, where δ ranges from 0% (no constraint) to 12% of the maximum distance between two objects. Table 3 (left) reports the WCSD value of an optimal solution and the total computation time. It shows that when the margin constraint is weak, the optimal WCSD value does not change. But the computation time decreases significantly when this additional constraint becomes stronger. The reason is that the total number of feasible solutions decreases and the search space is reduced. When the margin constraint is weak, propagating this constraint is more time-consuming than its benefits.

Table 3. margin and must-link constraint with dataset Iris

Margin Constraint	WCSD	Total time
no constraint	573.552	4174s
$\delta = 2\%$ max Dist	573.552	1452s
$\delta = 4\%$ max Dist	573.552	84.4s
$\delta = 6\%$ max Dist	573.552	0.3s
$\delta = 8\%$ max Dist	2169.21	0.1s
$\delta = 10\%$ max Dist	2412.43	0.04s
$\delta = 12\%$ max Dist	2451.32	0.04s

# must-link	WCSD	Total time
no constraint	573.552	4174s
0.2%	602.551	1275.1s
0.4%	602.551	35.6s
0.6%	617.012	16.1s
0.8%	622.5	3.5s
1%	622.5	1.6s
100%	646.045	0.04s

WCSD and must-link constraints. Let us now add must-link constraints, where the number of must-link constraints, generated from the true classes of objects, varies from 0.2 to 1% of the total number of pairs. Results are expressed in Table 3 (right), giving the WCSD value and the total computation time. In fact, the optimal value of WCSD, with no information on classes, does not correspond to the WCSD found when considering the partition of this dataset into the 3 defined classes. The more must-link constraints, the less computation time is needed for finding the optimal value, and the closer to the value of WCSD, when considering the 3 initial classes. The reduction of computation time can be easily explained, since when an object is instantiated, objects that must be linked to it are immediately instantiated too. Furthermore, with any kind of additional constraint, the total number of feasible solutions is always equal or less than the case without constraint.

Table 4. Example of appropriate combinations of user-constraints

Data set	User constraints	WCSD	Total time
Wine	margin: $\delta = 1.5\%$ max Distance minimum capacity: $\beta = 30$	1.40×10^6	11.2s
GR666	margin: $\delta = 1.5\%$ max Distance diameter: $\gamma = 50\%$ max Distance	1.79×10^8	12.4s
Letter Recognition	# must-link constraints $= 0.1\%$ total pairs # cannot-link constraints $= 0.1\%$ total pairs margin: $\delta = 10\%$ max Distance	5.84×10^6	11.5s
Vehicle	margin: $\delta = 3\%$ max Distance diameter: $\gamma = 40\%$ max Distance	1.93×10^9	1.6s

WCSD and appropriate user-constraints. Finding an exact solution minimizing the WCSD is difficult. However, with appropriate combination of user-constraints, the performance can be boosted. Table 4 presents some examples where our model can get an optimal solution with different user-constraints, which reduce significantly the search space.

5.5   Interest of the Model Flexibility

Our system finds an optimal solution when there exists one; otherwise no solution is returned. Let us show the interest of combining different kinds of constraints. Figure 2 presents 3 datasets in 2 dimensions, similar to those used in [7].

The first dataset is composed of 4 groups with different diameters. The second one is more difficult, since groups do not have the same shape. The third one contains outliers: outliers are not handled and are therefore integrated in classes.

When optimizing the maximal diameter, the solver tends to find rather homogeneous groups, as shown in Figure 3. Adding a min-margin constraint (with

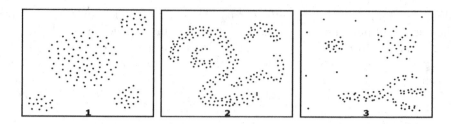

Fig. 2. Datasets

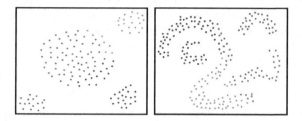

Fig. 3. Max-diameter optimization

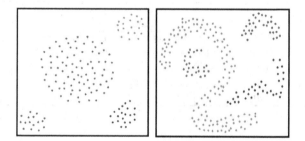

Fig. 4. Maximal diameter optimization + margin constraint

Fig. 5. Diameter opt. (left) Margin opt. (center) Margin opt. + density const. (right)

$\delta = 5\%$ of the maximum distance between pairs of points) improves the quality of the solution (see Figure 4). Let us notice that maximizing the minimum margin allows also to find this solution.

Concerning the third dataset, minimizing the maximum diameter or maximizing the minimum margin do not allow finding a good solution (see Figure 5). The quality of the solution is improved when a density constraint is added with $MintPts = 4$ and $\epsilon = 25\%$ from the maximal distance between pairs of points.

6 Conclusion

We have proposed a declarative framework for Constrained Clustering based on Constraint Programming. It allows to choose among different optimization criteria and to integrate various kinds of constraints. Besides the declarativity of CP, one of its advantage is that it allows to find an optimal solution, whereas most approaches find only local optima. On the other hand, complexity makes it difficult to handle very large databases. Nevertheless, integrating constraints enables to reduce the search space, depending on their ability to filter the domain of variables. Moreover, working on search strategies and on constraint propagation enables to increase the efficiency and to deal with larger problems. We plan to work on the search strategies and on the constraint propagators, thus being able to address larger databases. We do believe that global constraints adapted to the clustering tasks must be developed. From the Data Mining point of view, more optimization criteria should be added.

References

1. De Raedt, L., Guns, T., Nijssen, S.: Constraint programming for itemset mining. In: Proceedings of the 14th ACM SIGKDD International Conference on Knowledge Discovery and Data Mining, pp. 204–212 (2008)
2. De Raedt, L., Guns, T., Nijssen, S.: Constraint Programming for Data Mining and Machine Learning. In: Proc. of the 24th AAAI Conf. on Artificial Intelligence (2010)
3. Boizumault, P., Crémilleux, B., Khiari, M., Loudni, S., Métivier, J.P.: Discovering Knowledge using a Constraint-based Language. CoRR abs/1107.3407 (2011)
4. Dao, T.B.H., Duong, K.C., Vrain, C.: Une approche en PPC pour la classification non supervisée. In: 13e Conférence Francophone sur l'Extraction et la Gestion des Connaissances EGC (2013)
5. Brusco, M., Stahl, S.: Branch-and-Bound Applications in Combinatorial Data Analysis (Statistics and Computing), 1st edn. Springer (2005)
6. Gonzalez, T.: Clustering to minimize the maximum intercluster distance. Theoretical Computer Science 38, 293–306 (1985)
7. Ester, M., Kriegel, H.P., Sander, J., Xu, X.: A Density-Based Algorithm for Discovering Clusters in Large Spatial Databases with Noise. In: Second International Conference on Knowledge Discovery and Data Mining, pp. 226–231 (1996)
8. Wagstaff, K., Cardie, C.: Clustering with instance-level constraints. In: Proc. of the 17th International Conference on Machine Learning, pp. 1103–1110 (2000)

9. Davidson, I., Ravi, S.S.: Clustering with Constraints: Feasibility Issues and the k-Means Algorithm. In: Proc. 5th SIAM Data Mining Conference (2005)
10. Wagstaff, K., Cardie, C., Rogers, S., Schroedl, S.: Constrained k-means clustering with background knowledge. In: Proceedings of the Eighteenth International Conference on Machine Learning, pp. 577–584 (2001)
11. Bilenko, M., Basu, S., Mooney, R.J.: Integrating constraints and metric learning in semi-supervised clustering. In: Proceedings of the Twenty-First International Conference on Machine Learning, pp. 11–18 (2004)
12. Davidson, I., Ravi, S.S.: Agglomerative hierarchical clustering with constraints: Theoretical and empirical results. In: Jorge, A.M., Torgo, L., Brazdil, P.B., Camacho, R., Gama, J. (eds.) PKDD 2005. LNCS (LNAI), vol. 3721, pp. 59–70. Springer, Heidelberg (2005)
13. Lu, Z., Carreira-Perpinan, M.A.: Constrained spectral clustering through affinity propagation. In: 2008 IEEE Conference on Computer Vision and Pattern Recognition, pp. 1–8. IEEE (June 2008)
14. Wang, X., Davidson, I.: Flexible constrained spectral clustering. In: KDD 2010: Proceedings of the 16th ACM SIGKDD International Conference on Knowledge Discovery and Data Mining, pp. 563–572 (2010)
15. Rossi, F., van Beek, P., Walsh, T. (eds.): Handbook of Constraint Programming. Foundations of Artificial Intelligence. Elsevier B.V. (2006)
16. Guns, T., Nijssen, S., De Raedt, L.: k-Pattern set mining under constraints. IEEE Transactions on Knowledge and Data Engineering (2011)
17. Métivier, J.-P., Boizumault, P., Crémilleux, B., Khiari, M., Loudni, S.: Constrained Clustering Using SAT. In: Hollmén, J., Klawonn, F., Tucker, A. (eds.) IDA 2012. LNCS, vol. 7619, pp. 207–218. Springer, Heidelberg (2012)
18. Mueller, M., Kramer, S.: Integer linear programming models for constrained clustering. In: Pfahringer, B., Holmes, G., Hoffmann, A. (eds.) DS 2010. LNCS, vol. 6332, pp. 159–173. Springer, Heidelberg (2010)
19. Schmidt, J., Brändle, E.M., Kramer, S.: Clustering with attribute-level constraints. In: ICDM, pp. 1206–1211 (2011)
20. Davidson, I., Ravi, S.S., Shamis, L.: A SAT-based Framework for Efficient Constrained Clustering. In: SDM, pp. 94–105 (2010)
21. Aloise, D., Hansen, P., Liberti, L.: An improved column generation algorithm for minimum sum-of-squares clustering. Math. Program. 131(1-2), 195–220 (2012)
22. Dao, T.B.H., Duong, K.C., Vrain, C.: Constraint programming for constrained clustering. Technical Report 03, LIFO, Université d'Orléans (2013)
23. Bache, K., Lichman, M.: UCI machine learning repository (2013)
24. Reinelt, G.: TSPLIB - A t.s.p. library. Technical Report 250, Universität Augsburg, Institut für Mathematik, Augsburg (1990)
25. Grötschel, M., Holland, O.: Solution of large-scale symmetric travelling salesman problems. Math. Program. 51, 141–202 (1991)

SNNAP: Solver-Based Nearest Neighbor for Algorithm Portfolios

Marco Collautti, Yuri Malitsky, Deepak Mehta, and Barry O'Sullivan

Cork Constraint Computation Centre,
Department of Computer Science, University College Cork, Ireland
collautt@dei.unipd.it, {y.malitsky,d.mehta,b.osullivan}@4c.ucc.ie

Abstract. The success of portfolio algorithms in competitions in the area of combinatorial problem solving, as well as in practice, has motivated interest in the development of new approaches to determine the best solver for the problem at hand. Yet, although there are a number of ways in which this decision can be made, it always relies on a rich set of features to identify and distinguish the structure of the problem instances. In this paper, we show how one of the more successful portfolio approaches, ISAC, can be augmented by taking into account the past performance of solvers as part of the feature vector. Testing on a variety of SAT datasets, we show how our new formulation continuously outperforms an unmodified/standard version of ISAC.

1   Introduction

It is becoming increasingly recognized in the constraint programming (CP) and satisfiability (SAT) communities that there is no single best solver for all problems. Instead solvers tend to excel on a particular set of instances while offering subpar performance on everything else. This observation has led to the pursuit of approaches that, given a collection of existing solvers, attempt to select the most appropriate one for the problem instances at hand. The way in which these portfolio solvers make the selection, however, varies greatly. One approach can be to train a regression model to predict the performance of each solver, selecting the expected best one [16,15]. Alternatively, a ranking approach can be used over all solvers [6]. It is also possible to train a forest of trees, where each tree makes a decision for every pair of solvers, deciding which of the two is likely to be better, selecting the one voted upon most frequently [17]. Research has also been conducted on creating a schedule of solvers to call in sequence rather than committing to a single one [7,12]. An overview of many of these approaches is presented in [9]. Yet regardless of the implementation, portfolio-based approaches have been dominating the competitions in satisfiability (SAT) [7,16], constraint programming (CP) [12], and quantified Boolean formulae (QBF) [14].

One of the most successful portfolio techniques is referred to as Instance-Specific Algorithm Configuration (ISAC) [8]. Originally the approach was demonstrated to outperform the then reigning regression-based approach, SATzilla, in the SAT domain [16], as well as a number of other existing portfolios [11]. The

H. Blockeel et al. (Eds.): ECML PKDD 2013, Part III, LNAI 8190, pp. 435–450, 2013.
© Springer-Verlag Berlin Heidelberg 2013

approach was later embraced by the 3S solver which won 2 gold medals in the 2011 SAT competition [1]. ISAC's applicability has also been demonstrated in other domains such as set covering (SCP) and mixed integer (MIP) problems [8].

The guiding principle behind ISAC is to cluster the training instances based on a representative feature vector. It is assumed that if the feature vector is good, then the instances within the same cluster will all have similar underlying structure and will therefore yield to the same solver. Up to now, however, ISAC has relied on a pre-specified set of features and an objective-oblivious clustering. This means that any anomalous behavior of the features could result in an improper clustering. Similarly, it cannot be assumed the identified clusters group the instances in an optimal manner. In short, there has been a disconnect between the clustering objective and the performance objective.

For this reason we propose a new approach, SNNAP, in which we try to use the past performance of the solvers in the portfolio as part of the feature vector. This approach redefines the feature vector to automatically encode our desired objective of having similar instances in the same cluster, and is thus able to significantly improve the results of the original algorithm. However, unlike feature landmarking [13,3] where the evaluation of multiple simple algorithms provide insight into the success of more complex algorithms, SNNAP tackles a domain where a problem needs to only be evaluated once, but highly efficiently.

In this paper, we will first give an overview of the ISAC methodology and demonstrate that straightforward feature filtering is not enough to significantly improve performance. We will then show a number of modifications to the employed distance metric that improve the quality of the clusterings, and thus the overall performance. We will conclude by presenting SNAPP, an approach that combines predictive modeling as a way of generating features to be used by dynamic clustering. To demonstrate our results, we use a collection of SAT datasets: one that contains only randomly generated instances, one with only handcrafted instances, one with industrial instances, and finally a mixture of all three.

2   ISAC

The fundamental principle behind ISAC is that instances with similar features are likely to have commonalities in their structure, and that there exists at least one solver that is best at solving that particular structure. Therefore the approach works as presented in Algorithm 1. In the training phase, ISAC is provided with a list of training instances T, their corresponding feature vectors F, and a collection of solvers A.

First, the computed features are normalized so that every feature ranges in [-1,1]. This normalization helps keep all the features at the same order of magnitude, and thereby avoids the larger values being given more weight than smaller values. Using these normalized values, the instances are clustered. Although any clustering approach can be used, in practice *g-means* [5] is employed in order to avoid specifying the desired number of clusters. This clustering approach assumes

---

Algorithm 1. ISAC

1: **function** ISAC-TRAIN(T, F, A)
2:     $(\bar{F}) \leftarrow Normalize(F)$
3:     $(k, C, S) \leftarrow Cluster(T, \bar{F})$
4:     **for all** $i = 1, \ldots, k$ **do**
5:         $BS_i \leftarrow FindBestSolver(T, S_i, A)$
6:     **end for**
7:     **return** (k, C, BS)
8: **end function**

1: **function** ISAC-RUN(x, k, C, BS)
2:     $j \leftarrow FindClosestCluster(k, x, C)$
3:     **return** $BS_j(x)$
4: **end function**

---

that a good cluster is one in which the data follows a Gaussian distribution. Therefore, starting with all instances in the same cluster, g-$means$ iteratively applies 2-means clustering, accepting the partition only if the two new clusters are more Gaussian than their predecessor. Once the instances are clustered, we add an additional step that merges all clusters smaller than a predefined threshold into their neighboring clusters. The final result of the clustering is a set of k clusters S, and a set of cluster centers C. For each cluster we then determine a single solver, which is usually the one that has best average performance on all instances in its cluster. When the procedure is presented with a previously unseen instance x, ISAC assigns it to the nearest cluster and runs the solver designated to that cluster on the instance x.

In practice, this standard version of ISAC has been continuously shown to perform well, commonly outperforming the choice of a single solver on all instances. In many situations ISAC has even outperformed existing state-of-the-art regression-based portfolios [11]. However, the current version of ISAC also accepts the computed clustering on faith, even though it might not be optimal or might not best capture the relationship between problem instances and solvers. It also does not take into account that some of the features might be noisy or misleading. Therefore, the following sections will show the advantages of applying feature filtering. We will then show how the feature vector can be extended to include the performance of solvers on the training instances. This has the effect of increasing the chances that instances for which the same solver performs well are placed into the same cluster. Finally, we extend the approach to combine predictive modeling and dynamic clustering to find the best cluster for each new instance.

3   Experimental Setup

The satisfiability (SAT) domain was selected to test our proposed methodologies due to the large number of diverse instances and solvers that are available. We compiled four datasets using all instances from the 2006 SAT Competition to the present day 2012 SAT Challenge [2]. These four datasets contain 2140, 735, 1098 and 4243 instances, respectively, as follows:

RAND: instances have been generated at random;

HAND: instances are hand-crafted or are transformations from other NP-Complete problems;

INDU: instances come from industrial problems;

ALL: instances are the union of the previous datasets.

We rely on the standard set of 115 features that have been embraced by the SAT community [16]. Specifically, using the feature code made available by UBC,[1] we compute features with the following settings: *-base, -sp, -dia, -cl, -ls, and -lobjois*; but having observed previously that the features measuring computation time are unreliable, we discard those 9. These features cover information such as the number of variables, number of clauses, average number of literals per clause, proportion of positive to negative literals per clause, etc. Finally, we also run 29 of the most current SAT solvers, many of which have individually shown very good performance in past competitions. Specifically, we used:

− clasp-2.1.1_jumpy	− cryptominisat_2011	− sattimep
− clasp-2.1.1_trendy	− eagleup	− sparrow
− ebminisat	− gnoveltyp2	− tnm
− glueminisat	− march_rw	− cryptominisat295
− lingeling	− mphaseSAT	− minisatPSM
− lrglshr	− mphaseSATm	− sattime2011
− picosat	− precosat	− ccasat
− restartsat	− qutersat	− glucose_21
− circminisat	− sapperlot	− glucose_21_modified.
− clasp1	− sat4j-2.3.2	

Each of the solvers was run on every instance with a 5,000 second timeout. We then removed instances that could not be solved by any of the solvers within the allotted time limit. We further removed instances that we deemed too easy, i.e. those where every solver could solve the instance within 15 seconds. This resulted in our final datasets comprising of 1949 Random, 363 Crafted, and 805 industrial instances, i.e. 3117 instances in total.

All the experiments presented in this paper were evaluated using 10-fold cross validation, where we averaged the results over all folds. We then repeat each experiment 10 times to decrease the bias of our estimates of the performance. In our experiments we commonly present both the average and the PAR-10 performance; PAR-10 is a weighted average where every timeout is treated as having taken 10 times the timeout. We also present the percentage of instances not solved.

---

[1] http://www.cs.ubc.ca/labs/beta/Projects/SATzilla/

We evaluate all our results comparing them to three benchmark values:

- *Virtual Best Solver (VBS)*: This is the lower bound of what is achievable with a perfect portfolio that, for every instance, always chooses the solver that results in the best performance.
- *Best Single Solver (BSS)*: This is the desired upper bound, obtained by solving each instance with the solver whose average running time is the lowest on the entire dataset.
- *Instance-Specific Algorithm Configuration (ISAC)*: This is the pure ISAC methodology obtained with the normal set of features and clustering.

Our results will always lie between the *VBS* and the *BSS* with the ultimate goal to improve over *ISAC*. Results will be divided according to their dataset.

4 Applying Feature Filtering

The current standard version of ISAC assumes that all features are equally important. But as was shown in [10], this is often not the case, and it is possible to achieve comparable performance with a fraction of the features, usually also resulting in slightly better overall performance. The original work presented in [10] only considered a rudimentary feed-forward selection. In this paper we utilize four common and more powerful filtering techniques: Chi-squared, information gain, gain ratio, and symmetrical uncertainty. Because all these approaches depend on a classification for each instance, we use the name of the best solver for that purpose.

Chi-squared. The Chi-squared test is a correlation-based filter and makes use of "contingency tables". One advantage of this function is that it does not need the discretization of continuous features. It is defined as:

$$\chi^2 = \sum_{ij} (M_{ij} - m_{ij})^2 / m_{ij} \qquad \text{where } m_{ij} = M_{i.} M_{.j} / m$$

M_{ij} is the number of times objects with feature values $Y = y_j, X = x_i$ appear in a dataset, y_i are classes and x_j are features.

Information gain. Information gain is based on information theory and is often used in decision trees and is based on the calculation of entropy of the data as a whole and for each class. For this ranking function continuous features must be discretized in advance.

Gain ratio. This function is a modified version of the *information gain* and it takes into account the *mutual information* for giving equal weight to features with many values and features with few values. It is considered to be a stable evaluation.

Symmetrical uncertainty. The symmetrical uncertainty is built on top of the mutual information and entropy measures. It is particularly noted for its low bias for multivalued features.

Table 1. Results on the SAT benchmark, comparing the Virtual Best Solver (VBS), the Best Single Solver (BSS), the original ISAC approach (ISAC) and ISAC with different feature filtering techniques: "chi.squared", "information.gain", "symmetrical.uncertainty" and "gain.ratio".

RAND	Runtime - avg (std)	Par 10 - avg (std)	% not solved - avg (std)
BSS	1551 (0)	13154 (0)	25.28 (0)
ISAC	826.1 (6.6)	4584 (40.9)	8.1 (0.2)
chi.squared	1081 (42.23)	7318 (492.7)	14 (1)
information.gain	851.5 (32.33)	5161 (390)	8.7 (0.8)
symmetrical.uncertainty	840.2 (13.15)	4908 (189.5)	8.76 (0.4)
gain.ratio	830.3 (21.3)	4780 (210)	9 (0.4)
VBS	358 (0)	358 (0)	0 (0)

HAND	Runtime - avg (std)	Par 10 - avg (std)	% not solved - avg (std)
BSS	2080 (0)	15987 (0)	30.3 (0)
ISAC	1743 (34.4)	13994 (290.6)	26.5 (0.9)
chi.squared	1544 (37.8)	11771 (435)	23.5 (0.9)
information.gain	1641 (38.9)	12991 (443)	24.3 (0.9)
symmetrical.uncertainty	1686 (27.3)	13041 (336)	25.7 (0.7)
gain.ratio	1588 (43.7)	12092 (545)	22.4 (1)
VBS	400 (0)	400 (0)	0 (0)

INDU	Runtime - avg (std)	Par 10 - avg (std)	% not solved - avg (std)
BSS	871 (0)	4727 (0)	8.4 (0)
ISAC	763.4 (4.7)	3166 (155.6)	5.2 (0.7)
chi.squared	708.1 (25.3)	3252 (218)	5.8 (0.4)
information.gain	712.6 (7.24)	2578 (120)	4.3 (0.3)
symmetrical.uncertainty	716.4 (16.76)	2737 (150)	4.4 (0.3)
gain.ratio	705.4 (19.9)	2697 (284)	4.1 (0.6)
VBS	319 (0)	319 (0)	0 (0)

ALL	Runtime - avg (std)	Par 10 - avg (std)	% not solved
BSS	2015 (0)	4726 (0)	30.9 (0)
ISAC	1015 (10.3)	6447 (92.4)	11.8 (0.2)
chi.squared	1078 (29.7)	7051 (414)	11.79 (0.8)
information.gain	1157 (18.9)	7950 (208)	15 (0.4)
symmetrical.uncertainty	1195 (28.7)	8067 (341)	15.6 (0.7)
gain.ratio	1111 (17.4)	6678 (225)	13.39 (0.5)
VBS	353 (0)	353 (0)	0 (0)

Restricting our filtering approaches to finding the best 15 features, we can see in Table 1 that the results are highly dependent on the dataset taken into consideration. For the random dataset there is no major improvement due to using just a subset of the features. Yet we can achieve almost the same result as the original ISAC by just using a subset of the features calculated using the *gain.ratio* function, a sign that not all the features are needed. We can further observe the improvements are more pronounced in the hand-crafted and industrial datasets. For them, the functions that give the best results are, respectively, *chi.squared* and *gain.ratio*, but in the latter the result is almost identical to the one given by *chi.squared.*

These results show that not all features are necessary for the clustering, and that it is possible to improve performance through the careful selection of the features. However, we also observe that the improvements can sometimes be minor, and are highly dependent on the filtering approach used and the dataset it is evaluated on.

5 Extending the Feature Space

While the original version of ISAC employs Euclidean distance for clustering, there is no reason to believe that this is the best distance function. As an alternative one might learn a weighted distance metric, where the weights are tuned to match the desired similarity between two instances. For example, if two instances have the same best solver, then the distance between these two instances should be small. Alternatively, when a solver performs very well on one instance, but poorly on another, it might be desirable for these instances to be separated by a large distance.

In our initial experiments, we have trained a distance function that attempts to capture this desired behavior. Yet the resulting performance often times proved worse than the standard Euclidean distance. There are a number of reasons for this. First, while we know that some instances should be closer or farther from each other, the ideal magnitude of the distance cannot be readily determined. Second, the effectiveness of the distance function depends on near perfect accuracy since any mistake can distort the distance space. Third, the exact form of the distance function is not known. It is, for example, possible that even though two instances share the same best solver, they should nevertheless be allowed to be in opposite corners of the distance space. We do not necessarily want every instance preferring the same solver to be placed in the same cluster, but instead want to avoid contradictory preferences within the same cluster.

Due to these complications, we propose an alternate methodology for refining the feature vector. Specifically, we propose to add the normalized performance of all solvers as part of the features. In this setting, for each instance, the best performing solver is assigned a value of -1, while the worst performing is assigned to 1. Everything in between is scaled accordingly. The clustering is then done on both the set of the normal features and the new ones. During testing, however, we do not know the performance of any of the solvers beforehand, so we set all those features to 0.

We see in Table 2 that the performance of this approach (called NormTimes ISAC) was really poor: never comparable with the running times of the normal ISAC. The main reason was that we were taking into consideration too many solvers during the computation of the new features.

As an alternative, we consider that matching the performance of all solvers is too constraining. Implicitly ISAC assumes that a good cluster is one where

Table 2. Results on the SAT benchmark, comparing the Best Single Solver (BSS), the Virtual Best Solver (VBS), the original ISAC approach (ISAC), the ISAC approach with extra features coming from the running times: "NormTimes ISAC" has the normalized running times while "BestTwoSolv ISAC" takes into consideration just the best two solvers per each instance.

RAND	Runtime - avg (std)	Par 10 - avg (std)	% not solved - avg (std)
BSS	1551 (0)	13154 (0)	25.28 (0)
ISAC	826.1 (6.6)	4584 (40.9)	8.1 (0.2)
NormTimes ISAC	1940 (-)	15710 (-)	30 (-)
BestTwoSolv ISAC	825.6 (5.7)	4561 (87.8)	8.1 (0.2)
VBS	358 (0)	358 (0)	0 (0)

HAND	Runtime - avg (std)	Par 10 - avg (std)	% not solved - avg (std)
BSS	2080 (0)	15987 (0)	30.3 (0)
ISAC	1743 (34.4)	13994 (290.6)	26.5 (0.9)
NormTimes ISAC	1853 (-)	14842 (-)	28.3 (-)
BestTwoSolv ISAC	1725 (29.2)	13884 (124.4)	26.5 (0.8)
VBS	400 (0)	400 (0)	0 (0)

INDU	Runtime - avg (std)	Par 10 - avg (std)	% not solved - avg (std)
BSS	871 (0)	4727 (0)	8.4 (0)
ISAC	763.4 (4.7)	3166 (155.6)	5.2 (0.7)
NormTimes ISAC	934.3 (-)	5891 (-)	10.8 (-)
BestTwoSolv ISAC	750.5 (2.4)	2917 (157.3)	4.7 (0.4)
VBS	319 (0)	319 (0)	0 (0)

ALL	Runtime - avg (std)	Par 10 - avg (std)	% not solved - avg (std)
BSS	2015 (0)	4727 (0)	30.9 (0)
ISAC	1015 (10.3)	6447 (92.4)	11.8 (0.2)
NormTimes ISAC	1151 (-)	6923 (-)	12.5 (-)
BestTwoSolv ISAC	1019 (11.5)	6484 (172.3)	11.9 (0.3)
VBS	353 (0)	353(0)	0 (0)

the instances all prefer the same solver. For this reason we decided to take into account only the performance of the best two solvers per each instance. This was accomplished by extending the normal set of features with a vector of new features (one per each solver), and assigning a value of 1 to the components corresponding to the best two solvers and 0 to all the others. In the testing set, since we again do not know which are the best two solvers before hand, all the new features are set to the constant value of 0. As can be seen in Table 2, depending on which of the four datasets was used we got different results (this approach is called bestTwoSolv ISAC): we observed a small improvement in the hand-crafted and industrial datasets, while for the other two datasets the results were almost the same as the pure ISAC methodology.

The drawbacks of directly extending the feature vector with the performance of solvers are two-fold. First, the performance of the solvers is not available prior to solving a previously unseen test instance. Secondly, even if the new features are helpful in determining a better clustering, there is usually a large number of original features that might be resisting the desired clustering. Yet even though these extensions to the feature vector did not provide a compelling case to be used instead of the vanilla ISAC approach, they nonetheless supported the assumption that by considering solver performances on the training data it is possible to improve the quality of the overall clustering. This has inspired the Solver-based Nearest Neighbor Approach that we describe in the next section.

6 SNNAP

There are two main takeaway messages from extending the feature vector with solver performances. First, the addition of solver performances can be helpful, but the inclusion of the original features can be disruptive for finding the desired cluster. Second, it is not necessary for us to find instances where the relation of every solver is similar to the current instance. It is enough to just know the best two or three solvers for an instance. Using these two ideas we propose SNNAP which is presented as Algorithm 2.

During the training phase the algorithm is provided with a list of training instances T, their corresponding features vectors F and the running times R of every solver in our portfolio. We then train a single model PM for every solver to predict the expected runtime on a given instance. We have claimed previously that such models are difficult to train properly since any misclassification can result in the selection of the wrong solver. In fact, this was partly why the original version of ISAC outperformed these types of regression-based portfolios. Clusters provide better stability of the resulting prediction of which solver to choose. We are, however, not interested in using the trained model to predict the single best solver to be used on the instance. Instead, we just want to know which solvers are going to behave well on a particular instance.

For training the model, we scale the running times of the solvers on one instance so that the scaled vector will have a mean of 0 and unitary standard deviation. We saw that this kind of scaling is crucial in helping the following phase of prediction. Thus we are not training to predict runtime. We are learning to predict when a solver will perform much better than usual. Doing so, for every instance, every solver that behaves better than one standard deviation from the mean will receive a score less than -1, the solvers which behaves worse than one standard deviation from the mean a score greater of 1, and the others will lie in between. Random forests [4] were used as the prediction model.

In the prediction phase the procedure is presented with a previously unseen instance x, the prediction models PM (one per each solver), the training instances T, their running times R (and the scaled version $\bar{R}$), the portfolio of solvers A and the size of the desired neighborhood k. The procedure first uses the prediction models to infer the performances PR of the solvers on the instance x, using its originally known features. SNNAP then continues to use these performances to compute a distance between the new instance and every training instance, selecting the k nearest among them. The distance calculation takes into

Algorithm 2. Solver-based Nearest Neighbor for Algorithm Portfolios

---

1: **function** SNNAP-TRAIN(T, F, R)
2:     **for all** instances i in T **do**
3:         $\bar{R}_i \leftarrow Scaled(R_i)$
4:     **end for**
5:     **for all** solver j in the portfolio **do**
6:         $PM_j \leftarrow PredictionModel(T, F, \bar{R})$
7:     **end for**
8:     **return** PM
9: **end function**

1: **function** SNNAP-RUN($x, PM, T, R, \bar{R}, A, k$)
2:     $PR \leftarrow Predict(PM, x)$
3:     $dist \leftarrow CalculateDistance(PR, T, \bar{R})$
4:     $neighbors \leftarrow FindClosestInstances(dist, k)$
5:     $j \leftarrow FindBestSolver(neighbors, R)$
6:     **return** $A_j(x)$
7: **end function**

---

account only the scaled running time of the instances of the training set and the predicted performances PR of the different solvers on the instance x. At the end the instance x will be solved using the solver that behaves best (measured as the average running time) on the k neighbors previously chosen.

It is worth highlighting again that we are not trying to predict the running times of the solvers on the instances but, after scaling, we predict a ranking amongst the solvers on a particular instance: which will be the best, which the second best, etc. Moreover, as shown in the next section, we are not interested in learning a ranking among all the solvers, but just among a small subset of them, specifically for each instance which will be the best n solvers.

6.1   Choosing the Distance Metric

The k-nearest neighbors approach is usually used in conjunction with the weighted Euclidean distance; unfortunately the Euclidean distance does not take into account the performances of the solvers in a way that is helpful to us. What is needed is a distance metric that takes into account the performances of the solvers and that would allow the possibility of making some mistakes in the prediction phase without too much prejudice on the performances. Thus the metric should be trained with the goal that the k-nearest neighbors always prefer to be solved by the same solver while instances that prefer different solvers are separated by a large margin.

Given two instances a, b and the running times of the m algorithms in the portfolio A on both of them $R_{a_1}, \ldots R_{a_m}$ and $R_{b_1}, \ldots R_{b_m}$, we identify which are the best n solvers on each ($A_{a_1}, \ldots A_{a_n}$) and ($A_{b_1}, \ldots A_{b_n}$) and define their distance as a Jaccard distance:

$$1 - \frac{|intersection((A_{a_1}, \ldots A_{a_n}), (A_{b_1}, \ldots A_{b_n}))|}{|union((A_{a_1}, \ldots A_{a_n}), (A_{b_1}, \ldots A_{b_n}))|}$$

Using this definition two instances that will prefer the exact same n solvers will have a distance of 0, while instances which prefer completely different solvers will have a distance of 1. Moreover, using this kind of distance metric we are no longer concerned with making small mistakes in the prediction phase: even if we switch the ranking between the best n solvers the distance between two instances will remain the same. In our experiments, we focus on setting $n = 3$, as with higher values the performances degrades.

6.2   Numerical Results

In our approach, with the Jaccard distance metric, for each instance we are interested in knowing which are the n best solvers. In the prediction phase we used random forests which achieved high levels of accuracy: as stated in Table 3 we correctly made 91, 89, 91 and 91% (respectively RAND, HAND, INDU and ALL datasets) of the predictions. We compute these percentages in the following manner. There are 29 predictions made (one per each solver) per each instance, giving us a total of 5626, 1044, and 2320 predictions per category. We define accuracy as the percentage of matches between the predicted best n solvers and the true best n.

Table 3. Statistics of the four datasets used: instances generated at random "RAND", hand-crafted instances "HAND", industrial instances "INDU" and the union of them "ALL".

	RAND	HAND	INDU	ALL
Number of instances considered	1949	363	805	3117
Number of predictions	5626	1044	2320	9019
Accuracy in the prediction phase	91%	89%	91%	91%

Having tried different parameters we use the performance of just the $n = 3$ best solvers in the calculation of the distance metric and a neighborhood size of 60. Choosing a larger number of solvers degrades the results. This is most likely due to scenarios where one instance is solved well by a limited number of solvers, while all the others time out.

As we can see in Table 4 the best improvement, as compared to the standard ISAC, is achieved in the hand-crafted dataset. Not only are the performances improved by 60%, but also the number of unsolved instances is halved; this also has a great impact on the PAR10 evaluation.[2] It is interesting to note that

---

[2] PAR10 score is a penalized average of the runtimes: for each instance that is solved within 5000 seconds (the timeout threshold), the actual runtime in seconds denotes the penalty for that instance. For each instance that is not solved within the time limit, the penalty is set to 50000, which is 10 times the original timeout.

Table 4. Results on the SAT benchmark, comparing the Best Single Solver (BSS), the original ISAC approach (ISAC), our SNNAP approach (SNNAP) (also with feature filtering) and the Virtual Best Solver (VBS)

RAND	*Runtime - avg (std)*	*Par 10 - avg (std)*	*% not solved - avg (std)*
BSS	1551 (0)	13154 (0)	25.28 (0)
ISAC	826.1 (6.6)	4584 (40.9)	8.1 (0.2)
SNNAP	791.4 (15.7)	4119 (207)	7.3 (0.2)
SNNAP + Filtering	723 (9.27)	3138 (76.9)	5.28 (0.1)
VBS	358 (0)	358 (0)	0 (0)

HAND	*Runtime - avg (std)*	*Par 10 - avg (std)*	*% not solved - avg (std)*
BSS	2080 (0)	15987 (0)	30.3 (0)
ISAC	1743 (34.4)	13994 (290.6)	26.5 (0.9)
SNNAP	1063 (33.86)	6741 (405.5)	12.4 (0.4)
SNNAP + Filtering	995.5 (18.23)	6036 (449)	10.5 (0.4)
VBS	400 (0)	400 (0)	0 (0)

INDU	*Runtime - avg (std)*	*Par 10 - avg (std)*	*% not solved - avg (std)*
BSS	871 (0)	4727 (0)	8.4 (0)
ISAC	763.4 (4.7)	3166 (155.6)	5.2 (0.7)
SNNAP	577.6 (21.5)	1776 (220.8)	2.6 (0.4)
SNNAP + Filtering	540 (15.52)	1630 (149)	2.4 (0.4)
VBS	319 (0)	319 (0)	0 (0)

ALL	*Runtime - avg (std)*	*Par 10 - avg (std)*	*% not solved - avg (std)*
BSS	2015 (0)	4727 (0)	30.9 (0)
ISAC	1015 (10.3)	6447 (92.4)	11.8 (0.2)
SNNAP	744.2 (14)	3428 (141.2)	5.8 (0.2)
SNNAP + Filtering	692.9 (7.2)	2741 (211.9)	4.5 (0.1)
VBS	353 (0)	353 (0)	0 (0)

the hand-crafted dataset is the one that proves to be most difficult, in terms of solving time, while being the setting in which we achieved the most improvement.

We also achieved a significant improvement, although lower than that with the Hand-crafted dataset, on the Industrial and ALL ($\sim$ 25%) datasets. Here the number of unsolved instances was also halved. In the random dataset we achieved the lowest improvement but, yet, we were able to overtake significantly the standard ISAC approach.

We have also applied feature filtering to SNNAP and the results are shown in Table 4. Feature filtering is again proving beneficial, significantly improving the results for all our datasets and giving us a clue that not all 115 features are essential. Results in the table have been reported only for the more successful ranking function (gain.ratio for the Random dataset, chi.squared for Hand-crafted and Industrial and the overAll dataset).

These consistent results for SNNAP are encouraging. In particular, it is clear that the dramatic decrease in the number of unsolved instances is highly

Table 5. Matrix for comparing instances solved and not solved using SNNAP and ISAC for the four datasets: RAND, HAND, INDU and ALL. Values are in percentages

RAND				*HAND*		
SNNAP \ ISAC	Solved	Not Solved		SNNAP \ ISAC	Solved	Not Solved
Solved	89.4	3.3		Solved	70.2	17.4
Not Solved	2.5	4.8		Not Solved	3.3	9.1

INDU				*ALL*		
SNNAP \ ISAC	Solved	Not Solved		SNNAP \ ISAC	Solved	Not Solved
Solved	93.9	3.5		Solved	85.8	8.4
Not Solved	0.9	1.7		Not Solved	2.4	3.4

important, as they are key to lowering the average and the PAR10 scores. This result can also be observed in Table 5, where we can see the percentage of instances solved/not solved by each approach. In particular the most significant result is achieved, again, in the HAND dataset where the number of instances not solved by ISAC, but solved by our approach is 17.4% of the overall instances, while the number of instances not solved by our approach but solved by ISAC is only 3.3%. As we can see this difference is also considerable in the other three datasets. Deliberately, we chose to show this matrix only for the version of ISAC and SNNAP without feature filtering as it offers an unbiased comparison between the two approaches, as we have shown that ISAC does not improve after feature filtering.

Another useful statistic is represented by the number of times that an approach is able to select the best solver for a given instance. In the random dataset SNNAP is able to select the best solver for 39% of the instances, as compared with 35% for ISAC. For the Hand-crafted dataset those values are 25% and 17%, respectively, for the Industrial 29% and 21%, respectively, and for the ALL 32% and 26%, respectively. These values suggest that ISAC is already behaving well on the RAND dataset and, for this reason, the improvement achieved with our new approach is smaller in that case, while the improvement is more significant in the other three datasets. These results also show that there is room for further improvement.

The final thing we analyze are the frequencies with which the solvers are chosen by the three strategies: the Virtual Best Solver (VBS), ISAC and SNNAP. Table 6 presents the frequency with which each of the 29 solvers in our portfolio were selected by each strategy, highlighting the best single solver (BSS) for each category. In this table we can see that ISAC tends to favor selecting the Best Single Solver. This is particularly clear in the Hand-crafted dataset: the BSS (15) is chosen only in 8% of the instances by the VBS, while it is the more frequently chosen solver by ISAC. Note also that in the ALL dataset the VBS approach never chooses the BSS, while this solver is one of the top three chosen by ISAC. On the other hand, the more often a solver is chosen by the VBS, the more often

448   M. Collautti et al.

Table 6. Frequencies of solver selections for VBS, ISAC and SNNAP. Results are expressed as percentages and entries with value < 0.5 have been reported as '-'. In bold the top three solvers for each approach are identified. Reported are also the Best Single Solver (BSS) for each dataset. Solvers are: 1: clasp-2.1.1_jumpy, 2: clasp-2.1.1_trendy, 3: ebminisat, 4: glueminisat, 5: lingeling, 6: lrglshr, 7: picosat, 8: restartsat, 9: circminisat, 10: clasp1, 11: cryptominisat_2011, 12: eagleup, 13: gnoveltyp2, 14: march_rw, 15: mphaseSAT, 16: mphaseSATm, 17: precosat, 18: qutersat, 19: sapperlot, 20: sat4j-2.3.2, 21: sattimep, 22: sparrow, 23: tnm, 24: cryptominisat295, 25: minisatPSM, 26: sattime2011, 27: ccasat, 28: glucose_21, 29: glucose_21_modified.

	1	2	3	4	5	6	7	8	9	10	11	12	13	14	15	16	17	18	19	20	21	22	23	24	25	26	27	28	29
RAND																													
BSS	-	-	-	-	-	-	-	-	-	-	-	-	-	-	-	-	-	-	-	-	-	-	-	-	-	-	**100**	-	-
VBS	-	-	-	-	-	-	-	-	-	-	17	6	**19**	-	2	-	-	-	-	3	**18**	5	-	-	-	4	**24**	-	-
ISAC	-	-	6	1	-	-	-	-	-	-	8	-	**20**	-	-	**12**	-	-	-	-	-	-	-	-	-	4	**52**	-	-
SNNAP	-	-	5	-	-	-	-	-	-	-	-	3	1	**27**	2	2	-	-	-	0	**12**	4	-	-	-	2	**41**	-	-
HAND																													
BSS	-	-	-	-	-	-	-	-	-	-	-	-	-	-	**100**	-	-	-	-	-	-	-	-	-	-	-	-	-	-
VBS	-	3	12	-	2	-	**17**	1	**13**	-	4	-	5	7	8	1	3	2	-	-	7	**10**	5	3	1	6	3	-	2
ISAC	-	22	21	-	1	-	12	-	1	-	2	-	1	-	**40**	-	-	-	-	2	-	**26**	-	-	-	6	-	-	-
SNNAP	-	10	11	-	1	-	-	-	**32**	-	-	-	1	-	**25**	-	-	-	-	-	2	4	-	4	-	**19**	-	-	2
INDU																													
BSS	-	-	-	-	**100**	-	-	-	-	-	-	-	-	-	-	-	-	-	-	-	-	-	-	-	-	-	-	-	-
VBS	-	5	13	-	**15**	-	8	3	**25**	-	4	4	-	3	-	1	1	9	-	-	4	-	-	4	5	-	-	**20**	3
ISAC	-	1	-	1	**31**	-	5	-	6	-	**17**	-	2	-	-	-	5	**19**	-	-	-	-	-	2	-	-	**32**	-	-
SNNAP	-	1	-	1	**38**	-	-	-	2	-	-	-	-	-	4	1	-	**19**	-	2	-	-	4	-	-	**19**	-	**29**	-
ALL																													
BSS	-	-	-	-	-	-	-	-	-	-	-	-	-	-	-	**100**	-	-	-	-	-	-	-	-	-	-	-	-	-
VBS	-	2	-	1	4	-	2	2	1	1	3	11	-	**13**	3	-	1	-	-	-	3	**13**	4	2	1	3	**15**	5	1
ISAC	-	6	11	-	14	-	-	-	2	-	-	2	-	**12**	5	**12**	-	-	-	-	-	4	4	-	-	2	**31**	**32**	2
SNNAP	-	5	-	1	10	-	-	-	3	-	-	2	-	**18**	2	5	-	3	-	-	-	**11**	2	3	-	2	**24**	4	3

it is chosen by SNNAP. This big discrepancy between the VBS and BSS in this dataset is one of the reason for the poorer performance of ISAC and one of the reasons for the improvement observed when using SNNAP.

7   Conclusions

Instance-Specific Algorithm Configuration (ISAC) is a successful approach to tuning a wide range of solvers for SAT, MIP, set covering, and others. This approach assumes that the features describing an instance are enough to group instances so that all instances in the cluster prefer the same solver. Yet there is no fundamental reason why this hypothesis should hold. In this paper we show that the assumptions that ISAC makes can be strengthened. We show that not all employed features are useful and that it is possible to achieve similar performance with only a fraction of the features that are available. We then show that it is possible to extend the feature vector to include the past performances of solvers to help guide the clustering process. In the end, however, we introduce an alternative view of ISAC which uses the existing features to predict the best three solvers for a particular instance. Using k-nearest neighbors, the approach then scans the training data to find other instances that preferred the same solvers, and uses them as a dynamically formed training set to select the best solver to use. We call this methodology Solver-based Nearest Neighbors for Algorithm Portfolios (SNNAP).

The benefit of the SNNAP approach over ISAC is that the cluster formulated by the k-NN comprises of instances that are most similar to the new instance, something that ISAC assumes but has no way of enforcing. Additionally, the approach is not as sensitive to incorrect decisions by the predictive model. For example, it does not matter if the ranking of the top three solvers is incorrect, any permutation is acceptable. Furthermore, even if one of the solvers is incorrectly included in the top n, the k-NN generates a large enough training set to find a reasonable solver that is likely to work well in general.

This synergy between prediction and clustering that enforces the desired qualities of our clusters is the reason that SNNAP consistently and significantly outperforms the traditional ISAC methodology. Consequently, this paper presents a solver portfolio for combinatorial problem solving that out-performs the state-of-the-art in the area.

Acknowledgements. This work was supported by the European Commission through the ICON FET-Open project (Grant Agreement 284715) and by Science Foundation Ireland under Grant 10/IN.1/I3032.

References

1. SAT Competition (2011), http://www.cril.univ-artois.fr/SAT11/
2. SAT Competitions, http://www.satcompetition.org/

3. Bensusan, H., Giraud-Carrier, C.: Casa Batlo is in Passeig de Gracia or landmarking the expertise space. In: ECML Workshop on Meta-Learning: Building Automatic Advice Strategies for Model Selection and Method Combination (2000)
4. Breiman, L.: Random forests. Machine Learning 45(1), 5–32 (2001)
5. Hamerly, G., Elkan, C.: Learning the k in k-means. In: NIPS (2003)
6. Hurley, B., O'Sullivan, B.: Adaptation in a CBR-based solver portfolio for the satisfiability problem. In: Agudo, B.D., Watson, I. (eds.) ICCBR 2012. LNCS, vol. 7466, pp. 152–166. Springer, Heidelberg (2012)
7. Kadioglu, S., Malitsky, Y., Sabharwal, A., Samulowitz, H., Sellmann, M.: Algorithm selection and scheduling. In: Lee, J. (ed.) CP 2011. LNCS, vol. 6876, pp. 454–469. Springer, Heidelberg (2011)
8. Kadioglu, S., Malitsky, Y., Sellmann, M., Tierney, K.: ISAC – Instance-Specific Algorithm Configuration. In: ECAI, pp. 751–756 (2010)
9. Kotthoff, L., Gent, I., Miguel, I.P.: An evaluation of machine learning in algorithm selection for search problems. AI Communications (2012)
10. Kroer, C., Malitsky, Y.: Feature filtering for instance-specific algorithm configuration. In: ICTAI, pp. 849–855 (2011)
11. Malitsky, Y., Sellmann, M.: Instance-specific algorithm configuration as a method for non-model-based portfolio generation. In: Beldiceanu, N., Jussien, N., Pinson, É. (eds.) CPAIOR 2012. LNCS, vol. 7298, pp. 244–259. Springer, Heidelberg (2012)
12. O'Mahony, E., Hebrard, E., Holland, A., Nugent, C., O'Sullivan, B.: Using case-based reasoning in an algorithm portfolio for constraint solving. In: AICS (2008)
13. Pfahringer, B., Bensusan, H., Giraud-Carrier, C.: Meta-learning by landmarking various learning algorithms. In: ICML (2000)
14. Pulina, L., Tacchella, A.: A self-adaptive multi-engine solver for quantified boolean formulas. Constraints 14(1), 80–116 (2009)
15. Silverthorn, B., Miikkulainen, R.: Latent class models for algorithm portfolio methods. In: AAAI (2010)
16. Xu, L., Hutter, F., Hoos, H.H., Leyton-Brown, K.: Satzilla: Portfolio-based algorithm selection for SAT. CoRR (2011)
17. Xu, L., Hutter, F., Shen, J., Hoos, H.H., Leyton-Brown, K.: Satzilla2012: Improved algorithm selection based on cost-sensitive classification models (2012), SAT Competition

Area under the Precision-Recall Curve: Point Estimates and Confidence Intervals

Kendrick Boyd[1], Kevin H. Eng[2], and C. David Page[1]

[1] University of Wisconsin-Madison, Madison, WI
boyd@cs.wisc.edu,page@biostat.wisc.edu
[2] Roswell Park Cancer Institute, Buffalo, NY
Kevin.Eng@RoswellPark.org

Abstract. The area under the precision-recall curve (AUCPR) is a single number summary of the information in the precision-recall (PR) curve. Similar to the receiver operating characteristic curve, the PR curve has its own unique properties that make estimating its enclosed area challenging. Besides a point estimate of the area, an interval estimate is often required to express magnitude and uncertainty. In this paper we perform a computational analysis of common AUCPR estimators and their confidence intervals. We find both satisfactory estimates and invalid procedures and we recommend two simple intervals that are robust to a variety of assumptions.

1 Introduction

Precision-recall (PR) curves, like the closely-related receiver operating characteristic (ROC) curves, are an evaluation tool for binary classification that allows the visualization of performance at a range of thresholds. PR curves are increasingly used in the machine learning community, particularly for imbalanced data sets where one class is observed more frequently than the other class. On these imbalanced or skewed data sets, PR curves are a useful alternative to ROC curves that can highlight performance differences that are lost in ROC curves [1]. Besides visual inspection of a PR curve, algorithm assessment often uses the area under a PR curve (AUCPR) as a general measure of performance irrespective of any particular threshold or operating point (e.g., [2,3,4,5]).

Machine learning researchers build a PR curve by first plotting precision-recall pairs, or points, that are obtained using different thresholds on a probabilistic or other continuous-output classifier, in the same way an ROC curve is built by plotting true/false positive rate pairs obtained using different thresholds. Davis and Goadrich [6] showed that for any fixed data set, and hence fixed numbers of actual positive and negative examples, points can be translated between the two spaces. After plotting the points in PR space, we next seek to construct a curve and compute the AUCPR and to construct 95% (or other) confidence intervals (CIs) around the curve and the AUCPR.

However, the best method to construct the curve and calculate area is not readily apparent. The PR points from a small data set are shown in Fig. 1. Questions

H. Blockeel et al. (Eds.): ECML PKDD 2013, Part III, LNAI 8190, pp. 451–466, 2013.
© Springer-Verlag Berlin Heidelberg 2013

immediately arise about what to do with multiple points with the same x-value (recall), whether linear interpolation is appropriate, whether the maximum precision for each recall are representative, if convex hulls should be used as in ROC curves, and so on. There are at least four distinct methods (with several variations) that have been used in machine learning, statistics, and related areas to compute AUCPR, and four methods that have been used to construct CIs. The contribution of this paper is to discuss and analyze eight estimators and four CIs empirically. We provide evidence in favor of computing AUCPR using the *lower trapezoid, average precision*, or *interpolated median* estimators and using *binomial* or *logit* CIs rather than other methods that include the more widely-used (in machine learning) ten-fold *cross-validation*. The differences in results using these approaches are most striking when data are highly skewed, which is exactly the case when PR curves are most preferred over ROC curves.

Section 2 contains a review of PR curves, Section 3 describes the estimators and CIs we evaluate, and Section 4 presents case studies of the estimators and CIs in action.

2   Area Under the Precision-Recall Curve

Consider a binary classification task where models produce continuous outputs, denoted Z, for each example. Diverse applications are subsumed by this setup, e.g., a medical test to identify diseased and disease-free patients, a document ranker to distinguish relevant and non-relevant documents to a query, and generally any binary clas-

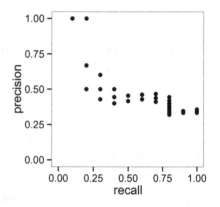

Fig. 1. Empirical PR points obtained from a small data set with 10 positive examples and 20 negative examples

sification task. The two categories are often naturally labelled as positive (e.g., diseased, relevant) or negative (e.g., disease-free, non-relevant). Following the literature on ROC curves [7,8], we denote the output values for the negative examples by X and the output values for the positive examples by Y (Z is a mixture of X and Y). These populations are assumed to be independent when the class is known. Larger output values are associated with positive examples, so for a given threshold c, an example is predicted positive if its value is greater than c. We represent the category (or class) with the indicator variable D where $D = 1$ corresponds to positive examples and $D = 0$ to negative examples. An important aspect of a task or data set is the class skew $\pi = P(D = 1)$. Skew is also known as prevalence or a prior class distribution.

Several techniques exist to assess the performance of binary classification across a range of thresholds. While ROC analysis is the most common, we are interested in the related PR curves. A PR curve may be defined as the set of points:

$$PR(\cdot) = \{(Recall(c), Prec(c)), -\infty < c < \infty\}$$

where $Recall(c) = P(Y > c)$ and $Prec(c) = P(D = 1|Z > c)$. Recall is equivalent to true positive rate or sensitivity (the y-axis in ROC curves), while precision is the same as positive predictive value. Since larger output values are assumed to be associated with positive examples, as c decreases, $Recall(c)$ increases to one and $Prec(c)$ decreases to π. As c increases, $Prec(c)$ reaches one as $Recall(c)$ approaches zero under the condition that "the first document retrieved is relevant" [9]. In other words, whether the example with the largest output value is positive or negative greatly changes the PR curve (approaching $(0, 1)$ if positive and $(0, 0)$ if negative). Similarly, estimates of precision for recall near 0 tend to have high variance, and this is a major difficulty in constructing PR curves.

It is often desirable to summarize the PR curve with a single scalar value. One summary is the area under the PR curve (AUCPR), which we will denote θ. Following the work of Bamber [7] on ROC curves, AUCPR is an average of the precision weighted by the probability of a given threshold.

$$\theta = \int_{-\infty}^{\infty} Prec(c)dP(Y \leq c) \tag{1}$$

$$= \int_{-\infty}^{\infty} P(D = 1|Z > c)dP(Y \leq c). \tag{2}$$

By Bayes' rule and using that Z is a mixture of X and Y,

$$P(D = 1|Z > c) = \frac{\pi P(Y > c)}{\pi P(Y > c) + (1 - \pi)P(X > c)}$$

and we note that $0 \leq \theta \leq 1$ since $Prec(c)$ and $P(Y \leq c)$ are bounded on the unit square. Therefore, θ might be viewed as a probability. If we consider Eq. (2) as an importance-sampled Monte Carlo integral, we may interpret θ as the fraction of positive examples among those examples whose output values exceed a randomly selected $c \sim Y$ threshold.

3   AUCPR Estimators

In this section we summarize point estimators for θ and then introduce CI methods.

3.1   Point Estimators

Let $X_1, \ldots, X_m$ and $Y_1, \ldots, Y_n$ represent observed output values from negative and positive examples, respectively. The skew π is assumed to be given or is set to $n/(n+m)$. An empirical estimate of the PR curve, $\widehat{PR}(\cdot)$, can be derived by the empirical estimates of each coordinate:

$$\widehat{Recall}(c) = n^{-1} \sum_{i=1}^{n} I(Y_i > c)$$

$$\widehat{Prec}(c) = \frac{\pi \widehat{Recall}(c)}{\pi \widehat{Recall}(c) + (1 - \pi) m^{-1} \sum_{j=1}^{m} I(X_j > c)}$$

where $I(A)$ is the indicator function for event A.

We review a number of possible estimators for θ.

Trapezoidal Estimators. For fixed $\widehat{Recall}(t)$, the estimated precision may not be constant (so $\widehat{PR}(\cdot)$ is often not one-to-one). This corresponds to cases where an observed $Y_{(i)} < X_j < Y_{(i+1)}$ for some i and j where $Y_{(i)}$ denotes the ith order statistic (ith largest value among the Y_i's). As the threshold is increased from $Y_{(i)}$ to X_j, recall remains constant while precision decreases. Let $r_i = \widehat{Recall}(Y_{(n-i)})$, so that $r_1 \leq r_2 \leq \cdots \leq r_n$, and p_i^{max} be the largest sample precision value corresponding to r_i. Likewise, let p_i^{min} be the smallest sample precision value corresponding to r_i. This leads immediately to a few choices for estimators based on the empirical curve using trapezoidal estimation [10].

$$\widehat{\theta}_{LT} = \sum_{i=1}^{n-1} \frac{p_i^{min} + p_{i+1}^{max}}{2} (r_{i+1} - r_i) \tag{3}$$

$$\widehat{\theta}_{UT} = \sum_{i=1}^{n-1} \frac{p_i^{max} + p_{i+1}^{max}}{2} (r_{i+1} - r_i) \tag{4}$$

corresponding to a *lower trapezoid* (Eq. (3)) and *upper trapezoid* (Eq. (4)) approximation. Note the *upper trapezoid* method uses an overly optimistic linear interpolation [6]; we include it for comparison as it is one of the first methods a non-expert is likely to use due to its similarity to estimating area under the ROC curve.

Interpolation Estimators. As suggested by Davis and Goadrich [6] and Goadrich et al. [1], we use PR space interpolation as the basis for several estimators. These methods use the non-linear interpolation between known points in PR space derived from a linear interpolation in ROC space.

Davis and Goadrich [6] and Goadrich et al. [1] examine the interpolation in terms of the number of true positives and false positives corresponding to each PR point. Here we perform the same interpolation, but use the recall and precision of the PR points directly, which leads to the surprising observation that the interpolation (from the same PR points) does not depend on π.

Theorem 1. *For two points in PR space (r_1, p_1) and (r_2, p_2) (assume WLOG $r_1 < r_2$), the interpolation for recall r' with $r_1 \leq r' \leq r_2$ is*

$$p' = \frac{r'}{ar' + b} \tag{5}$$

where

$$a = 1 + \frac{(1 - p_2)r_2}{p_2(r_2 - r_1)} - \frac{(1 - p_1)r_1}{p_1(r_2 - r_1)}$$

$$b = \frac{(1 - p_1)r_1}{p_1} - \frac{(1 - p_2)r_1 r_2}{p_2(r_2 - r_1)} + \frac{(1 - p_1)r_1^2}{p_1(r_2 - r_1)}$$

Proof. First, we convert the points to ROC space. Let s_1, s_2 be the false positive rates for the points (r_1, p_1) and (r_2, p_2), respectively. By definition of false positive rate,

$$s_i = \frac{(1 - p_i)\pi r_i}{p_i(1 - \pi)}. \tag{6}$$

A linear interpolation in ROC space for $r_1 \leq r' \leq r_2$ has a false positive rate of

$$s' = s_1 + \frac{r' - r_1}{r_2 - r_1}(s_2 - s_1). \tag{7}$$

Then convert back to PR space using

$$p' = \frac{\pi r'}{\pi r + (1 - \pi)s'}. \tag{8}$$

Substituting Eq. (7) into Eq. (8) and using Eq. (6) for s_1 and s_2, we have

$$p' = \pi r' \left[\pi r' + \frac{\pi(1 - p_1)r_1}{p_1} + \frac{\pi(r' - r_1)}{r_2 - r_1}\left(\frac{(1 - p_2)r_2}{p_2} - \frac{(1 - p_1)r_1}{p_1}\right) \right]^{-1}$$

$$= r' \left[r'\left(1 + \frac{(1 - p_2)r_2}{p_2(r_2 - r_1)} - \frac{(1 - p_1)r_1}{p_1(r_2 - r_1)}\right) + \right.$$

$$\left. \frac{(1 - p_1)r_1}{p_1} - \frac{(1 - p_2)r_1 r_2}{p_2(r_2 - r_1)} + \frac{(1 - p_1)r_1^2}{p_1(r_2 - r_1)} \right]^{-1}$$

$\square$

Thus, despite PR space being sensitive to π and the translation to and from ROC space depending on π, the interpolation in PR space *does not* depend on π. One explanation is that each particular PR space point inherently contains the information about π, primarily in the precision value, and no extra knowledge of π is required to perform the interpolation.

The area under the interpolated PR curve between these two points can be calculated analytically using the definite integral:

$$\int_{r_1}^{r_2} \frac{r'}{ar' + b} \, dr' = \frac{br' - a\log(a + br')}{b^2} \Big|_{r'=r_1}^{r'=r_2}$$

$$= \frac{br_2 - a\log(a + br_2) - br_1 + a\log(a + br_1)}{b^2}$$

where a and b are defined as in Theorem 1.

With the definite integral to calculate the area between two PR points, the question is: which points should be used. The achievable PR curve of Davis of Goadrich [6] uses only those points (translated into PR space) that are on the ROC convex hull. We also use three methods of summarizing from multiple PR points at the same recall to a single PR point to interpolate through. The summaries we use are the max, mean, and median of all p_i for a particular r_i. So we have four estimators using interpolation: *convex*, *max*, *mean*, and *median*.

Average Precision. Avoiding the empirical curve altogether, a plug-in estimate of θ, known in information retrieval as *average precision* [11], is:

$$\widehat{\theta}_A = \frac{1}{n} \sum_{i=1}^{n} \widehat{Prec}(Y_i) \tag{9}$$

which replaces the distribution function $P(Y \leq c)$ in Eq. (2) with its empirical cumulative distribution function.

Binormal Estimator. Conversely, a fully parametric estimator may be constructed by assuming that $X_j \sim \mathcal{N}(\mu_x, \sigma_x^2)$ and $Y_j \sim \mathcal{N}(\mu_y, \sigma_y^2)$. In this *binormal* model [12], the MLE of θ is

$$\widehat{\theta}_B = \int_0^1 \frac{\pi t}{\pi t + (1 - \pi)\Phi\left(\frac{\hat{\mu}_y - \hat{\mu}_x}{\sigma_x} + \frac{\hat{\sigma}_y}{\hat{\sigma}_x}\Phi^{-1}(t)\right)} \, dt \tag{10}$$

where $\hat{\mu}_x, \hat{\sigma}_x, \hat{\mu}_y, \hat{\sigma}_y$ are sample means and variances of X and Y and $\Phi(t)$ is the standard normal cumulative distribution function.

3.2   Confidence Interval Estimation

Having discussed AUCPR estimators, we now turn our attention to computing confidence intervals (CIs) for these estimators. Our goal is to determine a simple, accurate interval estimate that is logistically easy to implement. We will compare two computationally intensive methods against two simple statistical intervals.

Bootstrap Procedure. A common approach is to use a *bootstrap* procedure to estimate the variation in the data and to either extend a symmetric, normal-based interval about the point estimate or to take the empirical quantiles from resampled data as interval bounds [13]. Because the relationship between the number of positive examples n and negative examples m is crucial for estimating PR points and hence curves, we recommend using stratified bootstrap so π is preserved exactly in all replicates. In our simulations we chose to use empirical quantiles for the interval bounds and perform 1000 bootstrap replicates.

Cross-Validation Procedure. Similarly, a *cross-validation* approach is a wholly data driven method for simultaneously producing the train/test splits required for unbiased estimation of future performance and producing variance estimates. In k-fold cross-validation, the available data are partitioned into k folds. $k - 1$ folds are used for training while the remaining fold is used for testing. By performing evaluation on the results of each fold separately, k estimates of performance are obtained. A normal approximation of the interval can be constructed using the mean and variance of the k estimates. For more details and discussion of k-fold cross-validation, see Dietterich [14]. For our case studies we use the standard $k = 10$.

Binomial Interval. Recalling that $0 \leq \theta \leq 1$, we may interpret $\hat{\theta}$ as a probability associated with some binomial$(1, \theta)$ variable. If so, a CI for θ can be constructed through the standard normal approximation:

$$\hat{\theta} \pm \Phi_{1-\alpha/2}\sqrt{\frac{\hat{\theta}(1 - \hat{\theta})}{n}}$$

We use n for the sample size as opposed to $n + m$ because n specifies the (maximum) number of unique recall values in $\widehat{PR}(\cdot)$. The *binomial* method can be applied to any $\hat{\theta}$ estimate once it is derived. A weakness of this estimate is that it may produce bounds outside of $[0, 1]$, even though $0 \leq \theta \leq 1$.

Logit Interval. To obtain an interval which is guaranteed to produce endpoints in $[0, 1]$, we may use the logistic transformation $\hat{\eta} = \log \frac{\hat{\theta}}{(1-\hat{\theta})}$ where $\hat{\tau} = s.e.(\hat{\eta}) = (n\hat{\theta}(1 - \hat{\theta}))^{-1/2}$ by the delta method [15].

On the logistic scale, an interval for η is $\hat{\eta} \pm \Phi_{1-a/2}\hat{\tau}$. This can be converted pointwise to produce an asymmetric *logit* interval bounded in $[0, 1]$:

$$\left[\frac{e^{\hat{\eta}-\Phi(1-\alpha/2)\hat{\tau}}}{1 + e^{\hat{\eta}-\Phi(1-\alpha/2)\hat{\tau}}}, \frac{e^{\hat{\eta}+\Phi(1-\alpha/2)\hat{\tau}}}{1 + e^{\hat{\eta}+\Phi(1-\alpha/2)\hat{\tau}}} \right].$$

4   Case Studies

We use simulated data to evaluate the merits of the candidate point and interval estimates discussed in Section 3 with the goal of selecting a subset of desirable procedures.[1] The ideal point estimate would be unbiased, robust to various distributional assumptions on X and Y, and have good convergence as $n + m$ increases. A CI should have appropriate coverage, and smaller widths of the interval are preferred over larger widths.

---

[1] R code for the estimators and simulations may be found at
http://pages.cs.wisc.edu/~boyd/projects/2013ecml_aucprestimation/

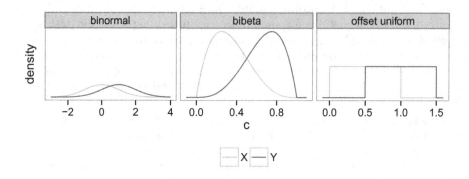

Fig. 2. Probability density functions for X (negative) and Y (positive) output values for binormal ($X \sim N(0,1), Y \sim N(1,1)$), bibeta ($X \sim B(2,5), Y \sim B(5,2)$), and offset uniform ($X \sim U(0,1), Y \sim U(0.5,1.5)$) case studies

We consider three scenarios for generating output values X and Y. Our intention is to cover representative but not exhaustive cases whose conclusions will be relevant generally. The densities for these scenarios are plotted in Fig. 2. The true PR curves (calculated using the cumulative distribution functions of X and Y) for $\pi = 0.1$ are shown in Fig. 3. Fig. 3 also contains sample empirical PR curves that result from drawing data from X and Y. These are the curves the estimators work from, attempting to recover the area under the true curve as accurately as possible.

For unbounded continuous outputs, the binormal scenario assumes that $X \sim \mathcal{N}(0,1)$ and $Y \sim \mathcal{N}(\mu,1)$ where $\mu > 0$. The distance between the two normal distributions, μ, controls the discriminative ability of the assumed model. For test values bounded on $[0,1]$ (such as probability outputs), we replace the normal distribution with a beta distribution. So the bibeta scenario has $X \sim B(a,b)$ and $Y \sim B(b,a)$ where $0 < a < b$. The larger the ratio between a and b, the better able to distinguish between positive and negative examples. Finally, we model an extreme scenario where the support of X and Y is not the same. This offset uniform scenario is given by $X \sim U(0,1)$ and $Y \sim U(\gamma, 1+\gamma)$ for $\gamma \geq 0$: that is X lies uniformly on $(0,1)$ while Y is bounded on $(\gamma, \gamma+1)$. If $\gamma = 0$ there is no ability to discriminate, while $\gamma > 1$ leads to perfect classification of positive and negative examples with a threshold of $c = 1$. All results in this paper use $\mu = 1, a = 2, b = 5,$ and $\gamma = 0.5$. These were chosen as representative examples of the distributions that produce reasonable PR curves.

This paper exclusively uses simulated data drawn from specific, known distributions because this allows calculation of the true PR curve (shown in Fig. 3) and the true AUCPR. Thus, we have a target value to compare the estimates against and are able to evaluate the bias of an estimator and the coverage of a CI. This would be difficult to impossible if we used a model's predictions on real data because the true PR curve and AUCPR are unknown.

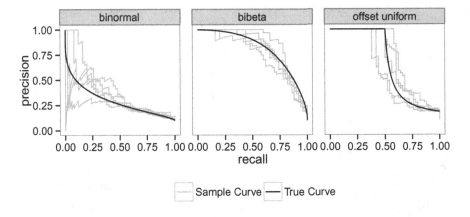

Fig. 3. True PR curves (calculated from the theoretical density functions) and sampled empirical PR curves, both at $\pi = 0.1$. Sampled PR curves use $n + m = 500$. The sampled PR curves were generated by connecting PR points corresponding to adjacent thresholds values.

4.1   Bias and Robustness in Point Estimates

For each scenario, we evaluate eight estimators: the non-parametric *average precision*, the parametric *binormal*, two trapezoidal estimates, and four interpolated estimates. Fig. 4 shows the bias ratio versus $n + m$ where $\pi = 0.1$ over 10,000 simulations, and Fig. 5 shows the bias ratio versus π where $n + m = 1000$. The bias ratio is the mean estimated AUCPR divided by the true AUCPR, so an unbiased estimator has a bias ratio of 1.0. Good point estimates of AUCPR should be unbiased as $n + m$ and π increase. That is, an estimator should have an expected value equal to the true AUCPR (calculated by numerically integrating Eq. 2).

As $n + m$ grows large, most estimators converge to the true AUCPR in every case. However, the *binormal* estimator shows the effect of model misspecification. When the data are truly binormal, it shows excellent performance but when the data are bibeta or offset uniform, the *binormal* estimator converges to the wrong value. Interestingly, the bias due to misspecification that we observe for the *binormal* estimate is lessened as the data become more balanced (π increases).

The *interpolated convex* estimate consistently overestimates AUCPR and appears far from the true value even at $n + m = 10000$. The poor performance of the *interpolated convex* estimator seems surprising given how it uses the popular convex hull ROC curve and then converts to PR space. Because the other interpolated estimators perform adequately, the problem may lie in evaluating the convex hull in ROC space. The convex hull chooses those particular points that give the best performance on the *test* set. Analogous to using the test set during training, the convex hull procedure may be overly optimistic and lead to the observed overestimation of AUCPR.

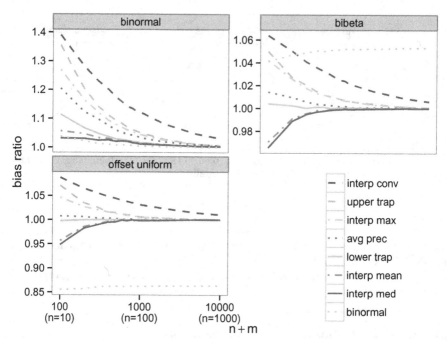

Fig. 4. Ratio of estimated AUCPR to true AUCPR (bias ratio) versus total number of examples $(n + m)$. $\pi = 0.1$ for all cases.

It is important to note that since $\pi = 0.1$ in Fig. 4, data are sparse at $n + m = 100$: there are $n = 10$ values of Y to evaluate the estimate. In these situations there is no clear winner across all three scenarios and estimators tend to overestimate AUCPR when n is small with a few exceptions. Among related estimators, *lower trapezoid* appears more accurate than the *upper trapezoid* method and the *mean* or *median interpolation* outperform the *convex* and *max interpolation*. Consequently, we will only consider the *average precision*, *interpolated median*, and *lower trapezoid* estimators since they are unbiased in the limit, less biased for small sample sizes, and robust to model misspecification.

4.2   Confidence Interval Evaluation

We use a two-step approach to evaluate confidence intervals (CIs) based on Chapter 7 of Shao [16]. In practice, interval estimates must come with a confidence guarantee: if we say an interval is an $(1-\alpha)\%$ CI, we should be assured that it covers the true value in at least $(1 - \alpha)\%$ of datasets [16,17,18]. It may be surprising to non-statisticians that an interval with slightly low coverage is ruled inadmissible, but this would invalidate the guarantee. Additionally, targeting an exact $(1 - \alpha)\%$ interval is often impractical for technical reasons, hence the *at least* $(1-\alpha)\%$. When an interval provides at least $(1-\alpha)\%$ coverage, it is considered a valid interval and this is the first criteria a potential interval must satisfy.

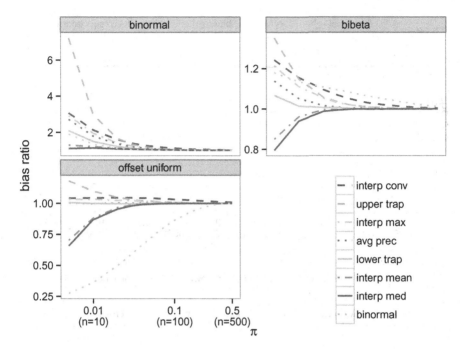

Fig. 5. Ratio of estimated AUCPR to true AUCPR (bias ratio) versus π. In all cases $n + m = 1000$.

After identifying valid methods for CIs, the second step is that we prefer the narrowest (or optimal) intervals among the valid methods. The trivial $[-\infty, +\infty]$ interval is a valid 95% CI because it always has at least 95% coverage (indeed, it has 100% coverage), but it conveys no useful information about the estimate. Thus we seek methods that produce the narrowest, valid intervals.

CI Coverage. The first step in CI evaluation is to identify valid CIs with coverage at least $(1 - \alpha)\%$. In Fig. 6, we show results over 10,000 simulations for the coverage of the four CI methods described in 3.2. These are 95% CIs, so the target coverage of 0.95 is denoted by the thick black line. As mentioned at the end of Section 4.1, we only consider the *average precision, interpolated median*, and *lower trapezoid* estimators for our CI evaluation.

A strong pattern emerges from Fig. 6 where the *bootstrap* and *cross-validation* intervals tend to have coverage below 0.95, though asymptotically approaching 0.95. Since the coverage is below 0.95, this makes the computational intervals technically invalid. The two formula-based intervals are consistently above the requisite 0.95 level. So *binomial* and *logit* produce valid confidence intervals.

Given the widespread use of *cross-validation* within machine learning, it is troubling that the CIs produced from that method fail to maintain the confidence guarantee. This is not an argument against *cross-validation* in general, only a

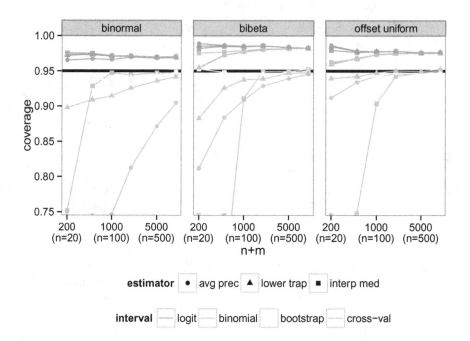

Fig. 6. Coverage for selected estimators and 95% CIs calculated using the four interval methods. Results for selected $n + m$ are shown for $\pi = 0.1$. To be valid 95% CIs, the coverage should be at least 0.95. Note that the coverage for a few of the *cross-validation* intervals is below 0.75. These points are represented as half-points along the bottom border.

caution against using it for AUCPR inference. Similarly, *bootstrap* is considered a rigorous (though computationally intensive) fall-back for non-parametrically evaluating variance, yet Fig. 6 shows it is only successful assymptotically as data size increases (and the data size needs to be fairly large before it nears 95% coverage).

CI Width. To better understand why *bootstrap* and *cross-validation* are failing, an initial question is: are the intervals *too* narrow? Since we have simulated 10,000 data sets and obtained AUCPR estimates on each using the various estimators, we have an empirical distribution from which we can calculate an ideal empirical width for the CIs. When creating a CI, only 1 data set is available, thus this empirical width is not available, but we can use it as a baseline to compare the mean width obtained by the various interval estimators. Fig. 7 shows coverage versus the ratio of mean width to empirically ideal width. As expected there is a positive correlation between coverage and the width of the intervals: wider intervals tend to provide higher coverage. For *cross-validation*, the widths tend to be slightly smaller than the *logit* and *binomial* intervals but still larger than the empirically ideal width. Coverage is frequently much lower though,

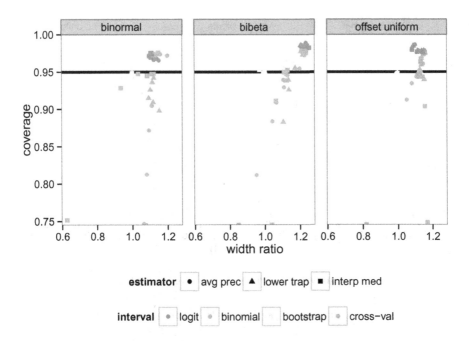

Fig. 7. Mean normalized width ratio versus coverage for *binomial, logit, cross-validation,* and *bootstrap* methods. Normalized width is the ratio of the CI width to the empirically ideal width. Width ratios below 1 suggest the intervals are overoptimistic. Results shown for $n + m \in 200, 500, 1000, 5000, 10000$ and $\pi = 0.1$. Note that the coverage for some of the *cross-validation* intervals is below 0.75. These points are represented as half-points along the bottom border.

suggesting the width of the interval is not the reason for the poor performance of *cross-validation*. However, interval width may be part of the issue with *bootstrap*. The *bootstrap* widths are either right at the empirically ideal width or even smaller.

CI Location. Another possible cause for poor coverage is that the intervals are for the wrong target value (i.e., the intervals are biased). To investigate this, we analyze the mean location of the intervals. We use the original estimate on the full data set as the location for the *binomial* and *logit* intervals since both are constructed around that estimate, the mid-point of the interval from *cross-validation*, and the median of the *bootstrap* replicates since we use the quantiles to calculate the interval. The ratio of the mean location to the true value (similar to Fig. 4) is presented in Fig. 8. The location of the *cross-validation* intervals is much farther from the true estimate than either the *bootstrap* or *binomial* locations, with *bootstrap* being a bit worse than *binomial*. This targeting of the wrong value for small $n + m$ is the primary explanation for the low coverages seen in Fig. 6.

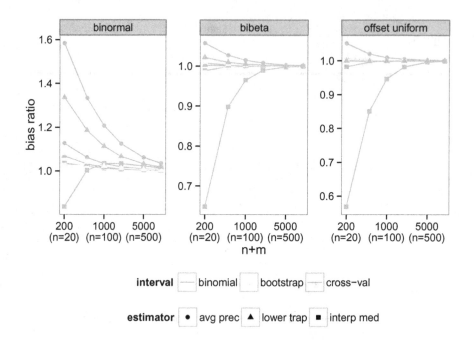

Fig. 8. Mean location of the intervals produced by the *binomial, bootstrap,* and *cross-validation* methods (*logit* is identical to *binomial*). As in Fig. 4, the y-axis is the bias ratio, the ratio of the location (essentially a point estimate based on the interval) to the true AUCPR. *Cross-validation* is considerably more biased than the other methods and *bootstrap* is slightly more biased than *binomial*.

Comments on *Bootstrap* and *Cross-validation* Intervals. The increased bias in the intervals produced by *bootstrap* and *cross-validation* occurs because these methods use many smaller data sets to produce a variance estimate. K-fold *cross-validation* reduces the effective data sets by a factor of k while *bootstrap* is less extreme but still reduces the effective data sets by a factor of 1.5. Since the estimators become more biased with smaller data sets (demonstrated in Fig. 4), the point estimates used to construct the *bootstrap* and *cross-validation* intervals are more biased, leading to the misplaced intervals and less than $(1 - \alpha)\%$ coverage.

Additionally, the *bootstrap* has no small sample theoretical justification and it is acknowledged it tends to break down for very small sample sizes [19]. When estimating AUCPR with skewed data, the critical number for this is the number of positive examples n, not the size of the data set $n + m$. Even when the data set itself seems reasonably large with $n + m = 200$, at $\pi = 0.1$ there are only $n = 20$ positive examples. With just 20 samples, it is difficult to get representative samples during the *bootstrap*. This also contributes to the lower than expected 95% coverage and is a possible explanation for the *bootstrap* widths being even smaller than the empirically ideal widths seen in Fig. 7.

We emphasize that both the *binomial* and *logit* intervals are valid and do not require the additional computation of *cross-validation* and *bootstrap*. For large sample sizes *bootstrap* approaches $(1 - \alpha)\%$ coverage, but it approaches from below, so care should be taken. *Cross-validation* is even more problematic, with proper coverage not obtained even at $n + m = 10,000$ for some of our case studies.

5 Conclusion

Our computational study has determined that simple estimators can achieve nearly ideal width intervals while maintaining valid coverage for AUCPR estimation. A key point is that these simple estimates are easily evaluated and do not require resampling or add to computational workload. Conversely, computationally expensive, empirical procedures (*bootstrap* and *cross-validation*) yield interval estimates that do not provide adequate coverage for small sample sizes and only asymptotically approach $(1 - \alpha)\%$ coverage.

We have also tested a variety of point estimates for AUCPR and determined that the parametric *binormal* estimate is extremely poor when the true generating distribution is not normal. Practically, data may be re-scaled (e.g., the Box-Cox transformation) to make this assumption fit better, but, with easily accessible nonparametric estimates that we have shown to be robust, this seems unnecessary.

The scenarios we studied are by no means exhaustive, but they are representative, and the conclusions can be further tested in specific cases if necessary. In summary, our investigation concludes that the *lower trapezoid, average precision,* and *interpolated median* point estimates are the most robust estimators and recommends the *binomial* and *logit* methods for constructing interval estimates.

Acknowledgments. We thank the anonymous reviewers for their detailed comments and suggestions. We gratefully acknowledge support from NIGMS grant R01GM097618, NLM grant R01LM011028, UW Carbone Cancer Center, ICTR NIH NCA TS grant UL1TR000427, CIBM Training Program grant 5T15LM007359, Roswell Park Cancer Institute, and NCI grant P30 CA016056.

References

1. Goadrich, M., Oliphant, L., Shavlik, J.: Gleaner: Creating ensembles of first-order clauses to improve recall-precision curves. Machine Learning 64, 231–262 (2006)
2. Richardson, M., Domingos, P.: Markov logic networks. Machine Learning 62(1-2), 107–136 (2006)
3. Liu, Y., Shriberg, E.: Comparing evaluation metrics for sentence boundary detection. In: IEEE International Conference on Acoustics, Speech and Signal Processing, ICASSP 2007, vol. 4, pp. IV–185. IEEE (2007)

4. Yue, Y., Finley, T., Radlinski, F., Joachims, T.: A support vector method for optimizing average precision. In: Proceedings of the 30th Annual International ACM SIGIR Conference on Research and Development in Information Retrieval, pp. 271–278. ACM (2007)
5. Natarajan, S., Khot, T., Kersting, K., Gutmann, B., Shavlik, J.: Gradient-based boosting for statistical relational learning: The relational dependency network case. Machine Learning 86(1), 25–56 (2012)
6. Davis, J., Goadrich, M.: The relationship between precision-recall and ROC curves. In: Proceedings of the 23rd International Conference on Machine learning, ICML 2006, pp. 233–240. ACM, New York (2006)
7. Bamber, D.: The area above the ordinal dominance graph and the area below the receiver operating characteristic graph. Journal of Mathematical Psychology 12(4), 387–415 (1975)
8. Pepe, M.S.: The statistical evaluation of medical tests for classification and prediction. Oxford University Press, USA (2004)
9. Gordon, M., Kochen, M.: Recall-precision trade-off: A derivation. Journal of the American Society for Information Science 40(3), 145–151 (1989)
10. Abeel, T., Van de Peer, Y., Saeys, Y.: Toward a gold standard for promoter prediction evaluation. Bioinformatics 25(12), i313–i320 (2009)
11. Manning, C.D., Raghavan, P., Schütze, H.: Introduction to Information Retrieval. Cambridge University Press, New York (2008)
12. Brodersen, K.H., Ong, C.S., Stephan, K.E., Buhmann, J.M.: The binormal assumption on precision-recall curves. In: 2010 20th International Conference on Pattern Recognition (ICPR), pp. 4263–4266. IEEE (2010)
13. Efron, B.: Bootstrap methods: Another look at the jackknife. The Annals of Statistics 7(1), 1–26 (1979)
14. Dietterich, T.G.: Approximate statistical tests for comparing supervised classification learning algorithms. Neural Computation 10, 1895–1923 (1998)
15. DeGroot, M.H., Schervish, M.J.: Probability and Statistics. Addison-Wesley (2001)
16. Shao, J.: Mathematical Statistics, 2nd edn. Springer (2003)
17. Wasserman, L.: All of statistics: A concise course in statistical inference. Springer (2004)
18. Lehmann, E.L., Casella, G.: Theory of point estimation, vol. 31. Springer (1998)
19. Efron, B.: Bootstrap confidence intervals: Good or bad? Psychological Bulletin 104(2), 293–296 (1988)

Incremental Sensor Placement Optimization on Water Network[*]

Xiaomin Xu[1], Yiqi Lu[1], Sheng Huang[2], Yanghua Xiao[1,**], and Wei Wang[1]

[1] School of computer science, Fudan University, Shanghai, China
{10210240041,10210240030,shawyh,weiwang1}@fudan.edu.cn
[2] IBM China Research Lab, Shanghai, China
huangssh@cn.ibm.com

Abstract. Sensor placement on water networks is critical for the detection of accidental or intentional contamination event. With the development and expansion of cities, the public water distribution systems in cities are continuously growing. As a result, the current sensor placement will lose its effectiveness in detecting contamination event. Hence, in many real applications, we need to solve the *incremental sensor placement* (ISP) problem. We expect to find a sensor placement solution that reuses existing sensor deployments as much as possible to reduce cost, while ensuring the effectiveness of contamination detection. In this paper, we propose scenario-cover model to formalize ISP and prove that ISP is NP-hard and propose our greedy approaches with provable quality bound. Extensive experiments show the effectiveness, robustness and efficiency of the proposed solutions.

1 Introduction

Monitoring water networks in cities for the safety of water quality is of critical importance for the living and development of societies. One of the efforts for water safety is building *early warning systems* (EWSs) with the aim to detect contamination event promptly by installing sensors in a water distribution network. When placing sensors in a water network, it is always desired to maximize the effectiveness of these sensors with minimal deployment cost. Hence, optimizing the sensor placement on water network so that the adverse impact of contaminant event on public health is minimized under the budget limit becomes one of the major concerns of researchers or engineers when building EWSs.

Problem Statement. With the development and expansion of cities, especially those cities in developing countries like China, India, the public municipal water distribution systems are ever-expanding. As a consequence, the current sensor placement of EWS generally will lose its effectiveness in detecting contamination events. See Figure 1 as an example. The sensor deployment in the original water network is shown in Figure 1(a). Sometime later, the water network may significantly expand (the expanded network is shown in Figure 1(b)). Clearly, if we keep the sensor deployment

---

[*] Supported by the National NSFC (No. 61003001, 61170006, 61171132, 61033010); Specialized Research Fund for the Doctoral Program of Higher Education No. 20100071120032; NSF of Jiangsu Province (No. BK2010280).
[**] Correspondence author.

H. Blockeel et al. (Eds.): ECML PKDD 2013, Part III, LNAI 8190, pp. 467–482, 2013.
© Springer-Verlag Berlin Heidelberg 2013

unchanged, it will be hard for the sensors in the original locations to detect the contamination events occurring in the expanded part of the network (marked by the red dotted line in Figure 1(b)).

To keep the effectiveness of EWS in detecting such contamination events, in most cases people needs to add more sensors or reinstall existing sensors. In the previous example, to ensure the effectiveness of contamination detection on the expanded network, we may move one sensor deployed in v_2 (the green node in Figure 1(a)) from the original network and reinstall it on a location (say v_9) of the expanded network. We may also need to buy two new sensors and deploy them on v_{13} and v_{16}.

Effectiveness of above redeployment comes at the cost of reinstalling and adding new sensors. However, in many cases, the budget is limited. To reduce the cost, the newly installed sensors and reinstalled sensors are always expected to be minimized. Consequently, in real applications of water network, an *incremental sensor placement* (ISP) problem usually arises, in which *we need to find a sensor placement that reuses existing sensor placement as much as possible, and meanwhile guarantees the effectiveness to detect contaminations.*

Despite of the importance of ISP, rare efforts can be found to solve it efficiently and effectively. Most of existing related works focus on sensor placement on a static water network. These solutions on static water networks can not be straightforwardly extended to solve ISP since reusing current sensor placement is a new objective of the problem. If we directly recalculate a sensor placement on the current water networks, the resulting sensor deployment may not have sufficient overlap with the current deployment, which will lead to high cost of redeployment. Furthermore, the water networks in real life are generally evolving in a complicated way, which also poses new challenges to solve ISP.

In summary, it is a great challenge to design a strategy to deploy sensors incrementally under limited budget without losing the effectiveness of contamination detection.

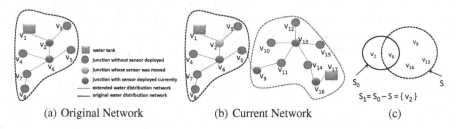

(a) Original Network   (b) Current Network   (c)

Fig. 1. Demonstration of a possible incremental sensor deployment solution. Originally, 2 sensors were deployed. When the network was expanded, 2 additional sensors would be deployed and one original sensor (marked by green node) was moved to the expanded part of the network.

Contribution and Organization. To overcome the difficulties stated above, we reduce the ISP problem to *maximal coverage* (MC) problem. Based on this model, we develop several heuristic algorithms to solve ISP. The contribution of this paper can be summarized as follows:

1. We reduce sensor placement optimization to *maximal coverage* problem.

2. We propose the problem of ISP. We show that ISP is NP-hard, and the objective function of ISP is *submodular*.
3. We propose a series of greedy algorithms with provable bound on the solution quality to solve static sensor placement and then extend them to solve ISP.
4. We conduct extensive experiments to show the effectiveness, robustness and efficiency of our proposed solutions.

The remainder is organized as follows. Sec 2 briefly reviews the related works. Sec 3 formalizes the ISP problem and builds the theory for this problem. In Sec 4, we present the solutions for the proposed problems. Sec 5 shows the experimental results. We close the paper with a brief conclusion in Sec 6.

2 Related Work

A large number of approaches have been proposed for optimizing water sensor networks. They differ from each other in the design and/or performance objective. [1] developed a formulation related to a set covering problem with an premise that sampling at a location supplied by upstream nodes provides information about water at the upstream nodes. Subsequently several researchers refined the model by greedy heuristic-based algorithm [2] and genetic algorithm [3]. [4,5] introduced a scenario in which the objective is to ensure a pre-specified maximum volume of contaminated water consumed prior to detection and also reduced this problem to set cover problem. [6,7] introduced an MIP(mixed integer programming) solution for the objective to minimize the expected fraction of population exposed to a contamination. The objective of [8,9] is to ensure that the expected impact of a contamination event is within a pre-specified level, and [8] introduced a formulation based on set cover and solved the problem using genetic algorithm while [9] use a MIP based solution. In order to achieve the objective defined in [10], [11,12,13,14,15,16,17,18,19,20,21] adopted multi-objective optimization by different methods such as heuristic, predator-prey model or local search method and [22] used the submodular property to achieve an approximation guarantee.

In our paper, we use the concept of submodularity [23,24] to solve the problem of sensor placement on dynamic water network. Submodularity was widely used in sensor placement optimization [22,25]. But these solutions are mostlybuilt for static water network. Besides sensor placement, submodularity has also been widely adopted in finding influencers [26], influence maximization [27] and network structural learning [28]. All the solutions mentioned above are not designed for incremental sensor placement, which is a more realistic problem in real-world. To the best of our knowledge, this paper is the first one addressing the incremental sensor placement on dynamic water networks.

3 Incremental Sensor Placement

In this section, we formalize ISP. We start this section with the introduction of preliminary concepts in Sec 3.1. Please refer to [6] for more background knowledge. Then, we propose a scenario-cover based model in Sec 3.3, upon which ISP is defined in Sec 3.2. Finally, we show that ISP is NP-hard in Sec 3.4 and the submodularity of the objective function used in ISP in Sec 3.5.

3.1 Preliminaries

A water distribution system is modeled as an undirected graph $G = (V, E)$, with vertices in V representing junctions, tanks, and other resources in the systems; edges in E representing pipes, pumps, and valves. Suppose we need to monitor a set of *contamination scenarios* $\mathcal{A}$. Each *contamination scenario* $c \in \mathcal{A}$ can be characterized by a quadruple of the form (v_x, t_s, t_f, X), where $v_x \in V$ is the origin of the contamination event, t_s and t_f are the start and stop times of a contamination event, respectively, and X is the contamination event profile (which describes the contamination material injected at a particular concentration or at a given rate).

Let $L \subseteq V$ be the set of all possible *sensor locations* (*e.g.* junctions in the water network). A placement of p sensors on $L \subseteq V$ is called a *sensor placement*. In general, we do not distinguish sensors placed at different locations. Thus, a subset of L holding the sensors can uniquely determine a sensor placement. In this paper, we always use the subset of L to describe a sensor placement.

One of key issues in water sensor placement is the measurement of the total impact of contamination scenario c when c is detected by a sensor deployed at vertex v in *a sensor placement S*. We use $d_{c,v}$ to denote such an impact. More specifically, $d_{c,v}$ represents the total damaging influence caused by contamination scenario c during the time period from the beginning of c to the time point at which c was detected by a sensor deployed at vertex v. $d_{c,v}$ can be defined from various aspects, including *volume of contaminated water consumed* [4], *population affected* [6], and *the time until detection* [2]. In this paper, we use *the time until detection* as the quantitative criteria to evaluate the adverse impact of each contamination scenario. Solutions proposed in this paper can be directly extended on other adverse impact measures.

The contamination scenarios can be simulated by water quality analysis software (*e.g.* EPANET [1]). Based on the simulation data, $d_{c,v}$ can be computed accordingly. In this paper, each $d_{c,v}$ can be considered as the input of the major problem that will be addressed. Given a set of contamination scenarios $\mathcal{A}$ and sensor locations L, we have a $|\mathcal{A}| \times |L|$ matrix with each element representing $d_{c,v}$. We call this matrix as *contamination scenario matrix* (CS for short) and denote it by $D_{\mathcal{A},L}$. We use Example 1 to illustrate above basic concepts.

Example 1 (Basic Concepts). We give in Table 1(a) a scenario set (i.e., $\mathcal{A}$) for water distribution network described in Figure 1(a). A matrix $D_{\mathcal{A},L}$ provided by contamination simulation is given in Table1(b), in which *the time until detection* is adopted as the measure of the adverse impact of each contamination scenario. As an example, $d_{c_1,v_1} = 7$ implies that placing a sensor on v_1 can detect contamination event c_1 within 7 minutes.

3.2 Scenario-Cover Based Modeling

In this subsection, we will reduce sensor placement optimization to maximal coverage problem. Based on this model, we will formalize the static sensor placement problem

---

[1] http://www.epa.gov/nrmrl/wswrd/dw/epanet.html

Table 1. Table 1(a): Scenario set $\mathcal{A}$. Table 1(b): Contamination scenario matrix $D_{\mathcal{A},L}$. Each element in the matrix denotes the the time (in minutes). Table 1(c): $R_{v,M}$ for each vertex shown in Figure 1(a).

(a)

scenarios	v_x	t_s	t_f	X(mg/min)
c_1	v_1	0:00	1:00	1000
c_2	v_2	0:20	1:00	1000
c_3	v_7	0:15	1:00	1000
c_4	v_5	0:12	1:00	1000

(b)

	v_1	v_2	v_3	v_4	v_5	v_6	v_7	v_8
c_1	7	9	12	18	14	13	23	14
c_2	12	5	8	16	12	12	15	17
c_3	14	12	16	15	12	7	5	11
c_4	26	18	17	13	5	7	14	15

(c)

V	$R_{v,M}$	V	$R_{v,M}$
v_1	$\{c_1\}$	v_5	$\{c_4\}$
v_2	$\{c_1,c_2\}$	v_6	$\{c_3,c_4\}$
v_3	$\{c_2\}$	v_7	$\{c_3\}$
v_4	$\emptyset$	v_8	$\emptyset$

and incremental sensor placement problem. After that, in subsection 3.4, we will show that problem ISP (Definition 4) is NP-hard.

In real applications, it is a reasonable objective to limit the worst damage caused by a potential contamination event under a certain level. For example, we usually expect that the contaminated population can not exceed a certain threshold when we choose a rational water sensor placement strategy. We use *credit M* to capture the worst damage that we can afford.

Definition 1 (Covered Scenario with Credit M). *Given a contamination scenario matrix $D_{\mathcal{A},L}$ and a sensor placement S, if a contamination scenario c can be detected by a sensor deployed at vertex $v \in S$ such that $d_{c,v} \leq M$, then we say that the contamination scenario c is* covered *by S with credit M.*

Given the contamination scenario matrix $D_{\mathcal{A},L}$, the set of scenarios in $\mathcal{A}$ that is covered by v with credit M can be uniquely determined. This set can be formally defined as:

$$R_{v,M} = \{c | d_{c,v} \in D_{\mathcal{A},L}, d_{c,v} \leq M\} \tag{1}$$

$R_{v,M}$ is the set of scenarios in $\mathcal{A}$ whose total harmful impact is within credit M if we place a sensor on vertex v. If we add vertex v into sensor placement S, then each contamination scenario in $R_{v,M}$ would be covered by S. In other words, $R_{v,M}$ represents the contamination detection performance when v is added into sensor placement.

Given the water network, $D_{\mathcal{A},L}$ and credit M, we hope that *the sensor placement can cover contamination scenarios with credit M as many as possible.* For this purpose, we use $F(S)$, an evaluation function defined on sensor placement, formally defined in Definition 2, to precisely quantify the *the number of scenarios* covered *by S with credit M.* Thus, our design objective is to maximize $F(S)$.

Definition 2 (Evaluation Function). *Given a credit M and a contamination scenario matrix $D_{\mathcal{A},L}$, the evaluation function $F(S)$ of a sensor placement $S \subseteq L$ is defined as:*

$$F(S) = | \bigcup_{v \in S} R_{v,M} | \tag{2}$$

Now the *static sensor placement optimization* on a water network can be rewritten in the scenario-cover based model. We illustrate key concepts in scenario-cover based modeling and the SP problem based on this modeling in Example 2.

Definition 3 (Sensor Placement (SP)). *Given a water network G with sensor locations L, scenario set $\mathcal{A}$, CS matrix $D_{\mathcal{A},L}$, a credit M and an integer p, finding a sensor placement $S \subseteq L$ such that $|S| \leq p$ and $F(S)$ is maximized.*

Example 2 (Scenario-cover Based Modeling). Continue the previous example and consider the water network shown in Figure 1(a). Suppose the detection time credit $M = 10$ minutes and contamination scenario set is $\mathcal{A}$ shown in Table 1(a). Table 1(c) shows the covered scenario set $R_{v,M}$ for each v. We find that $S = \{v_2, v_6\}$ can cover all the scenarios in $\mathcal{A}$. Hence, S is a solution of SP.

Note that sensor placement optimization can be also reduced to *integer linear programming* (ILP), which could be solved by the state-of-the-art *mixed integer programming* (MIP) solver such as ILOG's AMPL/CPLEX9.1 MIP solver that was widely used in the previous works [9]. However, in general, these solvers are less efficient and can not scale up to large water distribution networks.

3.3   Incremental Sensor Placement

Now, we are ready to formalize *incremental sensor placement* (ISP). In ISP, beside quality, *cost* is another major concern. Following the scenario-cover based modeling, the quality of a sensor placement can also be measured by $F(S)$. Next, we first give the cost constraint in ISP then give the formal definition of ISP.

Cost Constraint. Let S_0, S be the sensor placement in the original water network and the adjusted sensor placement in the expanded network, respectively. In general, when the network grows larger, we need more sensors. Hence, $|S| \geq |S_0|$. To reduce the cost of S, we may reuse S_0. There are two kinds of reuse:

1. First, the sensor placement on some location may be completely preserved. $S \cap S_0$ contains such sensor locations. No cost will be paid for this kind of reuse.
2. Second, we may move sensors to other locations. For such kind of reuse, we need to pay the uninstall and reinstall cost.

Let $S_1 \subseteq S$ be the set of sensor locations in which sensors are reused in the second case. The relationships among S_0, S_1 and S for the running example is illustrated in Figure 1(c).

Let C_1, C_2 be the cost of installing and uninstalling a sensor, respectively. The cost of replacing an existing sensor to a different location can be approximated by C_1 and C_2. Suppose the overall cost budget C is given. Then, we expect $|S_1|(C_1 + C_2) + (|S| - |S_0|)C_1 \leq C$. In general, C_1, C_2 can be considered as constants in a typical water network. Hence, limiting the cost is equivalent to limiting $|S_1|$ and $|S| - |S_0|$. Thus, we can use two input parameters k_1, k_2 to bound $|S_1|$ and $|S| - |S_0|$, respectively, to precisely model the cost constraint.

Problem Definition. Now, we can define the Incremental Sensor Placement (ISP) problem in Definition 4. We use Example 3 to illustrate ISP.

Definition 4 (Incremental Sensor Placement(ISP)). *Given a water network G with sensor locations L, scenario set $\mathcal{A}$, CS matrix $D_{\mathcal{A},L}$, a credit M and a sensor placement $S_0 \in L$ and two integers k_1, k_2, find a new sensor placement $S \subseteq L$ such that (1) $|S_1| \leq k_1, |S| - |S_0| = k_2$, where $S_1 = S_0 - S$ is the set of places at which sensors have been uninstalled from S_0 and replaced on a sensor location outside of S_0; and (2) $F(S)$ is maximized.*

Example 3 (ISP). Continue the previous example shown in Figure 1, where $S_0 = \{v_2, v_6\}$. Suppose $k_1 = 1, k_2 = 2$, which means that we could modify the location of at most 1 sensor deployed in S_0 and 2 additional sensors would be deployed on the network. Our goal is to find a sensor placement solution S such that $|S_0 - S| \leq 1$, $|S| - |S_0| = 2$ and $F(S)$ is maximized.

Notice that in our problem Definition 4(ISP), the sensor placement S cannot guarantee that all the contamination scenarios in scenario set $\mathcal{A}$ are covered. In fact, according to previous works [25,16,17], it requires an exponential number of sensors to detect all scenarios in $\mathcal{A}$, which is unaffordable in real world due to the limited budget. Hence, covering scenarios as many as possible under the budget limit is a more realistic objective for real applications.

3.4   NP-Hardness of ISP

In this subsection, we will show that ISP is NP-hard. Our proof consists of two steps. We first show that SP in our model is NP-hard by the reduction from the NP-Complete problem Maximal Coverage Problem (Definition 5) to SP. Then, we prove the NP-hardness of ISP by showing that SP is a special case of ISP.

Definition 5 (Maximal Coverage Problem (MC)). *Given a number k, universe V with n elements and a collection of sets $\mathcal{C} = \{C_1 ... C_m\}$ such that each $C_i \subseteq V$, find a subset $C' \subseteq \mathcal{C}$ such that $|C'| \leq k$ and $|\bigcup_{C_i \in C'} C_i|$ is maximized.*

Theorem 1 (NP-hardness of SP). *SP is NP-hard.*

Proof. It can be shown that for any instance of MC, i.e. $< V, \mathcal{C} = \{C_1, ..., C_m\}, k >$ we can construct an instance of SP accordingly such that there exists solution for this instance of MC if and only if the corresponding SP instance could be solved. For this purpose, let $V = \mathcal{A}$, $k = p$, and for each C_i we define a sensor location v_i such that $|L| = m$. Then, we just need to show that given $\mathcal{C} = \{C_1, ..., C_m\}$, we can always construct a pair $< M, D_{\mathcal{A},L} >$ such that each $R_{v_i, M} = C_i$. Clearly, it can be easily constructed. For any M (without loss of generality, we may assume it as an integer), we can create matrix $D_{\mathcal{A},L}$ as follows. For each $v_i \in L$ and each scenario $a \in \mathcal{A}$, we set $d_{a, v_i} = \infty$ at first. Then, for each $c \in C_i$, $i = 1, ..., m$, we set d_{c, v_i} to $M - 1$. As a result, we surely have $R_{v_i, M} = C_i$.   ∎

Theorem 2 (NP-hardness of ISP). *ISP is NP-hard.*

Proof. Let k be the number of senors to be placed in problem SP. We can simply set $S_0 = \emptyset$, and $k_1 = 0, k_2 = k$ to transform an instance of SP into a problem instance of ISP. Since SP is NP-hard, ISP is also NP-hard.   ∎

3.5   Submodularity of Evaluation Function

In this subsection, we will present the major properties of evaluation function $F(\cdot)$, especially the submodularity of this function, which underlies our major solution for ISP.

$F(S)$ has some obvious properties. First, it is *nonnegative*. It is obvious by definition. $F(\emptyset) = 0$, i.e., if we place no sensors, the evaluation function is 0. It is also *nondecreasing*, i.e., for placement placements $A \subseteq B \subseteq L$, it holds that $F(A) \leq F(B)$. The last but also the most important property is *submodularity*. Intuitively, adding a sensor to a large deployment brings less benefits than adding it to a small deployment. The diminishing returns are formalized by the combinatorial concept of *submodularity* [23] (given in Definition 6). In other words, adding sensor s to the smaller set A helps more than adding it to the larger set B. Theorem 3 gives the results.

Definition 6 (Submodularity). *Given a universe set S, a set function F is called submodular if for all subsets $A \subseteq B \subseteq S$ and an element $s \in S$, it holds that* $F(A \cup \{s\}) - F(A) \geq F(B \cup \{s\}) - F(B)$.

Theorem 3 (Submodularity of $F(\cdot)$). *Sensor placement evaluation function $F(\cdot)$ given in Definition 2 is submodular.*

Proof. For sensor placement $A \subseteq L$, $F(A \cup \{i\}) = |R_{i,M} \cup (\bigcup_{v \in A} R_{v,M})|$. Then, we have $F(A \cup \{i\}) - F(A) = |R_{i,M} \cup \bigcup_{v \in A} R_{v,M}| - |\bigcup_{v \in A} R_{v,M}| = |R_{i,M} - \bigcup_{v \in A} R_{v,M}|$. Given two sensor placements $A \subseteq B \subseteq L$. We have $\bigcup_{v \in A} R_{v,M} \subseteq \bigcup_{v \in B} R_{v,M}$. Thus, $R_{i,M} - (\bigcup_{v \in A} R_{v,M}) \supseteq R_{i,M} - (\bigcup_{v \in B} R_{v,M})$. Consequently, we have $F(A \cup \{i\}) - F(A) \geq F(B \cup \{i\}) - F(B)$. ∎

Thus, the SP problem on water distribution networks can be cast as a submodular optimization problem, which has been proven to be NP-hard and can be solved with bounded quality [22]. Next, we will present the detail of solutions based on submodularity of the evaluation function.

4 Algorithm Solutions

Let L be the set of locations that can hold sensors and q be the number of sensors to be installed. To solve SP, a brute-force solution needs to enumerate all $C_{|L|}^q$ sensor deployments, where C represents the binomial coefficient. For ISP, we should enumerate $\sum_{0 \leq i \leq k_1} C_{|S_0|}^i C_{|V|-|S_0|}^{i+k_2}$ possible deployments, where S_0 is the original sensor placement. In both cases, the search space is exponential. Hence, it is computationally prohibitive to find the global-optimal deployment for large water distribution network by an exhaustive enumeration approach. To overcome the complexity, we will first present a greedy approach to solve SP. Then, we will further extend it to solve ISP.

4.1 Greedy Algorithms for SP

The basic greedy heuristic algorithm starts from the empty placement $S = \emptyset$ and proceeds iteratively. In each iteration, a new place $v \in V$ which leads to the most increase of F, i.e.,

$$\delta_v = F(S \cup \{v\}) - F(S), \tag{3}$$

$v_c = \arg\max_{v \notin S}(\delta_v)$ would be added to S. In other words, at each iteration, we always select the place covering the largest number of uncovered scenarios. A fundamental result in [23] shows that the above greedy procedure will produce a *near optimal* solution for the class of *nondecreasing submodular functions*. More specifically, for any instance of SP, the greedy algorithm always return a sensor placement S such that $F(S) \geq (1 - \frac{1}{e})F(S^*)$, where S^* is the global optimal solution to this instance. Hence, the greedy solutions achieve an approximation ratio at least $1 - \frac{1}{e} \approx 63\%$ compared to the global optimal solution.

We present the greedy procedure in Algorithm 1. After the selection of v_c, we update $R_{v,M}$ by excluding the contamination scenarios that have been covered by v_c (line 6-8). Thus, when the procedure proceeds, the size of $R_{v,M}$ becomes progressively smaller. Note that the size of $R_{v,M}$ for each candidate vertex in the beginning of each iteration is equal to δ_v as defined in Equation 3.

The running time of the algorithm is proportional to the number of sensor locations $|L| = n$ of the water network, the number of sensors to be deployed p, the size of contamination scenarios $|\mathcal{A}| = m$ and the time taken to calculate the size of remaining uncovered scenarios for each $v \in L - S_G$. The update of $R_{v,M}$ needs set union operation, whose complexity is $O(m)$. In each iteration, $O(n)$ vertices need to be evaluated on its quality function. Hence, the total running time is $O(pnm)$.

Algorithm 1. GreedySP

Input: p, M, $R_{v,M}$ for each $v \in L$
Output: S_G
  1: $iter \leftarrow 1$
  2: $S_G \leftarrow \emptyset$
  3: **while** $iter \leq p$ **do**
  4:     $v_c \leftarrow \arg\max_{v \in L - S_G} |R_{v,M}|$
  5:     $S_G \leftarrow \{v_c\} \bigcup S_G$
  6:     **for** each $v \in L - S_G$ **do**
  7:         $R_{v,M} \leftarrow R_{v,M} - R_{v_c,M}$
  8:     **end for**
  9:     $iter \leftarrow iter + 1$
 10: **end while**
 11: **return** S_G

4.2 Algorithms for ISP

In this section, we will further apply the above greedy heuristic to solve ISP. Compared to SP, effective reuse of the original sensor placement S_0 is one of the distinctive concern of ISP. We use a function *Select* to decide the subset of the original sensor placement to be reused. And we will discuss different selection strategies used in *Select* and their effectiveness in detail later. According to our definition of ISP (see Definition 4), there are at least $|S_0| - k_1$ sensors remain unchanged in the new sensor placement.

The greedy algorithm to solve ISP is presented in Algorithm 2, which consists of three major steps. In the first step (line 1), we use *Select* to choose $|S_0| - k_1$ sensors in S_0 to be preserved (denoted by S_r). In the second step (line 2-5), the algorithm updates $R_{v,M}$ for every sensor location $v \in L - S_r$ at which no sensor is deployed by eliminating the scenarios *covered* by the $|S_0| - k_1$ sensors. In the last step (line 7), we directly call Algorithm 1 to calculate a solution for deploying $k_1 + k_2$ sensors on $L - S_r$. The union of the result of the last step and S_r will be returned as the final answer.

Obviously, the selection strategy used in *Select* function has a significant impact on the quality of the final sensor placement. We investigate three candidate strategies: *randomized, greedy* and *simulated annealing*. The effectiveness of three strategies will be tested in the experimental section.

Randomized Heuristic (RH). The straightforward strategy is choosing $|S_0| - k_1$ sensors from the original sensor placement S_0 randomly such that each sensor has the same probability to be selected.

Greedy Heuristic (GH). The greedy heuristic is identical to that used in Algorithm 1. Start with the empty placement, $S = \emptyset$ and proceeds $|S_0| - k_1$ times iteratively. In each iteration, select vertex $v \in S_0$ such that $\delta(v)$ is maximal from the remaining vertices in S_0.

Simulated Annealing (SA). First, we use Randomized Heuristic to choose $|S_0| - k_1$ sensors denoted as S_{RH}. Then, we start from S_{RH}, and perform a local search based approach called Simulated Annealing to get a local optimal solution. Simulated Annealing proceeds iteratively. Let S_{cur} be the solution to be optimized in current iteration. The simulated annealing proceeds as follows. At each round, it proposes an exchange of a selected vertex $s \in S_{cur}$ and an unselected vertex $s' \in S_0 - S_{cur}$ randomly, then computes the quality gain function for the exchange of s, s' by

$$\alpha(s, s') = F(S_{cur} \cup \{s'\} - \{s\}) - F(S_{cur}) \tag{4}$$

If $\alpha(s, s')$ is positive, i.e. the exchange operation improves the current solution, the proposal is accepted. Otherwise, the proposal is accepted with probability $\exp(\frac{\alpha(s,s')}{\vartheta_t})$, where ϑ_t is the *annealing temperature* at round t, and $\vartheta_t = Cq^t$ for some large constant C and small constant q $(0 < q < 1)$. We use exponential decay scheme for *annealing temperature*. Such exchanges are repeated until the number of iterations reach to the user-specified upper limit.

Algorithm 2. Greedy Algorithm for ISP

Input: $p, M, R_{v,M}$ for each $v \in V, S_0, k_1$
Output: S_G
  1: $S_r \leftarrow Select(S_0, |S_0| - k_1)$
  2: **for** each $u \in S_r$ **do**
  3:    **for** each $v \in L - S_r$ **do**
  4:       $R_{v,M} \leftarrow R_{v,M} - R_{u,M}$
  5:    **end for**
  6: **end for**
  7: $S' \leftarrow GreedySP(k_1 + k_2, M, \{R_{v,m} | v \in L - S_r\})$
  8: **return** $S_r \cup S'$

---

5   Experimental Study

In this section, we present our experimental study results. We show in Sec 5.1 the experiment setup. Then, in Sec 5.2 we justify our scenario-cover based modeling through comparisons of different solutions to SP. Sec 5.3 5.4 5.5 5.6 will study our solution to ISP, including the influence of parameters k_1, k_2, robustness and performance.

5.1 Experiment Setup

We use two real water networks provided by BWSN challenge [10] to test our algorithms. The first one is *BWSN1* containing 129 nodes. The second one is *BWSN2* containing 12,527 nodes. We run EPANET [1] hydraulic simulation and water quality simulation for the two water distribution networks and use the *time until detection* [10] as the criteria to evaluate contamination impact. For network *BWSN1*, 516 contamination scenarios were generated for each of the vertices in the water distribution network at 4 different attack time (*i.e. the start time of contamination* t_s): 6 A.M., 12 A.M., 6 P.M., 12 P.M.. The reason why we vary the start time is that during simulation of water network, parameters of resources in the water systems such as junction's water pressure or pipe's flow velocity would change over time, which will influence the propagation of contamination events. Each contamination scenario features 96-hour injection of a fictional contaminant (*i.e. stop time point* t_f *of each scenario is set to be the end time point of contamination simulation*) at strength $1000mg/min$ (using EPANET's 'MASS' injection type). We record the contaminant concentration at intervals of 5 *minutes* and use these time points as the time series.

For network *BWSN2*, we generate 4000 contamination scenarios originating from 1000 randomly selected vertices in the water distribution network with settings identical to that of *BWSN1*. In the following experiments, we fix M as 120 *min* and 150 *min* for *BWSN1* and *BWSN2*, respectively. For the water quality simulation on the two networks, we assume that a deployed sensor would alarm when the concentration of contaminant surpasses $10mg/L$. To compare the effectiveness of our solution, we use *detect ratio*, defined as $\frac{F(S)}{|\mathcal{A}|}$, *i.e*, the proportion of covered contamination scenario, to measure the quality of the sensor placement S, where $|\mathcal{A}|$ denotes the total number of contamination scenarios considered. We run all algorithms on a machine with 2G memory and 2.2GHZ AMD processor. All algorithms are implemented in C++.

5.2 Effectiveness and Efficiency of Solutions to SP

In this subsection, we will show the effectiveness and efficiency of our SP solution. The purpose is to justify the scenario-cover based model. Through the comparisons to other solutions, we will show that by modeling sensor placement optimization in the form of scenario-cover, our solution achieves good solution quality (comparable to the state-of-the-art solution) but consumes significantly less running time. With the growth of real water network, improving the scalability without sacrificing the quality will be a more and more critical concern.

We compare to the following approaches:

1. *Random placement.* In a random placement, we randomly select p sensor locations as a solution. We repeat it for 100 times. For each random placement, we calculate its *detect ratios*. Then, we summarize the the median, minimum, maximum, 10th, 25th, 75th, and 90th-percentiles over 100 random solutions in our experiment.

2. *Exhaustive search*. It's a brute-force solution by enumerating all possible sensor placements. Clearly it gets the optimal result but consumes the most time. Since the time cost is unaffordable, we estimate the entire running time by multiplying the time of enumerating one placement and the number of possible enumerations .

3. *MIP(mixed-integer programming)*. The state-of-the-art approach for SP uses mixed-integer programming modeling. We use LINDO [2], a state-of-the-art MIP solvers, to solve SP.

On large network We first compare random placement to our greedy approach on the large network *BWSN2*. MIP and exhaustive search can not scale to large networks, hence are omitted here. The result is shown in Figure 2(a). We can see that our method can detect significantly more scenarios than random placement. Even the optimal one in 100 random placements is worse than our greedy solution. Hence, in the following test, random placement will not compared.

On small network We also compared different approaches on the small network: *BWSN1*. The result is given in Figure 2(b) and Figure 2(c). From Figure 2(b),we can see that our greedy solution's quality is comparable to that produced by MIP. Their difference is less than 3.7%. However, MIP costs almost one order of magnitude more time than greedy solution. This can be observed from Figure 2(c).

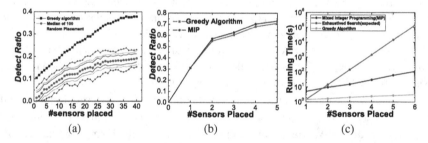

(a)   (b)   (c)

Fig. 2. Comparison of effectiveness and time. Figure 2(a) compares the results of greedy solution and random placement strategy on *BWSN2*. Figure 2(b) compares solution quality(*detect ratio*) of our greedy algorithm and MIP. Figure 2(c) compares running time of our greedy algorithm, exhaustive search and MIP.

5.3   Influence of k_1

In this experiment, we explore the influence of k_1, i.e., the maximal number of redeployed sensors, on the solution quality of ISP. By this experiment, we justify the motivation of ISP. We show that we only need to modify a relatively small part of the original sensor placement to keep the effectiveness of sensors while significantly reducing the deployment cost.

We first need to simulate the growth of a water network since no real evolving water network data is available. For the simulation, we first find a subregion of the entire water network and consider it as the network at the earlier time. We solve the SP problem on

---

[2] http://www.lindo.com/

this subregion and obtain a sensor placement S_0 on this subregion. Then, the entire water network can be regarded as the network after growth with S_0 as the original sensor placement.

Due to the expansion of water network, S_0 may fail to detect contamination events. In this experiment, we exclude the influence caused by deploying new sensors through setting k_2 as 0. Note that by setting $k_2 = 0$, no new sensors will be added in our solution. Hence, the *detect ratio* may be quite small for a large network (as indicated in Figure 3(c)). Then, we vary the upper bound of reinstalled sensors k_1 from 0 to $|S_0|$ and observe the evolution of *detect ratio* varying with k_1. For each network, we set $|S_0| = 5$ and $|S_0| = 10$ for *BWSN1* and *BWSN2*, respectively. We define *increase rate* of *detect ratio* as the difference of *detect ratio* between $k_1 = d$ and $k_1 = d + 1$.

The results are shown in Figure 3(a) and 3(c). It is clear that *detect ratio* increases with the growth of k_1, indicating that if we allow more sensors to be redeployed, we can cover more contamination scenarios. However, the *increase rate* gradually decreases when k_1 increases. It is interesting to see that there exist critical points for both two networks ($k_1 = 2$ for BWSN1 and $k_1 = 4$ for BWSN2, respectively), after which the *detect ratio* will increase very slowly.

Considering results given above, we find a good trade-off between *detect ratio* and deployment cost. Since after the critical point, the improvement of *detect ratio* is slower than the increase of cost, generally we can set k_1 at the critical point to trade quality for cost.

Note that k_1 is the upper limit of the actual redeployed sensors. We further summarize the actual number of redeployed sensors in Figure 3(b) and 3(d). We can see that the actual number of reinstalled sensors is always equal to k_1, which implies that the original sensor placement S_0 needs to be redeployed completely to enhance the *detect ratio* on current water network.

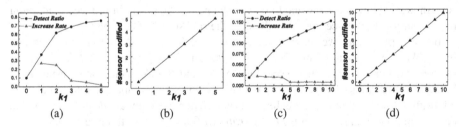

(a)   (b)   (c)   (d)

Fig. 3. (a)(c):*detect ratio* for *BWSN1* and *BWSN2* (b)(d): the actual number of redeployed sensors in *BWSN1* and *BWSN2*. The results show that solution quality generally increases with the growth of k_1 and there exists some critical point at which we can seek for a good tradeoff between solution quality and deployment cost.

5.4   Influence of k_2 and the Selection Strategies

In this experiment, we explore the influence of k_2 and compare the effectiveness of three strategies used in *Select* function to solve ISP. We set k_1 as 2 and 5 for *BWSN1* and *BWSN2*, respectively. Other parameters are the same as the previous experiment. We set iteration number to be 1000 for the simulated annealing strategy.

From Figure 4(a), i.e., the result on *BWSN1*, we can see that simulated annealing and random heuristic strategy show only minor priority in *detect ratio* over the greedy heuristic. However, on *BWSN2* (shown in Figure 4(b)), the performance of the three strategies are quite close to each other. Such results imply that our solution is generally independent on the selection strategy. For comparison, we also present median of *detect ratio* of 100 complete random placements in Figure 4(b).

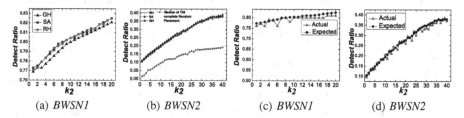

| (a) *BWSN1* | (b) *BWSN2* | (c) *BWSN1* | (d) *BWSN2* |

Fig. 4. Figure 4(a),4(b) show the influence of k_2 and compare different selection strategies used in *Select* function. It shows that solution quality increases with the growth of k_2, and three strategies shows similar effectiveness. Figure 4(c),4(d) present the robustness of proposed solutions. *Expected* represents the *detect ratio* on the training scenario set; *Actual* represents the average of *detect ratio* on test scenario sets. These figures show that our sensor placement solution is *robust* against newly introduced contamination scenarios.

5.5 Robustness of Solutions

In general, there may exist potentially infinite number of possible contamination scenarios. A placement strategy with high *detect ratio* on the training contamination scenario set may be ineffective to detect contamination scenarios not belonging in the training set. An ideal solution is expected to be *robust* against the newly introduced contamination scenarios. In this section, we will show the robustness of our solution.

In our experiments, we set $|S_0| = 5, 10$, $k_1 = 2, 5$ for *BWSN1* and *BWSN2*, respectively. We use simulated annealing with 1000 iterations for the *Select* function. We randomly generate four scenario sets for *BWSN1* and *BWSN2* as test sets, respectively. Each test set contains 100 scenarios. We choose one set as $\mathcal{A}$ and get solution S using Algorithm 2 with simulated annealing heuristic. Then, we compare *detect ratio* of S on the training set $\mathcal{A}$ and the average *detect ratio* of S on the other three sets.

The result is shown in Figure 4(c) and 4(d), where we vary the number of new sensors (k_2) and observe the evolution of *detect ratio*. It can be observed that when k_2 increases, *detect ratio* of test case on average is quite close to that of the training sets, implying that our sensor placement which is effective on training scenario set $\mathcal{A}$ works well on other scenario sets as well. Hence, our solution to ISP is *robust* against unknown contamination scenarios.

5.6 Performance of Our Solution for ISP

In this experiment, we test the performance of our solutions for ISP. MIP can not be easily extended on ISP problem. Random placement is certainly the fastest, but as shown in Sec 5.2, is of quite low quality. Hence, in this experiment, we only compare to exhaustive searching whose time is estimated as stated in Sec 5.2. We fix $|S_0| = 5$ and $|S_0| = 10$ for *BWSN1* and *BWSN2*, respectively.

The result is shown in Figure 5, from which we can see that our solution is faster than exhaustive search, and the speedup is almost two or three orders of magnitudes. Our algorithm generally linearly increases with the growth of k_2. Even on the large network with ten thousands of nodes, our solution finds a solution within 2-3 hours. The performance result implies that our approach can scale up to large water networks.

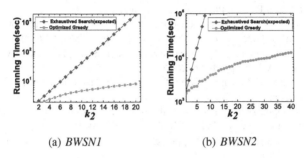

| (a) *BWSN1* | (b) *BWSN2* |

Fig. 5. Running time comparison of solutions to ISP.

6   Conclusion

In this paper, we propose a new problem: incremental sensor placement optimization problem (ISP), in which we need to find an optimal sensor placement for the dynamic-evolving water network with the following two objectives: (1) keeping the deployment cost limited and (2) maximizing the effectiveness of the new sensor placement. We show this problem is NP-hard. We prove that the objective function used in the definition of ISP is submodular. Based on this property, we propose several greedy algorithms to solve this problem. Experimental results verify the effectiveness, robustness and efficiency of proposed solutions. We will further consider more realistic constraints on our problem to solve more specific real sensor optimization problems.

References

1. Lee, B.H., Deininger, R.: Optimal locations of monitoring stations inwater distribution system. J. Environ. Eng. 118(1), 4–16 (1992)
2. Kumar, A., et al.: Identification of monitoring stations in water distribution system. J. Envir. Engin. 123(8), 746–752 (1997)
3. Al-Zahrani, M.A., Moied, K.: Locating optimum water quality monitoring stations in water distribution system. In: Proc. World Water and Environmental Resources Congress (2001)
4. Kessler, A., et al.: Detecting accidental contaminations in municipal water networks. J. Water Resour. Plann. Manage 124(4), 192–198 (1998)
5. Ostfeld, A., Kessler, A.: Protecting urban water distribution systems against accidental hazards intrusions. In: IWA Second Conference (2001)

6. Berry, J.W., Fleischer, L., et al.: Sensor placement in municipal water networks. J. Water Resour. Plann. Manage 131(3), 237–243 (2005)

7. Jean-paul, W., et al.: A multiple-objective analysis of sensor placement optimization in water networks. In: Critical Transitions In Water And Environmental Resources Management, pp. 1–10 (2006)

8. Ostfeld, A., Salomons, E.: Optimal layout of early warning detection stations for water distribution systems security. J. Water Resour. Plann. Manage 130(5), 377–385 (2004)

9. Berry, J., Hart, W.E., et al.: Sensor placement in municipal water networks with temporal integer programming models. J. Water Resour. Plann. Manage 132(4), 218–224 (2006)

10. Ostfeld, A., et al.: Battle of water sensor networks: A design challenge for engineers and algorithms. J. Water Resour. Plann. Manage 134(6), 556 (2008)

11. Ostfeld, A., Salomons, E.: Sensor network design proposal for the battle of the water sensor networks (bwsn). J. Water Resour. Plann. Manage 134(6), 556–568 (2008)

12. ZhengYi, W., Tom, W.: Multi objective optimization of sensor placement in water distribution systems. In: WDSA 2006, pp. 1–11 (2006)

13. Dorini, G., et al.: An efficient algorithm for sensor placement in water distribution systems. In: WDSA 2006, pp. 1–13 (2006)

14. Guan, J., et al.: Optimization model and algorithms for design of water sensor placement in water distribution systems. In: WDSA 2006, pp. 1–16 (2006)

15. Berry, J., et al.: A facility location approach to sensor placement optimization. In: WDSA 2006, pp. 1–4 (2006)

16. Huang, J.J., et al.: Multiobjective optimization for monitoring sensor placement in water distribution systems. In: WDSA 2006, pp. 1–14 (2006)

17. Preis, A., Ostfeld, A.: Multiobjective sensor design for water distribution systems security. In: WDSA 2006, pp. 1–17 (2006)

18. Ghimire, S.R., Barkdoll, B.D.: Heuristic method for the battle of the water network sensors: Demand-based approach. In: WDSA 2006, pp. 1–10 (2006)

19. Gueli, R.: Predator-prey model for discrete sensor placement. In: WDSA 2006, pp. 1–9 (2006)

20. Ghimire, S.R., Barkdol, B.D.: A heuristic method for water quality sensor location in a municipal water distribution system:mass related based approach. In: WDSA 2006, pp. 1–11 (2006)

21. Eliades, D., Polycarpou, M.: Iterative deepening of pareto solutions in water sensor networks. In: WDSA 2006, pp. 1–19 (2006)

22. Leskovec, J., et al.: Cost-effective outbreak detection in networks. In: KDD 2007, pp. 420–429 (2007)

23. Nemhauser, G.L., et al.: An analysis of approximations for maximizing submodular set functions. Mathematical Programming 14, 265–294 (1978)

24. Sviridenko, M.: A note on maximizing a submodular set function subject to a knapsack constraint. Operations Research Letters 32, 41–43 (2004)

25. Krause, A., et al.: Efficient sensor placement optimization for securing large water distribution networks. J. Water Resour. Plann. Manage (2008)

26. Goyal, A., et al.: Learning influence probabilities in social networks. In: Goyal, A., et al. (eds.) WSDM 2010, pp. 241–250. ACM (2010)

27. Chen, W., et al.: Efficient influence maximization in social networks. In: KDD 2009, pp. 199–208. ACM (2009)

28. Gomez Rodriguez, M., et al.: Inferring networks of diffusion and influence. In: KDD 2010, pp. 1019–1028. ACM (2010)

Detecting Marionette Microblog Users
for Improved Information Credibility

Xian Wu[1], Ziming Feng[1], Wei Fan[2], Jing Gao[3], and Yong Yu[1]

[1] Shanghai Jiao Tong University, Shanghai, 200240, P.R. China
{wuxian,fengzm,yyu}@apex.sjtu.edu.cn
[2] Huawei Noah's Ark Lab, Hong Kong
david.fanwei@huawei.com
[3] University at Buffalo, NY 14260, USA
jing@buffalo.edu

Abstract. In this paper, we mine a special group of microblog users: the "marionette" users, who are created or employed by backstage "puppeteers", either through programs or manually. Unlike normal users that access microblogs for information sharing or social communication, the marionette users perform specific tasks to earn financial profits. For example, they follow certain users to increase their "statistical popularity", or retweet some tweets to amplify their "statistical impact". The fabricated follower or retweet counts not only mislead normal users to wrong information, but also seriously impair microblog-based applications, such as popular tweets selection and expert finding. In this paper, we study the important problem of detecting marionette users on microblog platforms. This problem is challenging because puppeteers are employing complicated strategies to generate marionette users that present similar behaviors as normal ones. To tackle this challenge, we propose to take into account two types of discriminative information: (1) individual user tweeting behaviors and (2) the social interactions among users. By integrating both information into a semi-supervised probabilistic model, we can effectively distinguish marionette users from normal ones. By applying the proposed model to one of the most popular microblog platform (Sina Weibo) in China, we find that the model can detect marionette users with f-measure close to 0.9. In addition, we propose an application to measure the credibility of retweet counts.

Keywords: marionette microblog user, information credibility, fake followers and retweets.

1 Introduction

The flourish of Microblog services, such as Twitter, Sina Weibo and Tencent Weibo, has attracted enormous number of web users. According to recent statistics, the number of Twitter users has exceeded 500 million in July 2012,[1] and

---

[1] http://semiocast.com/publications

H. Blockeel et al. (Eds.): ECML PKDD 2013, Part III, LNAI 8190, pp. 483–498, 2013.
© Springer-Verlag Berlin Heidelberg 2013

Sina Weibo has more than 300 million users in Match 2012.[2] Such a large volume of participants has made microblog a new social phenomenon that attracts attention from a variety of domains, such as business intelligence, social science and life science.

In Microblog services, the messages (tweets) usually deliver time sensitive information, e.g., "What the user is doing now?". By following others, a user will be notified of all their posted tweets and thus keep track of what these people are doing or thinking about. Therefore, the number of followers measures someone's popularity, and can indicate how much influence someone has. For celebrities, a large number of followers shows their social impact and can increase their power in advertisement contract negotiations. As for normal users, a relatively large number of followers represents rich social connections and promotes one's position in social networks. Therefore, both celebrities and normal users are eager to get more followers.

Due to the retweet mechanism, information propagation is quite efficient in microblog services. Once a user posts a message, his followers will be notified immediately. If these followers further retweet this message, their followers can view it immediately as well. In this way, the number of audiences can grow at an exponential rate. Therefore, the retweet count of a message represents its popularity, as more users wish to share with their followers. On many microblog platforms (e.g., Sina Weibo), the retweet count is adopted as the key metric to select top stories.[3] As a result, some microblog users are willing to purchase more retweets to promote their messages for commercial purpose.

The desire for more followers and retweets triggers the emergence of a new microblog business: follower and retweet purchase. The backstage puppeteers maintain a large pool of marionette users. To purchase followers or retweets, the buyer first provides his user id or tweet id. Then the puppeteer activates certain number of marionette users to follow this buyer or retweet his message. The number of followers or retweets depends on the price paid. The fee is typically modest, 25 USD for 5,000 followers in Twitter, and 15 Yuan (i.e., 2.5 USD) for 10,000 followers in Sina Weibo. Moreover, the massive following process is quite efficient. For example, it only takes one night to add 10,000 followers in Sina Weibo, which can make someone become "famous" overnight.

From the perspectives of people who made the follower and retweet purchase, the marionette users can satisfy their needs to become famous and help in promoting commercial advertisements, but overall the fabrications conducted by marionette users can lead to serious damages:

- The purchased followers fabricate the social influence of users, and the purchased retweets amplify the public attention paid to the messages. As a result, the fake numbers can mislead real users and data mining applications based on microblog data, such as [1]
- Beside promoting advertisements, the marionette users are sometimes employed to distribute rumors [2]. It will not only mislead normal users but

---

also provide wrong evidence for business [3] and government's establishment of policies and strategies. Thus, this becomes a serious financial and political problem.

– To disguise as normal users, the marionette users are operated by puppeteers to perform some random actions, including following, retweeting and replying. Such actions can interfere operating and viewing experiences of normal users and result in unpleasant user experience.

Therefore, identifying marionette users is a key challenge for ensuring normal functioning of microblog services. However, marionette users are difficulty to detect. Monitoring the marionette users over the past two years, we find that the difficulty of detecting marionette users has gradually gone up. Back in November 2011, we purchased 2,000 followers from Taobao (China EBay) and all these fake followers were recognized by microblog services and deleted within two days. Such quick detection can be attributed to the several discriminative features. For example, the marionette accounts are usually created from the same IP address within a short period of time, many marionette users posted no original tweets but only performed massive following or retweeting, and so on. Therefore, the microblog services can employ simple rules to detect marionette users and delete their accounts. However, the marionette users are evolving and becoming more intelligent. Nowadays, the puppeteers hire people or use crowdsourcing to create marionette accounts manually. To make these accounts behave like normal users, the puppeteers develop highly sophisticated strategies that operate the marionette users to follow celebrities, reply to hot tweets, and conduct other complicated operations. These disguises can easily overcome the filtering strategies of microblog platforms and make marionette users much more difficult to be detected. In February 2012, we purchased another 4,000 followers. This time, 1,790 marionette users survived after five weeks, and around 1,000 marionette users are still active by Feb 2013.

After analyzing the behaviors of marionette users and comparing them with normal users, we find that the following two types of information are useful in detecting marionette users.

– Local Features: The features that describe individual user behaviors, which could be either textual or numerical. The local features can capture the different behaviors between normal and marionette users. For example, the following/follower counts are important features that distinguish a large portion of normal users from marionette users. The time interval between tweets and the tweet posting devices also serve as effective clues to detect marionette users.

– Social Relations: The following, retweeting or other relationships among users. Such relations provide important information for marionette user detection. For example, the marionette users will follow both normal users and other marionette users. They follow normal users to disguise or for profits, and follow other marionette users to help them to disguise. On the other hand, the normal users are less likely to follow marionette users. Therefore the neighboring users that are connected by the "following" relation can be used to recognize marionette users.

The two types of features provide complementary predictive powers for the task of marionette user detection. Therefore, we propose a probabilistic model that seamlessly takes both the rich local features as well as social relations among users into consideration to detect marionette users more effectively. On the dataset collected from Sina Weibo, the proposed model is able to detect marionette users with the f-measure close to 0.9. As a result, we are able to measure the true popularity of hot tweets. For example, given a hot tweet, we can first extract the users who retweet it and then evaluate whether these users are marionette users. The percentage of normal users can be used to measure the true popularity.

2   The Proposed Model

In this section, we describe the proposed probabilistic model that integrates local features and social relations in marionette user detection.

2.1   Notation Description

Let u_i denote a microblog user and let the vector x_i denote the features of u_i. Each dimension of x_i represents a local feature, which could be the follower count of u_i or the tweeting device u_i has used before. Let the binary variable y_i denote the label of u_i, 1 stands for the marionette user and 0 stands for the normal user. Let $V^{(i)} = \{v_1^{(i)}, v_2^{(i)}, \ldots, v_{M(i)}^{(i)}\}$ denote the $M(i)$ users who are related to u_i. In microblog services, the social relations between users can be either explicit or implicit. To be concrete, "followed" and "following" are explicit relations while retweeting one's tweet or "mention" someone in a tweet establish implicit social relations. In this paper, we target to predict the label y_i of u_i given its local features x_i and his social relations $V^{(i)}$.

2.2   Problem Formulation

We will first describe how to only use local features that describe user behavior on the microblogging platform, such as follower/following counts, the posting devices, to build a discriminative model. We will later describe how to incorporate social relations into this model to further improve the performance. If we only consider the local features, marionette user detection is a typical classification problem. A variety of classification models can be used, among which we choose Logistic Regression because it can be adapted to incorporate social relations which will be shown later in this paper. We first describe how to model local user features using Logistic Regression model.

We introduce the sigmoid function in Eq.(1) to represent the probability of belonging to marionette or normal class given feature values, i.e., $P(y_i|x_i)$, for each user.

$$P_\theta(y_i|x_i) = h_\theta(x_i)^{y_i}(1 - h_\theta(x_i))^{(1-y_i)} \qquad (1)$$

where $h_\theta(x_i) = \frac{1}{1+e^{-\theta^T x_i}}$ is equal to the probability that u_i is a marionette user. θ is the set of parameters that characterizes the sigmoid function. With Eq.(1), we can formulate the joint probability over N labeled users in Eq.(2), in which we try to find the parameter θ that maximizes this data likelihood.

$$\max_\theta \ \prod_{i=1}^{N} P_\theta(y_i|x_i) \qquad (2)$$

In the above formulation, each user is treated separately and the prediction of a marionette user only depends on one's local features. However, besides the local features, the relations between users are also discriminative for the task of predicting marionette users. To incorporate the social relations, we modify the objective function from Eq.(2) to Eq.(3).

$$\max_{\theta,\alpha} \ \prod_{i=1}^{N}\{P_\theta(y_i|x_i) \prod_{j=1}^{M(i)} P_\alpha(y_i|y_j^{(i)})^d\} \qquad (3)$$

In Eq.(3), we assume that, for each user u_i, the label of its $M(i)$ neighbors $y_0^{(i)}$, $y_1^{(i)}$, ..., $y_{M(i)}^{(i)}$ are known in advance. Then we can integrate the effect of local features and user connections together to predict marionette users. d is the coefficient that balances between the social relations and local features. The larger d is, the more biased the model is towards the social relations in making the predictions. Note that to simplify the presentation, we consider the case where only one type of social relations exists in Eq.(3). However, the proposed model is general enough and can be easily adapted to cover multi-type social relations. Take the microblog system for example, the common user relations include follower, following, mention, retweet and reply. We can introduce different parameter α to correspond to each kind of relation and model all relations in one unified framework.

In Eq.(3), $P_\theta(y_i|x_i)$ is formulated using the same sigmoid function shown in Eq.(1). $P_\alpha(y_i|y_j^{(i)})$ will be modeled using Bernoulli distribution and characterized by parameter α as shown in Eq.(4).

$$P_\alpha(y_i|y_j^{(i)} = k) = \alpha_k^{y_i}(1 - \alpha_k)^{(1-y_i)} \quad (k = 0, 1) \qquad (4)$$

As k is either 1 or 0, we can write down all the possible $P_\alpha(y_i|y_j^{(i)} = k)$ in Eq.(5).

$$\begin{bmatrix} P(y_i=0|y_j^{(i)}=0)=\alpha_0 & P(y_i=1|y_j^{(i)}=0)=1-\alpha_0 \\ P(y_i=0|y_j^{(i)}=1)=\alpha_1 & P(y_i=1|y_j^{(i)}=1)=1-\alpha_1 \end{bmatrix} \qquad (5)$$

For each user, the parameter α measures the influence received from his neighbors. α_0 indicates the chance of a user being a normal user if his neighbor is a normal user. If the neighbor is normal, the larger α_0 is, this user is more likely to be a normal user. Similarly, α_1 indicates the chance of a user being a normal user if his neighbor is a marionette user. If the neighbor is marionette, the larger α_1 is, this user is more likely to be a normal user. The logarithm of the joint probability in Eq.(3) can be represented in Eq.(6):

$$\ell(\theta, \alpha) = \sum_{i=1}^{N} y_i \log h_\theta(x_i) + (1 - y_i) \sum_{i=1}^{N} \log(1 - h_\theta(x_i))$$

$$+ d \sum_{i=1}^{N} \sum_{j=1}^{M(i)} \sum_{k=y_j^{(i)}} (y_i \log \alpha_k + (1 - y_i) \log(1 - \alpha_k)) \tag{6}$$

The model parameters θ and α will be inferred by maximizing the log-likelihood in Eq.(6). To solve this optimization problem, it is natural to apply gradient descent approaches. Notice that θ is only included in the first part of Eq.(6) and α is only included in the second part, we can maximize each part separately to infer θ and α. θ can be obtained via numerical optimization methods using the same procedure in the aforementioned Logistic Regression formulation. As for α, we can derive the following analytical solution by maximizing the following objective function.

$$\alpha_k = \frac{\sum_{i=1}^{N} \sum_{j=1}^{M(i)} \sum_{k=y_j^{(i)}} y_i}{\sum_{i=1}^{N} \sum_{j=1}^{M(i)} \sum_{k=y_j^{(i)}} 1} \tag{7}$$

The above model takes social relations into consideration, but it has several disadvantages that may prevent its usage in real practice: First, the model only works in a supervised scenario where the class labels of all the neighbors of each user are observed. This is a strong assumption and can only be achieved by spending huge amounts of time and labeling costs to get sufficient training data. Second, even if we acquire sufficient labeled data, the discriminative information hidden in the labeled data is not fully utilized in the model. As shown in Eq.(3), the labels on a user's neighbors are only used in modeling $P(y_i|y_j^{(i)})$ without considering the relationship between the labels of these neighbors and their local features. Intuitively, if two neighbors have the same class label but different local features, their effect on the target user's label should be different.

Therefore, we propose to adapt Eq.(3) to Eq.(8) by considering both class labels and local features of a user's neighbors:

$$\max_{\theta, \alpha} \prod_{i=1}^{N} \{ P_\theta(y_i|x_i) \prod_{j=1}^{M(i)} P_{\alpha,\theta}(y_i|x_j^{(i)})^d \} \tag{8}$$

The only difference between Eq.(3) and Eq.(8) is that we replace $P_\alpha(y_i|y_j^{(i)})$ with $P_{\alpha,\theta}(y_i|x_j^{(i)})$. In this way, the proposed model incorporates the local features of the neighbors and the model does not have the strong assumption that the neighbors' labels are fully observed.

In Eq.(8), we represent $P_\theta(y_i|x_i)$ using the same sigmoid function shown in Eq.(1). As for $P_{\alpha,\theta}(y_i|x_j^{(i)})$, its formulation can be inferred based on Eq.(9).

$$P_{\alpha,\theta}(y_i|x_j^{(i)}) = \sum_{k=0}^{1} P_{\alpha,\theta}(y_i, y_j^{(i)} = k|x_j^{(i)}) \qquad (9)$$

$$= \sum_{k=0}^{1} P_{\alpha,\theta}(y_i|y_j^{(i)} = k, x_j^{(i)}) P_\theta(y_j^{(i)} = k|x_j^{(i)})$$

We assume that the label of a user y_i is conditionally independent of the local features $x_j^{(i)}$ of his neighbor given the label of this neighbor $y_j^{(i)}$, and thus we have $P_{\alpha,\theta}(y_i|y_j^{(i)} = k, x_j^{(i)}) = P_\alpha(y_i|y_j^{(i)} = k)$. Hence, we modify Eq.(9) accordingly into Eq.(10).

$$P_{\alpha,\theta}(y_i|x_j^{(i)}) = \sum_{k=0}^{1} P_\alpha(y_i|y_j^{(i)} = k) P_\theta(y_j^{(i)} = k|x_j^{(i)}) \qquad (10)$$

By plugging the above definition of $P_{\alpha,\theta}(y_i|x_j^{(i)})$ into the proposed objective function in Eq.(8), we effectively integrate users and their neighbors' local features together with social relations in the discriminative model to distinguish marionette and normal users. Accordingly, the log-likelihood in Eq. (6) is modified to Eq. (11).

$$\ell(\theta, \alpha) = \sum_{i=1}^{N} y_i \log h_\theta(x_i) + (1 - y_i) \sum_{i=1}^{N} \log(1 - h_\theta(x_i))$$

$$+ d \sum_{i=1}^{N} \sum_{j=1}^{M(i)} \log \sum_{k=0}^{1} P_\alpha(y_i|y_j^{(i)} = k) P_\theta(y_j^{(i)} = k|x_j^{(i)}) \qquad (11)$$

2.3   Parameter Estimation

In the proposed model, two sets of parameters need to be estimated: θ in both $P_\theta(y_i|x_j)$ and $P_{\alpha,\theta}(y_i|x_j^{(i)})$, and α in $P_{\alpha,\theta}(y_i|x_j^{(i)})$. These parameters should be obtained by maximizing the logarithm of Eq.(11). As the class labels of one's neighbors are unknown, we treat them as latent hidden variables during the inference procedure. The following hidden variable $z_{jk}^{(i)}$ is introduced in Eq. (12).

$$z_{jk}^{(i)} \propto P_{\alpha,\theta}(y_i, y_j^{(i)} = k|x_j^{(i)}))$$

$$\propto P_\alpha(y_i|y_j^{(i)} = k) P_\theta(y_j^{(i)} = k|x_j^{(i)}) \qquad (12)$$

Based on this hidden variable, the objective function in Eq.(11) can be represented in Eq.(13):

$$\ell'(z_{jk}^{(i)}, \theta, \alpha) = \sum_{i=1}^{N} \log P_\theta(y_i|x_i) + d \sum_{i=1}^{N} \sum_{j=1}^{M(i)} \sum_{k=0}^{1} z_{jk}^{(i)} \log P_\alpha(y_i|y_j^{(i)} = k)$$

$$+ d \sum_{i=1}^{N} \sum_{j=1}^{M(i)} \sum_{k=0}^{1} z_{jk}^{(i)} \log P_\theta(y_j^{(i)} = k|x_j^{(i)}) \tag{13}$$

We propose to use EM method to iteratively update model parameters and hidden variables. At the E-Step, the hidden variable $z_{jk}^{(i)}$ can be calculated via Eq.(14):

$$z_{jk}^{(i)} = \frac{P_\alpha(y_i|y_j^{(i)} = k)P_\theta(y_j^{(i)} = k|x_j^{(i)})}{\sum_{k=0}^{1} P_\alpha(y_i|y_j^{(i)} = k)P_\theta(y_j^{(i)} = k|x_j^{(i)})} \tag{14}$$

At the M-Step, we maximize the parameter $\ell'(z_{jk}^{(i)}, \theta, \alpha)$ with respect to α and get the following solution of α in Eq.(15).

$$\alpha_k = \frac{\sum_{i=1}^{N} \sum_{j=1}^{M(i)} z_{jk}^{(i)} y_i}{\sum_{i=1}^{N} \sum_{j=1}^{M(i)} z_{jk}^{(i)}} \tag{15}$$

The estimation of θ can be transformed into the parameter estimation process of Logistic Regression by constructing a training set. Initially, the training data set only includes N labeled users $\{(x_1, y_1), \ldots, (x_N, y_N)\}$. Then for each neighbor of the users, two instances $(x_j^{(i)}, y_j^{(i)} = 0)$ and $(x_j^{(i)}, y_j^{(i)} = 1)$ are generated and added into the training data set. In total, there are $2 \sum_{i=1}^{N} M(i)$ new instances added. The weights of the newly added instances are different from those of the initial ones. For the initial training instance (x_i, y_i), its weight is 1, while the weight of the newly added instance $(x_j^{(i)}, y_j^{(i)} = k)$ is $d \times z_{jk}^{(i)}$. The detailed parameter estimation process is summarized in Algorithm 1.

After obtaining the values of α and θ using Algorithm 1 from data, we can now use the proposed model to predict the class label of a new user u_i. This user's label y_i can be predicted according to Eq.(16).

$$\arg\max_{y_i} \quad P_\theta(y_i|x_i) \prod_{j=1}^{M(i)} P_{\alpha,\theta}(y_i|x_j^{(i)})^d \tag{16}$$

where $P_\theta(y_i|x_i)$ can be calculated using Eq.(1) and $P_{\alpha,\theta}(y_i|x_j^{(i)})$ can be calculated using Eq.(10).

2.4   Time Complexity

Another perspective we want to discuss is the time complexity and the number of iterations needed to converge. As shown in Algorithm 1, the parameter estimation process basically consists of EM iterations. During each iteration, the

---

Algorithm 1. Parameter Estimation Process

Data: Training data set $D = \{(x_1, y_1), \ldots, (x_N, y_N)\}$ and their unlabeled
        neighbors.
Result: Value of θ and α

1 **while** *EM not converged* **do**
2   **E step:**
3   Update $z_{jk}^{(i)}$ according to Eq.(14).
4   **M step:**
5   Update α according to Eq.(15).
6   For each neighbor of each instance in D, add two instances
       $\{(x_j^{(i)}, y_j^{(i)} = k), k = 0, 1\}$ and assign weights $d \times z_{jk}^{(i)}$.
7   Apply parameter estimation of Logistic Regression to calculate θ.

---

value of the hidden variable $z_{jk}^{(i)}$, θ and α are updated. According to Eq.(14) and
Eq.(15), the time complexity for calculating $z_{jk}^{(i)}$ and α is $O(NM)$ where N is
the number of instances and M is the average of the number of neighbors. As
for θ, the calculation is the same as parameter estimation process of Logistic Regression, whose time complexity depends on the optimization method adopted.
In total, the time complexity for training is $O(TNM + TL)$ where T denotes the
number iterations and L represents the time complexity of Logistic Regression
optimization.

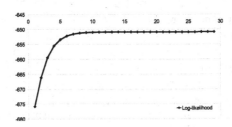

Fig. 1. The Log-likelihood Value with EM Iterations

We illustrate the convergence speed of the algorithm on the Weibo data set in
Figure 1. We calculate the log-likelihood after each round of iteration and plot
the values of log-likelihood with respect to each iteration. It can be observed
that Algorithm 1 converges quickly. After 8 rounds, the log-likelihood becomes
stable. Therefore, a small iteration number can achieve good performance. On
the training data set consisting of 12,000 users with 30 iterations, the proposed
approach only takes less than 10 seconds to converge on a commodity PC.

3   Experiments

In this section, we evaluate the proposed probabilistic model from two perspectives: (1) we calculate the classification accuracy and show that incorporating social relations can indeed improve the performance; (2) we demonstrate an application which measures the credibility of hot tweets with the suspicion of marionette user promotion.

3.1   Data Sets

Classification Corpus. We acquire a data set that consists of labeled marionette and normal users to evaluate the proposed model.

- **Marionette Users:** To collect the corpus of marionette users, we first created three phishing Sina Weibo accounts and bought followers from three Taobao shops for three times. Each time we purchased 2,000 followers and altogether there are 6,000 in total. The first purchase was made on November 2011 and the other two were made on February 2012. On Feb 2013, one year after the purchase, we re-examined these bought marionette users and found that around 1,000 are still active while the rest have already been deleted or blocked by Sina Weibo. Over 1/6 marionette users are not discovered by Sina weibo for over a year. To target a more challenging problem and compensate the existing detecting methods of Sina Weibo, we select these well hidden marionette users into our corpus.
- **Normal Users:** As for the normal users, we first select several seed users manually and crawl the users that they are following. After that, the crawled users are taken as new seeds to continue the crawling. Through this iterative procedure, we collect users whose identifications have been verified by Sina Weibo. As Sina requires the users to fax their ID copies for verification, we are confident these users are normal users. From these verified users, we randomly select 1,000 into our corpus that is the same amount as the marionette users. In real life, the distribution of normal and marionette users is usually imbalanced. However, to make the classifier more accurate, we decide to under sample the normal users and use a balanced training set to train the classifier which is commonly used in imbalanced classification [4].

For each obtained user, we further randomly select 5 users from all their followers into the data set. As a result, this data set consists of 2,000 labeled users and 10,000 unlabeled users. The profiles and posted tweets of all these 12,000 users are crawled.

Suspicious Hot Tweet Corpus. In Sina Weibo, the account named "social network analysis"[4] listed several hot tweets that were suspiciously promoted by marionette users. This account visualized the retweeting propagations of these

---

[4] http://weibo.com/dmonsns

suspicious tweets and identified the topological differences compared with the normal hot tweets. For each suspicious tweet mentioned by this account, we first retrieve the list of users who have retweeted this tweet and randomly select 200 users to crawl their profiles, posted tweets as well as that of their 5 neighbors.

3.2   Feature Description

In this subsection, we analyze some local features of microblog users whether they are discriminative in marionette user classification.

Number of Tweets/Followings/Followers. For each user, we extract the number of their posted tweets, the number of their followings and followers, and demonstrate the comparison results in three sub-figures of Figure 2 respectively. The x-axis represents different numbers of tweets, followings and followers, while the y-axis represents the number of users with the same number of tweets, followings and followers. Both axis are in the logarithmic scale.

In Figure 2(a), we find the marionettes are relatively inactive in tweeting, a large proportion marionettes post less than 20 tweets. On the contrary, the normal users are more active. The most "energetic" normal user posts more than 30,000 tweets. Therefore, a large number of tweets can be an effective feature to recognize normal users; In Figure 2(b), we find the number of followings of most marionettes lies between 100 to 1,000. One possible explanation to this range is that the puppeteer restricts the maximal following times to avoid being detected by microblog services.

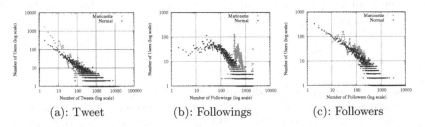

(a): Tweet   (b): Followings   (c): Followers

Fig. 2. User Number Distribution on Number of Tweets, Followings and Followers

Tweet Posting Device. Microblog services provide multiple access manners, including web interface, mobile clients, third party microblog applications, etc. Thus, we try to figure out whether there are any differences in posting devices between normals and marionettes. All the tweets are posted from 1,912 different sources, in which 1,707 different sources are used by normal users and 869 different sources are used by marionette users. Table 1 lists the top 5 mostly used sources for normals and marionettes respectively. We find that more than half tweets of normal users are posted via "Sina Web", thus the web interface remains the primary choice for accessing microblog and "iPhone" and "Android" are two most popular mobile clients. While for marionette users, most tweets are posted via "Sina Mobile" which denotes that majority tweets are posted via the web browsers of cell phones. In this case, if massive user accounts are created from some mobile IP address, the microblog service could not block this IP as it could be the real requests from normal users in the same district. Besides, the IP address can change when the puppeteer relocates.

Table 1. Top 5 Most Used Devices to Post Tweets

Normal		Marionette	
Device	#Tweet	Device	#Tweet
Sina Web	356,192	Sina Mobile	209,739
iPhone	59,996	Sina Web	29,365
Android	54,778	UC Browser	4,775
Sina Mobile	19,733	Android	2,577
S60	19,278	iPhone	2,112

Besides above local features, we also select: *the maximal, minimal, middle and average length of tweets; the maximal, minimal, middle and average time interval between tweets; the percentage of retweets.* We did not include the word bag features here, this is because we want make the model more generic. Since the marionette users owned by the same backstage puppeteer will retweet the same tweet, if the bag-of-word features are utilized as features, the trained model will incline to these word features and become over fit. To our knowledge, the bot detection of a popular search engine [5] only use behavior features, the words are used in blacklist for pre-filtering.

3.3   Classification Evaluation

To show the advantages of incorporating social relations, we compare with the baseline method which only applies Logistic Regression on the local features without considering social relations. When evaluating the proposed model, we set different values of d and different numbers of neighbors to illustrate the impact of social relations on the marionette user detection task. We implement the proposed method based on Weka [6] and the recorded accuracy is the average computed based on 5-fold cross validation.

- Baseline: The baseline model is a Logistic Regression classification model which adopts the local features introduced in previous sub section.
- Light-Neighbor: This model is the proposed model which adopts the same local features as the baseline model and incorporates the social relations with the setting of 5 neighbors and $d = 0.1$.
- Heavy-Neighbor: Similar to Light-Neighbor model, except this model biases more towards social relations with a higher degree setting $d = 0.5$.

Table 2. Classification Results on Three Models

	Precision	Recall	F-Measure
Baseline	0.884	0.875	0.872
Light-Neighbor	0.900	0.890	0.887
Heavy-Neighbor	**0.907**	**0.895**	**0.892**

Table 2 lists the weighted classification precision, recall and f-measure over three models. We can find that incorporating social relation increases the performance of detecting marionette users.

We also evaluate the proposed model with different settings of d and different number of neighbors and show the results in Figure 3. From this figure, we can find that in general when the number of neighbors increases, the classification accuracy increases. Similarly, when the value d increases, the accuracy improves as well. This clearly demonstrates the importance of casting social relations in the classification model. The more neighbors we included and the stronger influence we give to social relations, the better the performance is.

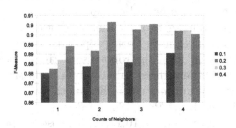

Fig. 3. Classification F-Measure with Different Neighbor and Degree Settings

3.4   Credibility Measurement

We further apply the proposed probabilistic model to detect the credibility of hot retweets. Firstly we apply the model learned from the *Classification Corpus* to classify the users in *Suspicious Hot Tweets Corpus*, and then we can obtain the percentage of marionette users who retweet the hot tweets.

Table 3 lists some Weibo accounts that post suspicious hot tweets, the possible promotion purpose and the percentage of marionette users. The percentages for the first four tweets are quite high, which suggests that most of their retweets are conducted by marionette users. Although the retweet of the last tweet shown in Table 3 involves more normal users, it might be attributed to the fact that marionette users attract the attention of many normal users and thus the goal of promotion is achieved through marionette user purchase.

Table 3. The Marionette User Percentage of Suspicious Hot Tweets

Tweet Author	Promotion Purpose	Marionette
A Web Site of Clothing Industry	Web Site Promotion	100.00%
A Famous Brand of Women's Dress	Weibo Account Promotion	98.61%
Provincial Culture Communication Co., LTD	Ceremony Advertisement	93.62%
A Anti-Worm Software for Mobile Device	A Security Issue Reminder	92.44%
A Famous China Smart Phone Manufacturer	The Advertisement of Sale Promotion	43.04%

4   Related Works

In this section, we describe related works from two perspectives: (1) the credibility issues of web data and corresponding solutions; (2) the credibility issues of microblog data.

4.1   Credibility of Web Data

Prior to the emergence of microblog services, the web has existed for over two decades. Many web services have been experiencing all kinds of malicious attacks. For example, the robot users submit specific queries or conduct fake clicks towards search engines, aiming to hack the ranking or auto-suggestion results [7]. The approaches like [5,8] have been proposed to detect and exclude such automated traffic. Different from the robot users, the marionette users possess social relations which can be utilized to build better classifiers.

Besides robot users and automated traffic, another web data issue is the link spam that tries to increase the PageRank of certain pages by creating a large number of links pointing to them. [9,10,11,12] propose to optimize search engine ranking and minimize the effects of the link spam. The marionette user detection is different from link spam detection, as the former is a classification problem which targets to separate the marionette users from normal users, while the latter is a ranking problem that targets to lower the rank of link spam web pages. Moreover, the link spam detection methods like [9] rely on the large link structure on the web, while the marionette detection only requires the local features and social connections of each user.

4.2   Credibility of Microblog Data

Due to the massive usage of microblog data, the credibility of microblog data becomes extremely important. [13] explored the information credibility of news propagated through Twitter and proposed to assess the credibility level of newsworthy topics. [14,15] identified the "Link Farmer" in microblog systems. This type of users try to acquire more followers and distribute spams. The main difference between the link farmers and marionette users is that the former one is seeking for followers and the latter one is providing followers. [16,17] analyzed the possible harm that link farmers could have done to microblog applications and [18] proposed several classifiers to detect the link farmers on Facebook. [19]

identified the cyber criminals. Different from marionette users, the cyber criminals generate direct harm to normal users by spreading phishing scams.

The SMFSR method proposed by [20] is related to the proposed approach in the sense that it combines user activities, social regularization and semi-supervised labeling in one framework. Specifically, it employed a matrix factorization based method to find spammers in social networks. Different from the proposed approach, this method is transductive rather than inductive. In other words, it is difficult to be used tof predict over new users not originally in the training set. Every time, new users are added, the entire matrix factorization needs to run again.

5 Conclusions

In the paper, we first discuss the business model of puppeteers and marionette users or how they make profits in microblog services. The following facts motivate the emergence of marionette user purchase: 1) to increase the number of followers and fake their popularity, some users purchase marionette users to follow them; and 2) to increase the retweet time and make promotion tweet to the front page story, the advertiser pays marionette users to retweet their tweets. Marionette users cheat in microblog services by manipulating fake retweets and following relations. Therefore, to ensure information trustworthiness and security guarantee, it is extremely important to detect marionette users in a timely manner. Facing the challenges posed by the complicated strategies adopted by marionette users, we propose an effective probabilistic model to fully utilize local user features and social relations in detecting marionette users. We propose an iterative EM procedure to infer model parameters from data and the model can then be used to predict whether a user is marionette or normal. Experiments on Sina Weibo data show that the proposed method achieves a very high f-measure close to 0.9, and the further analysis on some retweet examples demonstrates the effectiveness of the proposed model in measuring the true credibility of information on microblog platforms.

References

1. Sakaki, T., Okazaki, M., Matsuo, Y.: Earthquake shakes twitter users: real-time event detection by social sensors. In: Proceedings of the 19th International Conference on World Wide Web, WWW 2010, pp. 851–860 (2010)
2. Yu, L., Asur, S., Huberman, B.A.: Artificial inflation: The real story of trends in sina weibo (2012),
   http://www.hpl.hp.com/research/scl/papers/chinatrends/weibospam.pdf
3. Bollen, J., Mao, H., Zeng, X.J.: Twitter mood predicts the stock market. CoRR abs/1010.3003 (2010)
4. Chawla, N.V., Bowyer, K.W., Hall, L.O., Kegelmeyer, W.P.: Smote: synthetic minority over-sampling technique. J. Artif. Int. Res. 16(1), 321–357 (2002)
5. Kang, H., Wang, K., Soukal, D., Behr, F., Zheng, Z.: Large-scale bot detection for search engines. In: Proceedings of the 19th International Conference on World Wide Web, WWW 2010, pp. 501–510 (2010)

6. Hall, M., Frank, E., Holmes, G., Pfahringer, B., Reutemann, P., Witten, I.H.: The weka data mining software: an update. SIGKDD Explor. Newsl. 11(1), 10–18 (2009)
7. Buehrer, G., Stokes, J.W., Chellapilla, K.: A large-scale study of automated web search traffic. In: Proceedings of the 4th International Workshop on Adversarial Information Retrieval on the Web, AIRWeb 2008, pp. 1–8. ACM (2008)
8. Yu, F., Xie, Y., Ke, Q.: Sbotminer: large scale search bot detection. In: Proceedings of the Third ACM International Conference on Web Search and Data Mining, WSDM 2010 (2010)
9. Gyöngyi, Z., Garcia-Molina, H., Pedersen, J.: Combating web spam with trustrank. In: Proceedings of the Thirtieth International Conference on Very Large Data Bases, VLDB 2004, vol. 30, pp. 576–587. VLDB Endowment (2004)
10. Wu, B., Davison, B.D.: Identifying link farm spam pages. Special interest tracks and posters of the 14th International Conference on World Wide Web, WWW 2005, pp. 820–829 (2005)
11. Krishnan, V., Raj, R.: Web spam detection with anti-trust rank. In: AIRWeb 2006, pp. 37–40 (2006)
12. Benczur, A.A., Csalogany, K., Sarlos, T., Uher, M., Uher, M.: Spamrank - fully automatic link spam detection. In: Proceedings of the First International Workshop on Adversarial Information Retrieval on the Web, AIRWeb (2005)
13. Castillo, C., Mendoza, M., Poblete, B.: Information credibility on twitter. In: Proceedings of the 20th International Conference on World Wide Web, WWW 2011, pp. 675–684 (2011)
14. Ghosh, S., Viswanath, B., Kooti, F., Sharma, N.K., Korlam, G., Benevenuto, F., Ganguly, N., Gummadi, K.P.: Understanding and combating link farming in the twitter social network. In: Proceedings of the 21st International Conference on World Wide Web, WWW 2012, pp. 61–70 (2012)
15. Wagner, C., Mitter, S., Körner, C., Strohmaier, M.: When social bots attack: Modeling susceptibility of users in online social networks. In: 2nd Workshop on Making Sense of Microposts at WWW 2012 (2012)
16. Silvia Mitter, C.W., Strohmaier, M.: Understanding the impact of socialbot attacks in online social networks. In: WebSci, pp. 15–23 (2013)
17. Boshmaf, Y., Muslukhov, I., Beznosov, K., Ripeanu, M.: Design and analysis of a social botnet. Comput. Netw. 57(2), 556–578 (2013)
18. Reddy, R.N., Kumar, N.: Automatic detection of fake profiles in online social networks (2012), http://ethesis.nitrkl.ac.in/3578/1/thesis.pdf
19. Yang, C., Harkreader, R., Zhang, J., Shin, S., Gu, G.: Analyzing spammers' social networks for fun and profit: a case study of cyber criminal ecosystem on twitter. In: Proceedings of the 21st International Conference on World Wide Web, WWW 2012, pp. 71–80 (2012)
20. Zhu, Y., Wang, X., Zhong, E., Liu, N.N., Li, H., Yang, Q.: Discovering spammers in social networks. In: AAAI (2012)

Will My Question Be Answered?
Predicting "Question Answerability" in Community Question-Answering Sites

Gideon Dror, Yoelle Maarek, and Idan Szpektor

Yahoo! Labs, MATAM, Haifa 31905, Israel
{gideondr,yoelle,idan}@yahoo-inc.com

Abstract. All askers who post questions in Community-based Question Answering (CQA) sites such as Yahoo! Answers, Quora or Baidu's Zhidao, expect to receive an answer, and are frustrated when their questions remain unanswered. We propose to provide a type of "heads up" to askers by predicting how many answers, if at all, they will get. Giving a preemptive warning to the asker at posting time should reduce the frustration effect and hopefully allow askers to rephrase their questions if needed. To the best of our knowledge, this is the first attempt to predict the actual number of answers, in addition to predicting whether the question will be answered or not. To this effect, we introduce a new prediction model, specifically tailored to hierarchically structured CQA sites. We conducted extensive experiments on a large corpus comprising 1 year of answering activity on Yahoo! Answers, as opposed to a single day in previous studies. These experiments show that the F_1 we achieved is 24% better than in previous work, mostly due the structure built into the novel model.

1 Introduction

In spite of the huge progress of Web search engines in the last 20 years, many users' needs still remain unanswered. Query assistance tools such as query completion, and related queries, cannot, as of today, deal with complex, heterogeneous needs. In addition, there will always exist subjective and narrow needs for which content has little chance to have been authored prior to the query being issued.

Community-based Question Answering (CQA) sites, such as *Yahoo! Answers*, *Quora*, *Stack Overflow* or *Baidu Zhidao*, have been precisely devised to answer these different needs. These services differ from the extensively investigated Factoid Question Answering that focuses on questions such as *"When was Mozart born?"*, for which unambiguous answers typically exist, [1]. Though CQA sites also feature factoid questions, they typically address other needs, such as opinion seeking, recommendations, open-ended questions or very specific needs, *e.g.* *"What type of bird should I get?"* or *"What would you choose as your last meal?"*.

Questions not only reflect diverse needs but can be expressed in very different styles, yet, all askers expect to receive answers, and are disappointed otherwise. Unanswered questions are not a rare phenomenon, reaching 13% of the questions in the Yahoo! Answers dataset that we studied, as detailed later, and users whose questions remain

H. Blockeel et al. (Eds.): ECML PKDD 2013, Part III, LNAI 8190, pp. 499–514, 2013.

unanswered are considerably more prone to churning from the CQA service [2] . One way to reduce this frustration is to proactively recommend questions potential answerers, [3,4,5,6]. However, the asker has little or no influence on the answerers' behavior. Indeed, a question by itself may exhibit some characteristics that reduce its potential for answerability. Examples include a poor or ambiguous choice of words, a given type of underlying sentiment, the time of the day when the question was posted, as well as sheer semantic reasons if the question refers to a complex or rare need.

In this work, we focus on the askers, investigate a variety of features they can control, and attempt to predict, based on these features, the expected number of answers a new question might receive, even before it is posted. With such predictions, askers can be warned in advance, and adjust their expectations, if their questions have little chances to be answered. This work represents a first step towards the more ambitious goal of assisting askers in posting answerable questions, by not only indicating the expected number of answers but also suggesting adequate rephrasing. Furthemore, we can imagine additional usages of our prediction mechanism, depending on the site priorities. For instance, a CQA site such as Yahoo! Answers that attempts to satisfy all users might decide to promote questions with few predicted answers in order to achieve a higher answering rate. Alternatively a socially oriented site like Quora, might prefer to promote questions with many predicted answers in order to encourage social interaction between answerers.

We cast the problem of predicting the number of answers as a regression task, while the special case of predicting whether a question will receive any answer at all is viewed as a classification task. We focus on Yahoo! Answers, one of the most visited CQA sites with 30 millions questions and answers a month and 2.4 asked questions per second [7]. For each question in Yahoo! Answers, we generate a set of features that are extracted only from the question attributes and are available before question submission. These features capture asker's attributes, the textual content of the question, the category to which the question is assigned and the time of submission.

In spite of this rich feature set, off-the-shelf regression and classification models do not provide adequate predictions in our tasks. Therefore, we introduce a series of models that better address the unique attributes of our dataset. Our main contributions are threefold:

1. we introduce a novel task of predicting the number of expected answers for a question before it is posted,
2. we devise hierarchical learning models that consider the category-driven structure of Yahoo! Answers and reflect their associated heterogeneous communities, each with its own answering behavior, and finally,
3. we conduct the largest experiment to date on answerability, as we study a year-long question and answer activity on Yahoo! Answers, as opposed to a day-long dataset in previous work.

2 Background

With millions of active users, Yahoo! Answers hosts a very large amount of questions and answers on a wide variety of topics and in many languages. The system is content-centric, as users are socially interacting by engaging in multiple activities around a specific question. When a user asks a new question, she also assigns it to a specific category, within a predefined hierarchy of categories, which should best match the general topic of the question. For example, the question *"What can I do to fix my bumper?"* was assigned to the category *'Cars & Transportation > Maintenance & Repairs'*. Each new question remains "open" for four days (with an option for extension), or less if the asker chose a best answer within this period. Registered users may answer a question as long as it remains "open".

One of the main issues in Yahoo! Answers, and in community-based question answering in general, is the high variance in perceived question and answer quality. This problem drew a lot of research in recent years. Some studies attempted to assess the quality of answers [8,9,10,11], or questions [12,13], and rank them accordingly. Others looked at active users for various tasks such, scoring their "reliability" as a signal for high quality answers or votes [14,15,16], identifying spammers [17], predicting whether the asker of a question will be satisfied with the received answers [18,19] or matching questions to specific users [3,4,5].

Our research belongs to the same general school of work but focuses on estimating the number of answers a question will receive. Prior work that analyzes questions, did it in retrospect, either after the questions had been answered [9], or as a ranking task for a given collection of questions [12,13]. In contrast, we aim at predicting the number of answers for every new question *before* it is submitted.

In a related work, Richardson and White [20] studied whether a question will receive an answer or not. Yet, they conducted their study in a different environment, an IM-based synchronous system, in which potential answerers are known. Given this environment they could leverage features pertaining to the potential answerers, such as reputation. In addition, they considered the specific style of messages sent over IM, including whether a newline was entered and whether some polite words are added. Their experiment was of a small scale on 1,725 questions, for which they showed improvement over the majority baseline. We note that their dataset is less skewed than in Yahoo! Answers. Indeed their dataset counted about 42% of unanswered questions, while Yahoo! Answers datasets typically count about 13% of unanswered questions. We will later discuss the challenges involved in dealing with such a skewed dataset.

A more related prior work that investigated question answerability is Yang et al. [21], who addressed the same task of coarse (yes/no) answerability as above but in the same settings as ours, namely Yahoo! Answers. Yang et al. approached the task as a classification problem with various features ranging from content analysis, such as category matching, polite words and hidden topics, to asker reputation and time of day. They used a one-day dataset of Yahoo! Answers questions and observed the same ratio of unanswered questions as we did in our one-year dataset, namely 13%. Failing to construct a classifier for this heavily skewed dataset, Yang et al. resorted to learning from an artificially balanced training set, which resulted in improvements over the majority baseline. In this paper, we also address this classification task, with the same type of skewed dataset. However, unlike Yang et al., we attempt to improve over the majority baseline without artificially balancing the dataset.

Finally another major differentiator with the above previous work is that we do not stop at simply predicting whether a question will be answered or not, but predict the exact number of answers the question would receive.

3   Predicting Question Answerability

One key requirement of our work, as well as a differentiator with typical prior work on question analysis, is that we want to predict answerability before the question is posted. This imposes constraints on the type of data and signals we can leverage. Namely, we can only use data that is intrinsic to a new question before submission. In the case of Yahoo! Answers, this includes: (a) the title and the body of the question, (b) the category to which the question is assigned, (c) the identity of the user who asked the question and (d) the date and time the question is being posted.

We view the prediction of the expected number of answers as a regression problem, in which a target function (a.k.a the *model*) $\hat{y} = f(x)$ is learned, with x being a vector-space representation of a given question, and $\hat{y} \in \mathbb{R}$ an estimate for y, the number of answers this question will actually receive. All the different models we present in this section are learned from a training set of example questions and their known number of answers, $D = \{(x_i, y_i)\}$. The prediction task of whether a question will receive an answer at all is addressed as a classification task. It is similarly modeled by a target function $\hat{y} = f(x)$ and the same vector space representation of a question, yet, the training target is binary, with answered (unanswered) questions being the positive (negative) examples.

To fully present our models for the two tasks, we next specify how a question representation x is generated, and then introduce for each task novel models (*e.g.* $f(x)$) that address the unique properties of the dataset.

3.1   Question Features

In our approach, each question is represented by a feature vector. For any new question, we extract various attributes that belong to three main types of information: question meta data, question content, and user data. In the rest of this paper we use the term *feature family* to denote a single attribute extracted from the data. Question attributes may be numerical, categorical or set-valued (*e.g.* the set of word tokens in the title). Hence, in order to allow learning by gradient-based methods, we transformed all categorical attributes to binary features, and binned most of the numeric attributes. For example, the category of a question is represented as 1287 binary features and the hour it was posted is represented as 24 binary features. Tables 3.1, 2 and 3 describe the different feature families we extract, grouped according to their information source: the question text, the asker and question meta data.

3.2   Regression Models

Following the description of the features extracted from each question, we now introduce different models (by order of complexity) that use the question feature vector in

Table 1. Features extracted from title and body texts

Feature Family	Description	# Features
Title tokens	The tokens extracted from the title, not including stop words	45,011
Body tokens	The tokens extracted from the body, not including stop words	45,508
Title sentiment	The positive and negative sentiment scores of the title, calculated by the SentiStrength tool [22]	2
Body sentiment	The mean positive and negative sentiment scores of the sentences in the body	2
Supervised LDA	The number of answers estimated by supervised Latent Dirichlet Allocation (SLDA) [23], which was trained over a small subset of the training set	1
Title WH	WH-words (*what, when, where* ...) extracted from the question's title	11
Body WH	WH-words extracted from the question's body	11
Title length	The title length measured by the number of tokens after stopword removal, binned on a linear scale	10
Body length	The body length, binned on an exponential scale since this length is not constrainted	20
Title URL	The number of URLs that appear within the question title	1
Body URL	The number of URLs that appear within the question body	1

order to predict the number of answers. We remind the reader that our training set consists of pairs $D = \{(x_i, y_i)\}$, where $x_i \in R^F$ is the F dimensional feature vector representation of question q_i, and $y_i \in \{0, 1, 2 \ldots\}$ is the known number of answers for q_i.

Baseline Model. Yang et al. [21] compare the performance of several classifiers, linear and non-linear, on a similar dataset. They report that a linear SVM significantly outperforms all other classifiers. Given these findings, as well as the fact that a linear model is both robust [24] and can be trained very efficiently for large scale problems, we chose a linear model $f(x_i) = w^T x_i + b$ as our baseline model.

Feature Augmentation Model. One of the unique characteristics of the Yahoo! Answers site is that it consists of questions belonging to a variety of categories, each with its community of askers and answerers, temporal activity patterns, jargon etc., and that the categories are organized in a topical taxonomy. This structure, which is inherent to the data, suggests that more complex models might be useful in modeling the data. One effective way of incorporating the category structure of the data in a regression model is to enrich the features with category information. Specifically, we borrowed the idea from [25], which originally utilized such information for domain adaptation.

To formally describe this model, we consider the Yahoo! Answers category taxonomy as a rooted tree T with *"All Categories"* as its root. When referring to the category

Table 2. Features extracted based on the asker

Feature Family	Description	# Features
Asker ID	The identity of the asker, if it asked at least 50 questions in the training set. We ignore askers who asked fewer questions since their ID statistics are unreliable	175,714
Mean # of answers	The past mean number of answers the asker received for her questions, binned on an exponential scale	26
# of questions	The past number of questions asked by the asker, binned on a linear scale and on an exponential scale	26
Log # of questions	The logarithm of the total number of questions posted by the asker in the training set, and the square of the logarithm. For both features we add 1 to the argument of the logarithm to handle test users with no training questions.	2

Table 3. Features extracted from the question's meta data

Feature Family	Description	# Features
Category	The ID of the category that the question is assigned to	1,287
Parent Category	The ID of the parent category of the assigned category for the question, based on the category taxonomy	119
Hour	The hour at which the question was posted, capturing daily patterns	24
Day of week	The day-of-week in which the question was posted, capturing weekly patterns	7
Week of year	The week in the year in which the question was posted, capturing yearly patterns	51

tree we will use interchangeably the term *node* and *category*. We denote the category of a question q_i by $C(q_i)$. We further denote by $P(c)$ the set of all nodes on the path from the tree root to node c (including c and the root). For notational purposes, we use a binary representation for $P(c)$: $P(c) \in \{0,1\}^{|T|}$, where $|T|$ is the number of nodes in the category tree.

The feature augmentation model represents each question q_i by $\widehat{x}_i \in R^{F|T|}$ where $\widehat{x}_i = P(C(q_i)) \otimes x_i$ where $\otimes$ represents the Kronecker product. For example, given question q_i that is assigned to category '*Dogs*', the respective node path in T is '*All Questions/Pets/Dogs*'. The feature vector $\widehat{x}_i$ for q_i is all zeros except for three copies of x_i corresponding to each of the nodes '*All Questions*','*Pets*' and '*Dogs*'.

The rationale behind this representation is to allow a separate set of features for each category, thereby learning category specific patterns. These include learning patterns for leaf categories, but also learning lower resolution patterns for intermediate nodes in the tree, which correspond to parent and top categories in Yahoo! Answers. This permits a good tradeoff between high resolution modeling and robustness, obtained by the higher level category components shared by many examples.

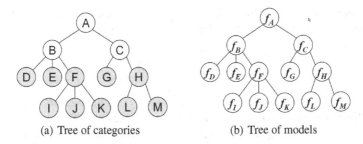

(a) Tree of categories   (b) Tree of models

Fig. 1. An illustration of the subtree model structure. Shaded nodes in (a) represent categories populated with questions, while unshaded nodes are purely navigational.

Subtree Model. An alternative to the feature augmentation model is to train several linear models, each specializing on a different subtree of T. Let us note a subset of the dataset as $D_c = \{(x_i, y_i) | q_i \in S(c)\}$, where $S(c)$ is the set of categories in the category subtree rooted at node c. We also note a model trained on D_c as f_c. Since there is a one-to-one correspondence between models and nodes in T, the set of models f_c can be organized as a tree isomorphic to T. Models in deeper levels of the tree are specialized on fewer categories than models closer to the root. Figure 1 illustrates a category tree and its corresponding model tree structure. The shaded nodes in 1(a) represent categories to which some training questions are assigned.

One simplistic way of using the model tree structure is to apply the root model (f_A in Figure 1(b)) to all test questions. Note that this is identical to the baseline model. Yet, there are many other ways to model the data using the model tree. Specifically, any set of nodes that also acts as a tree cut defines a regression model, in which the number of answers for a given question q_i is predicted by the first model in the set encountered when traversing from the category c_i, assigned to q_i, to the root of T. In this work, we shall limit ourselves to three such cuts:

TOP: q_i is predicted by model $f_{Top(c_i)}$, where $Top(c)$ is the category in $P(c)$ directly connected to the root.
PARENT: q_i is predicted by model $f_{Parent(c_i)}$
NODE: q_i is predicted by model f_{c_i}

In Figure 1 the TOP model refers to $\{f_B, f_C\}$, the PARENT model refers to $\{f_B, f_C, f_F, f_H\}$ and the NODE model refers to $\{f_D, f_E, \dots f_M\}$.

Ensemble of Subtree Models. In order to further exploit the structure of the category taxonomy in Yahoo! Answers, the questions in each category c are addressed by all models in the path between this category and the tree root, under the subtree framework described above. Each model in this path introduces a different balance between robustness and specificity. For example, the root model is the most robust, but also the least specific in terms of the idiomatic attributes of the target category c. At the other end of the spectrum, f_c is specifically trained for c, but it is more prone for over fitting the data, especially for categories with few training examples.

Instead of picking just one model on the path from c to the root, the ensemble model for c learns to combine all subtree models by training a meta linear model:

$$f(x_i) = \sum_{c' \in P(c)} \alpha_{cc'} f_{c'}(x_i) + b_c \tag{1}$$

where $f_{c'}$ are the subtree models described previously and the weights $\alpha_{cc'}$ and b_c are optimized over a validation set. For example, the ensemble model for questions assigned to category E in Figure 1(a) are modeled by a linear combination of models f_A, f_B and f_E, which are trained on training sets D, D_B and D_E respectively.

3.3   Classification Models

The task of predicting whether a question will be answered or not is an important special case of the regression task. In this classification task, we treat questions that were not answered as negative examples and questions that were answered as the positive examples. We emphasize that our dataset is skewed, with the negative class constituting only 12.68% of the dataset. Furthermore, as already noted in [26], the distribution of the number of answers per question is very skewed, with a long tail of questions having high number of answers.

As described in the background section, this task was studied by Yang et al. [21], who failed to provide a solution for the unbalanced dataset. Instead, they artificially balanced the classes in their training set by sampling, which may reduce the performance of the classifier on the still skewed test set. Unlike Yang et al., who used off-the-shelf classifiers for the task, we devised classifiers that specifically address the class imbalance attribute of the data. We noticed that a question that received one or two answers could have easily gone unanswered, while this is unlikely for questions with dozen answers or more. When projecting the number of answers y_i into two values, this difference between positive examples is lost and may produce inferior models. The following models attempt to deal with this issue.

Baseline Model. Yang et al. [21] found that linear SVM provides superior performance on this task. Accordingly, we choose as baseline a linear model, $f(x_i) = w^T x_i + b$ trained with hinge loss. We train the model on the binarized dataset $D_0 = \{(x_i, \text{sign}(y_i - 1/2))\}$ (see our experiment for more details).

Feature Augmentation Model. In this model, we train the same baseline classifier presented above. Yet the feature vector fed into the model is the augmented feature representation introduced for the regression models.

Ensemble Model. In order to capture the intuition that not "all positive examples are equal", we use an idea closely related to works based on Error Correcting Output Coding for multi-class classification [27]. Specifically, we construct a series of binary classification datasets $D_t = \{(x_i, z_i^t)\}$ where $z_i^t = \text{sign}(y_i - 1/2 - t)$ and $t = 0, 1, 2, \ldots$. In this series, D_0 is a dataset where questions with one or more answers are considered positive, while in D_{10} only examples with more than 10 answers are considered positive. We note that these datasets have varying degrees of imbalance between the positive and the negative classes.

Denoting by f_t the classifier trained on D_t, we construct the final ensemble classifier by using a logistic regression

$$f(x) = \sigma(\sum_t \alpha_t f_t(x) + b) \tag{2}$$

where $\sigma(u) = (1 + e^{-u})^{-1}$ and the coefficients α_t and b are learned by minimizing the log-likelihood loss on the validation set.

Ensemble of Feature Augmentation Models. In this model, we train the same ensemble classifier presented above. Yet the feature vector fed into the model is the augmented feature representation introduced for the regression models.

Classification Ensemble of Subtree Models. As our last classification model, we directly utilize regression predictions to differentiate between positive examples. We use the same model tree structure used in the regression by ensemble of subtree models. All models are linear regression models trained exactly as in the regression problem, in order to predict the number of answers for each question. The final ensemble model is a logistic regression function of the outputs of the individual regression models:

$$f(x_i) = \sigma(\sum_{c' \in P(c)} \alpha_{cc'} f_{c'}(x_i) + b_c) \tag{3}$$

where $f_{c'}$ are the subtree regression models and the weights $\alpha_{cc'}$ and b_c are trained using the validation set.

4 Experiments

We describe here the experiments we conducted to test our regression and classification models, starting with our experimental setup, then presenting our results and analyses.

4.1 Experimental Setup

Our dataset consists of a uniform sample of 10 million questions out of all non-spam English questions submitted to Yahoo! Answers in 2009. The questions in this dataset were asked by more than 3 million different users and were assigned to $1,287$ categories out of the $1,569$ categories. A significant fraction of the sampled questions (12.67%) remained unanswered. The average number of answers per question is 4.56 ($\sigma = 6.11$). The distribution of the number of answers follows approximately a geometric distribution.

The distributions of questions among users and among categories are extremely skewed, with a long tail of users who posted one or two questions and sparsely populated categories. These distributions are depicted in Figure 2, showing a power law

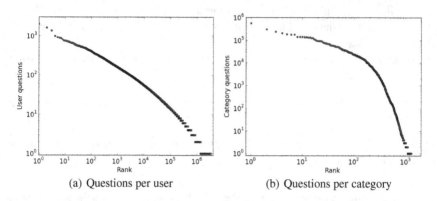

(a) Questions per user   (b) Questions per category

Fig. 2. Distribution of number of questions depicted as a function of ranks

behavior for the questions per asker distribution. A large fraction of the categories have quite a few questions, for example, about half of all categories in our dataset count less than 50 examples.

We randomly divided our dataset into three sets: 80% training, 15% test and 5% validation (for hyper-parameter tuning). Very few questions in the dataset attracted hundreds of answers. To eliminate the ill effect of these questions on model training, we modified the maximum number of answers per question to 64. This resulted in changing the target of about 0.03% of the questions.

Due to its speed, robustness and scalability, we used the Vowpal Wabbit tool[1] whenever possible. All regression models were trained with squared loss, except for ensemble of subtree models, Eq. 1, whose coefficients were learned by a least squares fit. All classification models were trained using Vowpal Wabbit with hinge loss, except for the ensemble models, Eq. 2 and 3, whose coefficients were learned by maximizing the log-likelihood of the validation set using Stochastic Gradient Descent. We note that for a node c, where ensemble models should have been trained based on less than 50 validation examples, we refrained from training the ensemble model and used the NODE subtree model of c as a single component of the ensemble model.

Table 4 compares between the various trained models with respect to the number of basic linear models used in composite models and the average number of features observed per linear model. The meta-parameters of the ensemble models (Eq. 1 and 3) were not included in the counting.

4.2   Results

The performance of the different regression models on our dataset was measured by Root Mean Square Error (RMSE) [28] and by the Pearson correlation between the predictions and the target. Table 5 presents these results. As can be seen, all our models outperform the baseline off-the-shelf linear regression model, with the best performing

---

[1] http://hunch.net/~vw/

Table 4. Details of the regression and classification models, including the number of linear models in each composite model, and the average number of features used by each linear model

Regression			Classification		
Model	# linear models	features per model	Model	# linear models	features per model
Baseline	1	267,781	Baseline	1	267,781
Feature augmentation	1	12,731,748	Feature augmentation	1	12,731,748
Subtree - TOP	26	88,358	Ensemble	7	267,781
Subtree - PARENT	119	26,986	Feature augmentation Ens.	7	12,731,748
Subtree - NODE	924	9,221	Ens. of subtree models	955	13,360
Ens. of subtree models	955	13,360			

Table 5. Test performance for the regression models

Model	RMSE	Pearson Correlation
Baseline	5.076	0.503
Feature augmentation	4.946	0.539
Subtree - TOP	4.905	0.550
Subtree - PARENT	4.894	0.552
Subtree - NODE	4.845	0.564
Ens. of subtree models	4.606	0.620

Table 6. Test performance for the classification models

Model	AUC
Baseline	0.619
Feature augmentation	0.646
Ensemble	0.725
Feature augmentation ensemble	0.739
Ensemble of subtree models	0.781

model achieving about 10% relative improvement. These results indicate the importance of explicitly modeling the different answering patterns within the heterogeneous communities in Yahoo! Answers, as captured by categories. Interestingly, the feature-augmentation model, which attempts to combine between categories and their ancestors, performs worse than any specific subtree model. One of the reasons for this is the huge number of parameters this model had to train (see Table 4), compared to the ensemble of separately trained subtree models, each requiring considerably fewer parameters to tune. A t-test based on the Pearson correlations shows that each model in Table 5 is significantly better than the preceding one, with P-values close to zero.

The performance of models for the classification task was measured by the area under the ROC Curve (AUC) [29]. AUC is a preferred performance measure when class distributions are skewed, since it measures the probability that a positive example is scored higher than a negative example. Specifically, the AUC of a majority model is always 0.5, independently of the distribution of the targets.

Inspecting the classification results in Table 6, we can see that all the novel models improve over the baseline classifier, with the best performing ensemble of subtrees classifier achieving an AUC of 0.781, a substantial relative improvement of 26% over the baseline's result of 0.619. A t-test based on the estimated variance of AUC [30] shows that each model in Table 6 is statistically significantly superior to its predecessor with P-values practically zero.

We next examine in more depth the performance of the ensemble of classifiers and the ensemble of subtree regressors (the third and fifth entries in Table 6 respectively). We see that the ensemble of classifiers explicitly models the differences between questions with many and few answers, significantly improving over the baseline. Yet, the ensemble of subtree regressors not only models this property of the data but also the differences in answering patterns within different categories. Its higher performance indicates that both attributes are key factors in prediction. Thus the task of predicting the actual number of answers has additional benefits, it allows for a better understanding of the structure of the dataset, which also helps for the classification task.

Finally, we compared our results to those of Yang et al. [21]. They measured the F_1 value on the predictions of the minority class of unanswered questions, for which their best classifier achieved an F_1 of 0.325. Our best model for this measure was again the ensemble of subtree models classifier, which achieved an F_1 of 0.403. This is a substantial increase of 24% over Yang et al.'s best result, showing again the benefits of a structured classifier.

4.3   Error Analysis

We investigated where our models err by measuring the average performance of our best performing models as a function of the number of answers per test question, see Figure 3. We split the test examples into disjoint sets characterized by a fixed number of answers per question and averaged the RMSE of our best regressor on each set (Figure 3(a)). Since our classifier is not optimized for the Accuracy measure, we set a specific threshold on the classifier output, choosing the 12.67^2 percentile of test examples with lowest scores as negatives. Figure 3(b) shows the error rate for this threshold. We note that the error rate for zero answers refers to false positives rate and for all other cases it refers to the false negatives rate.

Figure 3(a) exhibits a clear minimum in the region most populated with questions, which shows that the regressor is optimized for predicting values near 0. Although the RMSE increases substantially with the number of answers, it is still moderate. In general, the RMSE we obtained is approximately linear to the square root of the number of answers. Specifically, for questions with large number of answers, the RMSE is much smaller than the true number of answers. For example, for questions with more than 10 answers, which constitute about 13% of the dataset, the actual number of answers is approximately twice the RMSE on average. This shows the benefit of using the regression models as input to a answered/unanswered classifier, as we did in our best performing classifier. This is reflected, for example, in the very low error rates (0.0064 or less) for questions with more than 10 answers in Figure 3(b).

While the regression output effectively directs the classifier to the correct decision for questions with around 5 or more answers, Figure 3(b) still exhibits substantial error rates for questions with very few or no answers. This is due to the inherent randomness in the answering process, in which questions that received very few answers could have easily gone unanswered and vice versa and are thus difficult to predict accurately.

---

[2] This is the fraction of negative examples in our training set.

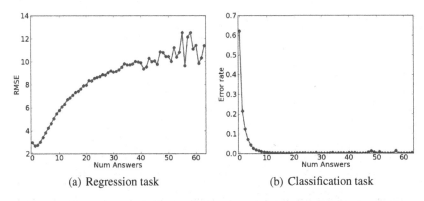

(a) Regression task   (b) Classification task

Fig. 3. Performance of the Subtree Ensemble models as a function of the number of answers

In future work, we want to improve these results by employing boosting approaches and constructing specialized classifiers for questions with very few answers.

4.4 Temporal Analysis

Intuitively, the time at which a question is posted should play a role in a social-media site, therefore, like Yang et al. [21], we use temporal features. Our dataset, which spans over one year, confirms their reported patterns of hourly answering: questions posted at night are most likely to be answered, while "afternoon questions" are about 40% more likely to remain unanswered.

To extend this analysis to longer time periods, we analyzed weekly and yearly patterns. We first calculated the mean number of answers per question and the fraction of unanswered questions as a function of the day of week, as shown in Figure 4. A clear pattern can be observed: questions are more often answered towards the end of the week, with a sharp peak on Fridays and a steep decline over the weekend. The differences between the days are highly statistically significant (t-test, two sided tests). The two graphs in Figure 4 exhibit extremely similar characteristics, indicating that the fraction of unanswered questions is negatively correlated with the average number of answers per question. This suggests that both phenomena are controlled by a supply and demand equilibrium. This can be explained by two hypotheses: (a) both phenomena are driven by an increase in questions (Yang et al.'s hypothesis) or (b) both phenomena are driven by a decrease in the number of answers.

To test the above two hypotheses, we extracted the number of questions, number of answers and fraction of unanswered questions on a daily basis. Each day is represented in Figure 5 as a single point, as we plot the daily fraction of unanswered questions as a function of the daily average number of answers per question (Figure 5(a)) and as a function of the total number of daily questions (Figure 5(b)). We note that while some answers are provided on a window of time longer than a day, this is a rare phenomenon. The vast majority of answers are obtained within about twenty minutes from the question posting time [5], hence our daily analysis.

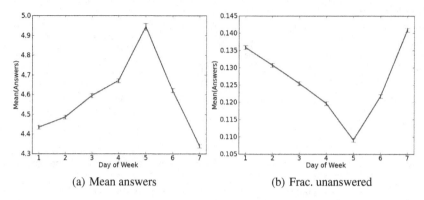

(a) Mean answers   (b) Frac. unanswered

Fig. 4. Mean number of answers and fraction of number of answers as a function of the day of the week, where '1' corresponds to Monday and '7' to Sunday

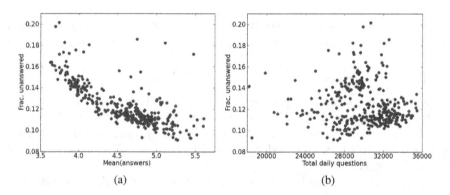

(a)   (b)

Fig. 5. The daily fraction of unanswered questions as a function of the daily mean number of answers and as a function of the total number of questions

Figure 5(a) exhibits a strong negative correlation (Pearson correlation $r = -0.631$), while almost no effect is observed in Figure 5(b) ($r = 0.010$). We further tested the correlation between the daily total number of answers and the fraction of fraction of unanswered questions, and here as well a significant negative correlation was observed ($r = -0.386$). These findings support the hypothesis that deficiency in answerers is the key factor affecting the fraction of unanswered questions, and not the overall number of questions, which was Yang et al's hypothesis. This result is important, because it implies that more questions in a community-based question answering site will not reduce the performance of the site, as long as an active community of answerers strives at its core.

5   Conclusions

In this paper, we investigated the answerability of questions in community-based question answering sites. We went beyond previous work that returned a binary result of whether or not the question will be answered. We focused on the novel task of predicting the actual number of expected answers for new questions in community-based question answering sites, so as to return feedback to askers before they post their questions. We introduced a series of novel regression and classification models explicitly designed for leveraging the unique attributes of category-organized community-based question answering sites. We observed that these categories host diverse communities with different answering patterns.

Our models were tested over a large set of questions from Yahoo! Answers, showing significant improvement over previous work and baseline models. Our results confirmed our intuition that predicting answerability at a finer grained level is beneficial. They also showed the strong effect of the different communities interacting with questions on the number of answers a question will receive. Finally, we discovered an important and somehow counter-intuitive fact, namely that an increased number of questions will not negatively impact answerability, as long as the community of answerers is maintained.

We constructed models that are performant at scale: even the ensemble models are extremely fast at inference time. In future work, we intend to increase response time even further and consider incremental aspects in order to return predictions as the asker types, thus providing, in real-time, dynamic feedback and a more engaging experience. To complement this scenario, we are also interested in providing question rephrasing suggestions for a full assistance solution.

References

1. Voorhees, E.M., Tice, D.M.: Building a question answering test collection. In: SIGIR (2000)
2. Dror, G., Pelleg, D., Rokhlenko, O., Szpektor, I.: Churn prediction in new users of yahoo! answers. In: WWW (Companion Volume), pp. 829–834 (2012)
3. Li, B., King, I.: Routing questions to appropriate answerers in community question answering services. In: CIKM, pp. 1585–1588 (2010)
4. Horowitz, D., Kamvar, S.D.: The anatomy of a large-scale social search engine. In: Proceedings of the 19th International Conference on World Wide Web, WWW 2010, pp. 431–440. ACM, New York (2010)
5. Dror, G., Koren, Y., Maarek, Y., Szpektor, I.: I want to answer; who has a question?: Yahoo! answers recommender system. In: KDD, pp. 1109–1117 (2011)
6. Szpektor, I., Maarek, Y., Pelleg, D.: When relevance is not enough: Promoting diversity and freshness in personalized question recommendation. In: WWW (2013)
7. Rao, L.: Yahoo mail and im users update their status 800 million times a month. TechCrunch (October 28, 2009)
8. Jeon, J., Croft, W.B., Lee, J.H., Park, S.: A framework to predict the quality of answers with non-textual features. In: Proceedings of the 29th Annual International ACM SIGIR Conference on Research and Development in Information Retrieval, SIGIR 2006, pp. 228–235. ACM, New York (2006)
9. Agichtein, E., Castillo, C., Donato, D., Gionis, A., Mishne, G.: Finding high quality content in social media, with an application to community-based question answering. In: Proceedings of ACM WSDM, WSDM 2008. ACM Press, Stanford (2008)
10. Shah, C., Pomerantz, J.: Evaluating and predicting answer quality in community qa. In: Proceedings of the 33rd International ACM SIGIR Conference on Research and Development in Information Retrieval, SIGIR 2010, pp. 411–418. ACM, New York (2010)

11. Surdeanu, M., Ciaramita, M., Zaragoza, H.: Learning to rank answers on large online qa collections. In: ACL, pp. 719–727 (2008)
12. Song, Y.I., Lin, C.Y., Cao, Y., Rim, H.C.: Question utility: A novel static ranking of question search. In: AAAI, pp. 1231–1236 (2008)
13. Sun, K., Cao, Y., Song, X., Song, Y.I., Wang, X., Lin, C.Y.: Learning to recommend questions based on user ratings. In: CIKM, pp. 751–758 (2009)
14. Jurczyk, P., Agichtein, E.: Discovering authorities in question answer communities by using link analysis. In: Proceedings of the Sixteenth ACM Conference on Information and Knowledge Management, CIKM 2007, pp. 919–922. ACM, New York (2007)
15. Bian, J., Liu, Y., Zhou, D., Agichtein, E., Zha, H.: Learning to recognize reliable users and content in social media with coupled mutual reinforcement. In: WWW, pp. 51–60 (2009)
16. Lee, C.T., Rodrigues, E.M., Kazai, G., Milic-Frayling, N., Ignjatovic, A.: Model for voter scoring and best answer selection in community q&a services. In: Web Intelligence, pp. 116–123 (2009)
17. Lee, K., Caverlee, J., Webb, S.: Uncovering social spammers: social honeypots + machine learning. In: SIGIR, pp. 435–442 (2010)
18. Liu, Y., Agichtein, E.: You've got answers: Towards personalized models for predicting success in community question answering. In: ACL (Short Papers), pp. 97–100 (2008)
19. Agichtein, E., Liu, Y., Bian, J.: Modeling information-seeker satisfaction in community question answering. ACM Transactions on Knowledge Discovery from Data 3(2), 10:1–10:27 (2009)
20. Richardson, M., White, R.W.: Supporting synchronous social q&a throughout the question lifecycle. In: WWW, pp. 755–764 (2011)
21. Yang, L., Bao, S., Lin, Q., Wu, X., Han, D., Su, Z., Yu, Y.: Analyzing and predicting not-answered questions in community-based question answering services. In: AAAI (2011)
22. Thelwall, M., Buckley, K., Paltoglou, G., Cai, D., Kappas, A.: Sentiment in short strength detection informal text. J. Am. Soc. Inf. Sci. Technol. 61(12), 2544–2558 (2010)
23. Blei, D., McAuliffe, J.: Supervised topic models. In: Platt, J., Koller, D., Singer, Y., Roweis, S. (eds.) Advances in Neural Information Processing Systems 20. MIT Press, Cambridge (2008)
24. Draper, N.R., Smith, H.: Applied Regression Analysis, 3rd edn. Wiley Series in Probability and Statistics. Wiley-Interscience (April 1998)
25. Daume III, H.: Frustratingly easy domain adaptation. In: Proceedings of the 45th Annual Meeting of the Association of Computational Linguistics, Prague, Czech Republic, pp. 256–263. Association for Computational Linguistics (June 2007)
26. Adamic, L.A., Zhang, J., Bakshy, E., Ackerman, M.S.: Knowledge sharing and yahoo answers: everyone knows something. In: Proceedings of the 17th International Conference on World Wide Web, WWW 2008, pp. 665–674. ACM, New York (2008)
27. Dietterich, T.G., Bakiri, G.: Solving multiclass learning problems via error-correcting output codes. Journal of Artificial Intelligence Research 2 (1995)
28. Bibby, J., Toutenburg, H.: Prediction and Improved Estimation in Linear Models. John Wiley & Sons, Inc., New York (1978)
29. Provost, F.J., Fawcett, T.: Analysis and visualization of classifier performance: Comparison under imprecise class and cost distributions. In: KDD, pp. 43–48 (1997)
30. Cortes, C., Mohri, M.: Confidence intervals for the area under the roc curve. In: NIPS (2004)

Learning to Detect Patterns of Crime

Tong Wang[1], Cynthia Rudin[1], Daniel Wagner[2], and Rich Sevieri[2]

[1] Massachusetts Institute of Technology, Cambridge, MA 02139, USA
{tongwang,rudin}@mit.edu
[2] Cambridge Police Department, Cambridge, MA 02139, USA
{dwagner,rsevieri}@cambridgepolice.org

Abstract. Our goal is to automatically detect patterns of crime. Among a large set of crimes that happen every year in a major city, it is challenging, time-consuming, and labor-intensive for crime analysts to determine which ones may have been committed by the same individual(s). If automated, data-driven tools for crime pattern detection are made available to assist analysts, these tools could help police to better understand patterns of crime, leading to more precise attribution of past crimes, and the apprehension of suspects. To do this, we propose a pattern detection algorithm called *Series Finder*, that grows a pattern of discovered crimes from within a database, starting from a "seed" of a few crimes. Series Finder incorporates both the common characteristics of all patterns and the unique aspects of each specific pattern, and has had promising results on a decade's worth of crime pattern data collected by the Crime Analysis Unit of the Cambridge Police Department.

Keywords: Pattern detection, crime data mining, predictive policing.

1 Introduction

The goal of crime data mining is to understand patterns in criminal behavior in order to predict crime, anticipate criminal activity and prevent it (e.g., see [1]). There is a recent movement in law enforcement towards more empirical, data driven approaches to *predictive policing*, and the National Institute of Justice has recently launched an initiative in support of predictive policing [2]. However, even with new data-driven approaches to crime prediction, the fundamental job of crime analysts still remains difficult and often manual; *specific* patterns of crime are not necessarily easy to find by way of automated tools, whereas larger-scale density-based trends comprised mainly of background crime levels are much easier for data-driven approaches and software to estimate. The most frequent (and most successful) method to identify specific crime patterns involves the review of crime reports each day and the comparison of those reports to past crimes [3], even though this process can be extraordinarily time-consuming. In making these comparisons, an analyst looks for enough commonalities between a past crime and a present crime to suggest a pattern. Even though automated detection of specific crime patterns can be a much more difficult problem than

H. Blockeel et al. (Eds.): ECML PKDD 2013, Part III, LNAI 8190, pp. 515–530, 2013.

estimating background crime levels, tools for solving this problem could be extremely valuable in assisting crime analysts, and could directly lead to actionable preventative measures. Locating these patterns automatically is a challenge that machine learning tools and data mining analysis may be able to handle in a way that directly complements the work of human crime analysts.

In this work, we take a machine learning approach to the problem of detecting specific patterns of crime that are committed by the same offender or group. Our learning algorithm processes information similarly to how crime analysts process information instinctively: the algorithm searches through the database looking for similarities between crimes in a growing pattern and in the rest of the database, and tries to identify the modus operandi (M.O.) of the particular offender. The M.O. is the set of habits that the offender follows, and is a type of motif used to characterize the pattern. As more crimes are added to the set, the M.O. becomes more well-defined. Our approach to pattern discovery captures several important aspects of patterns:

- *Each M.O. is different.* Criminals are somewhat self-consistent in the way they commit crimes. However, different criminals can have very different M.O.'s. Consider the problem of predicting housebreaks (break-ins): Some offenders operate during weekdays while the residents are at work; some operate stealthily at night, while the residents are sleeping. Some offenders favor large apartment buildings, where they can break into multiple units in one day; others favor single-family houses, where they might be able to steal more valuable items. Different combinations of crime attributes can be more important than others for characterizing different M.O's.
- *General commonalities in M.O. do exist.* Each pattern is different but, for instance, similarity in time and space are often important to any pattern and should generally by weighted highly. Our method incorporates both general trends in M.O. and also pattern-specific trends.
- *Patterns can be dynamic.* Sometimes the M.O. shifts during a pattern. For instance, a novice burglar might initially use bodily force to open a door. As he gains experience, he might bring a tool with him to pry the door open. Occasionally, offenders switch entirely from one neighborhood to another. Methods that consider an M.O. as stationary would not naturally be able to capture these dynamics.

2   Background and Related Work

In this work, we define a "pattern" as a series of crimes committed by the same offender or group of offenders. This is different from a "hotspot" which is a spatially localized area where many crimes occur, whether or not they are committed by the same offender. It is also different than a "near-repeat" effect which is localized in time and space, and does not require the crimes to be committed by the same offender. To identify true patterns, one would need to consider information beyond simply time and space, but also other features of the crimes, such as the type of premise and means of entry.

An example of a pattern of crime would be a series of housebreaks over the course of a season committed by the same person, around different parts of East Cambridge, in houses whose doors are left unlocked, between noon and 2pm on weekdays. For this pattern, sometimes the houses are ransacked and sometimes not, and sometimes the residents are inside and sometimes not. This pattern does not constitute a "hotspot" as it's not localized in space. These crimes are not "near-repeats" as they are not localized in time and space (see [4]).

We know of very few previous works aimed directly at detecting specific patterns of crime. One of these works is that of Dahbur and Muscarello [5],[1] who use a cascaded network of Kohonen neural networks followed by heuristic processing of the network outputs. However, feature grouping in the first step makes an implicit assumption that attributes manually selected to group together have the same importance, which is not necessarily the case: each crime series has a signature set of factors that are important for that specific series, which is one of the main points we highlighted in the introduction. Their method has serious flaws, for instance that crimes occurring before midnight and after midnight cannot be grouped together by the neural network regardless of how many similarities exists between them, hence the need for heuristics. Series Finder has no such serious modeling defect. Nath [6] uses a semi-supervised clustering algorithm to detect crime patterns. He developed a weighting scheme for attributes, but the weights are provided by detectives instead of learned from data, similar to the baseline comparison methods we use. Brown and Hagen [7] and Lin and Brown [8] use similarity metrics like we do, but do not learn parameters from past data.

Many classic data mining techniques have been successful for crime analysis generally, such as association rule mining [7–10], classification [11], and clustering [6]. We refer to the general overview of Chen et al. [12], in which the authors present a general framework for crime data mining, where many of these standard tools are available as part of the COPLINK [13] software package. Much recent work has focused on locating and studying hotspots, which are localized high-crime-density areas (e.g., [14–16], and for a review, see [17]).

Algorithmic work on semi-supervised clustering methods (e.g., [18, 19]) is slightly related to our approach, in the sense that the set of patterns previously labeled by the police can be used as constraints for learned clusters; on the other hand, each of our clusters has different properties corresponding to different M.O.'s, and most of the crimes in our database are not part of a pattern and do not belong to a cluster. Standard clustering methods that assume all points in a cluster are close to the cluster center would also not be appropriate for modeling dynamic patterns of crime. Also not suitable are clustering methods that use the same distance metric for different clusters, as this would ignore the pattern's M.O. Clustering is usually unsupervised, whereas our method is supervised. Work on (unsupervised) set expansion in information retrieval (e.g., [20,21]) is very relevant to ours. In set expansion, they (like us) start with

---

[1] Also see http://en.wikipedia.org/wiki/Classification_System_for_Serial_Criminal_Patterns

a small seed of instances, possess a sea of unlabeled entities (webpages), most of which are not relevant, and attempt to grow members of the same set as the seed. The algorithms for set expansion do not adapt to the set as it develops, which is important for crime pattern detection. The baseline algorithms we compare with are similar to methods like Bayesian Sets applied in the context of Growing a List [20, 21] in that they use a type of inner product as the distance metric.

3   Series Finder for Pattern Detection

Series Finder is a supervised learning method for detecting patterns of crime. It has two different types of coefficients: pattern-specific coefficients $\{\eta_{\hat{\mathcal{P}},j}\}_j$, and pattern-general coefficients $\{\lambda_j\}_j$. The attributes of each crime (indexed by j) capture elements of the M.O. such as the means of entry, time of day, etc. Patterns of crime are grown sequentially, starting from candidate crimes (the seed). As the pattern grows, the method adapts the pattern-specific coefficients in order to better capture the M.O. The algorithm stops when there are no more crimes within the database that are closely related to the pattern.

The crime-general coefficients are able to capture common characteristics of all patterns (bullet 2 in the introduction), and the pattern-specific coefficients adjust to each pattern's M.O. (bullet 1 in the introduction). Dynamically changing patterns (bullet 3 in the introduction) are captured by a similarity S, possessing a parameter d which controls the "degree of dynamics" of a pattern. We discuss the details within this section.

Let us define the following:

- $\mathscr{C}$ – A set of all crimes.
- $\mathscr{P}$ – A set of all patterns.
- $\mathcal{P}$ – A single pattern, which is a set of crimes. $\mathcal{P} \subseteq \mathscr{P}$.
- $\hat{\mathcal{P}}$ – A pattern grown from a seed of pattern $\mathcal{P}$. Ideally, if $\mathcal{P}$ is a true pattern and $\hat{\mathcal{P}}$ is a discovered pattern, then $\hat{\mathcal{P}}$ should equal $\mathcal{P}$ when it has been completed. Crimes in $\hat{\mathcal{P}}$ are represented by $C_1, C_2, ...C_{|\hat{\mathcal{P}}|}$.
- $\mathcal{C}_{\hat{\mathcal{P}}}$ – The set of candidate crimes we will consider when starting from $\hat{\mathcal{P}}$ as the seed. These are potentially part of pattern $\mathcal{P}$. In practice, $\mathcal{C}_{\hat{\mathcal{P}}}$ is usually a set of crimes occurring within a year of the seed of $\mathcal{P}$. $\mathcal{C}_{\hat{\mathcal{P}}} \subseteq \mathscr{C}$.
- $s_j(C_i, C_k)$ – Similarity between crime i and k in attribute j. There are a total of J attributes. These similarities are calculated from raw data.
- $\gamma_{\hat{\mathcal{P}}}(C_i, C_k)$ – The overall similarity between crime i and k. It is a weighted sum of all J attributes, and is pattern-specific.

3.1   Crime-Crime Similarity

The pairwise similarity γ measures how similar crimes C_i and C_k are in a pattern set $\hat{\mathcal{P}}$. We model it in the following form:

$$\gamma_{\hat{\mathcal{P}}}(C_i, C_k) = \frac{1}{\Gamma_{\hat{\mathcal{P}}}} \sum_{j=1}^{J} \lambda_j \eta_{\hat{\mathcal{P}},j} s_j(C_i, C_k),$$

where two types of coefficients are introduced:

1. λ_j – pattern-general weights. These weights consider the general importance of each attribute. They are trained on past patterns of crime that were previously labeled by crime analysts as discussed in Section 3.4.
2. $\eta_{\hat{\mathcal{P}},j}$ – pattern-specific weights. These weights capture characteristics of a specific pattern. All crimes in pattern $\hat{\mathcal{P}}$ are used to decide $\eta_{\hat{\mathcal{P}},j}$, and further, the defining characteristics of $\hat{\mathcal{P}}$ are assigned higher values. Specifically:

$$\eta_{\hat{\mathcal{P}},j} = \sum_{i=1}^{|\hat{\mathcal{P}}|} \sum_{k=1}^{|\hat{\mathcal{P}}|} s_j(C_i, C_k)$$

$\Gamma_{\hat{\mathcal{P}}}$ is the normalizing factor $\Gamma_{\hat{\mathcal{P}}} = \sum_{j=1}^{J} \lambda_j \eta_{\hat{\mathcal{P}},j}$. Two crimes have a high $\gamma_{\hat{\mathcal{P}}}$ if they are similar along attributes that are important specifically to that crime pattern, and generally to all patterns.

3.2 Pattern-Crime Similarity

Pattern-crime similarity S measures whether crime $\tilde{C}$ is similar enough to set $\hat{\mathcal{P}}$ that it should be potentially included in $\hat{\mathcal{P}}$. The pattern-crime similarity incorporates the dynamics in M.O. discussed in the introduction. The dynamic element is controlled by a parameter d, called the *degree of dynamics*. The pattern-crime similarity is defined as follows for pattern $\hat{\mathcal{P}}$ and crime $\tilde{C}$:

$$S(\hat{\mathcal{P}}, \tilde{C}) = \left(\frac{1}{|\hat{\mathcal{P}}|} \sum_{i=1}^{|\hat{\mathcal{P}}|} \gamma_{\hat{\mathcal{P}}}(\tilde{C}, C_i)^d \right)^{(1/d)}$$

where $d \geqslant 1$. This is a soft-max, that is, an ℓ_d norm over $i \in \hat{\mathcal{P}}$. Use of the soft-max allows the pattern $\hat{\mathcal{P}}$ to evolve: crime i needs only be very similar to a few crimes in $\hat{\mathcal{P}}$ to be considered for inclusion in $\hat{\mathcal{P}}$ when the degree of dynamics d is large. On the contrary, if d is 1, this forces patterns to be very stable and stationary, as new crimes would need to be similar to most or all of the crimes already in $\hat{\mathcal{P}}$ to be included. That is, if $d = 1$, the dynamics of the pattern are ignored. For our purpose, d is chosen appropriately to balance between including the dynamics (d large), and stability and compactness of the pattern (d small).

3.3 Sequential Pattern Building

Starting with the seed, crimes are added iteratively from $C_{\hat{\mathcal{P}}}$ to $\hat{\mathcal{P}}$. At each iteration, the candidate crime with the highest pattern-crime similarity to $\hat{\mathcal{P}}$ is tentatively added to $\hat{\mathcal{P}}$. Then $\hat{\mathcal{P}}$'s cohesion is evaluated, which measures the cohesiveness of $\hat{\mathcal{P}}$ as a pattern of crime: Cohesion$(\hat{\mathcal{P}}) = \frac{1}{|\hat{\mathcal{P}}|} \sum_{i \in \hat{\mathcal{P}}} S(\hat{\mathcal{P}} \backslash C_i, C_i)$. While the cohesion is large enough, we will proceed to grow $\hat{\mathcal{P}}$. If $\hat{\mathcal{P}}$'s cohesion is below a threshold, $\hat{\mathcal{P}}$ stops growing. Here is the formal algorithm:

1: **Initialization:** $\hat{\mathcal{P}} \leftarrow \{$Seed crimes$\}$
2: **repeat**
3:     $C_{\text{tentative}} = \arg\max_{C \in (\mathcal{C}_{\hat{\mathcal{P}}} \backslash \hat{\mathcal{P}})} S(\hat{\mathcal{P}}, C)$
4:     $\hat{\mathcal{P}} \leftarrow \hat{\mathcal{P}} \cup \{C_{\text{tentative}}\}$
5:     Update: $\eta_{\hat{\mathcal{P}},j}$ for $j \in \{1, 2, \ldots J\}$, and Cohesion$(\hat{\mathcal{P}})$
6: **until** Cohesion$(\hat{\mathcal{P}}) <$ cutoff
7: $\hat{\mathcal{P}}^{\text{final}} := \hat{\mathcal{P}} \backslash C_{\text{tentative}}$
8: **return** $\hat{\mathcal{P}}^{\text{final}}$

3.4 Learning the Pattern-General Weights λ

The pattern-general weights are trained on past pattern data, by optimizing a performance measure that is close to the performance measures we will use to evaluate the quality of the results. Note that an alternative approach would be to simply ask crime analysts what the optimal weighting should be, which was the approach taken by Nath [6]. (This simpler method will also be used in Section 5.2 as a baseline for comparison.) We care fundamentally about optimizing the following measures of quality for our returned results:

– The fraction of the true pattern $\mathcal{P}$ returned by the algorithm:

$$\text{Recall}(\mathcal{P}, \hat{\mathcal{P}}) = \frac{\sum_{C \in \mathcal{P}} \mathbb{1}(C \in \hat{\mathcal{P}})}{|\mathcal{P}|}.$$

– The fraction of the discovered crimes that are within pattern $\mathcal{P}$:

$$\text{Precision}(\mathcal{P}, \hat{\mathcal{P}}) = \frac{\sum_{C \in \hat{\mathcal{P}}} \mathbb{1}(C \in \mathcal{P})}{|\hat{\mathcal{P}}|}.$$

The training set consists of true patterns $\mathcal{P}_1, \mathcal{P}_2, \ldots \mathcal{P}_\ell, \ldots \mathcal{P}_{|\mathscr{P}|}$. For each pattern $\mathcal{P}_\ell$ and its corresponding $\hat{\mathcal{P}}_\ell$, we define a gain function $g(\hat{\mathcal{P}}_\ell, \mathcal{P}_\ell, \lambda)$ containing both precision and recall. The dependence on $\lambda = \{\lambda_j\}_{j=1}^J$ is implicit, as it was used to construct $\hat{\mathcal{P}}_\ell$.

$$g(\hat{\mathcal{P}}_\ell, \mathcal{P}_\ell, \lambda) = \text{Recall}(\mathcal{P}_\ell, \hat{\mathcal{P}}_\ell) + \beta \cdot \text{Precision}(\mathcal{P}_\ell, \hat{\mathcal{P}}_\ell)$$

where β is the trade-off coefficient between the two quality measures. We wish to choose λ to maximize the gain over all patterns in the training set.

$$\underset{\lambda}{\text{maximize}} \quad G(\lambda) = \sum_\ell g(\hat{\mathcal{P}}_\ell, \mathcal{P}_\ell, \lambda)$$

$$\text{subject to} \quad \lambda_j \geqslant 0, \ j = 1, \ldots, J,$$

$$\sum_{j=1}^J \lambda_j = 1.$$

The optimization problem is non-convex and non-linear. However we hypothesize that it is reasonably smooth: small changes in λ translate to small changes in G. We use coordinate ascent to approximately optimize the objective, starting from different random initial conditions to avoid returning local minima. The procedure works as follows:

1: Initialize λ randomly, Converged=0
2: **while** Converged=0 **do**
3:   **for** $j = 1 \rightarrow J$ **do**
4:     $\lambda_j^{\text{new}} \leftarrow \text{argmax}_{\lambda_j} \; G(\lambda)$ (using a linesearch for the optima)
5:   **end for**
6:   **if** $\lambda^{\text{new}} = \lambda$ **then**
7:     Converged= 1
8:   **else**
9:     $\lambda \leftarrow \lambda^{\text{new}}$
10: **end if**
11: **end while**
12: **return** λ

We now discuss the definition of each of the J similarity measures.

4   Attribute Similarity Measures

Each pairwise attribute similarity $s_j : \mathcal{C} \times \mathcal{C} \rightarrow [0,1]$ compares two crimes along attribute j. Attributes are either categorical or numerical, and by the nature of our data, we are required to design similarity measures of both kinds.

4.1   Similarity for Categorical Attributes

In the Cambridge Police database for housebreaks, categorical attributes include "type of premise" (apartment, single-family house, etc.), "ransacked" (indicating whether the house was ransacked) and several others. We wanted a measure of agreement between crimes for each categorical attribute that includes (i) whether the two crimes agree on the attribute (ii) how common that attribute is. If the crimes do not agree, the similarity is zero. If the crimes do agree, and agreement on that attribute is unusual, the similarity should be given a higher weight. For example, in residential burglaries, it is unusual for the resident to be at home during the burglary. Two crimes committed while the resident was in the home are more similar to each other than two crimes where the resident was not at home. To do this, we weight the similarity by the probability of the match occurring, as follows, denoting C_{ij} as the jth attribute for crime C_i:

$$s_j(C_i, C_k) = \begin{cases} 1 - \sum_{q \in Q} p_j^2(x) & \text{if} \quad C_{ij} = C_{kj} = x \\ 0 & \text{if} \quad C_{ij} \neq C_{kj} \end{cases}$$

where $p_j^2(x) = \frac{n_x(n_x-1)}{N(N-1)}$, with n_x the number of times x is observed in the collection of N crimes. This is a simplified version of Goodall's measure [22].

4.2   Similarity for Numerical Attributes

Two formats of data exist for numerical attributes, either exact values, such as time 3:26pm, or a time window, e.g., 9:45am - 4:30pm. Unlike other types of crime such as assault and street robbery, housebreaks usually happen when the resident is not present, and thus time windows are typical. In this case, we need a similarity measure that can handle both exact time information and range-of-time information. A simple way of dealing with a time window is to take the midpoint of it (e.g., [15]), which simplifies the problem but may introduce bias.

Time-of-day profiles. We divide data into two groups: exact data $(t_1, t_2, \ldots, t_{m_e})$, and time window data $(\tilde{t}_1, \tilde{t}_2, \ldots, \tilde{t}_{m_r})$ where each data point is a range, $\tilde{t}_i = [t_{i,1}, t_{i,2}]$, $i = 1, 2, \ldots m_r$. We first create a profile based only on crimes with exact time data using kernel density estimation: $\hat{p}_{\text{exact}}(t) \propto \frac{1}{m_e} \sum_{i=1}^{m_e} K(t - t_i)$ where the kernel $K(\cdot)$ is a symmetric function, in our case a gaussian with a chosen bandwidth (we chose one hour). Then we use this to obtain an approximate distribution incorporating the time window measurements, as follows:

$$p(t|\tilde{t}_1, \ldots, \tilde{t}_{m_r}) \propto p(t) \cdot p(\tilde{t}_1, \ldots, \tilde{t}_{m_r}|t)$$
$$\approx \hat{p}_{\text{exact}}(t) \cdot \hat{p}(\text{range includes } t|t).$$

The function $\hat{p}(\text{range includes } t|t)$ is a smoothed version of the empirical probability that the window includes t:

$$\hat{p}(\text{range includes } t|t) \propto \frac{1}{m_r} \sum_{i=1}^{m_r} \tilde{K}(t, \tilde{t}_i)$$

where $\tilde{t}_i = [t_{i,1}, t_{i,2}]$ and $\tilde{K}(t, \tilde{t}_i) := \int_\tau \mathbf{1}_{\tau \in [t_{i,1}, t_{i,2}]} K(t - \tau) d\tau$. K is again a gaussian with a selected bandwidth. Thus, we define:

$$\hat{p}_{\text{range}}(t) \propto \hat{p}_{\text{exact}}(t) \cdot \frac{1}{m_r} \sum_{i=1}^{m_r} \tilde{K}(t, \tilde{t}_i).$$

We combine the exact and range estimates in a weighted linear combination, weighted according to the amount of data we have from each category:

$$\hat{p}(t) \propto \frac{m_e}{m_e + m_r} \hat{p}_{\text{exact}}(t) + \frac{m_r}{m_e + m_r} \hat{p}_{\text{range}}(t).$$

We used the approach above to construct a time-of-day profile for residential burglaries in Cambridge, where $\hat{p}(t)$ and $\hat{p}_{\text{exact}}(t)$ are plotted in Figure 1(a). To independently verify the result, we compared it with residential burglaries in Portland between 1996 and 2011 (reproduced from [23]) shown in Figure 1(b).[2] The temporal pattern is similar, with a peak at around 1-2pm, a drop centered around 6-7am, and a smaller drop at around midnight, though the profile differs slightly in the evening between 6pm-2am.

---

[2] To design this plot for Portland, range-of-time information was incorporated by distributing the weight of each crime uniformly over its time window.

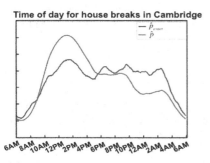
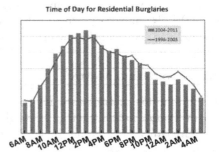

(a) Cambridge, from year 1997 to 2012   (b) Portland, from year 1996 to 2011

Fig. 1. Time of day profiling for house breaks in two different cities

A unified similarity measure for numeric attributes. We propose a similarity measure that is consistent for exact and range numerical data. The similarity decays exponentially with the distance between two data values, for either exact or range data. We use the expected average distance over the two ranges as the distance measure. For example, let crime i happen within $t_i := [t_{i,1}, t_{i,2}]$ and crime k happen within $t_k := [t_{k,1}, t_{k,2}]$. Then

$$\tilde{d}(t_i, t_k) = \int_{t_{i,1}}^{t_{i,2}} \int_{t_{k,1}}^{t_{k,2}} \hat{p}(\tau_i|t_i)\hat{p}(\tau_k|t_k)d(\tau_i, \tau_k)d\tau_i d\tau_k$$

where $\hat{p}$ was estimated in the previous subsection for times of the day, and $d(\tau_i, \tau_k)$ is the difference in time between τ_i and τ_k. The conditional probability is obtained by renormalizing $\hat{p}(\tau_i|t_i)$ to the interval (or exact value) t_i. The distance measure for exact numerical data can be viewed as a special case of this expected average distance where the conditional probability $\hat{p}(\tau_i|t_i)$ is 1.

The general similarity measure is thus:

$$s_j(C_i, C_k) := \exp\left(-\tilde{d}(z_i, z_k)/\Upsilon_j\right)$$

where Υ_j is a scaling factor (e.g, we chose $\Upsilon_j = 120$ minutes in the experiment), and z_i, z_k are values of attribute j for crimes i and k, which could be either exact values or ranges of values. We applied this form of similarity measure for all numerical (non-categorical) crime attributes.

5   Experiments

We used data from 4855 housebreaks in Cambridge between 1997 and 2006 recorded by the Crime Analysis Unit of the Cambridge Police Department. Crime attributes include geographic location, date, day of the week, time frame, location of entry, means of entry, an indicator for "ransacked," type of premise, an indicator for whether residents were present, and suspect and victim information. We also have 51 patterns collected over the same period of time that were curated and hand-labeled by crime analysts.

5.1   Evaluation Metrics

The evaluation metrics used for the experimental results are *average precision* and *reciprocal rank*. Denoting $\hat{\mathcal{P}}^i$ as the first i crimes in the discovered pattern, and $\Delta\mathrm{Recall}(\mathcal{P}, \hat{\mathcal{P}}^i)$ as the change in recall from $i - 1$ to i:

$$\mathrm{AveP}(\mathcal{P}, \hat{\mathcal{P}}) := \sum_{i=1}^{|\hat{\mathcal{P}}|} \mathrm{Precision}(\mathcal{P}, \hat{\mathcal{P}}^i)\Delta\mathrm{Recall}(\mathcal{P}, \hat{\mathcal{P}}^i).$$

To calculate reciprocal rank, again we index the crimes in $\hat{\mathcal{P}}$ by the order in which they were discovered, and compute

$$\mathrm{RR}(\mathcal{P}, \hat{\mathcal{P}}) := \frac{1}{\left(\sum_{r=1}^{|\hat{\mathcal{P}}|} \frac{1}{r}\right)} \sum_{C_i \in \mathcal{P}} \frac{1}{\mathrm{Rank}(C_i, \hat{\mathcal{P}})},$$

where $\mathrm{Rank}(C_i, \hat{\mathcal{P}})$ is the order in which C_i was added to $\hat{\mathcal{P}}$. If C_i was never added to $\hat{\mathcal{P}}$, then $\mathrm{Rank}(C_i, \hat{\mathcal{P}})$ is infinity and the term in the sum is zero.

5.2   Competing Models and Baselines

We compare with hierarchical agglomerative clustering and an iterative nearest neighbor approach as competing baseline methods. For each method, we use several different schemes to iteratively add discovered crimes, starting from the same seed given to Series Finder. The pairwise similarity γ is a weighted sum of the attribute similarities:

$$\gamma(C_i, C_k) = \sum_{j=1}^{J} \hat{\lambda}_j s_j(C_i, C_k).$$

where the similarity metrics $s_j(C_i, C_k)$ are the same as Series Finder used. The weights $\hat{\lambda}$ were provided by the Crime Analysis Unit of the Cambridge Police Department based on their experience. This will allow us to see the specific advantage of Series Finder, where the weights were learned from past data.

Hierarchical agglomerative clustering (HAC) begins with each crime as a singleton cluster. At each step, the most similar (according to the similarity criterion) two clusters are merged into a single cluster, producing one less cluster at the next level. *Iterative nearest neighbor classification* (NN) begins with the seed set. At each step, the nearest neighbor (according to the similarity criterion) of the set is added to the pattern, until the nearest neighbor is no longer sufficiently similar. HAC and NN were used with three different criteria for cluster-cluster or cluster-crime similarity: *Single Linkage* (SL), which considers the most similar

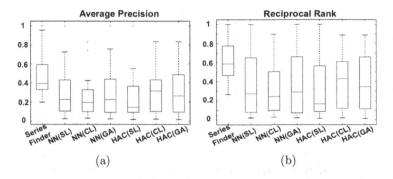

Fig. 2. Boxplot of evaluation metrics for out-of-sample patterns

pair of crimes; *Complete Linkage* (CL), which considers the most dissimilar pair of crimes, and *Group Average* (GA), which uses the averaged pairwise similarity [24]. The incremental nearest neighbor algorithm using the S_{GA} measure, with the weights provided by the crime analysts, becomes similar in spirit to the Bayesian Sets algorithm [20] and how it is used in information retrieval applications [21].

$$S_{SL}(R,T) := \max_{C_i \in R, C_k \in T} \gamma(C_i, C_k)$$

$$S_{CL}(R,T) := \min_{C_i \in R, C_k \in T} \gamma(C_i, C_k)$$

$$S_{GA}(R,T) := \frac{1}{|R||T|} \sum_{C_i \in R} \sum_{C_k \in T} \gamma(C_i, C_k).$$

5.3 Testing

We trained our models on two-thirds of the patterns from the Cambridge Police Department and tested the results on the remaining third. For all methods, pattern $\hat{\mathcal{P}}_\ell$ was grown until all crimes in $\mathcal{P}_\ell$ were discovered. Boxplots of the distribution of average precision and reciprocal ranks over the test patterns for Series Finder and six baselines are shown in Figure 2(a) and Figure 2(b). We remark that Series Finder has several advantages over the competing models: (i) Hierarchical agglomerative clustering does not use the similarity between seed crimes. Each seed grows a pattern independently, with possibly no interaction between seeds. (ii) The competing models do not have pattern-specific weights. One set of weights, which is pattern-general, is used for all patterns. (iii) The weights used by the competing models are provided by detectives based on their experience, while the weights of Series Finder are learned from data.

Since Series Finder's performance depends on pattern-specific weights that are calculated from seed crimes, we would like to understand how much each additional crime within the seed generally contributes to performance. The average precision and reciprocal rank for the 16 testing patterns grown from 2, 3

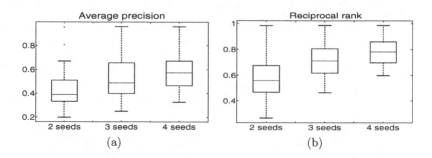

Fig. 3. Performance of Series Finder with 2, 3 and 4 seeds

and 4 seeds are plotted in Figure 3(a) and Figure 3(b). For both performance measures, the quality of the predictions increases consistently with the number of seed crimes. The additional crimes in the seed help to clarify the M.O.

5.4   Model Convergence and Sensitivity Analysis

In Section 3, when discussing the optimization procedure for learning the weights, we hypothesized that small changes in λ generally translate to small changes in the objective $G(\lambda)$. Our observations about convergence have been consistent with this hypothesis, in that the objective seems to change smoothly over the course of the optimization procedure. Figure 4(a) shows the optimal objective value at each iteration of training the algorithm on patterns collected by the Cambridge Police Department. In this run, convergence was achieved after 14 coordinate ascent iterations. This was the fastest converging run over the randomly chosen initial conditions used for the optimization procedure.

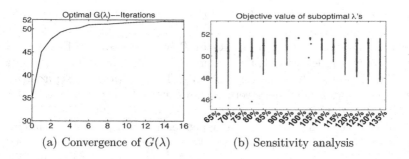

(a) Convergence of $G(\lambda)$ (b) Sensitivity analysis

Fig. 4. Performance analysis

We also performed a sensitivity analysis for the optimum. We varied each of the J coefficients λ_j from 65% to 135%, of its value at the optimum. As each coefficient was varied, the others were kept fixed. We recorded the value of $G(\lambda)$

at several points along this spectrum of percentages between 65% and 135%, for each of the λ_j's. This allows us to understand the sensitivity of $G(\lambda)$ to movement along any one of the axes of the J-dimensional space. We created box plots of $G(\lambda)$ at every 5th percentage between between 65% and 135%, shown in Figure 4(b). The number of elements in each box plot is the number of dimensions J. These plots provide additional evidence that the objective $G(\lambda)$ is somewhat smooth in λ; for instance the objective value varies by a maximum of approximately 5-6% when one of the λ_j's changes by 10-15%.

6   Expert Validation and Case Study

We wanted to see whether our data mining efforts could help crime analysts identify crimes within a pattern that they did not yet know about, or exclude crimes that were misidentified as part of a pattern. To do this, Series Finder was trained on all existing crime patterns from the database to get the pattern-general weights λ. Next, using two crimes in each pattern as a seed, Series Finder iteratively added candidate crimes to the pattern until the pattern cohesion dropped below 0.8 of the seed cohesion. Crime analysts then provided feedback on Series Finder's results for nine patterns.

There are now three versions of each pattern: $\mathcal{P}$ which is the original pattern in the database, $\hat{\mathcal{P}}$ which was discovered using Series Finder from two crimes in the pattern, and $\mathcal{P}_{\text{verified}}$ which came from crime experts after they viewed the union of $\hat{\mathcal{P}}$ and $\mathcal{P}$. Based on these, we counted different types of successes and failures for the 9 patterns, shown in Table 1. The mathematical definition of them is represented by the first 4 columns. For example, *correct finds* refer to crimes that are not in $\mathcal{P}$, but that are in $\hat{\mathcal{P}}$, and were verified by experts as belonging to the pattern, in $\mathcal{P}_{\text{verified}}$.

Table 1. Expert validation study results

Type of crimes	$\mathcal{P}$	$\hat{\mathcal{P}}$	$\mathcal{P}_{\text{verified}}$	$\mathcal{P}_1$	$\mathcal{P}_2$	$\mathcal{P}_3$	$\mathcal{P}_4$	$\mathcal{P}_5$	$\mathcal{P}_6$	$\mathcal{P}_7$	$\mathcal{P}_8$	$\mathcal{P}_9$
Correct hits	$\subseteq$	$\subseteq$	$\subseteq$	6	5	6	3	8	5	7	2	10
Correct finds	$\not\subseteq$	$\subseteq$	$\subseteq$	2	1	0	1	0	1	2	2	0
Correct exclusions	$\subseteq$	$\not\subseteq$	$\not\subseteq$	0	0	4	1	0	2	1	0	0
Incorrect exclusions	$\subseteq$	$\not\subseteq$	$\subseteq$	0	0	1	0	1	0	0	0	1
False hits	$\not\subseteq$	$\subseteq$	$\not\subseteq$	2	0	0	0	2	2	0	0	0

Correct hits, *correct finds* and *correct exclusions* count successes for Series Finder. Specifically, correct finds and correct exclusions capture Series Finder's improvements over the original database. Series Finder was able to discover 9 crimes that analysts had not previously matched to a pattern (the sum of the correct finds) and exclude 8 crimes that analysts agreed should be excluded (the sum of correct exclusions). *Incorrect exclusions* and *false hits* are not successes.

Table 2. Example: A 2004 Series

NO	Cri type	Date	Loc of entry	Mns of entry	Premises	Rans	Resid	Time of day	Day	Suspect	Victim
1	Seed	1/7/04	Front door	Pried	Aptment	No	Not in	8:45	Wed	null	White F
2	Corr hit	1/18/04	Rear door	Pried	Aptment	Yes	Not in	12:00	Sun	White M	White F
3	Corr hit	1/26/04	Grd window	Removed	Res Unk	No	Not in	7:30-12:15	Mon	null	Hisp F
4	Seed	1/27/04	Rear door	Popped Lock	Aptment	No	Not in	8:30-18:00	Tues	null	null
5	Corr exclu	1/31/04	Grd window	Pried	Res Unk	No	Not in	13:21	Sat	Black M	null
6	Corr hit	2/11/04	Front door	Pried	Aptment	No	Not in	8:30-12:30	Wed	null	Asian M
7	Corr hit	2/11/04	Front door	Pried	Aptment	No	Not in	8:00-14:10	Wed	null	null
8	Corr hit	2/17/04	Grd window	Unknown	Aptment	No	Not in	0:35	Tues	null	null
9	Corr find	2/19/04	Door: unkn	Pried	Aptment	No	Not in	10:00-16:10	Thur	null	White M
10	Corr find	2/19/04	Door: unkn	Pried	Aptment	No	Not in	7:30-16:10	Thur	null	White M
11	Corr hit	2/20/04	Front door	Broke	Aptment	No	Not in	8:00-17:55	Fri	null	null
12	Corr hit	2/25/04	Front door	Pried	Aptment	Yes	Not in	14:00	Wed	null	null

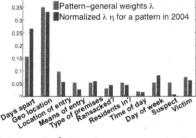

(a) Locations of crimes   (b) λ and $\frac{1}{\Gamma_{\mathcal{P}}}\lambda \cdot \eta$ for a pattern in 2004

Fig. 5. An example pattern in 2004

On the other hand, false hits that are similar to the crimes within the pattern may still be useful for crime analysts to consider when determining the M.O.

We now discuss a pattern in detail to demonstrate the type of result that Series Finder is producing. The example provided is Pattern 7 in Table 1, which is a series from 2004 in Mid-Cambridge covering a time range of two months. Crimes were usually committed on weekdays during working hours. The premises are all apartments (except two unknowns). Figure 5(a) shows geographically where these crimes were located. In Figure 5(a), four categories of crime within the 2004 pattern are marked with different colored dots: seed crimes are represented with blue dots, correct hits are represented with orange dots, the correct exclusion is represented with a red dot and the two correct finds are represented with green dots. Table 2 provides some details about the crimes within the series.

We visualize the M.O. of the pattern by displaying the weights in Figure 5(b). The red bars represent the pattern-general weights λ and the blue bars represent the total normalized weights obtained from the product of pattern-general weights and pattern-specific weights for this 2004 pattern. Notable observations about this pattern are that: the time between crimes is a (relatively) more important characteristic for this pattern than for general patterns, as the crimes in the pattern happen almost every week; the means and location of entry are relatively less important as they are not consistent; and the suspect information is also relatively less important. The suspect information is only present in one of the crimes found by Series Finder (a white male). Geographic closeness is less important for this series, as the crimes in the series are spread over a relatively large geographic distance.

Series Finder made a contribution to this pattern, in the sense that it detected two crimes that analysts had not previously considered as belonging to this pattern. It also correctly excluded one crime from the series. In this case, the correct exclusion is valuable since it had suspect information, which in this case could be very misleading. This exclusion of this crime indicates that the offender is a white male, rather than a black male.

7 Conclusion

Series Finder is designed to detect patterns of crime committed by the same individual(s). In Cambridge, it has been able to correctly match several crimes to patterns that were originally missed by analysts. The designer of the near-repeat calculator, Ratcliffe, has stated that the near-repeat calculator is not a "silver bullet" [25]. Series Finder also is not a magic bullet. On the other hand, Series Finder can be a useful tool: by using very detailed information about the crimes, and by tailoring the weights of the attributes to the specific M.O. of the pattern, we are able to correctly pinpoint patterns more accurately than similar methods. As we have shown through examples, the extensive data processing and learning that goes into characterizing the M.O. of each pattern leads to richer insights that were not available previously. Some analysts spend hours each day searching for crime series manually. By replicating the cumbersome process that analysts currently use to find patterns, Series Finder could have enormous implications for time management, and may allow analysts to find patterns that they would not otherwise be able to find.

Acknowledgements. Funding for this work was provided by C. Rudin's grants from MIT Lincoln Laboratory and NSF-CAREER IIS-1053407. We wish to thank Christopher Bruce, Julie Schnobrich-Davis and Richard Berk for helpful discussions.

References

1. Berk, R., Sherman, L., Barnes, G., Kurtz, E., Ahlman, L.: Forecasting murder within a population of probationers and parolees: a high stakes application of statistical learning. Journal of the Royal Statistical Society: Series A (Statistics in Society) 172(1), 191–211 (2009)
2. Pearsall, B.: Predictive policing: The future of law enforcement? National Institute of Justice Journal 266, 16–19 (2010)
3. Gwinn, S.L., Bruce, C., Cooper, J.P., Hick, S.: Exploring crime analysis. Readings on essential skills, 2nd edn. BookSurge, LLC (2008)
4. Ratcliffe, J.H., Rengert, G.F.: Near-repeat patterns in Philadelphia shootings. Security Journal 21(1), 58–76 (2008)

5. Dahbur, K., Muscarello, T.: Classification system for serial criminal patterns. Artificial Intelligence and Law 11(4), 251–269 (2003)
6. Nath, S.V.: Crime pattern detection using data mining. In: Proceedings of Web Intelligence and Intelligent Agent Technology Workshops, pp. 41–44 (2006)
7. Brown, D.E., Hagen, S.: Data association methods with applications to law enforcement. Decision Support Systems 34(4), 369–378 (2003)
8. Lin, S., Brown, D.E.: An outlier-based data association method for linking criminal incidents. In: Proceedings of the Third SIAM International Conference on Data Mining. (2003)
9. Ng, V., Chan, S., Lau, D., Ying, C.M.: Incremental mining for temporal association rules for crime pattern discoveries. In: Proceedings of the 18th Australasian Database Conference, vol. 63, pp. 123–132 (2007)
10. Buczak, A.L., Gifford, C.M.: Fuzzy association rule mining for community crime pattern discovery. In: ACM SIGKDD Workshop on Intelligence and Security Informatics (2010)
11. Wang, G., Chen, H., Atabakhsh, H.: Automatically detecting deceptive criminal identities. Communications of the ACM 47(3), 70–76 (2004)
12. Chen, H., Chung, W., Xu, J., Wang, G., Qin, Y., Chau, M.: Crime data mining: a general framework and some examples. Computer 37(4), 50–56 (2004)
13. Hauck, R.V., Atabakhsb, H., Ongvasith, P., Gupta, H., Chen, H.: Using COPLINK to analyze criminal-justice data. Computer 35(3), 30–37 (2002)
14. Short, M.B., D'Orsogna, M.R., Pasour, V.B., Tita, G.E., Brantingham, P.J., Bertozzi, A.L., Chayes, L.B.: A statistical model of criminal behavior. Mathematical Models and Methods in Applied Sciences 18, 1249–1267 (2008)
15. Mohler, G.O., Short, M.B., Brantingham, P.J., Schoenberg, F.P., Tita, G.E.: Self-exciting point process modeling of crime. Journal of the American Statistical Association 106(493) (2011)
16. Short, M.B., D'Orsogna, M., Brantingham, P., Tita, G.: Measuring and modeling repeat and near-repeat burglary effects. Journal of Quantitative Criminology 25(3), 325–339 (2009)
17. Eck, J., Chainey, S., Cameron, J., Wilson, R.: Mapping crime: Understanding hotspots. Technical report, National Institute of Justice, NIJ Special Report (August 2005)
18. Basu, S., Banerjee, A., Mooney, R.: Semi-supervised clustering by seeding. In: International Conference on Machine Learning, pp. 19–26 (2002)
19. Wagstaff, K., Cardie, C., Rogers, S., Schrödl, S.: Constrained k-means clustering with background knowledge. In: Int'l Conf. on Machine Learning, pp. 577–584 (2001)
20. Ghahramani, Z., Heller, K.: Bayesian sets. In: Proceedings of Neural Information Processing Systems (2005)
21. Letham, B., Rudin, C., Heller, K.: Growing a list. Data Mining and Knowledge Discovery (to appear, 2013)
22. Boriah, S., Chandola, V., Kumar, V.: Similarity measures for categorical data: A comparative evaluation. In: Proceedings of the Eighth SIAM International Conference on Data Mining, pp. 243–254 (2008)
23. Criminal Justice Policy Research Institute: Residential burglary in Portland, Oregon. Hatfield School of Government, Criminal Justice Policy Research Institute, http://www.pdx.edu/cjpri/time-of-dayday-of-week-0
24. Hastie, T., Tibshirani, R., Friedman, J., Franklin, J.: The elements of statistical learning: data mining, inference and prediction. Springer (2005)
25. National Law Enforcement and Corrections Technology Center: 'Calculate' repeat crime. TechBeat (Fall 2008)

Space Allocation in the Retail Industry: A Decision Support System Integrating Evolutionary Algorithms and Regression Models

Fábio Pinto and Carlos Soares

INESC TEC/Faculdade de Engenharia, Universidade do Porto*
Rua Dr. Roberto Frias, s/n
Porto, Portugal 4200-465
fhpinto@inescporto.pt, csoares@fe.up.pt

Abstract. One of the hardest resources to manage in retail is space. Retailers need to assign limited store space to a growing number of product categories such that sales and other performance metrics are maximized. Although this seems to be an ideal task for a data mining approach, there is one important barrier: the representativeness of the available data. In fact, changes to the layout of retail stores are infrequent. This means that very few values of the space variable are represented in the data, which makes it hard to generalize. In this paper, we describe a Decision Support System to assist retailers in this task. The system uses an Evolutionary Algorithm to optimize space allocation based on the estimated impact on sales caused by changes in the space assigned to product categories. We assess the quality of the system on a real case study, using different regression algorithms to generate the estimates. The system obtained very good results when compared with the recommendations made by the business experts. We also investigated the effect of the representativeness of the sample on the accuracy of the regression models. We selected a few product categories based on a heuristic assessment of their representativeness. The results indicate that the best regression models were obtained on products for which the sample was not the best. The reason for this unexpected results remains to be explained.

Keywords: Retail, Representativeness of Sample, Evolutionary Algorithms, Regression.

1 Introduction

This paper adresses the problem of assigning space to product categories in retail stores. According to the business specialists, space is one of the most expensive resources in retail [1]. This makes product category space allocation one of the

---

* Part of this work was carried out while the authors were at Faculdade de Economia, Universidade do Porto.

H. Blockeel et al. (Eds.): ECML PKDD 2013, Part III, LNAI 8190, pp. 531–546, 2013.

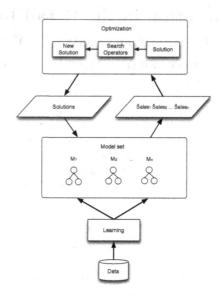

Fig. 1. Decision Support System architecture

most difficult decisions while defining the layout[1] of a retail store. Given that space is a limited resource, it is essential for the retailer to assess the effects of allocating more space to product category x instead of product category y. One approach is based on the estimated space elasticity of product categories [2]. *Space elasticity* is the impact on the sales of a given product or product category of varying (typically 1%) the space allocated to it.

We developed a Decision Support System (DSS) for space allocation that combines machine learning techniques with a meta-heuristic optimization method (Figure 1). The meta-heuristic searches the space of all admissible space allocations. For each solution considered, the corresponding total sales are estimated based on individual forecasts of the sales of each product category obtained using sales forecasting models. Besides the space allocated to the corresponding category in the solution, the inputs to these models include other variables characterizing the store and the product category. The models are induced using machine learning techniques on historical data.

This approach involves a number of challenges, the most important of which are 1) the representativeness of the data, 2) the evaluation of the individual models and 3) the evaluation of the whole system. The data collected by retail companies represents a tiny fraction of its domain because changes to the space assigned to a product category are not frequent. Therefore, it is hard to obtain models with good generalization capacity. To address this issue, we developed

---

[1] The layout is a schema combining text, graphics and photos to represent the physical distribution of products and product categories in a retail store, as well as their sizes and weights.

a measure of the space volatility of product categories. Surprisingly, however, the best models were not obtained on the product categories with more space volatility according to our measure.

The evaluation of both the individual models and the complete system are also challenging. The development of the individual models is hard because it is difficult to relate their accuracy with the quality of the global store layout. Therefore, knowing when to stop the development process is hard. We addressed this issue by setting thresholds, which were defined in collaboration with the business experts. Finally, the evaluation of the impact on sales of the layouts recommended by the system on real stores is not possible. On the other hand, evaluating them on historical data, even if a suitable resampling methodology is used, is not entirely convincing to the business users. We addressed this issue by using the system to make recommendations for a store that had its layout been recently makeover. We compared our recommendation to the new layout which was implemented, with very satisfactory results according to the experts.

We discuss these issues in this paper, which is organized as follows. Section 2 presents related work. In Section 3 we describe the data used as well as the measure to assess space volatility and the results of its application. Section 4 presents the methodology used to build the sales forecasting models. Sections 5 and 6 describe two experiments in modeling product categories sales: first, only four product categories were modeled, carefully selected according to the space volatility measure; secondly, we model all product categories. In Section 7 we specify how we combined the forecasting models with an optimization algorithm for our DSS. The case study is presented in Section 8, together with the results obtained. Finally, in Section 9 we present some conclusions and define future work.

2   Space Allocation in Retail Stores

Space allocation is done at multiple occasions. Retailers are obviously faced with this problem when opening new stores. Additionally, they must also make seasonal adjustments (e.g., add a camping section before summer) as well as temporary ones (e.g., to accommodate brand promotions). Finally, as the business evolves due to changes in the socio-economic conditions of the customer base (e.g., economic crisis) and new trends (e.g., inclusion of gourmet section), the layout must also be adapted.

Space allocation is done at multiple levels of granularity, namely the shop sections (e.g., vegetables) and all levels of the product hierarchy, including individual brands and products. Depending on the granularity level, decisions may involve location in 3D (i.e., not only the position in the store but also the height of the shelf in which the products are displayed) as well as size (e.g., length of shelf space assigned) and positioning relative to other products. Furthermore decisions are affected by a number of business constraints (i.e., contracts with product manufacturers), which makes the problem of space allocation even more complex.

Thus it comes as not surprise that product space allocation within a retail store is a common research topic. Several studies were conducted for studying the process of product allocation using Econometrics [3], Operations Research [4] and even Machine Learning, with the application of association rules [5] and Genetic Algorithms [6]. However, these papers are not concerned with product category space allocation. They focus on distributing products on previously assigned product category space.

Desmet and Renaudin [2] published the first paper concerning the problem of product category space allocation. They used Econometrics to model sales behaviour of product categories and estimated the respective space elasticity for each. Despite interesting, their results were partially questionable, with some estimated space elasticities with negative value. Castro [1] followed a very similar approach.

We believe that this is the first published work combining optimization and machine learning techniques on this problem.

3   Data

This Section details the data collected for the project and describes the process of identifying the categories with the most representative samples, including the space volatility measure.

3.1   Data Collection

The dataset comprises two years (2009-10) of data with monthly observations for 110 product categories. Overall, the dataset contains 332,885 observations.

The majority of the variables were provided by the retail company in which this project was developed. Due to confidentiality reasons, we can not give insights about their construction but we can motivate the purpose for their inclusion in our dataset. For all variables, i represents a product category, m represents a month and s a store: 1) $\mathbf{Sales}_{i,m,s}$ is the target variable; 2) $\mathbf{Area\_t}_{i,m,s}$ is the total area,[2] in square meters, assigned to a product category. [1] showed its significance on sales forecasting models for retail; 3) $\mathbf{Area\_pe}_{i,m,s}$ is the permanent area, in square meters, of a product category. Permanent area does not change due to seasonality factors. This type of area only changes during store layout restructuring; 4) $\mathbf{Area\_pr}_{i,m,s}$ is the promotional area, in square meters, of a product category. Promotional area changes mainly due to seasonality factors; 5) $\mathbf{m}$ is the month of the example. It is included for seasonality purposes as retail sales are highly seasonal and we expect to model that volatility with this nominal variable; 6) $\mathbf{Insignia}_s$ is the insignia of the store. The retail company has three different types (insignias) of stores. This nominal variable captures different sales behaviour among these insignias; 7) $\mathbf{Cluster}_s$ is the sales potential cluster of the store (nominal variable). The retail company divides its stores into

---

[2] The total area is the sum of the permanent and promotional areas.

four distinct clusters according to their sales potential; 8) **Cluster_Client**$_s$ is the client profile cluster of the store. The retail company divides its stores in four distinct clusters according to the profile of their customers. Again, the inclusion of this nominal variable seemed relevant given that it is expected that different customers will result in stores with different sales behaviour; 9) **PPI_County**$_s$ is the Purchasing Power Index of the region in which the store is located. It is expected that the larger the value of this variable, the larger the value of sales; 10) **N_W_Days**$_m$ is the number of non-working days of the month. Customers do most of their shopping on non-working days so it is expected that the larger the value of this variable, the larger the value of sales; 11) **C_P_Index**$_{i,s}$ is the category penetration index by store's client profile cluster. This is a discrete variable calculated for each product category within each customer cluster, so, there are 4 indexes by product category, one for each cluster. This variable can capture the impact that different customers have in product category sales.

Although there may have another important factors affecting sales and space, these are the variables that were available for this project.

3.2 Representativeness of Sample

The main goal of the project is to implement an optimization algorithm that maximizes sales given the decision variables **Area_t**$_{i,m,s}$ for all categories i. Therefore, we need models that predict sales accurately over a wide range of values of the latter variables. To achieve this, it is necessary to have data that covers a representative part of the space. However, this is very unlikely, as there are few changes to the space allocated to a category. Given the importance of space to retail, changes must be carefully motivated. The categories that are changed most often may be changed twice a year, while many have constant shelf area over much longer periods than that. Furthermore, most variations are relatively small.

This is an important issue because the quality of the results depends not only on the quality of the variables but also on the representativeness of the training sample. This is illustrated in Figure 2. Samples A and B contain examples in every region of the combined domain of both variables, x_1 and x_2. However, without further domain knowledge, sample A is better than sample B because it is more dense. Sample C is the worst of the three because only a small subset of the space is represented. The data in our case study is both sparse (as in sample B) and concentrated in a very small area of the domain space (as in sample C).

To understand how serious this problem is in our case study, we developed two measures to characterize shelf space volatility. These measures can be applied to all variables representing space allocation, namely $Area\_t_{i,m,s}$, $Area\_pe_{i,m,s}$ and $Area\_pr_{i,m,s}$, which are represented by $a_{i,m,s}$. The **Mean Number of Changes** (MNC_i) of the shelf space of category i is defined as

$$MNC_i = \frac{\sum_{s=1}^{n} f_{i,m,s}}{n} \tag{1}$$

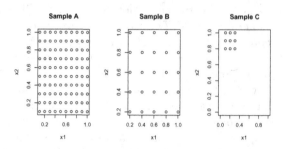

Fig. 2. Sample representativeness

where n is the number of observations and

$$f_{i,m,s} = \begin{cases} 1 & \text{if } a_{i,m,s} \neq a_{i,m-1,s} \\ 0 & \text{otherwise} \end{cases}$$

The **Absolute Mean Monthly Variation** $(AMMV_i)$ of the shelf space of category i is defined as (a percent value)

$$AMMV_i = \frac{\sum_{s=1}^{n} \frac{\sum_{m=2}^{k} \left| \frac{a_{i,m,s} - a_{i,m-1,s}}{a_{i,m-1,s}} \right|}{k}}{n} \times 100 \qquad (2)$$

The NMC_i and $AMMV_i$ measures are combined into a space volatility score, defined as $SV_i = MNC_i \times AMMV_i$.

Figure 3 shows the values of MNC_i and $AMMV_i$ for each product category in our dataset. The categories with a score value equal or greater than the 90th percentile are considered to have a high score level and are represented in red. As hypothesized, a great part of the product categories lie in the lower left side of the chart, meaning that those categories have reduced space volatility.

However, a high volatility score is not sufficient to ensure that the data sample for the corresponding category is representative of its domain. It is also important that the category has a homogeneous sales behavior across stores. The purpose of this requirement is to reduce the impact of the factor *store* in the relationship between product category space and sales. To do this, we compute the standard deviation of sales as a percentage of total sales by product category. The lower this value, the more homogeneous are the sales of the product category in the set of stores.

The graph of Figure 4 illustrates different scenarios that were found. It presents the indicators for the eleven product categories with a high volatility score: on the x-axis, the score for the permanent area of the categories; on the y-axis, the score for the promotional area; and the size of each point relates to the store homogeneity value of the respective product category, which are distinguished by different colours.

Interesting categories are: **3204**, with high volatility in terms of permanent area; above-average store homogeneity; **3101** with high volatility in terms of

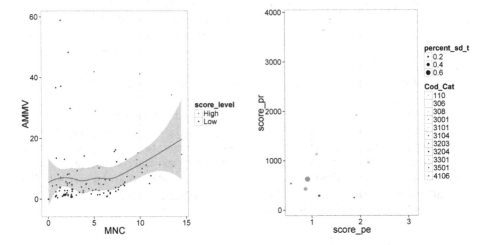

Fig. 3. Measures representing MNC_i and $AMMV_i$ for each category i. The categories with a high value of the score SV_i are represented in red color.

Fig. 4. Store homogeneity (size of points) for the categories with high sales volatility (SV_i) in permanent (y-axis) and promotional (x-axis) areas

promotional area and store homogeneity slightly below average; **3001** with considerable volatility in both area components and which is the product category with the best record of store homogeneity among the categories with a high volatility score; and **308**, with considerable volatility in both area components and which it is the product category with the worst record of store homogeneity among the selected categories. The latter category is the one with the worst data sample, according to our measures, while the other have the best samples.

4 Methodology

The problem is adressed as a regression task with $Sales_{i,m,s}$ as the dependent variable. Next, we describe how error is estimated and which regression algorithms were used.

4.1 Error Estimation

The error measures for the evaluation of the predictive models are: *Mean Relative Error*(MRE), defined as $\dfrac{\sum_{j=1}^{n} |\frac{y_j - \hat{y}_j}{y_j}|}{n} \times 100$; *Root Mean Squared Error* ($RMSE$), defined as $\sqrt{\dfrac{\sum_{j=1}^{n}(y_j - \hat{y}_j)^2}{n}}$; and *Variation Index*($varIndex$), defined as $\dfrac{RMSE}{\bar{y}}$. For all measures, y_j is the true value, $\hat{y}_j$ is the predicted value, and $\bar{y}$ is the mean of the output variable.

The performance of the regression algorithms will be compared with two benchmarks: a linear regression (LR), given that this technique was applied in past experiments on this problem [1][2]; and a baseline, whose predictions consist of the average of the target variable for the store which is being analyzed. These comparisons will help assess the difficulty of the phenomenon that we are modeling as well as how much useful knowledge the regression models are capturing.

The retail company defined 10% as the (maximum) target MRE value. We used this value as a treshold for sucess of our models.

Given that the dataset used in this work consisted of time series, the error was estimated using a hold-out strategy, ensuring that the test data was more recent than training data. The training set for each product category (and each model) consisted of one year and four months of observations; the remaining examples (eight months) were splited (4 months each) for a validation and a test set. The validation set was used for algorithm parameter tuning and the test set to assess the generalization error of the models.

4.2   Regression Algorithms

Several regression models were tested, namely: **Cubist**, an improvement of Quinlan's M5 regression trees [7]; **Artificial Neural Networks** (ANN), based on the empirical evidence of the capacity of ANN to successfully predict retail sales [8]; **Multivariate Adaptive Regression Splines** (MARS) [9]; **Support Vector Machines** (SVM) [10]; **Generalized Boosted Models** (GBM), R package implementation of boosting [11] models; and **Random Forests** (RF) [12].

The implementation of these algorithms available in the R [13] software was used in the experiments.

5   Modeling Categories with High Sales Volatility

The measures that were proposed in Section 3.2 to quantify space volatility were validated by the business experts. Nevertheless, we decided to analyze empirically if they are, in fact, good measures of the representativeness of the data samples. In case of positive results, we can define a strategy to systematically collect data for the product categories with data of insufficient quality based on the space volatility measures.

We focused on the four categories that were identified in Section 3.2, namely, **3204, 3101, 3001** and **308**. The results obtained on the category **308** are expected to be worse than on the others, given that the computed measures of space volatility and store homogeneity are clearly worst for this product category. Different parameters settings were tested and the best results are presented here.

Figure 5 shows the results for the four selected product categories in terms of MRE. Surprisingly, the MRE estimated for the models is far from the target value of 10%, except for product category **308**, which is actually quite close to the threshold. These results do not confirm our hypothesis. There seems to be no correlation between the variables calculated to characterize the representativeness of the data for each product category and the predictive performance

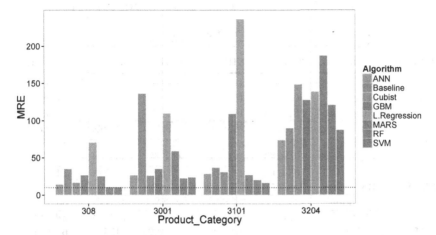

Fig. 5. Results for selected product categories, in terms of MRE. The dotted horizontal line represents the threshold defined by the client.

of the algorithms. Further analysis provided possible explanations for some of these results. For instance, the product category with the worst results is **3204**. Given that the models of this category do not include the variable m, since there was only one year of data for this category, we assume that the problem may be lack of representative data, as retail sales are seasonal. Nevertheless, at this point it became clear that we could not yet develop the data collection strategy based on the current version of these measures.

Table 1 presents more detailed results, according to three error measures. In terms of performance of the regression algorithms, SVM, RF and ANN obtain the best results. Overall, for these product categories, SVM is the best algorithm, according to the average of the error measures. Further observations can be made: 1) for all product categories, there is always at least one regression algorithm that is better than both benchmarks; 2) ANN presents reasonable performance from modeling retail sales, confirming previous results [8]; and 3) LR obtains poor performance for modeling retail sales. This fact is particularly important if we remember that the previous papers on the topic of product category space allocation [1][2] used this technique.

6   Modeling All Product Categories

Given the unexpected results in the preliminary experiments, we decided to run experiments on all product categories. Given that some of the 110 product categories of our dataset showed a very low number of observations, we only used the 89 categories with more than 1000 observations. Additionally, for 10 of these product categories, we only had one year of data.

We decided not to test ANN because of its computational cost. Additionally, the best SVM results in the previous section were obtained using two kernels,

Table 1. Detailed results on selected product categories results

Algorithm	Error Measure	Product Categories 3204	3101	3001	308	Average
Cubist	MRE	149.02%	31.10%	26.64%	16.97%	55.93%
	RMSE	300.57	4160.69	22118.84	6218.20	8199.58
	varIndex	182.33%	48.05%	44.12%	28.96%	75.86%
ANN	MRE	74.09%	28.97%	26.96%	14.49%	**36.13%**
	RMSE	208.57	2238.49	26816.24	5809.25	8768.14
	varIndex	135.97%	25.08%	60.33%	39.27%	65.16%
MARS	MRE	187.64%	27.16%	59.30%	25.70%	74.95%
	RMSE	348.37	2437.43	31767.91	7055.82	10402.38
	varIndex	211.33%	28.15%	63.36%	32.86%	83.93%
SVM	MRE	87.84%	16.70%	24.19%	11.22%	**34.98%**
	RMSE	94.81	2153.42	18680.24	5458.18	**6596.66**
	varIndex	57.52%	24.87%	37.26%	25.42%	**36.27%**
GBM	MRE	127.86%	109.4%	35.36%	26.93%	74.89%
	RMSE	220.68	6414.26	30985.78	15500	13280.18
	varIndex	127.78%	74.50%	59.28%	64.93%	81.62%
RF	MRE	121.28%	20.53%	22.92%	11.12%	43.96%
	RMSE	151.23	3044.51	19620.52	8373.84	**7797.53**
	varIndex	90.26%	35.20%	38.36%	38.28%	**50.53%**
LR	MRE	139.10%	236.87%	110.20%	71.13%	139.33%
	RMSE	264.26	5299.86	30761.88	18619.4	13736.35
	varIndex	160.31%	61.20%	61.35%	86.72%	92.40%
Baseline	MRE	90.40%	37.14%	136.72%	35.45%	74.93%
	RMSE	136.22	6485.44	78004.39	26796.21	27855.56
	varIndex	82.64%	74.90%	155.56%	124.80%	109.48%

therefore, we decided to use those two different kernels. So, we tested three algorithms: SVM with the radial and sigmoid kernels and RF.

6.1   Results

Figure 6 shows the results obtained for the three regression algorithms in terms of MRE on 89 product categories. The dotted horizontal line represents the threshold defined in Section 4. Surprisingly, given the results obtained earlier, several models are below or very close to the threshold. On one hand, this confirms that the measures used to assess space volatility need to be improved but, on the other, it indicates that the approach followed is viable, regardless of the apparent lack of representativeness of the data.

The statistical significance of the differences was based on the results of paired t-tests with $\alpha = 0.05$. Overall, SVM with the radial kernel obtained the best performance in 77 product categories; RF obtained the best performance in 10 product categories; and finally, SVM with sigmoid kernel presented superior

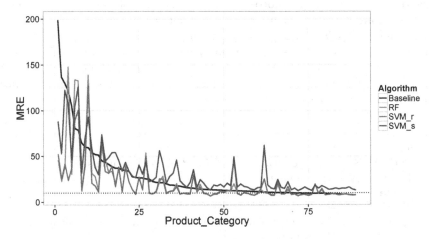

Fig. 6. Results of SVM - *radial kernel*, SVM - *sigmoid kernel* and RF for all product categories

performance in only two product categories. In several product categories, RF and SVM with radial kernel function showed a very similar performance and the statistical test did not distinguish them.

In Table 2 we compare the best model for each product category (of the three that were generated) with the respective baseline. On average, the regression algorithms show a better predictive performance than the baseline. More importantly, approximately half of them obtained an error that was very close or better than the threshold set by the retail company.

Table 2. Summary: best model *vs* Baseline (MRE)

Algorithm	Min	1stQ	Median	Mean	3rdQ	Max
Best model	6.51	8.76	11.01	19.05	22.95	126.11
Baseline	7.69	10.58	13.78	27.73	27.83	198.90

7 Decision Support System

Given that the models presented results that were very close to the goal set by the business experts, we decided to develop the method to optimize the assignment of shelf space to the product categories. As explained in more detail next, this method uses the models to predict the sales for the shelf space assigned by the solutions tested in the search process.

The optimization algorithm is an adapted version of a Genetic Algorithm (GA), an Evolutionary Computation (EC) framework. For further details on GAs, we refer the reader to [14].

Algorithm 1 shows the pseudocode of the algorithm implemented. The general EC method must be adapted to take into account specific issues of the problem at hand. In our case, solutions must not violate three fundamental constraints: a minimum value for each gene (the minimum space of a certain product category); a maximum value for each gene (the maximum space of a certain product category); and a maximum value for the sum of all genes (the maximum space of the sum of all product categories and layout).

begin
    INITIALISE population with historical and random solutions;
    EVALUATE each candidate;
    repeat
        SELECT parents;
        RECOMBINE pairs of parents with whole arithmetic recombination;
        MUTATE the resulting offspring;
        EVALUATE new candidates;
        SELECT top 30 individuals for the next generation;
    until *number of iterations* $= it$;
end

Algorithm 1. Optimization algorithm.

Representation of the solutions. A solution consists of the area of all product categories in a given store, $sol_h = g_{1,h}, g_{2,h}, ..., g_{i,h}$. Each solution has as many elements as the number of product categories.

The initial population consists of 30 solutions: 10 are historical assignments of shelf space to product categories in the store that is being optimized; and the remaining 20 are random solutions. The range of valid space values for each product category is defined by the corresponding maximum and minimum values that occurred in the most recent year available in the dataset for the store that is being optimized: $min_i \leq g_{i,h} \leq max_i$.

The fitness function that evaluates the solutions is based on the predictive models generated in Section 6. It uses those models to estimate the sales of the product categories, given the shelf size in the solutions together with the values of the remaining independent variables. The fitness of a solution is the sum of all sales forecasts, plus a penalty function ω_h for controlling the total space of the solution: $Fitness_h = sa\hat{l}es_1 + sa\hat{l}es_2 + ... + sa\hat{l}es_i + \omega_h$.

Parents selection. The selection of the solutions that will generate *offspring* is based on a probability associated with each, taking into account the fitness function output for each solution: $Prob(k) = Fitness(k) / \left(\sum_{k=1}^{30} Fitness \right)$. Then, from 30 solutions that constitute the population, 10 are selected for crossover.

Crossover. In this stage, the 10 selected solutions, generate 10 new solutions. In order not to violate the constraints that were imposed on all solutions, we applied an operator named whole arithmetic crossover [14]: given two selected solutions, sol_1 and sol_2, the average of the two is calculated, originating a new solution: $sol_3 = \frac{g_{1,1}+g_{1,2}}{2}, \frac{g_{2,1}+g_{2,2}}{2}, ..., \frac{g_{1,h}+g_{1,h}}{2}$.

Mutation. This operator randomly selects one or more solutions from the offspring. Two genes are randomly chosen from each of the selected solutions, and 1 unit (square meter) is transferred from the first category to the second: $sol_3 = g_{1,3}+1, g_{2,3}-1, ..., g_{h,3}$. However, this operator may disregard restrictions on the minimum and maximum area of each product category. In order to circumvent the problem, the mutation only occurs if the selected gene does not have a minimum or maximum value. The number of offspring selected for mutation is controlled by the mutation rate parameter.

Survival Selection. After the operators are applied, the offspring are added to the population. With a total population of 40 solutions, a new evaluation by the fitness function occurs. The top 30 are kept for new iteration. In this work we have opted for an elitist approach in the survivor selection mechanism [14], instead of the typical probabilistic one. We chose this approach given that the fitness values for our solutions were very close to each other, and a probabilistic selection lead the algorithm to the loss of good solutions.

The DSS performs optimization at the monthly level: given a store, a particular number of product categories and one month, the algorithm seeks the set of areas that maximizes monthly sales according to the predictive models generated in Section 6.

8   Case Study

The retail company proposed a test for a more realistic assessment of the system. In May 2011, a given store had a makeover of its logistics which included a new layout. The vast majority of the product categories that were available in this store underwent major changes in the shelf space. For this makeover, the analysts of the retail company based their space recommendations on data from 2009 to 2010, the very same that allowed us to build our dataset. Thus, it was proposed to test the developed DSS in this store and compare the results obtained with the recommendations of the business specialists.

8.1   Experimental Setup

In this experiment, we assumed that, except for the variables $Area\_t_{i,m,s}$ and $N\_W\_Days_m$, the other independent variables remained constant, given that it seems acceptable that from 2010 to 2011 they have not changed substantially. The predictive models generated for this experiment integrated all observations available in the dataset. Based on the experiments in Section 6, we selected the best algorithm for each product category.

Out of the 89 categories for which it is possible to build predictive models, 85 are part of the store layout, so, the solutions consisted of 85 genes. To evaluate the quality of the optimization and compare its results with recommendations from the business specialists, we used Pearson's correlation coefficient.

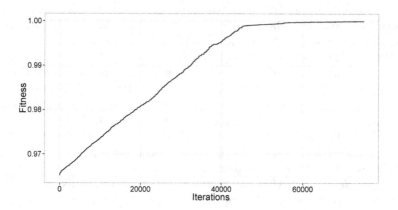

Fig. 7. DSS results for 75000 iterations, measured as the fitness of the best solution evaluated

8.2 Results

For 75,000 iterations, the optimization algorithm presents the performance showed in Figure 7. Due to confidentially issues, the fitness values were scaled between 0 and 1, 0 being the minimum fitness value and 1 being the maximum fitness achieved by the best solution in the population. There is a steep increase in fitness, which slows down as the system converges. A steady state is achieved near iteration 60,000.

Comparing the results obtained by the DSS with the recommendations made by the business specialists, we estimated a correlation of 0.66. Assuming that the space recommendations of the business specialists are ideal, these results are a good indicator of the quality of the developed DSS. We compared the predicted total sales according to our models of the two solutions.

Analysing the results in detail, some important questions arise. A first comparison between the output of the DSS and the real areas of the store in May 2011, shows that three product categories are clearly overvalued by the system. These categories have a common factor: they have very strong promotional campaigns at specific times of the year. It is clear that the models of these product categories (with MRE values of 11.17%, 24.90% and 16.80%) are failing to associate this boom in sales to seasonal factors. Those strong promotional campaigns also imply a significant, albeit temporary, increase in shelf space, which seems to be confusing the learning algorithms. Given that the m variable (month of the observation) varies systematically and the variable $Area\_t_{i,m,s}$ presents a large increase at those times just like the variable $Sales_{i,m,s}$, the forecasting models associated the sales boom to changes in the value of the variable $Area\_t_{i,m,s}$. This has implications in the DSS output: for these product categories, its recommendation is to have promotional campaigns at any time of the year. This clearly does not make sense in seasonal products.

Although, we assume that the recommendations made by the specialists are ideal, due to lack of better information, the product categories for which the system differs from those values are also interesting. For instance, for the adjacent product categories 1502 and 1503, the DSS suggests areas that are quite different from the recommendations of the business specialists. Given the excellent predictive accuracy of the models in question (6.92% and 12.17%, respectively), we found this information to be quite useful for the retail company.

9 Conclusions and Future Work

This paper presents the combination of sales forecasting models for individual product categories and a GA to develop a DSS for product category space allocation in retail stores. As far we know, this is the first time that machine learning and optimization techniques are applied to deal with this problem.

Given that not many changes are made to the layout of stores, it is expected that the data collected from the daily operation of retail stores is not adequate for modeling purposes. We developed two measures to assess the representativeness of the data associated with each product category. However, an empirical study indicated that the measures were not good predictors of the predictive accuracy of the models. Given the importance of collecting adequate data, an important goal is to improve these measures to serve as the basis for a systematic data collection strategy. This work can benefit from existing work on active learning [15].

Somewhat surprisingly, many of the models generated obtained satisfactory results and, thus, we developed the DSS. The system was evaluated with data from a major change in layout at a specific store, which had recently been performed. The results indicate that this system recommends space allocations in retail stores that are very similar to the recommendations made by business specialists. Furthermore, some of the differences in the recommendations were interesting from the business perspective.

We have developed models that are independent of each other. However, it is likely that there is some dependence between the sales of at least some of the product categories. We plan to test multi-target regression methods [16] to address this problem.

In this project, we have addressed only a very small part of the space allocation problem. An important development is the extension of the current system with recommendations concerning which categories to place next to each other. One approach to this problem is market basket analysis [17]. However, the challenge is how to combine the two types of recommendation.

Acknowledgements. This work was partially supported by Project Best-Case, which is co-financed by the North Portugal Regional Operational Programme (ON.2 - O Novo Norte), under the National Strategic Reference Framework (NSRF), through the European Regional Development Fund (ERDF) and by

National Funds through the FCT - Fundação para a Ciência e Tecnologia (Portuguese Foundation for Science and Technology) within project "Evolutionary algorithms for Decision Problems in Management Science" (PTDC/EGE-GES/099741/2008).

References

1. Castro, A., Brochado, A., Martins, F.: Supermarkets sales and retail area: a mixture regression model for segmentation. In: European Network for Business and Industrial Statistics (2007)
2. Desmet, P., Renaudin, V.: Estimation of product category sales responsiveness to allocated shelf space. International Journal of Research in Marketing (15), 443–457 (1998)
3. Gaur, V., Fisher, M., Raman, A.: An econometric analysis of inventory turnover performance in retail stores. Management Science 51, 181–193 (2005)
4. Dréze, X., Hoch, S.J., Purk, M.E.: Shelf management and space elasticity. Journal of Retailing 70(4), 301–326 (1994)
5. Nafari, M., Shahrabi, J.: A temporal data mining approach for shelf-space allocation with consideration of product price. Expert Systems with Applications (37), 4066–4072 (2010)
6. Hwang, H., Choi, B., Lee, G.: A genetic algorithm approach to an integrated problem of shelf space design and item allocation. Computers and Industrial Engineering (56), 809–820 (2009)
7. Quinlan, J.: Learning with continuous classes. In: Adams, Sterling (eds.) AI 1992, Singapore, pp. 343–348 (1992)
8. Alon, I., Qi, M., Sadowski, R.J.: Forecasting aggregate retail sales: a comparison of artificial neural networks and traditional methods. Journal of Retailing and Consumer Services 8(3), 147–156 (2001)
9. Friedman, J.: Multivariate adaptive regression splines. The Annals of Statistics 19(1), 1–141 (1991)
10. Vapnik, V., Cortes, C.: Support-vector networks. Machine Learning (20), 273–297 (1995)
11. Friedman, J.: Greedy function approximation: a gradient boosting machine. The Annals of Statistics 29(5), 1189–1232 (2001)
12. Breiman, L.: Random forests. Machine Learning 45(1), 5–32 (2001)
13. R Core Team: R: A Language and Environment for Statistical Computing. R Foundation for Statistical Computing, Vienna, Austria (2012)
14. Eiben, A., Simth, J.: Introduction to Evolutionary Computing, 1st edn. Natural Computing Series. Springer (2003)
15. Settles, B.: Active learning literature survey. Technical report, University of Wisconsin, Madison (2010)
16. Aho, T., Zenko, B., Dzeroski, S., Elomaa, T.: Multi-target regression with rule ensembles. Journal of Machine Learning Research 1, 1–48 (2012)
17. Russell, G.J., Petersen, A.: Analysis of cross category dependence in market basket selection. Journal of Retailing 76(3), 367–392 (2000)

Forest-Based Point Process for Event Prediction from Electronic Health Records

Jeremy C. Weiss and David Page

University of Wisconsin-Madison, Madison, WI, USA
jcweiss@cs.wisc.edu, page@biostat.wisc.edu

Abstract. Accurate prediction of future onset of disease from Electronic Health Records (EHRs) has important clinical and economic implications. In this domain the arrival of data comes at semi-irregular intervals and makes the prediction task challenging. We propose a method called multiplicative-forest point processes (MFPPs) that learns the rate of future events based on an event history. MFPPs join previous theory in multiplicative forest continuous-time Bayesian networks and piecewise-continuous conditional intensity models. We analyze the advantages of using MFPPs over previous methods and show that on synthetic and real EHR forecasting of heart attacks, MFPPs outperform earlier methods and augment off-the-shelf machine learning algorithms.

1 Introduction

Ballooning medical costs and an aging population are forcing governments and health organizations to critically examine ways of providing improved care while meeting budgetary constraints. A leading candidate to fulfill this mandate is the advancement of personalized medicine, the field surrounding the customization of healthcare to individuals. Predictive models for future onset of disease are the tools of choice here, though the application of existing models to existing data has had mixed results.

The research into improvements in predictive modeling has manifested in two main areas: better data and better models. Electronic health records (EHRs) now provide rich medical histories on individuals including diagnoses, medications, procedures, family history, genetic information, and so on. The individual may have regular check-ups interspersed with hospitalizations and medical emergencies, and the sequences of semi-irregular events can be considered as timelines.

Unlike timelines, the majority of models incorporating time use a time-series data representation. In these models data are assumed to arrive at regular intervals. Irregular arrivals of events violate this assumption and lead to missing data and/or aggregation, resulting in a loss of information. Experimentally, such methods have been shown to underperform analogous continuous-time models [1].

To address the irregularity of medical event arrivals, we develop a continuous-time model: multiplicative-forest point processes (MFPPs). MFPPs model the rate of event occurrences and assume that they are dependent on an event history

H. Blockeel et al. (Eds.): ECML PKDD 2013, Part III, LNAI 8190, pp. 547–562, 2013.
© Springer-Verlag Berlin Heidelberg 2013

in a piecewise-constant manner. For example, the event of aspirin consumption (or lack thereof) may affect the rate of myocardial infarction, or heart attack, which in turn affects the rate of thrombolytic therapy administration. Our goal is to learn a model that identifies such associations from data.

MFPPs build on previous work in piecewise-constant conditional intensity models (PCIMs) using ideas from multiplicative-forest continuous-time Bayesian networks (mfCTBNs) [2,3]. MFPPs extends the regression tree structure of PCIMs to regression forests. Unlike most forest learning algorithms, which minimize a classification loss through function gradient ascent or ensembling, MF-PPs are based on a multiplicative-forest technique developed in CTBNs. Here, a multiplicative assumption for combining regression tree values leads to optimal marginal log likelihood updates with changes in forest structure. The multiplicative representation allows MFPPs to concisely represent composite rates, yet also to have the flexibility to model rates with complicated dependencies. As the multiplicative forest model leads to representational and computational gains in mfCTBNs, we show that similar gains can be achieved in the point process domain. We conduct experiments to test two main hypotheses. First, we test for improvements in learning MFPPs over PCIMs, validating the usefulness of the multiplicative-forest concept. Second, we assess the ability of MFPPs to classify individuals for myocardial infarction from EHR data, compared to PCIMs and off-the-shelf machine learning algorithms.

Specifically we address two modeling scenarios for forecasting: *ex ante* (meaning "from the past") forecasting and supervised forecasting. An *ex ante* forecast is the traditional type of forecasting and occurs if no labels are available in the forecast region. An example of *ex ante* forecasting is the prediction of future disease onset from the present day forwards. Acquiring labels from the future is not possible, and labels from the past may introduce bias through a cohort effect. However, in some cases, labels may be used, and we call such forecasts "supervised". An example of supervised forecasting is the retrospective cohort study to predict the class of unlabeled examples as well as to identify risk factors leading to disease. The application of continuous-time models to the forecasting case is straightforward. When labels are available, however, we choose to apply MFPPs in a cascade learning framework, where the MFPP predictions contribute as features to supervised learning models.

In Section 2, we discuss point processes and contrast them from continuous-time Bayesian networks (CTBNs) noting their matching likelihood formulations given somewhat different problem setups. We show that multiplicative forest methods can be extended to point processes. We also introduce the problem of predicting myocardial infarction, discuss the various approaches to answering medical queries, and introduce our method of analysis. In Section 3, we present results on synthetic timelines and real health records data and show that MF-PPs outperform PCIMs on these tasks, and that the timeline analysis approach outperforms other standard machine learning approaches to the problem. We conclude in Section 4.

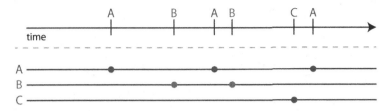

Fig. 1. A timeline (top) deconstructed into point processes (bottom)

2   Point Processes

Data that arrive at irregular intervals are aptly modeled with timelines. A time-line is a sequence of {event,time} pairs capturing the relative frequency and ordering of events. This representation arises in many domains, including neuron spike trains [4], high-frequency trading [5], and medical forecasting [6]. We describe and build upon one such model: the point process.

A point process treats each event type individually and specifies that it (re-)occurs according to the intensity (or rate) function $\lambda(t|h)$ over time t given an event history h. Figure 1 shows a sample timeline of events deconstructed into individual point processes. The conditional intensity model (CIM) is a probabilistic model formed by the composition of such processes. Our work will build on piecewise-constant conditional intensity models (PCIMs), which make the assumption that the intensity functions $\lambda(t|h)$ are constant over positive-length intervals. PCIMs represent the piecewise-constant conditional intensity functions with regression trees, and one is shown in Figure 2 (left).

The piecewise-constant intensity assumption is convenient for several reasons. For one, the likelihood can be computed in closed form. We can also compute the sufficient statistics by counting events and computing a weighted sum of constant-intensity durations. With these, we can directly estimate the maximum likelihood model parameters. Finally, we note that with this assumption the likelihood formulation becomes identical to the one used in continuous-time Bayesian networks (CTBNs). The shared likelihood formula lets us apply a recent advance in learning CTBNs: the use of multiplicative forests. Multiplicative forests produce intensities by taking the product of the regression values in active leaves. For example, a multiplicative forest equivalent to the tree described above is shown in Figure 2 (right). These models were shown to have large empirical gains for parameter and structure learning similar to those seen in the transition from tree models to random forests or boosted trees [3]. Our first goal is to show that a similar learning framework can be applied to point processes. We describe the model in fuller detail below.

2.1   Piecewise-Continuous Conditional Intensity Models (PCIMs)

Let us consider the finite set of event types $l \in \mathcal{L}$. An event sequence or trajectory x is an ordered set of {time, event} pairs $(t, l)_{i=1}^{n}$. A history h at time t is the

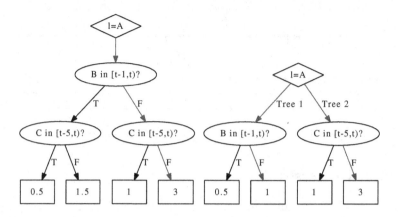

Fig. 2. A piecewise-constant conditional intensity tree for determining the rate of event type A (left). An equivalent multiplicative intensity forest (right). An example of active paths are shown in red. The active path in the tree corresponds to the intersection of active paths in the forest, and the output intensity is the same ($3 = 1 \times 3$).

subset of x whose times are less than t. Let l_0 denote the null event type, and use the null event pairs (l_0, t_0) and (l_0, t_{end}) to denote the start and end times of the trajectory. Then the likelihood of the trajectory given the CIM θ is:

$$p(x|\theta) = \prod_{l \in \mathcal{L}} \prod_{i=1}^{n} \lambda_l(t_i|h_i, \theta)^{\mathbb{1}(l=l_i)} e^{\int_{-\infty}^{t} \lambda_l(\tau|x,\theta) d\tau}$$

PCIMs introduce the assumption that the intensity functions are constant over intervals. As described in [2], let Σ_l be a set of discrete states so that we obtain the set of parameters λ_{ls} for $s \in \Sigma_l$. The active state s is determined by a mapping $\sigma_l(t, x)$ from time and trajectory to s. Let S_l hold the pair $(\Sigma_l, \sigma_l(t, x))$ and let $S = \{S_l\}_{l \in \mathcal{L}}$. Then the PCIM likelihood simplifies to:

$$p(x|S, \theta) = \prod_{l \in \mathcal{L}} \prod_{s \in \Sigma_l} \lambda_{ls}^{M_{ls}(x)} e^{-\lambda_{ls} T_{ls}(x)} \tag{1}$$

where $M_{ls}(x)$ is the count of events of type l while s is active in trajectory x, and $T_{ls}(x)$ is the total duration that s, for event type l, is active.

2.2 Continuous-Time Bayesian Networks (CTBNs)

Continuous-time Bayesian networks model a set of discrete random variables $x_1, x_2, \ldots, x_d = \mathcal{X}$ over continuous time, each with s_i number of discrete states for i in $\{1, \ldots, d\}$ [7]. CTBNs make the assumption that the probability of transition for variable x out of state x^j at time t is given by the exponential distribution $\lambda_{x^j|u} e^{-\lambda_{x^j|u} t}$ with rate parameter (intensity) $\lambda_{x^j|u}$ given parents setting u. The variable transitions from state x^j to x^k with probability $\Theta_{x^j x^k|u}$, where $\Theta_{x^j x^k|u}$

is an entry in state transition matrix Θ_u. The parents setting u is an element of the joint state U_x over parent variables of x, and the parent dependencies are provided in a directed possibly-cyclic graph. A complete CTBN model can be described by two components: a distribution $\mathcal{B}$ over the initial joint state, typically represented by a Bayesian network, and a directed graph over variables $\mathcal{X}$ with corresponding conditional intensity matrices (CIMs). The CIMs hold the intensities $\lambda_{x^j|u}$ and state transition probability matrices Θ_u.

The CTBN likelihood is defined as follows. A trajectory, or a timeline, is broken down into independent intervals of fixed state. For each interval $[t_0, t_{\text{end}})$, the duration $t = t_{\text{end}} - t_0$ passes and a variable x transitions at t_{end} from state x^j to x^k. All other variables $x_i \neq x$ rest during this interval in their active states x_i'. Then, the interval density is given by:

$$\underbrace{\lambda_{x^j|u}e^{-\lambda_{x^j|u}t}}_{x \text{ transitions}} \underbrace{\Theta_{x^j x^k|u}}_{\text{to state } x^k} \underbrace{\prod_{x_i':x_i \neq x} e^{-\lambda_{x_i'|u}t}}_{\text{while } x_i\text{'s rest}}$$

The trajectory likelihood is given by the product of intervals:

$$\prod_{x \in \mathcal{X}} \prod_{x^j \in x} \prod_{u \in U_x} \lambda_{x^j|u}^{M_{x^j|u}} e^{-\lambda_{x^j|u}T_{x^j|u}} \prod_{x^k \neq x^j} \Theta_{x^j x^k|u}^{M_{x^j x^k|u}} \tag{2}$$

where the $M_{x^j|u}$ (and $M_{x^j x^k|u}$) are the numbers of transitions out of state x^j (to state x^k), and where the $T_{x^j|u}$ are the amounts of time spent in x^j given parents settings u. Defining rate parameter $\lambda_{x^i x^j|u} = \lambda_{x^i|u}\Theta_{x^i x^j|u}$ and set element $p = x^j \times u$ (as in [3]), Equation 2 can be rewritten as:

$$\prod_{x \in \mathcal{X}} \prod_{x' \in x} \prod_{p} \lambda_{x'|p}^{M_{x'|p}} e^{-\lambda_{x'|p}T_p} \tag{3}$$

Note how the form of the likelihood in Equation 1 is identical to Equation 3.

2.3   Contrasting PCIMs and CTBNs

Despite the similarity in form, PCIMs and CTBNs model distinctly different types of continuous-time processes. Table 1 contrasts the two models. The primary difference is that, unlike point processes, CTBNs model a persistent, joint state over time. That is, a CTBN provides a distribution over the joint state for any time t. Additionally, CTBN variables must possess a 1-of-s_i state representation for $s_i > 1$ whereas point processes typically assume non-complementary event types. Furthermore, in CTBNs, observations are typically not of changes in state at particular times but instead probes of the state at a time point or interval. With persistent states, CTBNs can be used to answer interpolative queries, whereas CIMs are designed specifically for forecasting. Another notable difference is that CTBNs are Markovian: the intensities are determined entirely by the current state of the system. While more restrictive, this assumption allows for

Table 1. Contrasting piecewise-constant continuous intensity models (PCIMs) and multiplicative-forest continuous-time Bayesian networks (mfCTBNs). Key similarities are highlighted in blue.

	PCIM	mfCTBN
Model of:	event sequence	persistent state
Intensities	piecewise-constant	network-dependent constant
Dependence	event history	joint state (Markovian)
Labels	event types	variables
Emissions	events	states (x', 1 of s_i)
Structure	regression tree	multiplicative forest
Evidence	events	(partial) observations of states
Likelihood	$\prod_l \prod_s \lambda_{ls}^{M_{ls}} e^{-\lambda_{ls} T_{ls}}$	$\prod_{x'} \prod_p \lambda_{x'|p}^{M_{x'|p}} e^{-\lambda_{x'|p} T_p}$

variational and MCMC methods to be applied. On the other hand, PCIMs lend themselves to forecasting because the potentially prohibitive inference about the persistent state that CTBNs require is no longer necessary. This is because the rate of event occurrences depends on the event history instead of the current state.

2.4 Multiplicative-Forest Point Processes (MFPPs)

The similar likelihood forms allow us to extend the multiplicative-forest concept [3] to PCIMs. Following [2], we define the state Σ_l and mapping $\sigma_l(t, x)$ according to regression trees. Let $\mathcal{B}_l$ be the set of basis state functions $f(t, x)$ that maps to a basis state set Σ_f, akin to $\sigma(t, x)$ that maps to a single element s. As in [3], we can view the basis functions as set partitions of the space over $\Sigma = \Sigma_{l_1} \times \Sigma_{l_2} \times \ldots \Sigma_{l_{|\mathcal{L}|}}$. Each interior node in the regression tree is associated with a basis function f. Each leaf holds a non-negative real value: the intensity. Thus one path ρ through the regression tree for event type l corresponds to a recursive subpartition resulting in a set Σ_ρ, and every $(l, s) \in \Sigma_\rho$ corresponds to leaf intensity $\lambda_{l\rho}$, i.e., we set $\lambda_{ls} = \lambda_{l\rho}$. Figure 2 shows an example of the active path providing the intensity ($\lambda_{ls} = \lambda_{l\rho} = 3$).

MFPPs replace these trees with random forests. Given that each tree represents a partition, the intersection of trees, *i.e.* a forest, forms a finer partition. The subpartition corresponding to a single intensity is given by the intersection $\Sigma_\rho = \bigcap_{j=1}^k \Sigma_{\rho,j}$ of sets corresponding to the active paths through trees $1 \ldots k$. The intensity $\lambda_{l\rho}$ is given by the product of leaf intensities. Figure 2 (right) shows an example of the active paths in a tree, producing the forest intensity ($\lambda_{ls} = \lambda_{l\rho} = 1 \times 3$).

MFPPs use the PCIM generative framework. Forecasting is performed by forward sampling or importance sampling to generate an approximation to the distribution at future times. Learning MFPPs is analogous to learning mfCTBNs. A tree is learned iteratively by replacing a leaf with a branch with corresponding leaves. As in forest CTBNs, MFPPs have (1) a closed form marginal log likelihood update and (2) a simple maximum likelihood calculation for modification proposals. The intensities for the modification are the ratios between observed (M_{ls}) divided by expected ($\lambda_{ls}T_{ls}$) number of events prior to modification and while Σ_ρ is active. These two properties together provide the best greedy update to the forest model.

The use of multiplicative forest point processes has several advantages over previous methods.

– Compared to trees, forest models can represent more intensities per parameter, which is equal to the number of leaves in the model. For example, if a ground truth model has k stumps, that is, k single-split binary trees, then the forest can represent the model with $2k$ parameters. An equivalent tree would require 2^k parameters. This example arises whenever two risk factors are independent, i.e., their risks multiply.
– While forests can represent these independences when needed, they also can represent non-linear processes by increasing the depth of the tree beyond one. This advantage was established in previous work comparing trees to Poisson Networks [2,8], and forests possess advantages of both approaches.
– Unlike most forest models, multiplicative-forest trees may be learned in an order that is neither sequential nor simultaneous. The forest appends a stump to the end of its tree list when that modification improves the marginal likelihood the most. Otherwise it increases the depth of one tree. The data determines which expansion is selected.
– Multiplicative forests in CTBNs are restricted to learning from the current state (the Markovian assumption), whereas MFPPs learn from a basis set over some combination of the event history, deterministic, and constant features.
– Compared to the application of supervised classification methods to temporal data, the point process model identifies patterns of event sequences over time and uses them for forecasting. Figure 3 shows an example of the supervised forecasting setup. In this case, it may be harder to predict event B without using recurrent patterns of event sequences.

We hypothesize that these advantages will result in improved performance at forecasting, particularly in domains where risk factors are independent. As many established risk factors for cardiovascular disease are believed to contribute to the overall risk independently, we believe that MFPPs should outperform tree methods at this task. Because of their facility in modeling irregular series of events, we also believe that MFPPs should also outperform off-the-shelf machine learning methods.

2.5   Related Work

A rich literature exists on point processes focusing predominantly on spatial forecasting. In spatial domains, the point process is the temporal component of

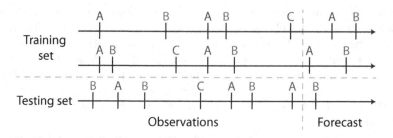

Fig. 3. Supervised forecasting. Labels are provided by the binary classification outcome: whether at least one event occurs in the forecasted region.

a model used to predict spatiotemporal patterns in data. The analysis of multivariate, spatial point processes is related to our work in its attempt to characterize the joint behavior of variables, for example, using Ripley's K function test for spatial homogeneity [9]. However, these methods do not learn dependency structures among variables; instead they seek to characterize cross-correlations observed in data. Generalized linear models for simple point processes are more closely related to our work. Here, a linear assumption for the intensity function is made, seen for example in Poisson networks [8]. PCIMs adopt a non-parametric approach and was shown to substantially improve upon previous methods in terms of model accuracy and learning time [2]. Our method builds on upon the PCIM framework.

Risk assessment for cardiovascular disease is also well studied. The primary outcome of most studies is the identification of one or a few risk factors and the quantification of the attributable risk. Our task is slightly different; we seek to predict from data the onset of future myocardial infarctions. The prediction task is closely related to risk stratification. For cardiovascular disease, the Framingham Heart Study is the landmark study for risk assessment [10]. They provide a 10-year risk of cardiovascular disease based on age, cholesterol (total and HDL), smoking status, and blood pressure. A number of studies have been since conducted purporting significant improvements over the Framingham Risk Score using different models or by collecting additional information [11]. In particular, the use of EHR data to predict heart attacks was previously addressed in [12]. However, in that work the temporal dependence of the outcome and its predictors was strictly logical and limited the success of their approach. We seek to show that, compared to standard approaches learning from features segmented in time, a point process naturally models timeline data and results in improved risk prediction.

3   Experiments

We evaluate MFPPs in two experiments. The first uses a model of myocardial infarction and stroke, and the goal is to learn MFPPs to recover the ground truth

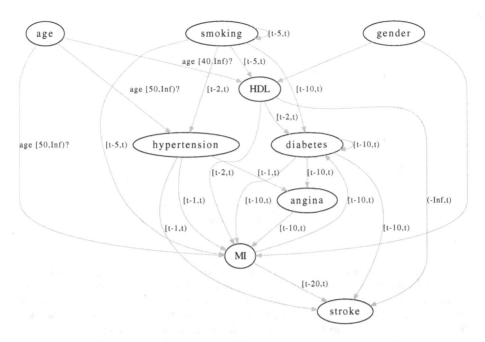

Fig. 4. Ground truth dependency structure of heart attack and stroke model. Labels on the edges determine the active duration of the dependency. Omitted in the graph is the age dependency for all non-deterministic nodes if the subject is older than 18.

model from sampled data. The second is an evaluation of MFPPs in predicting myocardial infarction diagnoses from real EHR data.

3.1   Model Experiment: Myocardial Infarction and Stroke

We introduce a ground truth PCIM model of myocardial infarction and stroke. The dependency structure of the model is shown in Figure 4. To compare MFPPs with PCIMs, we sample k trajectories from time 0 to 80 for $k = \{50, 100, 500, 1000, 5000, 10000\}$. We train each model with these samples and calculate the average log likelihood on a testing set of 1000 sampled trajectories. Each model used a BIC penalty to determine when to terminate learning. For features, we constructed a feature generator that uniformly at random selects an event type trigger and an active duration of one of $\{t-1, t-5, t-10, t-20, t-50\}$ to t. Note that the feature durations do not have a direct overlap with the dependency intervals shown in Figure 4. Our goal was to show that, even without being able to recover the exact ground truth model, we could get close with surrogate features. MFPPs were allowed to learn up to 10 trees each with 10 splitting features; PCIMs were allowed 1 tree with 100 splitting features. We also performed a two-tailed paired t-test to test for significant differences in MFPP and PCIM log likelihood. We ran each algorithm 250 times for each value of k.

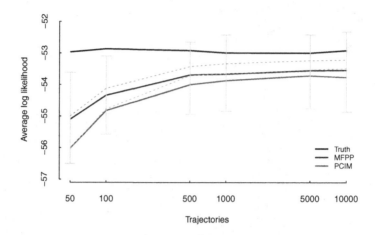

Fig. 5. Average log likelihoods for the {ground truth, MFPP, PCIM} model by the number of training set trajectories. Error bars in gray indicate the 95 percent confidence interval (omitted for the ground truth and PCIM models). Paired t-tests comparing MFPPs and PCIMs were significant at a p-value of 1-e20. Dotted lines show the likelihoods when ground truth features were made available to the models.

Figure 5 shows the average log likelihood results. Both MFPPs and PCIMs appear to converge to close to the ground truth model with increasing training set sizes. The lack of complete convergence is likely due to the mismatch in ground truth dependencies and the features available for learning. Error bars indicating the empirical 95 percent confidence intervals are also shown for MFPP. Similar error bars were observed for the ground truth and PCIM models but were omitted for clarity. The width of the interval is due to the variance in testing set log likelihoods. If we look at level average log likelihood lines in Figure 5, we observe that we only need a fraction of the data to learn a MFPP model equally good as the PCIM model. Both models completed all runs in under 15 minutes each.

We used a two-sided paired t-test to test for significant differences in the average log likelihood. For all numbers of trajectories k, the p-value was smaller than 1e-20. We conclude that the MFPP algorithm significantly outperformed the PCIM algorithm at recovering the ground truth model from data of this size.

3.2 EHR Prediction: Myocardial Infarction

In this section we describe the experiment on real EHR data. We define the task to be forecasting future onset of myocardial infarction between the years 2005 and 2010 given event data prior to 2005. We propose two forms of this experiment: *ex ante* and supervised forecasting. First, we test the ability of

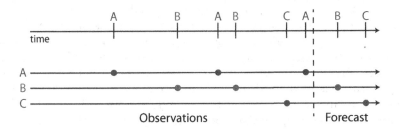

Fig. 6. *Ex ante* (traditional) forecasting. No labels for any example are available in the forecast region. The goal is to recover the events (*B* and *C*) from observations in the past.

MFPP to forecast events between 2005 and 2010 in all patients given the data leading up to 2005. Figure 6 depicts the *ex ante* forecasting setup.

Second, we split our data into training and testing sets to test MFPP in its ability to perform supervised forecasting. In this setup, we provide data between 2005 and 2010 for the training set in addition to all data prior to 2005 for both training and testing sets. We choose to focus on the outcome of whether a subject has at least one myocardial infarction event between the 2005 and 2010. Figure 3 shows the supervised forecasting setup.

We use EHR data from the Personalized Medicine Research Project (PMRP) cohort study run at the Marshfield Clinic Research Foundation [13]. The Marshfield Clinic has followed a patient population residing in northern Wisconsin and the outlying areas starting in the early 1960s up to the present. From this cohort, we include all subjects with at least one event between 1970 and 2005, and with at least one event after 2010 or a death record after 2005. Filtering with these inclusion criteria resulted in a study population of 15,446, with 428 identified individuals with a myocardial infarction event between 2005 and 2010.

To make learning and inference tractable, we selected additional event types from the EHR corresponding to risk factors identified in the Framingham Heart Study[10]: age, date, gender, LDL (critical low, low, normal, high, critical high, abnormal), blood pressure (normal, high), obesity, statin use, diabetes, stroke, angina, and bypass surgery. Because the level of detail specified in EHR event codes is fine, we use the above terms that represent aggregates over the terms in our database, *i.e.*, we map the event codes to one of the coarse terms. For example, an embolism lodged in the basilar artery is one type of stroke, and we code it simply as "stroke". The features we selected produced an event list with over 1.8 million events. As MFPPs require selecting active duration windows to learn, we used durations of size {0.25, 1, 2, 5, 10, 100 (ever)}, with more features focused on the recent past. Our intuition suggests that events occurring in the recent past are more informative than more distant events.

We compare MFPP against two sets of machine learning algorithms based on the experimental setup. For *ex ante* forecasting, we test against PCIMs [2] and homogeneous Poisson point processes, which assume independent and constant event rates. We assess their performance using the average log likelihood of the true events in the forecast region and precision-recall curves for our target event of interest: myocardial infarction. For supervised forecasting, we test against random forests and logistic regression [2,14]. As MFPP is not an inherently supervised learning algorithm, we also include a random forest learner using features corresponding to the intensity estimates based on the *ex ante* forecasting setup. We call this method MFPP-RF. We use modified bootstrapping to generate non-overlapping training and testing sets, and we train on 80 percent of the entire data. We compare the supervised forecasting methods only in terms of precision-recall due to the non-correspondence of the methods' likelihoods.

We also make a small modification to the MFPP and PCIM learning procedure when learning for modeling myocardial infarction, i.e., rare, events. On each iteration we expand one node in the forest of every event type instead of the forest of a single event type. The reason for this is that low intensity variables contribute less to the likelihood, so choosing the largest change in marginal log likelihood will tend to ignore modeling low intensity variables. By selecting an expansion for every event type each iteration, we ensure a rich modeling of myocardial infarction in the face of high frequency events such as blood pressure measurements and prescription refills. We note that because of the independence of likelihood components for each event type, this type of round-robin expansion is still guaranteed to increase the model likelihood. This statement would not hold, for example, in CTBNs, where a change in a variable intensity may change its latent state distribution, affecting the likelihood of another variable. Finally, for ease of implementation and sampling, we learn trees sequentially and limit the forest size to 40 total splits.

***Ex Ante* Forecasting Results.** Table 2 shows the average log likelihood results for *ex ante* forecasting for the MFPP, PCIM and homogeneous Poisson point process models. Both MFPPs and PCIMs perform much better than the baseline homogeneous model. MFPPs outperform PCIMs by a similar margin observed in the synthetic data set.

Figure 7 shows the precision-recall curve for predicting a myocardial infarction event between 2005 and 2010 given data on subjects prior to 2005. MFPPs and PCIMs perform similarly at this task. The high-recall region is of particular interest in the medical domain because it is more costly to miss a false negative (e.g. undiagnosed heart attack) than a false positive (false alarm). Simply put, clinical practice follows the "better safe than sorry" paradigm, so performance high-recall region is of highest concern. We plot the precision-recall curves between recalls of 0.5 and 1.0 for this reason. The absolute precision for all methods remains low and might exhibit the challenging nature of *ex ante* forecasting. Alternatively, the low precision results could be

Table 2. Log likelihood of {MFPP, PCIM, independent homogeneous Poisson processes} for forecasting patient medical events between 2005 and 2010.

Method	Log likelihood
MFPP	12.1
PCIM	10.3
Poisson	-54.8

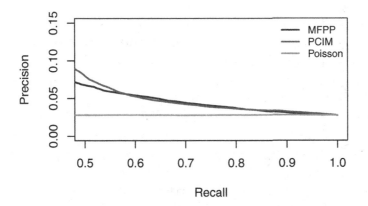

Fig. 7. Precision-recall curves for *ex ante* forecasting. MFPPs are compared against PCIMs and homogeneous Poisson point processes.

a result of potential incompatibility of the exponential waiting time assumption and medical event data. Since forecasting can be considered a type of extrapolative prediction, a violation of the model assumptions could lead to suboptimal predictions. Despite these limitations, compared to the baseline precision of $428/15{,}446 = 0.028$, the trained methods do provide utility in forecasting future MI events nonetheless.

Supervised Forecasting Results. Figure 8 provides the precision-recall curve for the supervised forecasting experiment predicting at least one myocardial infarction event between 2005 and 2010. As we see, MFPP underperforms compared to all supervised learning methods. However, the MFPP predicted intensities features boosts the MFPP-RF performance compared to the other classifiers. This suggests that while MFPP is a valuable model but may not be optimized for classification.

MFPPs also provide insight into the temporal progression of events. Figure 9 shows the first two trees of the forest learned for the rate of myocardial infarction. We observe the effects on increased risk: history of heart attack, elevated LDL cholesterol levels, abnormal blood pressure, and history of bypass surgery. While the whole forest is not shown (see http://cs.wisc.edu/~jcweiss/ecml2013/), the first two trees provide the main effects on the rate. As you progress through the forest, the range over intensity factors narrows towards 1. The tapering effect of relative tree "importance" is a consequence of experimental decision to learn the forest sequentially, and it provides for nice interpretation: the first few trees identify the main effects, and subsequent trees make fine adjustments for the contribution of additional risk factors.

As Figure 9 shows, the dominating factor of the rate is whether a recent myocardial infarction event was observed. In part, this may be due to an increased risk of recurrent disease, but also because some EHR events are "treated

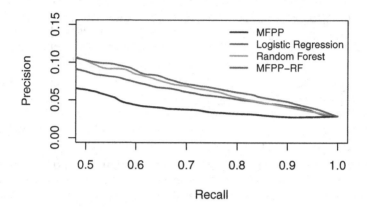

Fig. 8. Precision-recall curves for supervised forecasting. MFPPs are compared against random forests, logistic regression, and random forests augmented with MFPP intensity features.

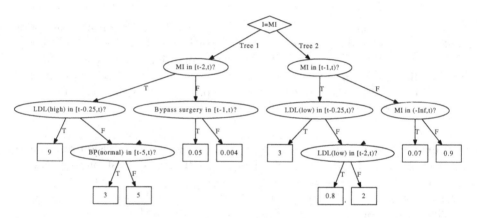

Fig. 9. First two trees in the MFPP forest. The model shows the rate predictions for myocardial infarction (MI) based on cholesterol (LDL), blood pressure (BP), previous MI, and bypass surgery. Time is in years; for example, [t-1,t) means "within the last year", and (-Inf, t) means "ever before".

for" events, meaning that the diagnosis is documented because care is provided. Care for incident heart attacks occurs over the following weeks, and so-called myocardial infarction events may recur over that time frame.

Despite the recurrence effect, the MFPP model provides an interpretable representation of risk factors and their interactions with other events. For example, Tree 1 shows that elevated cholesterol levels increase the rate of heart attack recurrence while normotensive blood pressure measurements decrease it. The findings corroborate established risk factors and their trends.

4   Conclusion

In this work we introduce an efficient multiplicative forest learning algorithm to the point process community. We developed this algorithm by combining elements of two continuous-time models taking advantage of their similar likelihood forms. We contrasted the differences between the two models and observed that the multiplicative forest extension of the CTBN framework would integrate cleanly into the PCIM framework. We showed that unlike CTBNs, MFPP forests can be learned independently because of the PCIM likelihood decomposition and intensity dependence on event history. We applied this model to two data sets: a synthetic model, where we showed significant improvements over the original PCIM model, and a cohort study, where we observed that MFPP-RFs outperformed standard machine learning algorithms at predicting future onset of myocardial infarctions. We provide multiplicative-forest point process code at http://cs.wisc.edu/~jcweiss/ecml2013/.

While our work has shown improved performance in two different comparisons, it would also be worthwhile to consider extensions of this framework to marked point processes. Marked point processes are ones where events contain additional information. The learning framework could leverage the information about the events to make better predictions. For example, this could mean the difference between reporting that a lab test was ordered and knowing the value of the lab test. The drawback of immediate extension to marked point processes is that the learning algorithm needs to be paired with a generative model of events in order to conduct accurate forecasting. Without the generative ability, sampled events would lack the information required for continued sampling. The integration of these methods with continuous-state representations would also help allow modeling of clinical events such as blood pressure to be more precise. Finally, we would like to be able to scale our methods and apply MFPPs to any disease. Because EHR systems are constantly updated, we can acquire new up-to-date information on both phenotype and risk factors. To fully automate the process in the present framework, we need to develop a way to address the scope of the EHR, selecting and aggregating the pertinent features for each disease of interest and identifying the meaningful time frames of interest.

Acknowledgments. We acknowledge Michael Caldwell at the Marshfield Clinic for his assistance and helpful comments, the anonymous reviewers for their insightful comments, and we are deeply grateful for the support provided by the CIBM Training Program grant 5T15LM007359, NIGMS grant R01GM097618, and NLM grant R01LM011028.

References

1. Nodelman, U., Shelton, C.R., Koller, D.: Learning continuous time Bayesian networks. In: Uncertainty in Artificial Intelligence (2003)
2. Gunawardana, A., Meek, C., Xu, P.: A model for temporal dependencies in event streams. In: Advances in Neural Information Processing Systems (2011)

3. Weiss, J., Natarajan, S., Page, D.: Multiplicative forests for continuous-time processes. In: Advances in Neural Information Processing Systems 25, pp. 467–475 (2012)
4. Brown, E.N., Kass, R.E., Mitra, P.P.: Multiple neural spike train data analysis: state-of-the-art and future challenges. Nature Neuroscience 7(5), 456–461 (2004)
5. Engle, R.: The econometrics of ultra-high-frequency data. Econometrica 68(1), 1–22 (2000)
6. Diggle, P., Rowlingson, B.: A conditional approach to point process modelling of elevated risk. Journal of the Royal Statistical Society. Series A (Statistics in Society), 433–440 (1994)
7. Nodelman, U., Shelton, C., Koller, D.: Continuous time Bayesian networks. In: Uncertainty in Artificial Intelligence, pp. 378–387. Morgan Kaufmann Publishers Inc. (2002)
8. Rajaram, S., Graepel, T., Herbrich, R.: Poisson-networks: A model for structured point processes. In: AI and Statistics, vol. 10 (2005)
9. Ripley, B.: The second-order analysis of stationary point processes. Journal of Applied Probability, 255–266 (1976)
10. Wilson, P., D'Agostino, R., Levy, D., Belanger, A., Silbershatz, H., Kannel, W.: Prediction of coronary heart disease using risk factor categories. Circulation 97(18), 1837–1847 (1998)
11. Tzoulaki, I., Liberopoulos, G., Ioannidis, J.: Assessment of claims of improved prediction beyond the Framingham risk score. JAMA 302(21), 2345 (2009)
12. Weiss, J., Natarajan, S., Peissig, P., McCarty, C., Page, D.: Machine learning for personalized medicine: predicting primary myocardial infarction from electronic health records. AI Magazine 33(4), 33 (2012)
13. McCarty, C.A., Wilke, R.A., Giampietro, P.F., Wesbrook, S.D., Caldwell, M.D.: Marshfield clinic personalized medicine research project (pmrp): design, methods and recruitment for a large population-based biobank. Personalized Medicine 2(1), 49–79 (2005)
14. Breiman, L.: Random forests. Machine Learning 45(1), 5–32 (2001)

On Discovering the Correlated Relationship between Static and Dynamic Data in Clinical Gait Analysis

Yin Song[1], Jian Zhang[1], Longbing Cao[1], and Morgan Sangeux[2,3,4]

[1] Advanced Analytics Institute (AAI), University of Technology, Sydney, Australia
[2] Hugh Williamson Gait Analysis Laboratory, The Royal Children's Hospital Melbourne, 50 Flemington Road, Parkville Victoria 3052, Australia
[3] Murdoch Childrens Research Institute, Victoria, Australia
[4] School of Engineering, The University of Melbourne, Victoria, Australia
yin.song@student.uts.edu.au, {jian.zhang,longbing.cao}@uts.edu.au,
morgan.sangeux@rch.org.au

Abstract. 'Gait' is a person's manner of walking. Patients may have an abnormal gait due to a range of physical impairment or brain damage. Clinical gait analysis (CGA) is a technique for identifying the underlying impairments that affect a patient's gait pattern. The CGA is critical for treatment planning. Essentially, CGA tries to use patients' physical examination results, known as *static* data, to interpret the dynamic characteristics in an abnormal gait, known as *dynamic* data. This process is carried out by gait analysis experts, mainly based on their experience which may lead to subjective diagnoses. To facilitate the automation of this process and form a relatively objective diagnosis, this paper proposes a new probabilistic correlated static-dynamic model (CSDM) to discover correlated relationships between the dynamic characteristics of gait and their root cause in the static data space. We propose an EM-based algorithm to learn the parameters of the CSDM. One of the main advantages of the CSDM is its ability to provide intuitive knowledge. For example, the CSDM can describe what kinds of static data will lead to what kinds of hidden gait patterns in the form of a decision tree, which helps us to infer dynamic characteristics based on static data. Our initial experiments indicate that the CSDM is promising for discovering the correlated relationship between physical examination (static) and gait (dynamic) data.

Keywords: Probabilistic graphical model, Correlated static-dynamic model (CSDM), Clinical gait analysis (CGA), EM algorithm, Decision tree.

1 Introduction

The past 20 years have witnessed a burgeoning interest in clinical gait analysis for children with cerebral palsy (CP). The aim of clinical gait analysis is to determine a patient's impairments to plan manageable treatment. Usually, two

H. Blockeel et al. (Eds.): ECML PKDD 2013, Part III, LNAI 8190, pp. 563–578, 2013.
© Springer-Verlag Berlin Heidelberg 2013

Table 1. An Excerpt Data Set from the Static Data

Subject	Internal_Rotation_r	Internal_Rotation_l	Anteversion_r	$\cdots$	Knee_Flexors_l
1	58	63	25	$\cdots$	3+
2	60	71	15	$\cdots$	4
3	53	52	29	$\cdots$	3
$\vdots$	$\vdots$	$\vdots$	$\vdots$	$\vdots$	$\vdots$

types of data are used in clinical gait analysis: *static* data, which is the physical examination data that is measured when the patient is not walking, such as the shape of the femur and the strength of the abductor muscles. Table 1 shows an excerpt data set from the static data. From the table, we can see that there are many attribute values for each subject. The other type of data is *dynamic* data, which records the dynamic characteristics that evolve during a gait trial and usually can be displayed in curves. Fig. 1 shows gait curve examples for one subject. Gait curves are recorded from multiple dimensions (i.e., from different parts of the body), such as the pelvis and hips. Since each subject has multiple trials, there are multiple curves for each dimension. In addition, each dimension has both the left and right side of the body. Thus, the total number of curves for each dimension is the number of trials multiplied by two. We use the red line to denote the dynamic of the left side and the blue line to denote the counterpart of the right side. Fig. 1(a)-(d) show 4 different dimensions of the dynamics. Each curve in each dimension represents the corresponding dynamics of one trial for the left or right part. The grey shaded area termed as *normal* describes the dynamic curve obtained from healthy people with a range of +/- 1 standard deviations for each observation point. From the example data shown above, we can see that describing the relationship between the *static* and *dynamic* data in the clinical gait data is not intuitive.

In practice, static data is used to explain abnormal features in dynamic data. In other words, gait analysis experts try to discover the correlated relationships between *static* and *dynamic* data for further clinical diagnosis. This process has been conducted empirically by clinical experts and thus is *qualitative*. In this paper, we make an initial exploration to discover the *quantitative* correlated relationships between the *static* data and *dynamic* curves.

The rest of the paper is organize as following: The next section reviews the work related to this paper and Section 3 follows by the problem formalization. Then, Section 4 proposes a probabilistic graphical model to simulate the generating process of the data and gives an EM-based recipe for learning the model given training data. Experimental results on both synthetic and real-world data sets are reported in Section 5 and Section 6 concludes this paper.

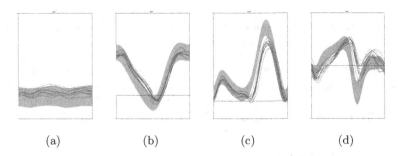

(a)   (b)   (c)   (d)

Fig. 1. Example Gait Curves for One Patient with 6 Trials: (a) The Pelvic Tilt Dimension; (b) The Hip Flexion Dimension; (c) The Knee Flexion Dimension; (d) The Dorsiflexion Dimension

2   Related Work

Recent research in CGA [3,5,13,12] have made initial attempts at the automatic discovery of correlated relationships in clinical gait data by machine learning methods such as multiple linear regression [5] and fuzzy decision trees [12]. However, previous researchers usually preprocessed the gait data and discarded the dynamic characteristics of that data, which fails to explore the correlated relationship between static data and dynamic curves. To the best of our knowledge, our work is a first attempt to explore this correlated relationship comprehensively.

Probabilistic models related to this paper exists, for example, hidden Markov models (HMMs) [11] and conditional random fields (CRFs) [6]. Since these models focus on modeling dynamic curves, they cannot be applied directly here. By contrast, the aim of this paper is to jointly model the static and dynamic data considering their correlated relationships.

3   Problem Statement

The following terms are defined:

- A *static profile* is a collection of static physical examination features of one subject denoted by $\mathbf{y} = (y_1, y_2, \cdots, y_L)$, where the subscript i ($1 \leq i \leq L$) denotes the ith attribute of the physical examination features, e.g., the Internal_Rotation_r attribute in Table 1.
- A *gait profile* is a collection of M gait trials made by one subject denoted by $\mathbf{X}_{1:M} = \{\mathbf{X}_1, \mathbf{X}_2, \cdots, \mathbf{X}_M\}$.
- A *gait trial* (cycle) is multivariate time series denoted by $\mathbf{X}_m = (\mathbf{x}_{m1}, \mathbf{x}_{m2}, \cdots, \mathbf{x}_{mN})$, where $\mathbf{x}_{mj}$ ($1 \leq m \leq M$ and $1 \leq j \leq N$) is the jth vector observation of the time series and $\mathbf{x}_{mj} = \begin{bmatrix} x_{m1j} & x_{m2j} & \cdots & x_{mDj} \end{bmatrix}^T$ (D is the number of the dimensions for dynamic data and N is the length of the time series). For example, one dimension of the multivariate time series

$(x_{mj1}, x_{mj2}, \cdots, x_{mjN})$ $(1 \leq j \leq D)$ can be plotted as one curve in Fig. 1(a) and represents the dynamics of that dimension for one trial. $\mathbf{X}_m$ can be seen as a collection of such curves in different dimensions.

Our goal was to develop a probabilistic model $p(\mathbf{X}_{1:M}, \mathbf{y})$ that considers the correlated relationships between the *static profile* (i.e., static data) and the corresponding *gait profile* (i.e., dynamic data). In other words, we aim to produce a probabilistic model that assigns 'similar' data high probability.

4   Proposed Model

4.1   Motivation

The basic idea is to construct the data generating process based on the domain knowledge gained by gait experts and model the process. Specifically, *static profile* $\mathbf{y}$ of a subject determines the generation of that subject's potential gait pattern. We denote this hidden gait pattern as a latent variable $\mathbf{h}$, a vector whose elements h_g $(1 \leq g \leq G)^1$ are 0 or 1 and sum to 1, where G is the number of hidden gait patterns. The generation of the corresponding *gait profile* $\mathbf{X}_{1:M}$ is then determined by this latent variable $\mathbf{h}$. In other words, the gait pattern is characterized by a distribution on the gait data. Due to the high dimensionality of $p(\mathbf{X}_{1:M}|\mathbf{h})$, the generating process of it is not intuitive. Thus, we need to consider the corresponding physical process. According to [8], a gait trial can usually be divided into a number of phases and each vector observation $\mathbf{x}_{mj}$ belongs to a certain state indicating its phase stage. These states are usually not labeled and we thus introduce latent variables $\mathbf{z}_{mj}$ $(1 \leq m \leq M, 1 \leq j \leq N_m)$ for each vector observation $\mathbf{x}_{mj}$ in each gait trial $\mathbf{X}_m$. We thus have two advantages: firstly, $p(\mathbf{X}_{1:M}|\mathbf{h})$ can be decomposed into a set of conditional probability distributions (CPDs) whose forms are intuitive to obtain; secondly, the dynamic process of the gait trials are captured by utilizing the domain knowledge.

4.2   The Correlated Static-Dynamic Model

We propose a novel correlated static-dynamic model (CSDM), which models the above conjectured data generating process. As mentioned before, existing models (e.g., HMMs and CRFs), cannot be directly used here. This is because HMMs only model the dynamic data $p(\mathbf{X}_m)$ and CRFs only model the relationship between $\mathbf{X}_m$ and $\mathbf{z}_m$, i.e., $p(\mathbf{z}_m|\mathbf{X}_m)$ $(1 \leq m \leq M)$, which is different to our goal of jointly modeling the *static* and *gait* profiles $p(\mathbf{X}_{1:M}, \mathbf{y})$. The graphical model for the CSDM is shown in Fig. 2 (subscript m is omitted for convenience). We use conventional notation to represent the graphical model [2]. In Fig. 2, each node represents a random variable (or group of random variables). For instance, a *static profile* is represented as a node $\mathbf{y}$. The directed links express

---

^{1} $h_g = 1$ denotes the gth hidden gait pattern.

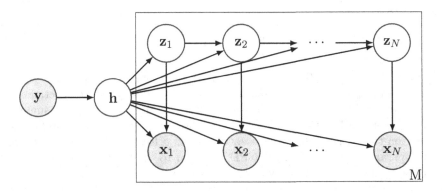

Fig. 2. The Graphical Model of the CSDM

probabilistic causal relationships between these variables. For example, the arrow from the *static profile* **y** to the hidden gait pattern variable **h** indicates their causal relationships. For multiple variables that are of the same kind, we draw a single representative node and then surround this with a plate, labeled with a number indicating that there are many such kinds of nodes. An example can be found in Fig. 2 in which M trials $\mathbf{Z}_{1:M}, \mathbf{X}_{1:M}$ are indicated by a plate label with M. Finally, we denote observed variables by shading the corresponding nodes and the observed *static profile* **y** is shown as shaded node in Fig. 2. To further illustrate the domain knowledge-driven data generating process in Fig. 2, the generative process for a *static profile* **y** to generate a *gait profile* $\mathbf{X}_{1:M}$ is described as follows:

1. Generate the static profile **y** by $p(\mathbf{y})$
2. Generate the latent gait pattern **h** by $p(\mathbf{h}|\mathbf{y})$
3. For each of the M trials
   (a) Generate the initial phase state $\mathbf{z}_{m1}$ from $p(\mathbf{z}_{m1}|\mathbf{h})$
   (b) Generate the corresponding gait observation $\mathbf{x}_{m1}$ by $p(\mathbf{x}_{m1}|\mathbf{z}_{m1}, \mathbf{h})$
   (c) For each of the gait observations $\mathbf{x}_{mn}$ $(2 \leq n \leq N)$
      i. Generate the phase state $\mathbf{z}_{mn}$ from $p(\mathbf{z}_{mn}|\mathbf{z}_{m,n-1}, \mathbf{h})$
      ii. Generate the the corresponding gait observation $\mathbf{x}_{mn}$ from $p(\mathbf{x}_{mn}|\mathbf{z}_{mn}, \mathbf{h})$

4.3 The Parameters of the CSDM

The parameters (i.e., the variables after the semicolon of each CPD) governing the CPDs of the CSDM are listed in the following[2]:

$$p(\mathbf{h}|\mathbf{y};\mathbf{d}) = \prod_{g=1}^{G} d_g(\mathbf{y})^{h_g} \tag{1}$$

where d_g $(1 \leq g \leq G)$ is a set of mapping functions $(\mathbf{y} \rightarrow d_g(\mathbf{y}) \equiv p(h_g = 1|\mathbf{y}))$ and h_g is the gth element of $\mathbf{h}$. Since the input $\mathbf{y}$ of the functions is a mixture of discrete and continuous values, it is not intuitive to assume the format of the functions. Thus, here we use the form of a probability estimation tree (PET) [9] to represent the CPD $p(\mathbf{h}|\mathbf{y};\mathbf{d})$. To be more specific, the parameters governing the CPD is similar to the form "if $\mathbf{y}$ in some value ranges, then the probability of $h_g = 1$ is $d_g(\mathbf{y})$".

$$p(\mathbf{z}_{m1}|h;\boldsymbol{\pi}) = \prod_{g=1}^{G} \prod_{k=1}^{K} \pi_{gk}^{h_g,\ z_{m1k}} \tag{2}$$

where $\boldsymbol{\pi}$ is a matrix of probabilities with elements $\pi_{gk} \equiv p(z_{m1k} = 1|h_g = 1)$.

$$p(\mathbf{z}_{mn}|\mathbf{z}_{m,n-1},h;\mathbf{A}) = \prod_{g=1}^{G} \prod_{k=1}^{K} \prod_{j=1}^{K} a_{gjk}^{h_g,\ z_{m,n-1,j},\ z_{mnk}} \tag{3}$$

where $\mathbf{A}$ is a matrix of probabilities with elements $a_{gjk} \equiv p(z_{mnk} = 1|z_{m,n-1,j} = 1, h_g = 1)$.

$$p(\mathbf{x}_{ml}|\mathbf{z}_{ml},h;\boldsymbol{\Phi}) = \prod_{g=1}^{G} \prod_{k=1}^{K} p(\mathbf{x}_{ml}|\boldsymbol{\phi}_{gk})^{h_g,z_{mlk}} \tag{4}$$

where $\boldsymbol{\Phi}$ is a matrix with elements $\boldsymbol{\phi}_{gk}$. For efficiency, in this paper, we assume that $p(\mathbf{x}_{ml};\boldsymbol{\phi}_{gk}) = \mathcal{N}(\mathbf{x}_{ml};\boldsymbol{\mu}_{gk},\boldsymbol{\sigma}_{gk})$, which is Gaussian distribution, and thus $\boldsymbol{\phi}_{gk} = (\boldsymbol{\mu}_{gk},\boldsymbol{\sigma}_{gk})$.

Thus, the CSDM can be represented by the parameters $\boldsymbol{\theta} = \{\mathbf{d},\boldsymbol{\pi},\mathbf{A},\boldsymbol{\mu},\boldsymbol{\sigma}\}$.

4.4   Learning the CSDM

In this section we present the algorithm for learning the parameters of the CSDM, given a collection of *gait profiles* $\mathbf{X}_{s,1:M}$ and corresponding *static profiles* $\mathbf{y}_s$ $(1 \leq s \leq S)$ for different subjects. We assume each pair of gait and static profiles are independent of every others since they are from different subjects and share the same set of model parameters. Our goal is to find parameters $\boldsymbol{\theta}$ that maximize the log likelihood of the observed data $\mathbf{X}_{1:S,1:M},\mathbf{y}_{1:S}$[3].

$$L(\boldsymbol{\theta}) = \sum_{s=1}^{S} \log p(\mathbf{X}_{s,1:M}|\mathbf{y}_s;\boldsymbol{\theta}) \tag{5}$$

---

[3] We add the subscript s for representing the sth profile in the rest of the paper.

Algorithm 1. The Learning Algorithm for the Proposed CSDM.

Input   : An initial setting for the parameters $\boldsymbol{\theta}^{old}$
Output: Learned parameters $\boldsymbol{\theta}^{new}$

1 **while** *the convergence criterion is not satisfied* **do**
2 | Estep();
3 | $\boldsymbol{\theta}^{new}$ = Mstep();
4 **end**

Directly optimizing the above function with respect to $\boldsymbol{\theta}$ is very difficult because of the involvement of latent variables [2]. We adopted an expectation-maximization (EM)-based algorithm [4] to learn the parameters, yielding the iterative method presented in Algorithm 1. First, the parameters $\boldsymbol{\theta}^{old}$ need to be initialized. Then in the E step, $p(\mathbf{z}_{s,1:M}, \mathbf{h}_s | \mathbf{X}_{s,1:M}, \mathbf{y}_s, \boldsymbol{\theta}^{old})$ $(1 \leq s \leq S)$ is inferred given the parameters $\boldsymbol{\theta}^{old}$ and will be used in M step. The M step then obtains the new parameters $\boldsymbol{\theta}^{new}$ that maximize the $Q(\boldsymbol{\theta}, \boldsymbol{\theta}^{old})$ function with respect to $\boldsymbol{\theta}$ as follows:

$$Q(\boldsymbol{\theta}, \boldsymbol{\theta}^{old}) = \sum_{s,\mathbf{h},\mathbf{z}} p(\mathbf{z}_{s,1:M}, \mathbf{h}_s | \mathbf{X}_{s,1:M}, \mathbf{y}_s; \boldsymbol{\theta}^{old}) \log p(\mathbf{h}_s, \mathbf{z}_{s,1:M}, \mathbf{X}_{s,1:M}, \mathbf{y}_s; \boldsymbol{\theta})$$

(6)

The E and M steps iterate until the convergence criterion is satisfied. In this manner, $L(\boldsymbol{\theta})$ is guaranteed to increase after each interaction.

Challenges of the Learning Algorithms. The challenges of the above algorithm is in the calculation of the E step and the M step. A standard forward-backward inference algorithm [11] cannot be directly used here for the E step because of the introduction of latent variables $\mathbf{h}_s$ $(1 \leq s \leq S)$. We provided a modified forward-backward inference algorithm in Algorithm 2 considering the involvement of $\mathbf{h}_s$ $(1 \leq s \leq S)$. In calculating the M step, it was difficult to find an analytic solution for $\mathbf{d}(\cdot)$. We utilized a heuristic algorithm to solve it in Procedure estimatePET. The details of the implementation for E and M steps are discussed in the following.

The E Step. Here we provide the detailed process of inferring the posterior distribution of the latent variables $\mathbf{h}_{1:S}, \mathbf{z}_{1:S,1:M}$ given the parameters of the model $\boldsymbol{\theta}^{old}$. Actually, we only infer some marginal posteriors instead of the joint posterior $p(\mathbf{z}_{s,1:M}, \mathbf{h}_s | \mathbf{X}_{s,1:M}, \mathbf{y}_s, \boldsymbol{\theta}^{old})$. This is because only these marginal posteriors will be used in the following M-step. We define the following notations for these marginal posteriors $\boldsymbol{\gamma}$ and $\boldsymbol{\xi}$ and auxiliary variables $\boldsymbol{\alpha}$ and $\boldsymbol{\beta}$ $(1 \leq s \leq S, 1 \leq m \leq M, 1 \leq n \leq N, 2 \leq n' \leq N, 1 \leq j \leq K, 1 \leq k \leq K, 1 \leq g \leq G)$:

$$\alpha_{sgmnk} = p(\mathbf{x}_{sm1}, \cdots, \mathbf{x}_{smn}, z_{smnk} | h_{sg}; \boldsymbol{\theta}^{old})$$

(7)

$$\beta_{sgmnk} = p(\mathbf{x}_{s,m,n+1}, \cdots, \mathbf{x}_{smN} | z_{smnk}, h_{sg}; \boldsymbol{\theta}^{old})$$

(8)

Procedure. forward

> **input** : A set of the parameters $\boldsymbol{\theta}$
> **output**: The variables $\boldsymbol{\alpha}$
>
> // Initialization;
> $\alpha_{sgm1k} = \pi_{gk}\mathcal{N}(\mathbf{x}_{sm1}; \boldsymbol{\mu}_{gk}, \boldsymbol{\sigma}_{gk})$ for all s, g, m and k;
>
> 1 **for** $s{=}1$ **to** S **do** // Induction
> 2   **for** $g{=}1$ **to** G **do**
> 3   **for** $m{=}1$ **to** M **do**
> 4   **for** $n{=}1$ **to** $N{-}1$ **do**
> 5   **for** $k{=}1$ **to** K **do**
> 6 $\alpha_{s,g,m,n+1,k} = \sum_{j=1}^{K} \alpha_{sgmnj}\, a_{gjk}\mathcal{N}(\mathbf{x}_{s,m,n+1}; \boldsymbol{\mu}_{gk}, \boldsymbol{\sigma}_{gk})$;
> 7   **end**
> 8   **end**
> 9   **end**
> 10   **end**
> 11 **end**

Procedure. backward

> **input** : A set of the parameters $\boldsymbol{\theta}$
> **output**: The variables $\boldsymbol{\beta}$
>
> // Initialization;
> $\beta_{sgmNk} = 1$ for all s, g, m and k;
>
> 1 **for** $s{=}1$ **to** S **do** // Induction
> 2   **for** $g{=}1$ **to** G **do**
> 3   **for** $m{=}1$ **to** M **do**
> 4   **for** $n{=}N{-}1$ **to** 1 **do**
> 5   **for** $j{=}1$ **to** K **do**
> 6 $\beta_{sgmnk} = \sum_{j=1}^{K} a_{gjk}\mathcal{N}(\mathbf{x}_{s,m,n+1}; \boldsymbol{\mu}_{gk}, \boldsymbol{\sigma}_{gk})\beta_{s,g,m,n+1,j}$;
> 7   **end**
> 8   **end**
> 9   **end**
> 10   **end**
> 11 **end**

$$\gamma_{sgmnk} = p(z_{smnk}, h_{sg}|\mathbf{X}_{sm}, \mathbf{y}_s; \theta^{old}) \tag{9}$$

$$\xi_{s,g,m,n'-1,j,n',k} = p(z_{s,m,n'-1,j}, z_{smn'k}|h_{sg}, \mathbf{X}_{sm}, \mathbf{y}_s; \theta^{old}) \tag{10}$$

The inference algorithm is presented in Algorithm 2. Specifically, line 1 calls Procedure forward to calculate the forward variables $\boldsymbol{\alpha}$, while line 2 calls Procedure backward to calculate the backward variables $\boldsymbol{\beta}$. Then line3-15 calculate the value of each element of the posteriors $\boldsymbol{\gamma}$ and $\boldsymbol{\xi}$ and the h_s^* $(1 \le s)$ on the

---

Algorithm 2. Estep()

---

input : An initial setting for the parameters $\boldsymbol{\theta}^{old}$
output: Inferred posterior distributions $\boldsymbol{\gamma}$, $\boldsymbol{\xi}$ and h_s^* ($1 \leq s \leq S$)

   /* Calculation of $\boldsymbol{\alpha}$, $\boldsymbol{\beta}$ */
1   Call Procedure forward using $\boldsymbol{\theta}^{old}$ as input;
2   Call Procedure backward using $\boldsymbol{\theta}^{old}$ as input;
   /* Calculation of $\boldsymbol{\gamma}$, $\boldsymbol{\xi}$ and h_s^* ($1 \leq s \leq S$)   */
3   **for** $s=1$ **to** S **do**
4   **for** $g=1$ **to** G **do**
5   **for** $m=1$ **to** M **do**
6              $p(\mathbf{X}_{sm}|h_{sg};\boldsymbol{\theta}^{old}) = \sum_{k=1}^{K} \alpha_{sgmNk}$;
7   **for** $n=1$ **to** N **do**
8                  $\gamma_{sgmnk} = \frac{\alpha_{sgmnk}\beta_{sgmnk}}{p(\mathbf{X}_{sm}|h_{sg};\boldsymbol{\theta}^{old})}$;
9                  $\xi_{s,g,m,n-1,j,n,k} = \frac{\alpha_{s,g,m,n-1,k}\mathcal{N}(\mathbf{x}_{smn}';\boldsymbol{\mu}_{gk},\boldsymbol{\sigma}_{gk})a_{gjk}\beta_{sgmnk}}{p(\mathbf{X}_{sm}|h_{sg};\boldsymbol{\theta}^{old})}$ $(n > 2)$;
10   **end**
11   **end**
12   **end**
13      $p(h_{sg}|\mathbf{y}_s;\boldsymbol{\theta}^{old}) = \prod_{m=1}^{M} p(\mathbf{X}_{sm}|h_{sg};\boldsymbol{\theta}^{old})$
        $p(h_{sg}|\mathbf{X}_{s,1:M},\mathbf{y}_s;\boldsymbol{\theta}^{old}) = \frac{p(h_{sg}|\mathbf{y}_s;\boldsymbol{\theta}^{old})p(h_{sg}|\mathbf{y}_s;\boldsymbol{\theta}^{old})}{\sum_{g=1}^{G} p(h_{sg}|\mathbf{y}_s;\boldsymbol{\theta}^{old})p(h_{sg}|\mathbf{y}_s;\boldsymbol{\theta}^{old})}$;
14      $h_s^* = \arg\max_{g} p(h_{sg}|\mathbf{X}_{s,1:M},\mathbf{y}_s;\boldsymbol{\theta}^{old})$;
15 **end**

---

basis of the $\boldsymbol{\alpha}$, $\boldsymbol{\beta}$ and $\boldsymbol{\theta}^{old}$. These posteriors will be used in the M-step for updating the parameters.

The M Step. Here we provide the detailed process for M step. Basically, it updates the parameters by maximizing the $Q(\theta, \theta^{old})$ with respect to them. If substituting the distributions with inferred marginal posteriors in the Q function, we can obtain

$$Q(\theta, \theta^{old}) = \sum_{s,\mathbf{h},\mathbf{z}_{s,1:M}} p(\mathbf{z}_{s,1:M},\mathbf{h}|\mathbf{X}_{s,1:M},\mathbf{y}_s;\boldsymbol{\theta}^{old}) \sum_{g=1}^{G} h_{sg} \log d_g(\mathbf{y})$$

$$+ \sum_{s,g,m,k} \gamma_{sgm1k} \log \pi_{gk}$$

$$+ \sum_{s,g,m,j,k} \sum_{n=2}^{N} \xi_{s,g,m,n-1,j,n,k} \log a_{gjk}$$

$$+ \sum_{s,g,m,n,k} \gamma_{sgmnk} \log \mathcal{N}(\mathbf{x}_{smn};\boldsymbol{\mu}_{gk},\boldsymbol{\sigma}_{gk}) \qquad (11)$$

Procedure. estimatePET

    input : The data tuple $(\mathbf{y}_s, h_s^*)$ $(1 \leq s \leq S)$
    output: The learned PET **d**

1 **while** *stopping rule is not satisfactory* **do**
2 | Examine all possible binary splits on every attribute of $\mathbf{y}_s$ $(1 \leq s \leq S)$;
3 | Select a split with best optimization criterion;
4 | Impose the split on the PET **d**;
5 | Repeat recursively for the two child nodes;
6 **end**
7 **for** *node in the PET* $\mathbf{d}(\cdot)$ **do**
8 | Do Laplace correction on each node;
9 **end**

Then the update formula for parameters $\mathbf{d}, \boldsymbol{\pi}, \mathbf{A}, \boldsymbol{\mu}, \boldsymbol{\sigma}$ can be obtained by maximizing the Q with respect to them, respectively:

- Updating of $\mathbf{d}$: Maximizing Q with respect to $\mathbf{d}$ is equivalent to maximizing the first item of Equation 11. However, $\mathbf{y}$ is a mixture of discrete and continuous values and it is impractical to find an analytic solution to $\mathbf{d}$. Here we consider a heuristic solution through the formation of probability estimation trees (PETs), which is a decision tree [7] with a Laplace estimation [10] of the probability on class memberships [9]. The heuristic algorithm for estimating the PET is described in Procedure estimatePET.
- Updating of $\boldsymbol{\pi}$, $\mathbf{A}$, $\boldsymbol{\mu}$ and $\boldsymbol{\sigma}$: Maximization Q with respect to $\boldsymbol{\pi}, \mathbf{A}, \boldsymbol{\mu}, \boldsymbol{\sigma}$ is easily achieved using appropriate Lagrange multipliers, respectively. The results are as follows:

$$\pi_{gk} = \frac{\sum\limits_{s,m,g} \gamma_{sgm1k}}{\sum\limits_{s,m,k,g} \gamma_{sgm1k}} \tag{12}$$

$$a_{gjk} = \frac{\sum\limits_{s,m,n,g} \xi_{s,g,m,n-1,j,n,k}}{\sum\limits_{s,m,l,n,g} \xi_{s,g,m,n-1,j,n,k}} \tag{13}$$

$$\mu_{gk} = \frac{\sum\limits_{s,m,g,n} \gamma_{sgmnk}\mathbf{x}_{smn}}{\sum\limits_{s,m,n,g} \gamma_{sgmnk}} \tag{14}$$

$$\sigma_{gk} = \frac{\sum\limits_{s,m,g,n} \gamma_{sgmnk}(\mathbf{x}_{smn} - \mu_{gk})(\mathbf{x}_{smn} - \mu_{gk})^T}{\sum\limits_{s,m,n,g} \gamma_{sgmnk}} \tag{15}$$

Algorithm 3 summarizes the whole process of the M step.

Algorithm 3. Mstep()

---

input : Inferred posterior distributions γ, ξ and h_s^* $(1 \leq s \leq S)$
output: The updated parameters θ^{new}

1 Call Procedure estimatePET to update $\mathbf{d}(\cdot)$;
2 Update $\pi, \mathbf{A}, \mu_{gk}, \sigma_{gk}$ according to Equation 12-15;

---

5   Empirical Study

The aim of this study is to test:

- The feasibility of the learning algorithm for the CSDM. Since we have proposed an iterative (i.e., EM-based) learning method, it is pivotal to show its convergence on the gait data set.
- The predictability of the CSDM. The aim of the CSDM is to discover the correlated relationship between the static and dynamic data. Thus, it is interesting to validate its predictive power on other data falling outside the scope of the training data set.
- The usability of the CSDM. Because the CSDM is designed to be used by gait experts, we need to demonstrate intuitive knowledge extracted by the CSDM.

5.1   Experimental Settings

we sampled the synthetic data from the true parameters listed in Table 2. We varied the s_0 for different sample sizes (e.g., $s_0 = 100, 500, 1500$) to represent relatively small, medium and large data sets. The real-world data set we used was provided by the gait lab at the Royal Children's Hospital, Melbourne[4]. We have collected a subset of static and dynamic data for 99 patients. The static data subset consisted of 8 attributes summarized in Table 3. There were at most 6 gait trials for each subject and each gait trial had 101 vector observations. In principle, curves for both left and right sides may be included. However, for simplicity and consistency, we only used the right side curves of the hip rotation dimension for analysis in this pilot study.

5.2   Experimental Results

Convergence of the Learning Process. For each iteration, we calculate the averaged log-likelihood as $\frac{1}{S}\sum_{s=1}^{S}\sum_{m=1}^{M}\log p(\mathbf{X}_{sm}, \mathbf{y}_s; \theta^{old})$, where θ^{old} is the parameters updated from last iteration. Fig. 3(a) shows the CSDM against the iteration numbers for different sample sizes of the synthetic data and Fig. 3(b) shows the results of the averaged log-likelihoods for CSDMs using different numbers (represented as G) of hidden gait patterns. As expected, the averaged log-likelihood is not monotonic all the time, since part of the learning process uses

---

[4] http://www.rch.org.au/gait/

Table 2. The Parameters for the Synthetic Data

d	if $-50 \leq y < -25$, $p(h_1 = 1\|y) = 1$, if $-25 \leq y < 0$, $p(h_2 = 1\|y) = 1$, if $0 \leq y < 25$, $p(h_1 = 1\|y) = 1$, if $25 \leq y < 50$, $p(h_2 = 1\|y) = 1$.
π	$\pi_{1,1:2} = \begin{bmatrix} 0.5 & 0.5 \end{bmatrix}$ $\pi_{2,1:2} = \begin{bmatrix} 0.5 & 0.5 \end{bmatrix}$
A	$a_{1,1:2,1:2} = \begin{bmatrix} 0.6 & 0.4 \\ 0.4 & 0.6 \end{bmatrix}$ $a_{2,1:2,1:2} = \begin{bmatrix} 0.4 & 0.6 \\ 0.6 & 0.4 \end{bmatrix}$
μ	$\mu_{1,1:2,1} = \begin{bmatrix} 0 \\ 3 \end{bmatrix}$ $\mu_{2,1:2,1} = \begin{bmatrix} 1 \\ 4 \end{bmatrix}$
σ	$\sigma_{1,1:2,1} = \begin{bmatrix} 1 & 1 \end{bmatrix}$ $\sigma_{2,1:2,1} = \begin{bmatrix} 1 & 1 \end{bmatrix}$

Table 3. Description of the Static Data

Name of Attributes	Data Type	Value Range
internalrotation_r (ir_r)	continuous	23 to 90
internalrotation_l (ir_l)	continuous	20 to 94
externalrotation_r (er_r)	continuous	-5 to 57
externalrotation_l (er_l)	continuous	-26 to 51
anteversion_r (a_r)	continuous	10 to 50
anteversion_l (a_l)	continuous	4 to 45
hipabductors_r (h_r)	discrete	-1 to 5
hipabductors_l (h_l)	discrete	-1 to 5

a heuristic algorithm. However, the best averaged log-likelihoods are usually achieved after at most 5 iterations, which proves the convergence of the proposed learning algorithm. It can be seen from Fig. 3(a), a larger sample size will lead to a higher log-likelihood for the learning algorithm. For the real-world data set, $G = 4$[5] shows the fastest convergence rate of the three settings for CSDMs.

Predictive Performance. We measured the CSDM predictive accuracy in terms of how well the future gait profile can be predicted given the static profile and learned parameters. Since the final prediction is a set of complex variables, we measure the predictive log-likelihood $\sum_{s'=1}^{S'} \log p(\mathbf{X}_{s',1:M}|\mathbf{y}_{s'};\boldsymbol{\theta})$ in the testing data with S' static and gait profiles, where $\boldsymbol{\theta}$ is learned from the training data. Then, the following can be obtained by using Bayes rule:

$$\log p(\mathbf{X}_{s',1:M}|\mathbf{y}_{s'};\boldsymbol{\theta}) = \log(\sum_g p(h_{s'g}|\mathbf{y}_{s'};\boldsymbol{\theta})p(\mathbf{X}_{s',1:M}|h_{s'g};\boldsymbol{\theta})) \qquad (16)$$

where $p(h_{s'g}|\mathbf{y}_{s'};\boldsymbol{\theta})$ and $p(\mathbf{X}_{s',1:M}|h_{s'g};\boldsymbol{\theta})$ can be calculated by using the line 13 and 14 of Algorithm 2 (i.e., E step).

---

[5] The number of G is suggested by gait experts not exceeding 4.

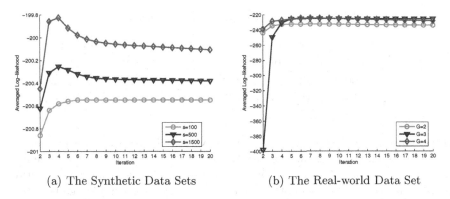

(a) The Synthetic Data Sets   (b) The Real-world Data Set

Fig. 3. Log-likelihood for the CSDM against the iteration numbers for different numbers of hidden gait pattern G

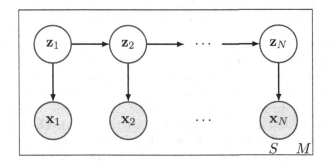

Fig. 4. The Graphical Model for the Baseline Algorithm

Without loss of generality, we proposed a baseline algorithm which ignored the static data for modeling and prediction to compare with our proposed method. The baseline model is a standard HMM with multiple observation sequences, whose graphical model is depicted in Fig. 4. It assumes all the gait trials are independently generated from an HMM. Using the standard algorithm provided in [1,11], we can learn the parameters of the baseline model, denoted as θ_0 from the training data. Accordingly, the predictive averaged log-likelihood for new gait trials can be calculated as $\sum_{s'=1}^{S'} \log p(\mathbf{X}_{s',1:M}; \theta_0)$.

We compare the CSDM with the alternating baseline scheme, an HMM with multiple sequences. We report on averages over 10 times 5-fold cross validations for the synthetic and real-world data, respectively. As shown in Table 4(a), all the CSDMs outperformed the baseline algorithm significantly. This may be because the proposed CSDM captures the correlated relationships existing in the data rather than ignoring them. Similarly, it can be observed from Table 4(b) that all the CSDMs achieved higher log-likelihoods than their counterparts of the baseline model. This proves the predictive power of our proposed CSDM on real-world data.

Table 4. The Comparison of the Log-likelihoods

(a) The Synthetic Data

	$s_0 = 100$	$s_0 = 500$	$s_0 = 1500$
CSDM	**-8016**	**-40090**	**-120310**
Baseline	-8025	-40132	-120420

(b) The Real-world Data

	$G = 2$	$G = 3$	$G = 4$
CSDM	**-1310**	**-1388**	**-1299**
Baseline	-1426	-1502	-1426

Extracting Knowledge from the CSDM. In this section, we provide an illustrative example of extracting intuitive knowledge from a CSDM on the gait data. Our real-world data are described in Section 5.1. We used the EM algorithm described in Section 4.4 to find the model parameters for a 4-hidden-gait-pattern CSDM as suggested by gait experts. Given the learned CSDM, we can extract the intuitive knowledge from the data set to answer the following questions:

- What kinds of static data will lead to what kinds of hidden gait patterns?
- What does the gait look like for each hidden gait pattern?

The first question is actually asking what is $p(\mathbf{h}|\mathbf{y}; \boldsymbol{\theta})$ (and subscript s is omitted since all s share the same parameters). Fig. 5(a) shows an answer to the first question in the form of a decision tree representation. This tree[6] decides hidden gait patterns based on the 8 features of the static data (e.g., ir_r, er_r and a_r) used in the data set. To decide the hidden gait patterns based on the static data, start at the top node, represented by a triangle ($\triangle$). The first decision is whether ir_r is smaller than 57. If so, follow the left branch, and see that the tree classifies the data as gait pattern 2. If, however, anteversion exceeds 57, then follow the right branch to the lower-right triangle node. Here the tree asks whether er_r is is smaller than 21.5. If so, then follow the right branch to see the question of next node until the tree classifies the data as ones of the gait patterns. For other nodes, the gait patterns can be decided in similar manners.

The second question is actually asking $\arg\max_g p(h_{sg}|\mathbf{X}_{s,1:M}, \mathbf{y}_s; \boldsymbol{\theta})$ ($1 \leq s \leq S$). In other words, we need to infer which gait trials belong to the corresponding hidden gait patterns in the corpus. We use line 14 described in Algorithm 2 to obtain the hidden gait pattern names of the gait trials. We can then plot representative gaits for each hidden gait pattern to answer the second question above, as shown in Figures 5(b)-5(e). Fig. 5(e) shows a collection of gaits for the hidden gait pattern 4. We can see that most of them fall into the normal area, which may indicate that these gaits are good. Fig. 5(c) shows a collection of gaits for the hidden gait pattern 2 and most of them are a little below the normal area, indicating that these gaits are not as good. By contrast, most of the gaits in Fig. 5(b) representing hidden gait pattern 1 fall outside the normal area and are abnormal gaits. Fig. 5(d) shows that the representative gaits for hidden gait pattern 3 are slightly above the normal area, which indicates these gaits are

---

[6] For simplicity, we do not display the fully tree and only display the gait pattern with the highest probability.

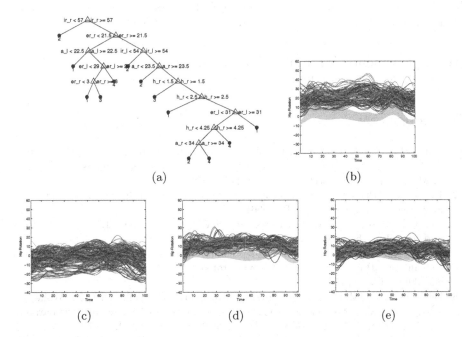

(a)   (b)

(c)   (d)   (e)

Fig. 5. Extracted Knowledge from the CSDM: (a) The Decision Tree to Predict Gait Patterns Given the Static Data, (b)-(e) Representative Gaits for Gait Pattern 1-4.

only slightly abnormal. Most subjects displaying pattern 1 and some subjects displaying pattern 3 would be susceptible to have surgery. By extracting the different paths that lead to those two patterns from the decision tree in Fig. 5(a), we can infer what combinations of static data may have clinical implications.

6   Conclusions and Future Work

This paper presents a new probabilistic graphical model (i.e., CSDM) for quantitatively discovering the correlated relationship between static physical examination data and dynamic gait data in clinical gait analysis. To learn the parameters of the CSDM on a training data set, we proposed an EM-based algorithm. One of the main advantages of the CSDM is its ability to provide intuitive knowledge. For example, the CSDM informs us what kinds of static data will lead to what kinds of hidden gait patterns and what the gaits look like for each hidden gait pattern. The experiments on both synthetic and real-world data (excerpted from patient records at the Royal Children's Hospital, Melbourne) showed promising results in terms of learning convergence, predictive performance and knowledge discovery. One direction for future work is to improve the CSDM with semi-supervised learning. Currently the CSDM is learned totally unsupervised, which may generate unexpected results due to its highly stochastic nature. Further collaboration with gait analysis experts may alleviate this problem through manual labeling of some examples. We also plan to collect more real-world data and include all static and dynamic outputs from clinical gait analysis.

Acknowledgments. We thank the four anonymous reviewers for their valuable comments on our manuscript!

References

1. Baum, L., Petrie, T., Soules, G., Weiss, N.: A maximization technique occurring in the statistical analysis of probabilistic functions of markov chains. The Annals of Mathematical Statistics 41(1), 164–171 (1970)
2. Bishop, C.: Pattern recognition and machine learning. Information Science and Statistics. Springer, New York (2006)
3. Chau, T.: A review of analytical techniques for gait data. part 1: fuzzy, statistical and fractal methods. Gait & Posture 13(1), 49–66 (2001)
4. Dempster, A., Laird, N., Rubin, D.: Maximum likelihood from incomplete data via the em algorithm. Journal of the Royal Statistical Society. Series B (Methodological) 39(1), 1–38 (1977)
5. Desloovere, K., Molenaers, G., Feys, H., Huenaerts, C., Callewaert, B., Walle, P.: Do dynamic and static clinical measurements correlate with gait analysis parameters in children with cerebral palsy? Gait & Posture 24(3), 302–313 (2006)
6. Lafferty, J., McCallum, A., Pereira, F.: Conditional random fields: Probabilistic models for segmenting and labeling sequence data. In: Proceedings of the Eighteenth International Conference on Machine Learning, pp. 282–289. Morgan Kaufmann Publishers Inc. (2001)
7. Olshen, L., Stone, C.: Classification and regression trees. Wadsworth International Group (1984)
8. Perry, J., Davids, J.: Gait analysis: normal and pathological function. Journal of Pediatric Orthopaedics 12(6), 815 (1992)
9. Provost, F., Domingos, P.: Tree induction for probability-based ranking. Machine Learning 52(3), 199–215 (2003)
10. Provost, F., Fawcett, T.: Robust classification for imprecise environments. Machine Learning 42(3), 203–231 (2001)
11. Rabiner, L.: A tutorial on hidden markov models and selected applications in speech recognition. Readings in Speech Recognition 53(3), 267–296 (1990)
12. Sagawa, Y., Watelain, E., De Coulon, G., Kaelin, A., Armand, S.: What are the most important clinical measurements affecting gait in patients with cerebral palsy? Gait & Posture 36, S11–S12 (2012)
13. Zhang, B.I., Zhang, Y., Begg, R.K.: Gait classification in children with cerebral palsy by bayesian approach. Pattern Recognition 42(4), 581–586 (2009)

Computational Drug Repositioning by Ranking and Integrating Multiple Data Sources

Ping Zhang[1], Pankaj Agarwal[2], and Zoran Obradovic[3]

[1] Healthcare Analytics Research, IBM T.J. Watson Research Center, USA
pzhang@us.ibm.com
[2] Computational Biology, GlaxoSmithKline R&D, USA
pankaj.agarwal@gsk.com
[3] Center for Data Analytics and Biomedical Informatics, Temple University, USA
zoran.obradovic@temple.edu

Abstract. Drug repositioning helps identify new indications for marketed drugs and clinical candidates. In this study, we proposed an integrative computational framework to predict novel drug indications for both approved drugs and clinical molecules by integrating chemical, biological and phenotypic data sources. We defined different similarity measures for each of these data sources and utilized a weighted k-nearest neighbor algorithm to transfer similarities of nearest neighbors to prediction scores for a given compound. A large margin method was used to combine individual metrics from multiple sources into a global metric. A large-scale study was conducted to repurpose 1007 drugs against 719 diseases. Experimental results showed that the proposed algorithm outperformed similar previously developed computational drug repositioning approaches. Moreover, the new algorithm also ranked drug information sources based on their contributions to the prediction, thus paving the way for prioritizing multiple data sources and building more reliable drug repositioning models.

Keywords: Drug Repositioning, Drug Indication Prediction, Multiple Data Sources, Metric Integration, Large Margin Method.

1 Introduction

In response to the high cost and risk in traditional *de novo* drug discovery, discovering potential uses for existing drugs, also known as drug repositioning, has attracted increasing interests from both the pharmaceutical industry and the research community [1]. Drug repositioning can reduce drug discovery and development time from 10-17 years to potentially 3-12 years [2].

Candidates for repositioning are usually either market drugs or drugs that have been discontinued in clinical trials for reasons other than safety concerns. Because the safety profiles of these drugs are known, clinical trials for alternative indications are cheaper, potentially faster and carry less risk than *de novo* drug development. Then, any newly identified indications can be quickly evaluated from phase II clinical trials.

H. Blockeel et al. (Eds.): ECML PKDD 2013, Part III, LNAI 8190, pp. 579–594, 2013.

Among the 51 new medicines and vaccines that were brought to market in 2009, new indications, new formulations, and new combinations of previously marketed products accounted for more than 30% [3]. Drug repositioning has drawn widespread attention from the pharmaceutical industry, government agencies, and academic institutes. However, current successes in drug repositioning have primarily been the result of serendipity or clinical observation. Systematic approaches are urgently needed to explore repositioning opportunities.

A reasonable systematic method for drug repositioning is the application of phenotypic screens by testing compounds with biomedical and cellular assays. However, this method also requires the additional wet bench work of developing appropriate screening assays for each disease being investigated, and it thus remains challenging in terms of cost and efficiency. Data mining and machine learning offer an unprecedented opportunity to develop computational methods to predict all possible drug repositioning using available data sources. Most of these methods have used chemical structure, protein targets, or phenotypic information (e.g., side-effect profiles, gene expression profiles) to build predictive models and some have shown promising results [4-11].

In this study, we proposed a new drug repositioning framework: Similarity-based LArge-margin learning of Multiple Sources (SLAMS), which ranks and integrates multiple drug information sources to facilitate the prediction task. In the experiment, we investigated three types of drug information: (1) chemical properties - compound fingerprints; (2) biological properties - protein targets; (3) phenotypic properties - side-effect profiles. The proposed framework is also extensible, and thus the SLAMS algorithm can incorporate additional types of drug information sources.

The rest of the paper is organized as follows. Section 2 presents the related work. Section 3 describes our SLAMS algorithm. Section 4 presents the conducted experiment and the achieved results. Finally, section 5 concludes the paper.

2   Related Work

Recent research has shown that computational approaches have the potential to offer systematic insights into the complex relationships among drugs, targets, and diseases for successful repositioning. Currently, there are five typical computational methods in drug repositioning: (1) predicting new drug indications on the basis of the chemical structure of the drug [4]; (2) inferring drug indications from protein targets interaction networks [5, 6]; (3) identifying relationships between drugs based on the similarity of their side-effects [7, 8]; (4) analyzing gene expression following drug treatment to infer new indications [9, 10]; (5) building a background chemical-protein interactome (CPI) using molecular docking [11]. All of these methods only focus on different aspects of drug-like activities and therefore result in biases in their predictions. Also, these methods suffer according to the noise in the given drug information source.

Li and Lu [12] developed a method for mining potential new drug indications by exploring both chemical and bipartite-graph-boosted molecular features in similar drugs. Gottlieb et al. [13] developed a method called PREDICT where the drug pairwise similarity was measured by similarities of chemical structures, side effects, and drug targets. These computed similarities were then used as features of a logistic regression classifier for predicting the novel associations between drugs and diseases.

This paper differs from the related studies in the following aspects:

1. We consider multiple chemical properties, biological properties, and phenotypic properties at the same time, unlike references [4-11]. Our SLAMS algorithm can also incorporate additional types of drug properties.
2. Li and Lu [12], tried all representative weights for multiple data sources in a brute-force way, but SLAMS assigns weights to all data sources without manual tuning.
3. We use a large margin method (i.e., minimize hinge-rank-loss) to integrate multiple sources, which is usually more optimal than a logistic regression method (i.e., minimize log-loss) [13] from the machine learning theory perspective. Also, the weight vector derived from a large margin method is more interpretable.
4. We use canonical correlation analysis (CCA) [14] to impute missing values of side-effect profiles. Then, we augmented known side-effect profiles with predicted side-effect profiles to build a new side-effect source.
5. We use multiple measures (e.g., precision, recall, F-score) to evaluate the results of drug repositioning experiments. Many previous methods used only area under the ROC curve (AUC) to evaluate their performance, but a high AUC score does not mean much in a highly imbalanced classification task [15] and unfortunately drug repositioning is such a task.

3   Method

In this section, we present the SLAMS algorithm for drug repositioning by integrating multiple data sources. First, we present the algorithmic framework. Second, we present a similarity-based scoring component for each data source. We also introduce the CCA to imputing missing side-effect profiles. Third, we present a large margin method to integrate multiple scoring components.

3.1   Algorithm Overview

The SLAMS algorithm is based on the observation that similar drugs are indicated for similar diseases. In this study, we identify a target drug d_x's potential new indications through similar drugs (e.g., d_y) as follows: If two drugs d_x and d_y are found to be similar, and d_y is used for treating disease s, then d_x is a repositioning candidate for disease s treatment. There are multiple metrics to measure the similarity between two drugs from different aspects of drug-like activities. The objective of SLAMS is to integrate individual metrics from multiple sources into a global metric.

The SLAMS process framework is illustrated in Fig. 1, where m data sources are involved in the integration process. Each candidate drug d_x queries i-th ($i=1,...,m$) data source and gets the prediction score for indicated disease s as $f^i(d_x,s)$. Then m

prediction scores from multiple data sources are combined into a single, final score $f^E(d_x,s)$. The details of scoring a single data source via k-nearest neighbor classifier and integrating multiple prediction scores via large margin method will be presented next.

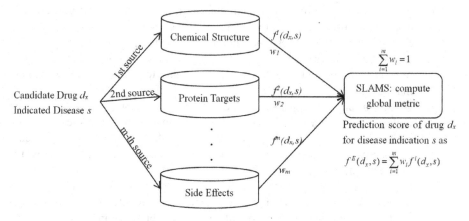

Fig. 1. Illustration of SLAMS Framework

3.2 Similarity Measures

A drug's chemical structure, protein targets, and side-effect profiles are important features in drug design, and evidently associated with its therapeutic use. Also these features are orthogonal to each other and so we consider them in the study.

Computing Similarity of Drug Chemical Structures. Our method for calculating the pairwise similarity $sim_{chem}(d_x,d_y)$ is based on the *2D* chemical fingerprint descriptor of each drug's chemical structure in PubChem [16]. We used chemistry development kit [1] (CDK) [17] to encode each chemical component into an 881-dimensional chemical substructure vector defined in PubChem. That is, each drug d is represented by a binary fingerprint $h(d)$ in which each bit indicates the presence of a predefined chemical structure fragment. The pairwise chemical similarity between two drugs d_x and d_y is computed as the Tanimoto coefficient of their fingerprints:

$$sim_{chem}(d_x,d_y) = \frac{h(d_x) \bullet h(d_y)}{|h(d_x)| + |h(d_y)| - h(d_x) \bullet h(d_y)}$$

where $|h(d_x)|$ and $|h(d_y)|$ are the counts of structure fragments in drugs d_x and d_y respectively. The dot product $h(d_x) \bullet h(d_y)$ represents the number of structure fragments shared by two drugs. The sim_{chem} score is in the [0, 1] range.

---

[1] Available at http://sourceforge.net/projects/cdk/.

Computing Similarity of Drug Protein Targets. A drug target is the protein in the human body whose activity is modified by a drug resulting in a desirable therapeutic effect. Our method for calculating the pairwise similarity $sim_{target}(d_x, d_y)$ is based on the average of sequence similarities of the two target protein sets:

$$sim_{target}(d_x, d_y) = \frac{1}{|P(d_x)| \, \| P(d_y)\|} \sum_{i=1}^{|P(d_x)|} \sum_{j=1}^{|P(d_y)|} g(P_i(d_x), P_j(d_y))$$

where given a drug d, we present its target protein set as $P(d)$; then $|P(d)|$ is the size of the target protein set of drug d. The sequence similarity function of two proteins g is calculated as a Smith-Waterman sequence alignment score [18]. The sim_{target} score is in the [0, 1] range.

Computing Similarity of Drug Side-Effect Profiles. Clinical side effects provide a human phenotypic profile for the drug, and this profile can suggest additional drug indications. In this subsection, we define the side-effect similarity first. Then, we introduce a method to predict drug side-effect profiles from chemical structure. There are two reasons for this: (1) the current side-effect dataset doesn't cover all drugs. By imputing missing side-effect profiles and using the predicted side-effect profiles with other known data sources, we have more data to train a predictive drug repositioning model; (2) in a real drug discovery pipeline, side-effect information is collected from phase I all the way through phase IV. The candidate drugs for repositioning may not have completed side-effect profiles in the early phases. It is easier to apply the predictive model to the candidate drug with predicted side-effect profiles with other known information.

Definition of Side-effect Similarity. Side-effect keywords were obtained from the SIDER database, which contains information about marketed medicines and their recorded adverse drug reactions [19]. Each drug d was represented by 1385-dimensional binary side-effect profile $e(d)$ whose elements encode for the presence or absence of each of the side-effect key words by 1 or 0 respectively. The pairwise side-effect similarity between two drugs d_x and d_y is computed as the Tanimoto coefficient of their fingerprints:

$$sim_{se}(d_x, d_y) = \frac{e(d_x) \bullet e(d_y)}{|e(d_x)| + |e(d_y)| - e(d_x) \bullet e(d_y)}$$

where $|e(d_x)|$ and $|e(d_y)|$ are the counts of side-effect keywords for drugs d_x and d_y respectively. The dot product $e(d_x) \bullet e(d_y)$ represents the number of side effects shared by two drugs. The sim_{se} score is in the [0, 1] range.

Predicting drug side-effect profiles. Suppose that we have a set of n drugs with p substructure features and q side-effect features. Each drug is represented by a chemical substructure feature vector $x=(x_1,...,x_p)^T$, and by a side-effect feature vector $y=(y_1,...,y_q)^T$. Consider two linear combinations for chemical substructures and side effects as $u_i=\alpha^T x_i$ and $v_i=\beta^T y_i$ ($i=1,2,...,n$), where $\alpha=(\alpha_1,..., \alpha_p)^T$ and $\beta=(\beta_1,..., \beta_q)^T$ are weight vectors. The goal of canonical correlation analysis is to find weight vectors α and β which maximize the following canonical correlation coefficient [14]:

$$\rho = corr(u,v) = \frac{\sum_{i=1}^{n} \alpha^T x_i \cdot \beta^T y_i}{\sqrt{\sum_{i=1}^{n} (\alpha^T x_i)^2} \sqrt{\sum_{i=1}^{n} (\beta^T y_i)^2}}$$

Let X denote the $n \times p$ matrix as $X=[x_1,...,x_n]^T$, and Y denote the $n \times q$ matrix as $Y=[y_1,...,y_n]^T$. Then consider the following optimization problem:

$$\max\{\alpha^T X^T Y \beta\} \text{ subject to } \| \alpha \|_2^2 \le 1, \| \beta \|_2^2 \le 1$$

Solving the problem, we obtain m pairs of weight vectors $\alpha_1,..., \alpha_m$ and $\beta_1,..., \beta_m$ (m is the counts of canonical components).

Given the profile of chemical substructure x_{new} for a drug of unknown side effects, we use the following prediction score for its potential side-effect profile y_{new} as:

$$y_{new} = \sum_{k=1}^{m} \beta_k \rho_k \alpha_k^T x_{new} = B \Lambda A^T x_{new}$$

where $A=[\alpha_1,..., \alpha_m]$, $B=[\beta_1,..., \beta_m]$ and Λ is the diagonal matrix whose elements are canonical correlation coefficients. If the j-th element in y_{new} has a high score, the new drug is predicted to have the j-th side-effect ($j=1,2,...,q$).

CCA was showed to be accurate and computationally efficient in prediction of the drug side-effect profiles [20]. Using CCA we augmented the drug-side-effect relationship list with side-effect predictions for drugs that are not included in SIDER, based on their chemical properties. We can use the similarity metric defined in the last subsection to calculate the side-effect similarity.

Computing Prediction Score from a Single Data Source. To calculate the likelihood that drug d_x has the indication s, we use a weighted variant of the k-nearest neighbor (k-NN) algorithm. The optimization of the model parameter k was done in a cross validation setting ($k=20$ in the study). For the i-th data source, the prediction score f of an indication s for the drug d_x is calculated as:

$$f^i(d_x,s) = \sum_{d_y \in N_k(d_x)} sim^i(d_x,d_y) \cdot C(s \in indications(d_y))$$

where $sim^i(d_x,d_y)$ denotes the similarity score between two drugs d_x and d_y from the i-th source, C is a characteristic function that return 1 if d_y has an indication s and 0 otherwise, and $N_k(d_x)$ are the k nearest neighbors of drug d_x according to the metric sim^i which is determined by the type of i-th data source. The metric sim^i can be one of the similarities defined in the previous subsections (i.e., chemical structure, protein targets, and side effects), or any additional types of drug information sources. Thus, our SLAMS algorithm is extensible. We propose a k-NN scoring component for drug repositioning tasks due to its simplicity of implementation on multiple data sources, straightforward use of multiple scores, and its competitive accuracy with more complex algorithms [21].

3.3   Combining Multiple Measures

We considered multiple data sources and obtained several prediction scores for each pair (d,s). Given m scores for a drug-disease pair (d,s) (i.e., there are m different data sources), we propose a large margin method to calculate final score f^E as a weighted average of individual scores:

$$f^E(d_x,s) = \sum_{i=1}^{m} w_i f^i(d_x,s)$$

where w_i is the corresponding weight for the i-th $(i=1,...,m)$ data source.

We learn the weights from training data using a large margin method as follows. Let us assume that we are given m data sources, $\{D_j, j = 1,...,m\}$, and n drugs $\{x_i, i = 1,...,n\}$. Each drug is assigned to several indications from the set of k indications. Let Y_i denote the set of indications that drug x_i is assigned to, and $\overline{Y}_i$ denote the set of indications that drug x_i is not assigned to. Then, let $f(x,y)$ be a vector of length m, whose j-th element is the score of drug x for indication y on the data source D_j. A weight vector w, used for integration of m prediction, is found by solving the following optimization problem:

$$\min_{w,\xi} \sum_{i} \sum_{y \in Y_i, \bar{y} \in \overline{Y}_i} \xi_i(y,\bar{y})$$
$$s.t\; w^T(f(x_i,y) - f(x_i,\bar{y})) \geq -\xi_i(y,\bar{y}), \forall i, y \in Y_i, \bar{y} \in \overline{Y}_i$$
$$\xi_i(y,\bar{y}) \geq 0, \forall i, y \in Y_i, \bar{y} \in \overline{Y}_i$$
$$w^T e = 1;\; w \geq 0$$

where e is a vector of ones. The resulting convex optimization problem can be solved using standard optimization tools, such as CVX[2]. With the trained weight vector w, the drug-indication scores from different data sources can be integrated by taking their weighted average as $w^T \cdot f^i(x,y)$.

4   Experimental Results

In this section we experimentally evaluate the proposed SLAMS algorithm on a drug repositioning task.

4.1   Data Description

In the experiment, we analyzed the approved drugs from DrugBank [22], which is a widely used public database of drug information. From DrugBank, we collected 1007 approved small-molecule drugs with their corresponding target protein information. Furthermore, we mapped these drugs to several other key drug resources including

---

[2] Available at http://cvxr.com/.

PubChem [16] and UMLS [23] in order to extract other drug related information. In the end, we extracted chemical structures of the 1007 drugs from PubChem. Each drug was represented by an 881-dimensional binary profile whose elements encode for the presence or absence of each PubChem substructure by 1 or 0, respectively. There are 122,022 associations between 1007 drugs and 881 PubChem substructures.

To facilitate collecting target protein information, we mapped target proteins to UniProt Knowledgebase [24], a central knowledgebase including most comprehensive and complete information on proteins. In the end, we extracted 3152 relationships between 1007 drugs and 775 proteins.

Side-effect keywords were obtained from the SIDER database [19]. SIDER presents an aggregate of dispersed public information on drug side effects. SIDER extracted information on marked medicines and their recorded side effects from public documents and package inserts, which resulted in a collection of 888 drugs and 1385 side-effect keywords. Merging these 888 SIDER drugs to the 1007 DrugBank approved drugs, we obtained 40,974 relationships between 613 drugs and 1385 side effects. A total number of 394 drugs from DrugBank approved list could not be mapped to SIDER drug names. We used the method described in subsection 3.2 to predict their side-effect profiles. Finally we obtained 19,385 predicted relationships between these 394 drugs and 1385 side effects.

We obtained a drug's known use(s) through extracting treatment relationships between drugs and diseases from the National Drug File - Reference Terminology[3] (NDF-RT), which is part of the UMLS [23]. The drug-disease treatment relationship list is also used by Li and Lu [12] as the gold standard set of drug repositioning task. We normalized various drug names in NDF-RT to their active ingredients. From the normalized NDF-RT data set, we were able to extract therapeutic uses for 799 drugs out of the 1007 drugs, which constructed a gold standard set of 3250 treatment relationships between 799 drugs and 719 diseases. We plotted the statistics of the gold drug-disease relationship in Fig. 2. Most of drugs (75%) treat <5 indicated diseases; 18% of drugs treat 5 to 10 diseases; only 7% of drugs treat >10 diseases (Fig. 2(a)). Although the disease *hypertension* has 78 related drugs, 80% of diseases have only <5 drugs; 10% of diseases have 5-10 drugs; and remaining 10% of diseases have >10 drugs (Fig. 2(b)).

All the data used in our experiments are available at our website[4].

4.2   Evaluation Measures

In the study, we modeled the drug repositioning task as a binary classification problem where each drug either treats or does not treat a particular disease. We measure the final classification performance using four criteria: precision, recall, F-score, and area under the ROC curve. In order to provide the definitions of these four criteria, we first define the classification confusion table for binary classification

---

[3] NDF-RT found at http://www.nlm.nih.gov/research/umls/sourcereleasedocs/current/NDFRT/.

[4] Available at http://astro.temple.edu/~tua87106/drugreposition.html.

problems where the two classes are indicated as positive and negative, which is constructed by comparing the actual data labels and predicted outcomes (see Table 1).

Then we can define the classification evaluation metrics as: True Positive Rate = $TP / (TP+FN)$, False Positive Rate = $FP / (FP+TN)$, Precision = $TP / (TP+FP)$, Recall = $TP / (TP+FN)$, and F-Score = $2 \cdot Precision \cdot Recall / (Precision+Recall)$.

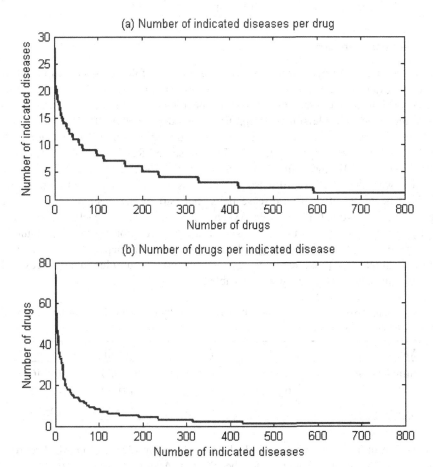

Fig. 2. Statistics of the drug-disease relationship dataset. (a) The number of indicated diseases per drug. (b) The number of drugs per indicated disease.

Table 1. Confusion matrix

	Actual Value	
Predicted	True Positive (TP)	False Positive (FP)
Value	False Negative (FN)	True Negative (TN)

The confusion matrix can be used to construct a point in the ROC curve, which is a graphical plot of true positive rate against false positive rate. The whole ROC curve can be plotted by varying threshold value for prediction score, above which the output is predicted as positive, and negative otherwise. Then we can use the area under the ROC curve (AUC) as a measure. The other three measures (precision, recall, and F-score) require setting the prediction threshold. In the experiment, a threshold was selected according to the maximum F-score of the predictions. Finally the precision, recall, and F-score are calculated over this specific threshold.

4.3   Method Comparison

To evaluate our SLAMS algorithm, we applied it in a 10-fold cross-validation setting. To avoid easy prediction cases, we hid all the associations involved with 10% of the drugs in each fold, rather than hiding 10% of the associations. In our comparisons, we considered three multiple source integration methods: (1) **PREDICT** [13] that uses similarity measures as features, and learns a logistic regression classifier that weighs the different features to yield a classification score. Replicating the settings of Gottlieb *et al.* [13], the training set used for the PREDICT logistic regression classifier was the true drug-disease associations (positive set), and a randomly generated negative set of drug-disease pairs (not part of the positive set), twice as large as the positive set. (2) **Simple Average** that assumes that each data source is equally informative, thus simply averages all k-NN prediction scores from multiple data sources. (3) The **SLAMS** algorithm proposed in this paper that uses a large margin method to automatically weigh and integrate multiple data sources. All evaluation measures are summarized in Table 2.

Table 2. Comparison of SLAMS vs. alternative integration methods according to AUC, precision, recall, and F-score

Method	AUC	Precision	Recall	F-score
Simple Average	0.8662	0.3144	0.6085	0.4146
PREDICT	0.8740	0.3228	0.5987	0.4194
SLAMS	0.8949	0.3452	0.6505	0.4510

As shown in Table 2, our proposed SLAMS algorithm obtained an AUC score of 0.8949. The score was superior to the Simple Average (AUC = 0.8662) and PREDICT (AUC = 0.8740). Also our proposed SLAMS algorithm produced a higher precision of 34.52% and a recall of 65.05% compared with Simple Average (31.44% for precision and 60.85% for recall) and PREDICT (32.28% for precision and 59.87% for recall). The results showed that our proposed SLAMS algorithm, a large-margin method, is better at integrating multiple drug information sources than simple average and logistic regression strategies.

An interesting observation is that for all methods, the AUC score is quite large (around 0.9 in the experiment), but the actual ability to detect and predict positive samples (i.e., the new drug-disease pairs) is low: even for the best method in the experiment - SLAMS, on average 34.52% of its predicted indications will be correct and about 65.05% of the true indications will be revealed for the previously unseen drugs. The reason for this is that the drug repositioning task is a highly imbalanced problem where the dataset has an approximate 1:176 positive to negative ratio. Consequently, a large change in the number of false positives can lead to a small change in the false positive rate used in ROC analysis. Therefore, AUC scores can present an overly optimistic view of an algorithm's performance for the drug repositioning task. Unlike Li and Lu [12] and Gottlieb *et al.* [13], we reported precision, recall, and F-score in addition to AUC.

4.4   Data Source Comparison

In the study, the three data sources we used reveal three different aspects of a drug: (1) chemical properties - compound fingerprints; (2) biological properties - protein targets; (3) phenotypic properties - side-effect profiles. The weight vector w derived from SLAMS is interpretable: the i-th element of w corresponds to the i-th data source, and the sum of all elements of w is 1. The SLAMS weights of each data source and standard deviation during the 10-fold cross-validation are plotted in Fig. 3.

To further characterize the abilities of different data sources and/or their combinations to predict new drug-disease relationships (i.e., drug repositioning), we used SLAMS through a 10-fold validation with different data-source combinations. To conduct a fair and accurate comparison across different data sources, the same experimental conditions were maintained by using the same training drugs and test drugs for each fold. Fig. 4 shows the ROC curves for different data sources based on cross-validation experiments, and Table 3 summarizes the evaluation results.

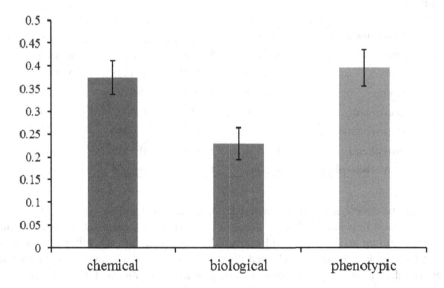

Fig. 3. Distribution of SLAMS weights and standard deviation for chemical, biological and phenotypic data sources in 10-fold cross-validation experiments

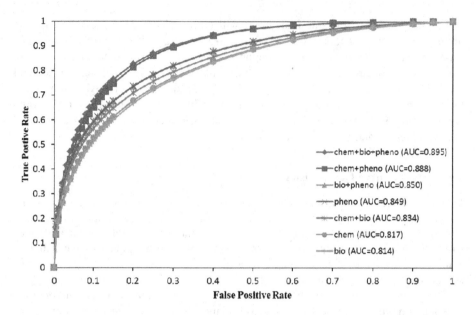

Fig. 4. The ROC comparison in 10-fold cross validation for various data-source combinations using SLAMS. chem: chemical properties; bio: biological properties; pheno: phenotypic properties. Data sources are sorted in the legend of the figure according to their AUC score.

Table 3. Comparison of various data-source combinations according to AUC, precision, recall, and F-score

Data Source	AUC	Precision	Recall	F-score
chem	0.8171	0.2232	0.4633	0.3013
bio	0.8139	0.2166	0.4592	0.2944
pheno	0.8492	0.2685	0.5117	0.3522
chem+bio	0.8339	0.2366	0.5012	0.3215
chem+pheno	0.8876	0.3281	0.6244	0.4302
bio+pheno	0.8503	0.2733	0.5119	0.3563
chem+bio+pheno	0.8949	0.3452	0.6505	0.4510

When the data sources were compared independently, the phenotypic data appeared to be the most informative (highest AUC of 0.8492), and chemical and biological data achieved similar AUC. This could be partially explained with the following reasons. Drug indications (i.e., drug's indicated diseases) and side effects are both measureable behavioral or physiological changes in response to the treatment. Intuitively, if drugs treating a disease share the same side-effects, this may be manifestation of some underlying mechanism-of-action (MOA) linking the indicated disease and the side-effect. Furthermore, both drug indications and side-effects are observations on human in the clinical stage, so there is less of a translational issue. Therefore, phenotypic data is a much more important drug information source with regard to predicting drug indications.

In the experiment while combing any two data sources will improve the AUC, the increase obtained by adding chemical structures on top of phenotypic properties (from 0.8492 to 0.8876) is much more significant than adding biological targets information on it (from 0.8492 to 0.8503). It seems that chemical properties and phenotypic properties are complementary. Combing all three data sources, we obtained the highest AUC score. On the other hand, if we focus on precision and recall, adding chemical properties to phenotypic properties yielded a dramatic increase (~22% in precision and recall). However, in our experiments there was no significant improvement when adding biological properties to phenotypic properties.

4.5   Analysis of Novel Predictions

During the 10-fold cross-validation, our SLAMS method with all chemical, biological, and phenotypic properties produced 3870 false-positive drug-disease associations. In other words, these associations were predicted by our method but they were not present in the gold standard. Some of these associations could be false, but a few associations could be true and can be considered as drug repositioning candidates in the real-world drug discovery. Taking the disease *Rheumatoid Arthritis* as an example, in Table 4 our SLAMS method found 10 drugs to treat it. These 10 drugs don't have associations with *Rheumatoid Arthritis* in the gold standard, and they have their own indications other than *Rheumatoid Arthritis*.

In order to test whether our predictions are in accordance with current experimental knowledge, we checked the extent to which they appear in current clinical trials. In Table 4, the drugs *Ramipril, Meloxicam,* and *Imatinib* have been tested for treating the disease *Rheumatoid Arthritis* in some clinical trials. In other words, pharmaceutical investigators have been aware of the associations of the drugs and *Rheumatoid Arthritis*, although they are still in the experimental stage. We downloaded all drug-disease data from registry of federally and privately supported clinical trials conducted around the world[5]. Overall, we acquired 18,392 unique drug-disease associations that are being investigated in clinical trials (phases I-IV). In all, 4798 of these associations involve drugs and diseases that are present in our data set with the exact names, spanning 4066 associations that are not part of our gold standard. Of these 4066 associations, our 3870 false-positive drug-disease associations cover 21% (i.e., 854 associations). It was highly unlikely that our false-positive predictions identified this set of 854 experimental drug-disease associations by chance (p < 0.0001, Fisher's exact test [25]). Hence, we conclude that false-positive novel uses predicted by our method attained significant coverage of drug-disease associations tested in clinical trials. All predicted drug-disease associations in our experiments are available at our website[6].

---

[5] Clinical trials found at http://clinicaltrials.gov/.
[6] Available at http://astro.temple.edu/~tua87106/drugreposition.html.

Table 4. Repositioned drugs for Rheumatoid Arthritis predicted by our method

Drug Name	Original Uses	Treat Rheumatoid Arthritis in clinical trial
Ramipril	Hypertension	NCT00273533 proposed in Jan 2006
	Diabetic Nephropathies	
	Heart Failure	
Lisinopril	Hypertension	N/A
	Heart Failure	
Mercaptopurine	Lymphoma	N/A
Meloxicam	Osteoarthritis	NCT00042068 proposed in July 2002
Mefenamic Acid	Menorrhagia	N/A
	Fever	
	Dysmenorrhea	
Zileuton	Asthma	N/A
Imatinib	Gastrointestinal Neoplasms	NCT00154336 proposed in Sept 2005
	Leukemia, Myeloid	
	Blast Crisis	
Allopurinol	Gout	N/A
Imiquimod	Condylomata Acuminata	N/A
Masoprocol	Keratosis	N/A

5   Conclusion

In response to the high cost and risk in traditional *de novo* drug discovery, discovering potential uses for existing drugs, also known as drug repositioning, has attracted increasing interests from both the pharmaceutical industry and the research community. From a serendipitous drug repositioning to systematic or rational ways, a variety of computational approaches using single source have been developed. However, the complexity of the problem clearly needs methods to integrate drug information from multiple sources for better solutions.

In this paper, we proposed SLAMS, a new drug repositioning framework by integrating chemical (i.e., compound signatures), biological (i.e., protein targets), and phenotypic (i.e., side effects) properties. Experimental results showed that our method is superior to a few existing computational drug repositioning methods. Furthermore, our predictions statistically overlap drug-disease associations tested in clinical trials, suggesting that the predicted drugs may be regarded as valuable repositioning candidates for further drug discovery research. An important property of our method is that it allows easy integration of additional drug information sources. Moreover, the method ranked multiple drug information sources based on their contributions to the prediction, thus paving the way for prioritizing multiple data sources and building more reliable drug repositioning models.

Acknowledgment. This project was funded in part under a grant with the GlaxoSmithKline LLC.

References

1. Hurle, M.R., Yang, L., Xie, Q., Rajpal, D.K., Sanseau, P., Agarwal, P.: Computational drug repositioning: from data to therapeutics. Clin. Pharmacol. Ther. 93(4), 335–341 (2013)
2. Ashburn, T.T., Thor, K.B.: Drug Repositioning: Identifying and Developing New Uses for Existing Drugs. Nature Reviews Drug Discovery 3, 645–646 (2004)
3. Sardana, D., Zhu, C., Zhang, M., Gudivada, R.C., Yang, L., Jegga, A.G.: Drug repositioning for orphan diseases. Brief Bioinform 12(4), 346–356 (2011)
4. Keiser, M.J., Setola, V., Irwin, J.J., Laggner, C., Abbas, A.I., Hufeisen, S.J., Jensen, N.H., Kuijer, M.B., Matos, R.C., Tran, T.B., Whaley, R., Glennon, R.A., Hert, J., Thomas, K.L., Edwards, D.D., Shoichet, B.K., Roth, B.L.: Predicting new molecular targets for known drugs. Nature 462, 175–181 (2009)
5. Li, J., Zhu, X., Chen, J.Y.: Building Disease-Specific Drug-Protein Connectivity Maps from Molecular Interaction Networks and PubMed Abstracts. PLoS Comput. Biol. 5(7), e1000450 (2009)
6. Kotelnikova, E., Yuryev, A., Mazo, I., Daraselia, N.: Computational approaches for drug repositioning and combination therapy design. J. Bioinform Comput. Biol. 8(3), 593–606 (2010)
7. Campillos, M., Kuhn, M., Gavin, A.C., Jensen, L.J., Bork, P.: Drug target identification using side-effect similarity. Science 321, 263–266 (2008)
8. Yang, L., Agarwal, P.: Systematic Drug Repositioning Based on Clinical Side-Effects. PLoS ONE 6(12), e28025 (2011)
9. Hu, G., Agarwal, P.: Human Disease-Drug Network Based on Genomic Expression Profiles. PLoS ONE 4(8), e6536 (2009)
10. Sirota, M., Dudley, J.T., Kim, J., Chiang, A.P., Morgan, A.A., Sweet-Cordero, A., Sage, J., Butte, A.J.: Discovery and preclinical validation of drug indications using compendia of public gene expression data. Sci. Transl. Med. 3(96), 96ra77 (2011)
11. Luo, H., Chen, J., Shi, L., Mikailov, M., Zhu, H., Wang, K., He, L., Yang, L.: DRAR-CPI: a server for identifying drug repositioning potential and adverse drug reactions via the chemical-protein interactome. Nucleic Acids Res. 39(Web Server Issue), W492–W498 (2011)
12. Li, J., Lu, Z.: A New Method for Computational Drug Repositioning Using Drug Pairwise Similarity. In: IEEE International Conference on Bioinformatics and Biomedicine (2012)
13. Gottlieb, A., Stein, G.Y., Ruppin, E., Sharan, R.: PREDICT: a method for inferring novel drug indications with application to personalized medicine. Mol. Syst. Biol. 7, 496 (2011)
14. Hotelling, H.: Relations between two sets of variates. Biometrika 28, 321–377 (1936)
15. Davis, J., Goadrich, M.: The Relationship Between Precision-Recall and ROC Curves. In: International Conference on Machine Learning (2006)
16. Wang, Y., Xiao, J., Suzek, T.O., Zhang, J., Wang, J., Bryant, S.H.: PubChem: a public information system for analyzing bioactivities of small molecules. Nucleic Acids Res. 37(Web Server Issue), W623–W633 (2009)
17. Steinbeck, C., Han, Y., Kuhn, S., Horlacher, O., Luttmann, E., Willighagen, E.: The Chemistry Development Kit (CDK): an open-source Java library for Chemo- and Bioinformatics. J. Chem. Inf. Comput. Sci. 43(2), 493–500 (2003)

18. Smith, T.F., Waterman, M.S., Burks, C.: The statistical distribution of nucleic acid similarities. Nucleic Acids Res. 13, 645–656 (1985)
19. Kuhn, M., Campillos, M., Letunic, I., Jensen, L.J., Bork, P.: A side effect resource to capture phenotypic effects of drugs. Molecular Systems Biology 6, 343 (2010)
20. Pauwels, E., Stoven, V., Yamanishi, Y.: Predicting drug side-effect profiles: a chemical fragment-based approach. BMC Bioinformatics 12, 169 (2011)
21. Pandey, G., Myers, C.L., Kumar, V.: Incorporating functional inter-relationships into protein function prediction algorithms. BMC Bioinformatics 10, 142 (2009)
22. Wishart, D.S., Knox, C., Guo, A.C., Cheng, D., Shrivastava, S., Tzur, D., Gautam, B., Hassanali, M.: DrugBank: a knowledgebase for drugs, drug actions and drug targets. Nucleic Acids Res. 36(Database Issue), D901–D906 (2008)
23. Olivier, B.: The Unified Medical Language System (UMLS): integrating biomedical terminology. Nucleic Acids Res. 32(Database Issue), D267–D270 (2004)
24. Apweiler, R., Bairoch, A., Wu, C.H., Barker, W.C., Boeckmann, B., Ferro, S., Gasteiger, E., Huang, H., Lopez, R., Magrane, M., Martin, M.J., Natale, D.A., O'Donovan, C., Redaschi, N., Yeh, L.S.: UniProt: the Universal Protein knowledgebase. Nucleic Acids Res. 32(Database Issue), D115–D119 (2004)
25. Upton, G.: Fisher's exact test. Journal of the Royal Statistical Society, Series A 155(3), 395–402 (1992)

Score As You Lift (SAYL): A Statistical Relational Learning Approach to Uplift Modeling

Houssam Nassif[1], Finn Kuusisto[1], Elizabeth S. Burnside[1], David Page[1], Jude Shavlik[1], and Vítor Santos Costa[1,2]

[1] University of Wisconsin, Madison, USA
nassif@wisc.edu, {finn,shavlik}@cs.wisc.edu, EBurnside@uwhealth.org,
page@biostat.wisc.edu, vsc@dcc.fc.up.pt
[2] CRACS-INESC TEC and DCC-FCUP, University of Porto, Portugal

Abstract. We introduce Score As You Lift (SAYL), a novel Statistical Relational Learning (SRL) algorithm, and apply it to an important task in the diagnosis of breast cancer. SAYL combines SRL with the marketing concept of uplift modeling, uses the area under the uplift curve to direct clause construction and final theory evaluation, integrates rule learning and probability assignment, and conditions the addition of each new theory rule to existing ones.

Breast cancer, the most common type of cancer among women, is categorized into two subtypes: an earlier in situ stage where cancer cells are still confined, and a subsequent invasive stage. Currently older women with in situ cancer are treated to prevent cancer progression, regardless of the fact that treatment may generate undesirable side-effects, and the woman may die of other causes. Younger women tend to have more aggressive cancers, while older women tend to have more indolent tumors. Therefore older women whose in situ tumors show significant dissimilarity with in situ cancer in younger women are less likely to progress, and can thus be considered for watchful waiting.

Motivated by this important problem, this work makes two main contributions. First, we present the first multi-relational uplift modeling system, and introduce, implement and evaluate a novel method to guide search in an SRL framework. Second, we compare our algorithm to previous approaches, and demonstrate that the system can indeed obtain differential rules of interest to an expert on real data, while significantly improving the data uplift.

1 Introduction

Breast cancer is the most common type of cancer among women, with a 12% incidence in a lifetime [2]. Breast cancer has two basic categories: an earlier *in situ* stage where cancer cells are still confined to where they developed, and a subsequent *invasive* stage where cancer cells infiltrate surrounding tissue. Since nearly all in situ cases can be cured [1], current practice is to treat in situ occurrences in order to avoid progression into invasive tumors [2]. Nevertheless, the

H. Blockeel et al. (Eds.): ECML PKDD 2013, Part III, LNAI 8190, pp. 595–611, 2013.

time required for an in situ tumor to reach invasive stage may be sufficiently long for an older woman to die of other causes, raising the possibility that treatment may not have been necessary.

Cancer occurrence and stage are determined through biopsy, a costly, invasive, and potentially painful procedure. Treatment is also costly and may generate undesirable side-effects. Hence there is a need for pre-biopsy methods that can accurately identify patient subgroups that would benefit most from treatment, and especially, those who do not need treatment. For the latter, the risk of progression would be low enough to employ watchful waiting (mammographic evaluation at short term intervals) rather than biopsy [26].

Fortunately, the literature confirms that the pre-biopsy mammographic appearance as described by radiologists can predict breast cancer stage [28,29]. Furthermore, based on age, different pre-biopsy mammographic features can be used to classify cancer stage [18]. A set of mammography features is *differentially-predictive* if it is significantly more predictive of cancer in one age group as compared to another. We may be able to use such differentially-predictive features to recommend watchful waiting for older in situ patients accurately enough to safely avoid additional tests and treatment.

In fact, younger women tend to have more aggressive cancers that rapidly proliferate, while older women tend to have more indolent cancers [8,13]. We assume that younger in situ patients should always be treated, due to the longer potential time-span for cancer progression. We also assume that older in situ patients whose mammography features are similar to in situ in younger patients should also be treated, because the more aggressive nature of cancer in younger patients may be conditioned on those features. On the other hand, older in situ patients whose mammography features are significantly different from features observed in younger in situ patients are less likely to experience rapid proliferation, and can thus be recommended for watchful waiting.

The general task of identifying differentially predictive features occurs naturally in diverse fields. Psychologists initially assessed for differential prediction using linear regression, defining it as the case where a common regression equation results in systematic nonzero errors of prediction for given subgroups [6]. The absence of differential prediction over different groups of examinees was an indicator of the fairness of a cognitive or educational test [31].

Psychologists aim to decrease differential prediction on their tests. This is not the case in the closely related concept of *uplift modeling*, a modeling and classification method used in marketing to determine the incremental impact of an advertising campaign on a given population. Uplift modeling is effectively a differential prediction approach aimed at maximizing uplift [11,16,23]. Uplift is defined as the difference in a model or intervention M's lift scores over the subject and control sets:

$$Uplift_M = Lift_M(subject) - Lift_M(control). \tag{1}$$

Given a fraction ρ such that $0 \leq \rho \leq 1$, a model M's lift is defined as the number of positive examples amongst the model's ρ-highest ranking examples.

Uplift thus captures the additional number of positive examples obtained due to the intervention. We generate an uplift curve by ranging ρ from 0 to 1 and plotting $Uplift_M$. The higher the uplift curve, the more profitable a marketing model/intervention is.

The motivating problem at hand can readily be cast as an uplift modeling problem (see Table 1). Even though we are not actively altering the cancer stage as a marketing intervention would alter the subject population behavior, one may argue that time is altering the cancer stage. Our subject and control sets are respectively older and younger patients with confirmed breast cancer —where time, as an intervention, has altered the cancer stage— and we want to predict in situ versus invasive cancer based on mammography features. By maximizing the in situ cases' uplift, which is the difference between a model's in situ lift on the older and younger patients, we are identifying the older in situ cases that are most different from younger in situ cases, and thus are the best candidates for watchful waiting. Exactly like a marketing campaign would want to target consumers who are the most prone to respond, we want to target the ones that differ the most from the control group.

Table 1. Casting mammography problem in uplift modeling terms

Intervention	Subject Group	Control Group	Positive Class	Negative Class
Time	Older cohort	Younger cohort	In Situ	Invasive

In recent work, Nassif et al. inferred older-specific differentially-predictive in situ mammography rules [20]. They used Inductive Logic Programming (ILP) [14], but defined a differential-prediction-sensitive clause evaluation function that compares performance over age-subgroups during search-space exploration and rule construction. To assess the resulting theory (final set of rules), they constructed a TAN classifier [9] using the learned rules and assigned a probability to each example. They finally used the generated probabilities to construct the uplift curve to assess the validity of their model.

The ILP-based differential prediction model [20] had several shortcomings. First, this algorithm used a differential scoring function based on m-estimates during clause construction, and then evaluated the resulting theory using the area under the uplift curve. This may result in sub-optimal performance, since rules with a high differential m-estimate score may not generate high uplift curves. Second, it decoupled clause construction and probability estimation: after rules are learned, a TAN model is built to compute example probabilities. Coupling these two processes together may generate a different theory with a lower ILP-score, but with a more accurate probability assignment. Finally, rules were added to the theory independently of each other, resulting in redundancies. Having the addition of newer rules be conditioned on the prior theory rules is likely to improve the quality and coverage of the theory.

In this work, we present a novel relational uplift modeling Statistical Relational Learning (SRL) algorithm that addresses all the above shortcomings. Our method, Score As You Lift (SAYL), uses the area under the uplift curve score during clause construction and final theory evaluation, integrates rule learning and probability assignment, and conditions the addition of new theory rules to existing ones. This work makes two main contributions. First, we present the first multi-relational uplift modeling system, and introduce, implement and evaluate a novel method to guide search in an SRL framework. Second, we compare our algorithm to previous approaches, and demonstrate that the system can indeed obtain differential rules of interest to an expert on real data, while significantly improving the data uplift.

2   Background: The SAYU Algorithm

Score As You Use (SAYU) [7] is a Statistical Relational Learner [10] that integrates search for relational rules and classification. It starts from the well known observation that a clause or rule r can be mapped to a binary attribute b, by having $b(e) = 1$ for an example e if the rule r explains e, and $b(e) = 0$ otherwise.

This makes it possible to construct classifiers by using rules as attributes, an approach known as *propositionalization* [32]. One limitation, though, is that often the propositional learner has to consider a very large number of possible rules. Moreover, these rules tend to be very correlated, making it particularly hard to select a subset of rules that can be used to construct a good classifier.

SAYU addresses this problem by evaluating the contribution of rules to a classifier as soon as the rule is generated. Thus, SAYU generates rules using a traditional ILP algorithm, such as Aleph [27], but instead of scoring the rules individually, as Aleph does, every rule SAYU generates is immediately used to construct a statistical classifier. If this new classifier improves performance over the current set of rules, the rule is added as an extra attribute.

Algorithm 1. SAYU

$Rs \leftarrow \{\}; M_0 \leftarrow InitClassifier(Rs)$
while $DoSearch()$ **do**
    $e^+ \leftarrow RandomSeed();$
    $\perp_{e^+} \leftarrow saturate(e);$
    while $c \leftarrow reduce(\perp_{e^+})$ **do**
        $M \leftarrow LearnClassifier(Rs \cup \{c\});$
        if $Better(M, M_0)$ **then**
            $Rs \leftarrow Rs \cup \{c\}; M_0 \leftarrow M;$
            break
        end if
    end while
end while

Algorithm 1 shows SAYU in more detail. SAYU maintains a current set of clauses, Rs, and a current reference classifier, M_0. SAYU extends the Aleph [27] implementation of Progol's MDIE algorithm [17]. Thus, it starts search by randomly selecting a positive example as seed, e^+, generating the corresponding bottom clause, $\perp_{e^+}$, and then generating clauses that subsume $\perp_{e^+}$. For every new such clause c, it constructs a classifier M and compares M with the current M_0. If better, it accepts c by adding it to Rs and making M the default classifier. SAYU can terminate search when all examples have been tried without adding new clauses. In practice, termination is often controlled by a time limit.

Quite often, most execution time will be spent learning classifiers. Therefore, it is important that the classifier can be learned in a reasonable time. Further, the classifier should cope well with many related attributes. We use the TAN classifier, a Bayesian network that extends naive Bayes with at most one other edge per attribute [9]. TAN has quadratic learning time, which is acceptable for SAYU, and compensates well for highly dependent attributes.

Second, comparing two classifiers is not trivial. SAYU reserves a tuning set for this task: if the classifier M has a better score on both the initial training and tuning sets, the new rule is accepted. The scoring function depends on the problem at hand. Most often SAYU has been used in skewed domains, where the area under the precision-recall curve is regarded as a good measure [5], but the algorithm allows for any metric.

The original SAYU algorithm accepts a logical clause as soon as it improves the network. It may be the case that a later clause would be even better. Unfortunately, SAYU will switch seeds after selecting a clause, so the better clause may be ignored. One solution is to make SAYU less greedy by *exploring* the search space for each seed, up to some limit on the number of clauses, before accepting a clause. We call this version of SAYU *exploration SAYU*: we will refer to it as *e-SAYU*, and to the original algorithm as *greedy SAYU*, or *g-SAYU*.

Algorithm 2. e-SAYU

$Rs \leftarrow \{\}; M_0 \leftarrow InitClassifier(Rs)$
while $DoSearch()$ **do**
    $e^+ \leftarrow RandomSeed()$;
    $\perp_{e^+} \leftarrow saturate(e^+)$;
    $c_{e^+} \leftarrow \top; M_{e^+} \leftarrow M_0$;
    while $c \leftarrow reduce(\perp_{e^+})$ **do**
        $M \leftarrow LearnClassifier(Rs \cup \{c\})$;
        if $Better(M, M_e)$ **then**
            $c_{e^+} \leftarrow c; M_{e^+} \leftarrow M$;
        end if
    end while
    if $c_{e^+} \neq \top$ **then**
        $Rs \leftarrow Rs \cup \{c_{e^+}\}; M_0 \leftarrow M_{e^+}$;
    end if
end while

Algorithm 2 details e-SAYU. It differs from g-SAYU in that it keeps track, for each seed, of the current best classifier M_{e^+} and best clause c_{e^+}. At the end, if a clause c_{e^+} was found, we commit to that clause and update the classifier.

3   Background: Uplift Modeling

Next we discuss uplift in more detail and compare it to related measures.

3.1   Uplift

Let P be the number of positive examples and N the number of negative examples in a given dataset D. Lift represents the number of true positives detected by model m amongst the top-ranked fraction ρ. Varying $\rho \in [0, 1]$ produces a lift curve. The area under the lift curve AUL for a given model and data becomes:

$$AUL = \int Lift(D, \rho)d\rho \approx \frac{1}{2} \sum_{k=1}^{P+N} (\rho_{k+1} - \rho_k)(Lift(D, \rho_{k+1}) + Lift(D, \rho_k)) \quad (2)$$

Uplift compares the difference between the model M over two groups, subjects s and controls c. It is obtained by:

$$Uplift(M_s, M_c, \rho) = Lift_{M_s}(S, \rho) - Lift_{M_c}(C, \rho). \quad (3)$$

Since uplift is a function of a single value for ρ, the area under the uplift curve is the difference between the areas under the lift curves of the two models, $\Delta(AUL)$.

It is interesting to note the correspondence of the uplift model to the differential prediction framework [20]. The subjects and controls groups are disjoint subsets, and thus form a 2-strata dataset. $Lift_M$ is a differential predictive concept, since maximizing $Uplift(M_s, M_c, \rho)$ requires $Lift_{M_s}(S, \rho) \gg Lift_{M_c}(C, \rho)$. Finally, $Uplift$ is a differential-prediction-sensitive scoring function, since it is positively correlated with $Lift_{M_s}(S, \rho)$ and negatively correlated with $Lift_{M_c}(C, \rho)$.

3.2   Lift AUC and ROC AUC

In order to obtain more insight into this measure it is interesting to compare uplift and lift curves with receiver operating characteristic (ROC) curves. We define AUL as the area under the lift curve, and AUR as the area under the ROC curve. There is a strong connection between the lift curve and the ROC curve: Let $\pi = \frac{P}{P+N}$ be the prior probability for the positive class or skew, then:

$$AUL = P * (\frac{\pi}{2} + (1 - \pi)\,AUR) \,\, [30, \text{p. } 549]. \quad (4)$$

In uplift modeling we aim to optimize for uplift over two sets, that is we aim at obtaining new classifiers such that $\Delta(AUL^*) > \Delta(AUL)$, where $\Delta(AUL) = AUL_s - AUL_c$, subscripts s and c referring to the subject and control groups, respectively. The equation $\Delta(AUL^*) > \Delta(AUL)$ can be expanded into:

$$AUL_s^* - AUL_c^* > AUL_s - AUL_c. \quad (5)$$

Further expanding and simplifying we have:

$$P_s(\frac{\pi_s}{2} + (1 - \pi_s)AUR_s^*) - P_c(\frac{\pi_c}{2} - (1 - \pi_c)AUR_c^*) >$$

$$P_s(\frac{\pi_s}{2} + (1 - \pi_s)AUR_s) - P_c(\frac{\pi_c}{2} - (1 - \pi_c)AUR_c)$$

$$P_s(1 - \pi_s)AUR_s^* - P_c(1 - \pi_c)AUR_c^* > P_s(1 - \pi_s)AUR_s - P_c(1 - \pi_c)AUR_c$$

$$P_s(1 - \pi_s)(AUR_s^* - AUR_s) > P_s(1 - \pi_s)(AUR_c^* - AUR_c)$$

and finally

$$\frac{AUR_s^* - AUR_s}{AUR_c^* - AUR_c} > \frac{P_c}{P_s}\frac{1 - \pi_c}{1 - \pi_s}. \tag{6}$$

In a balanced dataset, we have $\pi_c = \pi_s = \frac{1}{2}$ and $P_c = P_s$, so we have that $\frac{1-\pi_c}{1-\pi_s} = 1$. In fact, if the subject and control datasets have the same skew we can conclude that $\Delta(AUL^*) > \Delta(AUL)$ implies $\Delta(AUR^*) > \Delta(AUR)$.

In the mammography dataset, the skews are $P_s = 132$, $\pi_s = \frac{132}{132+401}$ (older), and $P_c = 110$, $\pi_c = \frac{110}{110+264}$ (younger). Thus equation 6 becomes:

$$\frac{AUR_s^* - AUR_s}{AUR_c^* - AUR_c} > 0.86. \tag{7}$$

Therefore we cannot guarantee that $\Delta(AUL^*) > \Delta(AUL)$ implies $\Delta(AUR^*) > \Delta(AUR)$ on this data, as we can increase uplift with rules that have similar accuracy but cover more cases in the older cohort, and there are more cases to cover in the older cohort. On the other hand, breast cancer is more prevalent in older women [1], so uplift is measuring the true impact of the model.

In general, we can conclude that the two tests are related, but that uplift is sensitive to variations of dataset size and skew. In other words, uplift is more sensitive to variations in coverage when the two groups have different size. In our motivating domain, this is particularly important in that it allows capturing information related to the larger prevalence of breast cancer in older populations.

4   SAYL: Integrating SAYU and Uplift Modeling

SAYL is a Statistical Relational Learner based on SAYU that integrates search for relational rules and *uplift modeling*. Similar to SAYU, every valid rule generated is used for classifier construction via propositionalization, but instead of constructing a single classifier, SAYL constructs two classifiers; one for each of the subject and control groups. Both classifiers use the same set of attributes, but are trained only on examples from their respective groups. If a rule improves the area under the uplift curve (uplift AUC) by threshold θ, the rule is added to the attribute set. Otherwise, SAYL continues the search.

The SAYL algorithm is shown as Algorithm 3. Like SAYU, SAYL maintains separate training and tuning example sets, accepting rules only when the classifiers produce a better score on both sets. This requirement is often extended

Algorithm 3. SAYL

$Rs \leftarrow \{\}; M_0^s, M_0^c \leftarrow InitClassifiers(Rs)$
while $DoSearch()$ **do**
    $e_s^+ \leftarrow RandomSeed();$
    $\perp_{e_s^+} \leftarrow saturate(e);$
    while $c \leftarrow reduce(\perp_{e_s^+})$ **do**
        $M^s, M^c \leftarrow LearnClassifiers(Rs \cup \{c\});$
        if $Better(M^s, M^c, M_0^s, M_0^c)$ **then**
            $Rs \leftarrow Rs \cup \{c\}; M_0^s, M_0^c \leftarrow M^s, M^c;$
            break
        end if
    end while
end while

with a specified threshold of improvement θ, or a minimal rule coverage requirement *minpos*. Additionally, SAYL also has a greedy (g-SAYL) and exploratory (e-SAYL) versions that operate in the same fashion as they do for SAYU.

The key difference between SAYL and SAYU, then, is that SAYL maintains a distinction between the groups of interest by using two separate classifiers. This is what allows SAYL to demonstrate differential performance as opposed to standard metrics, such as the area under a precision-recall curve. To compute uplift AUC, SAYL simply computes the area under the lift curve for each of the groups using the two classifiers and returns the difference.

SAYL and SAYU also differ in selecting a seed example to saturate. Instead of selecting from the entire set of positive examples, SAYL only selects seed examples from the positive examples in the subject group. This is not necessary, but makes intuitive sense as clauses produced from examples in the subject set are more likely to produce greater lift on the subject set in the first place.

5   Experimental Results

Our motivating application is to detect differential older-specific in situ breast cancer by maximizing the area under the uplift curve (uplift AUC). We apply SAYL to the breast cancer data used in Nassif *et al.* [20]. The data consists of two cohorts: patients younger than 50 years form the *younger* cohort, while patients aged 65 and above form the *older* cohort. The older cohort has 132 in situ and 401 invasive cases, while the younger one has 110 in situ and 264 invasive.

The data is organized in 20 extensional relations that describe the mammogram, and 35 intensional relations that connect a mammogram with related mammograms, discovered at the same or in prior visits. Some of the extensional features have been mined from free text [19]. The background knowledge also maintains information on prior surgeries. The data is fully described in [18].

We use 10-fold cross-validation, making sure all records pertaining to the same patient are in the same fold. We run SAYL with a time limit of one hour per fold. We run folds in parallel. On top of the ILP memory requirements, SAYL requires an extra 0.5 gigabyte of memory for the Java Virtual Machine. For each cross-validated run, we use 4 training, 5 tuning and 1 testing folds. For each fold, we used the best combination of parameters according to a 9-fold internal cross-validation using 4 training, 4 tuning and 1 testing folds. We try both e-SAYL and g-SAYL search modes, vary the minimum number *minpos* of positive examples that a rule is required to cover between 7 and 13 (respectively 5% and 10% of older in situ examples), and set the threshold θ to add a clause to the theory if its addition improves the uplift AUC to 1%, 5% and 10%. We concatenate the results of each testing set to generate the final uplift curve.

Table 2. 10-fold cross-validated SAYL performance. AUC is Area Under the Curve. Rule number averaged over the 10 folds of theories. For comparison, we include results of Differential Prediction Search (DPS) and Model Filtering (MF) methods [20]. We compute the *p*-value comparing each method to DPS, * indicating significance.

Algorithm	Uplift AUC	Lift(older) AUC	Lift(younger) AUC	Rules Avg #	DPS *p*-value
SAYL	58.10	97.24	39.15	9.3	0.002 *
DPS	27.83	101.01	73.17	37.1	-
MF	20.90	100.89	80.99	19.9	0.0039 *
Baseline	11.00	66.00	55.00	-	0.0020 *

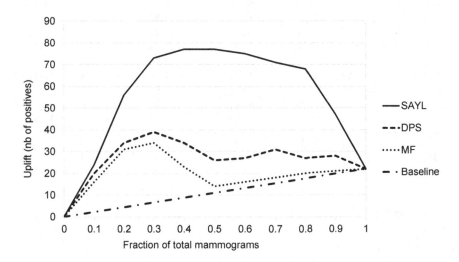

Fig. 1. Uplift curves for the ILP-based methods (Differential Prediction Search (DPS) and Model Filtering (MF), both with *minpos* = 13 [20]), a baseline random classifier, and SAYL with cross-validated paramters. Uplift curves start at 0 and end at 22, the difference between older (132) and younger (110) total in situ cases. The higher the curve, the better the uplift.

Table 2 compares SAYL with the Differential Prediction Search (DPS) and Model Filtering (MF) ILP methods [20], both of which had *minpos* = 13 (10% of older in situ). A baseline random classifier achieves an uplift AUC of 11. We use the Mann-Whitney test at the 95% confidence level to compare two sets of experiments. We show the *p*-value of the 10-fold uplift AUC paired Mann-Whitney of each method as compared to DPS, DPS being the state-of-the-art in relational differential prediction. We also plot the uplift curves in Figure 1.

SAYL 10-fold cross-validation chose g-SAYL in 9 folds and e-SAYL in 1, while *minpos* was 13 (10% of older in situ) in 5 folds, and 7 (5%) in the remaining 5 folds. θ was selected to be 1% in 4 folds, 5% in 3 folds, and 10% in the remaining 3 folds. Table 3 shows how sensitive SAYL is to those different parameters.

Table 3. 10-fold cross-validated SAYL performance under various parameters. *minpos* is the minimum number of positive examples that a rule is required to cover. θ is the uplift AUC improvement threshold for adding a rule to the theory. We also include results of SAYL using cross-validated parameters and Differential Prediction Search (DPS). We compute the *p*-value comparing each method to DPS, * indicating significance.

minpos	θ (%)	search mode	Uplift AUC	Lift(older) AUC	Lift(younger) AUC	Rules Avg #	DPS *p*-value	
13	1	g-SAYL	63.29	96.79	33.50	16.4	0.002	*
13	1	e-SAYL	43.51	83.82	40.31	2.0	0.049	*
13	5	g-SAYL	58.06	96.14	38.07	5.9	0.002	*
13	5	e-SAYL	53.37	85.66	32.29	1.8	0.027	*
13	10	g-SAYL	61.68	96.26	34.58	3.6	0.002	*
13	10	e-SAYL	65.36	90.50	25.14	1.1	0.002	*
7	1	g-SAYL	**65.48**	98.82	33.34	18.3	0.002	*
7	1	e-SAYL	25.50	74.39	48.90	3.0	0.695	
7	5	g-SAYL	58.91	96.67	37.76	5.8	0.002	*
7	5	e-SAYL	32.71	79.52	46.81	2.5	0.557	
7	10	g-SAYL	61.98	96.87	34.89	3.6	0.002	*
7	10	e-SAYL	52.35	83.64	31.29	1.6	0.002	*
-	-	SAYL	58.10	97.24	39.15	9.3	0.002	*
13	-	DPS	27.83	**101.01**	**73.17**	**37.1**	-	

6   Discussion

6.1   Model Performance

SAYL significantly outperforms DPS (Table 2, Figure 1), while ILP-based runs have the highest older and younger lift AUC (Tables 2, 3). This is because ILP methods use different metrics during clause construction and theory evaluation, and decouple clause construction from probability estimation. SAYL builds models that are slightly less predictive of in situ vs. invasive over the younger subset, as measured by the slightly lower older lift AUC, but on the other hand it effectively maximizes uplift. In fact, increasing lift on one subset will most often increase lift on the other subset, since both sets share similar properties. SAYL avoids this pitfall by selecting rules that generate a high differential lift, ignoring rules with good subject lift that are equally good on the controls. These results confirm the limitations of a pure ILP approach, demonstrating significantly higher uplift using SAYL.

e-SAYL explores a larger search space for a given seed before selecting a rule to add to the theory. This results in smaller theories than greedy g-SAYL. Increasing θ, the uplift AUC improvement threshold for adding a rule to the theory, also results in smaller theories, as expected. Ranging *minpos* between 7 and 13 doesn't seem to have a sizable effect on rule number.

g-SAYL's performance remains constant across all parameters, its uplift AUC varying between 58.06 and 65.48. At the same time, its theory size ranges from 3.6 to 18.3. This indicates that the number of rules is not correlated with uplift AUC. Another indication comes from e-SAYL, whose theory size changes little $(1.1 - 3.0)$, while its performance tends to increase with increasing *minpos* and θ. Its uplift AUC jumps from the lowest score of 25.50, where it is significantly worse than g-SAYL, to nearly the highest score of 65.36. In fact, g-SAYL outperforms e-SAYL on all runs except *minpos* = 13 and $\theta = 10\%$.

e-SAYL is more prone to over fitting, since it explores a larger search space and is thus more likely to find rules tailored to the training set with a poor generalization. By increasing *minpos* and θ, we are restricting potential candidate rules to the more robust ones, which decreases the chances of converging to a local minima and overfitting. This explains why e-SAYL had the worst performances with lowest *minpos* and θ values, and why it achieved the second highest score of all runs at the highest *minpos* and θ values. These limited results seem to suggest using e-SAYL with *minpos* and θ equal to 10%.

6.2 Model Interpretation

SAYL returns two TAN Bayes-net models, one for the older and one for the younger, with first-order logic rules as the nodes. Each model includes the classifier node, presented top-most, and the same rules. All rules depend directly on the classifier and have at least one other parent. Although both models have the same rules as nodes, TAN learns the structure of each model on its corresponding data subset separately, resulting in different networks. SAYL identifies the features that best differentiate amongst subject and control positive examples, while TAN uses these features to create the best classifier over each set.

To generate the final model and inspect the resulting rules, we run SAYL with 5 folds for training and 5 for tuning. As an example, Figures 2 and 3 respectively show the older and younger cases TAN models of g-SAYL with *minpos* = 13 and $\theta = 5\%$. The older cohort graph shows that the increase in the combined BI-RADS score is a key differential attribute. The BI-RADS score is a number that summarizes the examining radiologist's opinion and findings concerning the mammogram [3]. We then can see two sub-graphs: the left-hand side sub-graph focuses on the patient's history (prior biopsy, surgery and family history), whereas the right-hand side sub-graph focuses on the examined breast (BI-RADS score, mass size). In contrast, the younger cohort graph is very different: the graph has a shorter depth, and the combined BI-RADS increase node is linked to different nodes...

As the number of rules increases, it becomes harder for humans to interpret the cohort models, let alone their uplift interaction. In ILP-based differential prediction methods [20], theory rules are independent and each rule is an older in situ differential rule. In SAYL, theory rules are dependent on each other, whereas a rule can be modulating another rule in the TAN graph. This is advantageous because such modulated rule combinations can not be expressed in ILP-theory, and therefore might not be learnable. On the other hand, SAYL individual rules are not required to be older in situ specific. A SAYL rule can predict invasive, or be younger specific, as long as the resulting model is uplifting older in situ. Which decreases clinical rule interpretability.

The average number of rules returned by SAYL is lower than ILP-based methods (Table 2), SAYL effectively removes redundant rules by conditioning the

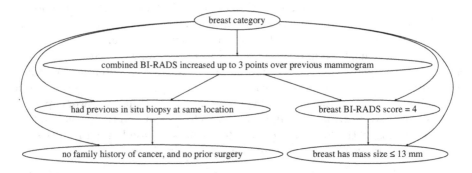

Fig. 2. TAN model constructed by SAYL over the older cases: the topmost node is the classifier node, and the other nodes represent rules inserted as attributes to the classifier. Edges represent the main dependencies inferred by the model.

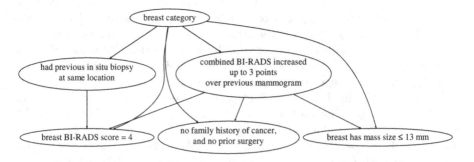

Fig. 3. TAN model constructed by SAYL over the younger cases. Notice that it has the same nodes but with a different structure than its older counterpart.

addition of a new rule on previous ones. We also note that SAYL, like SAYU, tends to like short rules [7]. DPS found five themes amongst its older in situ rules with a significantly better precision and recall: calcification, prior in situ biopsy, BI-RADS score increase, screening visit, and low breast density [20].

For SAYL runs returning small theories, the resulting rules tend to be differential and fall within these 5 themes. For example, g-SAYL with $minpos = 13$ and $\theta = 10\%$ returns 3 rules:

1. Current study combined BI-RADS increased up to 3 points over previous mammogram.
2. Had previous in situ biopsy at same location.
3. Breast BI-RADS score = 4.

These rules cover two of the 5 DPS themes, namely prior in situ biopsy and BI-RADS score increase.

As the number of SAYL returned rules increases, rule interactions become more complex, individual rules tend not to remain older in situ differential, and rules are no longer confined to the above themes. In the Figures 2 and 3 example, we recover the prior in situ biopsy and BI-RADS score increase themes, but we also have non-thematic rules like "no family history of cancer, and no prior surgery". In the two runs returning the largest theories, g-SAYL with $\theta = 1\%$ and $minpos = 7$ and 13, we recover 4 of the themes, only missing calcification. Note that, as the graph size increases, medical interpretation of the rules becomes more difficult, as well as identifying novel differential themes, since rules are conditioned on each other.

Although the SAYL rules may not be differential when viewed individually, the SAYL final model is differential, significantly outperforming DPS in uplift AUC. DPS, on the other hand, is optimized for mining differential rules, but performs poorly as a differential classifier. SAYL returns a TAN Bayes net whose nodes are logical rules, a model that is human interpretable and that offers insight into the underlying differential process. Greedy g-SAYL's performance depended little on the parameters, while exploratory e-SAYL's performance increased when requiring more robust rules.

7   Related Work

Differential prediction was first used in psychology to assess the fairness of cognitive and educational tests, where it is defined as the case where consistent nonzero errors of prediction are made for members of a given subgroup [6]. In this context, differential prediction is usually detected by either fitting a common regression equation and checking for systematic prediction discrepancies for given subgroups, or by building regression models for each subgroup and testing for differences between the resulting models [15,31]. If the predictive models differ in terms of slope or intercept, it implies that bias exists because systematic errors of prediction would be made on the basis of group membership. An example is assessing how college admission test scores predict first year cumulative grades for males and females. For each gender group, we fit a regression model. We then compare the slope, intercept and/or standard errors for both models. If they differ, the test exhibits differential prediction and may be considered unfair.

In contrast to most studies of *differential prediction* in psychology, marketing's *uplift modeling* assumes an active agent. Uplift modeling is used to understand the best targets for an advertising campaign. Seminal work includes Radcliffe and Surry's true response modeling [23], Lo's true lift model [16], and Hansotia and Rukstales' incremental value modeling [11]. As an example, Hansotia and Rukstales construct a regression and a decision tree, or CHART, model to identify customers for whom direct marketing has sufficiently large impact. The splitting criterion is obtained by computing the difference between the estimated probability increase for the attribute on the subject set and the estimated probability increase on the control set.

In some applications, especially medical decision support systems, gaining insight into the underlying classification logic can be as important as system performance. Insight into the classification logic in medical problems can be an important method to discover disease patterns that may not be known or easily otherwise gleaned from the data. Recent developments include tree-based approaches to uplift modeling [24,25], although ease-of-interpretation was not an objective in their motivating applications. Wanting to maximize rule interpretability, Nassif *et al.* [20] opted for ILP-based rule learning instead of decision-trees because the latter is a special case of the former [4].

To the best of our knowledge, the first application of uplift modeling in medical domains is due to Jaśkowski and Jaroszewicz [12], who adapt standard classifiers by using a simple class variable transformation. Their transformation avoids using two models by assuming that both sets have the same size and combining the examples into a single set. They also propose an approach where two classifiers are learned separately but they help each other by labeling extra examples. Instead, SAYL directly optimizes an uplift measure.

Finally, we observe that the task of discriminating between two dataset strata is closely related to the problem of Relational Subgroup Discovery (RSD), that is, "given a population of individuals with some properties, find subgroups that are statistically interesting" [32]. In the context of multi-relational learning systems, RSD applies a first propositionalization step and then applies a weighted covering algorithm to search for rules that can be considered to define a sub-group in the data. Although the weighting function is defined to focus on unexplored data by decreasing the weight of covered examples, RSD does not explicitly aim at discovering the differences between given partitions.

8 Future Work

A key contribution of this work is constructing a relational classifier that maximizes uplift. SAYL effectively identifies older in situ patients with mammography features that are significantly different from those observed in the younger in situ cases. But one may argue that, for a model to be clinically relevant, we should take into account all mammography features when staging an uplift comparison. We can start the SAYL TAN model with the initial set of attributes, and then learn additional rules, composed of relational features or a combinations of attributes, to maximize uplift [21]. This could potentially increase the achievable lift on both the subject and control groups, making the uplift task harder.

Given the demonstrated theoretical similarity between lift and ROC curves (Section 3.2), and the fact that ROC curves are more widely used especially in the medical literature, it is interesting to compare our approach with a SAYL version that optimizes for ROC AUC.

Finally, we are in the process of applying SAYL to different problems. For example, working on uncovering adverse drug effects, SAYL can be used to construct a model identifying patient subgroups that have a differential prediction before and after drug administration [22].

9   Conclusion

In this work, we present Score As You Lift (SAYL), a novel Statistical Relational Learning algorithm and the first multi-relational uplift modeling system. Our algorithm maximizes the area under the uplift curve, uses this measure during clause construction and final theory evaluation, integrates rule learning and probability assignment, and conditions the addition of new theory rules to existing ones. SAYL significantly outperforms previous approaches on a mammography application ($p = 0.002$ with similar parameters), while still producing human interpretable models. We plan on further investigating the clinical relevance of our model, and to apply SAYL to additional differential problems.

Acknowledgment. This work is supported by US National Institute of Health (NIH) grant R01-CA165229. VSC was funded by the ERDF through the Progr. COMPETE, the Portuguese Gov. through FCT, proj. ABLe ref. PTDC/EEI-SII/2094/2012, ADE (PTDC/ EIA-EIA/121686/2010), and by FEDER/ON2 and FCT project NORTE-07-124-FEDER-000059.

References

1. American Cancer Society: Breast Cancer Facts & Figures 2009-2010. American Cancer Society, Atlanta, USA (2009)
2. American Cancer Society: Cancer Facts & Figures 2009. American Cancer Society, Atlanta, USA (2009)
3. American College of Radiology, Reston, VA, USA: Breast Imaging Reporting and Data System (BI-RADS^{TM}), 3rd edn. (1998)
4. Blockeel, H., De Raedt, L.: Top-down induction of first-order logical decision trees. Artificial Intelligence 101, 285–297 (1998)
5. Boyd, K., Davis, J., Page, D., Santos Costa, V.: Unachievable region in precision-recall space and its effect on empirical evaluation. In: Proceedings of the 29th International Conference on Machine Learning, ICML 2012, Edinburgh, Scotland (2012)

6. Cleary, T.A.: Test bias: Prediction of grades of negro and white students in integrated colleges. Journal of Educational Measurement 5(2), 115–124 (1968)
7. Davis, J., Burnside, E., de Castro Dutra, I., Page, D.L., Santos Costa, V.: An integrated approach to learning bayesian networks of rules. In: Gama, J., Camacho, R., Brazdil, P.B., Jorge, A.M., Torgo, L. (eds.) ECML 2005. LNCS (LNAI), vol. 3720, pp. 84–95. Springer, Heidelberg (2005)
8. Fowble, B.L., Schultz, D.J., Overmoyer, B., Solin, L.J., Fox, K., Jardines, L., Orel, S., Glick, J.H.: The influence of young age on outcome in early stage breast cancer. Int. J. Radiat. Oncol. Biol. Phys. 30(1), 23–33 (1994)
9. Friedman, N., Geiger, D., Goldszmidt, M.: Bayesian network classifiers. Machine Learning 29, 131–163 (1997)
10. Getoor, L., Taskar, B. (eds.): An Introduction to Statistical Relational Learning. MIT Press (2007)
11. Hansotia, B., Rukstales, B.: Incremental value modeling. Journal of Interactive Marketing 16(3), 35–46 (2002)
12. Jaśkowski, M., Jaroszewicz, S.: Uplift modeling for clinical trial data. In: ICML 2012 Workshop on Clinical Data Analysis, Edinburgh, Scotland (2012)
13. Jayasinghe, U.W., Taylor, R., Boyages, J.: Is age at diagnosis an independent prognostic factor for survival following breast cancer? ANZ J. Surg. 75(9), 762–767 (2005)
14. Lavrac, N., Dzeroski, S.: Inductive Logic Programming: Techniques and Applications, Ellis Horwood, New York (1994)
15. Linn, R.L.: Single-group validity, differential validity, and differential prediction. Journal of Applied Psychology 63, 507–512 (1978)
16. Lo, V.S.: The true lift model - a novel data mining approach to response modeling in database marketing. SIGKDD Explorations 4(2), 78–86 (2002)
17. Muggleton, S.H.: Inverse entailment and Progol. New Generation Computing 13, 245–286 (1995)
18. Nassif, H., Page, D., Ayvaci, M., Shavlik, J., Burnside, E.S.: Uncovering age-specific invasive and DCIS breast cancer rules using Inductive Logic Programming. In: ACM International Health Informatics Symposium (IHI), Arlington, VA, pp. 76–82 (2010)
19. Nassif, H., Woods, R., Burnside, E.S., Ayvaci, M., Shavlik, J., Page, D.: Information extraction for clinical data mining: A mammography case study. In: IEEE International Conference on Data Mining (ICDM) Workshops, Miami, Florida, pp. 37–42 (2009)
20. Nassif, H., Santos Costa, V., Burnside, E.S., Page, D.: Relational differential prediction. In: Flach, P.A., De Bie, T., Cristianini, N. (eds.) ECML PKDD 2012, Part I. LNCS, vol. 7523, pp. 617–632. Springer, Heidelberg (2012)
21. Nassif, H., Wu, Y., Page, D., Burnside, E.S.: Logical Differential Prediction Bayes Net, improving breast cancer diagnosis for older women. In: American Medical Informatics Association Symposium (AMIA), Chicago, pp. 1330–1339 (2012)
22. Page, D., Santos Costa, V., Natarajan, S., Barnard, A., Peissig, P., Caldwell, M.: Identifying adverse drug events by relational learning. In: AAAI 2012, Toronto, pp. 1599–1605 (2012)
23. Radcliffe, N.J., Surry, P.D.: Differential response analysis: Modeling true response by isolating the effect of a single action. In: Credit Scoring and Credit Control VI, Edinburgh, Scotland (1999)
24. Radcliffe, N.J., Surry, P.D.: Real-world uplift modelling with significance-based uplift trees. White Paper TR-2011-1, Stochastic Solutions (2011)

25. Rzepakowski, P., Jaroszewicz, S.: Decision trees for uplift modeling with single and multiple treatments. Knowledge and Information Systems 32, 303–327 (2012)
26. Schnitt, S.J.: Local outcomes in ductal carcinoma in situ based on patient and tumor characteristics. J. Natl. Cancer Inst. Monogr. 2010(41), 158–161 (2010)
27. Srinivasan, A.: The Aleph Manual, 4th edn. (2007),
    http://www.comlab.ox.ac.uk/activities/machinelearning/
    Aleph/aleph.html
28. Tabar, L., Tony Chen, H.H., Amy Yen, M.F., Tot, T., Tung, T.H., Chen, L.S., Chiu, Y.H., Duffy, S.W., Smith, R.A.: Mammographic tumor features can predict long-term outcomes reliably in women with 1-14-mm invasive breast carcinoma. Cancer 101(8), 1745–1759 (2004)
29. Thurfjell, M.G., Lindgren, A., Thurfjell, E.: Nonpalpable breast cancer: Mammographic appearance as predictor of histologic type. Radiology 222(1), 165–170 (2002)
30. Tufféry, S.: Data Mining and Statistics for Decision Making, 2nd edn. John Wiley & Sons (2011)
31. Young, J.W.: Differential validity, differential prediction, and college admissions testing: A comprehensive review and analysis. Research Report 2001-6, The College Board, New York (2001)
32. Zelezný, F., Lavrac, N.: Propositionalization-based relational subgroup discovery with RSD. Machine Learning 62(1-2), 33–66 (2006)

A Theoretical Framework for Exploratory Data Mining: Recent Insights and Challenges Ahead

Tijl De Bie and Eirini Spyropoulou

Intelligent Systems Lab, University of Bristol, UK
tijl.debie@gmail.com, enxes@bristol.ac.uk

Abstract. Exploratory Data Mining (EDM), the contemporary heir of Exploratory Data Analysis (EDA) pioneered by Tukey in the seventies, is the task of facilitating the extraction of interesting nuggets of information from possibly large and complexly structured data. Major conceptual challenges in EDM research are the understanding of how one can formalise a nugget of information (given the diversity of types of data of interest), and how one can formalise how interesting such a nugget of information is to a particular user (given the diversity of types of users and intended purposes). In this Nectar paper we briefly survey a number of recent contributions made by us and collaborators towards a theoretically motivated and practically usable resolution of these challenges.

1 Exploratory Data Mining

From the seventies of the previous century, Tukey, Friedman, and collaborators advocated complementing research into statistical tools for *confirmatory* analysis of hypotheses with the development of tools that allow the interactive and *exploratory* analysis of data [24]. The sort of techniques they proposed for this ranged from the very simple (the use of summary statistics for data, and simple visual data summarisation techniques including the box plot as well as now largely obsolete techniques such as the stem-and-leaf plot), to advanced techniques for dimensionality reduction such as projection pursuit and its variants [6,10]. While recognising the development of confirmatory analysis techniques (such as hypothesis tests and confidence intervals, allowing us to infer population properties from a sample) as one of the greatest achievements of the twentieth century, Tukey complained that "Anything to which a confirmatory procedure was not explicitly attached was decried as 'mere descriptive statistics', no matter how much we learned from it."

Since then, data has evolved in size and complexity, and the techniques developed in the past century for EDA are only rarely applicable in their basic unaltered form. Nevertheless, we argue that the problem identified by Tukey is greater than ever. Today's data size and complexity more often than not demand an extensive exploration stage by means of capable and intuitive EDM techniques, before predictive modelling or confirmatory analysis can realistically and usefully be applied.

H. Blockeel et al. (Eds.): ECML PKDD 2013, Part III, LNAI 8190, pp. 612–616, 2013.

There are however a few important research challenges that need resolving before EDM techniques can optimally fulfil this need:

- The concept of a 'nugget of information' found in data needs to be formalised. We will refer to such a nugget of information as a *pattern*.
- To allow for automating the search for interesting patterns in data, the concept of *interestingness* of a pattern needs to be formalised mathematically. Clearly, interestingness is a subjective concept, such that the formalisation must depend on the user's perspective.
- These theoretical insights need to be turned into practical methods and eventually a toolbox for EDM 'in the wild'.

Given the nature of this Nectar paper track, in most of the remainder of this short note we will focus on our own contributions towards the resolution of these challenges. Here we only briefly mention a very incomplete list of works that have influenced our thinking or that have otherwise impacted on EDM research: In 2000 Mannila wrote a highly insightful letter in SIGKDD Explorations about frameworks for data mining [19]; Several prominent researchers advocate data compression as the key operation in the data mining process [5,20]; Recent influential work from Mannila and others on swap randomizations has advocated the use of empirical hypothesis testing in the development of interestingness measures [8,9,18]; The work on tiling databases [7] has been inspirational to our earliest work on this topic. For a more comprehensive overview of data mining interestingness measures based on novelty we refer the reader to our survey paper [14]. Finally, much of our work was also inspired by applied bioinformatics research where exploratory analysis was required, and where we found that current techniques fell short [17,16].

2 Patterns and Their Interestingness

Let us start by clarifying the key terminology. Let Ω be the (measurable) space to which the *data*, denoted as x, is known to belong. We will refer to Ω as the *data domain*. Then, in our work we defined the notion of a *pattern* by means of a subset Ω' of the data domain, saying that a pattern defined by $\Omega' \subseteq \Omega$ is *present* in the data x iff $x \in \Omega'$. This definition is as expressive as it is simple. Most, if not all, types of data mining patterns can be expressed in this way, including the results of frequent pattern miners, dimensionality reduction methods, clustering algorithms, methods for community detection in networks, and more.

The simplicity of this definition further allows us to reason about the interestingness of a pattern in terms of how it affects a user's beliefs about the data. To achieve this, we have opted to represent the beliefs of a user by means of a probability measure P defined over the data domain Ω, to which we refer as the *background distribution*. The interestingness of a pattern is then related to how the background distribution is affected by revealing a pattern to a user, i.e. the degree to which revealing a pattern enhances the user's belief attached to the actual value of the data under investigation.

To do this, several issues need to be studied, such as how to come up with a sensible background distribution without putting too large a burden on the user, how the revealing of a pattern affects the background distribution, how a change in background distribution should be translated into interestingness, and the cost (e.g. in terms of mental energy or processing capacity) presented to a user when processing the revealed pattern.

In answer to these questions, in [2,1,4] we presented formal arguments demonstrating that a robust approach to quantifying interestingness is based on three elements: (1) inferring the background distribution as the one of maximum entropy subject to constraints that formalise the user's prior beliefs about the data; (2) the quantification of the *information content* of the pattern, as minus the logarithm of the probability $P(x \in \Omega')$ under this background distribution that the data belongs to the restricted domain Ω' defined by the pattern; and (3) trading off this information content with the *length of the description* required to communicate the pattern to the user.

Most commonly the purpose of the data miner is to obtain as good an understanding of the data (overall information content of the set of patterns revealed) within specific bounded resource constraints (overall description length of all the patterns revealed). Initially in [2] and later more formally in [1], we argued that this amounts to solving a weighted budgeted set coverage problem. While this problem does not allow for an efficient optimal solution, it can be approximated provably well in a greedy way, iteratively selecting the next best pattern. Hereby, the next best pattern is defined as the one that maximizes the ratio of its information content (given the current background distribution) to its description length. Thus, matching this common usage setting, we proposed to formalize the interestingness of a pattern as the ratio of its information content and its description length, called its *information ratio* (or compression ratio). It represents how densely information is compressed in the description of the pattern.

3   Data and Users in the Real World

Initially we demonstrated our theoretical results on the particular data mining problem of frequent itemset mining [2] for a relatively simple type of prior beliefs (namely the row and column sums), and for a simple type of pattern (namely a tile [7]). In our later work we extended it in the following directions:

- Using more complex types of pattern (in casu noisy tiles) [11] as well as allowing more complex types of prior beliefs to be taken into account on simple binary databases, such as tile densities and itemset frequencies [12].
- Expanding these ideas toward real-valued data, for local pattern types [15] as well as global clustering pattern types [3,13].
- The development of a new expressive pattern syntax for multi-relational data with binary and n-ary relationships, the formalisation of subjective interestingness for a certain type of prior information, and the development of efficient algorithms to mine these patterns [21,22,23].

4   An Encompassing Toolbox for Exploratory Data Mining?

We believe there is significant value to be gained by further expanding these theoretical insights as well as the practical instantiations of the framework. We hope and anticipate that this may ultimately result in a modular and expandable toolbox for EDM that can be applied to data as it presents itself in real-life, and that is effectively usable by experts and lay users alike.

Most real-world structured data is multi-relational data in some way, including simple binary and attribute-value tables, traditional (relational) databases, (annotated) graphs, as well as RDF data and the semantic web. We therefore believe that a general EDM toolbox could most easily be built upon our recent work on multi-relational data mining. In this work we developed a new pattern syntax for multi-relational data with categorical attribute values, an associated interestingness measure along the lines of the advocated framework (demonstrated for a simple but important type of prior beliefs), as well as efficient mining algorithms [21,22,23].

Of course, in order to mature into a fully fledged EDM toolbox, this starting point requires a number of advances. Some of these, however, we have already partially developed for simpler types of data. For example, the resulting toolbox will need to be able to deal with real-valued data, which requires the definition of a new multi-relational pattern syntax and the adaptation of the prior belief types for real-valued data developed in [15,3,13] to the multi-relational case. Another required step will be the incorporation of more complex types of prior information also for categorical data, along the lines of our previous work on single-relational data [12].

Acknowledgements. Most importantly we are grateful to Kleanthis-Nikolaos Kontonasios, as well as to Jilles Vreeken, Mario Boley, and Matthijs van Leeuwen, all of whom have made important contributions to the development of the vision advocated in this Nectar paper. This work was partially funded by EPSRC grant EP/G056447/1.

References

1. De Bie, T.: An information-theoretic framework for data mining. In: Proc. of the 17th ACM SIGKDD International Conference on Knowledge Discovery and Data Mining, KDD 2011 (2011)
2. De Bie, T.: Maximum entropy models and subjective interestingness: an application to tiles in binary databases. Data Mining and Knowledge Discovery 23(3), 407–446 (2011)
3. De Bie, T.: Subjectively interesting alternative clusters. In: Proceedings of the 2nd MultiClust Workshop: Discovering, Summarizing and Using Multiple Clusterings (2011)
4. De Bie, T., Kontonasios, K.-N., Spyropoulou, E.: A framework for mining interesting pattern sets. SIGKDD Explorations 12(2) (December 2010)

5. Faloutsos, C., Megalooikonomou, V.: On data mining, compression, and kolmogorov complexity. Data Mining and Knowledge Discovery 15, 3–20 (2007)
6. Friedman, J., Tukey, J.: A projection pursuit algorithm for exploratory data analysis. IEEE Transactions on Computers 100(9), 881–890 (1974)
7. Geerts, F., Goethals, B., Mielikäinen, T.: Tiling databases. In: Suzuki, E., Arikawa, S. (eds.) DS 2004. LNCS (LNAI), vol. 3245, pp. 278–289. Springer, Heidelberg (2004)
8. Gionis, A., Mannila, H., Mielikäinen, T., Tsaparas, P.: Assessing data mining results via swap randomization. ACM Transactions on Knowledge Discovery from Data 1(3), 14 (2007)
9. Hanhijarvi, S., Ojala, M., Vuokko, N., Puolamäki, K., Tatti, N., Mannila, H.: Tell me something I don't know: Randomization strategies for iterative data mining. In: Proc. of the 15th ACM SIGKDD International Conference on Knowledge Discovery and Data Mining (KDD 2009), pp. 379–388 (2009)
10. Huber, P.: Projection pursuit. The annals of Statistics, 435–475 (1985)
11. Kontonasios, K.-N., De Bie, T.: An information-theoretic approach to finding informative noisy tiles in binary databases. In: Proceedings of the 2010 SIAM International Conference on Data Mining (2010)
12. Kontonasios, K.-N., De Bie, T.: Formalizing complex prior information to quantify subjective interestingness of frequent pattern sets. In: Proc. of the 11th International Symposium on Intelligent Data Analysis, IDA (2012)
13. Kontonasios, K.-N., De Bie, T.: Subjectively interesting alternative clusterings. Machine Learning (2013)
14. Kontonasios, K.-N., Spyropoulou, E., De Bie, T.: Knowledge discovery interestingness measures based on unexpectedness. WIREs Data Mining and Knowledge Discovery 2(5), 386–399 (2012)
15. Kontonasios, K.-N., Vreeken, J., De Bie, T.: Maximum entropy modelling for assessing results on real-valued data. In: Proceedings of the IEEE International Conference on Data Mining, ICDM (2011)
16. Lemmens, K., De Bie, T., Dhollander, T., Keersmaecker, S.D., Thijs, I., Schoofs, G., De Weerdt, A., De Moor, B., Vanderleyden, J., Collado-Vides, J., Engelen, K., Marchal, K.: DISTILLER: a data integration framework to reveal condition dependency of complex regulons in escherichia coli. Genome Biology 10(R27) (2009)
17. Lemmens, K., Dhollander, T., De Bie, T., Monsieurs, P., Engelen, K., Winderickx, J., De Moor, B., Marchal, K.: Inferring transcriptional module networks from ChIP-chip-, motif- and microarray data. Genome Biology 7(R37) (2006)
18. Lijffijt, J., Papapetrou, P., Puolamki, K.: A statistical significance testing approach to mining the most informative set of patterns. In: Data Mining and Knowledge Discovery (December 2012)
19. Mannila, H.: Theoretical frameworks for data mining. SIGKDD Explorations (2000)
20. Siebes, A., Vreeken, J., van Leeuwen, M.: Item sets that compress. In: SIAM Conference on Data Mining (2006)
21. Spyropoulou, E., De Bie, T.: Interesting multi-relational patterns. In: Proceedings of the IEEE International Conference on Data Mining, ICDM (2011)
22. Spyropoulou, E., De Bie, T., Boley, M.: Mining interesting patterns in multi-relational data. In: Data Min. Knowl. Discov. (2013)
23. Spyropoulou, E., De Bie, T., Boley, M.: Mining interesting patterns in multi-relational data with n-ary relationships. In: Proceedings of the International Conference on Discovery Science, DS (2013)
24. Tukey, J.: Exploratory data analysis, Reading, MA, vol. 231 (1977)

Tensor Factorization
for Multi-relational Learning

Maximilian Nickel[1] and Volker Tresp[2]

[1] Ludwig Maximilian University, Oettingenstr. 67, Munich, Germany
nickel@dbs.ifi.lmu.de
[2] Siemens AG, Corporate Technology, Otto-Hahn-Ring 6, Munich, Germany
volker.tresp@siemens.com

Abstract. Tensor factorization has emerged as a promising approach for solving relational learning tasks. Here we review recent results on a particular tensor factorization approach, i.e. RESCAL, which has demonstrated state-of-the-art relational learning results, while scaling to knowledge bases with millions of entities and billions of known facts.

1 Introduction

Exploiting the information contained in the relationships between entities has been essential for solving a number of important machine learning tasks. For instance, social network analysis, bioinformatics, and artificial intelligence all make extensive use of relational information, as do large knowledge bases such as Google's Knowledge Graph or the Semantic Web. It is well-known that, in these and similar domains, statistical relational learning (SRL) can improve learning results significantly over non-relational methods. However, despite the success of SRL in specific applications, wider adoption has been hindered by multiple factors: without extensive prior knowledge about a domain, existing SRL methods often have to resort to structure learning for their functioning; a process that is both time consuming and error prone. Moreover, inference is often based on methods such as MCMC and variational inference which introduce additional scalability issues. Recently, tensor factorization has been explored as an approach that overcomes some of these problems and that leads to highly scalable solutions. Tensor factorizations realize multi-linear latent factor models and contain commonly used matrix factorizations as the special case of bilinear models. We will discuss tensor factorization for relational learning by the means of RESCAL [6,7,5], which is based on the factorization of a third-order tensor and which has shown excellent learning results; outperforming state-of-the-art SRL methods and related tensor-based approaches on benchmark data sets. Moreover, RESCAL is highly scalable such that large knowledge bases can be factorized, which is currently out of scope for most SRL methods. In our review of this model, we will also exemplify the general benefits of tensor factorization for relational learning, as considered recently in approaches like [10,8,1,4,2]. In the following, we will mostly follow the notation outlined in [3]. We will also assume that all relationships are of dyadic form.

H. Blockeel et al. (Eds.): ECML PKDD 2013, Part III, LNAI 8190, pp. 617–621, 2013.
© Springer-Verlag Berlin Heidelberg 2013

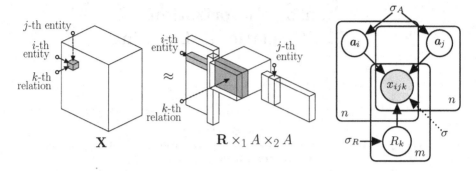

Fig. 1. Factorization of an adjacency tensor **X** using the RESCAL model

Fig. 2. Graphical plate model of RESCAL

2   Relational Learning via Tensor Factorization

Dyadic relational data has a natural representation as an adjacency tensor $\mathbf{X} \in \mathbb{R}^{n \times n \times m}$ whose entries x_{ijk} correspond to all possible relationships between n entities over m different relations. The entries of **X** are set to

$$x_{ijk} = \begin{cases} 1, & \text{if the relationship } relation_k(entity_i, entity_j) \text{ exists} \\ 0, & \text{otherwise.} \end{cases}$$

RESCAL [6] is a latent factor model for relational learning, which factorizes an adjacency tensor **X** into a core tensor $\mathbf{R} \in \mathbb{R}^{r \times r \times m}$ and a factor matrix $A \in \mathbb{R}^{n \times r}$ such that

$$\mathbf{X} \approx \mathbf{R} \times_1 A \times_2 A. \tag{1}$$

Equation (1) can be equivalently specified as $x_{ijk} \approx \boldsymbol{a}_i^T R_k \boldsymbol{a}_j$, where the column vector $\boldsymbol{a}_i \in \mathbb{R}^r$ denotes the i-th row of A and the matrix $R_k \in \mathbb{R}^{r \times r}$ denotes the k-th frontal slice of **R**. Consequently, $\boldsymbol{a}_i$ corresponds to the latent representation of $entity_i$, while R_k models the interactions of the latent variables for $relation_k$. The dimensionality r of the latent space A is a user-given parameter which specifies the complexity of the model. The symbol "$\approx$" denotes the approximation under a given loss function. Figure 1 shows a visualization of the factorization. Probabilistically, eq. (1) can be interpreted as estimating the joint distribution over *all possible* relationships

$$P(X|A, \mathbf{R}) = \prod_{i=1}^{n} \prod_{j=1}^{n} \prod_{k=1}^{m} P(x_{ijk} \mid \boldsymbol{a}_i^T R_k \boldsymbol{a}_j). \tag{2}$$

Hence, a RESCAL factorization of an adjacency tensor **X** computes a complete model of a relational domain where the state of a relationship x_{ijk} depends on the matrix-vector product $\boldsymbol{a}_i^T R_k \boldsymbol{a}_j$. Here, a Gaussian likelihood model would imply a least squares loss function, while a Bernoulli likelihood model would imply a logistic loss function [6,5]. To model attributes of entities efficiently, coupled tensor factorization can be employed [7,11], where simultaneously to eq. (1) an attribute matrix $F \in \mathbb{R}^{n \times \ell}$ is factorized such that $F \approx AW$ and where the latent factor A is shared between the factorization of $\mathbf{X}$ and F. RESCAL and other tensor factorizations feature a number of important properties that can be exploited for tasks like link prediction, entity resolution or link-based clustering:

Efficient Inference. The latent variable structure of RESCAL decouples inference such that global dependencies are captured during learning, whereas prediction relies only on a typically small number of latent variables. It can be seen from eq. (2) that a variable x_{ijk} is conditionally independent from all other variables given the expression $a_i^T R_k a_j$. The computational complexity of these matrix-vector multiplications depends only on the dimensionality of the latent space A, what enables, for instance, fast query answering on knowledge bases. It is important to note that this locality of computation does not imply that the likelihood of a relationship is only influenced by local information. On the contrary, the conditional independence assumptions depicted in fig. 2 show that information is propagated globally when computing the factorization. Due to the colliders in fig. 2, latent variables (a_i, a_j, R_k) are not d-separated from other variables such that they are possibly dependent on all remaining variables. Therefore, as the variable x_{ijk} depends on its associated latent variables $\{a_i, a_j, R_k\}$, it depends *indirectly* on the state of any other variable such that global dependencies between relationships can be captured. Similar arguments apply to tensor factorizations such as the TUCKER decomposition and CP, which explains the strong relational learning results of RESCAL and CP compared to state-of-the-art methods such as MLN or IRM [6,7,2,5].

Unique Representation. A distinctive feature of RESCAL is the unique representation of entities via the latent space A. Standard tensor factorization models such as CP and TUCKER compute a bipartite model of relational data, meaning that entities have different latent representations as subjects or objects in a relationship. For instance, a TUCKER-2 model would factorize the frontal slices of an adjacency tensor $\mathbf{X}$ as $X_k \approx AR_k B^T$ such that entities are represented as subjects via the latent factor A and as objects via the latent factor B. However, relations are usually not bipartite and in these cases a bipartite modeling would effectively break the flow of information from subjects to objects, as it does not account for the fact that the latent variables a_i and b_i refer to the identical entity. In contrast, RESCAL uses a unique latent representation a_i for each entity in the data set, what enables efficient information propagation via the dependency structure shown in fig. 2 and what has been demonstrated to be critical for capturing correlations over relational chains. For instance, consider the task to predict the party membership of presidents of the United States of America. When the party membership of a president's vice president is known, this can be done with high accuracy, as both persons have usually been members of the same party, meaning that the formula *vicePresident*$(x, y) \wedge party(y, z) \Rightarrow party(x, z)$ holds with high probability. For this and similar examples, it has been shown that bipartite models such as CP and TUCKER fail to capture the necessary correlations, as, for instance, the object representation b_y does not reflect that person y is in a relationship to party z as a subject. RESCAL, on the other hand, is able to propagate the required information, e.g. the party membership of y, via the unique latent representations of the involved entities [6,7].

Latent Representation. In relational data, the similarity of entities is determined by the similarity of their relationships, following the intuition that "if two objects are in the same relation to the same object, this is evidence that they may be the same object" [9]. This notion of similarity is reflected in RESCAL via the latent space A. For the i-th entity, all possible occurrences as a subject are grouped in the slice $X_{i,:,:}$ of an adjacency tensor, while all possible occurrences as an object are grouped in the slice $X_{:,i,:}$ (see figs. 3 and 4). According to the RESCAL model, these slices are computed by $\text{vec}(X_{i,:,:}) \approx a_i R_{(1)}(I \otimes A)^T$ and $\text{vec}(X_{:,i,:}) \approx a_i R_{(2)}(I \otimes A)^T$. Since the terms $R_{(1)}(I \otimes A)^T$ and $R_{(2)}(I \otimes A)^T$ are constant for different values of i, it is sufficient to consider only the similarity of a_p and a_q to determine the *relational* similarity of $entity_p$ and $entity_q$. As this measure of similarity is based on the latent representations of entities, it is not only based on counting identical relationships of identical entities, but it also considers the similarity of the entities that are involved in a relationship. The previous intuition could therefore be restated as *if two objects are in similar relations to similar objects, this is evidence that they may be the same object.* Latent representations of entities have been exploited very successfully for entity resolution and also enabled large-scale hierarchical clustering on relational data [6,7]. Moreover, since the matrix A is a vector space representation of entities, non-relational machine learning algorithms such as k-means or kernel methods can be conveniently applied to any of these tasks.

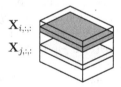
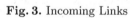

Fig. 3. Incoming Links   **Fig. 4.** Outgoing Links

High Scalability. The scalability of algorithms has become of utmost importance as relational data is generated in an unprecedented amount and the size of knowledge bases grows rapidly. RESCAL-ALS is a highly scalable algorithm to compute the RESCAL model under a least-squares loss. It has been shown that it can efficiently exploit the sparsity of relational data as well as the structure of the factorization such that it features *linear* runtime complexity with regard to the number of entities n, the number of relations m, and the number of known relationships $\text{nnz}(X)$, while being cubic in the model complexity r. This property allowed, for instance, to predict various high-level classes of entities in the YAGO 2 ontology, which consists of over three million entities, over 80 relations or attributes, and over 33 million existing relationships, by computing low-rank factorizations of its adjacency tensor on a single desktop computer [7].

3 Conclusion and Outlook

RESCAL has shown state-of-the-art relational learning results, while scaling up to the size of complete knowledge bases. Due to its latent variable structure, RESCAL does not require deep domain knowledge and therefore can be easily applied to most domains. Its latent representation of entities enables the application of non-relational algorithms to relational data for a wide range of tasks such as cluster analysis or entity resolution. RESCAL is applicable if latent factors are suitable for capturing the essential information in a domain. In ongoing research, we explore situations where plain latent factor models are not a very efficient approach to relational learning and examine how to overcome the underlying causes of these situations.

References

1. Bordes, A., Weston, J., Collobert, R., Bengio, Y.: Learning structured embeddings of knowledge bases. In: Proc. of the 25th Conference on Artificial Intelligence, SF, USA (2011)
2. Jenatton, R., Le Roux, N., Bordes, A., Obozinski, G.: A latent factor model for highly multi-relational data. In: Advances in Neural Information Processing Systems, vol. 25 (2012)
3. Kolda, T.G., Bader, B.W.: Tensor decompositions and applications. SIAM Review 51(3), 455–500 (2009)
4. Kolda, T.G., Bader, B.W., Kenny, J.P.: Higher-order web link analysis using multilinear algebra. In: Proc. of the Fifth International Conference on Data Mining, pp. 242–249 (2005)
5. Nickel, M., Tresp, V.: Logistic tensor-factorization for multi-relational data. In: ICML Workshop - Structured Learning: Inferring Graphs from Structured and Unstructured Inputs. Atlanta, GA, USA (2013)
6. Nickel, M., Tresp, V., Kriegel, H.P.: A three-way model for collective learning on multi-relational data. In: Proc. of the 28th International Conference on Machine Learning. pp. 809—816. Bellevue, WA, USA (2011)
7. Nickel, M., Tresp, V., Kriegel, H.P.: Factorizing YAGO: scalable machine learning for linked data. In: Proc. of the 21st International World Wide Web Conference, Lyon, France (2012)
8. Rendle, S., Freudenthaler, C., Schmidt-Thieme, L.: Factorizing personalized markov chains for next-basket recommendation. In: Proc. of the 19th International Conference on World Wide Web, pp. 811–820 (2010)
9. Singla, P., Domingos, P.: Entity resolution with markov logic. In: Proc. of the Sixth International Conference on Data Mining, Washington, DC, USA, pp. 572–582 (2006)
10. Sutskever, I., Salakhutdinov, R., Tenenbaum, J.B.: Modelling relational data using bayesian clustered tensor factorization. In: Advances in Neural Information Processing Systems 22 (2009)
11. Ylmaz, Y.K., Cemgil, A.T., Simsekli, U.: Generalised coupled tensor factorisation. In: Advances in Neural Information Processing Systems (2011)

MONIC and Followups on Modeling
and Monitoring Cluster Transitions

Myra Spiliopoulou[1,*], Eirini Ntoutsi[2], Yannis Theodoridis[3], and Rene Schult[4]

[1] Otto-von-Guericke University of Magdeburg, Germany
myra@iti.cs.uni-magdeburg.de
[2] Ludwig-Maximilians-University of Munich, Germany (work done while with[3])
ntoutsi@dbs.ifi.lmu.de
[3] Department of Informatics, University of Piraeus, Greece
ytheod@unipi.gr
[4] Mercateo AG, Germany (work done while with[1])
rene.schult@mercateo.com

Abstract. There is much recent discussion on data streams and big data, which except of their volume and velocity are also characterized by volatility. Next to detecting change, it is also important to interpret it. Consider customer profiling as an example: Is a cluster corresponding to a group of customers simply disappearing or are its members migrating to other clusters? Does a new cluster reflect a new type of customers or does it rather consist of old customers whose preferences shift? To answer such questions, we have proposed the framework MONIC [20] for modeling and tracking cluster transitions. MONIC has been re-discovered some years after publication and is enjoying a large citation record from papers on community evolution, cluster evolution, change prediction and topic evolution.

Keywords: cluster monitoring, change detection, dynamic data, data streams, big data.

1   Motivation

MONIC stands for MONItoring Clusters. It has appeared in [20] with following motivation: "In recent years, it has been recognized that the clusters discovered in many real applications are affected by changes in the underlying population. Much of the research in this area has focused on the *adaptation* of the clusters, so that they always reflect the current state of the population. Lately, research has expanded to encompass *tracing and understanding* of the changes themselves, as means of gaining valuable insights on the population and supporting strategic decisions. For example, consider a business analyst who studies customer profiles. Understanding how such profiles change over time would allow for a long-term proactive portfolio design instead of reactive portfolio adaptation."

Seven years later, this motivation holds unchanged and the need for evolution monitoring is exarcebated through the proliferation of big data. Next to

---

* Work on model monitoring (2013) partially supported by the German Research Foundation, SP 572/11-1 *IMPRINT: Incremental Mining for Perennial Objects*.

H. Blockeel et al. (Eds.): ECML PKDD 2013, Part III, LNAI 8190, pp. 622–626, 2013.

"Volume" and "Velocity", "Volatility" is a core characteristic of big data. Data mining methods should not only adapt to change but also describe change.

Citations to MONIC (according to Google scholar) follow a rather unusual distribution. There has been one citation to the article in 2006 and 11 in 2007. From 2008 on though, 15-30 new citations are added every year, reaching 132 in 2012 and achieving 141 in 2013. This means that the article has been re-discovered in 2008, reaching a peak in popularity increase in 2011 (31 citations) and remaining stable thereafter. We associate these values with the intensive investigation of the role of time in data mining and the study of drift in stream mining. The topics associated with MONIC are multifaceted. MONIC is cited by papers on community evolution, on evolution of topics in text streams, on cluster evolution in general (for different families of cluster algorithms) and on frameworks for change modeling and change prediction. An important followup in terms of specifying a research agenda is the work of Boettcher et al [3] on "exploiting the power of time in data mining".

2   MONIC at a Glance

MONIC models and traces the transitions of clusters built upon an accumulating dataset. The data are clustered at discrete timepoints, cluster transition between consecutive time points are detected and "projected" upon the whole stream so as to draw conclusions regarding the evolution of the underlying population.

To build a monitoring framework for streaming data, the following challenges should be addressed: (a) what is a cluster? (b) how do we find the "same" cluster at a later timepoint? (c) what transitions can a cluster experience and (d) how to detect these transitions?

Regarding challenge (a), in MONIC a cluster is described by the set of points it contains. MONIC is thus not restricted to a specific cluster definition and can be used equally with partitioning, density-based and hierarchical clustering methods. More importantly, this cluster description allows for different forgetting strategies over a data stream, and even allows for changes in the feature space. Feature space evolution may occur in document streams, where new features/words may show up. Hence, MONIC is flexible enough for change monitoring on different clustering algorithms with various data forgetting models.

Challenge (b), i.e. tracing a cluster at a later timepoint is solved by computing the overlap between the original cluster and the clusters built at the later timepoint. The overlap is a set intersection, except of considering only those cluster members that exist at both timepoints. Hence, the impact of data decay on cluster evolution is accounted for.

Cluster evolution, challenge (c), refers to the transitions that may be observed in a cluster's "lifetime". The simplest transitions are disappearance of an existing cluster, and the emerging of a new one. MONIC proposes a typification of cluster transitions, distinguishing between *internal transitions* that affect the cluster itself and *external transitions* that concern the relationship of a cluster to other clusters. Cluster merging, the absorption of a cluster by another, the split of a cluster into more clusters, as well as cluster survival, appearance and disappearance are external transitions. An internal transition is a change in the cluster's properties, such as its cardinality (shrink, expand), compactness or location (repositioning of its center, change in the descriptors of its distribution).

Next to specifying a list of transitions, there is need for a mechanism detecting them. Challenge (d) is dealt through *transition indicators*. These are heuristics which may be tailored to a specific algorithm, e.g. by tracing movements of a cluster's centroid, or algorithm-independent, e.g. by concentrating only on the cardinality of the cluster and the similarity among its members. The transition indicators are incorporated in a transition detection algorithm that takes as input the clusterings between two consecutive timepoints and outputs the experienced cluster transitions. The algorithm first detects external transitions corresponding to "cluster population movements" between consecutive timepoints and for clusters that "survive" at a later timepoint, internal transitions are detected. Based on the detected transitions, temporal properties of clusters and clusterings, such as lifetime and stability, are derived and conclusions about the evolution of the underlying population are drawn.

The low complexity of the algorithm, $\mathcal{O}(K^2)$, where K is the maximum number of clusters in the compared clusterings, makes MONIC applicable to real big volatile data applications. The memory consumption is also low, only the clusters at the compared timepoints are required as input to the clustering algorithm, whereas any previous clustering results is removed.

In the original paper, we investigated evolution over a document collection, namely ACM publications on Database Applications (archive H2.8) from 1998 to 2004 and studied particularly the dominant themes in the most prominent subarea of database applications, namely "data mining". In [19,14], we extended MONIC into MONIC +, which in contrast to MONIC that is a generic cluster transition modeling and detection framework, encompasses a typification of clusters and cluster-type-specific transition indicators, by exploiting cluster topology and cluster statistics for transition detection. Transition specifications and transition indicators form the basis for monitoring clusters over time. This includes detecting and studying their changes, summarizing them in an *evolution graph*, as done in [16,15], or predicting change, as done in [17].

3   MONIC and Beyond

MONIC has been considered in a variety of different areas and applications. Rather than elaborating on each paper, we explain very briefly the role of evolution in each area, picking one or two papers as examples.

Evolution in social networks: There are two aspects of evolution in social networks [18], best contrasted by the approaches [18,21,4]. In [21] the problem of community evolution is investigated from the perspective of migration of individuals. They observe communities as stable formations and assume that an individuum is mostly observed with other members of the own community and rarely with members of other communities, i.e. movement from one community/cluster to another is rare. This can be contrasted to the concept of community evolution, as investigated in [4], where it is asserted that the individuals define the clusters, and hence the clusters may change as individuals migrate. The two aspects of evolution are complementary: the one aspect concentrates on the clusters as stable formations, the other on the clusters as the result of individuals' movements.

Frameworks for change prediction & stream mining: Due to the high volatile nature of modern data, several generic works for change detection or incorporation of time in the mining process have been proposed. In [7] the problem of temporal relationship between clusters is studied and the TRACDS framework is proposed which "adds" to the stream clustering algorithms the temporal ordering information in the form of a dynamically changing Marcov Chain. In [6] the problem of cluster tracing in high dimensional feature spaces is considered. Subspace clustering is applied instead of full dimensional clustering thus associating each cluster with a certain set of dimensions. In [17] the MEC framework for cluster transition detection and visualization of the monitoring process is proposed. Other citations in this area include [2,16,15].

Spatiotemporal data: With the wide spread usage of location aware devices and applications, analyzing movement data and detecting trends is an important task nowadays. The problem of convoy discovery in spatiotemporal data is studied in [10], a convoy being a group of objects that have traveled together for some time. The discovery of flock patterns is considered in [22], defined as all groups of trajectories that stay "together" for the duration of a given time interval. Continuous clustering of moving objects is studied in [9]. Other citations in this area include [1,13].

Topic evolution: Topic monitoring is a necessity nowadays due to the continuous stream of published documents. In [11] a topic evolution graph is constructed by identifying topics as significant changes in the content evolution and connecting each topic with the previous topics that provided the context. [8] studies how topics in scientific literature evolve by using except for the words in the documents the impact of one document to the other in terms of citations. In [23] a method is proposed for detecting, tracking and updating large and small bursts in a stream of news. Recently [24], the method has been coupled with classifier counterparts for each topic in order to study dynamics of product features and their associated sentiment based on customer reviews. Other citations are [5,12].

4   Tracing Evolution Today

Scholars become increasingly aware on the importance of understanding evolution. This is reflected in the increasing number of citations on MONIC and in the diversity of the areas it is cited from. Modeling evolution, summarizing it and predicting it are cornerstone subjects in learning from data. Big data, characterized by volatility, will contribute further to the trend.

References

1. Aung, H.H., Tan, K.-L.: Discovery of evolving convoys. In: SSDBM, pp. 196–213 (2010)
2. Bifet, A., Holmes, G., Pfahringer, B., Kranen, P., Kremer, H., Jansen, T., Seidl, T.: Moa: Massive online analysis, a framework for stream classification and clustering. Journal of Machine Learning Research - Proceedings Track 11, 44–50 (2010)

3. Böttcher, M., Höppner, F., Spiliopoulou, M.: On exploiting the power of time in data mining. SIGKDD Explorations 10(2), 3–11 (2008)
4. Falkowski, T., Bartelheimer, J., Spiliopoulou, M.: Mining and visualizing the evolution of subgroups in social networks. In: Web Intelligence, pp. 52–58 (2006)
5. Gohr, A., Hinneburg, A., Schult, R., Spiliopoulou, M.: Topic evolution in a stream of documents. In: SDM (2009)
6. Günnemann, S., Kremer, H., Laufkötter, C., Seidl, T.: Tracing evolving clusters by subspace and value similarity. In: Huang, J.Z., Cao, L., Srivastava, J. (eds.) PAKDD 2011, Part II. LNCS, vol. 6635, pp. 444–456. Springer, Heidelberg (2011)
7. Hahsler, M., Dunham, M.H.: Temporal structure learning for clustering massive data streams in real-time. In: SDM, pp. 664–675 (2011)
8. He, Q., Chen, B., Pei, J., Qiu, B., Mitra, P., Giles, C.L.: Detecting topic evolution in scientific literature: how can citations help? In: CIKM, pp. 957–966 (2009)
9. Jensen, C.S., Lin, D., Ooi, B.C.: Continuous clustering of moving objects. TKDE 19(9), 1161–1174 (2007)
10. Jeung, H., Yiu, M.L., Zhou, X., Jensen, C.S., Shen, H.T.: Discovery of convoys in trajectory databases. CoRR, abs/1002.0963 (2010)
11. Jo, Y., Hopcroft, J.E., Lagoze, C.: The web of topics: discovering the topology of topic evolution in a corpus. In: WWW, pp. 257–266 (2011)
12. Lauschke, C., Ntoutsi, E.: Monitoring user evolution in twitter. In: BASNA Workshop, co-located with ASONAM (2012)
13. Ntoutsi, E., Mitsou, N., Marketos, G.: Traffic mining in a road-network: How does the traffic flow? IJBIDM 3(1), 82–98 (2008)
14. Ntoutsi, E., Spiliopoulou, M., Theodoridis, Y.: Tracing cluster transitions for different cluster types. Control & Cybernetics Journal 38(1), 239–260 (2009)
15. Ntoutsi, I., Spiliopoulou, M., Theodoridis, Y.: Summarizing cluster evolution in dynamic environments. In: Murgante, B., Gervasi, O., Iglesias, A., Taniar, D., Apduhan, B.O. (eds.) ICCSA 2011, Part II. LNCS, vol. 6783, pp. 562–577. Springer, Heidelberg (2011)
16. Ntoutsi, E., Spiliopoulou, M., Theodoridis, Y.: FINGERPRINT summarizing cluster evolution in dynamic environments. IJDWM (2012)
17. Oliveira, M.D.B., Gama, J.: A framework to monitor clusters evolution applied to economy and finance problems. Intell. Data Anal. 16(1), 93–111 (2012)
18. Spiliopoulou, M.: Evolution in social networks: A survey. In: Aggarwal, C.C. (ed.) Social Network Data Analytics, pp. 149–175. Springer (2011)
19. Spiliopoulou, M., Ntoutsi, E., Theodoridis, Y.: Tracing cluster transitions for different cluster types. In: ADMKD Workshop, co-located with ADBIS (2007)
20. Spiliopoulou, M., Ntoutsi, E., Theodoridis, Y., Schult, R.: MONIC – modeling and monitoring cluster transitions. In: KDD, pp. 706–711 (2006)
21. Tantipathananandh, C., Berger-Wolf, T.Y.: Finding communities in dynamic social networks. In: ICDM, pp. 1236–1241 (2011)
22. Vieira, M.R., Bakalov, P., Tsotras, V.J.: On-line discovery of flock patterns in spatio-temporal data. In: GIS, pp. 286–295 (2009)
23. Zimmermann, M., Ntoutsi, E., Siddiqui, Z.F., Spiliopoulou, M., Kriegel, H.-P.: Discovering global and local bursts in a stream of news. In: SAC, pp. 807–812 (2012)
24. Zimmermann, M., Ntoutsi, E., Spiliopoulou, M.: Extracting opinionated (sub)features from a stream of product reviews. In: DS (2013)

Towards Robot Skill Learning:
From Simple Skills to Table Tennis

Jan Peters[1,2], Jens Kober[1,2], Katharina Mülling[1,2], Oliver Krömer[1,2],
and Gerhard Neumann[2]

[1] Technische Universität Darmstadt, 64293 Darmstadt, Germany
mail@jan-peters.net, muelling@tuebingen.mpg.de, je.kober@gmail.com,
oli@robot-learning.de
[2] Max Planck Institute for Intelligent Systems, 72076 Tübingen, Germany
neumann@ias.tu-darmstadt.de

Abstract. Learning robots that can acquire new motor skills and refine existing ones have been a long-standing vision of both robotics, and
machine learning. However, off-the-shelf machine learning appears not
to be adequate for robot skill learning, as it neither scales to anthropomorphic robotics nor do fulfills the crucial real-time requirements. As
an alternative, we propose to divide the generic skill learning problem
into parts that can be well-understood from a robotics point of view. In
this context, we have developed machine learning methods applicable to
robot skill learning. This paper discusses recent progress ranging from
simple skill learning problems to a game of robot table tennis.

1 Introduction

Despite the many impressive motor skills exhibited by anthropomorphic robots,
the generation of motor behaviors has changed little since classical robotics. The
roboticist models the task dynamics as accurately as possible while using human
insight to create the desired robot behavior, as well as to eliminate all uncertain-
ties of the environment. In most cases, such a process boils down to recording a
desired trajectory in a pre-structured environment with precisely placed objects.
Such highly engineered approaches are feasible in highly structured industrial or
research environments. However, for robots to leave factory floors and research
environments, the strong reliance on hand-crafted models of the environment
and the robots needs to be reduced. Instead, a general framework is needed
for allowing robots to learn their tasks with minimal programming and in less
structured, uncertain environments. Such an approach clearly has to be based on
machine learning *combined* with robotics insights to make the high-dimensional
domain of anthropomorphic robots accessible. To accomplish this aim, three
major questions need to be addressed:

1. How can we develop efficient motor learning methods?
2. How can anthropomorphic robots learn basic skills similar to humans?
3. Can complex skills be composed with these elements?

H. Blockeel et al. (Eds.): ECML PKDD 2013, Part III, LNAI 8190, pp. 627–631, 2013.
© Springer-Verlag Berlin Heidelberg 2013

In the next sections, we will address these questions. We focus on model-free methods, which do not maintain an internal behavior simulator (i.e., a forward model) but operate directly on the data. Note that most methods transfer straightforwardly to model-based approaches.

2   Motor Learning Methods

After formalizing the necessary assumptions on robotics from a machine learning perspective, we show the concepts behind the our learning methods.

2.1   Modeling Assumptions

For addressing these questions, we focus on anthropomorphic robot systems which always are in a state $\mathbf{x} \in \mathfrak{R}^n$ that includes both the internal state of the robot (e.g., joint angles, velocities, acceleration in Fig. 1, but also internal variables) as well as external state variables (e.g., ball position and velocity), and execute motor commands $\mathbf{u} \in \mathfrak{R}^m$ at a high frequency (usually 500–1000Hz). The actions are taken in accordance to a parametrized, stationary, stochastic policy, i.e., a set of rules with exploration $\mathbf{u} \sim \pi_\theta(\mathbf{u}|\mathbf{x}) = p(\mathbf{u}|\mathbf{x}, \boldsymbol{\theta})$ where the parameters $\boldsymbol{\theta} \in \mathfrak{R}^N$ allow for learning. The

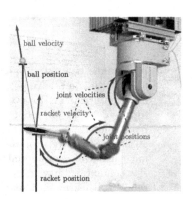

Fig. 1. Modeling of the learning task of paddling a ball

stochasticity in the policy allows capturing the variance of the teacher, can ease algorithm design, and there exist well-known problems where the optimal stationary policy is stochastic. Frequently used policies are linear in state feature $\phi(\mathbf{x})$ and have Gaussian exploration, i.e., $\pi_\theta(\mathbf{u}|\mathbf{x}) = \mathcal{N}(\mathbf{u}|\phi^T(\mathbf{x})\boldsymbol{\theta}, \sigma^2)$. After every motor command, the system transfers to a next state $\mathbf{x}' \sim p(\mathbf{x}'|\mathbf{x}, \mathbf{u})$, and receives a learning signal $r(\mathbf{x}, \mathbf{u})$. The learning signal can be a general reward (i.e., in full reinforcement learning), but can also contain substantially more structure (e.g., prediction errors in model learning or proximity to a demonstration in imitation), see [1].

During experiments, the system obtains a stream of data consisting of episodes $\boldsymbol{\tau} = [\mathbf{x}_1, \mathbf{u}_1, \mathbf{x}_2, \mathbf{u}_2, \ldots, \mathbf{x}_{T-1}, \mathbf{u}_{T-1}, \mathbf{x}_T]$ of length T, often also called trajectories or paths. These paths are obviously distributed according to

$$p_\theta(\boldsymbol{\tau}) = p(\mathbf{x}_1)\prod_{t=1}^{T} p(\mathbf{x}_{t+1}|\mathbf{x}_t, \mathbf{u}_t)\pi_\theta(\mathbf{u}|\mathbf{x}), \tag{1}$$

where $p(\mathbf{x}_1)$ denotes the start-state distribution. We will refer to the distribution of teacher's demonstrations or past data $p(\boldsymbol{\tau})$ by simply omitting $\boldsymbol{\theta}$. The rewards of a path can be formulated as a weighted sum of immediate rewards $r(\boldsymbol{\tau}) = \sum_{t=1}^{T} a_t r(\mathbf{x}_t, \mathbf{u}_t)$. Most motor skill learning problems can be phrased as optimizing the expected returns $J(\boldsymbol{\theta}) = E_\theta\{r(\boldsymbol{\tau})\} = \int p_\theta(\boldsymbol{\tau})r(\boldsymbol{\tau})d\boldsymbol{\tau}$.

2.2   Method Development Approach

The problem of learning robot motor skills can be modeled as follows: (1) The robots starts with an initial training data set obtained from demonstrations from which it learns an initial policy. (2) It subsequently learns how to improve this policy by repetitive training over multiple episodes. The first goal is accomplished by imitation learning while the second requires reinforcement learning. In addition, model learning is often needed for improved execution [2].

Imitation Learning. The goal of imitation learning is to successfully reproduce the policy of the teacher $\pi(\mathbf{u}|\mathbf{x})$. Many approaches exist in the literature [3,4]. However, this problem can be well-understood for stochastic policies: How can we reproduce the stochastic policy π given a demonstrated path distribution $p(\boldsymbol{\tau})$? The path distribution $p_{\boldsymbol{\theta}}(\boldsymbol{\tau})$ generated by the policy $\pi_{\boldsymbol{\theta}}$ should be as close as possible to the teacher's, i.e., it minimizes the Kullback-Leibler Divergence

$$D(p(\boldsymbol{\tau})\|p_{\boldsymbol{\theta}}(\boldsymbol{\tau})) = \int p(\tau)\log\frac{p(\boldsymbol{\tau})}{p_{\boldsymbol{\theta}}(\boldsymbol{\tau})}d\tau = \int p(\tau)\sum_{t=1}^{T}\log\frac{\pi(\mathbf{u}_t|\mathbf{x}_t)}{\pi_{\boldsymbol{\theta}}(\mathbf{u}_t|\mathbf{x}_t)}d\tau,$$

where the model of the system and the start-state distribution naturally cancel out. As $\log\pi(\mathbf{u}_t|\mathbf{x}_t)$ is an additive constant, the path rewards become

$$r(\boldsymbol{\tau}) \propto -\sum_{t=1}^{T}\log\pi_{\boldsymbol{\theta}}(\mathbf{u}_t|\mathbf{x}_t) = -\sum_{t=1}^{T}\left\|\mathbf{u} - \boldsymbol{\phi}^T(\mathbf{x})\boldsymbol{\theta}\right\|^2,$$

where the second part only holds true for our policy which is linear in the features and has Gaussian exploration. Clearly, the model-free imitation learning problem can be solved in one shot in this way [4].

Reinforcement Learning.   For general rewards, the problem is not straightforward as the expected return has no notion of data. Instead, for such a brute-force problem, learning can only happen indirectly as in value function methods [1] or using small steps in the policy space, as in policy gradient methods [5]. Instead of circumventing this problem, we realized that there exits a tight lower bound

$$J(\boldsymbol{\theta}) = \int p_{\boldsymbol{\theta}}(\boldsymbol{\tau})r(\boldsymbol{\tau})d\tau \geq D(p(\boldsymbol{\tau})r(\boldsymbol{\tau})\|p_{\boldsymbol{\theta}}(\boldsymbol{\tau})).$$

Hence, reinforcement learning becomes a series of reward-weighted self-imitation steps (Intuitively: "*Do what you are but better*") with the resulting policy update

$$\boldsymbol{\theta}' = \text{argmax}_{\boldsymbol{\theta}'}D(R(\boldsymbol{\tau})p_{\boldsymbol{\theta}}(\boldsymbol{\tau})\|p_{\boldsymbol{\theta}'}(\boldsymbol{\tau}))$$

which is guaranteed to converge to a local optimum. Taking such an approach, which stays close to the training data is often crucial for robot reinforcement learning as the robot avoids trying arbitrary, potentially destructive actions. The resulting methods have led to a series of highly successful robot reinforcement learning methods such as reward-weighted regression [5], LAWER [6], PoWER [4], and Cost-regularized Kernel Regression [7].

3 Application in Robot Skill Learning

The imitation and reinforcement learning approaches have so far been general, despite being geared for the robotics scenario. To apply these methods in robotics, we need appropriate policy representations. Such representation are needed both for simple and complex tasks.

Fig. 2. Swing the ball into the cup

3.1 Learning Simple Tasks with Motor Primitives

We chose policy features based on dynamical systems, which are an extension the ground-breaking work of Ijspeert, Nakanishi & Schaal refined in [4]. We will use these features to represent elementary movements, or Movement Primitives (MP).

The methods above are straightforward to apply by using a single motor primitive as a parametrized policy. Such elementary policies $\pi_\theta(\mathbf{u}|\mathbf{x})$ have both shape parameters $\boldsymbol{w}$ as well as task parameters $\boldsymbol{\gamma}$ where $\boldsymbol{\theta} = [\boldsymbol{w}, \boldsymbol{\gamma}]$. For example, an elementary policy can be used to learn a dart throwing movement by learning the shape parameters $\boldsymbol{w}$ without considering the task parameters $\boldsymbol{\gamma}$. However, when playing a dart game (e.g., around the clock), the robot has to adapt the elementary policy (which represents the throwing movement) to new fields on the dart board. In this case, the shape parameters $\boldsymbol{w}$ can be kept at fixed value and the goal-adaptation happens purely through the task parameters $\boldsymbol{\gamma}$.

Learning only the shape parameters of rhythmic motor primitives using just imitation learning, we have been able to learn ball paddling [4] as shown in Fig. 1. Using the combination of imitation and reinforcement learning, our robot managed to learn ball-in-a-cup in Fig. 2 to perfection within less than a hundred trials using only shape parameters [4]. By learning dart throwing with the shape parameters, and, subsequently, adapting the dart throwing movement to the context, we have managed to learn dart games based on context as well as another, black-jack-style sequential throwing game [7]. The latter two have been accomplished by learning a task parameter policy $\boldsymbol{\gamma} \sim \hat{\pi}(\boldsymbol{\gamma}|\mathbf{x})$.

3.2 Learning a Complex Task with Many Motor Primitives

When single primitives no longer suffice, a robot learning system does not only need context but also multiple motor primitives, as for example, in *robot table tennis*, see Fig. 3. A combination of primitives allows the robot to deal with many situations where only few primitives are activated in the same context [8]. The new policy combines multiple primitives as follows

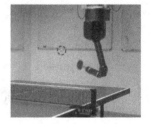

Fig. 3. Learning to Play Robot Table Tennis

$$\mathbf{u} \sim \pi_\theta(\mathbf{u}|\mathbf{x}) = \sum_{i=1}^{K} \pi_{\theta_0}(i|\mathbf{x}) \pi_{\theta_i}(\mathbf{u}|\mathbf{x}).$$

The policy $\pi_{\theta_0}(i|\mathbf{x})$ represents the probability of selecting primitive i, represented by $\pi_{\theta_i}(\mathbf{u}|\mathbf{x})$, based on the incoming ball and the opponent's position. The resulting system learned to return 69% of all balls after imitation learning, and could self-improve against a ball gun to up to 94% successful returns.

4 Conclusion

In this paper, we reviewed the imitation and reinforcement learning methods used to learn a large variety of motor skills. These range from simple tasks, such as ball-paddling, ball-in-a-cup, dart games, etc up to playing robot table tennis.

References

1. Kober, J., Bagnell, D., Peters, J.: Reinforcement learning in robotics: A survey. International Journal of Robotics Research, IJRR (2013)
2. Nguyen-Tuong, D., Peters, J.: Model learning in robot control: a survey. Cognitive Processing (4) (2011)
3. Argall, B., Chernova, S., Veloso, M., Browning, B.: A survey of robot learning from demonstration. Robotics and Autonomous Systems (2009)
4. Kober, J., Peters, J.: Imitation and reinforcement learning. IEEE Robotics and Automation Magazine (2010)
5. Peters, J.: Machine Learning of Motor Skills for Robotics. PhD thesis (2007)
6. Neumann, G., Peters, J.: Fitted Q-iteration by Advantage Weighted Regression. In: Advances in Neural Information Processing Systems 22, NIPS (2009)
7. Kober, J., Wilhelm, A., Oztop, E., Peters, J.: Reinforcement learning to adjust parametrized motor primitives to new situations. Autonomous Robots (2012)
8. Mülling, K., Kober, J., Krömer, O., Peters, J.: Learning to select and generalize striking movements in robot table tennis. IJRR (2013)

Functional MRI Analysis with Sparse Models

Irina Rish

IBM T.J. Watson Research Center
Computational Biology Center - Neuroscience
1101 Kitchawan Rd. Yorktown Heights, NY 10598
rish@us.ibm.com

Abstract. Sparse models embed variable selection into model learning (e.g., by using l_1-norm regularizer). In small-sample high-dimensional problems, this leads to improved generalization accuracy combined with interpretability, which is important in scientific applications such as biology. In this paper, we summarize our recent work on sparse models, including both sparse regression and sparse Gaussian Markov Random Fields (GMRF), in neuroimaging applications, such as functional MRI data analysis, where the central objective is to gain a better insight into brain functioning, besides just learning predictive models of mental states from imaging data.

Keywords: neuroimaging, fMRI, l_1-norm regularization, Lasso, Elastic Net, sparse Gaussian Markov Random Fields (GMRF).

1 Introduction

Predicting person's mental state based on his or her brain imaging data, such as functional MRI (fMRI), is an exciting and rapidly growing research area on the intersection of neuroscience and machine learning. A mental state can be cognitive, such as viewing a picture or reading a sentence [8], emotional, such as feeling happy, anxious, or annoyed while playing a virtual-reality videogame [1], reflect person's perception of pain [11,12,3], or person's mental disorder, such as schizophrenia [2,10], drug addiction [6], and so on.

In fMRI, an MR scanner non-invasively records a subject's blood-oxygenation-level dependent (BOLD) signal, known to be related to neural activity, as a subject performs a task of interest or is exposed to a particular stimulus. Such scans produce a sequence of 3D images, where each image typically has on the order of 10,000-100,000 subvolumes, or *voxels*, and the sequence typically contains a few hundreds of time points, or TRs (time repetitions). Thus, each voxel is associated with a time series representing the average BOLD signal within that subvolume (i.e., voxel activity) at each TR; a task or a stimulus is associated with the corresponding time series over the same set of TRs.

2 Sparse Regression

Our work is motivated by the traditional fMRI goal of discovering task-relevant brain areas. However, we wish to avoid limitations of traditional mass-univariate

H. Blockeel et al. (Eds.): ECML PKDD 2013, Part III, LNAI 8190, pp. 632–636, 2013.

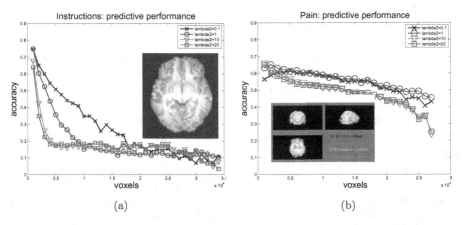

Fig. 1. Predictive accuracy of the Elastic Net for the task of predicting (a) "Instructions" task in PBAIC and (b) thermal pain perception. The insets visualize the sparse solutions found by the Elastic Net.

approaches such as GLM [4] that essentially performs *filter-based* variable selection based on individual voxel correlations with the task, and thus can miss important multivariate interactions, as noted by [5] and others. Thus, we focus instead on **sparse** multivariate models capable of identifying a relatively small subset of variables (voxels) that (*jointly*) predict the task well. In [1], we were among the first ones to apply sparse methods to fMRI, presenting our analysis of the PBAIC 2007 competition data [9] that we obtained using the Elastic Net (EN) approach [15]. EN improves upon the basic LASSO [14] by using a convex combination of l_1- and l_2-norm regularizers instead of just l_1. The effect of such combined penalty is that, on top of sparsity (voxel selection), a *grouping effect* is encouraged, i.e. joint inclusion (or exclusion) of groups of highly correlated variables (such as spatial clusters of voxels). The grouping property is particularly important from the interpretability perspective, since we hope to discover relevant brain areas rather than their single-voxel representatives sufficient for accurate prediction, as the basic LASSO does. We investigate the effects of both l_1 and l_2 regularization parameters on the predictive accuracy and stability, measured here as a support overlap between the regression coefficients learned for different runs of the experiment. We conclude that, (a) EN can be highly predictive about various mental states, achieving $0.7 - 0.9$ correlation between the true and predicted response variables (see Figure 1a), and (b) even among equally predictive models, increasing the l_2-regularization weight can help to improve the model stability.

Furthermore, our subsequent work presented in [11], demonstrates that the Elastic Net can be highly predictive about such subjective and seemingly hard-to-quantify experience as pain, achieving up to $0.75 - 0.8$ correlation between the predicted and actual pain perception ratings, drastically outperforming un-regularized linear regression, and identifying novel areas undiscovered by GLM[1].

However, given a brain map of task-relevant voxels, does this imply that the rest of the brain voxels is irrelevant? Not necessarily, since multiple sparse predic-tive solutions are possible in presence of highly-correlated predictors. In [12], we explore the space of sparse solutions, by first finding the best EN solution with 1000 voxels, removing those voxels from the set of predictors, and repeating the procedure until there are no more voxels left. Interestingly, for multiple tasks we considered, including pain perception and others, *no clear separation between rel-evant and irrelevant areas was observed*, as shown in Figure 1b, suggesting highly non-localized, "holographic" distribution of task information in the brain. The only task which demonstrated fast (exponential) performance degradation, and clear separation of relevant vs irrelevant areas, was a relatively simple auditory task from PBAIC (Figure 1a)[2]. Note that standard GLM method does not reveal such phenomenon, since, as shown in [12], individual voxel-task correlations al-ways seem to decay exponentially, and for many reasonably predictive (but not best) sparse solutions, their voxel would not even pass 0.1 correlation thresh-old. Thus, multivariate sparse regression is a much better tool than GLM for exploring actual distribution of task-relevant information in the brain.

3   Sparse Gaussian Markov Random Fields (GMRFs)

Though task-relevant brain areas are still the most common type of patterns considered in fMRI analysis, they have obvious limitations, since the brain is an interconnected, dynamical system, whose behavior may be better captured by modeling interactions across different area. For example, in our recent study of schizophrenia [2,10], task-based voxel activations are dramatically outperformed by network-based features, extracted from the voxel-level correlation matrices ("functional networks"), and yielding from 86% to 93% classification accuracy on schizophrenic vs. control subjects. Furthermore, we investigate structural dif-ferences of sparse Gaussian Markov Random Fields, or GMRFs, constructed from fMRI data via l_1-regularized maximum likelihood (inverse covariance esti-mation). Used as probabilistic classifiers, GMRFs often outperform state-of-art classifiers such as SVM (e.g., see Figure 2a from [2]). In [13], we proposed a simple, easily parallelizable greedy algorithm SINCO, for Sparse INverse CO-variance estimation.

Next, we developed a *variable-selection structure learning* approach for GM-RFs in [7]. A combination of (ℓ_1, ℓ_p) group-Lasso penalty with the l_1-penalty selects the most-important variables/nodes, besides simply sparsifying the set of

---

[1] The predictive accuracy can be further improved by combining EN for predicting the intensity of the painful stimulus (e.g., the temperature) from fMRI data with a novel analytic, differential-equation model that links temperature and pain perception [3].

[2] A possible hypothesis is that, while "simple" tasks are localized, more complex tasks/experiences (such as pain) tend to involve much more distributed brain areas (most of the brain, potentially).

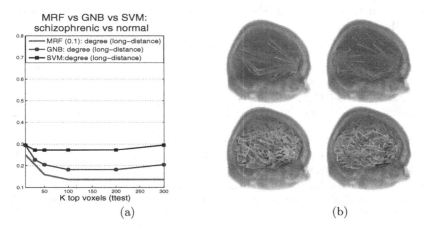

(a)   (b)

Fig. 2. (a) Gaussian MRF classifier predicts schizophrenia with 86% accuracy using just 100 top-ranked (most-discriminative) features, such as voxel degrees in a functional network. (b) Structures learnt for cocaine addicted (left) and control subjects (right), for sparse Markov net learning method with variable-selection via $\ell_{1,2}$ method (top) and without variable-selection, i.e., standard graphical lasso approach (bottom). Positive interactions are in blue, negative – in red. Structure density on top is 0.0016, while on the bottom it is 0.023 (number of edges in a complete graph is $\approx$378000).

edges. Our main advantage is a better interpretability of the resulting networks due to elimination of noisy variables (see below), combined with improvements in model likelihood and more accurate recovery of ground-truth structure. From an algorithmic point of view, we show that a block coordinate descent method generates a sequence of positive definite solutions. Thus, we reduce the original problem into a sequence of strictly convex (ℓ_1, ℓ_p) regularized quadratic minimization subproblems for $p \in \{2, \infty\}$. Our algorithm is well founded since the optimal solution of the maximization problem is unique and bounded. Figure 2b shows the network structures learnt for cocaine addicted vs healthy control subjects, comparing the two methods. The disconnected variables are not shown. The variable-selection sparse Markov network approach yields much fewer connected variables but a higher log-likelihood than graphical lasso, as reported in [7], which suggests that the discarded edges from the disconnected nodes are not important for accurate modeling of this dataset. Moreover, removal of a large number of nuisance variables (voxels) results into a much more interpretable model, clearly demonstrating brain areas involved in structural model differences that discriminate cocaine addicts from healthy control subjects. Cocaine addicts show increased interactions between the visual cortex (back of the brain, on the left here) and the prefrontal cortex (front of the brain image, on the right), while at the same time decreased density of interactions between the visual cortex with other brain areas. Given that the trigger for reward in this experiments was a visual stimulus, and that the prefrontal cortex is involved in higher-order cognitive functions such as decision making and reward processing, the alteration in this pathway in the addict group is highly significant from a neuroscientific perspective.

References

1. Carroll, M.K., Cecchi, G.A., Rish, I., Garg, R., Rao, A.R.: Prediction and interpretation of distributed neural activity with sparse models. Neuroimage 44(1), 112–122 (2009)
2. Cecchi, G., Rish, I., Thyreau, B., Thirion, B., Plaze, M., Paillere-Martinot, M.L., Martelli, C., Martinot, J.L., Poline, J.B.: Discriminiative network models of schizophrenia. In: Proc. of NIPS 2009 (2009)
3. Cecchi, G.A., Huang, L., Hashmi, J.A., Baliki, M., Centeno, M.V., Rish, I., Apkarian, A.V.: Predictive dynamics of human pain perception. PLoS Computational Biology 8(10) (2012)
4. Friston, K.J., et al.: Statistical parametric maps in functional imaging - a general linear approach. Human Brain Mapping 2, 189–210 (1995)
5. Haxby, J.V., Gobbini, M.I., Furey, M.L., Ishai, A., Schouten, J.L., Pietrini, P.: Distributed and Overlapping Representations of Faces and Objects in Ventral Temporal Cortex. Science 293(5539), 2425–2430 (2001)
6. Honorio, J., Ortiz, L., Samaras, D., Paragios, N., Goldstein, R.: Sparse and locally constant gaussian graphical models. In: Bengio, Y., Schuurmans, D., Lafferty, J., Williams, C.K.I., Culotta, A. (eds.) Advances in Neural Information Processing Systems 22, pp. 745–753 (2009)
7. Honorio, J., Samaras, D., Rish, I., Cecchi, G.A.: Variable selection for gaussian graphical models. In: Proceedings of the Fifteenth International Conference on Artificial Intelligence and Statistics, AISTATS 2012 (2012)
8. Mitchell, T.M., Hutchinson, R., Niculescu, R.S., Pereira, F., Wang, X., Just, M., Newman, S.: Learning to decode cognitive states from brain images. Machine Learning 57, 145–175 (2004)
9. Pittsburgh EBC Group. PBAIC Homepage (2007), http://www.ebc.pitt.edu/2007/competition.html
10. Rish, I., Cecchi, G., Thyreau, B., Thirion, B., Plaze, M., Paillere-Martinot, M.L., Martelli, C., Martinot, J.L., Poline, J.B.: Schizophrenia as a network disease: Disruption of emergent brain function in patients with auditory hallucinations. PLoS ONE 8(1) (2013)
11. Rish, I., Cecchi, G.A., Baliki, M.N., Apkarian, A.V.: Sparse regression models of pain perception. In: Proc. of Brain Informatics (BI 2010) (August 2010)
12. Rish, I., Cecchi, G.A., Heuton, K., Baliki, M.N., Apkarian, A.V.: Sparse regression analysis of task-relevant information distribution in the brain. In: Proc. of SPIE Medical Imaging 2012 (February 2012)
13. Scheinberg, K., Rish, I.: Learning sparse gaussian markov networks using a greedy coordinate ascent approach. In: Balcázar, J.L., Bonchi, F., Gionis, A., Sebag, M. (eds.) ECML PKDD 2010, Part III. LNCS, vol. 6323, pp. 196–212. Springer, Heidelberg (2010)
14. Tibshirani, R.: Regression shrinkage and selection via the lasso. Journal of the Royal Statistical Society, Series B 58(1), 267–288 (1996)
15. Zou, H., Hastie, T.: Regularization and variable selection via the elastic net. Journal of the Royal Statistical Society, Series B 67(2), 301–320 (2005)

Image Hub Explorer: Evaluating Representations and Metrics for Content-Based Image Retrieval and Object Recognition

Nenad Tomašev and Dunja Mladenić

Institute Jožef Stefan
Artificial Intelligence Laboratory
Jamova 39, 1000 Ljubljana, Slovenia
{nenad.tomasev,dunja.mladenic}@ijs.si

Abstract. Large quantities of image data are generated daily and visualizing large image datasets is an important task. We present a novel tool for image data visualization and analysis, Image Hub Explorer. The integrated analytic functionality is centered around dealing with the recently described phenomenon of *hubness* and evaluating its impact on the image retrieval, recognition and recommendation process. Hubness is reflected in that some images (*hubs*) end up being very frequently retrieved in 'top k' result sets, regardless of their labels and target semantics. Image Hub Explorer offers many methods that help in visualizing the influence of major image hubs, as well as state-of-the-art metric learning and hubness-aware classification methods that help in reducing the overall impact of extremely frequent neighbor points. The system also helps in visualizing both beneficial and detrimental visual words in individual images. Search functionality is supported, along with the recently developed hubness-aware result set re-ranking procedure.

Keywords: image retrieval, object recognition, visualization, k-nearest neighbors, metric learning, re-ranking, hubs, high-dimensional data.

1 Introduction

Image Hub Explorer is the first image collection visualization tool aimed at understanding the underlying *hubness* [1] of the k-nearest neighbor data structure. Hubness is a recently described aspect of the well known *curse of dimensionality* that arises in various sorts of intrinsically high-dimensional data types, such as text [1], images [2] and audio [3]. Its implications were most thoroughly examined in the context of music retrieval and recommendation [4]. Comparatively little attention was given to emerging hubs and the skewed distribution of influence in image data. One of the main goals of the Image Hub Explorer was to enable other researchers and practitioners to easily detect hubs in their datasets, as well as test and apply the built-in state-of-the-art hubness-aware data mining methods.

H. Blockeel et al. (Eds.): ECML PKDD 2013, Part III, LNAI 8190, pp. 637–640, 2013.

2   Resources

Image Hub Explorer is an analytics and visualization tool that is built on top of the recently developed *Hub Miner* Java data mining library that is focused on evaluating various types of *k*NN methods. Additional resources on the use of Image Hub Explorer (http://ailab.ijs.si/tools/image-hub-explorer/) and the Hub Miner library (http://ailab.ijs.si/nenad_tomasev/hub-miner-library/) are available online. This includes the demo video: http://youtu.be/LB9ZWuvmOqw.

3   Related Work

3.1   Hubs in High-Dimensional Data

The concentration of distances [5] in intrinsically high-dimensional data affects the distribution of neighbor occurrences and causes *hubs* to emerge as centers of influence in form of very frequent neighbor points. The *k*NN hubs often act as semantic singularities and are detrimental for the analysis [6]. Different representations and metrics exhibit different degrees of neighbor occurrence distribution skewness [2]. Selecting an appropriate feature representation paired with an appropriate distance measure is a nontrivial task and very important for improving system performance. This is what Image Hub Explorer was designed to help with.

Hubness-aware methods have recently been proposed for instance selection [7], clustering [8], metric learning [9][4], information retrieval [10], classification [11] and re-ranking [12]. Most of these methods are implemented and available in Image Hub Explorer.

3.2   Visualizing Image Collections

Visualization plays an essential role in examining large image databases. Several similarity-based visualization approaches have been proposed [13][14] and ImagePlot (http://flowingdata.com/2011/09/18/explore-large-image-collections-with-imageplot/) is a typical example. What these systems have in common is that they mostly focus on different ways of similarity-preserving projections of the data onto the plane, as well as selection strategies that determine which images are to be shown. Some hierarchical systems are also available [15]. These systems allow for quick browsing through large collections, but they offer no support for examining the distribution of influence and detection of emerging hub images.

4   System Components and Functions

Image Hub Explorer implements several views of the data, to facilitate easier analysis and interpretation. All images in all views can be selected and the information is shared among the views and updated automatically. The desired neighborhood size k is controlled by a slider and its value can be changed at any time.

   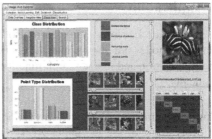

(a) MDS Data Overview   (b) Class Comparisons

Fig. 1. Screenshots of some selected Image Hub Explorer functions

Metric Learning plays an important role in the analysis. For any loaded data representation, many different metrics can be employed. Image Hub Explorer supports 7 primary metrics and 5 secondary metrics that are learned from the primary distance matrices. This includes two recently proposed hubness-aware distance measures: $simhub_s$ [9] and mutual proximity [4].

Data Overview gives a quick insight into the hub-structure of the data (Fig. 1(a)). The most influential image hubs are projected onto a visualization panel via multidimensional scaling (MDS) and the background coloring is determined based on the nature of their occurrences - green denotes the beneficial influences, red the detrimental ones. The occurrence frequency distribution is shown, followed by a set of statistics describing various aspects of neighbor occurrences and kNN set purity.

Neighbor View offers a deeper insight into the local neighbor structure. For each selected image, its k-neighbors and reverse k-neighbors are listed and any selected image can be inserted into the local subgraph visualization panel. This allows the user to visualize all kNN relations among a selected set of images as a graph, with distance values displayed on the edges.

Class View allows the user to compare the point type distributions among different classes, as well as a way to quickly select and examine the top hubs, good hubs and bad hubs for each class separately(Fig. 1(b)). Additionally, the global class-to-class occurrence matrix can be used to determine which classes cause most label mismatches in k-neighbor sets and which classes these mismatches are directed at.

Search, Re-ranking and Classification : Apart from simple browsing, image search is an important function in examining large image databases. Image Hub Explorer supports image queries, for which a set of k most similar images from the database is retrieved. Image Hub Explorer implements 8 different kNN classification methods to help with image labeling, as well as a hubness-aware result set re-ranking procedure [12].

Feature Assessment for quantized feature representations can easily be performed in Image Hub Explorer, as it calculates the occurrence profile for each visual word (codebook vector) and determines which features help in increasing the intra-class similarity and which increase the inter-class similarity. Beneficial and detrimental features and texture regions can be visualized on each image separately.

5 Applicability

The Image Hub Explorer system can also be used to visualize other data types, when rectangular nodes are shown instead of the loaded image thumbnails. Only the feature visualization and image search functions are restricted to working with image data specifically.

Acknowledgements. This work was supported by the Slovenian Research Agency, the ICT Programme of the EC under XLike (ICT-STREP-288342), and RENDER (ICT-257790-STREP).

References

1. Radovanović, M., Nanopoulos, A., Ivanović, M.: Hubs in space: Popular nearest neighbors in high-dimensional data. Journal of Machine Learning Research 11, 2487–2531 (2010)
2. Tomašev, N., Brehar, R., Mladenić, D., Nedevschi, S.: The influence of hubness on nearest-neighbor methods in object recognition. In: Proceedings of the 7th IEEE International Conference on Intelligent Computer Communication and Processing (ICCP), pp. 367–374 (2011)
3. Aucouturier, J., Pachet, F.: Improving timbre similarity: How high is the sky? Journal of Negative Results in Speech and Audio Sciences 1 (2004)
4. Schnitzer, D., Flexer, A., Schedl, M., Widmer, G.: Local and global scaling reduce hubs in space. Journal of Machine Learning Research, 2871–2902 (2012)
5. François, D., Wertz, V., Verleysen, M.: The concentration of fractional distances. IEEE Transactions on Knowledge and Data Engineering 19(7), 873–886 (2007)
6. Radovanović, M., Nanopoulos, A., Ivanović, M.: Nearest neighbors in high-dimensional data: The emergence and influence of hubs. In: Proc. 26th Int. Conf. on Machine Learning (ICML), pp. 865–872 (2009)
7. Buza, K., Nanopoulos, A., Schmidt-Thieme, L.: INSIGHT: Efficient and effective instance selection for time-series classification. In: Huang, J.Z., Cao, L., Srivastava, J. (eds.) PAKDD 2011, Part II. LNCS, vol. 6635, pp. 149–160. Springer, Heidelberg (2011)
8. Tomašev, N., Radovanović, M., Mladenić, D., Ivanović, M.: The role of hubness in clustering high-dimensional data. IEEE Transactions on Knowledge and Data Engineering 99, 1 (2013) (PrePrints)
9. Tomašev, N., Mladenić, D.: Hubness-aware shared neighbor distances for high-dimensional k-nearest neighbor classification. Knowledge and Information Systems (2013)
10. Tomašev, N., Rupnik, J., Mladenić, D.: The role of hubs in cross-lingual supervised document retrieval. In: Pei, J., Tseng, V.S., Cao, L., Motoda, H., Xu, G. (eds.) PAKDD 2013, Part II. LNCS, vol. 7819, pp. 185–196. Springer, Heidelberg (2013)
11. Tomašev, N., Mladenić, D.: Nearest neighbor voting in high dimensional data: Learning from past occurrences. Computer Science and Information Systems 9, 691–712 (2012)
12. Tomašev, N., Leban, G., Mladenić, D.: Exploiting hubs for self-adaptive secondary re-ranking in bug report duplicate detection. In: Proceedings of the ITI Conference, ITI 2013 (2013)
13. Nguyen, G.P., Worring, M.: Similarity based visualization of image collections. In: Int'l Worksh. Audio-Visual Content and Information Visualization in Digital Libraries (2005)
14. Nguyen, G.P., Worring, M.: Interactive access to large image collections using similarity-based visualization. J. Vis. Lang. Comput. 19(2), 203–224 (2008)
15. Tomašev, N., Fortuna, B., Mladenić, D., Nedevschi, S.: Ontogen extension for exploring image collections. In: Proceedings of the 7th IEEE International Conference on Intelligent Computer Communication and Processing (ICCP) (2011)

Ipseity – A Laboratory for Synthesizing and Validating Artificial Cognitive Systems in Multi-agent Systems

Fabrice Lauri, Nicolas Gaud, Stéphane Galland, and Vincent Hilaire

IRTES-SET, UTBM, 90010 Belfort cedex, France
{fabrice.lauri,nicolas.gaud,stephane.galland,vincent.hilaire}@utbm.fr

Abstract. This article presents an overview on IPSEITY [1], an open-source rich-client platform developed in C++ with the Qt [2] framework. IPSEITY facilitates the synthesis of artificial cognitive systems in multi-agent systems. The current version of the platform includes a set of plugins based on the classical reinforcement learning techniques like Q-Learning and Sarsa. IPSEITY is targeted at a broad range of users interested in artificial intelligence in general, including industrial practitioners, as well as machine learning researchers, students and teachers. It is daily used as a course support in Artificial Intelligence and Reinforcement Learning and it has been used successfully to manage power flows in simulated microgrids using multi-agent reinforcement learning [4].

Keywords: Multi-Agent Systems, Reinforcement Learning.

1 Introduction

Multi-agent systems constitute a fitted paradigm for solving various kinds of problems in a distributed way or for simulating complex phenomena emerging from the interactions of several autonomous entities. A multi-agent system (MAS) consists of a collection of *agents* that interact within a common environment. Every agent perceives some information extracted from its environment and acts upon it based on these perceptions.

The individual behaviors of the agents composing a MAS can be defined by using many decision making mechanisms and many programming languages according to the objective at hand. For instance, a planification mechanism can be used to fulfill the agent goals. A powerful alternative is to implement Reinforcement Learning (RL) algorithms that allow the agents to learn how to behave rationally. In this context, a learning agent tries to achieve a given task by continuously interacting with its environment. At each time step, the agent perceives the environment state and performs an action, which causes the environment to transit to a new state. A scalar reward evaluates the quality of each transition, allowing the agent to observe the cumulative reward along sequences of interactions. By trials and errors, such agents can manage to find a policy, that is a mapping from states to actions, which maximizes the cumulative reward.

To our knowledge, there is currently no multi-agent platform that allow users interested in multi-agent RL in particular to easily study the influence of some

H. Blockeel et al. (Eds.): ECML PKDD 2013, Part III, LNAI 8190, pp. 641–644, 2013.

(learning) parameters on the performance and the results obtained by different dedicated algorithms using accepted benchmarks. Indeed, RL-Glue[1], CLSquare[2], PIQLE[3], RL Toolbox[4], JRLF[5], LibPG[6]only support single-agent RL techniques. The MARL Toolbox[7] supports multi-agent reinforcement learning under Matlab, but unlike IPSEITY, it is not possible to remotely control other systems.

2   Overview

IPSEITY is a rich-client platform especially dedicated to facilitating the implementation and the experimental validation of different kinds of behaviors for cooperative or competitive agents in MASs.

2.1   Kernel Concepts

In IPSEITY, a set $\mathcal{A}$ of *agents* interact within a given *environment*. A set $\mathcal{G} = \{\mathcal{G}_1, \cdots, \mathcal{G}_N\}$ of agent groups, called *taxons* in IPSEITY, can be defined. Agents grouped together into the same taxon are likely to behave similarly (they share the same decision making mechanism). The behavior of an agent is exhibited according to its *cognitive system*. A cognitive system implements the decision process that allows an agent to carry out actions based on its perceptions. It can be plugged directly to a given agent or to a taxon. In the latter case, all the agents associated to the same taxon use the same decision process. If a cognitive system is based on RL techniques, plugging it to a taxon shared by several agents may speed up their learning in some cases, as any agent can immediately benefit from the experiences of the others.

The agents interact within the environment from different possible initial configurations, or *scenarios*. Scenarios allow the user to study the quality of the decisions taken individually or collectively by the agents under some initial environmental conditions. During a *simulation*, the agents evolve within the environment by considering several predefined scenarios whose order is handled by a *supervisor*, who can be the user himself or an agent.

2.2   Platform Architecture

IPSEITY simulates discrete-time or continuous-time multi-agent environments inherating from *AbstractEnvironment* (see Fig. 1a). Several agents (inherating from *AbstractAgent*) interact in a multi-agent environment on the basis of their cognitive system inherating from *AbstractCognitiveSystem*. A set of scenarii has to be defined in order to carry out the simulations. These are selected by a *AbstractSupervisionModule*.

---

[8] http://glue.rl-community.org/wiki/Main_Page

[9] http://ml.informatik.uni-freiburg.de/research/clsquare

[10] http://piqle.sourceforge.net

[11] http://www.igi.tu-graz.ac.at/gerhard/ril-toolbox/general/overview.html

[12] http://mykel.kochenderfer.com/jrlf

[13] https://code.google.com/p/libpgrl

[14] http://busoniu.net/repository.php

The cognitive systems in IPSEITY can be decomposed into several plugins. Each of them takes part in the decision process. As shown in Fig. 1b, the reinforcement learning cognitive system is currently built from three classes of plugins: a *behavior module*, that selects actions from states, a *memory*, that stores the Q-values of the state-action pairs, and a *learning module*, that updates the contents of the memory from data obtained after environmental transitions. Currently, *Epsilon-Greedy* and *Softmax* are predefined plugins that can be used as behavior modules, *Q-Learning* and *Sarsa* have been implemented and can be used as learning modules. The Q-value memory can be instanciated by a plugin implementing either a static or a dynamic lookup table, or a linear function approximator using a *feature extraction module* like CMAC for example. More details about the kernel concepts and about the software architecture are available on the web site[15].

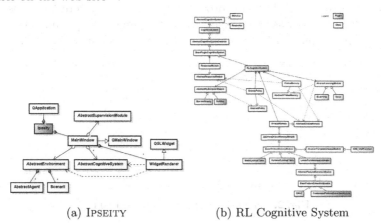

(a) IPSEITY   (b) RL Cognitive System

Fig. 1. Architecture of IPSEITY and architecture of a RL cognitive system

2.3   Properties

IPSEITY was designed to possess the following properties:

Flexibility: IPSEITY uses kernel concepts and components (i.e. data representations and algorithms) that are as flexible as possible. For example, the perceptions and the responses of agents are represented by 64-bit float vectors, allowing agents to be immerged either in discrete environments or in continuous environments.

Modularity: IPSEITY uses modules, or *plugin*, to implement kernel concepts and components. For example, the environments, the cognitive systems, the agent scheduling, and the selection of the simulation scenarios are all defined as plugins.

Easy Integration: These plugins (and in particular those related to cognitive systems) can be easily integrated in other systems and applications. For example, IPSEITY can be used to learn the behaviors of some agents. Once

---

[15] At http://www.ipseity-project.com/docs/IpseityTechnicalGuide.pdf.

the learning phase is finished, agents can perceive information from a remote environment and act according to the learnt behavior. Such integration has been realized between IPSEITY and JANUS [3]: remote cognitive systems with behaviors learnt using IPSEITY communicate using TCP-IP sockets with agents of a microgrid simulator developed under JANUS.

Extensibility: IPSEITY can easily be extended by specialized plugins for the target application area. Customized extensions include new environments, algorithms that take part in the decision processes of some cognitive systems or rendering modules for some predefined environments.

System Analysis: IPSEITY supports the user in keeping track of all the data, performed actions and results of a RL task. This data set is linked to a session, allowing the user to easily and rapidly compare the results obtained from different algorithms.

Graphical Interface: IPSEITY provides a user-friendly interface (see Fig.2) with informative icons and widgets for setting up all the parameters involved in the simulation of a MAS, including those of a cognitive system.

Fig. 2. Screenshots of some tabs in IPSEITY

3   Conclusion

An overview of IPSEITY has been presented in this article, focusing on RL. IPSEITY is highly modular and broadly extensible. It can be freely downloaded from http://www.ipseity-project.com under a GNU *GPLv3* open-source licence. It is intended to be enriched with state-of-the-art RL techniques very soon. Persons who want to contribute to this project are cordially encouraged to contact the first author.

References

1. http://www.ipseity-project.com
2. http://www.qt-project.org
3. Gaud, N., Galland, S., Hilaire, V., Koukam, A.: An organisational platform for holonic and multiagent systems. In: Hindriks, K.V., Pokahr, A., Sardina, S. (eds.) ProMAS 2008. LNCS, vol. 5442, pp. 104–119. Springer, Heidelberg (2009)
4. Lauri, F., Basso, G., Zhu, J., Roche, R., Hilaire, V., Koukam, A.: Managing Power Flows in Microgrids using Multi-Agent Reinforcement Learning. In: Agent Technologies in Energy Systems (ATES) (2013)

OpenML: A Collaborative Science Platform

Jan N. van Rijn[1], Bernd Bischl[2], Luis Torgo[3], Bo Gao[4],
Venkatesh Umaashankar[5], Simon Fischer[5], Patrick Winter[6], Bernd Wiswedel[6],
Michael R. Berthold[7], and Joaquin Vanschoren[1]

[1] Leiden University, Leiden, Netherlands
{jvrijn,joaquin}@liacs.nl
[2] TU Dortmund, Dortmund, Germany
bischl@statistik.tu-dortmund.de
[3] University of Porto, Porto, Portugal
ltorgo@inescporto.pt
[4] KU Leuven, Leuven, Belgium
bo.gao@cs.kuleuven.be
[5] Rapid-I GmbH, Dortmund, Germany
{venkatesh,fischer}@rapid-i.com
[6] KNIME.com AG
{patrick.winter,Bernd.Wiswedel}@knime.com
[7] University of Konstanz, Konstanz, Germany
Michael.Berthold@uni-konstanz.de

Abstract. We present OpenML, a novel open science platform that provides easy access to machine learning data, software and results to encourage further study and application. It organizes all submitted results online
so they can be easily found and reused, and features a web API which is
being integrated in popular machine learning tools such as Weka, KNIME,
RapidMiner and R packages, so that experiments can be shared easily.

Keywords: Experimental Methodology, Machine Learning, Databases,
Meta-Learning.

1 Introduction

Research in machine learning and data mining can be speeded up tremendously
by moving empirical research results "out of people's heads and labs, onto the
network and into tools that help us structure and alter the information" [3].
The massive streams of experiments that are being executed to benchmark new
algorithms, test hypotheses or model new datasets have many more uses beyond
their original intent, but are often discarded or their details are lost over time.
In this paper, we present OpenML[1], an open science platform for machine learning research. OpenML is a website where researchers can share their datasets,
implementations and experiments in such a way that they can easily be found
and reused by others. It offers a web API through which new resources and results can be submitted, and is being integrated in a number of popular machine

---

[1] OpenML can be found at http://www.openml.org.

H. Blockeel et al. (Eds.): ECML PKDD 2013, Part III, LNAI 8190, pp. 645–649, 2013.

learning and data mining platforms, such as Weka, RapidMiner, KNIME, and data mining packages in R, so that new results can be submitted automatically. Vice versa, it enables researchers to easily search for certain results (e.g., evaluations on a certain dataset), to directly compare certain techniques against each other, and to use all submitted data in advanced queries. An overview of the key components of OpenML is provided in Figure 1.

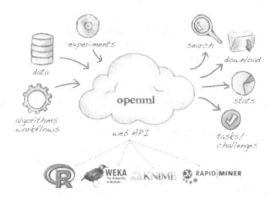

Fig. 1. Components of the OpenML platform

OpenML engenders a novel, collaborative approach to experimentation with important benefits. First, many questions about machine learning algorithms won't require the laborious setup of new experiments: they can be answered on the fly by querying the combined results of thousands of studies on all available datasets. OpenML also keeps track of experimentation details, ensuring that we can easily reproduce experiments later on, and confidently build upon earlier work [2]. Reusing experiments also allows us to run large-scale machine learning studies, yielding more generalizable results [1] with less effort. Finally, beyond the traditional publication of algorithms in journals, often in a highly summarized form, OpenML allows researchers to share all code and results that are possibly of interest to others, which may boost their visibility, speed up further research and applications, and engender new collaborations.

2   Sharing Experiments

To make experiments from different researchers comparable, OpenML uses *tasks*, well-described problems to be solved by a machine learning algorithm or workflow. A typical task would be: *Predict (target) attribute X of dataset Y with maximal predictive accuracy.* Somewhat similar to a data mining challenge, researchers are thus challenged to build algorithms that solve these tasks. Different tasks can be defined, e.g., parameter optimization, feature selection and clustering. They can be searched online, and will be automatically generated for newly submitted datasets. OpenML provides all necessary information to complete the task, such as a URL to download the input dataset, and what information should be submitted to the server. For some tasks, e.g., predictive tasks, it offers more structured input and output, such as exact train and test splits for cross-validation, and a submission format for all predictions. The server will then evaluate the predictions and store the scores for various evaluation metrics.

An attempt to solve a task is called a *run*, and includes the task itself, the algorithm or workflow (i.e., implementation) used, and a file detailing the obtained results. These are all submitted to the server, where new implementations will be registered. Workflows are represented as a set of algorithms, and can be downloaded into the workbenches for detailed inspection. For each implementation, an overview page will be generated containing data about all tasks on which this algorithm was run. This will detail the performance of the algorithm over a potentially wide range of datasets, with various parameter settings. For each dataset, a similar page is created, containing a ranking of algorithms that were run on tasks with that dataset as input.

OpenML provides a RESTful API for downloading tasks and uploading datasets, implementations and results. This API is currently being integrated in various machine learning platforms such as Weka, R packages, RapidMiner and KNIME. For instance, in WEKA[2], OpenML is integrated as part of the Weka Experimenter. Given a task, it automatically downloads all associated input data, runs the experiment, and uploads the results to OpenML.

3 Searching OpenML

OpenML links various bits of information together in a single database. All results for different algorithms on the same tasks are stored in such a way that algorithms can directly be compared against each other (using various evaluation measures), and parameter settings are stored so that the impact of individual parameters can be tracked. Moreover, for all datasets, it calculates meta-data concerning their features (e.g., type, distinct values or mean and standard deviation) and their distributions, such as the number of features, instances, missing values, default accuracy, class entropy and landmarking results [4]. Likewise, for algorithms, it includes information about the (hyper)parameters and properties, such as whether the algorithm can handle nominal/numeric features, whether it can perform classification and/or regression and a bias-variance profile.

OpenML allows users to easily search for results of interest. First, it stores textual descriptions for datasets, algorithms and implementations so that they can be found through simple keyword searches, linked to overview pages that detail all related results. Second, runs can be searched to directly compare many algorithms over many datasets (e.g., for benchmarking). Furthermore, the database can be queried directly through an SQL editor, or through pre-defined advanced queries such as "Show the effect of a parameter P on algorithm A on dataset D" and "Draw the learning curve for algorithm A on dataset D".[3] The results of such queries are displayed as data tables, scatterplots or line plots, which can be downloaded directly.

---

[2] A beta version can be downloaded from the OpenML website.
[3] See the 'Advanced' tab on http://www.openml.org/search.

4   Related Work

OpenML builds on previous work on *experiment databases* [5], but also enhances it by markedly facilitating the sharing of new experiments through the web API and by making results much easier to find and compare.

In terms of sharing algorithms or workflows, it is somewhat similar to MyExperiment[4], an online repository where users can search and share workflows so that interesting workflows can be reused. However, MyExperiment offers little support for storing the *results* of workflows, or comparing workflows based on performance metrics.

On the other hand, MLComp[5] is a platform on which users can run their algorithms on known datasets (or vice versa). These runs are performed on the servers of MLComp, which saves the user resources. Although very useful, especially for comparing runtimes, OpenML differs from MLComp in two key aspects: First, OpenML allows users much more flexibility in running experiments: new tasks can easily be introduced for novel types of experiments and experiments can be run in any environment. It is also being integrated in data mining platforms that researchers already use in daily research. Finally, OpenML allows more advanced search and query capabilities that allow researchers to reuse results in many ways beyond direct comparisons, such as meta-learning studies [5].

5   Conclusions and Future Work

OpenML aims to engender an open, collaborative approach to machine learning research. Experiments can be shared in full detail, which generates a large amount of reproducible results available for everyone. Moreover, integration with popular data mining tools will make it very easy for researchers to share experiments with OpenML and the community at large.

Future work includes support for a broader range of task types, e.g., time series analyses, graph mining and text mining.

Acknowledgments. This work is supported by grant 600.065.120.12N150 from the Dutch Fund for Scientific Research (NWO), and by the IST Programme of the European Community, under the Harvest Programme of the PASCAL2 Network of Excellence, IST-2007-216886.

References

1. Hand, D.: Classifier technology and the illusion of progress. Statistical Science (January 2006)
2. Hirsh, H.: Data mining research: Current status and future opportunities. Statistical Analysis and Data Mining 1(2), 104–107 (2008)

---

[4] http://www.myexperiment.org/
[5] http://www.mlcomp.org/

3. Nielsen, M.A.: The future of science: Building a better collective memory. APS Physics 17(10) (2008)
4. Peng, Y.H., Flach, P.A., Soares, C., Brazdil, P.B.: Improved dataset characterisation for meta-learning. In: Lange, S., Satoh, K., Smith, C.H. (eds.) DS 2002. LNCS, vol. 2534, pp. 141–152. Springer, Heidelberg (2002)
5. Vanschoren, J., Blockeel, H., Pfahringer, B., Holmes, G.: Experiment databases. A new way to share, organize and learn from experiments. Machine Learning 87(2), 127–158 (2012), https://lirias.kuleuven.be/handle/123456789/297378

ViperCharts:
Visual Performance Evaluation Platform

Borut Sluban and Nada Lavrač

Department of Knowledge Technologies, Jožef Stefan Institute,
Jamova 39, 1000 Ljubljana, Slovenia
{borut.sluban,nada.lavrac}@ijs.si

Abstract. The paper presents the ViperCharts web-based platform for visual performance evaluation of classification, prediction, and information retrieval algorithms. The platform enables to create interactive charts for easy and intuitive evaluation of performance results. It includes standard visualizations and extends them by offering alternative evaluation methods like F-isolines, and by establishing relations between corresponding presentations like Precision-Recall and ROC curves. Additionally, the interactive performance charts can be saved, exported to several formats, and shared via unique web addresses. A web API to the service is also available.

Keywords: classifiers, performance evaluation, web application.

1   Introduction and Related Work

Empirical evaluation of classification algorithms is mainly focused on their ability to correctly predict the desired class of data instances. Hence, several performance measures are derived from the confusion matrix, which provides the distribution of the predicted classes over the actual classes of a dataset. The choice of an evaluation measure depends on the task to be solved. Typically, classification is evaluated by accuracy which is the average of correct predictions over all the classes, however in the case of unbalanced class distributions, medical diagnostic or information retrieval tasks other evaluation measures which focus on a certain class, like precision, recall, false positive rate or specificity, may be more desirable. This paper presents the ViperCharts web-based platform for visual performance evaluation of classification, prediction, or information retrieval algorithms used in machine learning and data/text mining. Our goal is to provide a web environment which enables intuitive visualization of results and sharing of performance charts that are of interest to machine learning and data mining practitioners.

Tools for graphical representation of numerical data are provided by different software applications for calculation, numerical computing, statistical computing and statistical analysis, like Microsoft Excel[1], MATLAB[2], R[3] and SPSS[4].

---

[1] http://office.microsoft.com/en-us/excel/
[2] http://www.mathworks.com/products/matlab/
[3] http://www.r-project.org/
[4] http://www.ibm.com/software/analytics/spss/

H. Blockeel et al. (Eds.): ECML PKDD 2013, Part III, LNAI 8190, pp. 650–653, 2013.

Also several JavaScript, SVG or Flash based charting libraries and charting software, such as Highcharts[5], D3[6] or Google Chart Tools[7], offer data visualizations which can be embedded in web pages. However, among the usually supported chart types only specific line and scatter charts are interesting from the point of visualizing algorithm performance, since they depict the relation between different variables and/or performance measures of algorithms.

Common visualizations of algorithm performance in machine learning include the Lift curves [11], ROC curves [5], Precision-Recall Curves [8], and Cost curves [4], for probabilistic classifiers or ranking algorithms, and scatter charts in the ROC space or Precision-Recall space (PR space) for discrete classification algorithms. Existing environments for machine learning and data mining, like Weka [6], RapidMiner [9], KNIME [1] and Orange [3], as well as MATLAB and R, only offer support for the computation and visualization of ROC and Lift curves, whereas support for other performance visualizations may be available as third-party packages for different programing languages.

In this paper we present the ViperCharts platform, which aims to provide charting support for the evaluation of machine learning, data/text mining and information retrieval algorithms. It offers a range of performance visualizations and reveals the relations among different performance measures. Its main distinctive feature is its web-based design which requires no installation on the user's system, enables sharing of performance results, and offers enhanced visual performance evaluation through interactive charts and additional advanced functionality. Furthermore, it provides a unique environment for computing, visualizing and comparing performance results of different algorithms for various evaluation measures. The ViperCharts platform is accessible online at http://viper.ijs.si.

2   The ViperCharts Platform

The platform is designed to serve as a tool for data mining practitioners with the goal to produce visualizations of performance results achieved by their algorithms. ViperCharts is a web application running in the client's Web browser. The server side is written in Python[8] and uses the Django Web Framework[9].

Creating a chart requires two simple steps. First, the user selects the desired chart type, choosing among Scatter chart, Curve chart and Column chart. Where Scatter charts visualize the performance of discrete prediction algorithms in the ROC or PR space, Curve charts visualize Lift, ROC, PR and Cost curves, whereas Column charts visualize arbitrary values for a selection of algorithms. Second, the required data for the specific chart is copy-pasted into the provided form. Different data input formats are supported: TAB delimited data for copy-pasting from spreadsheet applications, CSV data and JSON formatted data. Finally, a *Draw chart* button triggers the chart visualization.

---

[5] http://www.highcharts.com
[6] http://d3js.org
[7] https://developers.google.com/chart/
[8] http://www.python.org/
[9] http://www.djangoproject.com

The desired performance measures are calculated from the provided data and stored in the platforms' database. In this way the visualization components have access to the data of a specific chart whenever the chart needs to be displayed.

Charts are drawn with the help of a JavaScript charting library, called Highcharts[5]. For each chart type we created a specific template including individual functionality available for a certain type of performance visualization. For example, in PR space charts we included the novel *F-isoline* evaluation approach [10] which enables to simultaneously visually evaluate algorithm performance in terms of recall, precision and the *F*-measure. As additional novelty, for ROC curve charts a corresponding PR curve chart can be created (and vice versa), since PR curves give a more informative picture of the algorithm's performance when dealing with highly skewed datasets, which provides additional insight for algorithms design, as discussed in [2]. A screenshot of the ViperCharts platform showing a chart of the PR space with *F*-isolines can be found in Figure 1.

The ViperCharts platform enables users to make their charts public and share them with research colleagues, export and publish them in research papers, or include them in their web sites. The sharing and embedding of charts is made easy by a unique URL address assigned to each chart. Exporting the charts into SVG, JPG, PNG or PDF formats is taken care of by the exporting module of the charting library that we use for chart visualization. A web API provides direct access to our performance visualization services, which were also integrated in the cloud based data mining platform, called ClowdFlows [7].

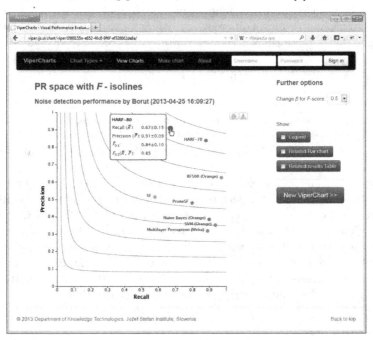

Fig. 1. A screenshot of the ViperCharts platform showing the performance of discrete prediction algorithms in the PR space enhanced by *F*-isolines.

3   Conclusion and Further Work

We have developed a novel web-based platform for visual performance evaluation of classification, prediction and information retrieval algorithms. The web-based design of the platform allows users to create, save and export their results' visualizations, without any installation needed. Users can easily publish and share their charts, since public charts can be simply accessed through unique web addresses. The performance visualizations can also be accesses through a web API or used as integrated components in the ClowdFlows data mining platform. We are working on covering even a wider range of performance visualizations used by machine learning practitioners, as well as including support for statistical significance tests. We will continue the development on the support for different input data formats and on the relations and conversions between different performance visualizations. We also plan to extend our API to enable integration into several existing data mining environments.

Acknowledgments. This work was partially funded by the Slovenian Research Agency and by the European Commission in the context of the FP7 project FIRST under the grant agreement n. 257928.

References

1. Berthold, M.R., Cebron, N., Dill, F., Gabriel, T.R., Kötter, T., Meinl, T., Ohl, P., Thiel, K., Wiswedel, B.: KNIME - the Konstanz information miner: version 2.0 and beyond. SIGKDD Explor. Newsl. 11(1), 26–31 (2009)
2. Davis, J., Goadrich, M.: The relationship between Precision-Recall and ROC curves. In: Proceedings of the 23rd ICML Conference, pp. 233–240 (2006)
3. Demšar, J., Zupan, B., Leban, G., Curk, T.: Orange: From experimental machine learning to interactive data mining. In: Boulicaut, J.-F., Esposito, F., Giannotti, F., Pedreschi, D. (eds.) PKDD 2004. LNCS (LNAI), vol. 3202, pp. 537–539. Springer, Heidelberg (2004)
4. Drummond, C., Holte, R.C.: Cost curves: An improved method for visualizing classifier performance. Machine Learning 65(1), 95–130 (2006)
5. Flach, P.A.: ROC Analysis. In: Encyclopedia of Machine Learning, pp. 869–875 (2010)
6. Hall, M., Frank, E., Holmes, G., Pfahringer, B., Reutemann, P., Witten, I.H.: The WEKA data mining software. SIGKDD Explorations Newsletter 11(1), 10–18 (2009)
7. Kranjc, J., Podpečan, V., Lavrač, N.: ClowdFlows: A cloud based scientific workflow platform. In: Flach, P.A., De Bie, T., Cristianini, N. (eds.) ECML PKDD 2012, Part II. LNCS, vol. 7524, pp. 816–819. Springer, Heidelberg (2012)
8. Manning, C.D., Raghavan, P., Schütze, H.: Introduction to Information Retrieval. Cambridge University Press (2008)
9. Mierswa, I., Wurst, M., Klinkenberg, R., Scholz, M., Euler, T.: YALE: Rapid prototyping for complex data mining tasks. In: ACM SIGKDD, pp. 935–940 (2006)
10. Sluban, B., Gamberger, D., Lavrač, N.: Ensemble-based noise detection: noise ranking and visual performance evaluation. In: Data Mining and Knowledge Discovery (2013), doi: 10.1007/s10618-012-0299-1
11. Witten, I.H., Frank, E.: Data Mining: Practical Machine Learning Tools and Techniques, 2nd edn. Morgan Kaufmann Publishers Inc. (2005)

Entityclassifier.eu: Real-Time Classification of Entities in Text with Wikipedia

Milan Dojchinovski[1,2] and Tomáš Kliegr[2]

[1] Web Engineering Group
Faculty of Information Technology
Czech Technical University in Prague
milan.dojchinovski@fit.cvut.cz
[2] Department of Information and Knowledge Engineering
Faculty of Informatics and Statistics
University of Economics, Prague, Czech Republic
tomas.kliegr@vse.cz

Abstract. Targeted Hypernym Discovery (THD) performs unsupervised classification of entities appearing in text. A hypernym mined from the free-text of the Wikipedia article describing the entity is used as a class. The type as well as the entity are cross-linked with their representation in DBpedia, and enriched with additional types from DBpedia and YAGO knowledge bases providing a semantic web interoperability. The system, available as a web application and web service at entityclassifier.eu, currently supports English, German and Dutch.

1 Introduction

One of the most significant challenges in text mining is the dimensionality and sparseness of the textual data. In this paper, we introduce Targeted Hypernym Discovery (THD), a Wikipedia-based entity classification system which identifies salient words in the input text and attaches them with a list of more generic words and concepts at varying levels of granularity. These can be used as a lower dimensional representation of the input text.

In contrast to the commonly used dimensionality reduction techniques, such as PCA or LDA, which are sensitive to the amount of data, THD provides the same quality of output for all sizes of input text, starting from just one word. Since THD extracts these types from Wikipedia, it can also process infrequent, but often information-rich words, such as named entities. Support for live Wikipedia mining is a unique THD feature allowing coverage of "zeitgeist" entities which had their Wikipedia article just established or updated.

THD is a fully unsupervised algorithm. A class is chosen for a specific entity as the one word (concept) that best describes its type according to the consensus of Wikipedia editors. Since the class (so as the entity) is mapped to DBpedia, the semantic knowledge base, one can traverse up the taxonomy to the desired class granularity. Additionally, the machine-readable information obtainable on the disambiguated entity and class from DBpedia and YAGO can be used for feature enrichment.

H. Blockeel et al. (Eds.): ECML PKDD 2013, Part III, LNAI 8190, pp. 654–658, 2013.

2   Architecture

THD is implemented in Java on top of the open source GATE framework[1].

Entity extraction module identifies entity candidates (noun phrases) in the input text. Depending on setting, entities can be restricted to named entities ("Diego Maradona") or common entities ("football").

Disambiguation module assigns entity candidate with a Wikipedia entry describing it. This module combines textual similarity between the entity candidate and article title with the importance of the article.

Entity classification module assigns each entity with one or more hypernyms. The hypernyms are mined with the THD algorithm (see Sec. 3) from the Wikipedia articles identified by the Disambiguation module. This mining is performed either on-line from live Wikipedia or from a Wikipedia mirror. The default option is to use the Linked Hypernyms Dataset, which contains 2.5 million article-hypernym pairs precomputed from a Wikipedia mirror.

Semantization module maps the entity as well as the class to DBpedia.org concepts. A "semantic enrichment" is also performed: once the entity is mapped, additional types are attached from DBpedia [1] and YAGO [2], the two prominent semantic knowledge bases. The final set of types returned for an entity thus contains the "linked hypernym" (hypernym mapped to DBpedia obtained with THD), and a set of DBpedia and YAGO types.

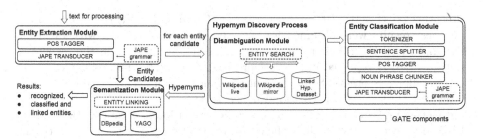

Fig. 1. Architecture overview

3   Hypernym Discovery Algorithm and Benchmark

Hypernym discovery is performed with hand-crafted lexico-syntactic patterns. These were in the past primarily used on larger text corpora with the intent to discover all word-hypernym pairs in the collection [7]. With *Targeted* Hypernym Discovery we apply lexico-syntactic patterns on a *suitable document* (Wikipedia article) with the intent to extract *one hypernym* at a time (details in [3,4]).

THD performance was measured on the following benchmarks independent on the input text: a) discovering correct hypernym given a Wikipedia article, b) linking hypernym to a semantic web identifier. The outcome of the evaluation[2]

---

[1] http://gate.ac.uk
[2] The results and the "High accuracy dataset" are available at
   http://ner.vse.cz/datasets/linkedhypernyms/.

Fig. 2. Screenshot of the system (edited to fit the page)

altogether on 16.500 entity articles (English, German, Dutch) is reported in [3]. The best results were obtained for the German person subset, with precision 0.98 and recall 0.95. This is on par with the the best results in the respective metrics recently reported in [5]: 0.97 precision for lexico-syntactic patterns and 0.94 recall for Syntactic-Semantic Tagger. The overall accuracy of discovering plain text (linked) hypernyms for English is 0.95 (0.85), for Dutch 0.93 (0.88) and German 0.95 (0.77). These numbers provide a lower bound on the error of THD, since they do not include the entity recognition error and particularly the disambiguation error (matching entity with a Wikipedia article).

4   Comparison with Related Systems

While techniques for Named Entity Recognition and classification (NER) are well-researched, NER classifiers typically need to be trained on large labeled document corpora, which generally involve only several labels, making them unsuitable for dimensionality reduction. Replacement of "Maradona" with "Person" loses too much meaning for most applications. The recent shift from human-annotated corpora to Wikipedia in some systems allows to provide types with finer granularity, and also broadening of the scope to "common" entities. In this section (and accompanying screencasts), we present a comparison with two best-known academic systems DBpedia Spotlight [6] and AIDA [8].

Real-time Mining. THD directly incorporates a text mining algorithm. Once an entity is disambiguated to a Wikipedia article, the system retrieves the article from Wikipedia and extracts the hypernym from its free text. The mining speed is about 1 second per entity including network overhead. This allows to discover types for entities, which had their article only recently added to Wikipedia, or adapt to changes in Wikipedia. The authors are not aware of any other system that incorporates query-time Wikipedia mining. AIDA and DBpedia Spotlight lookup the disambiguated entity in a database of types.

Complementarity to other Systems. Since THD extracts the types from *free text*, the results are largely complementary to types returned by other Wikipedia-based systems. These typically rely on DBpedia or YAGO knowledge-bases, which are populated from article *categories* and *"infoboxes"*, the semistructured information in Wikipedia. As a convenience, THD returns types from DBpedia and YAGO in addition to the mined hypernym. The complementary character of the results can be utilized for classifier fusion.

Right Granularity. For many entities DBpedia and YAGO-based systems provide a long list of possible types. For example, DBpedia assigns Diego Maradona with 40 types including `dbpedia-owl:SoccerManager`, `foaf:Person` as well as the highly specific `yago:1982FIFAWorldCupPlayers`. THD aids the selection of the "right granularity" by providing the most frequent type, as selected by Wikipedia editors for inclusion into the article's first sentence. For Maradona, as of time of writing, THD returns "manager".[3]

Multilinguality. System currently supports English, Dutch and German, extensibility to a new language requires only providing two JAPE grammars and plugging in correct POS tagger (ref. to Fig. 2). DBpedia Spotlight and AIDA support only English.

Acknowledgements. This research was supported by the European Union's 7th Framework Programme via the LinkedTV project (FP7-287911) and CTU in Prague grant (SGS13/100/OHK3/1T/18).

References

1. Bizer, C., et al.: DBpedia - a crystallization point for the web of data. Web Semant 7(3), 154–165 (2009)
2. Hoffart, J., Suchanek, F.M., Berberich, K., Weikum, G.: YAGO2: A spatially and temporally enhanced knowledge base from Wikipedia. Artificial Intelligence 194, 28–61 (2013)
3. Kliegr, T., Dojchinovski, M.: Linked hypernyms: Enriching DBpedia with Targeted Hypernym Discovery (Submitted)
4. Kliegr, T., et al.: Combining captions and visual analysis for image concept classification. In: MDM/KDD 2008. ACM (2008)
5. Litz, B., Langer, H., Malaka, R.: Sequential supervised learning for hypernym discovery from Wikipedia. In: Fred, A., Dietz, J.L.G., Liu, K., Filipe, J. (eds.) IC3K 2009. CCIS, vol. 128, pp. 68–80. Springer, Heidelberg (2011)

---

[3] As demonstrated in [4], the algorithm used can also return multi-word hypernyms ("soccer manager"). This feature is not yet available in THD.

6. Mendes, P.N., Jakob, M., Garcia-Silva, A., Bizer, C.: DBpedia spotlight: Shedding light on the web of documents. In: I-Semantics (2011)
7. Snow, R., Jurafsky, D., Ng, A.Y.: Learning syntactic patterns for automatic hypernym discovery. In: Advances in Neural Information Processing Systems, vol. 17, pp. 1297–1304. MIT Press, Cambridge (2005)
8. Yosef, M.A., et al.: AIDA: An online tool for accurate disambiguation of named entities in text and tables. PVLDB 4(12), 1450–1453 (2011)

Hermoupolis: A Trajectory Generator
for Simulating Generalized Mobility Patterns

Nikos Pelekis[1], Christos Ntrigkogias[2], Panagiotis Tampakis[2],
Stylianos Sideridis[2], and Yannis Theodoridis[2]

[1] Dept. of Statistics and Insurance Science, Univ. of Piraeus, Greece
npelekis@unipi.gr
[2] Dept. of Informatics, Univ. of Piraeus, Greece
xdrigog@gmail.com, {ptampak,ytheod}@unipi.gr, siderste@yahoo.gr

Abstract. During the last decade, the domain of mobility data mining has emerged providing many effective methods for the discovery of intuitive patterns representing collective behavior of trajectories of moving objects. Although a few real-world trajectory datasets have been made available recently, these are not sufficient for experimentally evaluating the various proposals, therefore, researchers look to synthetic trajectory generators. This case is problematic because, on the one hand, real datasets are usually small, which compromises scalability experiments, and, on the other hand, synthetic dataset generators have not been designed to produce mobility pattern driven trajectories. Motivated by this observation, we present *Hermoupolis*, an effective generator of synthetic trajectories of moving objects that has the main objective that the resulting datasets support various types of mobility patterns (clusters, flocks, convoys, etc.), as such producing datasets with available ground truth information.

Keywords: Mobility Data Mining, Trajectory Patterns, Synthetic Generators.

1 Introduction

The explosion of mobile devices and positioning technologies has now made possible and easier the collection of trajectory data of moving objects. The rapid growth of these technologies has increased the interest for data analysis upon trajectory datasets. As such, the field of mobility data mining has already many success stories to narrate, as these are described by works that identify various types of patterns, including, among others, clusters of entire trajectories [11] or of sub-trajectories [9][14], moving clusters [7], flocks [8][5], sequential trajectory patterns [3], convoys [6], swarms [10], and top-k representative trajectory samples [12].

The effectiveness of most of the afore-mentioned methods has been evaluated with the use of small datasets w.r.t. potential sizes of real-world datasets, which however, are not available, usually due to privacy issues. Even when datasets are available, the ground truth for such kind of patterns is absent, thus researchers have to evaluate their proposals with general-purpose validation metrics (e.g. intra vs. inter cluster distance). On the other hand, utilizing synthetic generators is a typical approach for

H. Blockeel et al. (Eds.): ECML PKDD 2013, Part III, LNAI 8190, pp. 659–662, 2013.
© Springer-Verlag Berlin Heidelberg 2013

researchers since it can support scalability experiments, however, synthetic datasets cannot guarantee the cardinality (or even the existence) of patterns within the synthetic population. For instance, experimentation of density-based clustering algorithms, like T-OPTICS [11] and TRACLUS [9], may be biased if the distribution of the data under experimentation does not include a sufficient number of density-connected groups of objects; similarly for other mobility patterns. We argue that the effectiveness, efficiency and scalability experiments should not be applied independently and over different datasets. Like efficiency and scalability, effectiveness should be tested in very large datasets; like effectiveness, efficiency and scalability should be tested in datasets that include patterns of varying, known cardinality. Only this way, experimental results are interpretable and useful.

To meet the above-described requirements, we present *Hermoupolis*[1], a *pattern-aware synthetic trajectory generator*, which produces annotated trajectories of moving objects following given mobility patterns. These mobility patterns imply the different profiles of movement that we want to reproduce, covering many of the examples cited earlier. On the other hand, related work includes data generators simulating either movement in free space, including GSTD [15], CENTRE [4] and C4C [11], or network-constrained movement, including Brinkhoff [1] and BerlinMOD [2]. Although the above generators present very interesting features, they cannot be considered pattern-aware, in the sense that we described earlier.

In what follows, Section 2 presents the details about *Hermoupolis* methodology and Section 3 describes demo specifications.

2   Generating Pattern-Aware Synthetic Trajectories

Hermoupolis takes as input a set of *Generalized Mobility Patterns* (GMP) along with a road network and a set of *Points of Interest* (PoI), and generates a set of network-constrained trajectories conforming to the requirements posed by *GMP*.

More formally, a *GMP* is a time ordered sequence of pairs <AMP, c>, where c is the cardinality of the trajectories that will be simulated and *AMP* is an *Atomic Mobility Pattern*. In turn, a *AMP* is a triplet <MBB, MFV, AFV>, where *MBB* is a *Minimum Bounding Box* that approximates the spatio-temporal space where the motion of simulated trajectories takes place, *MFV* is a *Movement Feature Vector* containing parameters that affect movement (distribution of speed, duration, agility, etc.) and *AFV* is an *Annotation Feature Vector* that contains textual information that is used to annotate the simulated recordings. Having in mind the recent advances in *semantic trajectory* modeling [13], *AFV* includes meta-information about a Stop at a *PoI*, where the user performed e.g. a leisure activity, or a Move (or trip) between two *PoI*, in-between which was performed e.g. by foot for a fitness activity.

In order to perform the trajectory generation task, *Hermoupolis* simulates all concurrent *AMP* from the set of *GMP*, while at the same time it interprets the sequence of each *GMP* and appropriately simulates Stops and Moves. Stop could be either stillness at or jerky movement around a *PoI*; Move could be a navigation from one Stop to another.

The simulation of the trips between Stops is actually performed by the well-known Brinkhoff data generator [1], which has been appropriately adapted in order to

---

[1] Hermoupolis, polis (=town) of Hermes, is the capital of Cyclades prefecture in Greece.

constrain movements according to the spatio-temporal restrictions set by *MBB* and *MFV*. Moreover, road networks and PoI databases that are public available at OpenStreetMap open source repository can be automatically loaded in *Hermoupolis*.

3   Demo Specifications

Hermoupolis design allows for the simulation of an extensive range of mobility patterns. For instance, Fig. 1(a) illustrates the generation of a set of trajectories representing the mobility behaviour of six different profiles of people (each depicted with a different colour) throughout a weekday. Although non-visible, each of these profiles corresponds to a *GMP* and is accompanied by a respective activity pattern (e.g. the green profile reflects young, single, working men following Home – Campus – Leisure - Home pattern). Obviously, such a dataset is appropriate for evaluating clustering techniques, such as [11] and [9].

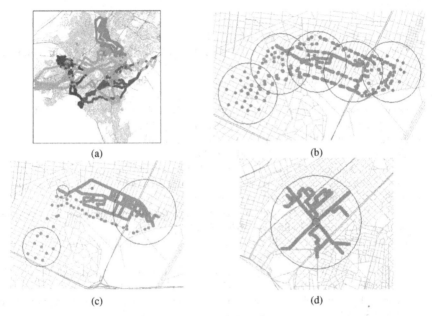

(a)   (b)

(c)   (d)

Fig. 1. Hermoupolis in action

Moreover, as a food for thought, in Fig. 1(b-d), we illustrate three simple different *GMP* aiming to simulate mobility patterns already available in the literature. In particular, Fig. 1(b) illustrates a *GMP* consisting of 9 *AMP* (5 Stops, for which we visualize their bounding circles, and 4 Moves). As the respective *MBB* of the 5 Stops are overlapping, one expects that some kind of flock or convoy pattern be hidden in the resulted trajectories. Then, in Fig. 1(c), we depict a *GMP*, composed of 3 Stops with varying spatial extent (indicated by the circles' radiuses). The first Move (between the first and second Stop) has large speed and small agility, while the opposite is true for the second Move (between the second and third Stop). We argue that such a simulation is appropriate for evaluating both sub-trajectory clustering algorithms, as well as moving object clusters like convoys, swarms, even patterns that can capture the thickness of the mobility patterns. Finally, in Fig. 1(d), we exhibit a *GMP* that simulates objects starting from a wide Stop (the large circle) routed towards a short Stop (the small circle) that lies inside the first! Intuitively, this is a simple way of simulating that several people from a region converge to a "meeting place".

Throughout the demonstration[2], users will be able to interact with *Hermoupolis* and generate the volumes of synthetic trajectories they wish by simulating various mobility patterns, such as the ones discussed above.

Acknowledgements. This work has been supported by the European Union's FP7-DATASIM (grant 270833) and IRSES-SEEK (grant 295179) projects.

References

1. Brinkhoff, T.: Generating network-based moving objects. GeoInformatica 6(2), 153–180 (2002)
2. Duntgen, C., Behr, T., Guting, R.H.: BerlinMOD: a benchmark for moving object databases. The VLDB Journal 18(6), 34 (2008)
3. Giannotti, F., Nanni, M., Pedreschi, D., Pinelli, F.: Trajectory Pattern Mining. In: Proc. of SIGKDD (2007)
4. Giannotti, F., Mazzoni, A., Puntoni, S., Renso, C.: Synthetic generation of cellular network positioning data. In: Proc. of ACM GIS, pp. 12–20 (2005)
5. Gudmundsson, J., Kreveld, M.J., Speckmann, B.: Efficient detection of patterns in 2d trajectories of moving points. GeoInformatica 11(2), 195–215 (2007)
6. Jeung, H., Yiu, M.L., Zhou, X., Jensen, C.S., Shen, H.T.: Discovery of Convoys in Trajectory Databases. In: Proc. of VLDB (2008)
7. Kalnis, P., Mamoulis, N., Bakiras, S.: On discovering moving clusters in spatio-temporal data. In: Medeiros, C.B., Egenhofer, M., Bertino, E. (eds.) SSTD 2005. LNCS, vol. 3633, pp. 364–381. Springer, Heidelberg (2005)
8. Laube, P., Imfeld, S., Weibel, R.: Discovering relative motion patterns in groups of moving point objects. IJGIS 19(6), 639–668 (2005)
9. Lee, J.G., Han, J., Whang, K.Y.: Trajectory clustering: A partition-and-group framework. In: Proc. of SIGMOD, pp. 593–604 (2007)
10. Li, Z., Ding, B., Han, J., Kays, R.: Swarm: mining relaxed temporal moving object clusters. In: Proc. of PVLDB, vol. 3(1-2), pp. 723–734 (2010)
11. Nanni, M., Pedreschi, D.: Time-focused clustering of trajectories of moving objects. JIIS 27(3) (2006)
12. Panagiotakis, C., Pelekis, N., Kopanakis, I., Ramasso, E., Theodoridis, Y.: Segmentation and Sampling of Moving Object Trajectories based on Representativeness. In: TKDE (2011)
13. Parent, C., Spaccapietra, S., Renso, C., Andrienko, G., Andrienko, N., Bogorny, V., Damiani, M.L., Gkoulalas, D.A., Macedo, J.A., Pelekis, N., Theodoridis, Y., Yan, Z.: Semantic trajectories modeling and analysis. ACM Computing Surveys 35(4) (2013)
14. Pelekis, N., Kopanakis, I., Kotsifakos, E., Frentzos, E., Theodoridis, Y.: Clustering Uncertain Trajectories. KAIS 28(1), 117–147 (2011)
15. Theodoridis, Y., Silva, J.R.O., Nascimento, M.A.: On the generation of spatiotemporal datasets. In: Proc. of SSD (1999)

---

[2] A preview of *Hermoupolis* including screenshots from the GUI, and an accompanying video is available at: http://infolab.cs.unipi.gr/pubs/pkdd2013/.

AllAboard: A System for Exploring Urban Mobility and Optimizing Public Transport Using Cellphone Data

Michele Berlingerio, Francesco Calabrese, Giusy Di Lorenzo, Rahul Nair,
Fabio Pinelli, and Marco Luca Sbodio

IBM Research, Dublin, Ireland
{mberling,fcalabre,giusydil,rahul.nair,fabiopin,marco.sbodio}@ie.ibm.com

1 Introduction

The deep penetration of mobile phones offers cities the ability to opportunistically monitor citizens' interactions and use data-driven insights to better plan and manage services. In this context, transit operators can leverage pervasive mobile sensing to better match observed demand for travel with their service offerings. With large scale data on mobility patterns, operators can move away from the costly and resource intensive transportation planning processes prevalent in the West, to a more data-centric view, that places the instrumented user at the center of development. In this framework, using mobile phone data to perform transit analysis and optimization represents a new frontier with significant societal impact, especially in developing countries.

In this demo, we present AllAboard, a system for optimizing public transport using cellphone data. Our system uses mobile phone location data to infer origin-destination flows in the city, which is then converted to ridership on the existing transit network. Sequential travel patterns extracted from individual call location data are used to propose new candidate transit routes. An optimization model evaluates which new routes would best improve the existing transit network to increase ridership and user satisfaction, both in terms of reduced travel and wait time. The system provides also a User Interface that allows the interaction with results and the data themselves. The system in its whole is intended to be used by city authorities for improving their public transport systems, using cell phone data, which have a large penetration even in developing countries, and provide a cheaper, faster, alternative to costly surveys.

The system has been tested using Call Detail Record data from Orange for the city of Abidjan, Ivory Coast, with the focus to improve the existing transit network. Four new routes have been proposed by the optimization system, resulting in an expected reduction of 10% city-wide travel times.

Several projects deal with the analysis of mobility and mobile phone data [2,1]. They present powerful mining engines but do not provide direct interaction with data and results. Other projects providing visualization and interaction, on the other hand, do not integrate the optimisation based on analytical results [3,4]. Our system integrates all these modules: the mobility analysis engine, the

H. Blockeel et al. (Eds.): ECML PKDD 2013, Part III, LNAI 8190, pp. 663–666, 2013.

optimisation, and an interactive user interface, providing a new environment to extract and use information from mobile phone data.

2   The AllAboard Platform

The system is implemented with a modular architecture. We isolate our data models using an abstraction layer that separates algorithm implementations from data stores. The current version of the system includes two core modules: *Mobility Mining* and *Optimizer*. These modules are implemented as components within an extensible framework, and other components can easily be added in the future. Each component provides a lightweight REST service exposing its functionalities. The REST services are also used to implement the AJAX-based Web user interface. The main algorithms for the current two cores are described in Sections 2.1, and 2.2 respectively.

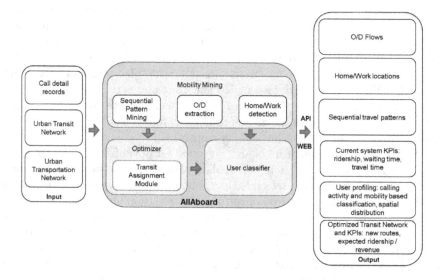

Fig. 1. Architecture of the AllAboard platform

2.1   Mobility Mining

This module is able to process mobile phone data in the form of records: (userID, timestamp, cellID), where a location is associated to each cellID, and extract information about users' stops, trajectories, Origin/Destination (O/D) flows (i.e. number of people moving from the origin O to the destination D in a given 1 hour time interval), frequent travel sequences, and home and work locations for each user.

The methodology to process the data follows these steps:

1. we extract the location of the stops performed by users
2. based on those, we estimate the O/D flows, used to feed the optimization module presented in Section 2.2
3. we exploit the data to better understand the mobility of the users, and use the results as additional input to the optimization module. For this, we extract frequent sequential patterns from the sequences of stops
4. we identify for each user, when possible, most likely home and work locations based on both stops and mobile phone activity patterns

The results of this process are visualised by the framework, allowing the user to interact with them. For instance, the user can select a particular antenna and then the system visualises its O/D flows and the relative temporal profile. The system is also able to show the sequential patterns emerging over time from the mobile phone data, highlighting the most frequent ones. An example of the two visualisations is represented in Figure 2.

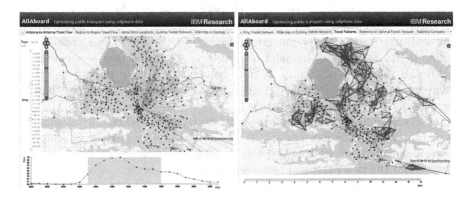

Fig. 2. Exploring mobility with AllAboard: O/D flows with temporal profiles (left); sequential patterns emerging from the data (right)

2.2 Optimal Transit Design

Given (a) an existing transit network, (b) O/D flows derived from mobile data representing travel demand, (c) a set of frequent sequences that serve as candidate new routes, (d) travel time estimates across the network, and (e) a resource budget in terms of fleet size, this module is able to determine an optimal set of new routes and their associated service frequencies, such that passenger journey times city-wide are minimized. A new route is defined by a sequence of transit stops and has an associated frequency.

The problem is strategic in nature as it represents a longer-term decision on the part of a public transport operator. The addition of new routes to the service network are intended to match current supply with revealed demand. From a

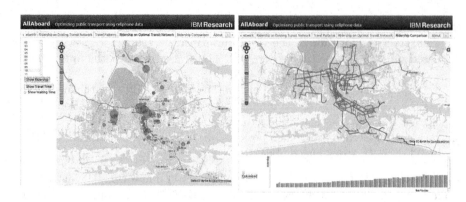

Fig. 3. Exploring optimization results with AllAboard: waiting times (left); comparing ridership on optimized network (right)

demand perspective, new routes will attract riders only if they offer competitive service to existing routes. The model therefore considers the user perspective, in terms of optimal strategies. From a supply side, new routes should fill service gaps and map user activity patterns observed in the data. The optimization routine therefore includes frequent sequences as potential new services. Taken together, a potential new route is only recommended if it directly addresses under-serviced demand, and does so by offering shorter journey times than the ones possible on the existing network.

The system offers an interactive User Interface that allows to visualise the results of the optimisation module, in particular in terms of ridership improvements, and decrease in waiting times, comparing the results between the existing network and the new extended one (see Figure 3).

Acknowledgements. We wish to thank the Orange D4D Challenge (http://www.d4d.orange.com) organizers for releasing the data we used for building and testing our prototype.

References

1. Calabrese, F., Ratti, C.: Real time rome. Networks and Communications Studies 20(3-4), 247–258 (2006)
2. Eagle, N., Sandy Pentland, A.: Reality mining: sensing complex social systems. Personal Ubiquitous Comput. 10(4), 255–268 (2006)
3. Giannotti, F., Nanni, M., Pedreschi, D., Pinelli, F., Renso, C., Rinzivillo, S., Trasarti, R.: Unveiling the complexity of human mobility by querying and mining massive trajectory data. The VLDB Journal 20(5), 695–719 (2011)
4. Zheng, Y., Zhou, X. (eds.): Computing with Spatial Trajectories. Springer (2011)

SciénScan – An Efficient Visualization and Browsing Tool for Academic Search

Daniil Mirylenka and Andrea Passerini

Department of Information Engineering and Computer Science
University of Trento, Via Sommarive 5, 38123, Trento, Italy
{dmirylenka,passerini}@disi.unitn.it

Abstract. In this paper we present ScienScan[1] – a browsing and visualization tool for academic search. The tool operates in real time by post-processing the query results returned by an academic search engine. ScienScan discovers topics in the search results and summarizes them in the form of a concise hierarchical topic map. The produced topical summary informatively represents the results in a visual way and provides an additional filtering control. We demonstrate the operation of ScienScan deploying it on top of the search API of Microsoft Academic Search.

1 Introduction

We often use academic search engines for exploratory tasks, such as reviewing related work or investigating unfamiliar topics, when the goal is not to retrieve specific publications but rather to improve our understanding of the topic in question. Performing exploratory search, we tightly interact with the search engine, refining our queries based on the retrieved results. Presented as endless unstructured lists, typical search results are difficult to examine and interpret, often making our exploratory search task tedious and even frustrating.

In this paper we describe ScienScan – an academic search tool that provides concise visual summaries of the query results. These summaries convey useful information about the topical structure of the result set in an intuitive way, without the user having to sift through individual items. In addition, the produced summaries serve as a filtering control, allowing the user to focus on the relevant subtopics of the query, and thus find papers more efficiently.

ScienScan presents a novel approach to visualization and browsing of academic search results. It employs state-of-the-art external tools and services as well as newly developed methods, and can run on top of third-party search engines. Based on the practical solutions, ScienScan is, to the best of our knowledge, the only available prototype tool providing this type of functionality.

2 Web Interface

ScienScan is a Web tool with an interface of a typical search engine (see Figure 2). Users type queries into the search box and obtain the list of search results. In

---

[1] http://scienscan.disi.unitn.it/

H. Blockeel et al. (Eds.): ECML PKDD 2013, Part III, LNAI 8190, pp. 667–671, 2013.

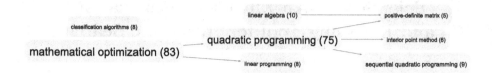

Fig. 1. Example summary of 100 search results for the query "quadratic programming"

addition to the standard controls, ScienScan displays a topic map of the retrieved results. The topic map is a small taxonomy of topics built so as to summarize the search results in the most informative way (see Figure 1 for an example).

The topic map is a directed acyclic graph, in which the child nodes represent the subtopics of the parent nodes. The nodes in the topic map represent topics relevant to the search results. Each topic covers a subset of the results in such a way that a parent topic always covers all the results covered by its child topics and, possibly, some additional results. The number of covered results is displayed in the parentheses near the topic title, with the font size of the title being proportional to this number. When the user clicks on a topic in the hierarchy, the topic and its subtopics become highlighted, and the displayed results get restricted to those covered by the selected topic.

The user can control the number of nodes in the topic map by moving the slider. ScienScan builds multiple instances of the topic maps of various sizes, and moving the slider switches between these instances, making the displayed topic map grow or shrink visually. The algorithms of ScienScan are implemented in such a way that bigger topic maps are built incrementally from smaller ones. This makes the computations efficient, and ensures that no topic disappears from the map when the map size is increased.

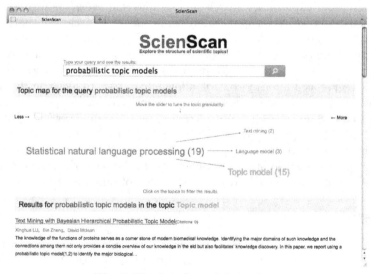

Fig. 2. The interface of ScienScan

3   Architecture

When the user submits the search query, ScienScan performs the following steps:

1. forward the query to an existing search engine and collect the results;
2. link the results to Wikipedia articles with the help of a topic annotator;
3. build the topic graph based on the retrieved Wikipedia articles, their categories and hierarchical relations between them;
4. summarize the constructed topic graph into a concise topic map;
5. visualize the topic map using a graph-drawing tool.

A detailed description of the steps 2-5 can be found in [4].

Search Engine. The current version of ScienScan relies on the search API of `Microsoft Academic Search`. The API restricts usage to 200 requests per minute and returns up to 100 search results per request. The latter restriction is not too limiting, as users seldom look beyond the first one hundred results.

Topic Annotation. During this step the search result titles and abstracts are mapped to Wikipedia articles, which can be viewed as fine-grained meaningfully labeled topics. Recently, there have been proposed multiple methods and tools for annotating texts with links to Wikipedia articles. Some of the implementations have publicly available demos, restricted web service APIs, and standalone tools requiring a snapshot of Wikipedia database (see, for example, TAGME, Machine Linking, DBPediaSpotlight and Wikipedia Miner[3]). We deployed an instance of the Wikipedia Miner web service for the purpose of topic annotation.

Building the Topic Graph. The following steps expand the set of article topics into a large topic graph based on the network of Wikipedia categories:

1. retrieve the parent categories of the discovered articles (transforming the set of articles into a bipartite article-category graph);
2. connect the categories according to their taxonomic relations in Wikipedia (transforming the bipartite graph into a general directed topic graph);
3. merge similar topics and break the cycles (making the topic graph acyclic);
4. detect and extend the main topic of the query (that with the most search results, making the topic graph more detailed in the area of the main topic).

Summarizing and Displaying the Topic Graph. At this step we have to reduce the graph containing about three hundred nodes to a concise taxonomy, such as shown in Figure 1. A good taxonomy must possess a number of important properties, such as high coverage of the search results, relevance, high frequency and low redundancy of the included topics. The current version of ScienScan applies a frequency-based heuristic algorithm to select the most informative set of nodes from the topic graph. A new version being under development uses a more advanced summarization algorithm based on structured-output prediction [4]. The mentioned algorithms prescribe which topics from the original topic graph should be included into the summary. In order to completely define the summary, we connect the topics with the minimum number of links that still maintain the hierarchical relations induced by the original graph. After the topic map is built, we submit it to the `graphviz` package [1] for visualization. The `dot` algorithm used in `graphviz` for drawing directed graphs produces the layered layout appropriate for displaying topic hierarchies.

4   Related Work and Discussion

To the best of our knowledge, there exist no other online academic search tools providing structured visual representations of the query results. Current popular scholarly services include publishers' digital libraries (such as ACM or IEEE), search engines (Google Scholar, Microsoft Academic Search or CiteSeer) and social networking sites (Mendeley, CiteULike, ResearchGate). These services typically provide browsing based on metadata, such as publication venue, authors or year, or a predefined topic categorization scheme. As categorization schemes are independent of the current query and rather coarse-grained, no visualization of the search results is provided based on them, nor on the metadata attributes.

In contrast to available tools, in the literature there have been proposed numerous sophisticated methods for detecting and visualizing research topics. The main approaches to this problem include frequent keyword-based methods, analyses of citation graph, and probabilistic topic models, the latter probably representing the most developed class of methods (see [2] for an example). In the context of search result visualization, these methods have the following shortcomings: *a)* they typically require access to the whole corpus of papers rather than only current results, *b)* (except for keyword-based methods) they do not provide short meaningful labels for discovered topics. Keyword-based methods have an additional shortcoming in that the topics correspond to verbatim keywords. ScienScan avoids these shortcomings by relying on Wikipedia-based topics.

TAG MY SEARCH [5] is an example of Wikipedia-based topic discovery applied to a related task of *general Web search* result clustering. Unlike ScienScan, TAG MY SEARCH uses only articles but not categories of Wikipedia to represent topics, and thus performs flat rather then hierarchical grouping of the search results. Action Science Explorer and Sci^2 represent publication collections as networks (for instance, citation-based), and provide visualization and exploration tools typical for network analysis, such as clustering and filtering based on metadata and network statistics. In contrast, ScienScan builds a higher-level view that is focused on the explicit labeled semantic topics and their hierarchical relations.

Acknowledgments. This research was partially supported by grant PRIN 2009LNP494 (Statistical Relational Learning: Algorithms and Applications) from Italian Ministry of University and Research.

References

1. Gansner, E.R., North, S.C.: An open graph visualization system and its applications to software engineering. Software – Practice & Experience 30(11) (2000)
2. He, Q., Chen, B., Pei, J., Qiu, B., Mitra, P., Giles, L.: Detecting topic evolution in scientific literature: how can citations help? In: CIKM, pp. 957–966. ACM (2009)

3. Milne, D., Witten, I.H.: An open-source toolkit for mining wikipedia. Artificial Intelligence 194, 222–239 (2013)
4. Mirylenka, D., Passerini, A.: Learning to grow structured visual summaries for document collections. In: ICML Workshop on Structured Learning (2013)
5. Scaiella, U., Ferragina, P., Marino, A., Ciaramita, M.: Topical clustering of search results. In: WSDM, pp. 223–232. ACM (2012)

InVis: A Tool for Interactive Visual Data Analysis

Daniel Paurat and Thomas Gärtner

University of Bonn and Fraunhofer IAIS, Sankt Augustin, Germany
{daniel.paurat,thomas.gaertner}@uni-bonn.de
http://kdml-bonn.de

Abstract. We present **InVis**, a tool to visually analyse data by interactively shaping a two dimensional embedding of it. Traditionally, embedding techniques focus on finding one fixed embedding, which emphasizes a single aspects of the data. In contrast, our application enables the user to explore the structures of a dataset by observing and controlling a projection of it. Ultimately it provides a way to search and find an embedding, emphasizing aspects that the user desires to highlight.

1 Introduction

We present an application[1] that enables the user to layout a two dimensional embedding of a possibly higher dimensional dataset by selecting and rearranging some of the embedded data points as *control points*. Working with our application resembles observing the shadow of a higher dimensional object from different angles and actively reshaping it. As the constellation of control points and the projection angle are dependent, specifying where the shadow of the chosen control points falls to, enforces the rest of the embedding to follow, see Figure 1. Gradually rearranging the constellation also changes the shadow gradually. An example for this is depicted on a real world dataset in Figure 2.

Fig. 1. The projection can be controlled by arranging the control points (dataset from [7])

Fig. 2. Re-positioning the control point *gin and tonic* (green) influences the location of related, gin containing, cocktails (dark)

Embedding data into a lower dimensional space for visual analysis is a wide field that is approached by a lot of different techniques. Many of them are unsupervised, like the well known *principle component analysis* (PCA) [5], *Isomap*

---

[1] The tool can be downloaded under:
http://www-kd.iai.uni-bonn.de/index.php?page=software_details&id=31

H. Blockeel et al. (Eds.): ECML PKDD 2013, Part III, LNAI 8190, pp. 672–676, 2013.

[11], *Locally linear embedding* [10], *non-negative matrix factorization* [6], *archetypal analysis* [3] and *CUR decomposition* [4].

Apart from these unsupervised embedding techniques, there are methods that take supervision into account, like *guided locally linear embedding* [1] 'and *supervised PCA* [2]. Many of the classic embedding methods also have a semi supervised extensions [12]. One particularly interesting setting is utilizing *must-link* and *cannot-link* constraints [13]. In this paper we employ the semi-supervised *least squares projections* (LSP) [8, 9] method, which computes an embedding based on a set of exemplary embedded data points.

In contrast to other authors applying semi-supervised embedding techniques, our aim is not a fixed one-time-embedding. Our application rather exploits the influence of the control points in order to enable the user to shape and steer a life-updating embedding. This active layout approach ultimately empowers the user to highlight aspects of the dataset that he considers interesting. This is illustrated in Figure 3 on a selection of four persons from the *CMU Face Images* dataset. While a regular PCA embedding does not directly convey insights, arranging a few control points in different constellations, can highlight different semantic aspects of the data.

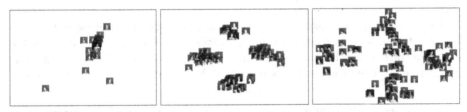

Fig. 3. A dataset of facial images embedded in different ways. The left figure shows a plain PCA embedding, while the other two figures use LSP to group the control points by person and by pose (*looking-straight, -up, -left* and *-right*), respectively.

2   Method

Consider a dataset X with n data records $x_1, ..., x_n$ from an instance space $\mathcal{X} \subseteq \mathbb{R}^d$ and the general task to map $\{x_1, ..., x_n\}$ into an embedding space $\mathcal{Y} \subseteq \mathbb{R}^2$, yielding $\{y_1, ...y_n\}$. To determine this mapping, the user chooses a set of k data records from X, denoted by $\hat{X}$, and fixes their coordinates in the embedding space, providing $\hat{Y}$. For the purpose of our application, we consider the desired projection $P : \mathcal{X} \to \mathcal{Y}$ to be the linear projection matrix with the least squared error in mapping $\hat{X}$ to $\hat{Y}$. Regarding $\hat{X}$ and $\hat{Y}$ as data matrices of shape $d \times k$ and $2 \times k$ we can formulate the system of linear equations $P\hat{X} \approx \hat{Y}$, which can be solved for P with least squared error efficiently, especially since the calculation only depends on k and not all n data points. The least squares projection matrix P is then used to determine the final embedding Y of all n data points X by matrix multiplication $PX = Y$. Note, that every time $\hat{Y}$

changes P has to be recalculated. P can be derived by right multiplying the pseudo inverse of $\hat{X}$ (given by $\hat{X}^\dagger = \hat{X}^T(\hat{X}\hat{X}^T)^\dagger$) to $\hat{Y}$. As long as the user only relocates the k control points, $\hat{X}^\dagger$ does not change and P can be determined by matrix multiplication with a time complexity of $\mathcal{O}(d^2 \cdot k)$. However, if the user alters the selection of the control points, the pseudo inverse $\hat{X}^\dagger$ has to be recalculated, which leads to an additional calculation with a time complexity of $\mathcal{O}(d^3 + d^2 \cdot k)$.

3   User Interface

Figure 4 shows the user interface of our application running an exemplary analysis of a cocktail ingredient dataset. The core of our tool is the interactive canvas on the left side, displaying the embedding. The initial control points are provided by five randomly chosen points, placed according to their coordinates of a PCA-embedding on the whole dataset. From here the user can interact with the canvas in the usual way by clicking and dragging. The user can select, or de-select a control point by middle-clicking it and he can reposition the point simply by left-clicking and dragging it to the new location. While relocating a control point, the embedding is constantly updated to provide the user with a "hands on" sensation.

To support practical usability of the application, we also provide some extra features that can help a user in the exploration process. The user can shift the center of the displayed data by *Ctrl*-dragging on the canvas and zoom in and out of different regions by using the mouse wheel. He can also search for a data record

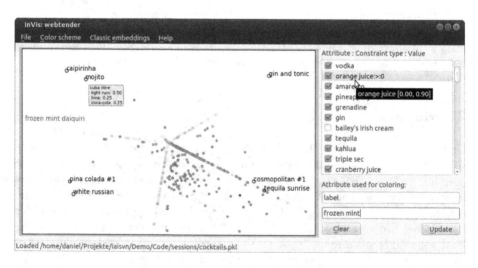

Fig. 4. A screenshot of InVis. The embedding shows some control points (black), a search result (red) and ingredient information (gray box). Interaction with the embedding is done by directly clicking and dragging. Setting the constraint "*orange juice:>:0*" fades the points not satisfying the restriction out.

by its name and highlight it in the embedding, or request additional information on any data point by right-clicking the according point in the embedding. In case of the cocktail dataset, ingredients and amounts of the particular cocktail are displayed. In case of an image database a thumbnail picture is rendered into the embedding. For deeper inspection of an attribute, we offer the option to colorize the data points according to the attribute-value and to fade out data points that do not satisfy a $\{>, <, =\}$-constraint. In addition, unwanted attributes can be excluded from any calculation and running sessions can be saved and restored.

4   Conclusion

We present a tool that encourages the user to explore a dataset in a "hands on" manner, by directly interacting with an embedding of it. In contrast to traditional one-time-embeddings our approach enables the user to develop a feeling for the underlying structure of the dataset by browsing it from different angles and layout the embedding in such a way that user desired aspects are emphasized.

Acknowledgment. Part of this work was supported by the German Science Foundation (DFG) under the reference number 'GA 1615/1-1'.

References

1. Alipanahi, B., Ghodsi, A.: Guided locally linear embedding. Pattern Recognition Letters 32(7), 1029–1035 (2011)
2. Barshan, E., Ghodsi, A., Azimifar, Z., Zolghadri Jahromi, M.: Supervised principal component analysis: Visualization, classification and regression on subspaces and submanifolds. Pattern Recognition 44(7), 1357–1371 (2011)
3. Cutler, A., Breiman, L.: Archetypal analysis. Technometrics 36(4), 338–347 (1994)
4. Drineas, P., Kannan, R., Mahoney, M.W.: Fast monte carlo algorithms for matrices iii: Computing a compressed approximate matrix decomposition. SIAM Journal on Computing 36(1), 184–206 (2006)
5. Hastie, T., Tibshirani, R., Friedman, J.: The Elements of Statistical Learning. Springer Series in Statistics. Springer New York Inc. (2001)
6. Lee, D.D., Seung, H.S., et al.: Learning the parts of objects by non-negative matrix factorization. Nature 401(6755), 788–791 (1999)
7. Neumann, M., Garnett, R., Moreno, P., Patricia, N., Kersting, K.: Propagation kernels for partially labeled graphs. In: ICML 2012 Workshop on Mining and Learning with Graphs (MLG 2012), Edinburgh, UK (2012)
8. Paiva, J.G.S., Schwartz, W.R., Pedrini, H., Minghim, R.: Semi-supervised dimensionality reduction based on partial least squares for visual analysis of high dimensional data. In: Computer Graphics Forum, vol. 31, pp. 1345–1354. Wiley Online Library (2012)
9. Paulovich, F.V., Nonato, L.G., Minghim, R., Levkowitz, H.: Least square projection: A fast high-precision multidimensional projection technique and its application to document mapping. IEEE Transactions on Visualization and Computer Graphics 14(3), 564–575 (2008)

10. Roweis, S.T., Saul, L.K.: Nonlinear dimensionality reduction by locally linear embedding. Science 290(5500), 2323–2326 (2000)
11. Tenenbaum, J.B., De Silva, V., Langford, J.C.: A global geometric framework for nonlinear dimensionality reduction. Science 290(5500), 2319–2323 (2000)
12. Yang, X., Fu, H., Zha, H., Barlow, J.: Semi-supervised nonlinear dimensionality reduction. In: Proceedings of the 23rd International Conference on Machine Learning, pp. 1065–1072. ACM (2006)
13. Zhang, D., Zhou, Z.H., Chen, S.: Semi-supervised dimensionality reduction. In: Proceedings of the 7th SIAM International Conference on Data Mining, pp. 629–634 (2007)

Kanopy: Analysing the Semantic Network around Document Topics

Ioana Hulpuş[1], Conor Hayes[2], Marcel Karnstedt[1], Derek Greene[3], and Marek Jozwowicz[1]

[1] Digital Enterprise Research Institute, NUI-Galway, Ireland
{ioana.hulpus,marcel.karnstedt,marek.jozwowicz}@deri.org
[2] College of Engineering and Informatics, NUI-Galway, Ireland
conor.hayes@nuigalway.ie
[3] School of Computer Science and Informatics, University College Dublin, Ireland
derek.greene@ucd.ie

Abstract. External knowledge bases, both generic and domain specific, available on the Web of Data have the potential of enriching the content of text documents with structured information. We present the Kanopy system that makes explicit use of this potential. Besides the common task of semantic annotation of documents, Kanopy analyses the semantic network that resides in DBpedia around extracted concepts. The system's main novelty lies in the translation of social network analysis measures to semantic networks in order to find suitable topic labels. Moreover, Kanopy extracts advanced knowledge in the form of subgraphs that capture the relationships between the concepts.

1 Introduction

Recent research has made progress in interlinking text documents with the Web of Data. One of the main benefits of this linkage is that the external knowledge can be used as a complementary source of information, enriching the content of the original documents and revealing semantic relations between them. There exist several popular systems for this task. Zemanta [13] focuses on enriching blog posts by recommending the authors publicly available content that can be added to the post, for example, images or links. OpenCalais [11] aims at annotating text with semantic entities and events that are extracted from it. WikipediaMiner [9] provides links to corresponding pages in Wikipedia [12] and related concepts, while DBpedia Spotlight [5] focuses on linking entities from text to those in DBpedia [6].

However, most of these systems just annotate the text with the disambiguated concepts extracted from external data sources. While this does add value to the user experience, we argue that this leaves most of the potential unexploited. DBpedia and other semi-structured knowledge bases offer a wealth of exploration options available with comparatively low processing costs. The relations between concepts can be analysed together with the graph structure around them, resulting in the discovery of rich knowledge that is not necessarily obvious from

H. Blockeel et al. (Eds.): ECML PKDD 2013, Part III, LNAI 8190, pp. 677–680, 2013.
© Springer-Verlag Berlin Heidelberg 2013

the text itself. Advantages of such an analysis are manifold. Besides concept linking, it can serve to: (i) provide explanations for concept linkage; (ii) enrich texts with a wealth of background knowledge; (iii) provide starting points for further knowledge exploration; (iv) provide insights in the quality/quantity of information the knowledge base contains about the topics discussed in the text reveling possible knowledge gaps.

Our system *Kanopy* demonstrates these advantages. It extracts a list of relevant topics from an input document, where each topic consists of a set of related words. The main objective of Kanopy is to automatically label each topic with a concept that captures its essence. Thus, for each topic Kanopy returns the topics, the top k recommended labels as well as a *topic signature graph*. This graph shows the relations between the topic words and the suggested labels, as well as other strongly related concepts. In order to achieve this, Kanopy tackles a number of difficult problems: keyphrase extraction from text, topic finding, concept linking and disambiguation, graph extraction and topic labelling.

An important approach related to our work is the REX system [1]. Given a pair of entities and a knowledge base (i.e, DBpedia), REX extracts a ranked list of semantic paths that explain the relationship between the two concepts. MING [4] is another related system that, given a group of concepts and a knowledge base, returns the most informative subgraph that contains all the given concepts and the relations between them. These systems are focused on extracting the relations between the seed concepts. Kanopy's main contribution lies in applying graph-based centralities to rank related concepts and extract the ones suitable for labelling the topic. The resulted concept must be central, from a semantic-graph perspective, with respect to all the topic words. In order to measure this *semantic centrality*, we analyse the semantic network that interconnects the concepts behind the words. When combined with a convenient user interface, Kanopy offers knowledge discovery beyond that offered by simple topic labelling.

2   The Kanopy System

In this section, we overview the key stages of the Kanopy processing pipeline, as illustrated in Figure 1.

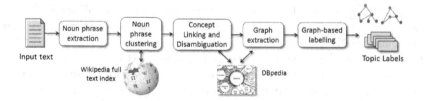

Fig. 1. Diagram illustrating the complete Kanopy processing pipeline

1. Noun Phrase Extraction. The main noun phrases are extracted from the raw input using the Stanford CoreNLP library [8]. These phrases are weighted and then filtered according to their TF-IDF score, computed with respect to the Wikipedia full-text corpus.

2. Noun Phrase Clustering. As in probabilistic topic models, we assume that a text document contains one or more sets of related "meaning-bearing words", where each set corresponds to a different topic. To identify these topics, we cluster the noun phrases obtained from the previous stage using agglomerative hierarchical clustering. Positive Pointwise Mutual Information (PPMI), with respect to the Wikipedia full-text corpus, is used as similarity metric. The applied linkage strategy and cut-off point can be adjusted from the user interface. After this stage, we obtain several clusters of noun phrases.

3. Concept linking & disambiguation. This stage links and, if necessary, disambiguates every noun phrase in each obtained topic to concepts from DBpedia. While many algorithms exists for word-sense disambiguation (WSD), we use the Eigenvalue-based WSD [2]. It is unsupervised, does not need preprocessing and it supports simultaneous disambiguation of a group of related words. It achieved approximately 10% better accuracy at disambiguating topic models than the state-of-art unsupervised WSD algorithms [2]. Its drawback lies in the higher computational complexity, but the parallel implementation inside Kanopy meets the requirements of an online demonstration.

4. Graph extraction. At the beginning of this stage, each remaining noun phrase is linked to exactly one DBpedia concept, which we refer to as *seed concepts*. A semantic network is extracted by activating the concepts and their relations within a distance of two hops from each seed. In the majority of cases, this is sufficient to connect the single concept subgraphs into one larger graph, which provides the input for the next stage [3].

5. Topic labelling. From this extracted topic graph we finally identify the most relevant concepts. We apply the *focused random walk betweenness* and *focused information centrality*, as defined in [3], which rank the nodes in the topic graph with respect to their *semantic centrality* to the seed concepts.

At this stage, each previously identified topic is represented by the set of seed concepts, the ranked label nodes, and the semantic network extracted at Step 4. This network contains some hundreds or thousands of nodes and edges. In the user interface, we present compressed topic graphs, consisting of only the shortest paths between seeds and labels as well as the nodes on these paths. Colour coding is used to differentiate the types of nodes and their relevance to the topic. Users can dynamically select how many of the top labels they want to inspect, resulting in graphs of different complexity.

3   Kanopy in Action

Kanopy is deployed as a web application. Users are encouraged to input any text in the user interface. Let us assume they copy and paste the body of a text related to research about the indian tigers endangered by extinction due to poor genetic diversity [10]. In the following, we exemplify some key insights that can be gained with the current version of Kanopy [7].

Some topics that Kanopy identifies with the default settings are about locality (India), scientific research, and genetics. The third one is of particular interest. As the column Extracted Concepts shows, it brings together different concepts found in the text, such as "Preservation breeding", "DNA", "Gene pool", "Extinction" and "Genetic structure". Opening the topic graph and setting the Top Labels slider to 1 reveals that the top candidates for labelling are either *Genetics* or *Biology*, depending on the chosen Centrality Measure. Hovering over the topic graph highlights that *Genetics* is directly connected to the text mentions of "Genetic structure" and "DNA", a fact that explains its high score. The graph also clarifies that *Population Genetics*, the second-ranked label, is part of *Genetics* – which in turn is part of *Biology*. None of these recommended labels occured in the original text. Setting the Topic granularity to "Fine" in the user interface, Kanopy splits this topic into two more focused ones, labelled *Population Genetics* and *Conservation Biology*. Besides these multi-concept topics, Kanopy also displays single concepts that remained isolated, such as *Tiger*, together with the DBpedia categories and classes they belong to.

We plan to demonstrate Kanopy's research along the lines outlined above. A future extension we envisage for the system is that of corpus analysis and exploration. Here, Kanopy can be used to extract an interconnected network of concepts, topics and documents. This use case would bring value for a range of domains, such as online journalism, education, and knowledge management.

Acknowledgements. This work was supported by Science Foundation Ireland (SFI) partly under Grant No. 08/CE/I1380 (Lion-2) and partly under Grant No. 08/SRC/I1407 (Clique: Graph and Network Analysis Cluster).

References

1. Fang, L., Sarma, A.D., Yu, C., Bohannon, P.: Rex: explaining relationships between entity pairs. Proc. VLDB Endow. 5(3), 241–252 (2011)
2. Hulpus, I., Hayes, C., Karnstedt, M., Greene, D.: An Eigenvalue-Based Measure for Word-Sense Disambiguation. In: FLAIRS 2012 (2012)
3. Hulpus, I., Hayes, C., Karnstedt, M., Greene, D.: Unsupervised Graph-based Topic Labelling using DBpedia. In: WSDM 2013, ACM (2013)
4. Kasneci, G., Elbassuoni, S., Weikum, G.: Ming: mining informative entity relationship subgraphs. In: CIKM 2009, pp. 1653–1656. ACM (2009)
5. Mendes, P.N., Jakob, M., García-Silva, A., Bizer, C.: DBpedia spotlight: shedding light on the web of documents. In: I-Semantics 2011, pp. 1–8 (2011)
6. http://dbpedia.org/
7. http://kanopy.deri.ie
8. http://nlp.stanford.edu/software/corenlp.shtml
9. http://wikipedia-miner.cms.waikato.ac.nz/
10. http://www.bbc.co.uk/news/uk-wales-22536571
11. http://www.opencalais.com/
12. http://www.wikipedia.org/
13. http://www.zemanta.com

SCCQL : A Constraint-Based Clustering System

Antoine Adam[1], Hendrik Blockeel[1], Sander Govers[2], and Abram Aertsen[2]

[1] KU Leuven, Departement of Computer Science,
Celestijnenlaan 200A, 3001 Leuven, Belgium
{antoine.adam,hendrik.blockeel}@cs.kuleuven.be
[2] KU Leuven, Department of Microbial and Molecular Systems,
Kasteelpark Arenberg 22, 3001 Leuven, Belgium
{sander.govers,abram.aertsen}@biw.kuleuven.be

Abstract. This paper presents the first version of a new inductive database system called SCCQL. The system performs constraint-based clustering on a relational database. Clustering problems are formulated with a query language, an extension of SQL for clustering that includes must-link and cannot-link constraints. The functioning of the system is explained. As an example of use of this system, an application in the context of microbiology has been developed that is presented here.

Keywords: inductive database, inductive query language, clustering.

1 Introduction

Data analysis is a non-trivial task: many methods require knowledge of advanced mathematics and statistics to be used correctly, and among those methods that do not, there is still the task of choosing among the many implementations that are available. Data mining environments such as Weka, Orange, RapidMiner, KNIME, etc., facilitate data analysis by allowing the user to construct workflows from predefined building blocks, helping the user choose among alternative techniques, etc. While this provides much support and flexibility, full flexibility is only achieved by allowing scripting or programming in addition to this.

Inductive databases go one step further. Based on the principle that there should be no inherent difference between querying and mining, they offer "data mining query languages", in which data mining tasks can be expressed as queries, and the results are again queriable (the "closure principle"). They set the stage for a more declarative approach to data mining: Just like SQL made it possible to query complex databases without having to program data navigation, inductive query languages should make it possible to formulate complex mining problems without having to choose or compose the optimal mining algorithm. Examples of such systems are SINDBAD/SiQL [6], ATLAS [8], DMX [5].

Constraint-based clustering is an example of a mining task where flexibility is desirable. It is a generalization of standard clustering in which the user can impose constraints on the clustering to be found, such as must-link and cannot-link constraints. Note that also classical parameters of clustering algorithms, such

H. Blockeel et al. (Eds.): ECML PKDD 2013, Part III, LNAI 8190, pp. 681–684, 2013.

as the number of clusters to be constructed, can be seen as constraints. One could then think of a language that allows the user to naturally formulate "clustering queries", which may involve a variety of constraints, and of an inductive database system that can execute such queries.

Adam et al. [1] recently proposed such a language. It extends SQL with the CLUSTER statement. The basic structure of this statement is as follows:

```
CLUSTER attributes FROM data [WITH constraints]
```

The currently available constraints are the number of clusters wanted and must-link and cannot-link constraints. These last ones are specified as follows:

```
[SOFT] MUST|CANNOT LINK (cdata) [BY attribute]
```

The complete grammar definition of the language and examples of queries can be found in the aforesaid paper. Note that, while some existing inductive database systems offer clustering queries, none of them offer constraint-based clustering in this manner.

The purpose of this demo is to exhibit a system that allows the user to run this type of query, and to demonstrate the ease with which such a system can be used to solve practical data mining questions. In the remainder of this paper, we briefly describe the architecture of the system and some of its features.

2   The SCCQL System

Figure 1 shows the architecture of the SCCQL ("Structured Constraint-based Clustering Query Language") system, which allows for clustering tuples in a relational database using the mentioned query language. When a cluster query is parsed, the parser does not interpret the **data** and **cdata** parts; these are sent to the actual SQL database, which retrieves the data and sends it to the cluster engine. For soft equivalence constraints, the engine learns a Mahalanobis distance as in [2]. The engine next chooses the algorithm to execute: CopKMeans [7] if there are hard equivalence constraints; the Weka [3] implementation of EM otherwise. Respecting the closure principle, which states that the result of a query should be queriable, the result returned is a table. It is a copy of the **data** table with an extra column holding the cluster assignment of each instance.

The system we developed includes an interface that provides two ways of building a query. On one hand, the user can make the query step by step: select the data to cluster, choose the attributes to use for the clustering, specify the number of clusters and add equivalence constraints. The query is then built from the different elements. On the other hand, the user can directly type in the query he wants to execute. The interface also includes some representation of the result table to help visualize it.

3   Application

The SCCQL system is being developed in a project that groups scientists from microbiology and computer science. One of the goals is to build a platform

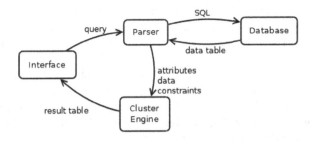

Fig. 1. Architecture scheme

that will help microbiologists to analyse data using data mining techniques. The SCCQL system will be part of this platform.

In the application, clustering has two purposes: finding groups of similar instances, and identifying outliers (instances that are far from any cluster, or in isolated clusters). Such outliers can help microbiologists better understand microbial behavior and growth characteristics. The relational database of the application stores different microbiological parameters of cells, such as the descendancy and physiological states within a population. For instance, one can be interested in clustering cells according to a number of static and/or dynamic parameters. This can be formulated by the following query:

```
CLUSTER LengthMean, WidthMean
FROM (SELECT c.Id, l.Mutant, AVG(s.Length) AS LengthMean,
            AVG(s.Width) AS WidthMean
      FROM stateovertime s, cell c, lineage l
      WHERE l.ExperimentId=5 AND c.LineageId = l.Id
        AND s.CellId = c.Id
      GROUP BY c.id) AS data
WITH SOFT MUST LINK WHERE data.Mutant=0 BY Mutant
```

More examples of queries can be found in [1] and on `http://people.cs.kuleuven.be/~antoine.adam/`.

4   Discussion and Conclusion

A first advantage of the SCCQL system is that it seamlessly integrates the specification of clustering constraints and of the data to be clustered. The latter allows for non-trivial data preprocessing. Not all kinds of preprocessing are easily expressed in SQL, but one could also integrate the system with a language such as SiQL, which does allow more advanced preprocessing. A second advantage is that it is goal-oriented: the user can tell the system what data to cluster, and under what constraints, without stating which clustering method should be used. Although the current implementation uses standard algorithms in its clustering engine, one could also use general solvers for those cases where no standard algorithm will suffice. The intelligence to chose the system correctly can be built into the system. No other system that we know of combines these two advantages.

For now, SCCQL focuses specifically on constraint-based clustering. Within this setting, its most important limitation is that it can exploit only predefined types of constraints. More types could be introduced (e.g., imposing a minimal cluster size, or requiring balanced clusters), but this will raise the problem of how to solve clustering tasks that combine certain types of constraints. SCCQL is not extensible in the sense of allowing the user to specify any constraint using a general-purpose constraint language. Finally, in the current implementation, the clustering process is not integrated in the database management system: the data is first retrieved from the database, then clustering is performed externally. This is transparent for the user, who just sees the resulting table, but a closer integration may have efficiency advantages.

The SCCQL system is work in progress. We have presented a first version that executes constraint-based clustering queries on a database. An application in the field of microbiology has been shown, but the system is essentially domain independent and can be combined with any SQL database.

Acknowledgements. We thank Tias Guns for his useful comments on this paper. This work is funded by the KU Leuven Research Fund (project IDO/10/012).

References

1. Adam, A., Blockeel, H.: A Query Language for Constraint-based Clustering. In: Proceedings of the 2013 Belgium-Netherlands Conference on Machine Learning, Benelearn (to appear, 2013)
2. Bar-Hillel, A., Hertz, T., Shental, N., Weinshall, D.: Learning a Mahalanobis Metric from Equivalence Constraints. Journal of Machine Learning Research 6(1), 937–965 (2005)
3. Hall, M., Frank, E., Holmes, G., Pfahringer, B., Reutemann, P., Witten, I.H.: The WEKA data mining software: an update. ACM SIGKDD Explorations Newsletter 11(1), 10–18 (2009)
4. Imieliński, T., Mannila, H.: A Database Perspective on Knowledge Discovery. Communication of the ACM 39(11), 58–64 (1996)
5. Data Mining eXtensions DMX,
   http://msdn.microsoft.com/en-us/library/ms132058.aspx
6. Wicker, J., Richter, L., Kessler, K., Kramer, S.: SINDBAD and SiQL: An Inductive Database and Query Language in the Relational Model. In: Daelemans, W., Goethals, B., Morik, K. (eds.) ECML PKDD 2008, Part II. LNCS (LNAI), vol. 5212, pp. 690–694. Springer, Heidelberg (2008)
7. Wagstaff, K.L.: Intelligent Clustering with Instance-level Constraints. PhD diss., Cornell University (2002)
8. Wang, H., Zaniolo, C.: Atlas: A Native Extension of SQL for Data Mining. In: Proceedings of the 3rd SIAM International Conference on Data Mining (2003)

Author Index